W9-CGP-414

Fundamentals of Cheese Science

Patrick F. Fox, PhD, DSc
Professor, Food Chemistry
Food Science and Technology
University College, Cork
Cork, Ireland

Timothy P. Guinee, PhD
Senior Research Officer
Dairy Products Research Centre
Teagasc
Cork, Ireland

Timothy M. Cogan, PhD
Senior Principal Research Officer
Dairy Products Research Centre
Teagasc
Visiting Professor
University College, Cork
Cork, Ireland

Paul L. H. McSweeney, PhD
Statutory Lecturer, Food Chemistry
Food Science and Technology
University College, Cork
Cork, Ireland

AN ASPEN PUBLICATION
Aspen Publishers, Inc.
Gaithersburg, Maryland
2000

The authors have made every effort to ensure the accuracy of the information herein. However, appropriate information sources should be consulted, especially for new or unfamiliar procedures. It is the responsibility of every practitioner to evaluate the appropriateness of a particular opinion in the context of actual clinical situations and with due considerations to new developments. The author, editors, and the publisher cannot be held responsible for any typographical or other errors found in this book.

Aspen Publishers, Inc., is not affiliated with the American Society of Parenteral and Enteral Nutrition.

Library of Congress Cataloging-in-Publication Data

Fundamentals of Cheese Science
Patrick F. Fox. . . [et al.].
p. cm.
Includes bibliographical references.
ISBN 0-8342-1260-9
1. Cheese. I. Fox, P.F.
SF271.F86 2000
637'.3—dc21
99-053386

Orders: (800) 638-8437
Customer Service: (800) 234-1660

Editorial Services: Jane Colilla
Library of Congress Catalog Card Number: 99-053386
ISBN: 0-8342-1260-9
Printed in the United States of America

1 2 3 4 5

Table of Contents

Preface

Cheese, which has been produced for about 5,000 years, is one of the classical fabricated foods in the human diet. During its long history, the volume and diversity of cheese production have increased such that today annual production is about 15×10^6 tonnes (representing about 35% of total world milk production) in at least 500 varieties. Cheese is one of our most complex and dynamic food products, and its study involves a wide range of disciplines, especially analytical and physical chemistry, biochemistry, microbiology, rheology, and sensory science. Not surprisingly, a large and diverse literature on the science and technology of cheese has accumulated over the past 100 years. Like in the case of the other great fermented foods, wine and beer, the epicurean attributes of cheese attract the attention of consumers and endow it with a certain social status. Cheese is a highly nutritious food with a very positive image. It is the quintessential consumer-ready food, yet it is one of the most flexible food ingredients. In many respects, cheese is the ideal food: nutritious, flexible in use and application, and sensorially appealing to a wide range of consumers.

Cheese has been the subject of a considerable number of books, but most of these were written with the general reader in mind (see "Suggested Readings" at the end of Chapter 1). There are at least three books on cheese technology, but the scientific aspects of cheese have been less well covered, the only exception being the two-volume set *Cheese: Chemistry, Physics and Microbiology*, edited by P.F. Fox. That set, though, assumes a substantial background knowledge on the part of its readers.

Fundamentals of Cheese Science provides comprehensive coverage of the scientific aspects of cheese, appropriate for anyone working with cheese, from researchers and professionals to undergraduate and graduate students in food science and technology. The book assumes familiarity with biochemistry, microbiology, and dairy chemistry, and it emphasizes fundamental principles rather than technological aspects.

The book is divided into 23 chapters that deal with the chemistry and microbiology of milk for cheesemaking, starter cultures, coagulation of milk by enzymes or by acidification, the microbiology and biochemistry of cheese ripening, the flavor and rheology of cheese, processed cheese, cheese as a food ingredient, public health and nutritional aspects of cheese, and various methods used for the analysis of cheese. The book contains copious references to other texts and review articles, but references to the primary literature are kept to a minimum to facilitate easy presentation.

Finally, the authors would like to express their appreciation for the highly skilled and enthusiastic assistance of Ms. Anne Cahalane in the preparation of the manuscript.

CHAPTER 1

Cheese: Historical Aspects

1.1 INTRODUCTION

Cheese is the generic name for a group of fermented milk-based food products produced throughout the world in a great diversity of flavors, textures, and forms. Sandine and Elliker (1970) suggest that there are more than 1,000 varieties of cheese. Walter and Hargrove (1972) describe about 400 varieties and list the names of a further 400, while Burkhalter (1981) classifies 510 varieties.

It is commonly believed that cheese evolved in the Fertile Crescent between the Tigris and Euphrates rivers, in Iraq, some 8,000 years ago, during the so-called Agricultural Revolution, when certain plants and animals were domesticated as sources of food. Among the earliest animals domesticated were goats and sheep; being small, gregarious, and easily herded, these were used to supply meat, milk, hides, and wool. Cattle were more difficult to domesticate; wild cattle were much larger and more ferocious than modern cattle and were also less well adjusted to the arid Middle East than goats and sheep. Apparently, cattle were used mainly as work animals (as they still are) and did not become a major source of milk until relatively recently. Man soon recognized the nutritive value of milk produced by domesticated animals, and milk and dairy products became important components of the human diet.

Milk is also a rich source of nutrients for bacteria that contaminate the milk and grow well under ambient conditions. Some contaminating bacteria utilize milk sugar, lactose, as a source of energy, producing lactic acid as a byproduct; these bacteria, known as *lactic acid bacteria (LAB),* include the genera *Lactococcus*, *Lactobacillus*, *Streptococcus*, *Enterococcus*, *Leuconostoc*, and *Pediococcus*. LAB are used in the production of a wide range of fermented milk, meat, and vegetable products. They are generally considered to be beneficial to human health and have been studied extensively (see Chapter 5).

Bacterial growth and acid production would have occurred in milk during storage or during attempts to dry milk in the prevailing warm, dry climate of the Middle East to produce a more stable product; air-drying of meat, and probably fruits and vegetables, appears to have been practiced as a primitive form of food preservation at this period of human evolution. When sufficient acid is produced, the principal proteins in milk, the caseins, coagulate at ambient temperature (21°C) in the region of their isoelectric points (≈ pH 4.6) to form a gel in which the fat and aqueous phases of milk are entrapped. Thus, the first fermented dairy foods were probably produced accidentally. Numerous, basically similar products are produced in various regions of the world by artisanal methods probably little different from those used several thousand years ago. Some descendants of these ancient fermented milks are now produced by scientifically based technology in sophisticated factories.

The first fermented dairy foods were produced by a fortuitous combination of events—the growth in milk of a group of lactic acid bacteria produced just enough acid to reduce the pH of the milk to the isoelectric point of the caseins, causing these proteins to coagulate. Neither the lactic acid bacteria nor the caseins were "designed" for this function. The caseins were designed to be coagulated enzymatically in the stomach of neonatal mammals, the gastric pH of which is around 6 (i.e., very much higher than the isoelectric point of the caseins). The ability of LAB to ferment lactose, a sugar specific to milk, is frequently encoded on plasmids, suggesting that this characteristic was acquired relatively recently in the evolution of these bacteria. Their natural habitats are vegetation, from which they presumably colonized the teats of mammals contaminated with lactose-containing milk.

An acid-induced milk gel is quite stable if left undisturbed, but if it is broken, either accidentally (e.g., by movement of the storage vessels) or intentionally, it separates into curds and whey. It would have soon been realized that the acid whey is a pleasant, refreshing drink for immediate consumption, whereas the curds could be consumed fresh or stored for future use. It was probably also realized that the shelf-life of the curds could be greatly extended by dehydration and/or by adding salt; heavily salted cheese varieties (e.g., Feta and Domiati) are still widespread throughout the Middle East and the Balkans, where the ambient temperature is high. Air- or sun-dried varieties of cheese are less common today, but numerous examples survive throughout the hot, dry areas of North Africa and the Middle East.

Today, acid-coagulated cheeses, which include Cottage cheese, Cream cheese, Quarg, Fromage frais, and some varieties of Queso blanco, represent approximately 25% of total cheese production and in some countries are the principal varieties. They are consumed fresh (not ripened) and are widely used in other products (e.g., cheesecake, cheese-based dips, and sauces).

An alternative mechanism for coagulating milk was also discovered at an early date. Many proteolytic enzymes can modify the milk protein system, causing it to coagulate under certain circumstances. Enzymes capable of causing this transformation are widespread and are found in bacteria, molds, plants, and animal tissues, but the most obvious source would have been animal stomachs. It would have been observed that the stomachs of young slaughtered animals frequently contained curds, especially if the animals had suckled shortly before slaughter; curds would also have been observed in the vomit of human infants. Before the development of pottery (about 5,000 B.C.), storage of milk in bags made from animal skins was probably common (it is still practiced in many countries). Stomachs from slaughtered animals provided ready-made, easily sealed containers; if stored in such containers, milk would extract coagulating enzymes (referred to as rennets) from the stomach tissue, leading to its coagulation during storage. The properties of rennet-coagulated milk curds are very different from those of curds produced by isoelectric (acid) precipitation. For example, they have better syneretic (curd-contracting) properties, making it possible to produce low-moisture cheese curd without hardening. Rennet-coagulated curds can, therefore, be converted to more stable low-moisture products than can acid curd, and rennet coagulation has become the principal mechanism for milk coagulation in cheese manufacture. Most modern cheese varieties and approximately 75% of total world production of cheese are produced by this mechanism.

During the storage of rennet-coagulated curds, various bacteria grow, and the enzymes in rennet continue to act. Thus, the flavor and texture of the cheese curds change during storage. When controlled, this process is referred to as ripening (maturation), during which a great diversity of characteristic flavors and textures develop. Although animal rennets were probably the first enzyme coagulants used, rennets produced from a range of plant species (e.g., figs and thistle) appear to have been common in Roman times. However, plant rennets are not suitable for the manufacture of long-ripened cheese varieties, and gastric proteinases from young animals became the standard rennets until a shortage of supply made it necessary to introduce rennet substitutes, which are discussed in Chapter 6.

While the coagulation of milk by the in situ production of lactic acid was, presumably, accidental, the use of rennets to coagulate milk was intentional. It was, in fact, quite an ingenious development—if the conversion of milk to cheese by the use of rennets was discovered today, it would be hailed as a major biotechnological discovery! The use of rennets in cheese manufacture is probably the oldest and is still one of the principal industrial applications of enzymes.

The advantages accruing from the ability to convert the principal constituents of milk to cheese would have been readily apparent: storage stability, ease of transport, and diversification of the human diet. Cheese manufacture accompanied the spread of civilization throughout the Middle East, Egypt, Greece, and Rome. There are several references in the Old Testament to cheese, such as in Job (1520 B.C.) and Samuel (1170–1017 B.C.); on the walls of Ancient Egyptian tombs; and in classical Greek literature, including Homer (12th century B.C.), Herodotus (484–408 B.C.), and Aristotle (384–322 B.C.). Cheese manufacture was well established at the time of the Roman Empire, and cheese was included in the rations of Roman soldiers. The demand for cheese in Rome must have exceeded supply, since the Emperor Diocletian (A.D. 284–305) fixed a maximum price for cheese. Many Roman writers, including Cato (about 150 B.C.), Varro (about 40 B.C.), Columella (A.D. 50), and Pliny (A.D. 23–89), described the manufacture, quality attributes, and culinary uses of cheese. Columella, in particular, gave a detailed account of cheese manufacture in his treatise on agriculture, *De Re Rustica*.

Movements of Roman armies and administrators contributed to the spread of cheese throughout the then known world. Although archeological evidence suggests that cheese may have been manufactured in pre-Roman Britain, the first unequivocal evidence credits the Romans with the establishment of cheesemaking in Britain. Palladius wrote a treatise on Roman-British farming in the 4th century A.D., which included a description of and advice on cheesemaking. Cheesemaking practice appears to have changed little from the time of Columella and Palladius until the 19th century.

The great migrations of peoples throughout Europe after the fall of the Roman Empire probably promoted the spread of cheese manufacture, as did the Crusaders and pilgrims of the Middle Ages. However, the most important contributors to the development of cheese technology and to the evolution of cheese varieties during the Middle Ages were the monasteries and feudal estates. In addition to their roles in the spread of Christianity and in the preservation and expansion of knowledge during the Dark Ages, the monasteries were major contributors to the advancement of agriculture in Europe and to the development and improvement of food commodities, notably wine, beer, and cheese. Many current cheese varieties were developed in monasteries, such as Wenslydale (Rievaulx Abbey, Yorkshire), Port du Salut or St. Paulin (Monastery of Notre Dame du Port du Salut, Laval, France), Fromage de Tamie (Abbey of Tamie, Lac d'Annecy, France), Maroilles (Abbey of Maroilles, Avesnes, France), and Trappist (Maria Stern Monastery, Banja Luka, Bosnia). The intermonastery movement of monks probably contributed to the spread of cheese varieties and to the development of new hybrid cheeses.

The great feudal estates of the Middle Ages were self-contained communities that, in the absence of an effective transport system, relied on a supply of locally produced foods. Surplus food was produced in summer and preserved to meet the requirements of the community throughout the year. Especially in cool, wet Europe, fermentation and salting were the most effective methods of food preservation. Well-known examples of products preserved by these methods include fermented and salted meat, salted fish, beer, wine, fermented vegetables, and cheese (the manufacture of which exploits both fermentation and salting). Cheese probably represented an item of trade when amounts beyond local requirements were available.

Within large estates, individuals acquired special skills that were passed on to succeeding generations. The feudal estates evolved into villages and some into larger communities. Because monasteries and feudal estates were essentially self-contained communities with limited intercommunity travel, it is readily apparent how several hundred distinct varieties of cheese could have evolved from essentially the same raw material. Traditionally, cheese varieties were produced in quite limited geographical regions, especially in mountainous areas. The localized production of certain varieties is still apparent and indeed is preserved through the designation of Appellation d'Origine Contrôlée. The regionalization of certain cheese varieties is still particularly marked in Spain, Italy, and France, where the production of many varieties is restricted to very limited, sometimes legally defined regions. Almost certainly, most cheese varieties evolved by accident because of particular local circumstances (e.g., a local species or breed of dairy animal or a peculiarity of the local milk supply with respect to chemical composition or microflora) or because of an unintended event during the manufacture or storage of the cheese (e.g., growth of molds or other microorganisms). Presumably, the accidents that led to desirable changes in the quality of the cheese were incorporated into the manufacturing protocol; each variety would thus have undergone a series of evolutionary changes and refinements.

The final chapter in the spread of cheese throughout the world resulted from the colonization of North and South America, Oceania, and Africa by European settlers who carried their cheesemaking skills with them. Cheese has become an item of major economic importance in some of these non-European countries, notably the United States, Canada, Australia, and New Zealand, but the varieties produced are mainly of European origin, modified in some cases to meet local conditions and requirements. There is no evidence that cheese was produced in the Americas or Oceania prior to colonization; in fact, animals had not been domesticated for milk production in these countries.

Cheesemaking remained a craft until relatively recently. With the gradual acquisition of knowledge about the chemistry and microbiology of milk and cheese, it became possible to gain more control over the cheesemaking process. Few new varieties have evolved as a result of the increased knowledge, but existing varieties have become better defined and their quality has become more consistent. Although the names of many current varieties were introduced several hundred years ago (Table 1–1), it is very likely that those cheeses were not comparable to their modern counterparts. Cheesemaking was not standardized until relatively recently. For example, the first attempt to standardize the well-known English varieties, Cheddar and Cheshire, was made by John Harding in the middle of the 19th century. Prior to that, "Cheddar cheese" was cheese produced around the village of Cheddar, in Somerset, England, and

Table 1–1 First Recorded Date for Some Major Cheese Varieties

Variety	*Year*
Gorgonzola	897
Schabzieger	1000
Roquefort	1070
Maroilles	1174
Schwangenkase	1178
Grana	1200
Taleggio	1282
Cheddar	1500
Parmesan	1579
Gouda	1697
Gloucester	1783
Stilton	1785
Camembert	1791
St. Paulin	1816

probably varied considerably depending on the cheesemaker and other factors. Cheese manufacture during most of its history was a farmstead enterprise. The first cheese factory in the United States was established near Rome, New York, in 1851, and the first in Britain at Longford, Derbyshire, in 1870. There were thousands of small-scale cheese manufacturers, and there must have been great variation within any one general type of cheese. When one considers the very considerable interfactory and indeed intrafactory variation in quality and characteristics that still occurs today in well-defined varieties (e.g., Cheddar) in spite of the very considerable scientific and technological advances, one can readily appreciate the variation that must have existed in earlier times.

Some major new varieties, notably Jarlsberg and Maasdamer, have been developed recently as a consequence of scientific research. Many other varieties have evolved considerably, even to the extent of becoming new varieties, as a consequence of scientific research and the development of new technology. Notable examples are Queso blanco as produced in the United States, Feta-type cheese produced from ultrafiltered milk, and various forms of Quarg. There has been a marked resurgence of farmhouse cheesemaking in recent years. Many of the cheeses being produced on farms are not standard varieties, and it will be interesting to see if some of these evolve to become major new varieties.

A main cause of variation in the characteristics of cheese is the species from which the milk is produced. Although milks from several species are used by humans, the cow is by far the principal producer. Worldwide, 85% of commercial milk is bovine. However, goats, sheep, and water buffalo are significant producers of milk in certain regions (e.g., the Mediterranean basin and India). Goats and sheep are especially important in cheese production, since the milks of these species are used mainly for the production of fermented milks and cheese. Many world-famous cheeses are produced from sheep's milk (e.g., Roquefort, Feta, Romano, and Manchego). Traditional Mozzarella is made from the milk of the water buffalo. As discussed in Chapter 3, there are very significant interspecies differences in the composition of milk that are reflected in the characteristics of the cheeses produced from them. There are also significant differences in milk composition between breeds of cattle, and these influence cheese quality, as do variations due to seasonal, lactational, and nutritional factors and of course the methods of milk production, storage, and collection.

1.2 CHEESE PRODUCTION AND CONSUMPTION

World production of cheese is roughly 15×10^6 tonnes per annum (about 35% of total milk production) and has increased at an average annual rate of about 4% over the past 30 years. Europe, with a production of roughly 8×10^6 tonnes per annum, is by far the largest producing block (Table 1–2). Thus, while cheese manufacture is practiced worldwide, it is apparent from Table 1–2 that cheese is primarily a product of European countries and those populated mostly by European immigrants.

Cheese consumption varies widely between countries, even within Europe (Table 1–3). Cheese consumption in most countries for which data are available has increased consistently over many years. Along with fermented milks, cheese is the principal growth product within the dairy sector. There are many reasons for the increased consumption of cheese, including a positive dietary image, convenience and flexibility in use, and the great diversity of flavors and textures. Cheese can be regarded as the quintessential convenience food: it can be used as a major component of a meal, as a dessert, as a component of other foods, or as a food ingredient; it can be consumed without preparation or subjected to various cooking processes. The most rapid growth in cheese consumption has occurred in its use as a food component or ingredient; these applications are discussed in Chapter 19.

Table 1–2 World Production of Cheese, 1994

Country	*Cheese Production (1,000 Tonnes)*	*Country*	*Cheese Production (1,000 Tonnes)*	*Country*	*Cheese Production (1,000 Tonnes)*
World	**15,084**	Colombia	51	Bulgaria	72
		Ecuador	7	Croatia	21
Africa	**511**	Peru	6	Czech Republic	139
Algeria	1	Uruguay	23	Denmark	291
Angola	1	Venezuela	70	Estonia	18
Botswana	2			Finland	89
Egypt	349	**Asia**	**1,018**	France	1,605
Ethiopia	3	Afghanistan	16	Germany	1,569
Mauritania	2	Armenia	9	Greece	216
Morocco	7	Azerbaijan	7	Hungary	88
Niger	14	Bangladesh	1	Iceland	3
Nigeria	7	China	206	Ireland	91
South Africa	37	Cyprus	5	Italy	1,017
Sudan	76	Georgia	3	Latvia	11
Tanzania	2	Iran	197	Lithuania	27
Tunisia	6	Iraq	17	Macedonia,	
Zambia	1	Israel	92	FYR of	1
Zimbabwe	2	Japan	114	Moldova	
		Jordan	4	Republic	3
North and Central		Kazakhstan	34	Netherlands	688
America	**4,130**	Kyrgyzstan	5	Norway	89
Canada	323	Lebanon	15	Poland	397
Costa Rica	6	Mongolia	1	Portugal	65
Cuba	15	Myanmar	29	Romania	42
Dominican		Syria	86	Russian	
Republic	3	Tajikistan	1	Federation	477
El Salvador	3	Turkey	139	Slovakia	44
Guatemala	11	Turkmenistan	13	Slovania	16
Honduras	8	Uzbekistan	14	Spain	163
Mexico	123	Yemen	10	Sweden	128
Nicaragua	6			Switzerland	133
Panama	7	**Europe**	**8,201**	United Kingdom	385
United States	3,627	Albania	1	Ukraine	71
		Austria	102	Yugoslavia, FR	14
South America	**677**	Belarus	39		
Argentina	405	Belgium-		**Oceania**	**544**
Bolivia	7	Luxembourg	75	Australia	270
Brasil	60	Bosnia-		New Zealand	274
Chile	51	Herzgovina	14		

Note: The following countries were included by the Food and Agriculture Organization in 1997, but no data for cheese production are available: Burkina Faso, Madagascar, Somalia, Jamaica, Trinidad and Tobago, India, Indonesia, Republic of Korea, Malaysia, Eritrea, Kenya, Namibia, Bhutan, Oman, Malta, Nepal, Pakistan, Philippines, Saudi Arabia, Thailand, Sri Lanka, United Arab Emirates, and Fiji.

Table 1–3 Consumption of Cheese, 1993 (Kilograms per Person per Year)

Country	*Fresh*	*Ripened*	*Total*
France	7.5	15.5	22.8
Greece	0.2	21.8	22.0
Italy	6.7	13.4	20.1
Belgium	4.7	15.1	19.8
Germany	8.0	10.5	18.5
Lithuania	11.6	6.8	18.4
Iceland	5.2	11.9	17.1
Switzerland	2.8	13.6	16.4
Sweden	0.9	15.5	16.4
Luxembourg	5.0	11.3	16.3
Netherlands	1.7	14.1	15.8
Denmark	0.9	14.5	15.4
Finland	2.3	12.0	14.3
Norway	0.2	14.0	14.2
Canada	0.9	12.4	13.3
United States	1.3	11.9	13.2
Austria	3.9	7.5	11.4
Czech and Slovak Republics	4.0	6.6	10.6
Estonia	5.6	4.4	10.0
Australia	–	–	8.8
United Kingdom	–	–	8.3
New Zealand	–	–	8.1
Hungary	3.3	4.6	7.9
Russia	2.8	4.9	7.7
Spain	–	–	7.0
Ireland	–	–	5.6
Chile	2.0	2.0	4.0
South Africa	0.1	1.5	1.6
Japan	0.2	1.2	1.4
India	0.2	–	0.2

1.3 CHEESE SCIENCE AND TECHNOLOGY

Cheese is the most diverse group of dairy products and is, arguably, the most academically interesting and challenging. While many dairy products, if properly manufactured and stored, are biologically, biochemically, and chemically stable, cheeses are, in contrast, biologically and biochemically active and consequently undergo changes in flavor, texture, and functionality during storage. Throughout manufacture and ripening, cheese production represents a finely orchestrated series of consecutive and concomitant biochemical events that, if synchronized and balanced, lead to products with highly desirable aromas and flavors, but, if unbalanced, result in off-flavors and off-odors. Considering that a basically similar raw material (milk from a very limited number of species) is subjected to a manufacturing protocol, the general principles of which are common to most cheese varieties, it is fascinating that such a diverse range of products can be produced. No two batches of the same variety are identical.

A further important facet of cheese production is the range of scientific disciplines involved: the study of cheese manufacture and ripening encompasses the chemistry and biochemistry of milk constituents, the chemical characterization of cheese constituents, microbiology, enzymology, molecular genetics, flavor chemistry, rheology, and chemical engineering. It is not surprising, therefore, that many scientists have become involved in the study of cheese manufacture and ripening. A voluminous scientific and technological literature has accumulated, including several textbooks (see "Suggested Readings") and chapters in many others. Many of these textbooks deal mainly with cheese technology or assume an overall knowledge of cheese. The present book is intended to provide a fairly comprehensive treatment of the scientific aspects of cheese.

REFERENCES

Burkhalter, G. (1981). *Catalogue of cheese* (Bulletin 141). Brussels: International Dairy Federation.

Sandine, W.E., & Elliker, P.R. (1970). Microbially induced flavors and fermented foods: Flavor in fermented dairy products. *Journal of Agricultural and Food Chemistry, 18*, 557–566.

Walter, H.E., & Hargrove, R.C. (1972). *Cheeses of the world*. New York: Dover.

SUGGESTED READINGS

Andrews, A.T., & Varley, J. (1994). *Biochemistry of milk and milk products*. Cambridge: Royal Society of Chemistry.

Anifantakis, E.M. (1991). *Greek cheeses: A tradition of centuries*. Athens: National Dairy Committee of Greece.

Berger, W., Klostermeyer, H., Merkenich, K., & Uhlmann, G. (1989). *Die Schmelzkäselerstellung*. Ledenburg, Germany: Benckiser-Knapsack GmbH.

Cantin, C. (1976). *Guide pratique des fromages*. Paris: Solar Editeur.

Cheke, V. (1959). *The story of cheesemaking in Britain*. London: Routledge & Kegan Paul.

Davis, J.G. (1965–1967). *Cheese* (Vols. 1–4). London: Churchill Livingstone.

DOC cheeses of Italy: A great heritage. (1992). Milan: Franco Angeli.

Eck, A., & Gilles, J.C. (1997). *Le fromage* (2d ed.). Paris: Technique et Documentation (Lavoisier).

Eekhof-Stork, N. (1976). *World atlas of cheese*. London: Paddington Press.

Fox, P.F. (1993). *Cheese: Chemistry, physics and microbiology* (2d ed., Vols. 1, 2). London: Chapman & Hall.

Glynn Christian's world guide to cheese. (1984). (S. Harris, trans.). London: Ebury Press.

Gonzalez, M.A., & del Cerro, C.G. (1988). *Quesos de Espana*. Madrid: Espasa-Calpe.

Kosikowski, F.V., & Mistry, V.V. (1997). *Cheese and fermented milks* (3d ed., Vols. 1, 2). Westport, CT: F.V. Kosikowski LLC.

Kosikowski, F.V., & Mocquot, G. (1958). *Advances in cheese technology*. (Food & Agriculture Organization Study No. 38). Rome: Food & Agriculture Organization.

Law, B.A. (1997). *Microbiology and biochemistry of cheese and fermented milks* (2d ed.). London: Chapman & Hall.

Layton, J.A. (1973). *The cheese handbook*. New York: Dover.

Lembo, P., & Spedicato, E. (1992). *I prodotti caseari del mezzogiorno*. Rome: Consiglio Nazionale delle Ricerche.

Mair-Waldburg, H. (1974). *Handbook of cheese; Cheeses of the world A to Z*. Kempten Allgan, Germany: Volkwertschaftlecher Verlag GmbH.

Malin, E.L., & Tunick, M.H. (1995). *Chemistry of structure-function relationships in cheese*. New York: Plenum Press.

Masui, K., & Yamada, T. (1996). *French cheeses*. London: Dorling Kindersley.

Meyer, A. (1973). *Processed cheese manufacture*. London: Food Trade Press.

Robinson, R.K. (1993). *Modern dairy technology* (2d ed., Vol. 2). London: Elsevier Applied Science.

Robinson, R.K. (1995). *A colour guide to cheese and fermented milks*. London: Chapman & Hall.

Robinson, R.K., & Tamime, A.Y. (1991). *Feta and related cheeses*. London: Ellis Horwood.

Robinson, R.K., & Wilbey, R.A. (1998). *Cheesemaking practice* (3d ed.). Gaithersburg, MD: Aspen Publishers.

Sammis, J.L. (1948). *Cheesemaking*. Madison, WI: Cheesemaker Book Co.

Scott, R. (1986). *Cheesemaking practice* (2d ed.). London: Elsevier Applied Science.

Simon, A.L. (1956). *Cheeses of the world*. London: Faber & Faber.

Squire, E.H. (1937). *Cheddar gorge: A book of English cheeses*. London: Collins.

van Slyke, L.L., & Price, W.V. (1949). *Cheese*. New York: Orange Judd.

CHAPTER 2

Overview of Cheese Manufacture

The production of all varieties of cheese involves a generally similar protocol (Figure 2–1), various steps of which are modified to give a product with the desired characteristics. The principal steps will be described in individual chapters. The objective of this chapter is to present a very brief description of the principal operations so that the operations described in the following chapters can be seen in an overall context.

2.1 SELECTION OF MILK

The composition of cheese is strongly influenced by the composition of the cheese milk, especially the content of fat, protein, calcium, and pH. The constituents of milk, which are described in Chapter 3, are influenced by several factors, including the species, breed, individuality, nutritional status, health, and stage of lactation of the producing animal. Owing to major compositional abnormalities, milk from cows in the very early or late stages of lactation and those suffering from mastitis should be excluded. Somatic cell (leucocyte) count is a useful index of quality. Some genetic polymorphs of the milk proteins have a significant effect on cheese yield and quality, and there is increasing interest in breeding for certain polymorphs. The milk should be free of chemical taints and free fatty acids, which cause off-flavors in the cheese, and antibiotics, which inhibit bacterial cultures.

The milk should be of good microbiological quality, as contaminating bacteria will be concentrated in the cheese curd and may cause defects or public health problems (see Chapter 4).

2.2 STANDARDIZATION OF MILK COMPOSITION

Milk for cheese is subjected to a number of pretreatments, with various objectives.

Different cheese varieties have a certain fat-in-dry-matter content, in effect, a certain fat:protein ratio, and this content has legal status in the "Standards of Identity" for many cheese varieties. While the moisture content of cheese, and hence the level of fat plus protein, is determined mainly by the manufacturing protocol, the fat:protein ratio is determined mainly by the fat:casein ratio in the cheese milk. Depending on the ratio required, it can be modified by

- removing some fat by natural creaming, as in the manufacture of Parmigiano-Reggiano, or centrifugation
- adding skim milk
- adding cream
- adding milk powder, evaporated milk, or ultrafiltration retentate (such additions also increase the total solids content of the milk and hence cheese yield, as discussed in Chapter 9)

Calcium plays a major role in the coagulation of milk by rennet and the subsequent processing

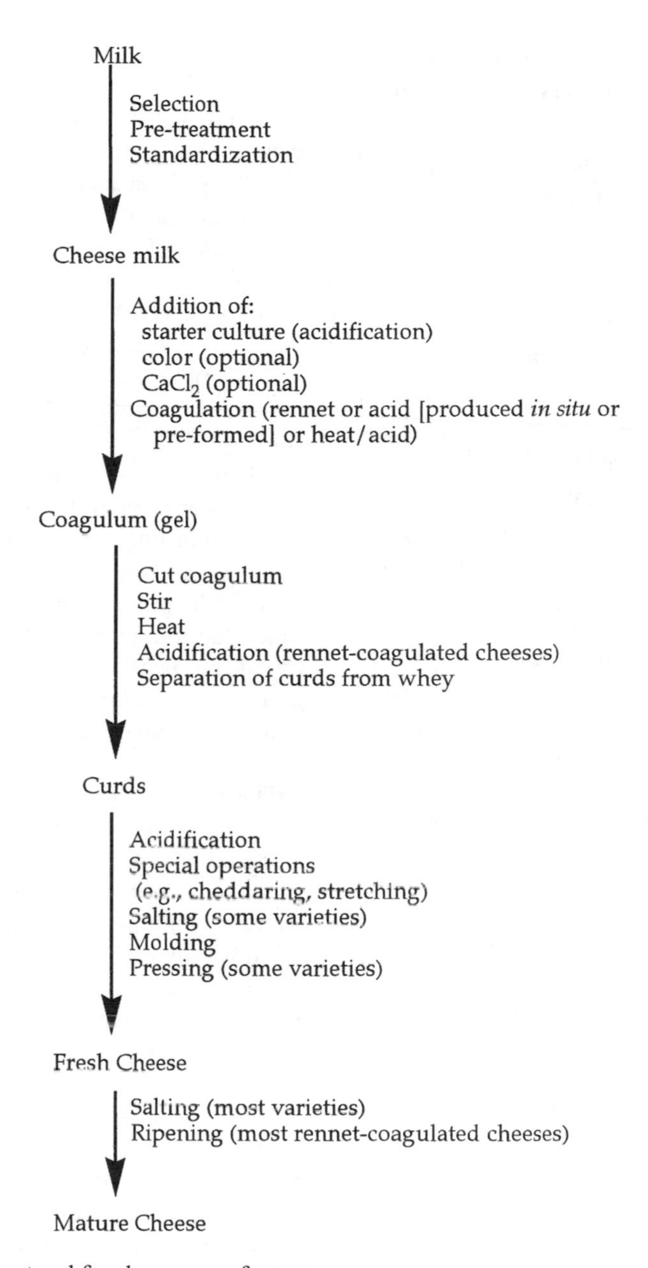

Figure 2–1 General protocol for cheese manufacture.

of the coagulum, and hence it is common practice to add $CaCl_2$ (e.g., 0.01%) to cheese milk.

The pH of milk is a critical factor in cheesemaking. The pH is inadvertently adjusted by the addition of 1.5–2% starter culture, which reduces the pH of the milk immediately by about 0.1 unit. Starter concentrates (sometimes called direct-to-vat starters), which are sometimes used, have no immediate acidifying effect.

Previously, it was standard practice to add the starter to the cheese milk 30–60 min before rennet addition. During this period, the starter began to grow and produce acid, a process referred to as *ripening*. Ripening served a number of functions:

- It allowed the starter bacteria to enter their exponential growth phase and hence to be highly active during cheesemaking; this is not necessary with modern high-quality starters.
- The lower pH was more favorable for rennet action and gel formation.

However, the practice increases the risk of bacteriophage infection of the starter as phage become distributed throughout the liquid milk but are locked in position after it has coagulated (see Chapter 5). Although ripening is still practiced for some varieties, it has been discontinued for most varieties.

The pH of milk on reception at the dairy is higher today than it was previously owing to improved hygiene during milking and the wider use of refrigeration at the farm and factory. In the absence of acid production by contaminating bacteria, the pH of milk increases slightly during storage due to the loss of CO_2 to the atmosphere. The natural pH of milk is about 6.6 but varies somewhat (e.g., it increases in late lactation and during mastitic infection).

To offset these variations and to reduce the pH as an alternative to ripening, the preacidification of milk by 0.1–0.2 pH units is recommended, either through the use of the acidogen gluconic acid-δ-lactone, or by limited growth of a lactic acid starter, followed by pasteurization (referred to as *prematuration*). Such a practice results in a gel with more uniform characteristics, reflected in the production of cheese of more uniform quality.

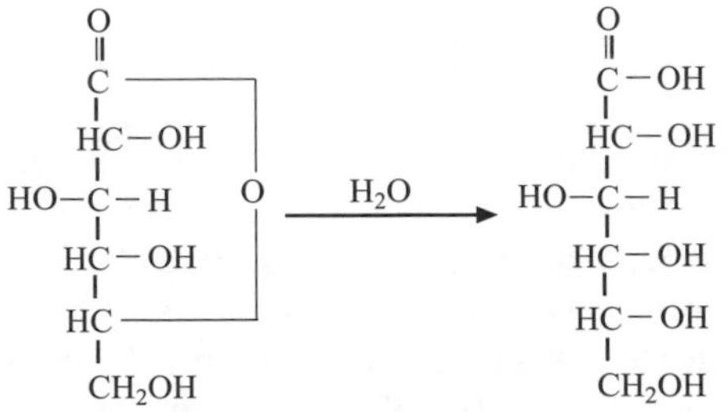

O
||
C
|
HC−OH
|
HO−C−H   O $\xrightarrow{H_2O}$
|
HC−OH
|
HC
|
CH_2OH

Gluconic acid δ-lactone

O
||
C−OH
|
HC−OH
|
HO−C−H
|
HC−OH
|
HC−OH
|
CH_2OH

Gluconic acid

2.3 HEAT TREATMENT OF MILK

Traditionally, all cheese was made from raw milk, a practice that remained widespread until the 1940s. Even today, significant amounts of cheese are made from raw milk in Europe. The use of raw milk is undesirable for two reasons:

1. dangers to public health
2. the presence of undesirable microorganisms, which may cause defects in flavor and/or texture

When cheese was produced from fresh milk on farms or in small local factories, the growth of contaminating microorganisms was minimal, but as cheese factories became larger, storage of milk for longer periods became necessary, and hence the microbiological quality of the milk varied. For public health reasons, it became increasingly popular from the beginning of this century to pasteurize milk for liquid consumption.

The pasteurization of cheese milk became widespread about 1940, primarily for public health reasons but also to provide a milk supply of more uniform bacteriological quality and to improve its keeping quality. Although a considerable amount of cheese is still produced from raw milk, on both an artisanal and factory scale, especially in southern Europe (including such famous varieties as Swiss Emmental, Gruyère, Comté, Parmigiano-Reggiano, and Grano Padano), pasteurized milk is now generally used, especially in large factories. Aspects of pasteurization are discussed in Chapter 4.

There are four alternatives to pasteurization for reducing the number of microorganisms in milk:

1. treatment with H_2O_2
2. activation of the lactoperoxidase-H_2O_2-thiocyanate system
3. bactofugation
4. microfiltration

These processes are also discussed briefly in Chapter 4.

2.4 CHEESE COLOR

Color is a very important attribute of foods and serves as an index of quality, although in some cases, it is merely cosmetic. The principal indigenous pigments in milk are carotenoids which are obtained from the animal's diet, especially from fresh grass and clover.

The carotenoids are secondary pigments involved in photosynthesis (the structure of β-carotene is shown in Figure 2–2). Owing to the conjugated double-bond system, they absorb ultraviolet and visible light, giving them colors ranging from yellow to red. They are responsible for the color of many foods (e.g., carrots, squashes, peppers, and corn). They are also present in the leaves of plants, in which their color is masked by the green chlorophylls. Some carotenoids have pro–vitamin A activity and may be converted to retinol (Figure 2–2) in the body.

Animals do not synthesize carotenoids but absorb them from plant materials in their diet. In addition to serving as pro-vitamin A, some animals store carotenoids in their tissues, which then acquire a color (e.g., salmon, cooked lobster, and egg yolk). Cattle transfer carotenoids to adipose tissue and milk but goats, sheep, and buffalo do not. Therefore, bovine milk and milk products are yellow to an extent dependent on the carotenoid content of the animal's diet. Products such as butter and cheese made from sheep, goat, or buffalo milk are very white in comparison with counterparts made from bovine milk. This yellowish color may make products produced from cow milk less acceptable than products produced from sheep, goat, or buffalo milk in Mediterranean countries, where the latter are traditional.

The carotenoids in bovine milk can be bleached by treatment with H_2O_2 or benzoyl peroxide or masked by chlorophyll or titanium oxide (TiO_2), although such practices are not permitted in all countries.

At the other end of the spectrum are individuals who prefer highly colored cheese, butter, or egg yolk. Intense colors may be obtained by adding carotenoids (synthetic or natural extracts). In the case of cheese and dairy products, annatto, extracted from the pericarp of the seeds of the annatto plant (*Bixa orellana*), a native of Brasil, is used most widely. Annatto contains two apocarotenoid pigments, bixin and norbixin (Figure 2–3). By suitable modification, the annatto pigments can be made fat soluble, for use in butter or margarine, or water soluble, for use in cheese.

Initially, annatto may have been used in cheese manufacture to give the impression of a high fat content in partially skimmed cheese, but some people believe that colored ("red") cheese tastes better than its white counterpart of equivalent quality.

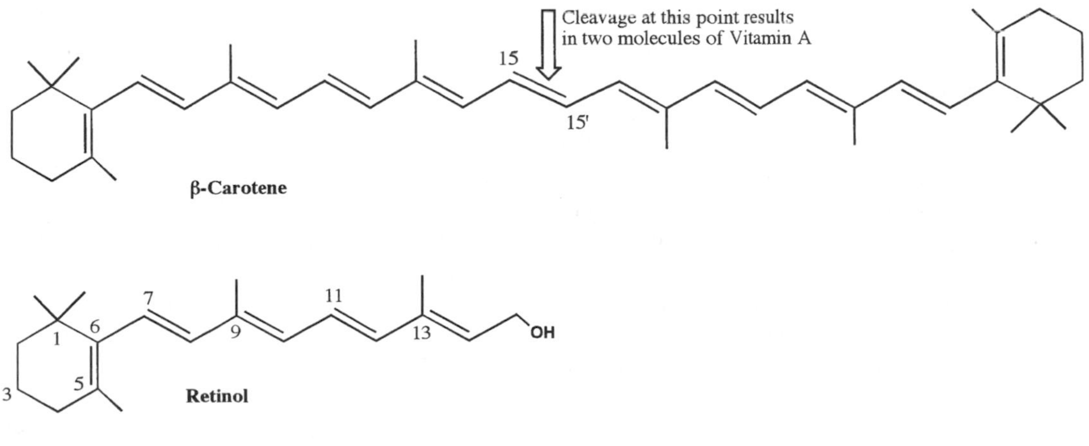

Figure 2–2 Structures of β-carotene and retinol.

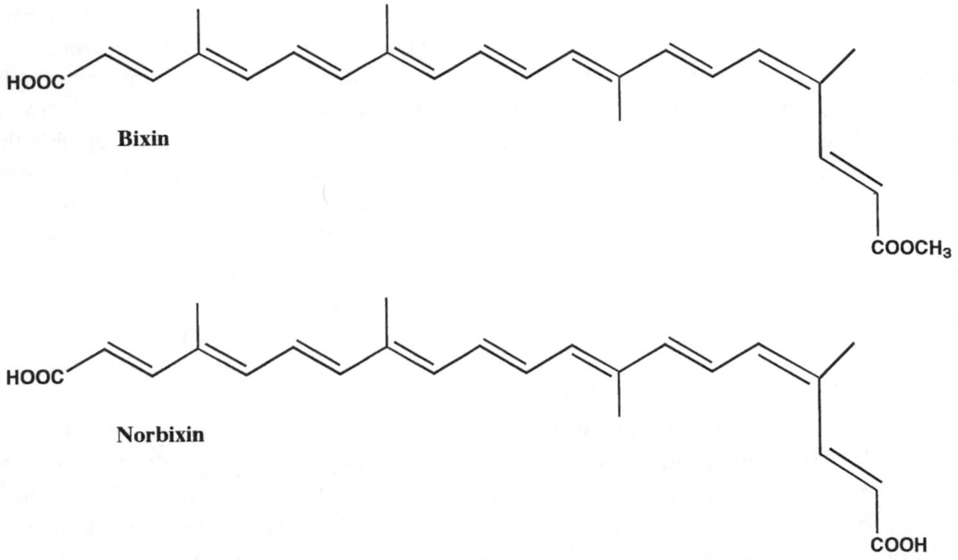

HOOC Bixin $COOCH_3$

HOOC Norbixin COOH

Figure 2–3 Structures of *cis*-bixin and norbixin, the apocarotenoid pigments in annatto.

2.5 CONVERSION OF MILK TO CHEESE CURD

After the milk has been standardized, pasteurized, or otherwise treated, it is transferred to vats (or kettles). These vats are of various shapes (hemispherical, rectangular, vertical cylindrical, and horizontal cylindrical), may be open or closed, and may range in size from a few hundred liters to 30,000 liters. Here, it is converted to cheese curd, a process that involves three basic operations: acidification, coagulation, and dehydration.

2.5.1 Acidification

Acidification is usually achieved through the in situ production of lactic acid through the fermentation of the milk sugar lactose by lactic acid bacteria. Initially, the indigenous milk microflora was relied upon to produce acid, but since this microflora was variable, the rate and extent of acidification were variable, resulting in cheese of variable quality. Cultures of lactic acid bacteria for cheesemaking were introduced commercially about 100 years ago, and these have become increasingly improved and refined. The science and technology of starters is described in Chapter 5. The acidification of curd for some artisanal cheeses still relies on the indigenous microflora.

Direct acidification using an acid (usually lactic or hydrochloric acid) or an acidogen (usually gluconic acid-δ-lactone) may be used as an alternative to biological acidification. It is used commercially to a significant extent in the manufacture of Cottage cheese, Quarg, and Feta-type cheese from ultrafiltration-concentrated milk and Mozzarella.

Direct acidification is more controllable than biological acidification, and, unlike starters, it is not susceptible to phage infection. However, in addition to acidification, the starter bacteria serve very important functions in cheese ripening (see Chapters 10 and 11), and hence chemical acidification is used mainly for cheese varieties for which texture is more important than flavor.

The rate of acidification is fairly characteristic of the variety, and its duration ranges from 5–6 hr for Cheddar and Cottage cheese to 10–12 hr for Dutch and Swiss types. The rate of acidification, which depends on the amount of starter used and on the temperature profile of the curd, has a major effect on the texture of cheese, mainly through its solubilizing effect on colloidal calcium phosphate (see Chapter 13).

Regardless of the rate of acidification, the ultimate pH of the curd for most hard cheese varieties is in the range 5.0–5.3 but it is 4.6 for the soft, acid-coagulated varieties (e.g., Cottage cheese, Quarg, and Cream cheese) and some rennet-coagulated varieties (e.g., Camembert and Brie).

The production of acid at the appropriate rate and time is critical for the manufacture of good quality cheese. Acid production affects several aspects of cheese manufacture, many of which are discussed in more detail in later chapters:

- coagulant activity during coagulation
- denaturation and retention of the coagulant in the curd during manufacture and hence the level of residual coagulant in the curd (this influences the rate of proteolysis during ripening and may affect cheese quality)
- curd strength, which influences cheese yield
- gel syneresis, which controls cheese moisture and hence regulates the growth of bacteria and the activity of enzymes in the cheese (consequently, it strongly influences the rate and pattern of ripening and the quality of the finished cheese)
- the extent of dissolution of colloidal calcium phosphate, which modifies the susceptibility of the caseins to proteolysis during ripening and influences the rheological properties of the cheese (e.g., compare the texture of Emmental, Gouda, Cheddar, and Cheshire cheese) (see Chapter 13)
- the growth of many nonstarter bacteria in cheese, including pathogenic, food-poisoning, and gas-producing microorganisms (properly made cheese is a very safe product from a public health standpoint) (see Chapter 20)

The level and time of salting have a major influence on pH changes in cheese. The concentration of NaCl in cheese (commonly 0.7–4%, which is equivalent to 2–10% salt in the moisture phase) is sufficient to halt the growth of starter bacteria. Some varieties, mostly of British origin, are salted by mixing dry salt with the curd toward the end of manufacture, and hence the pH of curd for these varieties must be close to the ultimate value (≈ pH 5.1) at salting. However, most varieties are salted by immersion in brine or by surface application of dry salt. Salt diffusion in cheese moisture is a relatively slow process and thus there is ample time for the pH to decrease to about 5.0 before the salt concentration becomes inhibitory throughout the interior of the cheese. The pH of the curd for most cheese varieties (e.g., Swiss, Dutch, Tilsit, and Blue cheese) is 6.2–6.5 at molding and pressing but decreases to around 5–5.2 during or shortly after pressing and before salting. The significance of various aspects of the concentration and distribution of NaCl in cheese is discussed in Chapter 8.

In a few special cases (e.g., Domiati), a high level of NaCl (10–12%) is added to the cheese milk, traditionally to control the growth of the indigenous microflora. This concentration of NaCl has a major influence not only on acid development but also on rennet coagulation, gel strength, and curd syneresis.

2.5.2 Coagulation

The essential step in the manufacture of all cheese varieties involves coagulation of the casein component of the milk protein system to form a gel that entraps the fat, if present. Coagulation may be achieved by

- limited proteolysis by selected proteinases (rennets)
- acidification to pH 4.6
- acidification to a pH value greater than 4.6 (perhaps ≈ 5.2) in combination with heating to roughly 90°C

The vast majority of cheese varieties (representing about 75% of total production) are produced by rennet coagulation, but some acid-coagulated varieties, such as Quarg and Cottage cheese, are of major importance. The acid-heat–coagulated cheeses are of relatively minor importance. They are usually produced from whey or a blend of whey and skim milk and probably evolved as a useful means for recovering the nutritionally valuable whey proteins. Their properties are very different from those of rennet- or acid-coagulated cheeses, and they are usually used as food ingredients. Important varieties are Ricotta and related varieties (indigenous to Italy), Anari (Cyprus), and Manouri (Greece) (see Chapters 17 and 19).

The coagulation of milk by rennets and the coagulation of milk by acid are discussed in Chapters 6 and 16, respectively.

A fourth (minor) group of cheeses is produced, not by coagulation, but by thermal evaporation of water from a mixture of whey and skim milk, whole milk, or cream and crystallization of lactose. Varietal names include Mysost and Gjetost. These cheeses, which are almost exclusive to Norway, bear little resemblance to rennet- or acid-coagulated cheeses and probably should be classified as whey products rather than cheese.

2.5.3 Postcoagulation Operations

Rennet- or acid-coagulated milk gels are quite stable if maintained under quiescent conditions, but if cut or broken, they rapidly undergo syneresis, expelling whey. Syneresis essentially concentrates the fat and casein of milk by a factor of 6–12, depending on the variety. In the dairy industry, concentration is normally achieved through thermal evaporation of water and more recently by removing water through semipermeable membranes. The syneresis of rennet- or acid-coagulated milk gels is thus an unusual method of dehydration.

The rate and extent of syneresis are influenced, inter alia, by the milk composition, especially the concentrations of Ca^{2+} and casein; the pH of the whey; the cooking temperature; the rate of stirring of the curd-whey mixture; and, of course, time (see Chapter 7). The composition of the finished cheese is, to a very large degree, determined by the extent of syneresis, and since this is readily under the control of the cheesemaker, it is here that the differentiation of the individual cheese varieties really begins, although the type and composition of the milk, the amount and type of starter, and the amount and type of rennet are also significant in this regard.

A more or less unique protocol has been developed for the manufacture of each cheese variety. Such protocols differ mainly with respect to the syneresis process. The protocols for the manufacture of the principal families of cheese are summarized in Chapter 17.

2.5.4 Salting

Salting is the last manufacturing operation. Salting promotes syneresis, but it is not a satisfactory method for controlling the moisture content of cheese curd, which is best achieved by ensuring that the degree of acidification, heating, and stirring in the cheese vat are appropriate to the particular variety. Salt has several functions in cheese, and these are described in Chapter 8. Although salting should be a very simple operation, quite frequently it is not performed properly, with adverse effects.

2.5.5 Ultrafiltration

As indicated previously, cheese manufacture is essentially a dehydration process. With the development of ultrafiltration as a concentration process, it was obvious that this process would have applications in cheese manufacture, not only for standardizing cheese milk with respect to fat and casein but more importantly for the preparation of a concentrate with the composition of the finished cheese, commonly referred to as *pre-cheese*. Standardization of cheese milk by adding ultrafiltration concentrate (retentate) is now common in some countries, but the manufacture of pre-cheese has to date been successful commercially for only certain cheese varieties, most notably ultrafiltration Feta and Quarg. Undoubtedly, the use of ultrafiltration will become much more widespread in cheese manufacture (see Chapter 17).

2.6 RIPENING

Fresh cheeses constitute a major proportion of the cheese consumed in some countries (see Table 1–3). Most of these cheeses are produced by acid coagulation and are described in Chapter 16. Although rennet-coagulated cheese varieties may be consumed at the end of manufacture and a few are (e.g., Burgos), most rennet-coagulated cheeses are ripened (cured, matured) for a period ranging from about 3 weeks to more than 2 years. Generally, the duration of ripening is inversely related to the moisture content of the cheese. Many varieties may be consumed at any of several stages of maturity, depending on the flavor preferences of consumers and economic factors.

Although curds for different cheese varieties are recognizably different at the end of manufacture (mainly as a result of compositional and textural differences arising from differences in milk composition and processing factors), the unique characteristics of the individual cheeses develop during ripening as a result of a complex set of biochemical reactions. The changes that occur during ripening—and hence the flavor, aroma, and texture of the mature cheese—are largely predetermined by the manufacturing process, that is, by the composition (especially moisture, NaCl, and pH), by the level of residual coagulant activity, by the type of starter, and in many cases by secondary inocula added to or gaining access to the milk or curd.

The biochemical changes that occur during ripening are caused by one or more of the following agents:

- the coagulant
- indigenous milk enzymes, especially proteinase and lipase, which are particularly important in cheese made from raw milk
- starter bacteria and their enzymes
- secondary microorganisms and their enzymes

The secondary microflora may arise from indigenous microorganisms that survive pasteurization or gain entry to the milk after pasteurization (e.g., *Lactobacillus*, *Pediococcus*, and *Micrococcus*). They may also be added as a secondary starter, such as citrate-positive *Lactococcus* or *Leuconostoc* spp. in Dutch-type cheese, *Propionibacterium* in Swiss cheese, *Penicillium roqueforti* in Blue varieties, *P. camemberti* in Camembert or Brie, or *Brevibacterium linens* in surface smear-ripened varieties (e.g., Tilsit and Limburger). In many cases, the characteristics of the finished cheese are dominated by the metabolic activity of these secondary microorganisms.

The primary biochemical changes involve glycolysis, lipolysis, and proteolysis, but these are followed and overlapped by a host of secondary catabolic changes to the compounds produced in these primary pathways, including deamination, decarboxylation, and desulfurylation of amino acids, β-oxidation of fatty acids, and even some synthetic reactions (e.g., esterification).

Although it is not yet possible to fully describe the biochemistry of cheese ripening, very considerable progress has been made on elucidating the primary reactions, and these are discussed in Chapter 11.

2.7 PROCESSED CHEESE PRODUCTS

Depending on culinary traditions, a variable proportion of mature cheese is consumed as cheese (such cheese is often referred to as *table cheese*). A considerable amount of natural cheese is used as an ingredient in other foods (e.g., Parmesan or Grana on pasta products, Mozzarella on pizza, Quarg in cheesecake, Ricotta in ravioli). A third major outlet for cheese is in the production of a broad range of processed cheese products, which in turn have a range of applications, especially as spreads, sandwich fillers, or food ingredients. These products are discussed in Chapters 18 and 19.

2.8 WHEY AND WHEY PRODUCTS

Only about 50% of the solids in milk are incorporated into cheese; the remainder (90% of the lactose, ≈ 20% of the protein, and ≈ 10% of the fat) are present in the whey. Until recently, whey was regarded as an essentially useless byproduct to be disposed of as cheaply as possible. However, in the interest of reducing environmental pollution, but also because it is now possible to produce valuable food products from whey, whey processing has become a major facet of the total cheese industry. The principal aspects of whey processing are discussed in Chapter 22.

CHAPTER 3

Chemistry of Milk Constituents

3.1 INTRODUCTION

Milk is a fluid secreted by the female of all mammals, of which there are more than 4,000 species, for the primary function of meeting the complete nutritional requirements of the neonate of the species. It must supply energy (mainly from fats and sugar [lactose]), amino acids (from proteins), vitamins, and atomic elements (commonly but inaccurately referred to as minerals). In addition, several physiological functions are performed by milk constituents, including antimicrobial substances (immunoglobulins, lactoperoxidase, and lactotransferrin), enzymes and enzyme inhibitors, vitamin-binding carrier proteins, and cell growth and control factors. Because the nutritional and physiological requirements of each species are more or less unique, the composition of milk shows very marked interspecies differences. The milks of only about 180 species have been analyzed, and the data for only about 50 of these species are considered to be reliable (sufficient number of samples, representative sampling, and adequate coverage of the lactation period). Not surprisingly, the milks of the principal dairying species (i.e., cow, goat, sheep, and buffalo) and *Homo sapiens* are among those that are well characterized. The gross composition of milks from selected species is summarized in Table 3–1. Very extensive data on the composition of bovine and human milks are compiled by Jensen (1995).

In addition to the principal constituents listed in Table 3–1, milk contains several hundred minor constituents, many of which have a major impact on the nutritional, technological, and sensoric properties of milk and dairy products (e.g., vitamins, small inorganic and organic ions, and flavor compounds).

Milk is a very variable biological fluid. In addition to interspecies differences, the milk of any particular species varies with the breed (in the case of commercial dairying species), health, nutritional status, stage of lactation, and age of the animal, the interval between milkings, and so on. In a bulked factory milk supply, variability due to many of these factors is reduced, but some variability persists, and it can even be quite large in situations where milk production is seasonal. In addition to variations in the concentrations of the principal and minor constituents due to the above factors, the chemistry of some of the constituents also varies (e.g., the fatty acid profile is strongly influenced by diet). Some of the variability in the composition and constituents of milk can be adjusted or counteracted by processing technology but some differences may persist. As will become apparent in later chapters, the variability of milk composition poses major problems in cheese production.

Physicochemically, milk is a very complex fluid. The constituents of milk occur in three phases. Most of the mass of milk is an aqueous solution of lactose, organic and inorganic salts,

Table 3–1 Composition of the Milks of Some Species

Species	*Total Solids (%)*	*Fat (%)*	*Protein (%)*	*Lactose (%)*	*Ash (%)*
Human	12.2	3.8	1.0	7.0	0.2
Cow	12.7	3.7	3.4	4.8	0.7
Goat	12.3	4.5	2.9	4.1	0.8
Sheep	19.3	7.4	4.5	4.8	1.0
Pig	18.8	6.8	4.8	5.5	–
Horse	11.2	1.9	2.5	6.2	0.5
Donkey	11.7	1.4	2.0	7.4	0.5
Reindeer	33.1	16.9	11.5	2.8	–
Domestic rabbit	32.8	18.3	11.9	2.1	1.8
Bison	14.6	3.5	4.5	5.1	0.8
Indian elephant	31.9	11.6	4.9	4.7	0.7
Polar bear	47.6	33.1	10.9	0.3	1.4
Grey seal	67.7	53.1	11.2	0.7	–

vitamins, and other small molecules. In this aqueous solution are dispersed proteins, some at the molecular level (whey proteins), others as large colloidal aggregates ranging in diameter from 50 to 600 nm (the caseins), and lipids, which exist in an emulsified state as globules ranging in diameter from 0.1 to 20 μm. Thus, colloidal chemistry is important in the study of milk, for example, in the context of surface chemistry, light scattering, and rheology.

Milk is a dynamic system owing to the instability of many of its structures (e.g., the milk fat globule membrane); changes in the solubility of many constituents, especially the inorganic salts and proteins, with temperature and pH; the presence of various enzymes that can modify constituents through lipolysis, proteolysis, or oxidation-reduction; the growth of microorganisms, which can cause major changes either directly through their growth (e.g., changes in pH or redox potential [E_h]), or indirectly through enzymes they excrete; and the interchange of gases with the atmosphere (e.g., CO_2). Milk was intended to be consumed directly from the mammary gland and to be expressed from the gland at frequent intervals. However, in dairy operations, milk is stored for various periods, ranging from a few hours to several days, during which it is cooled (and perhaps heated) and agitated to various degrees. These treatments will cause some physical changes and permit some enzymatic and microbiological changes that may alter the processing properties of milk. It may be possible to counteract some of these changes.

Although many of the minor constituents of milk are important from a nutritional standpoint, the technological properties of milk are determined mainly by its macroconstituents (proteins, lipids, and lactose) and some of its low molecular mass species, especially calcium, phosphate, citrate, and pH. The properties of these constituents, with emphasis on their significance in cheesemaking, are discussed briefly in this chapter. For a more thorough discussion, the reader is referred to Cayot and Lorient (1998); Fox (1982, 1983, 1985, 1989, 1992, 1995, 1997); Fox and McSweeney (1998); Jenness and Patton (1959); Walstra and Jenness (1984); Webb and Johnson (1965); Webb, Johnson, and Alford (1974); and Wong, Jenness, Keeney, and Marth (1988).

3.2 LACTOSE

Lactose is the principal carbohydrate in the milk of all mammals, which is the only source. Milk contains only trace amounts of other sugars, including glucose, fructose, glucosamine, galactosamine, neuraminic acid, and neutral and acidic oligosaccharides.

The concentration of lactose in milk varies widely between species (Table 3–2). The lactose content of cow milk varies with the breed of cow, individuality factors, udder infection, and especially stage of lactation. The concentration of lactose decreases progressively and significantly during lactation (Figure 3–1); this trend contrasts with the lactational trends for lipids and proteins, which, after decreasing during early lactation, increase strongly during the second half of lactation. Lactose and soluble ions (e.g., Na^+, K^+, and Cl^-) are the compounds mainly responsible for the osmotic pressure of milk. During mastitis, the concentration of NaCl in milk increases, resulting in an increase in osmotic pressure. This increase is compensated for by a decrease in the lactose content; that is, there is an inverse relationship between the concentration of NaCl and lactose in milk, which partly explains why certain milks with a high lactose content have a low ash content and vice versa (see Table 3–1). The inverse relationship between the concentration of lactose and chloride is the basis of the Koestler's chloride-lactose test for abnormal milk:

$$\text{Koestler number} = \frac{\text{percentage of chloride} \times 100}{\text{percentage of lactose}}$$

A Koestler number less than 2 indicates normal milk while a value greater than 3 is considered abnormal.

Lactose plays an important role in milk and milk products:

- It is essential in the production of fermented dairy products, including cheese.
- It contributes to the nutritive value of milk and its products. However, many non-Europeans have limited or zero ability to digest lactose in adulthood, leading to a syndrome known as *lactose intolerance*. Mature cheese is free of lactose, and hence cheese is suitable for inclusion in the diet of lactose-intolerant individuals.
- It affects the texture of certain concentrated and frozen products.
- It is involved in heat-induced changes in the color and flavor of highly heated milk products.

Table 3–2 Concentration of Lactose in the Milks of Selected Species

Species	*Lactose (%)*
California sea lion	0.0
Hooded seal	0.0
Black bear	0.4
Dolphin	0.6
Echidna	0.9
Blue whale	1.3
Rabbit	2.1
Red deer	2.6
Grey seal	2.6
Rat (Norwegian)	2.6
Mouse (house)	3.0
Guinea pig	3.0
Dog (domestic)	3.1
Sika deer	3.4
Goat	4.1
Elephant (Indian)	4.7
Cow	4.8
Sheep	4.8
Water buffalo	4.8
Cat (domestic)	4.8
Pig	5.5
Horse	6.2
Chimpanzee	7.0
Rhesus monkey	7.0
Human	7.0
Donkey	7.4
Zebra	7.4
Green monkey	10.2

3.2.1 Structure of Lactose

Lactose is a disaccharide consisting of galactose and glucose, linked by a β1–4 glycosidic bond (Figure 3–2). Its systematic name is *O*-β-D-galactopyranosyl-(1–4)-α-D-glucopyranose (α-lactose) or *O*-β-D-galactopyranosyl-(1–4)-β-D-glucopyranose (β-lactose). The hemiacetal group of the glucose moiety is potentially free

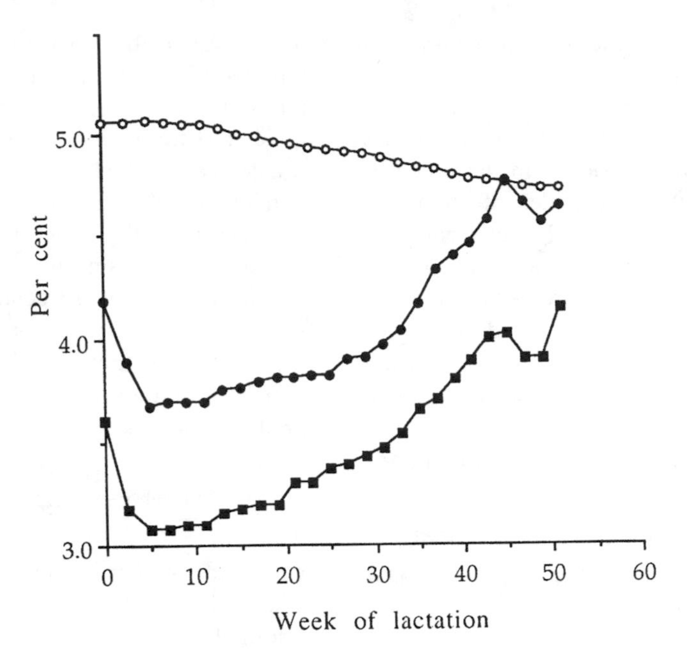

Figure 3–1 Changes in the concentrations of fat (●), protein (■), and lactose (○) in milk during lactation.

(i.e., lactose is a *reducing* sugar) and may exist as an α- or β-anomer. In the structural formula of the α-form, the hydroxyl group on the C_1 of glucose is cis to the hydroxyl group at C_2 (oriented downward) and vice versa for the β-form (oriented upward).

3.2.2 Biosynthesis of Lactose

Lactose is essentially unique to mammary secretions. It is synthesized from glucose absorbed from blood. One molecule of glucose is converted to UDP-galactose via the 4-enzyme Leloir pathway (Figure 3–3). UDP-galactose is then linked to another molecule of glucose in a reaction catalyzed by the enzyme lactose synthetase, a 2-component enzyme. Component A is a nonspecific galactosyl transferase that transfers the galactose from UDP-galactose to a number of acceptors. In the presence of the B component, which is the whey protein α-lactalbumin, the transferase becomes highly specific for glucose (its K_M decreases 1,000-fold), leading to the synthesis of lactose. Thus, α-lactalbumin is an enzyme modifier, and its concentration in the milk of several species is directly related to the concentration of lactose in those milks; the milks of some marine mammals contain neither α-lactalbumin nor lactose.

The presumed significance of this control mechanism is to enable mammals to terminate the synthesis of lactose when necessary, that is, to regulate and control osmotic pressure when there is an influx of NaCl, such as during mastitis or in late lactation (milk is isotonic with blood, the osmotic pressure of which is essentially constant). The ability to control osmotic pressure is sufficiently important to justify an elaborate control mechanism and "wastage" of the enzyme modifier.

3.2.3 Lactose Equilibrium in Solution

The configuration around the anomeric C_1 of the glucose moiety of lactose is not stable and can readily change (mutarotate) from the α- to the β-form and vice versa when the sugar is in solution as a consequence of the fact that the hemiac-

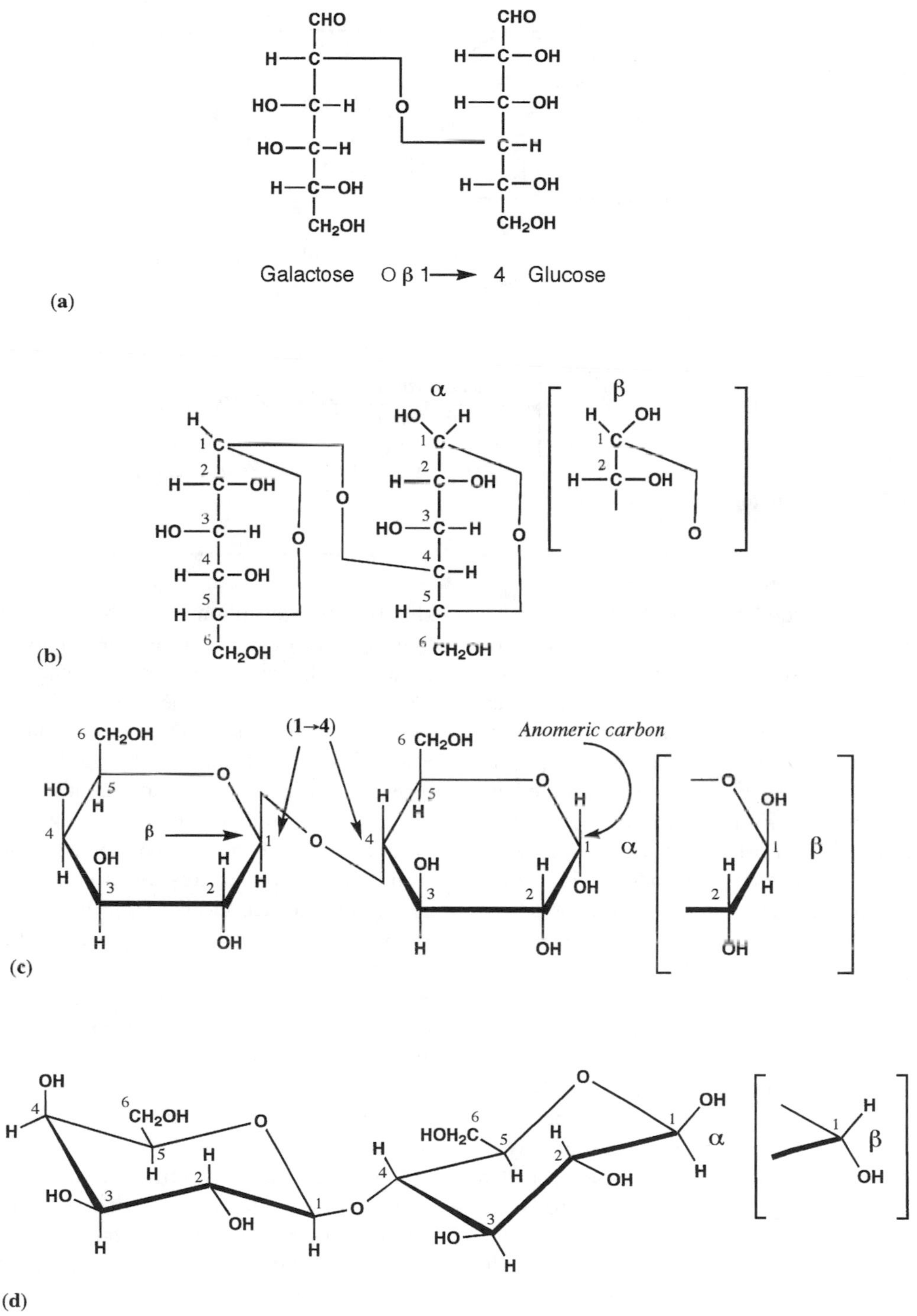

(a)

Galactose O β 1→ 4 Glucose

(b)

(c)

(d)

Figure 3–2 Structural formulae of α- and β-lactose: (a) open chains, (b) Fischer projection, (c) Haworth projection, and (d) conformational formula.

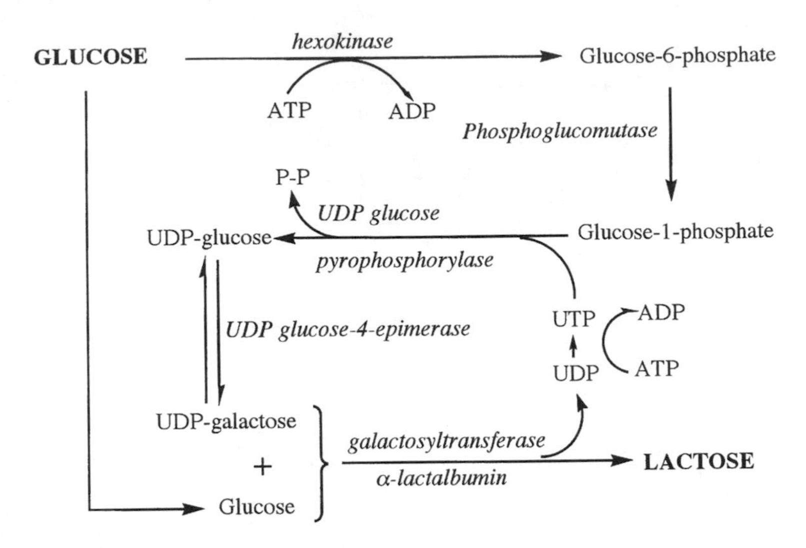

Figure 3–3 Pathway for lactose synthesis.

etal form is in equilibrium with the open-chain aldehyde form, which can be converted to either of the two isomeric forms (see Figure 3–2). When either isomer is dissolved in water, there is a gradual change from one form to the other until equilibrium is established (i.e., mutarotation). These changes are reflected by changes in optical rotation from +89.4° for α-lactose or +35° for β-lactose to a value of +55.4° at equilibrium. These values for specific rotation indicate that at equilibrium, a solution of lactose consists of 62.7% β anomer and 37.3% α anomer.

The α and β anomers of lactose differ markedly with respect to solubility, crystal shape, hydration of the crystals, hygroscopicity, specific rotation, and sweetness.

α-Lactose is soluble to around 7 g/100 ml H_2O at 20°C, and the solubility of β-lactose is around 50 g/100 ml. However, the solubility of α-lactose is more temperature dependent than that of β-lactose, and their solubility curves intersect at about 94°C (Figure 3–4). Thus, α-lactose is the form normally produced by crystallization. α-Lactose crystallizes as a monohydrate, whereas crystals of β-lactose are anhydrous. Although lactose has low solubility in comparison with other sugars, once in solution it crystallizes slowly, and precautions must be taken in the manufacture of concentrated and dehydrated products; otherwise, hygroscopicity, caking, and a grainy texture (due to the slow growth of lactose crystals to a size greater than 15 μm) will ensue. These physicochemical properties of lactose are of major concern to manufacturers of concentrated, dehydrated, and frozen dairy products, but problems can be avoided by proper manufacturing procedures. Such properties are of no consequence in cheese, in which all the lactose is utilized either during manufacture or early ripening; fresh curd contains about 1% lactose. The behavior of lactose is of major concern in the manufacture of whey powders, since around 70% of the total solids in whey are lactose, and hence the properties of whey concentrates and powders are strongly influenced by the properties of lactose.

In cheese, lactose is fermented to lactic acid by lactic acid bacteria, a process which has major, indeed vital, consequences for the manufacture and quality of cheese, as is discussed in Chapters 5, 10, and 11.

For further information on the properties of and the problems caused by lactose, the reader is referred to Fox (1985, 1997), Fox and

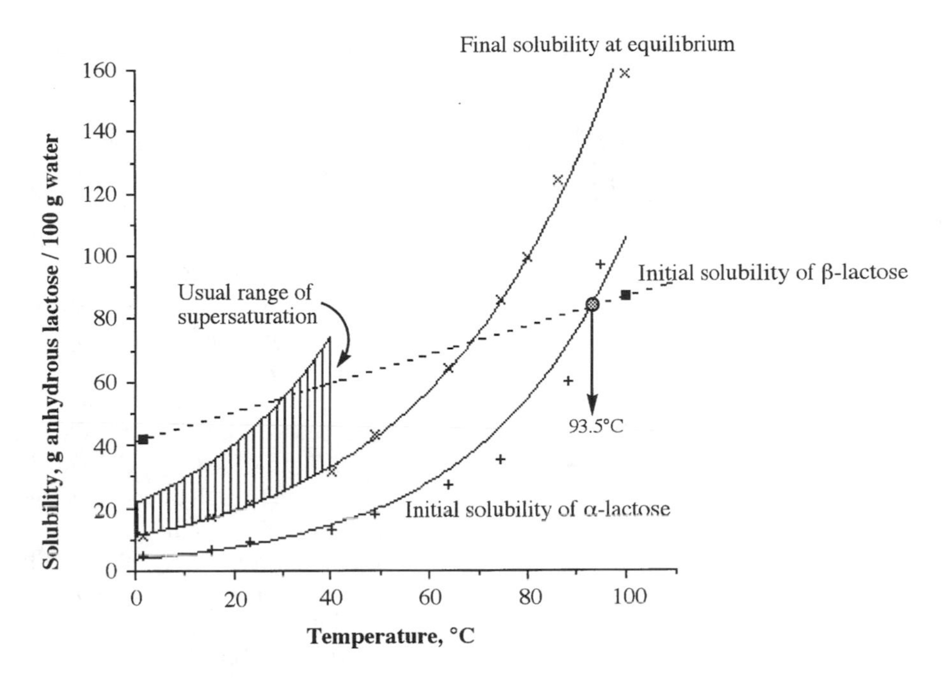

Figure 3–4 Solubility of lactose in water.

McSweeney (1998), Walstra and Jenness (1984), and Wong et al. (1988).

3.3 MILK LIPIDS

The lipid content of milk varies more widely than any other constituent; concentrations range from around 2% to more than 50% (Table 3–3). The average fat content of cow, goat, sheep, and buffalo milk is 3.5, 3.5, 6.5, and 7 g/L, respectively. Within any particular species, there are considerable variations due to breed, individuality, stage of lactation, age, animal health, nutritional status, interval between milking, and so on. Among the common breeds of dairy cattle, Jersey cows produce milk with the highest fat content (6–7%) and Holstein/Friesian, the lowest. Within any breed, there are considerable individual cow variations. The fat content of milk decreases for several weeks after parturition and then increases, especially toward the end of lactation (see Figure 3–1). If the interval between milkings is not equal, the milk obtained after the shorter interval has the higher fat content. The synthesis of all milk constituents, including fat, decreases during a mastitic infection, and the fat content of milk decreases slightly as the animal ages.

The lipids in milk are predominantly triglycerides (triacylglycerols), which make up about 98% of the total lipid fraction; the remaining 2% comprises diglycerides, monoglycerides, fatty acids, phospholipids, sterols (principally cholesterol), and trace amounts of fat-soluble vitamins (A, D, E, and K). Typical values for the concentration of the various lipids in milk are given in Table 3–4.

3.3.1 Fatty Acid Composition

Ruminant milk fats contain a greater diversity of fatty acids than other fats; about 400 fatty acids have been identified in bovine milk fat. The predominant fatty acids have a straight carbon chain with an even number of carbon atoms and may be saturated or unsaturated (1, 2, or 3 C = C

Table 3–3 Fat Content of Milks of Various Species

Species	*Fat Content (g/L)*
Cow	33–47
Buffalo	47
Sheep	40–99
Goat	41–45
Musk ox	109
Dall sheep	32–206
Moose	39–105
Antelope	93
Elephant	85–190
Human	38
Horse	19
Monkeys	10–51
Lemurs	8–33
Pig	68
Marmoset	77
Rabbit	183
Guinea pig	39
Snowshoe hare	71
Muskrat	110
Mink	134
Chinchilla	117
Rat	103
Red kangaroo	9–119
Dolphin	62–330
Manatee	55–215
Pygmy sperm whale	153
Harp seal	502–532
Bear (four species)	108–331

double bonds). There are smaller amounts of fatty acids with an uneven number of carbon atoms, branched or cyclic hydrocarbon chains, or hydroxyl or keto groups. The principal fatty acids in the milk fat of a selection of species are listed in Table 3–5.

The fatty acid profile of ruminant milk fats has a number of interesting features:

- Ruminant milk fats contain a considerable amount of butanoic acid ($C_{4:0}$) and are in fact the only fats that contain this acid. The high content of butanoic acid is due to the synthesis of 3-hydroxybutanoic acid (β-hydroxybutyric acid) and its reduction to butanoic acid by bacteria in the rumen. The high concentration of butanoic acid in ruminant milk fats provides the basis for the method commonly used to detect and quantify the adulteration of milk fat with other fats, that is, the Reichert-Meissl number (ml of 0.1 M KOH required to neutralize the volatile water-soluble fatty acids released from 5 g fat upon hydrolysis).
- Ruminant milk fats, in general, and ovine milk fat in particular, contain relatively high concentrations of middle chain fatty acids (hexanoic [$C_{6:0}$] to decanoic [$C_{10:0}$]). This is due to high thioacylhydrolase activity in the fatty acid synthetase complex, which causes the early release of fatty acids during the chain elongation process.
- The short and middle chain acids ($C_{4:0}$–$C_{10:0}$) are relatively volatile and water soluble and have a relatively low flavor threshold. They are esterified predominantly at the *sn*3 position of glycerol and hence are selectively released by lipases, especially by the indigenous lipoprotein lipase in milk. In milk and butter, the release of these highly flavored short chain fatty acids gives rise to off-flavors, referred to as *hydrolytic rancidity*. However, when present at an appropriate level, these short chain acids contribute positively to the flavor of cheese, especially hard Italian and blue-mold varieties.
- Ruminant milk fats contain low levels of polyunsaturated fatty acids (PUFAs; $C_{18:2}$, $C_{18:3}$), which are considered to be nutritionally desirable. However, the low level of PUFAs makes milk fat relatively resistant to *oxidative rancidity*. The low concentration of PUFAs in ruminant milk fats is due to the hydrogenation of dietary fatty acids by bacteria in the rumen, although ruminant feed usually contains high levels of PUFAs. On the positive side, biohydrogenation of PUFAs results in lower levels of *trans* isomers than chemical hydrogenation, such as is practiced in the processing of vegetable

Table 3–4 Composition of Individual Simple Lipids and Total Phospholipids in Milks of Various Species (Percentage of Total Lipids by Weight)

Lipid Class	*Cow*	*Buffalo*	*Human*	*Pig*	*Rat*	*Mink*
Triacylglycerols	97.5	98.6	98.2	96.8	87.5	81.3
Diacylglycerols	0.36	–	0.7	0.7	2.9	1.7
Monoacylglycerols	0.027	–	T*	0.1	0.4	T
Cholesteryl esters	T	0.1	T	0.06	–	T
Cholesterol	0.31	0.3	0.25	0.6	1.6	T
Free fatty acids	0.027	0.5	0.4	0.2	3.1	1.3
Phospholipids	0.6	0.5	0.26	1.6	0.7	15.3

* T = trace

oils. Trans fatty acids are considered to be nutritionally undesirable.

The concentration of PUFAs in ruminant milk fats can be increased by including protected lipids in the animal's diet. This involves encapsulating the dietary lipids in a layer of polymerized protein or using crushed vegetable seeds. Encapsulation protects the PUFAs against hydrogenation in the rumen, but the capsule is digested in the abomasum, liberating the encapsulated lipids, which are then metabolized, as in nonruminants. Fat has a major effect on the rheological properties of cheese. Polyunsaturated lipids, which have a low melting point, have an undesirable effect on cheese texture, but a low level is acceptable.

Although phospholipids are present at very low concentrations in milk, they play an important role in the emulsification of fat in milk. Milk contains a relatively low concentration of cholesterol, a high level of which in the diet is considered to be nutritionally undesirable. The cow transfers dietary carotenoids to its milk, and hence its milk fat has a yellow color, the intensity of which depends on the concentration of carotenoids in the animal's feed—fresh grass and especially clover and lucerne are rich in carotenoids (see Fox and McSweeney, 1998, for the structures of the principal phospholipids, cholesterol, and fat-soluble vitamins). Sheep and goats do not transfer dietary carotenoids to their milk, and consequently their milk fat and fat-containing products (including cheese) made from ovine or caprine milk are much whiter than their bovine counterparts. Products traditionally made from ovine or caprine milk may be unacceptable when made from bovine milk, owing to their yellow color, especially if the cows are fed on fresh grass. However, it is possible to bleach or mask the color of carotenoids (e.g., using H_2O_2, benzoyl peroxide, TiO_2, or chlorophyll).

Some carotenoids are precursors of vitamin A (retinol). Milk contains low levels of vitamin D, and liquid milk products are commonly fortified with vitamin D. Milk contains a substantial amount of vitamin E (tocopherols), which is a potent antioxidant. The tocopherol content of milk may be increased by supplementing the animal's diet with tocopherols; this may be done for nutritional or stability reasons. However, lipid oxidation is not a problem in cheese, probably because of its low redox potential (E_h: –150 mV).

3.3.2 Milk Fat as an Emulsion

Lipids are insoluble in and less dense than water (the specific gravity of fat and skim milk is around 0.9 and 1.036, respectively), and hence they would be expected to form a layer on the surface of milk. Lipids in general can be made compatible with water by forming an emulsion through homogenization, in which the fat is dispersed as small globules, each of which is surrounded by a layer of emulsifier. An emulsion is defined as a two-phase system, one phase (the

Table 3–5 Principal Fatty Acids in Milk Triacylglycerols or Total Lipids of Various Species (Percentage of Total by Weight)

Species	*4:0*	*6:0*	*8:0*	*10:0*	*12:0*	*14:0*	*16:0*	*16:1*	*18:0*	*18:1*	*18:2*	*18:3*	C_{20}–C_{22}
Cow	3.3	1.6	1.3	3.0	3.1	9.5	26.3	2.3	14.6	29.8	2.4	0.8	T*
Buffalo	3.6	1.6	1.1	1.9	2.0	8.7	30.4	3.4	10.1	28.7	2.5	2.5	T
Sheep	4.0	2.8	2.7	9.0	5.4	11.8	25.4	3.4	9.0	20.0	2.1	1.4	–
Goat	2.6	2.9	2.7	8.4	3.3	10.3	24.6	2.2	12.5	28.5	2.2	–	–
Musk ox	T	0.9	1.9	4.7	2.3	6.2	19.5	1.7	23.0	27.2	2.7	3.0	0.4
Dall sheep	0.6	0.3	0.2	4.9	1.8	10.6	23.0	2.4	15.5	23.1	4.0	4.1	2.6
Moose	0.4	T	8.4	5.5	0.6	2.0	28.4	4.3	4.5	21.2	20.2	3.7	–
Blackbuck antelope	6.7	6.0	2.7	6.5	3.5	11.5	39.3	5.7	5.5	19.2	3.3	–	–
Elephant	7.4	–	0.3	29.4	18.3	5.3	12.6	3.0	0.5	17.3	3.0	0.7	–
Human	–	T	T	1.3	3.1	5.1	20.2	5.7	5.9	46.4	13.0	1.4	T
Monkey (mean of six species)	0.4	0.6	5.9	11.0	4.4	2.8	21.4	6.7	4.9	26.0	14.5	1.3	–
Baboon	–	0.4	5.1	7.9	2.3	1.3	16.5	1.2	4.2	22.7	37.6	0.6	–
Lemur macaco	–	–	0.2	1.9	10.5	15.0	27.1	9.6	1.0	25.7	6.6	0.5	–
Horse	–	T	1.8	5.1	6.2	5.7	23.8	7.8	2.3	20.9	14.9	12.6	–
Pig	–	–	–	0.7	0.5	4.0	32.9	11.3	3.5	35.2	11.9	0.7	–
Rat	–	–	1.1	7.0	7.5	8.2	22.6	1.9	6.5	26.7	16.3	0.8	1.1
Guinea pig	–	T	–	–	–	2.6	31.3	2.4	2.9	33.6	18.4	5.7	T
Marmoset	–	–	–	8.0	8.5	7.7	18.1	5.5	3.4	29.6	10.9	0.9	7.0
Rabbit	–	T	22.4	20.1	2.9	1.7	14.2	2.0	3.8	13.6	14.0	4.4	T
Cottontail rabbit	–	–	9.6	14.3	3.8	2.0	18.7	1.0	3.0	12.7	24.7	9.8	0.4
European hare	–	T	10.9	17.7	5.5	5.3	24.8	5.0	2.9	14.4	10.6	1.7	T
Mink	–	–	–	–	0.5	3.3	26.1	5.2	10.9	36.1	14.9	1.5	–
Chinchilla	–	–	–	–	T	3.0	30.0	–	–	35.2	26.8	2.9	–
Red kangaroo	–	–	–	–	0.1	2.7	31.2	6.8	6.3	37.2	10.4	2.1	0.1
Platypus	–	–	–	–	–	1.6	19.8	13.9	3.9	22.7	5.4	7.6	12.2
Numbat	–	–	–	–	0.1	0.9	14.1	3.4	7.0	57.7	7.9	0.1	0.2
Bottle–nosed dolphin	–	–	–	–	0.3	3.2	21.1	13.3	3.3	23.1	1.2	0.2	17.3
Manatee	–	–	0.6	3.5	4.0	6.3	20.2	11.6	0.5	47.0	1.8	2.2	0.4
Pygmy sperm whale	–	–	–	–	–	3.6	27.6	9.1	7.4	46.6	0.6	0.6	4.5
Harp seal	–	–	–	–	–	5.3	13.6	17.4	4.9	21.5	1.2	0.9	31.2
Northern elephant seal	–	–	–	–	–	2.6	14.2	5.7	3.6	41.6	1.9	–	29.3
Polar bear	–	T	–	T	0.5	3.9	18.5	16.8	13.9	30.1	1.2	0.4	11.3
Grizzly bear	–	T	–	–	0.1	2.7	16.4	3.2	20.4	30.2	5.6	2.3	9.5

* T = trace.

discontinuous, dispersed, phase) being dispersed in the other (the continuous phase) and separated by a layer of emulsifier. In milk and cream, fat is the emulsified phase and water (or, more correctly, skim milk) is the continuous phase (i.e., milk is an oil-in-water emulsion). In butter (and margarine), the situation is reversed: water droplets are dispersed in a continuous oil/fat phase (i.e., butter is a water-in-oil emulsion).

Emulsifiers are amphipathic molecules, with hydrophobic (lipophilic, fat-loving) and hydrophilic (water-loving) domains. The principal natural emulsifiers are polar lipids and proteins; in addition, numerous synthetic emulsifiers are available and are used widely in the manufacture of high-fat foods.

In milk, the fat exists as globules, 0.1–20 μm in diameter (mean diameter, 3–4 μm). Numerically, most of the globules have a diameter less than 1 μm, but these small globules represent only a small fraction of the mass of milk fat. The globules are surrounded by a structured membrane, referred to as the milk fat globule membrane (MFGM), consisting mainly of phospholipids and proteins; the approximate composition of the MFGM is summarized in Table 3–6. The inner layers of the membrane are acquired within the secretory cell (mammocyte) as the fat globules move from the site of biosynthesis (i.e., the rough endoplasmic reticulum located toward the base of the cell) toward the apical membrane, through which they are expressed into the lumen of the mammary alveoli by exocytosis. During exocytosis, the fat globules become surrounded by the apical cell membrane, which therefore forms the outer layer of the MFGM of freshly secreted milk fat globules. However, much of this membrane, which has a typical trilaminar fluid mosaic structure, is lost as the milk ages, and much of it accumulates as lipoprotein particles, sometimes referred to as *microsomes*, in the skim milk phase.

Many of the indigenous enzymes in milk are constituents of the MFGM; consequently, isolated membrane (prepared by de-emulsification [churning] and washing) serves as the source material for the isolation of many indigenous milk enzymes. Xanthine oxidase is one of the principal proteins of the MFGM. Two notable exceptions are the principal indigenous proteinase plasmin and lipoprotein lipase (LPL), which are associated mainly with the casein micelles. The MFGM isolates and protects the triglycerides from LPL but if the membrane is damaged (e.g., by agitation), the enzyme and its substrate come into contact, and lipolysis and hydrolytic rancidity ensue, with undesirable consequences for the organoleptic quality of milk and many dairy products.

Although the milk fat emulsion is stable to phase separation, it does exhibit rapid creaming owing to the difference in density between the phases (i.e., the fat globules rise to the surface but remain discrete and can be redispersed by gentle agitation). The rate of creaming is governed by Stokes' law:

$$v = \frac{2r^2(\rho^1 - \rho^2)g}{9\eta}$$

Table 3–6 Gross Composition of the Milk Fat Globule Membrane

Component	*Amount in Fat Globule (mg 100 g⁻¹)*	*Amount in Fat Globule Surface (mg m⁻²)*	*Percentage of Total Membrane by Weight*
Protein	900	4.5	41
Phospholipid	600	3.0	27
Cerebrosides	80	0.4	3
Cholesterol	40	0.2	2
Neutral glycerides	300	1.5	14
Water	280	1.4	13
Total	2,200	11.0	100

where v is the velocity of particle movement; r is the radius of the globules; ρ^1 and ρ^2 are the densities of the continuous and dispersed phases, respectively; g is the acceleration due to gravity (9.8 m s^{-2}); and η is the viscosity coefficient of the emulsion.

For milk, the parameters of Stokes' law would suggest that a cream layer would form after about 60 hr, but in fact it forms in about 30 min. This large discrepancy between the actual and the predicted rate of creaming is due to flocculation of the fat globules: the large globules rise faster than and collide with smaller globules, and the globules form clusters owing to the agglutinating action of immunoglobulin M. This protein is referred to as a *cryoglobulin*, since it adsorbs onto the fat globules as the temperature is reduced. The cluster then rises as a unit, colliding with other globules as it does so and therefore rising at an accelerating rate. The cryoglobulins solubilize as the temperature is increased and are fully soluble above 37°C. Consequently, creaming is promoted by low temperatures and is very slow above 37°C. Cryoglobulins are denatured and inactivated by heating at time-temperature treatments greater than 74°C × 15 s; hence, severely pasteurized milk creams poorly or not at all. Sheep, goat, and buffalo milks are devoid of cryoglobulins and hence cream very slowly.

If the MFGM is physically damaged by high temperatures and/or agitation, the globules coalesce, and eventually phase inversion will occur (i.e., an oil-in-water emulsion is converted to a water-in-oil emulsion). Free (nonglobular) fat will float on the surface. Such damage occurs to at least some extent during cheesemaking; the free fat is not incorporated into the coagulum and floats as quite large masses on the surface of the whey and is lost to the cheese. About 10% of the fat in milk is normally lost in this way. It can be recovered from the whey by centrifugation and made into whey butter or other products.

Milk for many dairy products is "homogenized," usually by using a valve homogenizer. Homogenization reduces the size of the fat globules (average diameter, less than 1 μm) and denatures the cryoglobulins, and homogenized milk does not cream owing to the combined effects of globule size reduction and denaturation of cryoglobulins. The membrane on the fat globules in homogenized milk is mainly casein and does not protect the triglycerides against lipolysis. Homogenized milk must therefore be pasteurized before or immediately after homogenization to prevent the occurrence of hydrolytic rancidity.

Milk for cheesemaking is not normally homogenized, because homogenized milk forms a rennet coagulum (gel) with a lower tendency to undergo syneresis upon cutting or stirring than that from nonhomogenized milk. Homogenization results in cheese with a higher moisture content. This situation arises because the casein-coated fat globules behave somewhat like casein micelles, but they limit the contraction of the casein matrix. It may be advantageous to homogenize milk for low-fat cheese so as to obtain a higher moisture content and thus soften the texture of the cheese. In some cases, milk for Blue cheese is separated and the cream homogenized to promote lipolysis (which is desirable in Blue cheese). The lipolysed cream and skim milk are then combined and pasteurized before cheese manufacture. Milk for yogurt and cream cheese is also homogenized to

- prevent creaming during the relatively long gelation period
- increase the effective protein concentration by converting the fat globules to pseudo-protein particles, thereby giving a firmer gel for a given level of protein
- minimize syneresis

Fat plays an essential role in cheese quality:

- It acts as a plasticizer and affects cheese texture (low-fat cheese has a hard, crumbly texture).
- It serves as a source of fatty acids, which have a direct effect on cheese flavor and are changed to other flavor compounds (e.g., carbonyls, lactones, esters, and thioesters).
- It acts as a solvent for flavor compounds produced from lipids, proteins, or lactose.

With the objective of reducing the calorific content of cheese, there is considerable commer-

Table 3–7 Protein Content in the Milk of Various Species

Species	*Casein (%)*	*Whey Protein (%)*	*Total (%)*
Bison	3.7	0.8	4.5
Black bear	8.8	5.7	14.5
Camel (bactrian)	2.9	1.0	3.9
Cat	–	–	11.1
Cow	2.8	0.6	3.4
Domestic rabbit	9.3	4.6	13.9
Donkey	1.0	1.0	2.0
Echidna	7.3	5.2	12.5
Goat	2.5	0.4	2.9
Grey seal	–	–	11.2
Guinea pig	6.6	1.5	8.1
Hare	–	–	19.5
Horse	1.3	1.2	2.5
House mouse	7.0	2.0	9.0
Human	0.4	0.6	1.0
Indian elephant	1.9	3.0	4.9
Pig	2.8	2.0	4.8
Polar bear	7.1	3.8	10.9
Red kangaroo	2.3	2.3	4.6
Reindeer	8.6	1.5	10.1
Rhesus monkey	1.1	0.5	1.6
Sheep	4.6	0.9	5.5
White–tailed jack rabbit	19.7	4.0	23.7

cial interest in the production of low-fat cheeses, but the quality of such cheeses is reduced, and consequently they have had only limited marketability.

3.4 MILK PROTEINS

From a cheesemaking standpoint, the proteins of milk are its most important constituents. The protein content of milk shows large interspecies differences, ranging from about 1% for human milk to more than 20% for the milk of small mammals such as mice and rats (Table 3–7). There is a good correlation between the protein content of milk and the growth rate of the neonate of that species (Figure 3–5).

The proteins of milk belong to two main categories that can be separated based on their solubility at pH 4.6 at 20°C. Under these conditions, one of the groups precipitates; these are known as caseins. The proteins that remain soluble under these conditions are known as serum or whey proteins. Approximately 80% of the total nitrogen in bovine, ovine, caprine, and buffalo milks is casein, but casein constitutes only about 40% of the protein in human milk. Both caseins and whey proteins are heterogeneous and have very different molecular and physicochemical properties.

3.4.1 Caseins

Bovine casein consists of four types of protein with substantially different properties: α_{s1}–, α_{s2}–, β–, and κ–; these make up approximately 38%, 10%, 34%, and 15%, respectively, of whole casein. The caseins are well characterized at the molecular level (some of the major properties are summarized in Table 3–8), and the amino acid sequences are known (Figures 3–6 to 3–9). Some of the more important properties of the caseins are as follows:

- They are quite small molecules, with molecular masses of 20–25 kDa.
- All are phosphorylated. Most molecules of α_{s1}-casein contain 8 mol PO_4/mol of protein, but some contain 9 mol PO_4/mol. β-Casein molecules usually contain 5 mol PO_4/mol, but some contain 4 mol PO_4/mol. α_{s2}-Casein molecules contain 10, 11, 12, or 13 mol PO_4/mol. Most molecules of κ-casein contain 1 mol PO_4/mol, but some contain 2 or perhaps 3 mol PO_4/mol.
- The phosphate groups are esterified as monoesters of serine and most occur as clusters. The phosphate groups bind polyvalent cations strongly, causing charge neutralization and precipitation of α_{s1}-, α_{s2}-, and β-caseins at greater than 6 mM Ca^{2+} at 30°C. κ-Casein, which usually contains only 1 mol PO_4/mol, binds cations weakly and is not precipitated by them. It can stabilize up to 10 times its weight of calcium-sensitive caseins via the formation of micelles (see Section 3.4.2). In milk, the principal cation bound is calcium.

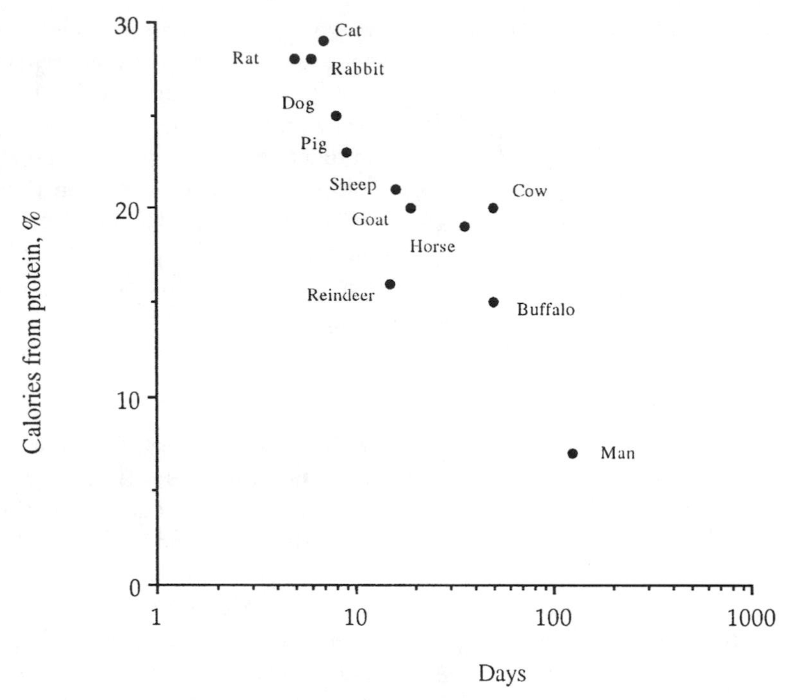

Figure 3–5 Relationship between the growth rate (days to double birthweight) of the young of some species of mammal and the protein content (expressed as the percentage of total calories derived from protein) of the milk of that species.

- Only α_{s2}- and κ-caseins contain cysteine, which normally exists as intermolecular disulphide bonds. α_{s2}-Casein usually occurs as disulphide-linked dimers, but up to at least 10 κ-casein molecules may be disulphide linked. The absence of cysteine or cystine in α_{s1}- and β-caseins increases the flexibility of these molecules.
- All the caseins, especially β-casein, contain relatively high levels of proline. In β-casein, 35 of the 209 residues are proline, and these are uniformly distributed throughout the molecule. The presence of a high level of proline prevents the formation of secondary structures (α-helices, β-sheets).
- Experimental techniques indicate that the caseins have low levels of secondary and tertiary structures, although theoretical calculations indicate that they do have some degree of higher structure. It has been suggested that, rather than lacking secondary structures, the caseins have very flexible structures, which have been described as *rheomorphic*. The lack of stable secondary and tertiary structures renders the caseins stable to denaturing agents (e.g., heat or urea); contributes to their surface activity properties; and makes them readily susceptible to proteolysis, which is important in cheese ripening.
- The caseins are relatively hydrophobic but have high surface hydrophobicity (owing to their open structures) rather than a high total hydrophobicity. The hydrophobic, polar, and charged residues are not uniformly distributed throughout the molecular sequences but occur as hydrophobic or hydrophilic patches (Figure 3–10), giving the caseins strongly amphiphatic structures that make them highly surface active. The N-terminal 2/3 of κ-casein, which is particularly significant in cheese manufacture,

Table 3–8 Amino Acid Composition of the Principal Proteins in Milk

Amino Acid	α_{s1}- *Casein B*	α_{s2}- *Casein A*	β- *Casein A^2*	κ- *Casein B*	γ^1- *Casein A^2*	γ^2- *Casein A^2*	γ^3- *Casein A*	*β–Lacto Globulin A*	*α–Lact Albumin B*
Asp	7	4	4	4	4	2	2	11	9
Asn	8	14	5	7	3	1	1	5	12
Thr	5	15	9	14	8	4	4	8	7
Ser	8	6	11	12	10	7	7	7	7
SerP	8	11	5	1	1	0	0	0	0
Glu	24	25	18	12	11	4	4	16	8
Gln	15	15	21	14	21	11	11	9	5
Pro	17	10	35	20	34	21	21	8	2
Gly	9	2	5	2	4	2	2	3	6
Ala	9	8	5	15	5	2	2	14	3
1/2 Cys	0	2	0	2	0	0	0	5	8
Val	11	14	19	11	17	10	10	10	6
Met	5	5	6	2	6	4	4	4	1
Ile	11	11	10	13	7	3	3	10	8
Leu	17	13	22	8	19	14	14	22	13
Tyr	10	12	4	9	4	3	3	4	4
Phe	8	6	9	4	9	5	5	4	4
Trp	2	2	1	1	1	1	1	2	4
Lys	14	24	11	9	10	4	3	15	12
His	5	3	5	3	5	4	3	2	3
Arg	6	6	4	5	2	2	2	3	1
PyroGlu	0	0	0	1	0	0	0	0	0
Total residues	199	207	169	209	181	104	102	162	123
Molecular weight	23,612	25,228	19,005	23,980	20,520	11,822	11,557	18,362	14,174
$H\Phi_{ave}$ (kJ/residue)	4.89	4.64	5.12	5.58	5.85	6.23	6.29	5.03	4.68

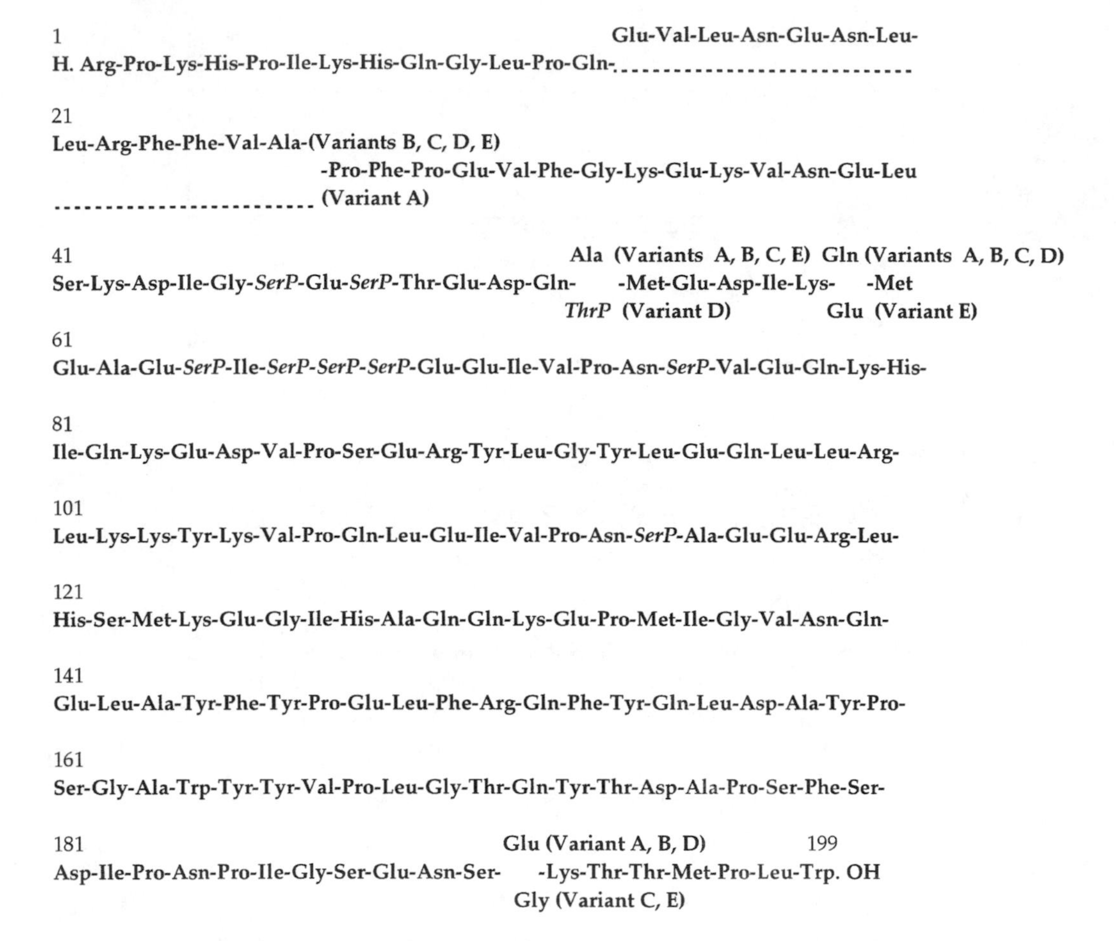

Figure 3–6 Amino acid sequence of α_{s1}-casein, showing the amino acid substitutions or deletions in the principal genetic variants.

is strongly hydrophobic, whereas the C-terminal 1/3 is strongly hydrophilic. The hydrophobicity of the caseins explains why their hydrolysates have a high propensity to bitterness, which is one of the principal defects in many cheese varieties.

- κ-Casein is glycosylated (α_{s1}-, α_{s2}-, and β-caseins are not). It contains galactose, galactosamine, and *N*-acetylneuraminic acid (sialic acid), which occur as either trisaccharides or tetrasaccharides attached to threonine residues in the C-terminal region. κ-Casein may contain 0 to 4 tri- or tetrasaccharides moieties (i.e., 10 variants of κ-casein exist). The presence of oligosaccharides attached to the C-terminal of κ-casein increases the hydrophilicity of that region.
- All the caseins exhibit genetic polymorphism that involves the substitution of 1 or 2 amino acids and rarely the deletion of a segment. The variant or variants present in milk are determined by simple Mendelian genetics. The presence of certain genetic variants in milk has a significant effect on the cheesemaking properties of the milk.

The preceding indicates that the casein system is extremely heterogeneous and that a logical nomenclature system is necessary. The follow-

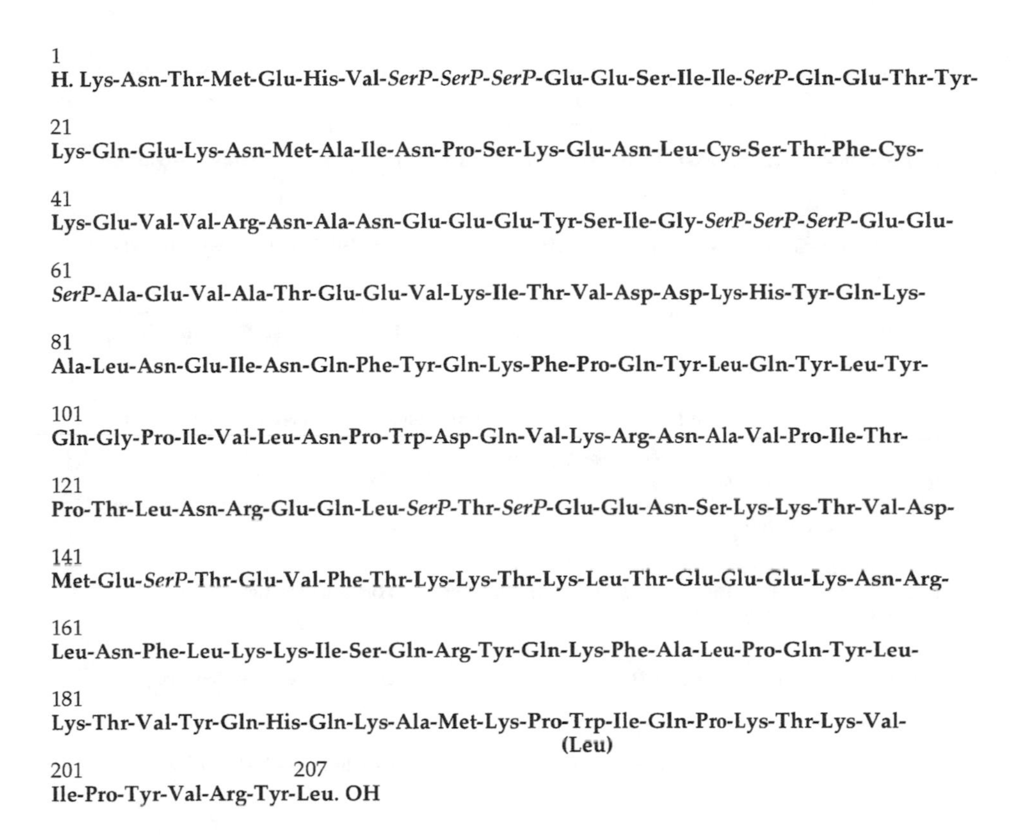
1
H. Lys-Asn-Thr-Met-Glu-His-Val-*SerP*-*SerP*-*SerP*-Glu-Glu-Ser-Ile-Ile-*SerP*-Gln-Glu-Thr-Tyr-

21
Lys-Gln-Glu-Lys-Asn-Met-Ala-Ile-Asn-Pro-Ser-Lys-Glu-Asn-Leu-Cys-Ser-Thr-Phe-Cys-

41
Lys-Glu-Val-Val-Arg-Asn-Ala-Asn-Glu-Glu-Glu-Tyr-Ser-Ile-Gly-*SerP*-*SerP*-*SerP*-Glu-Glu-

61
SerP-Ala-Glu-Val-Ala-Thr-Glu-Glu-Val-Lys-Ile-Thr-Val-Asp-Asp-Lys-His-Tyr-Gln-Lys-

81
Ala-Leu-Asn-Glu-Ile-Asn-Gln-Phe-Tyr-Gln-Lys-Phe-Pro-Gln-Tyr-Leu-Gln-Tyr-Leu-Tyr-

101
Gln-Gly-Pro-Ile-Val-Leu-Asn-Pro-Trp-Asp-Gln-Val-Lys-Arg-Asn-Ala-Val-Pro-Ile-Thr-

121
Pro-Thr-Leu-Asn-Arg-Glu-Gln-Leu-*SerP*-Thr-*SerP*-Glu-Glu-Asn-Ser-Lys-Lys-Thr-Val-Asp-

141
Met-Glu-*SerP*-Thr-Glu-Val-Phe-Thr-Lys-Lys-Thr-Lys-Leu-Thr-Glu-Glu-Glu-Lys-Asn-Arg-

161
Leu-Asn-Phe-Leu-Lys-Lys-Ile-Ser-Gln-Arg-Tyr-Gln-Lys-Phe-Ala-Leu-Pro-Gln-Tyr-Leu-

181
Lys-Thr-Val-Tyr-Gln-His-Gln-Lys-Ala-Met-Lys-Pro-Trp-Ile-Gln-Pro-Lys-Thr-Lys-Val-
(Leu)

201 207
Ile-Pro-Tyr-Val-Arg-Tyr-Leu. OH

Figure 3–7 Amino acid sequence of bovine α_{s2}-casein A, showing 11 of the 13 potential phosphorylation sites.

ing nomenclature has been adopted: The casein family is indicated by a Greek letter with a subscript, if necessary: α_{s1}-, α_{s2}-, β-, κ-. The Greek letter is followed by CN, and the genetic variant is indicated by a Latin letter (A, B, C, etc.) with a superscript, if necessary: α_{s1}-CN B, β-CN A^1. Finally, the number of phosphate residues is indicated: α_{s1}-CN B-8P, β-CN A^1-5P.

Minor components of the casein system are the γ-caseins, which are C-terminal fragments of β-casein produced through the action of the indigenous proteinase plasmin. The N-terminal fragments are included in the so-called proteose peptone fraction of milk protein. These peptides are summarized in Figure 3–11.

Ovine, caprine, and buffalo caseins generally exhibit heterogeneity similar to that of the bovine caseins.

3.4.2 Casein Micelles

As mentioned earlier, α_{s1}-, α_{s2}-, and β-caseins, which together constitute about 85% of whole casein, are precipitated by concentrations of Ca greater than 6 mM. Since bovine milk contains about 30 mM Ca, it might be expected that these caseins would precipitate in milk. However, κ-casein, which contains only 1 mol PO_4/mol, is insensitive to Ca^{2+} and, moreover, can stabilize up to 10 times its weight of the Ca-sensitive caseins against precipitation by Ca^{2+}. It does this via the formation of a type of quaternary structure, referred to as the *casein micelle*.

The principal properties of the casein micelle are summarized in Table 3–9. Many attempts have been made to elucidate the structure of the micelle. The most widely accepted view is that

1
H.Arg-Glu-Leu-Glu-Glu-LeuAsn-Val-Pro-Gly-Glu-Ile-Val-Glu-*SerP*-Leu*SerP*-*SerP*-*SerP*-Glu-

21 → γ_1-caseins (Variant C)
Glu-Ser-Ile-Thr-Arg-Ile-Asn-Lys-Lys-Ile-Glu-Lys-Phe-Gln- Ser/*SerP* -Glu- Lys/Glu -Gln-Gln-Gln-
(Variants A, B)

41
Thr-Glu-Asp-Glu-Leu-Gln-Asp-Lys-Ile-His-Pro-Phe-Ala-Gln-Thr-Gln-Ser-Leu-Val-Tyr-

61
Pro (Variants A^2, A^3)
Pro-Phe-Pro-Gly-Pro-Ile- -Asn-Ser-Leu-Pro-Gln-Asn-Ile-Pro-Pro-Leu-Thr-Gln-Thr
His (Variants C, A^1, and B)

81
Pro-Val-Val-Val-Pro-Pro-Phe-Leu-Gln-Pro-Glu-Val-Met-Gly-Val-Ser-Lys-Val-Lys-Glu-

→ γ_3–caseins
101(Variants A^1, A^2, B, C) His
Ala-Met-Ala-Pro-Lys- -Lys-Glu-Met-Pro-Phe-Pro-Lys-Tyr-Pro-Val-Glu-Pro-Phe-Thr-
(Variant A^3) Gln

121 Ser (Variants A, C) → γ_2–caseins
Glu- -Gln-Ser-Leu-Thr-Leu-Thr-Asp-Val-Glu-Asn-Leu-His-Leu-Pro-Leu-Pro-Leu-Leu-
Arg (Variant B)

141
Gln-Ser-Trp-Met-His-Gln-Pro-His-Gln-Pro-Leu-Pro-Pro-Thr-Val-Met-Phe-Pro-Pro-Gln-

161
Ser-Val-Leu-Ser-Leu-Ser-Gln-Ser-Lys-Val-Leu-Pro-Val-Pro-Gln-Lys-Ala-Val-Pro-Tyr-

181
Pro-Gln-Arg-Asp-Met-Pro-Ile-Gln-Ala-Phe-Leu-Leu-Tyr-Gln-Glu-Pro-Val-Leu-Gly-Pro-

201 209
Val-Arg-Gly-Pro-Phe-Pro-Ile-Ile-Val.OH

Figure 3–8 Amino acid sequence of bovine β-casein, showing the amino acid substitutions in the genetic variants and the principal plasmin cleavage sites (▼).

---

the micelles are composed of submicelles of mass around 5 × 10^6 kDa. The core of the submicelles is considered to consist of the Ca-sensitive α_{s1}-, α_{s2}-, and β-caseins, with variable amounts of κ-casein located principally on the surface of the submicelles. The κ-casein-deficient submicelles are located in the center of the micelles, and the κ-casein-rich submicelles are concentrated at the surface. The hydrophobic N-terminal segment of κ-casein is considered to interact hydrophobically with the Ca-sensitive caseins, with the hydrophilic C-terminal segment protruding from the surface, giving the whole micelle a hairy appearance (Figures 3–12 and 3–13). The colloidal stability of the micelles is attributed to a zeta potential of about -20 mV at 20°C and the steric stabilization provided by the protruding hairs. The submicelles are considered to be held together by microcrystals of calcium phosphate and perhaps hydrophobic and hydrogen bonds.

Although this model of the casein micelle is not universally accepted, it is adequate to explain many of the technologically important properties of the micelles, including rennet coagulation, which follows the specific hydrolysis

1
Pyro- Glu-Glu-Gln-Asn-Gln-Glu-Gln-Pro-Ile-Arg-Cys-Glu-Lys-Asp-Glu-Arg-Phe-Phe-Ser-Asp-

21
Lys-Ile-Ala-Lys-Tyr-Ile-Pro-Ile-Gln-Tyr-Val-Leu-Ser-Arg-Tyr-Pro-Ser-Tyr-Gly-Leu-

41
Asn-Tyr-Tyr-Gln-Gln-Lys-Pro-Val-Ala-Leu-Ile-Asn-Asn-Gln-Phe-Leu-Pro-Tyr-Pro-Tyr-

61
Tyr-Ala-Lys-Pro-Ala-Ala-Val-Arg-Ser-Pro-Ala-Gln-Ile-Leu-Gln-Trp-Gln-Val-Leu-Ser-

81
Asn-Thr-Val-Pro-Ala-Lys-Ser-Cys-Gln-Ala-Gln-Pro-Thr-Thr-Met-Ala-Arg-His-Pro-His-

101 105▼106
Pro-His-Leu-Ser-Phe▼Met-Ala-Ile-Pro-Pro-Lys-Lys-Asn-Gln-Asp-Lys-Thr-Glu-Ile-Pro-

121 Ile (Variant B)
Thr-Ile-Asn-Thr-Ile-Ala-Ser-Gly-Glu-Pro-*Thr*-Ser-*Thr*-Pro-*Thr*- -Glu-Ala-Val-Glu-
Thr (Variant A)

141 Ala (Variant B)
Ser-Thr-Val-Ala-Thr-Leu-Glu- - *SerP*-Pro-Glu-Val-Ile-Glu-Ser-Pro-Pro-Glu-Ile-Asn-
Asp (Variant A)

161 169
Thr-Val-Gln-Val-Thr-Ser-Thr-Ala-Val. OH

Figure 3–9 Amino acid sequence of bovine κ-casein, showing the amino acid substitutions in genetic polymorphs A and B and the chymosin cleavage site, (▼). The sites of posttransitional phosphorylation or glycosylation are italicized.

---

of the micelle-stabilizing κ-casein, as a result of which the stabilizing surface layer is lost.

As far as is known, the structure of the casein micelles in bovine, ovine, caprine, and buffalo milks is essentially similar.

3.4.3 Whey Proteins

The whey protein fraction of bovine, ovine, caprine, and buffalo milk contains four main proteins: β-lactoglobulin (β-lg, 50%), α-lactalbumin (α-la, 20%), blood serum albumin (BSA, 10%), and immunoglobulins (Ig, 10%; mainly IgG_1, with lesser amounts of IgG_2, IgA, and IgM). Human milk contains no β-lg and the principal Ig is IgA.

The principal properties of the whey proteins are listed in Table 3–8. In contrast to the caseins, the whey proteins possess high levels of secondary, tertiary, and quaternary structures. They are typical globular proteins and are denatured upon heating (e.g., completely at 90°C × 10 min). They are not phosphorylated and are insensitive to Ca^{2+}. All whey proteins contain intramolecular disulfide bonds that stabilize their structure. β-Lg contains one sulfydryl group that under certain conditions can undergo sulfydryl-disulfide interactions with other proteins; the most important of these interactions, with κ-casein, occurs upon heating at about 75°C × 15 s. The latter can markedly impair the rennet coagulation properties of milk and alter the gel structure and rheological and synertic properties of acid gel–based products such as yogurt and fresh cheeses.

The whey proteins are not directly involved in cheese manufacture. However, they are indirectly involved, as in these examples:

- Heat-induced interaction of whey proteins with κ-casein has undesirable effects on rennet coagulation.

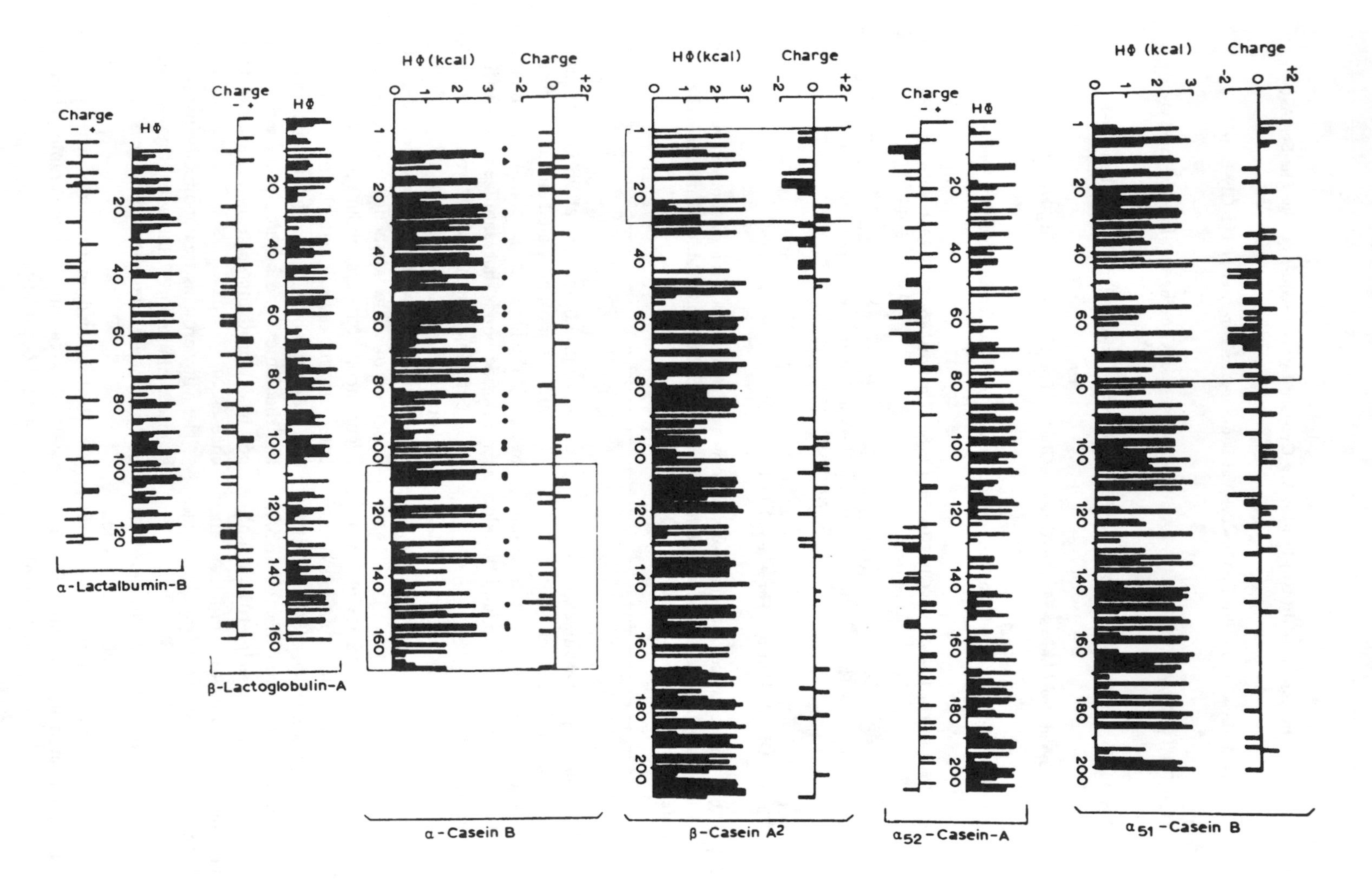

Figure 3–10 Schematic representation of the distribution of hydrophobic and charged residues in the principal milk proteins.

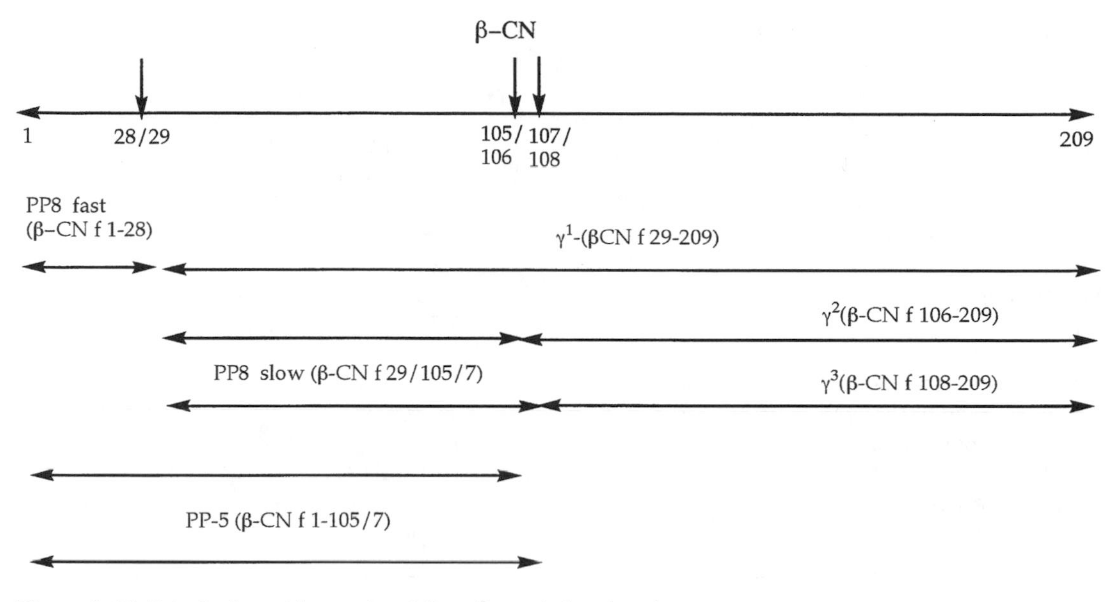

Figure 3–11 Principal peptides produced from β-casein by plasmin.

- Whey proteins are incorporated into cheese made from milk concentrated by ultrafiltration.
- Whey proteins are heat-denatured in the manufacture of some Quarg products.
- Valuable functional proteins are recovered from whey.

3.4.4 Minor Proteins

Milk contains numerous minor proteins. These are found mainly in the whey, but some are also found in the fat globule membrane. These minor proteins include enzymes (perhaps 60), enzyme inhibitors, metal-binding proteins (especially lactoferrin and osteopontin), vitamin-binding proteins, and several growth factors. As far as is known, most of these are of no consequence in cheese. Some of the indigenous enzymes are active in cheese during ripening, especially plasmin and xanthine oxidase and possibly acid phosphatase. Lipoprotein lipase is probably quite important in raw milk cheese and perhaps even in pasteurized milk cheese, since some probably partially survives HTST (high temperature, short time) pasteurization. The significance of other indigenous enzymes in cheese has not been investigated and perhaps warrants study.

3.5 MILK SALTS

After milk has been heated in a muffle furnace at around 600°C for 5 hr, a residue (ash), representing roughly 0.7 g/100 ml of the mass of the milk sample, remains. The ash contains the inorganic salts present in the original milk plus some elements, especially phosphorus, present originally in organic molecules, especially proteins and phospholipids, and lesser amounts of sugar phosphates and high-energy phosphates. The elements in the ash are changed from their original form; they are present, not as their original salts, but as oxides and carbonates. Organic salts, the most important of which is citrate, are lost on ashing. Fresh milk does not contain lactic acid, but lactic acid may be present in stored milk as a result of microbial growth. Although the salts of milk are quantitatively minor constituents, they are of major significance to its technological properties.

The typical concentration of the principal elements or compounds that constitute the salts of

Table 3–9 Average Characteristics of Casein Micelles

Characteristic	*Value*
Diameter	120 nm (range: 50–500 nm)
Surface area	8×10^{-10} cm^2
Volume	2.1×10^{-15} cm^3
Density (hydrated)	1.0632 g cm^{-3}
Mass	2.2×10^{-15}g
Water content	63%
Hydration	3.7 g H_2O g^{-1} protein
Voluminosity	4.4 cm^3 g^{-1}
Molecular weight (hydrated)	1.3×10^9 Da
Molecular weight (dehydrated)	5×10^8 Da
Number of peptide chains	10^4
Number of particles per milliliter of milk	10^{14}–10^{16}
Surface area of micelles per milliliter of milk	5×10^4 cm^2
Mean free distance	240 nm

milk are summarized in Table 3–10. Some of the salts are present in milk at concentrations below their solubility limit and are therefore fully soluble. However, others, especially calcium phosphate, exceed their solubility and occur partly in solution and partly in the colloidal phase, associated mainly with the casein micelles. These salts are collectively referred to as micellar or colloidal calcium phosphate (CCP), although several other elements or ions are present also. Several elements are also present in the MFGM, mainly as constituents of enzymes. There are several techniques for partitioning the colloidal and soluble salts (see Fox & McSweeney, 1998). Typical distributions are indicated in Table 3–10.

It is possible to either determine experimentally or to calculate (after making certain assumptions) the concentration of the principal ions in milk; these are also indicated in Table 3–10.

From a cheesemaking viewpoint, the most important salts or ions are calcium, phosphate, and, to a lesser extent, citrate. As shown in Table 3–10, bovine milk contains about 1200 mg Ca/L

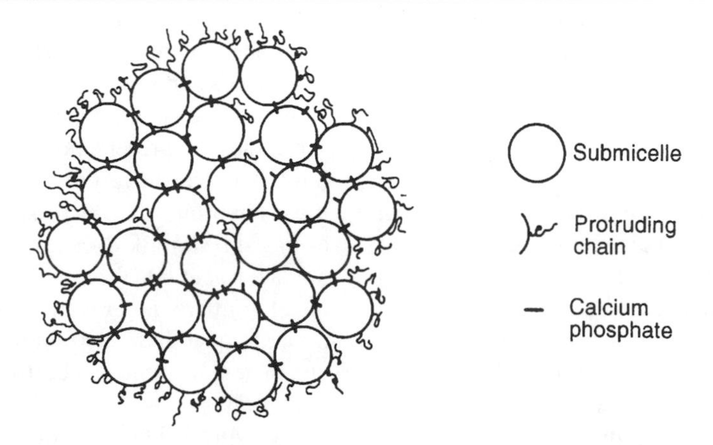

Figure 3–12 Submicelle model of the casein micelle.

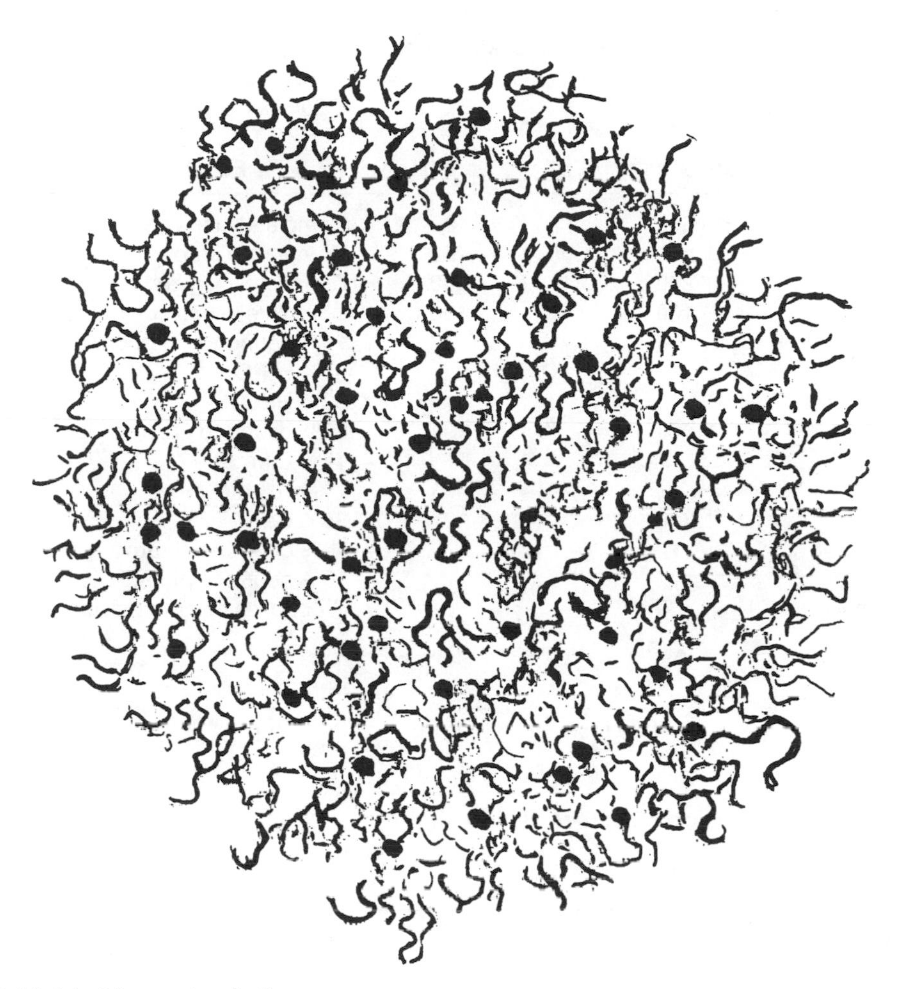

Figure 3–13 Model of the casein micelle.

(i.e., 30 mM). About 30% is soluble, most of which occurs as un-ionized salts of citrate, but about 30% exists as Ca^{2+}, which means that 10% of the total calcium exists as Ca^{2+} (2–3 mM). Although present at low concentrations, Ca^{2+} are of major significance in various aspects of the rennet coagulation of milk (see Chapter 6). The $[Ca^{2+}]$ is inversely related to the citrate concentration.

The insoluble calcium occurs mainly associated with the casein micelles, either as colloidal calcium phosphate (CCP) or casein Ca. CCP plays a major role in micellar integrity and has a very significant role in rennet coagulation. The precise composition and structure of CCP are not known. The simplest possible structure is tertiary phosphate, $Ca_3(PO_4)_2$, but the form for which the best experimental evidence exists is brushite, $CaHPO_4.2H_2O$, which forms microcrystals with organic casein phosphate.

3.6 pH OF MILK

As will become apparent in subsequent chapters, pH is a critical factor in several aspects of the manufacture and ripening of cheese curd.

Table 3–10 Concentration and Partition of Milk Salts

		Soluble		
Species	*Concentration (mg/L)*	*Percentage*	*Form*	*Colloidal (%)*
Sodium	500	92	Completely ionized	8
Potassium	1,450	92	Completely ionized	8
Chloride	1,200	100	Completely ionized	–
Sulphate	100	100	Completely ionized	–
Phosphate	750	43	10% bound to Ca and Mg 51% $H_2PO_4^-$ 39% HPO_4^{2-}	57
Citrate	1,750	94	85% bound to Ca and Mg 14% $Citrate^{3-}$ 1% $H.citrate^{2-}$	
Calcium	1,200	34	35% Ca^{2+} 55% bound to citrate 10% bound to phosphate	66
Magnesium	130	67	Probably similar to calcium	33

Table 3–11 Some Physical Properties of Milk

Property	*Value*
Osmotic pressure	≈ 700 kPa
Water activity, a_w	≈ 0.993
Boiling point	≈ 100.15°C
Freezing point	–0.522°C (approximately)
Redox potential, E_h (in equilibrium with air at 25°C and pH 6.6)	+0.25 to +0.35 V
Refractive index, n_D^{20}	1.3440 to 1.3485
Specific refractive index	≈ 0.2075
Density (20°C)	≈1030 kg × m^{-3}
Specific gravity (20°C)	≈ 1.0321
Specific conductance	≈ 0.0050 ohm^{-1} cm^{-1}
Ionic strength	≈ 0.07 M
Surface tension (20°C)	≈ 52 N m^{-1}
Coefficient of viscosity	2.127 mPa × s
Thermal conductivity (2.9% fat)	≈ 0.559 W • m^{-1} • K^{-1}
Thermal diffusivity (15–20°C)	≈ 1.25 × 10^{-7} m^2 • s^{-1}
Specific heat	≈ 3.931 kJ • kg^{-1} • K^{-1}
pH (at 25°C)	≈ 6.6
Titratable acidity	1.3–2.0 meq OH^- per 100 ml (0.14–0.16% as lactic acid)
Coefficient of cubic expansion (273–333 K)	0.0008 m^3 • m^{-3} • K^{-1}

The pH of milk at 25°C is usually in the range 6.5 to 7.0, with a mean value of 6.6. pH increases with advancing lactation and may exceed 7.0 in very late lactation; colostrum can have a pH as low as 6.0. The pH increases during mastitic infection owing to the increased permeability of the mammary gland membranes, which permits greater influx of blood constituents into the milk (the pH of cow's blood is 7.4). The difference in pH between blood and milk results from the active transport of various ions into the milk; precipitation of CCP, which results in the release of H^+ during the synthesis of casein micelles; higher concentrations of acidic groups in milk; and the relatively low buffering capacity of milk between pH 6.0 and 8.0.

One of the key events during the manufacture of cheese is the production of lactic acid from lactose by lactic acid bacteria (see Chapter 5). Consequently, the pH decreases to about 5.0. While lactic acid is primarily responsible for the decrease in pH, the actual pH attained is strongly affected by the buffering capacity of the milk and curd.

Milk contains a range of groups that are effective in buffering over a wide pH range. The principal buffering compounds in milk are its salts (particularly soluble phosphate, citrate, and bicarbonate) and acidic and basic amino acid side-chains of proteins (particularly the caseins). The contribution of these components to the buffering of milk is discussed in detail by Singh, McCarthy, and Lucey (1997).

The buffering capacity of milk and curd is of significance during cheesemaking, since it is the factor that determines the rate of decrease in pH caused by the production of lactic acid by the starter. The buffering capacity of milk is low near its natural pH but increases rapidly to a maximum at about pH 5.1. This means that, given a steady rate of acidification by the starter, the pH of milk decreases rapidly initially and later slows down. Since all of the soluble and some of the colloidal calcium phosphate are lost in the whey, it is not surprising that the buffering properties of cheese differ from those of milk. Cheddar and Emmental cheeses have maximum buffering capacities at around pH 4.8.

3.7 PHYSICOCHEMICAL PROPERTIES OF MILK

Information on the physicochemical properties of milk is important when developing and processing dairy products, designing processing equipment, and using dairy products in food products. Some of the principal physicochemical properties are summarized in Table 3–11.

REFERENCES

Cayot, P., & Lorient, D. (1998). *Structures et techno-fonctions des proteins du lait.* Paris: Lavoisier.

Fox, P.F. (Ed.). (1982). *Developments in dairy chemistry: Vol. 1. Proteins.* London: Applied Science Publishers.

Fox, P.F. (Ed.). (1983). *Developments in dairy chemistry: Vol. 2. Lipids.* London: Applied Science Publishers.

Fox, P.F. (Ed.). (1985). *Developments in dairy chemistry: Vol. 3. Lactose and minor constituents.* London: Elsevier Applied Science Publishers.

Fox, P.F. (Ed.). (1989). *Developments in dairy chemistry: Vol. 4. Functional proteins.* London: Elsevier Applied Science Publishers.

Fox, P.F. (Ed.). (1992). *Advanced dairy chemistry, Vol. 1. Milk proteins.* London: Elsevier Applied Science Publishers.

Fox, P.F. (Ed.). (1995). *Advanced dairy chemistry, Vol. 2. Lipids* (2d ed.). London: Chapman & Hall.

Fox, P.F. (Ed.). (1997). *Advanced dairy chemistry, Vol. 3. Lactose, water, salts and vitamins.* London: Chapman & Hall.

Fox, P.F., & McSweeney, P.L.H. (1998). *Dairy chemistry and biochemistry.* London: Chapman & Hall.

Jenness, R., & Patton, S. (1959). *Principles of dairy chemistry.* New York: Wiley.

Jensen, R.G. (Ed.). (1995). *Handbook of milk composition.* San Diego, CA: Academic Press.

Singh, H., McCarthy, O.J., & Lucey, J.A. (1997). Physico-chemical properties of milk. In P.F. Fox (Ed.), *Advanced dairy chemistry: Vol. 3. Lactose, water, salts and vitamins.* London: Chapman & Hall.

Walstra, P., & Jenness, R. (1984). *Dairy chemistry and physics.* New York: Wiley.

Webb, B.H., & Johnson, A.H. (Eds.). (1965). *Fundamentals of dairy chemistry*. Westport, CT: AVI Publishing.

Webb, B.H., Johnson, A.H., & Alford, J.A. (Eds.). (1974). *Fundamentals of dairy chemistry* (2d ed.). Westport, CT: AVI Publishing.

Wong, N.P., Jenness, R., Keeney, M., & Marth, E.H. (Eds.). (1988). *Fundamentals of dairy chemistry* (3d ed.). Westport, CT: AVI Publishing.

CHAPTER 4

Bacteriology of Cheese Milk

4.1 CONTAMINATION OF RAW MILK

The pH of milk (around 6.6), its temperature in the udder (around 38°C), and its high nutritional value are ideal for the growth of bacteria. However, bacteria growth does not usually occur because milk in the udder is sterile, unless the udder is infected. Bacteria can colonize the teat canal but are expelled in the first few squirts of milk. However, during milking, the milk becomes contaminated with microorganisms, mainly from the milking equipment, and it will, if maintained at a temperature above 15°C for several hours, coagulate due to the production of acid by adventitious bacteria, such as lactic acid bacteria (LAB) and coliforms. Therefore, great care must be taken to ensure that milk is produced hygienically. Today, there is no difficulty in producing milk with less than 5,000 colony-forming units (cfu) per ml, whereas 30 years ago it was difficult to produce it with less than 100,000 cfu/ml. These improvements in the microbial quality of raw milk are due to better hygiene during milking; improved design of milking equipment, making it easier to clean; cooling of the milk to a temperature less than 5°C within a few hours of production; and holding the milk at less than 5°C in easily cleaned, stainless steel bulk-storage tanks until it is collected and transported to the factory.

The sources of microorganisms in milk include the udder (unhealthy animals and the teat canal), the exterior of the udder (outside surfaces of the teats and udder), the bedding on which the cow lies, food eaten by the cow, the milker, the air, water used to wash the udder, and the milking and storage equipment.

Cows suffering from diseases like salmonellosis, tuberculosis, and brucellosis may shed the bacteria that cause these diseases into their milk. Normally, milk from cows infected with these diseases is not a major source of bacteria in raw milk. Mastitis is a bacterial infection of the mammary gland and is common in dairy cows. It can occur in subclinical (the majority of outbreaks) or clinical forms and is caused mainly by *Staphylococcus aureus*, although *Streptococcus dysgalactiae, Sc. agalactiae*, *Escherichia coli*, and *Corynebacterium* spp. may also be responsible. *S. aureus* is a Gram-positive coccus, and many strains produce heat-stable toxins, termed *enterotoxins*, which can cause food poisoning. Generally, growth to around 10^6 cfu/ml is necessary before sufficient toxin is produced to cause food poisoning. In subclinical mastitis, no physical change or abnormality is evident in the milk, whereas in clinical mastitis, large clots consisting of a mixture of milk, somatic cells, and bacteria are produced. Subclinical mastitis is generally manifested by increased numbers of polymorphonuclear leucocyte (PMNs) cells in the milk. In clinical mastitis, the number of bacteria in the milk varies depending on whether phagocytosis (i.e., engulfment by the PMNs) has occurred. At the beginning of infection, before phagocytosis has occurred, several million bacteria per ml may be present, but as phagocytosis develops, the numbers present will decrease rapidly to perhaps less than 1,000 bacteria per ml. In other words, bacterial numbers are high at the beginning of infection but very low as the infection progresses.

Small numbers of microorganisms can enter the teat canal of healthy animals from the outside of the teat but are generally washed out in the first squirts of milk. Less than 200 cfu/ml are probably added to the milk from the teat canal during milking.

During milking, milk can also become contaminated with bacteria from the air, the outside of the udder, the bedding, the feed, and the milker, but these are generally minor sources of contamination. However, extremely dirty udders may contaminate milk with up to 10^5 cfu/ml. Dirty udders are more likely to occur in winter, when the cows are housed, than in summer, when the cows are on pasture. Therefore, it is important to wash the udders and teats thoroughly before milking.

The major source of contamination of raw milk is improperly cleaned milking equipment. For this reason, considerable emphasis is placed on the satisfactory cleaning of the milking machine, its associated rubber hoses and pipework, and the bulk-storage tank. The machine should be cleaned after each milking, and the bulk-storage tank after it has been emptied. Hot and cold detergent washes are used and generally a hot acid rinse is given once a week to prevent the build up of "milk stone," which can harbor bacteria and make the equipment difficult to clean. Milk stone is composed mainly of calcium phosphate, but sufficient nutrients may also be present to allow significant microbial growth between milkings if the ambient temperature is greater than 15°C. Only very heavily contaminated milking equipment will cause a marked increase in the bacterial count in the raw milk. For example, 1 million organisms are required to increase the bacterial count of 1,000 L of milk by 1 bacterium per ml; therefore, to increase the count by 10,000/ml would require the addition of 10^{10} bacteria. The milking machine and its associated pipe lines and rubber hoses have a large surface area and may harbor such large numbers of bacteria if they are not adequately cleaned. Residues of milk left on the equipment after inadequate cleaning may contain sufficient nutrients to sustain bacterial growth at ambient temperatures. Immediately after milking, good quality milk produced using properly cleaned milking machines and bulk-storage tanks should have a count of less than 5,000 cfu/ml.

Gram-positive bacteria (e.g., *Micrococcus*, *Corynebacterium*, *Microbacterium*, *Lactobacillus*, *Lactococcus*, *Enterococcus*, etc.) and Gram-negative bacteria (*Pseudomonas*, *Achromobacter*, *Enterobacter*, *Escherichia*, *Flavobacterium*, etc.) are found in milk immediately after milking. In the past, milk was either not cooled at all or cooled to ambient temperature with water. Under these conditions, the growth of Gram-positive bacteria, particularly LAB, such as *Lactobacillus*, *Lactococcus*, *Enterococcus*, and *Streptococcus* spp., was more common than the growth of Gram-negative microorganisms. Many of the Gram-positive genera include cheese starter bacteria. Nowadays, milk is normally cooled to less than 5°C within 1–2 hr of milking, and the flora has changed from one dominated by Gram-positive bacteria to one dominated by Gram-negative, psychrotrophic bacteria, particularly of the genera *Pseudomonas* and *Achromobacter*. Psychrotrophs are generally defined as bacteria capable of growing at temperatures under 7°C.

Cooling significantly slows down the rate of multiplication of bacteria in raw milk (Figure 4–1). However, slow growth of bacteria, particularly psychrotrophs, still occurs at 4°C, and significant numbers (e.g., 10^6 or 10^7 cfu/ml) can often be reached in 3 or 4 days' storage on the farm. Raw milk may also be stored in silos for 1 or 2 days at the factory before use, during which further growth of psychrotrophs will occur. It is more important to use properly cleaned milking equipment than to cool the milk rapidly. In other words, rapid cooling of milk will not compensate

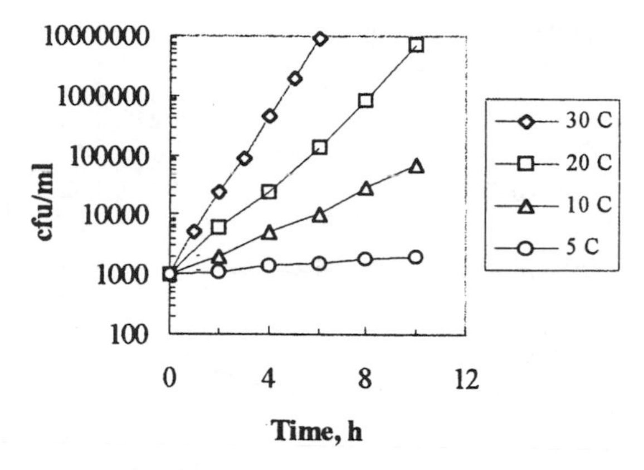

Figure 4–1 Effect of temperature on the growth of bacteria in a sample of raw milk.

for improperly cleaned milking machines and storage tanks, either on the farm or at the factory.

The growth of bacteria in four raw milks during storage at 5°C is shown in Figure 4–2. Little or no growth occurred during the first 2 days of storage, after which two of the milks showed a significant increase in bacterial numbers and the other two did not. This difference was probably due to differences in the species of organisms present and their ability to grow at 5°C. It is also noticeable that the initial level of contamination had little effect on the subsequent rates of bacterial growth. After 4 days at 5°C, counts in some cases exceeded 10^7 cfu/ml. Counts of this magnitude would be totally unacceptable in milk for cheesemaking—or indeed for making any dairy product.

Today, most raw milk received by cheese factories in developed countries has a viable count of less than 50,000 cfu/ml. The milk is often stored in large silos for perhaps 24 hr or longer, and therefore further growth and contamination from improperly cleaned silos can occur, so that milk for cheesemaking may have counts in excess of 10^5 cfu/ml. Such counts, while high, will not have a major effect on cheese quality. However, counts greater than 10^6 cfu/ml in the milk before pasteurization could affect cheese quality, because many psychrotrophs, especially *Pseudomonas* spp., produce heat-stable lipases and proteinases, which withstand heating to 100°C for 30 min, although the bacteria that produce these enzymes are killed. These enzymes may be retained in the curd during cheesemaking and cause off-flavors to develop during ripening, especially in semi-hard and hard varieties, which are ripened for a long time (e.g., Cheddar, Gouda, Comté, etc.).

4.2 PASTEURIZATION

Pasteurization of milk for cheesemaking became widespread after about 1940, primarily for public health reasons (*Mycobacterium tuberculosis*, the organism that causes tuberculosis, is killed by pasteurization) but also to provide a milk supply of more uniform bacteriological quality and to increase the keeping quality of the cheese. Batch pasteurization (low temperature, long time [LTLT] pasteurization; 63–65°C × 30 min) was used initially but was replaced by continuous high temperature, short time (HTST) pasteurization (72°C × 15 s). A line diagram of an HTST pasteurizer is shown in Figure 4–3.

Most (> 99.9%) of the bacteria found in raw milk are heat labile and are killed by pasteurization at 72°C for 15 s; most milk for cheesemaking is subjected to this heat treatment. Pasteurization kills all potential pathogens that

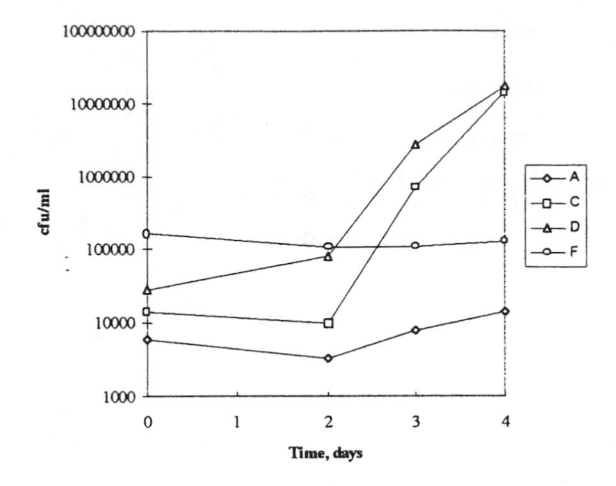

Figure 4–2 Growth of bacteria in four raw milks incubated at 5°C.

might be present in the milk, but spores of *Clostridium* and *Bacillus* are not killed by this treatment. In addition, organisms like *Micrococcus*, *Microbacterium,* and *Enterococcus* spp., which can withstand this heat treatment, are found in raw milk. These are called thermoduric bacteria and invariably come from improperly cleaned equipment. Generally, thermoduric bac-

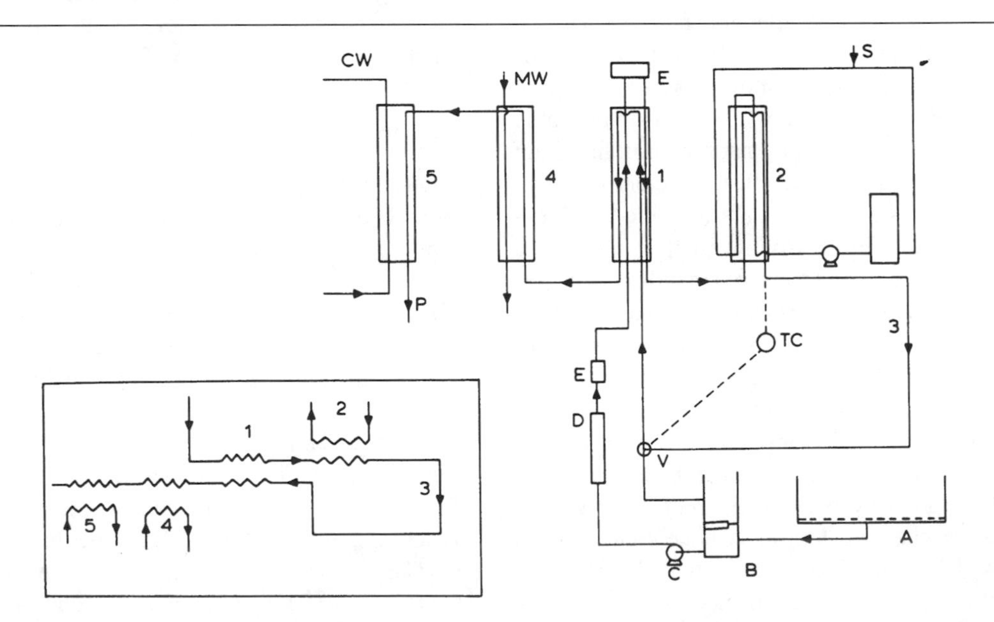

Figure 4–3 Layout of HTST pasteurizer (the insert shows a schematic diagram of the heat exchange sections): (A) feed tank, (B) balance tank, (C) feed pump, (D) flow controller, (E) filter, (P) product, (S) steam injection (hot water section), (V) flow diversion valve, (MW) mains water cooling, (CW) chilled water, (TC) temperature controller, (1) regeneration, (2) hot water section, (3) holding tube, (4) mains cooling water, and (5) chilled water cooling.

teria grow only slowly, if at all, in raw milk (*Enterococcus* spp. are exceptions), so that counts of thermoduric bacteria, even within 24 hr of milking, are a useful indicator of how well the milking equipment was cleaned. These bacteria can be enumerated by heat-treating the milk at 63°C for 30 min before plating. In cheese factories, another thermoduric bacterium, *Streptococcus thermophilus*, can grow as a biofilm in the regeneration section of the pasteurizer during long runs. This organism grows rapidly in milk and is also used as a starter culture (see Chapter 5).

Pasteurization also inactivates several enzymes in the milk, including lipase and alkaline phosphatase. Lack of alkaline phosphatase activity in milk indicates that the milk has been properly pasteurized.

In some countries, such as Canada, significant amounts of cheese are made from milk heat-treated to a temperature lower than pasteurization. This treatment is called *thermization* and generally involves heating the milk to 63°C for 10–15 s. This treatment results in less inactivation of enzymes and nonstarter lactic acid bacteria (NSLAB), which may be important in developing cheese flavor. Only some pathogenic and food-poisoning microorganisms are killed by thermization, and the milk must be subsequently fully pasteurized to meet public health regulations. The purpose of thermization is to kill psychrotrophs, which dominate the microflora of refrigerated milk and excrete potent proteinases and lipases, which may cause flavor and textural defects in cheese. Such treatments are also used in Europe to reduce the number of microorganisms in the raw milk as it is taken into the factory and thereby to prolong the keeping quality of raw milk, but in these cases the milk is subsequently pasteurized before cheesemaking.

While pasteurization reduces the risk of producing low-quality cheese resulting from the growth of undesirable bacteria and kills pathogens and food-poisoning microorganisms, pasteurization of cheese milk may damage its cheesemaking properties if the heat treatment is too severe (owing to heat-denaturation of the whey proteins and their interaction with κ-casein; see Chapter 6). The extent of this damage is negligible under HTST pasteurization conditions. The flavor of cheese made from pasteurized milk develops more slowly and is less intense than that made from raw milk, apparently because certain components of the microflora of raw milk contribute positively to cheese flavor. To overcome this deficiency, adjunct cultures of selected NSLAB are being recommended for use in the manufacture of long-ripened, low-moisture cheese made from pasteurized milk. This topic is discussed fully in Chapters 11 and 15.

Many cheeses, including such famous varieties as Emmental, Gruyère, Comté, and Parmigiano-Reggiano, are still produced, at both factory and farmhouse level, from raw milk. One of the important safety factors in these cheeses is that all are cooked to a high temperature (> 50°C) for up to 1 hr, which kills some of the bacteria in the raw milk. Legislation in many countries requires that cheese be made from HTST pasteurized milk; that it be aged for at least 60 days, during which food-poisoning or pathogenic bacteria die; or the cheese itself be pasteurized (i.e., converted to processed cheese). The public health aspects of cheese are discussed in Chapter 20.

4.3 ALTERNATIVES TO HEAT TREATMENT

There are at least four alternatives to heat treatment for reducing the numbers of bacteria in cheese milk:

- treatment with hydrogen peroxide (H_2O_2)
- activation of the lactoperoxidase-H_2O_2-thiocyanate system
- bactofugation
- microfiltration

4.3.1 Treatment with Hydrogen Peroxide

Hydrogen peroxide (H_2O_2) is a very effective bactericidal agent, the use of which is permitted in some countries, including the United States. The excess H_2O_2 is usually destroyed by adding catalase. The use of H_2O_2 to treat cheese milk is not practiced commercially to any great extent.

4.3.2 Lactoperoxidase-H_2O_2-Thiocyanate

Lactoperoxidase (LPO), an indigenous enzyme in milk, reduces H_2O_2 in the presence of a suitable reducing agent:

$$H_2O_2 + 2HA \rightarrow 2H_2O + 2A$$

One such reducing agent is the thiocyanate anion SCN^-, which is oxidized to various species (e.g., $OSCN^-$) that are strongly bactericidal. Milk contains a low concentration of indigenous SCN^-, arising from the catabolism of glucosinolates (from members of the *Cruciferae* family) by bacteria in the rumen (Figure 4–4). Milk contains no H_2O_2, which must be added or produced in situ via oxidation of glucose by glucose oxidase (Figure 4–5) or from xanthine by xanthine oxidase (Figure 4–6); or by starter cultures grown in the presence of O_2.

The LPO system is very effective for the "cold pasteurization" of milk. The process has been patented but has attracted limited, if any, interest in developed dairy countries, possibly for economic reasons. It is practiced to a small extent in developing countries.

4.3.3 Bactofugation

A high percentage (98–99%) of somatic and bacterial cells and bacterial spores in milk can be removed by centrifugation at high gravitational forces using a special centrifuge called a bactofuge. The cells and spores are more dense than milk serum and are concentrated in the sludge during bactofugation. The sludge is subsequently sterilized to kill the spores and bacteria and is then added back to the milk. Bactofugation is not widely used in general cheesemaking but is commonly used to remove

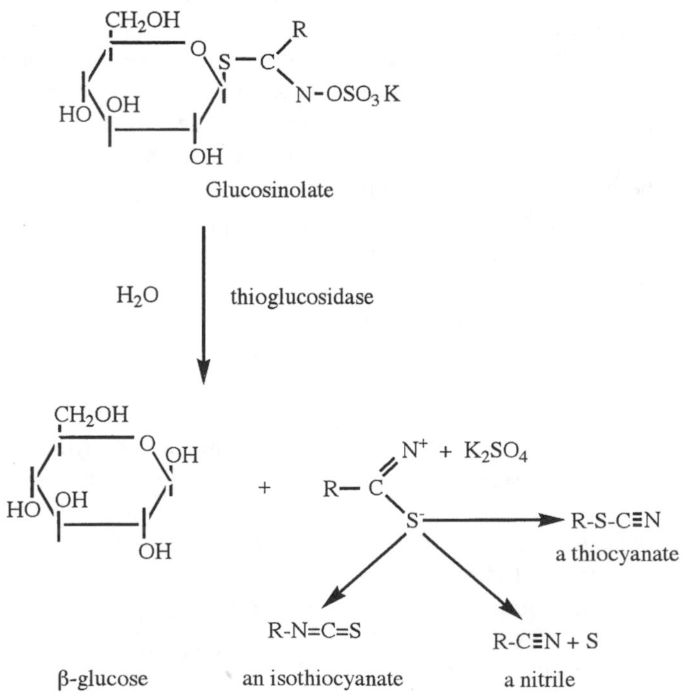

Figure 4–4 Enzymatic degradation of a glucosinolate.

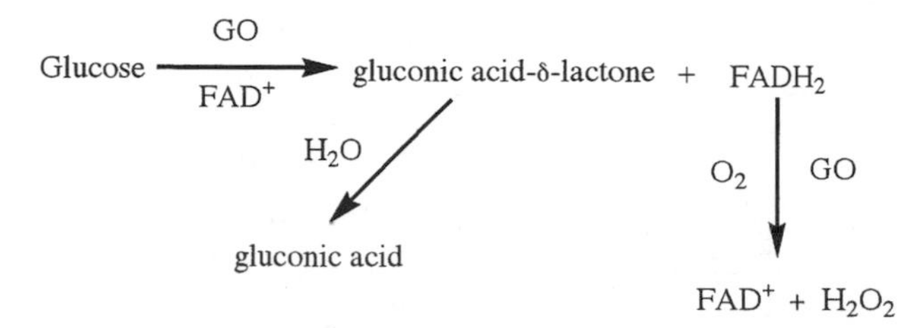

Figure 4–5 Formation of H_2O_2 upon oxidation of glucose by glucose oxidase (GO).

Clostridium tyrobutyricum from milk intended for Dutch- and Swiss-type cheeses; the latter undergo a propionic acid fermentation. Cheeses that undergo a propionic acid fermentation are generally ripened at below 13°C for a few weeks, after which the temperature is increased to around 22°C for 3–4 weeks to promote the fermentation. Clostridia, especially *Cl. tyrobutyricum*, can also grow under these circumstances and produce late gas (see Chapter 11), but bactofugation is very effective in eliminating spores of *Cl. tyrobutyricum* from the milk.

Cl. tyrobutyricum is an obligate anaerobe that can ferment lactic acid, the principal acid in cheese, and the anaerobic cheese environment is ideal for its growth. The major source of the organism in raw milk is improperly fermented silage. For this reason, feeding of silage to cows is prohibited in the areas of Switzerland where Emmental cheese is made. Contamination with spores is much greater in winter, when the cows are fed indoors with silage, than in summer, when they are on pasture. The vegetative cells are probably killed by pasteurization (although there is no proof of this), but the spores are heat resistant, requiring several minutes at 100°C to kill them.

When silage contaminated with clostridia is eaten by cows, the spores pass through their gastrointestinal tract. The teats and udders of cows lying in their own feces become contaminated with fecal material containing clostridial spores, which then contaminate the milk. Proper cleaning of the teats and udders will reduce contamination from this source. However, less than 10 spores per 100 ml of milk are sufficient to cause late gas production in Dutch-type cheeses. The critical number is lower in cheeses that undergo a propionic acid fermentation (i.e., Swiss-type cheeses) because of the higher ripening temperature used for these cheeses.

The design of the bactofuge (Figure 4–7) is essentially similar to that of separators used to separate the fat from milk but is modified in such a way that only bacteria and spores, which are more dense than skim milk, are forced outward and move down along the lower side of the upper member of a pair of disks and eventually through orifices in the bowl of the centrifuge as a bacterial concentrate called the bactofugate (this represents about 3%, v/v, of milk). Some large, dense casein micelles, perhaps as much as 6% of the total casein, are also removed by this process. The loss of casein will cause a decrease in cheese yield that may be avoided by heat-sterilizing the bactofugate and returning it to the milk or by otherwise supplementing the casein content (e.g., by adding ultrafiltration retentate).

4.3.4 Microfiltration

Microfiltration is a membrane separation process, in principle like reverse osmosis, nanofiltration or ultrafiltration, except that large pore-size membranes (0.8–1.4 μm) are used. The semipermeable membranes used retain the bacteria but allow milk constituents, including most of the casein micelles, to pass through in the permeate. The process can only be applied to skim milk, as the fat-globules in whole milk block the pores of the membrane and reduce its efficiency. Therefore, the cream must be separated from the milk, pasteurized, and added back to the microfiltered skim milk before cheesemaking.

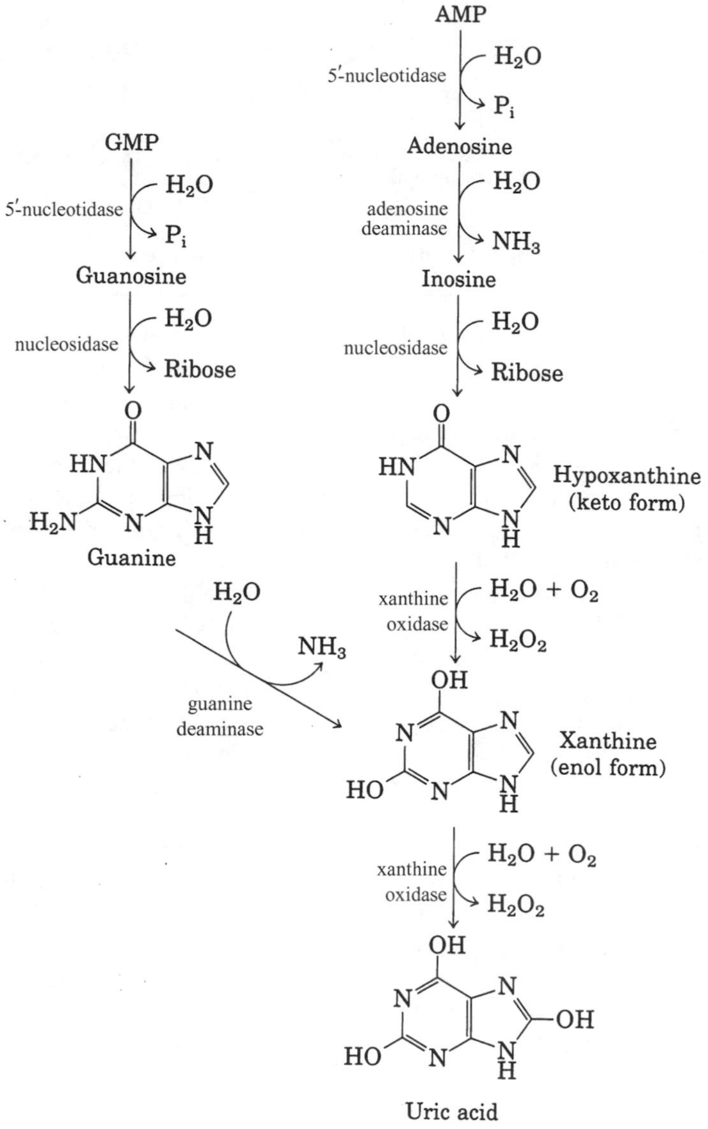

Figure 4–6 Diagram of catabolism of purine nucleotides, showing the production of H_2O_2 from hypoxanthine and xanthine by xanthine oxidase.

Microfiltration is very efficient at removing bacterial cells (> 99%) and is being used increasingly in the dairy industry, such as in the production of extra–long life pasteurized milk. It is not yet widely used for cheese milk except for the removal of spores from milk for Swiss and similar cheeses. The technique has been very useful in studying the effect of the indigenous raw-milk microflora and of enzymes inactivated by pasteurization on cheese flavor. The quality of Cheddar and Comté cheeses made from microfiltered milk is similar to that made from pasteurized milk and is different from that made from raw milk, which indicates that the differences in flavor between raw and pasteurized milk cheeses are due principally to the in-

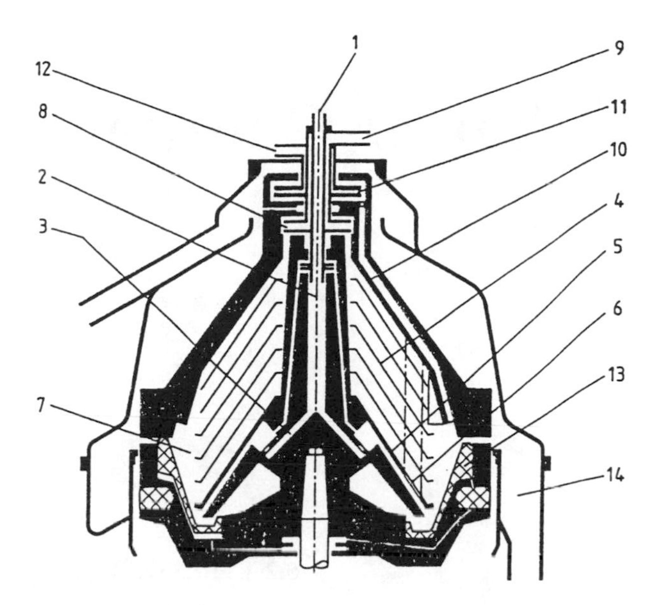

Figure 4–7 A cutaway of a Westfalia bacteria clarifier showing milk inlet (1), milk outlet (9), and bacteria stream (concentrate) outlet (12).

digenous microorganisms that are efficiently removed by microfiltration or killed by pasteurization rather than to the inactivation of indigenous enzymes or other heat-induced changes.

A microfiltration system known as the Bactocatch System has been developed by the Alfa Laval Company (Sweden) for the decontamination of milk as an alternative to pasteurization.

4.4 PREMATURATION

In some countries, particularly France, lactic cultures are added to the cheese milk, which is then incubated at 8–10°C for 12–15 hr (overnight), during which time a slight drop in pH occurs (ΔpH of 0.1 units). This process is called *prematuration,* and often the milk is pasteurized before the cheesemaking starter is added. Prematuration may suppress the growth of psychrotrophs during storage and/or produce compounds that stimulate the growth of the cheesemaking starter. In addition, the drop in pH stimulates rennet action. Alternatively, a proportion of the milk is inoculated with culture and grown until a pH of 4.6 is attained, and a sufficient volume is mixed with fresh bulk milk to reduce the pH of the fresh milk by 0.1 pH units.

SUGGESTED READINGS

Bramley, A.J., & McKinnon, C.H. (1990). The microbiology of raw milk. In R.K. Robinson (Ed.), *Dairy microbiology* (2d ed., Vol. 1). London: Elsevier Applied Science.

International Dairy Federation. (1996). *Symposium on Bacteriological Quality of Raw Milk.* Wolfpassing, Austria, 13–15 March 1996. Brussels: International Dairy Federation.

Jelen, P. (1985). *Introduction to food processing*. Englewood Cliffs, NJ: Prentice Hall.

Ledford, R.A. (1998). Raw milk and fluid milk products. In E.H. Marth & J.H. Steele (Eds.), *Applied dairy microbiology*. New York: Marcel Dekker.

Palmer, J. (No date). *Hygienic milk production and equipment cleaning*. Dublin: Teagasc.

Palmer, J. (1980). Contamination of milk from the milking environment. *International Dairy Federation Bulletin, 120*, 16–21.

Robinson, R.K. (Ed.). (1994). *Modern dairy technology* (2d ed., Vol. 1). London: Chapman & Hall.

CHAPTER 5

Starter Cultures

5.1 INTRODUCTION

In the manufacture of most cheeses, carefully selected strains of different species of lactic acid bacteria (LAB) are added to the milk shortly before renneting. Their major function is to produce lactic acid and, in some cases, flavor compounds, particularly acetic acid, acetaldehyde, and diacetyl. Acid production, in turn, has three functions: it promotes rennet activity; aids the expulsion of whey from the curd, thus reducing the moisture content of the cheese; and helps to prevent the growth of undesirable bacteria in the cheese. These cultures are called starters because they initiate (start) the production of acid. They are also called lactic cultures because they produce lactic acid. If raw milk is incubated at a temperature in the range 20–40°C, it will coagulate within 10–24 hr. This physical transformation is due to the growth of and acid production by adventitious LAB present in the raw milk. In the past, such coagulated milks were used as starter cultures for cheese manufacture and are probably the source of many starter strains in use today. Starter cultures are not used for many cheeses made in Mediterranean countries (e.g., La Serena and artisanal production of Manchego cheese); instead, the cheesemaker relies on the adventitious LAB present in the milk to grow during cheesemaking and produce the necessary acid.

5.2 TYPES OF CULTURES

Starter cultures are commonly divided into mesophilic cultures (with an optimum temperature of about 30°C) and thermophilic cultures (with an optimum temperature of about 42°C). Each group of starters can be further subdivided into defined- and mixed-strain cultures. Defined-strain cultures are pure cultures, the physiological characteristics of which are known and identifiable. Such cultures are used commercially in most large cheesemaking plants in Australia, New Zealand, the United States, the United Kingdom, and Ireland. They have been isolated mainly from mixed-strain cultures (see below) but also from fermented products made by indigenous LAB and plants. Before being used commercially, the strains are screened for important technological properties, such as their salt tolerance and their ability to grow and produce lactic acid in milk, resist attack by bacteriophage, utilize citrate, and produce good-quality cheese.

Mixed-strain cultures contain unknown numbers of strains of the same species. In addition, many of them also often contain bacteria from different genera of LAB, including *Lactococcus* and *Leuconostoc* spp. in the case of mesophilic, mixed-strain cultures and *Sc. thermophilus* and *Lb. delbrueckii* subsp. *bulgaricus*, *Lb. delbrueckii* subsp. *lactis*, and *Lb. helveticus* in the case of thermophilic, mixed-strain cultures. Many mixed cultures in use today are subcultures of coagulated milks that produced good-quality cheese at the turn of the century when cheese was beginning to be produced on a large scale. A further complication is that defined cultures are used rarely as pure cultures or single strains but as mixtures of 2 to 6 strains, which means that they can also, with some degree of justification, be called (defined) mixed cultures. Mixed-strain cultures generally contain phage (see later) and are generally used by small-scale producers, while defined cultures are usually used by large-scale producers. Some of the important distinguishing characteristics of the bacteria found in defined and mixed cultures are summarized in Table 5–1.

The main species of *Lactococcus* in mixed-strain mesophilic cultures is *Lc. lactis* subsp. *cremoris*, although *Lc. lactis* subsp. *lactis* is also found. These two subspecies are differentiated from each other by their growth response at 40°C and the ability to produce NH_3, ornithine, and citrulline from arginine (Table 5–1). In addition, *Lc. lactis* subsp. *lactis* contains glutamate decarboxylase, which produces γ-aminobutyric acid from glutamate, while *Lc. lactis* subsp. *cremoris* does not (Nomura, Kimoto, Someya, & Suzuki, 1999). The exact species of *Leuconostoc* found in starter cultures is not clear, but it is likely that both *Ln. mesenteroides* subsp. *cremoris* and *Ln. lactis* are present. The function of the *Leuconostoc* spp. is to metabolize citrate to CO_2, diacetyl, and acetate. CO_2 is responsible for eye formation in Edam and Gouda cheeses, while diacetyl and acetate are important flavor components of Quarg, Fromage frais, and Cottage cheese. For this reason, the citrate utilizers are often called aroma producers. Some of the *Lactococcus* strains found in mixed cultures are able to utilize citrate (Cit^+), and some are not (Cit^-). Cit^- lactococci dominate these cultures and generally make up about 90% of the organisms present, whereas the Cit^+ lactococci and leuconostoc make up the rest. Depending on the aroma producers present in the culture, mesophilic, mixed-strain starters containing only Cit^+ leuconostoc are called L cultures (L, the first letter of *Leuconostoc*), and those containing only Cit^+ lactococci are called D cultures (D, from *Streptococcus diacetilactis*, the old name for Cit^+ lactococcus). Cultures containing both Cit^+ lactococci and leuconostoc are called DL cultures, and those containing no aroma producer (i.e., only Cit^- lactococci are present) are called O cultures.

How do we know that mixed cultures contain different strains? This question is not easy to answer. Isolates from mixed cultures produce acid at widely different rates, e.g., pH values within the range 5–6.5 in milk after 6 hr of incubation are common (this may be related to lack of proteinase activity), have different phage-host patterns, and show several different plasmid profiles. Recently, a detailed analysis of 113 isolates from a mixed-strain DL culture (Flora Danica) was reported (Lodics & Steenson, 1990). This is a well-known commercial starter, commonly used in the manufacture of soft cheeses. Seventy isolates were identified as Lc. *lactis* subsp. *cremoris*, 2 as Cit^- *Lc. lactis* subsp. *lactis*, 21 as Cit^+ *Lc. lactis* subsp. *lactis*, 18 as *Ln. mesenteroides* subsp. *cremoris*, and 1 isolate could not be classified. Twenty different plasmid profiles were found, examples of 17 of which are shown in Figure 5–1. Most of the *Lc. lactis* subsp. *cremoris* isolates (58) fell into 12 groups (lanes A to L inclusive), while the 25 isolates of Cit^+ *Lc. lactis* subsp. *lactis* fell into 3 groups (lanes M, N, and O), and all of the *Ln. mesenteroides* subsp. *cremoris* isolates (18) belonged to the same plasmid group (lane Q). Twenty strains coagulated reconstituted skim milk (11% solids) in less than 8 hr at 21°C, 2 strains in 18 to 24 hr, and the remainder in more than 24 hr. Seven phages were isolated from the starter, and 51 isolates were insensitive to these phages. The remaining isolates showed various sensitivities to the 7 phages isolated from the culture itself and to 10 relatively well-known phages. These data support the view that genuinely different strains are present in mixed cultures.

Thermophilic cultures almost always consist of two organisms, *Streptococcus thermophilus*

Table 5–1 Some Distinguishing Characteristics of the Lactic Acid Bacteria Found in Commercial and Natural or Artisanal Starter Cultures

Name	Type[a]	Shape	Percentage of Lactic Acid Produced in Milk[b]	Metabolism of Citrate	NH_3 from Arginine	Growth at 10°C	Growth at 15°C	Growth at 40°C	Growth at 45°C
Commercial									
Streptococcus thermophilus	T	Coccus	0.6	–	–	–		+	+
Lactobacillus helveticus	T	Rod	2.0	–	–	–	–	+	+
Lactobacillus delbrueckii subsp. *bulgaricus*	T	Rod	1.8	–	±	–	–	+	+
Lactobaccilus delbrueckii subsp. *lactis*	T	Rod	1.8	–	±	–	–	+	+
Lactococcus lactis subsp. *cremoris*	M	Coccus	0.8	±	–	+	+	–	–
Lactococcus lactis subsp. *lactis*	M	Coccus	0.8	±	+	+	+	+	–
Leuconostoc lactis	M	Coccus	<0.5	+	–	+	+	–	–
Leuconostoc mesenteroides subsp. *cremoris*	M	Coccus	0.2	+	–	+	+	–	–
Natural or artisanal, above plus									
Lactobacillus casei subsp. *casei*		Rod		±	–		+		–
Lactobacillus paracasei subsp. *paracasei*		Rod			–	+	+	+	±
Lactobacillus paracasei subsp. *tolerans*		Rod			–		+		–
Lactobacillus rhamnosus		Rod			–	±	+	+	+
Lactobacillus plantarum		Rod		±	–		+		–
Lactobacillus curvatus		Rod			–		+		–
Lactobacillus fermentum		Rod			+		+	+	+
Enterococcus faecalis		Coccus		+	+	+	+	+	+
Enterococcus faecium		Coccus			+	+	+	+	+

Name	Fermentation of Sugar	Isomer of Lactate	Allosteric Lactate Dehydrogenase[c]	Fermentation of[d]						
				Glu	Gal	Lac	Rhm	Man	Raf	Mtl
Commercial										
Streptococcus thermophilus	Homo	L		+ –	+	–	+		–	
Lactobacillus helveticus	Homo	DL	–	+	+	+	–	±	–	–
Lactobacillus delbrueckii subsp. *bulgaricus*	Homo	D	–	+	–	+	–	+	–	–
Lactobaccilus delbrueckii subsp. *lactis*	Homo	D	–	+	–	+	–	+	–	–
Lactococcus lactis subsp. *cremoris*	Homo	L	+	+	+	+	–		–	
Lactococcus lactis subsp. *lactis*	Homo	L	+	+	+	+	–		–	
Leuconostoc lactis	Hetero	D	–	+	+	+		±	±	–
Leuconostoc mesenteroides subsp. *cremoris*	Hetero	D	–	+	±	+		–	–	–
Natural or artisanal, above plus										
Lactobacillus casei subsp. *casei*	Homo	L	+	+	+	±	–	+	–	+
Lactobacillus paracasei subsp. *paracasei*	Homo	L	+	+	+	+	–	+	–	+
Lactobacillus paracasei subsp. *tolerans*[e]	Homo	L	+	+	+	+	–	–	–	–
Lactobacillus rhamnosus	Homo	L	+	+	+	+	+	+	–	+
Lactobacillus plantarum	Homo	DL	–	+	+	+	–	+	+	±
Lactobacillus curvatus	Homo	DL	+	+	+	±	–	–	–	–
Lactobacillus fermentum	Hetero	DL	–	+	+	+	–	–	+	–
Enterococcus faecalis	Homo	L		+	+	+		+	–	+
Enterococcus faecium	Homo	L		+	+	+		+	±	±

[a] T = thermophilic; M = mesophilic.
[b] Approximate values; individual strains vary.
[c] Activated by fructose-1,6-phosphate and in lactobacilli also by Mn^{2+}.
[d] Glu = glucose; Gal = galactose; Lac = lactose; Rhm = rhamnose; Man = mannose; Raf = raffinose; Mtl = manitol.
[e] Survives heating at 72°C for 40 s.

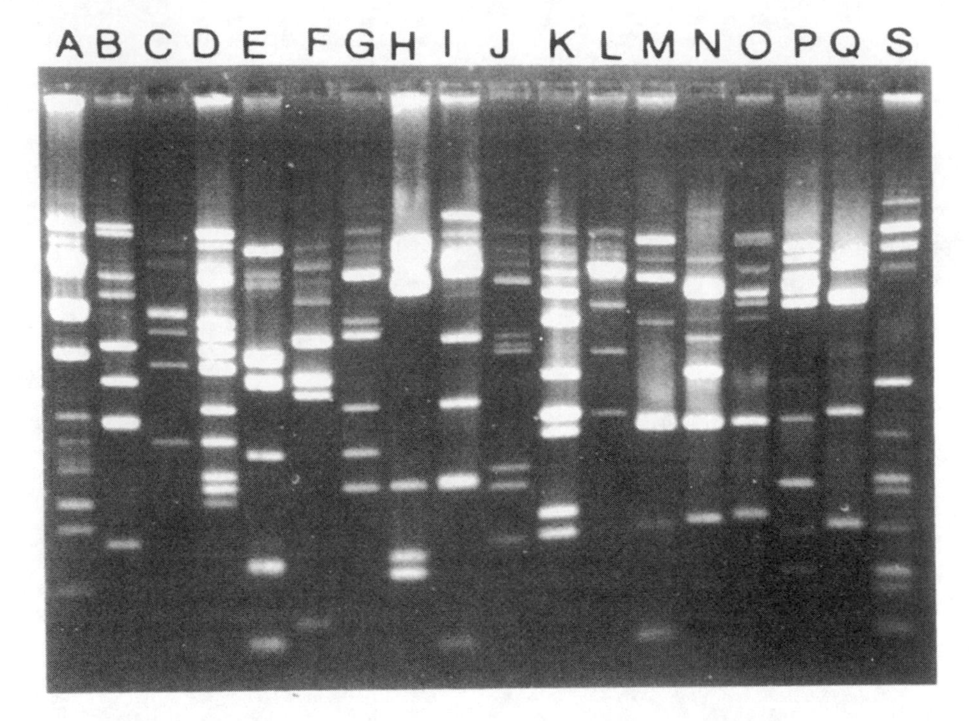

Figure 5–1 Plasmid profiles of representatives of different plasmid groups (lanes A to Q) isolated from a Flora Danica culture. Lane S, plasmid standard containing *Salmonella typhimurium* LT2 (60 MDa), *Escherichia coli* V517 (32, 5.2, 3.5, 3.0, 2.2, 1.7, 1.5. 1.2 MDa), pSA (23 MDa), and pSA3 (6.8 MDa) DNA.

and either *Lactobacillus helveticus*, *Lb. delbrueckii* subsp. *lactis*, or *Lb. delbrueckii* subsp. *bulgaricus*. These are often referred to as the coccus and rod, respectively. Bulk starters of the rod and coccus are generally grown individually for cheese manufacture, but they are grown together for yogurt production. In the latter case, they are genuinely mixed cultures, but defined-strain thermophilic cultures are also used. Like mesophilic mixed cultures, thermophilic mixed cultures may contain several strains of each species. For some products (e.g., Mozzarella cheese), the rod:coccus ratio is important, and it is much easier to control this ratio by growing the cultures separately.

The rod and coccus produce much more acid when they are grown together than when they are grown separately (Figure 5–2). Acid production is an excellent indicator of the growth of LAB (see Section 5.10). The improved growth is due to the production of amino acids, particularly leucine, isoleucine, and valine, from casein in the milk by the proteolytic system of the *Lactobacillus,* which stimulates the growth of *Sc. thermophilus*. The *Streptococcus,* in turn, produces small amounts of CO_2 and formic acid from lactose, which stimulate the growth of the *Lactobacillus*. Thus, the relationship is genuinely symbiotic. The CO_2 produced is not sufficient to be apparent in the milk. However, the growth of the rod and coccus are more difficult to control when they are grown together.

Whether symbiosis occurs between the *Leuconostoc* and the *Lactococcus* in mesophilic cultures is not clear; there is some evidence in the old literature that a certain amount of symbiosis occurs but the exact nature of the interaction has not been determined. A certain amount of symbiosis must occur, since many lactococci in mixed cultures do not have a proteinase system

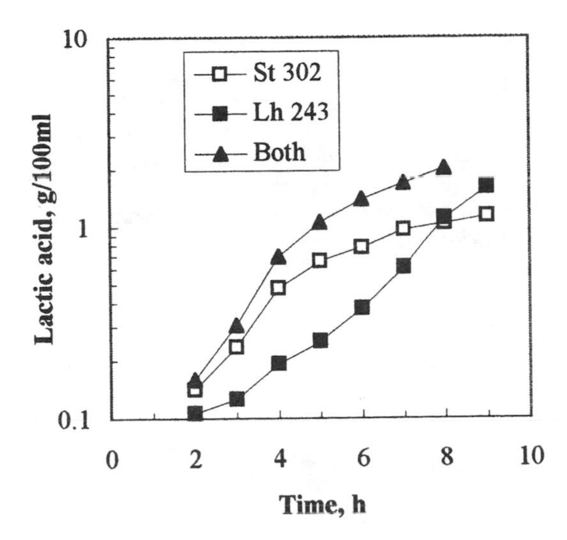

Figure 5–2 Acid production by *Lactobacillus helveticus* 243, *Streptococcus thermophilus* 302, or their combination in milk at 42°C.

(Prt⁻) and depend on the Prt⁺ strains to produce the amino acids they require for growth (discussed later). Examples of the cultures used in the production of different cheeses are shown in Table 5–2.

5.2.1 Artisanal (Natural) Cultures

In many countries, especially Italy, France, Switzerland, and Greece, other types of mixed cultures, called artisanal cultures, are used. These are derived mainly from the practice of "back-slopping," where some of the previous batch of cheese is used as the inoculum for the new batch. An example of this is the Greek cheese Kopanisti. More commonly, whey and milk from today's cheesemaking are incubated at a high temperature (45–52°C) for an extended period (4.5–18 hr), depending on the culture, for use in tomorrow's cheesemaking. The cheeses in which these cultures are used are made from raw milk. Such cultures depend on the presence of LAB in the raw milk, and the cultures are called *whey cultures* or *natural milk cultures*. The temperature of incubation and the pH exert selective pressure on the types of bacteria that grow under these conditions. The composition of these cultures is extremely complex and very variable and can include *Lb. delbrueckii* subsp. *lactis, Lb. delbrueckii* subsp. *bulgaricus, Lb. helveticus, Lb. plantarum, Lb. casei, Lb. paracasei, Lc. lactis, Sc. thermophilus, Enterococcus faecalis, Ec. faecium*, and *Leuconostoc* spp. (see Table 5–1). These cultures have many of the attributes of mixtures of both mesophilic and thermophilic cultures, and many of the bacteria in them are lysogenic (see Section 5.7.2) and, as such, are a potential source of phage in factories; however, they can be very inconsistent in performance.

5.2.2 Adjunct Cultures

Much hard cheese made commercially today is thought to lack flavor. The probable reasons for this are very low bacterial numbers in the raw milk and, more importantly, vastly improved hygiene in cheese factories. Because of this, various methods for improving flavor have been developed. Traditionally, only mesophilic cultures were used in the production of Cheddar and other low-cooked cheeses. However, in recent years, thermophilic starters, particularly *Sc. thermophilus* and *Lb. helveticus*, have been included in the starter and are thought to improve the flavor of the cheese. These organisms are very resistant to the cook temperature (≈38°C) used for those cheeses. However, their growth in such cheeses is limited to temperatures above 25°C. At temperatures below this, little growth and consequently little acid production occurs.

Carefully selected strains of mesophilic lactobacilli, particularly *Lb. paracasei* and *Lb. casei*, are also used to improve the flavor of some cheeses, especially Cheddar. The basis for their use is the fact that large numbers of these bacteria (10^8 cfu/g) are found in ripened cheese, and it is assumed they must have some role. However, despite extensive research over several decades, the role of these bacteria in cheese flavor development is still unclear (see Chapter 10).

Table 5–2 Starter Cultures Used in the Manufacture of Different Cheeses

Cheese	*Starter Cultures*	*Other Cultures*	*Important Products Other Than Lactic Acid*
Emmental cheese	*Sc. thermophilus* and *Lb. helveticus*; galactose-positive *Lb. delbrueckii* subsp. *lactis* may also be used	*Propionibacterium freudenreichii*	CO_2, propionate, and acetate
Mozzarella and other Italian cheese	*Sc. thermophilus* and *Lb. helveticus* or a mixed-strain thermophilic culture		
Cheddar cheese	Defined strains of *Lc. lactis* subsp. *lactis* or O, L, or DL mesophilic mixed cultures; sometimes thermophilic cultures are included		
Edam and Gouda cheese	Mainly DL mesophilic mixed cultures		CO_2 and acetate
Camembert and Brie cheese	O, L, or DL mesophilic mixed cultures	*Penicillium camemberti, Geotrichum candidum, Candida utilis*	
Tilset, Limburger, and Munster cheese	O, L, or DL mesophilic mixed cultures	*Brevibacterium linens, Geotrichum candidum, Candida utilis*	Sulphur compounds (e.g., methional)
Yogurt	Mainly thermophilic mixed cultures; defined strains of *Sc. thermophilus*, *Lb. delbrueckii* subsp. *bulgaricus*, and *Lb. delbrueckii* subsp. *lactis* may also be used		Acetaldehyde
Fromage frais and Quarg	O, L, or DL mesophilic mixed cultures		Diacetyl and acetate

5.2.3 Effect of Temperature on the Growth of Cultures

Temperature has a major effect on the growth of starter bacteria. In Figure 5–3, two different measurements, the generation time and the decrease in pH after 5.5 hr, are used as indicators of growth. The latter is quite acceptable, as the decrease in pH is a measure of the amount of acid produced by the culture, and acid production is directly proportional to the increase in cell numbers or cell mass (see Section 5.10). The optimum temperatures for the growth of *Lactococcus*, *Leuconostoc*, *Sc. thermophilus,* and *Lb. helveticus* are 30°C, 25°C, 42°C, and 42°C, respectively. The behavior of these starters at the cooking temperature of cheese is also important. *Lc. lactis* subsp. *lactis* grows slowly at 38°C, which is the cooking temperature for Cheddar cheese, whereas the growth of most strains of *Lc. lactis* subsp. *cremoris* is markedly inhibited at this temperature. A common cooking temperature for Swiss-type cheese is 54°C, and strains of *Sc. thermophilus* and *Lb. helveticus* produce little acid at this temperature (Figure 5–3). However, they can withstand this temperature and begin to grow again when the temperature decreases to 47°C and below.

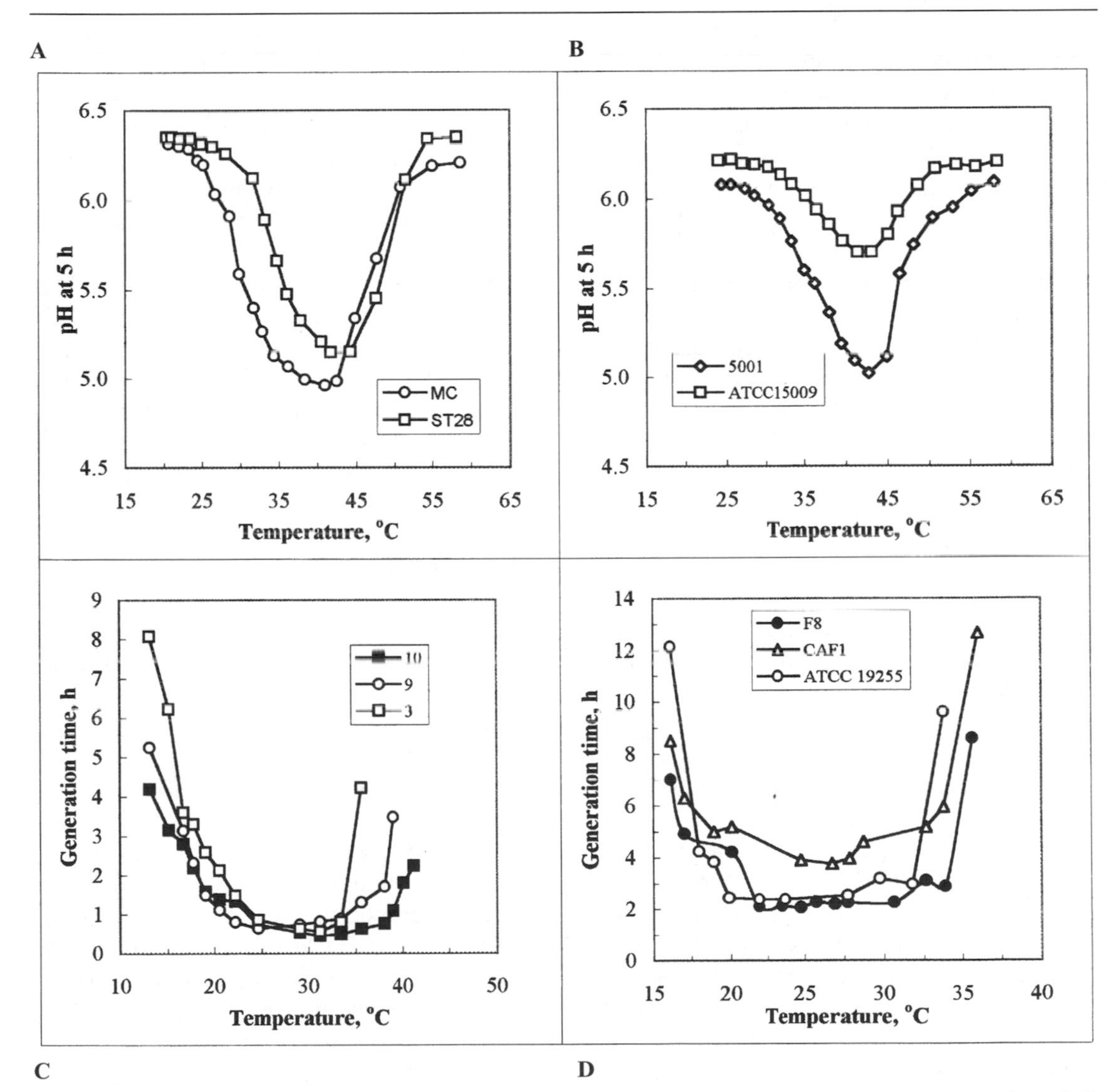

Figure 5–3 Effect of temperature on the growth of (A) two strains of *Sc. thermophilus*, (B) two strains of *Lb. helveticus* in milk, (C) *Lc. lactis* subsp. *lactis* 10 and *Lc. lactis* subsp. *cremoris* 3 and 9, and (D) three strains of *Ln. mesenteroides* subsp. *cremoris* in complex medium. Note differences in *y* axes.

5.3 TAXONOMY

There are 12 genera of lactic acid bacteria: *Aerococcus, Alloiococcus, Carnobacterium, Enterococcus, Lactobacillus, Lactococcus, Leuconostoc, Pediococcus Streptococcus, Tetragenococcus*, *Vagococcus*, and *Weissella.* Only five genera, however, contain organisms used as cheese starter cultures: *Lactococcus, Enterococcus*, *Leuconostoc, Streptococcus,* and *Lactobacillus.* All the lactic acid bacteria used in starter cultures are Gram-positive, catalase-negative, nonmotile, non-spore-forming bacteria.

The taxonomy of lactic acid bacteria, including those used in starter cultures, has gone through several major revisions during the past 30 years as a result of more detailed and sophisticated analyses. These include DNA:DNA and DNA:RNA hybridizations, comparative oligonucleotide cataloging and sequencing of the16S rDNA gene, and serological studies with superoxide dismutase. The current names are used in Table 5–1. The lactococci have probably gone through the greatest number of changes; their taxonomy is reviewed by Schleifer and Kilpper-Balz (1987).

5.3.1 Lactococcus

Lc. lactis was isolated from soured milk in 1873 by Lister and named *Bacterium lactis* (Latin for bacterium of milk). It was the first bacterium isolated in pure culture. In 1909, it was renamed *Streptococcus lactis* by Lohnis and placed in the genus *Streptococcus.* In the 1890s, Storch isolated a very similar organism from cream, which Orla-Jensen called *Sc. cremoris* (of cream). In 1937, Sherman divided the genus *Streptococcus* into four groups—pyogenic, lactic, faecal and viridans—and placed *Sc. lactis* and *Sc. cremoris* in the lactic group. The groups were relatively easily distinguished from each other on the basis of growth under different conditions (Table 5–3). Serology (Lancefield grouping), based on the presence of different carbohydrates in the cell wall, was also used at that time to separate the streptococci into different groups; for example, the lactic group reacted only with Group N antiserum and the faecal group only with Group D antisera. However, later studies showed that serology was not especially useful, since many streptococci isolated subsequently did not possess group-specific antigens, and the group D antigen was also present in organisms from the viridans group (e.g., *Sc. bovis*).

Homology studies of the DNA of *Sc. lactis* and *Sc. cremoris* have shown that these two organisms are closely related to each other, and in 1982 they were reclassified as subspecies of *Sc. lactis* and named *Sc. lactis* subsp. *lactis* and *Sc. lactis* subsp. *cremoris*, respectively (Garvie & Farrow, 1982). In 1985, it was realized that these organisms were only distantly related to the genus *Streptococcus*, and therefore they were transferred to a new genus, *Lactococcus*, as *Lc. lactis* subsp. *lactis* and *Lc. lactis* subsp. *cremoris*, respectively (Schleifer et al., 1985). The genus also includes four new species, *Lc. raffinolactis, Lc. garvieae, Lc. plantarum*, and *Lc. piscium,* which were isolated from spontaneously soured raw milk, a cow suffering from mastitis, frozen peas, and fish, respectively, and a subspecies, *Lc. lactis* subsp. *hordniae*, isolated from the leaf hopper insect. None of the new species grows well in milk and hence they are of little value as starter cultures.

In 1936, an organism similar to *Sc. lactis* was isolated from fermenting potatoes and named *Sc. diacetilactis.* The latter organism was later renamed *Sc. lactis* subsp. *diacetylactis. Diacetilactis* was the spelling used when the organism was considered to be a species, but the spelling changed to *diacetylactis* when the organism was given subspecies status. It differed from *Sc. lactis* only in its ability to metabolize citrate, and since the transport of citrate is plasmid encoded, it was renamed *Lc. lactis* subsp. *lactis* biovar. *diacetylactis.* Since *biovar* is not an accepted taxonomic epithet, this organism is now called citrate-utilizing (Cit^+) *Lc. lactis* subsp. *lactis* or Cit^+ *Lactococcus* to distinguish it from the vast majority of lactococci unable to metabolize citrate (Cit^-).

Table 5–3 Some Distinguishing Characteristics of the Different Groups of Streptococci

	Growth at or in the Presence of				
Group	*10°C*	*45°C*	*pH 9.6*	*6.5% NaCl*	*0.1% Methylene Blue*
Pyogenic	–	–	–	–	–
Viridans	–	+	–	–	–
Lactic	+	–	–	–	+
Faecal	+	+	+	+	+

Lc. lactis subsp. *lactis* and *Lc. lactis* subsp. *cremoris* are the main species isolated from mesophilic starters and raw milk soured at 18–30°C, and it is generally believed that *Lc. lactis* subsp. *cremoris* gives a better flavored cheese than *Lc. lactis* subsp. *lactis*. *Lc. lactis* subsp. *lactis* is able to grow at 40°C and, in the presence of 4% NaCl, produce NH_3 from arginine and ferment maltose, whereas *Lc. lactis* subsp. *cremoris* cannot (see Table 5–1). However, the latter organism can grow in the presence of 2% NaCl. The concentration of salt in cheese varies from 4% to 6% in the moisture of the cheese and is therefore inhibitory to most strains of lactococci. *Lc. lactis* subsp. *hordniae* is similar to *Lc. lactis* subsp. *cremoris*, except that it does not ferment lactose, galactose, maltose, or ribose. *Lc. raffinolactis* is also like *Lc. lactis* subsp. *lactis* but ferments melibiose and raffinose and does not grow at 40°C or produce NH_3 from arginine. Lactococci ferment sugars by the glycolytic pathway to L lactate.

Lactococci are spherical or ovoid cells that occur singly, in pairs, or in chains elongated in the direction of the chain, which can sometimes cause them to be misidentified as *Leuconostoc* spp. They grow at 10°C but not at 45°C or in the presence of 6.5% NaCl. They are nonmotile. Motile strains of *Lc. lactis* that have the Group N antigen have recently been transferred to the genus *Vagococcus*.

5.3.2 Enterococcus

There is considerable debate as to whether enterococci should be considered as starter cultures, since the main source of many of them is fecal material. Because of this, they are often used as indicators of fecal pollution. Occasionally, they cause endocarditis and urinary tract infections. Some of them, especially *Ec. faecalis*, are promiscuous and can easily pick up antibiotic resistance genes, especially for vancomycin, from plasmids or transposons. They are included here because they are common in artisanal cultures and are considered to impart desirable flavors to cheeses made with these cultures. Enterococci occur in pairs or chains and are salt and heat tolerant and generally grow in the presence of 6.5% NaCl and at 45°C. These properties make them ideal starters for cheesemaking, but their use has been questioned because they are used as indicators of fecal contamination of food. Unlike lactococci, enterococci are not killed by pasteurization to any great extent and can be found in large numbers in many cheeses.

In 1984, DNA hybridization studies showed that *Sc. faecalis* and *Sc. faecium*, which were then classified in the faecal group of *Streptococcus*, were only distantly related to the streptococci, and they were transferred to the new genus, *Enterococcus*, as *Ec. faecalis* and *Ec. faecium*, respectively (for a review, see Schleifer & Kilpper-Balz, 1987). Since then, 17 species have been added, and these are divided into individual species and 3 "species" groups, based on the sequences of the 16S rRNA. Group 1 contains *Ec. durans, Ec. faecium, Ec. mundtii,* and *Ec. hirae* (98.7–99.7% similarity); group 2 contains *Ec. avium, Ec. raffinosus, Ec. malodoratus*, and *Ec. pseudoavium* (99.3–99.7% similarity); and group 3 contains *Ec. gallinarum* and *Ec.*

casseliflavus (99.8% similarity). *Ec. dispar, Ec. saccharolyticus, Ec. sulfureus, Ec. columbae, Ec. cecorum,* and *Ec. faecalis* form individual lines of descent. Not all enterococci are of fecal origin. For example, *Ec. mundtii, Ec. sulfureus*, and *Ec. casseliflavus* have been isolated from plants; *Ec. malodoratus* from Gouda cheese (in which it caused off-flavor development); *Ec. durans* from milk and meat; *Ec. pseudoavium* from a cow with mastitis but not from bovine feces; and *Ec. raffinosus* from a clinical source (its habitat is not known). Therefore, their usefulness as indicators of fecal pollution is questionable.

Unfortunately, there are no simple biochemical or physiological tests that will categorically separate *Lactococcus* from *Enterococcus*. This can be done only by sophisticated techniques, such as protein profiling of extracts from whole cells by SDS-PAGE or genus-specific DNA probes. *Ec. faecalis* and *Ec faecium* are commonly found in the feces of humans and animals. These two species are easily separated from *Lactococcus* by their ability to grow at pH 9.6, at 10°C and at 45°C, and in the presence of 6.5% NaCl. The more recently recognized *Enterococcus* species do not give positive results with some of these tests. For example, *Ec. dispar* and *Ec. sulfureus* do not grow at 45°C, and *Ec. cecorum* and *Ec. columbae* do not grow at 10°C or in the presence of 6.5% salt and therefore may be confused with *Lactococcus* spp. In addition, a few, mainly from humans, isolates of *Lactococcus* grow at 45°C and in the presence of 6.5% NaCl. However, the likelihood of isolating these species from starters and cheese is small. Enterococci ferment sugars by the glycolytic pathway to L lactate.

In the 1930s, Lancefield introduced a method, based on the antigenic structure of the cell wall, to help in identifying what were then *Streptococcus* spp. Prior to the division of *Streptococcus* into *Lactococcus, Enterococcus*, and *Streptococcus*, the terms *fecal streptococci*, *Group D streptococci,* and *enterococci* were used interchangeably. Group D species are found in both *Enterococcus* (e.g., *Ec. faecalis, Ec. faecium, Ec. durans*) and *Streptococcus* (e.g., *Sc. bovis* and *Sc. equinus*), and therefore the Group D descriptor is illogical and the term *fecal streptococci* is now defunct. *Lactococcus* spp. reacts with Group N antiserum. The newer species of *Enterococcus,* e.g., *Ec. dispar*, *Ec. cecorum*, and *Ec. columbae*, do not have a Lancefield antigen. Based on these findings, Lancefield groupings are probably of little relevance today.

5.3.3 Streptococcus

These are also spherical to ovoid cells, arranged in pairs or chains. Currently, 39 species of *Streptococcus* are recognized but only one of them, *Sc. thermophilus,* is used as a starter culture. It grows at 45°C but not at 10°C and in the presence of 2.5% but not 4% NaCl, and it was included in the viridans group by Sherman (1937). It is closely related to *Sc. salivarius*, an inhabitant of the mouth. A few years ago, it was renamed *Sc. salivarius* subsp. *thermophilus*, but it has recently been restored to full species status. It ferments sugars by the glycolytic pathway to L lactate and does not have a Lancefield antigen.

5.3.4 Leuconostoc

These are spherical or lenticular cells that occur in pairs and chains and are commonly found in mesophilic cultures. Therefore, they can be confused with lactococci and (heterofermentative) lactobacilli. They differ from lactococci in three fundamental respects:

1. They ferment sugars heterofermentatively rather than homofermentatively, producing equimolar amounts of lactate, ethanol, and CO_2. Small amounts of acetate may also be produced.
2. They produce the D rather than the L isomer of lactate.
3. With the exception of *Ln. lactis*, they show no visual evidence of growth in litmus milk, unless yeast extract (0.3g/100 ml) is added.

Currently, the following species of *Leuconostoc* are recognized: *Ln. lactis*, *Ln. citreum*, *Ln. pseudomesenteroides*, *Ln. argentinum*, *Ln. fallax*, *Ln. amelibiosum*, *Ln. gelidum*, *Ln. carnosum*, *Ln. mesenteroides* subsp. *mesenteroides*, *Ln. mesenteroides* subsp. *dextranicum*, and *Ln. mesenteroides* subsp. *cremoris*. Despite the fact that leuconostocs were first identified in mixed-strain starter cultures in 1919, the exact species found in starters is still not clear. In the interim, different species have been implicated, but they have never been identified taxonomically, as researchers were more interested in how they behaved in milk fermentations than in their taxonomy. *Ln. mesenteroides* subsp. *cremoris* is certainly involved, since starter cultures are the only known source of this microorganism. It is also likely that *Ln. lactis* is involved. *Ln. mesenteroides* subsp. *cremoris* is unusual in that it ferments only lactose and its component monosaccharides, glucose, and galactose. *Leuconostoc* spp. do not hydrolyze arginine (Table 5–1).

5.3.5 Lactobacillus

These are rod-shaped cells that may be long and slender or short and sometimes bent, and they often occur in chains. Currently, about 64 different species of *Lactobacillus* are recognized. The genus is divided into three groups: obligately homofermentative, facultatively heterofermentative, and obligately heterofermentative, depending on whether they contain aldolase and phosphoketolase. The obligate heterofermenters can be coccobacillary in shape and may be confused with *Leuconostoc*.

The obligate homofermenters (Group 1) contain aldolase but not phosphoketolase and hence cannot ferment pentoses or gluconate; they ferment hexoses exclusively by the glycolytic (homofermentative) pathway to DL, L, and/or D lactate. This group includes all the thermophilic lactobacilli found in starter cultures (*Lb. helveticus*, *Lb. delbrueckii* subsp. *bulgaricus*, and *Lb. delbrueckii* subsp. *lactis*). In addition, all strains of *Lb. delbrueckii* subsp. *bulgaricus* and most strains of *Lb. delbrueckii* subsp. *lactis* excrete galactose in proportion to the amount of lactose taken up by the cell (see Section 5.4.3). *Lb. helveticus* produces DL lactate while *Lb. delbrueckii* produces only the D isomer.

The facultative heterofermenters (Group 2) contain both aldolase and phosphoketolase, and therefore they ferment hexoses homofermentatively to lactate and pentoses and gluconate heterofermentatively to lactate and acetate. Growth on glucose represses the formation of phosphoketolase. This group includes several of the lactobacilli found in artisanal cultures and mature cheeses (e.g., *Lb. casei, Lb. paracasei, Lb. plantarum*, and *Lb. curvatus*). These are generally referred to as the *nonstarter lactic acid bacteria* and are also called *mesophilic lactobacilli*. Nonstarter LAB counts of 10^7 to 10^9 cfu/g can be found in Cheddar cheese within 2 months of ripening (see Chapters 10 and 11).

The obligate heterofermenters (Group 3) possess phosphoketolase but not aldolase, and hence, like *Leuconostoc*, they ferment sugars heterofermentatively to equimolar concentrations of lactate, ethanol, and CO_2. Small amounts of acetate may be produced also. Almost invariably, members of this group produce NH_3 from arginine. The only Group 3 lactobacilli reported in cheese are *Lb. brevis* and *Lb. fermentum*, and these are also considered to be nonstarter bacteria.

Generally, Group 1 lactobacilli do not grow at 15°C but do grow at 45°C, while those in Groups 2 and 3 grow at 15°C but not at 45°C. This is not an absolute rule but generally applies to most of the lactobacilli found in starters and cheese. *Lb. fermentum* is an exception, as it is the only Group 3 *Lactobacillus* that grows at 45°C. The lactobacilli found in starter cultures are often called thermophilic because their optimum growth temperature is around 42°C. They are not true thermophiles, as they do not grow at 55°C. They are, however, able to withstand the cooking temperature (54°C) used in Swiss cheese manufacture.

5.3.6 Differentiation of Lactic Acid Bacteria in Starters

A relatively simple scheme involving only a few criteria including Gram reaction, catalase, shape, growth at different temperatures and in the presence of different concentrations of NaCl, and the pathway by which sugar is fermented can be used to identify the LAB found in starters (Table 5–4). The scheme is not definitive but provides good practical guidance.

During the past 20 years, several new techniques for identifying LAB have been developed, including methods for analyzing the amino acid and menaquione content of the cell wall, PAGE of the whole cell proteins, 16S rDNA sequencing, and randomly amplified polymorphic DNA (RAPD) techniques. All these techniques are very sophisticated and are either too slow or require standardized conditions or elaborate equipment for routine use. Species-specific probes based on 16S rDNA sequences and amplification by polymerase chain reaction (PCR) have been or are being developed for all LAB. These methods are relatively simple to use and will probably become routine tools for identification within a few years.

Recently, Taillez, Tremblay, Ehrlich, and Chopin (1998) used a RAPD technique with three different primers to study the relationships between 113 strains of lactococci from a culture collection. The strains could be divided into two groups, G1 and G2 plus G3, based on the genomic analysis, and they showed excellent agreement with the phenotypic analysis. Groups G1 and G3 contained only the *lactis* subspecies, whereas group G2 contained almost exclusively dairy strains of the *cremoris* subspecies. It was suggested that Group G2 evolved from the G3 strains through the loss of specific characteristics in the dairy environment.

Scanning electron micrographs of some species of LAB found in starter cultures are shown in Figure 5–4. The considerable variation that occurs in the shape of these bacteria is evident. A microcolony of *Lb. helveticus* in Grana cheese is also included.

5.3.7 Phylogeny

There is sound scientific evidence that all living organisms, including animals, plants, and microorganisms, evolved from a common ancestor. The study of the evolutionary history of bacteria is called phylogeny and it has received much attention during the past 20 years in attempts to understand the relationships between different microorganisms. To study phylogeny, one selects and compares the sequence of a macromolecule that is present and has the same function in all cells. Such molecules have been called evolutionary chronometers. The 16S rDNA gene is probably most widely used for this purpose, because it is a relatively large molecule (≈ 15,000 nucleotides) and some sequences are highly conserved while others are variable. A comparison of the sequences has allowed scientists to construct the universal tree of life. Bacteria form one of the 3 major domains (Archea and Eucarya are the others), and within the bacterial domain there are at least 12 distinct phylogenetic lineages, of which the Gram-positive bacteria are one. Gram-negative bacteria do not form a single group. The Gram-positive bacteria are divided into two branches, the clostridial branch, with a % mol guanine + cytosine (GC) content of less than 50 and the actinomycete branch, with a % mol GC greater than 55. All starter bacteria and the bacteria found in ripening cheese are Gram-positive. *Sc. thermophilus* and the species of starter LAB found in the genera *Lactococcus, Lactobacillus, Enterococcus*, and *Leuconosoc* belong to the clostridial subdivision, whereas other bacteria commonly found in ripening cheese *(Micrococcus*, *Corynebacterium, Brevibacterium*, and *Arthrobacter)* belong to the actinomycete branch of the Gram-positive bacteria (see Chapter 10).

Computer analysis of the 16S rRNA molecules has revealed short oligonucleotide sequences, called signature sequences, that are unique to each bacterial species and that enable the construction of species-specific nucleic acid probes usable for identifying microorganisms. These complement the biochemical and physi-

Table 5–4 Criteria for Discriminating between the Different Groups of Lactic Acid Bacteria Found in Cultures

					Growth in NaCl			*Growth at*		
	Shape	*Catalase*	*Gram's Stain*	*Fermentation of Glucose*	*20 g/L*	*40 g/L*	*65 g/L*	*10°C*	*15°C*	*45°C*
Lactococcus	Cocci	–	+	Homo	+	+/–	–	+	+	–
Leuconostoc	Cocci	–	+	Hetero	+	+/–	–	+	+	–
Enterococcus	Cocci	–	+	Homo	+	+	+	+	+	+
Sc. thermophilus	Cocci	–	+	Homo	+	–	–	–	?	+
Thermophilic (Group 1) *Lactobacillus*	Rods	–	+	Homo	+	–	–	?	–	+
Mesophilic (Group 2 & 3) *Lactobacillus*	Rods	–	+	Hetero or homo	+	+/–	+/–	?	+	–

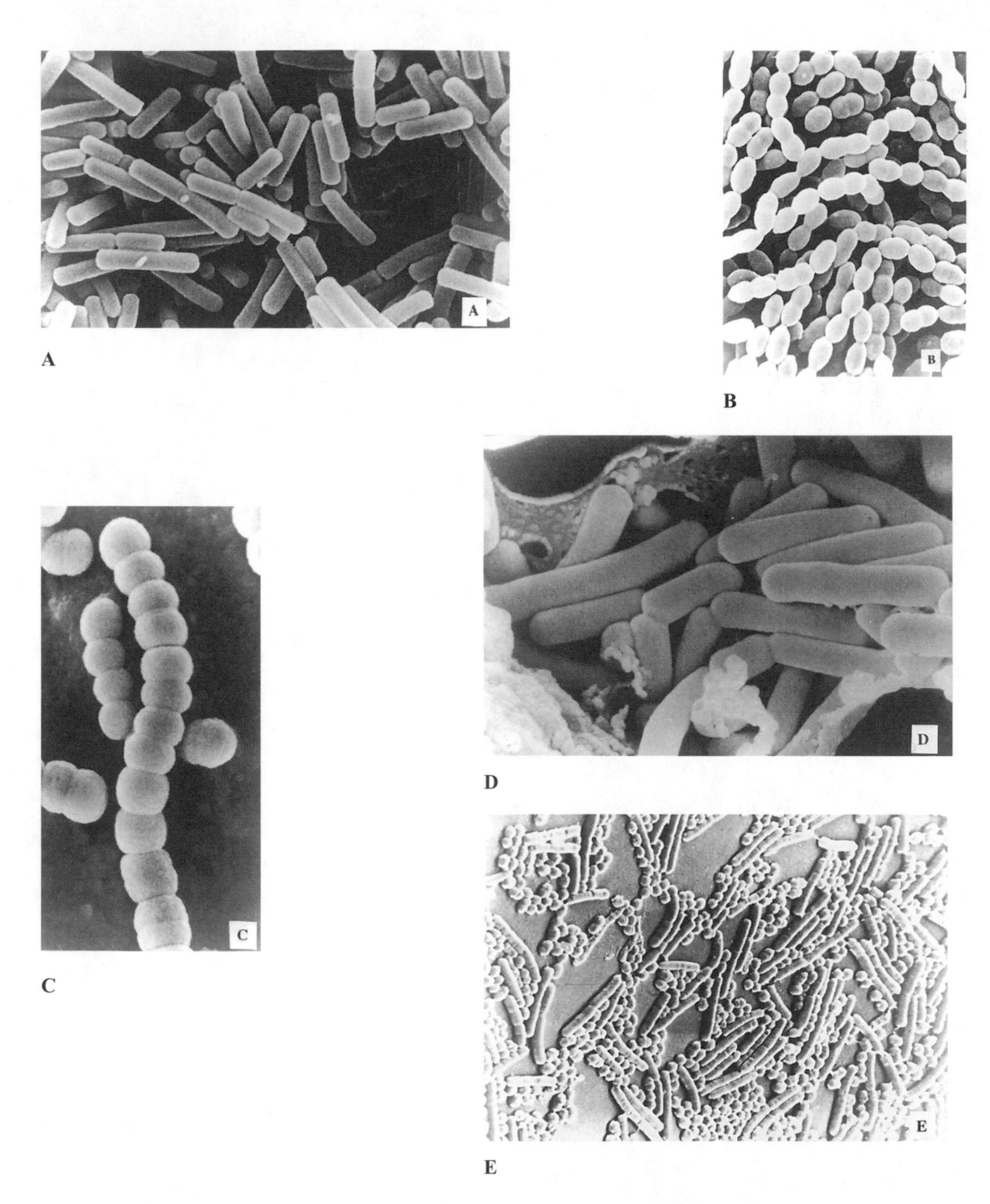

Figure 5–4 Scanning electron micrographs of some bacteria found in starter cultures. (A) *Lactobacillus helveticus* (13,000×); (B) *Streptococcus thermophilus* (9,000×), (C) *Lactococcus lactis* (13,000×); (D) a microcolony of *Lactobacillus helveticus* in Grana cheese (20,000×); (E) a mixed culture of *Streptococcus thermophilus* and *Lactobacillus delbrueckii* subsp. *bulgaricus*, used in yogurt manufacture (3,750×). The degree of magnification is shown in parenthesis.

ological tests normally used to identify bacteria.

The phylogenetic relationships of starter and some other LAB are outlined in Figure 5–5. Such analyses show interesting relationships. Both *Ec. faecalis* and *Str. bovis* are found in bovine feces and both react with Group D antigen, yet it is clear that *Enterococcus* is only distantly related to *Streptococcus* and *Lactococcus*. Although not shown, *Pediococcus* spp. are found in the *Lb. casei/Lb. parcacasei* group even though pediococci (tetrads) are morphologically quite distinct from lactobacilli (rods). In addition, the heterofermentative thermophile, *Lb. fermentum*, is relatively closely related to the facultatively heterofermentative mesophile, *Lb. casei,* and not to the other thermophilic lactobacilli. Recently, several heterofermentative lactobacilli, *Lb. viridescens*, *Lb. confusus,* and *Lb. halotolerans*, which are coccobacillary in shape, and the heterofermentative *Leuc. paramesenteroides* have been transferred to a new genus called *Weisella*. These considerations suggest that neither shape, nor type of fermentation, nor growth at 10°C, 15°C, or 45°C give absolute information on the relationship of LAB with each other.

The 16S rDNA sequences have also shown that *Enterococci* are more closely related to *Carnobacterium* and *Vagococcus* than to *Lactococcus* and *Streptococcus*.

5.4 METABOLISM OF STARTERS

5.4.1 Proteolysis

All LAB are auxotrophic and require several amino acids and vitamins for growth. Specific strains of LAB are still used to assay foods for vitamins (e.g., *Lb. delbruckii* ATCC 7830 for vitamin B_{12} and *Ec. faecalis* ATCC 8043 for folic acid). The requirements for amino acids are strain specific and vary from as few as 4 to perhaps 12 or more. Glutamic acid, methionine, valine, leucine, isoleucine, and histidine are required by most lactococci, and many strains have additional requirements for phenylalanine, tyrosine, lysine, and alanine. The amino acid requirements of *Sc. thermophilus* and *Leuconostoc* spp. are similar to those of the lactococci. Only one strain of *Lb. helveticus* has been studied; it required all the amino acids except glycine, alanine, serine, and cysteine. Lactococci possess many of the genes of the amino acid biosynthetic pathways in their chromosomes, and the amino acid requirements are probably the result of minor deletions in the nucleotide sequences. This is probably also true for the other starter LAB, but the question has not been studied.

Fully grown milk cultures of starter bacteria contain around 10^9 cfu/ml. The concentrations of amino acids and peptides in milk are low and sufficient to sustain only about 25% of the maximum number of starter cells present in a fully grown starter culture. Therefore, starter bacteria must have a proteolytic system to hydrolyze the milk proteins to the amino acids required for good growth in milk. The proteolytic system of *Lactococcus* involves a cell wall–associated proteinase, amino acid and peptide transport systems, and peptidases (for a review see Kunji, Mierau, Hagting, Poolman, & Konings, 1996). The generally accepted view of the system is shown in Figure 5–6.

The lactococcal proteinase (PrtP) is one of the most intensively studied enzymes of starter bacteria, but it is likely that other starter bacteria have similar systems. It is a serine proteinase that is synthesized in the cell as a pre-pro-proteinase and that is transformed into the mature, active proteinase by a process not yet completely understood. It is not a truly extracellular enzyme but is anchored to the cell membrane by its extremely hydrophobic C-terminal region. The mature proteinase, called PrtP, contains about 1,800 amino acid residues, has a molecular mass of about 185 kDa, and has an optimum pH of about 6. The gene encoding the proteinase (*prtP*) in several strains of lactococci has been cloned and sequenced. The nucleotide sequences are very similar (98% identical), implying that there is only one proteinase, but the amino acid sequences are sufficiently different to result in different activities on the various caseins. Two dif-

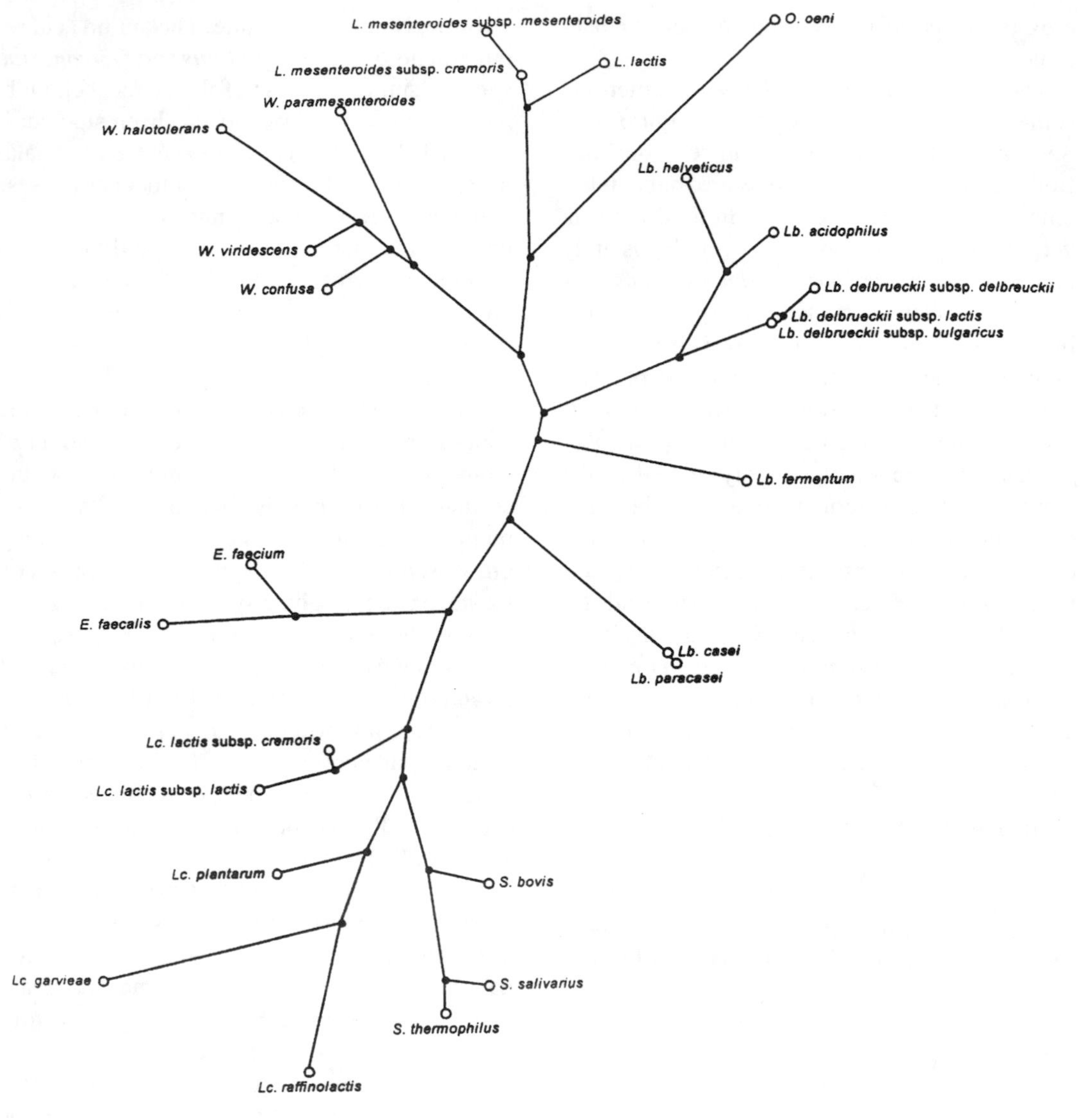

Figure 5–5 Phylogenetic tree showing the relationships among some starter lactic acid bacteria. *E, Enterococcus; L, Lactobacillus; Lc, Lactococcus; L, Leuconostoc; S, Streptococcus; O, Oenoccos; W, Weissella.*

ferent specificities are found, one of which, PI proteinase, hydrolyzes principally β-casein and to some extent κ-casein, while the other, PIII proteinase, acts efficiently on α_{s1}-, β-, and κ-caseins. PI-type proteinases hydrolyze the C-terminal region of β-casein, which is quite hydrophobic, producing bitter peptides that may be responsible for the development of bitter-flavored cheese. One way of preventing this is to use only strains that produce PIII-type proteinase as starters.

Only limited information is available on the proteinases of the other starter bacteria, but the available data indicate that they are similar to the lactococcal proteinases (e.g., the PrtP of *Lc. lactis* and *Lb. pararcasei* are 95% similar). Unusually, no proteolytic activity has been detected in *Sc. thermophilus* except in the so-called Asian

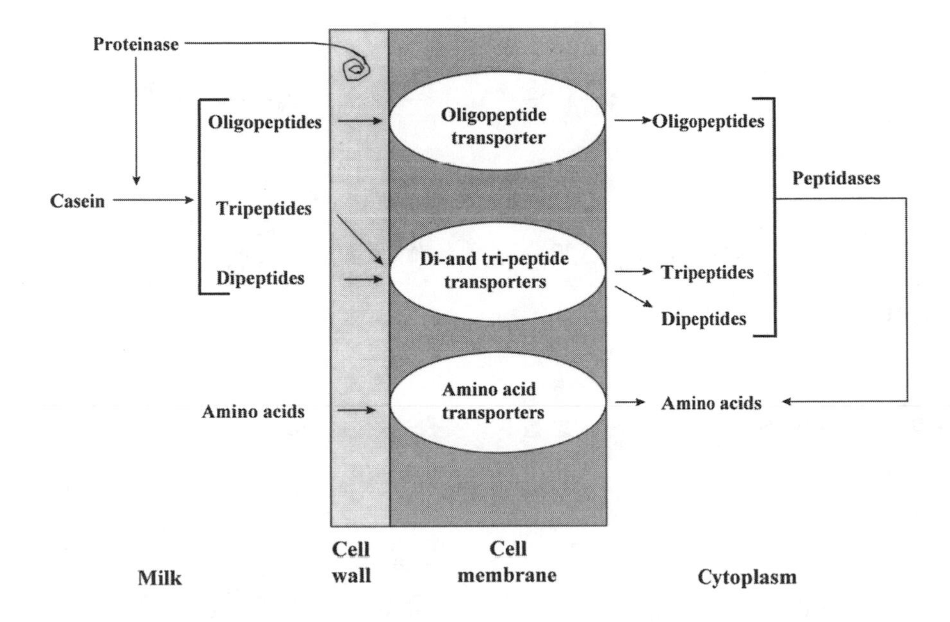

Figure 5–6 The proteolytic system of *Lactococcus.*

strains, which were isolated in outer Mongolia, India, and Japan. This lack of proteolytic activity may explain the symbiotic relationship between *Lb. delbrueckii* subsp. *bulgaricus* and *Sc. thermophilus* in yogurt cultures (Figure 5–2).

There are four different caseins in milk, α_{s1}-, α_{s2}-, β-, and κ-, which occur in the ratio of roughly 4:1:3:1 and make up about 80% of the total protein in milk (see Chapter 3). Hydrolysis of these proteins by lactococcal proteinases results in production of numerous oligopeptides of different sizes. For example, hydrolysis of β-casein, which contains 209 amino acid residues, by the PI proteinase results in the production of 100 peptides, the majority of which contain between 4 and 30 amino acid residues. Peptides containing up to 8 amino acid residues can be transported across the cell membrane into the cell. In the lactococci, various transport systems, including an oligopeptide transport system, a di- and tripetide transport system, and at least 10 amino acid transport systems, which have a high specificity for structurally related amino acids (e.g., Glu/Gln, Leu/Ile/Val) have been described. The driving forces for transport include the proton motive force, antiport and symport systems, and ATP hydrolysis (see Kunji et al., 1996, for a review).

Inside the cell, the peptides are hydrolyzed by peptidases to the individual amino acids necessary for the synthesis of the proteins required for cell growth. Numerous peptidases have been identified in starter LAB, including at least three aminopeptidases (Pep N, Pep A, and Pep C), two tripeptidases (Pep T and Pep 53), and two dipeptidases (Pep V and Pep D) that release single amino acids from the N-terminal end of the relevant substrates. In *Lc. lactis* subsp. *cremoris*, two different endopeptidases (Pep O and Pep F) have been identified that hydrolyze internal peptide bonds in peptides but not in the intact caseins. The proline content of casein is quite high, and, because of its structure, specific peptidases, called prolidases (Pep Q), aminopeptidase P (Pep P), X-prolyl-dipeptidyl aminopeptidase (Pep X), prolinase (Pep R), and proline iminopeptidase (Pep I), are required to hydrolyze it from peptides. Some of the important

Table 5–5 Properties of the Various Peptidases Found in Starter Lactic Acid Bacteria

Peptidase	*Name*	*Substrate n = 1,2,3...*	*Organism*	*Mol. Wt. (kDa)*	*Type*[a]	*Optimum pH*	*Location*[b]
General							
Aminopeptidase N	Pep N	X $\downarrow$ $(X)_n$	*Lc. lactis*	95	M	7	I
			Lb. delbrueckii	95	M	7	I
			Lb. helveticus	97	M	6.5	CW, I
Aminopeptidase C	Pep C	X $\downarrow$ $(X)_n$	*Lc. lactis*	50	T	7	I
			Lb. delbrueckii	52	T	7	I
			Lb. helveticus	50	T		I
Tripeptidase	Pep T	X $\downarrow$ X-X	*Lc. lactis*	46–52	M	7.5	I
	Pep 53		*Lc. lactis*	>23	M	5.8	CW
Dipeptidase	Pep V	X $\downarrow$ X	*Lc. lactis*	50	M	8	I
			Lb. delbrueckii	51	M	7.5	
			Lb. helveticus	50	M	8	I
	Pep D	X $\downarrow$ X	*Lb. helveticus*	54	T	6	
Proline Specific							
Prolidase	Pep Q	X $\downarrow$ Pro	*Lc. lactis*	42	M	7–8	I
			Lb. delbrueckii	41	M		I
Aminopeptidase P	Pep P	X $\downarrow$ Pro-$(X)_n$	*Lc. lactis*	43	M	8	I
X-prolyl-dipeptidyl aminopeptidase	Pep X	X Pro-$\downarrow$ $(X)_n$	*Lc. lactis*	59–90	S	7–8.5	CW, I
			Lb. delbrueckii	82–95	S	6.5 –7	I
			Lb. helveticus	88–95	S	6.5	I
Prolinase	Pep R	Pro $\downarrow$ X	*Lb. helveticus*	35		7.5	I
Proline iminopeptidase	Pep I	Pro $\downarrow$ X-$(X)_n$	*Lc. lactis*	30–50	S		I
			Lb. delbrueckii	33	S	6.5	CW
			Lb. helveticus	34	S		I

Note: Many of the enzymes were isolated from several strains of the species, and the data presented are a summary.

[a] M = metalloenzyme; S = serine peptidase; T = thiol peptidase.

[b] I = intracellular; CW = cell wall.

properties of the peptidases are summarized in Table 5–5. Generally, the peptidases were isolated from several strains of the same species, and the data shown are a summary. Pep R has been found only in *Lb. helevticus*. No carboxypeptidase has been found in LAB.

The peptidases are either serine-, metallo-, or thiol-enzymes and have pH optima in the range 6.0 to 8.0. All of them are located inside the cell and, acting together, they hydrolyze the peptides transported into the cell by the oligopeptide and the di- and tripeptide transport systems to their constituent amino acids for use in protein synthesis. They are also important in the ripening of cheese. During ripening, the starter bacteria gradually die and lyse and release their intracellular peptidases, which then act on any peptides present around the cell. The amino acids produced are considered to be the precursors of the flavor compounds necessary for the development of good-flavored cheese (see Chapter 11).

5.4.2 Arginine Metabolism

Many LAB produce NH_3 from arginine (Table 5–1) by the arginine deiminase pathway and simultaneously produce 1 mol of ATP per mol of arginine metabolized (Figure 5–7). Arginine is first hydrolyzed to NH_3 and citrulline by arginine deiminase. Ornithine carbamyltransferase then catalyses the phosphorolysis of citrulline to ornithine and carbamyl phosphate; the latter is then hydrolyzed to NH_3 and CO_2 by carbamate kinase, with the concomitant production of ATP. The uptake of arginine is driven by an antiport transport system in which ornithine is exchanged for arginine.

Lc. lactis subsp. *lactis* produces NH_3 from arginine via this pathway, whereas *Lc. lactis* subsp. *cremoris* does not, owing to the lack of at least one of the three enzymes of the pathway.

5.4.3 Lactose Metabolism

Lactose, the principal sugar found in milk, is a disaccharide composed of one glucose and one galactose residue linked by a β1-4 bond (see Chapter 3). The major function of starter LAB in cheesemaking is the production of lactic acid from the fermentation of lactose. Since starter bacteria do not contain a functional cytochrome system, their metabolism of sugars is fermentative rather than respirative, and ATP is produced by substrate-level phosphorylation rather than by oxidative phosphorylation. Despite their lack of a cytochrome system, most LAB are aerotolerant and grow quite well in air. They cannot grow without a sugar that serves as an energy source. Fermentation of sugars occurs by the glycolytic (Figure 5–8) or phosphoketolase (PK) (Figure 5–9) pathways. Lactate is the end-product of glycolysis, while lactate, ethanol, and CO_2 are the end products of the PK pathway.

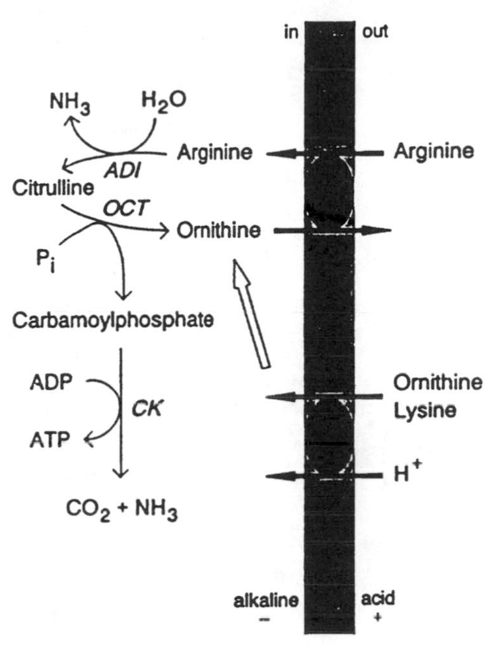

Figure 5–7 Arginine/ornithine antiport and the arginine deiminase pathway. Accumulation of ornithine (lysine) via the Dp-driven lysine transport system is also shown. ADI, arginine deiminase; OCT, ornithine carbamoyltransferase; CK, carbamate kinase.

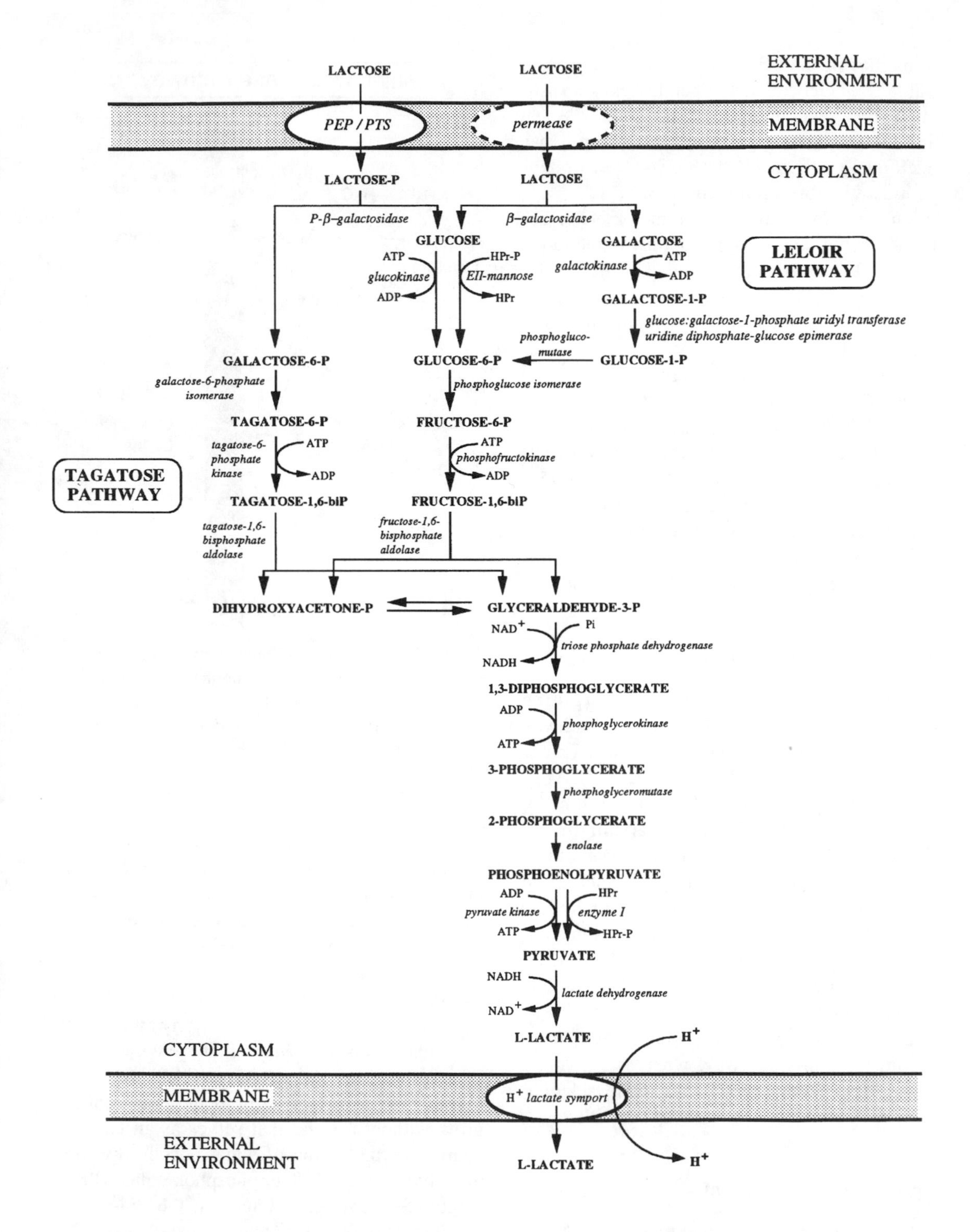

Figure 5–8 Glycolytic pathway of lactose metabolism in lactic acid bacteria.

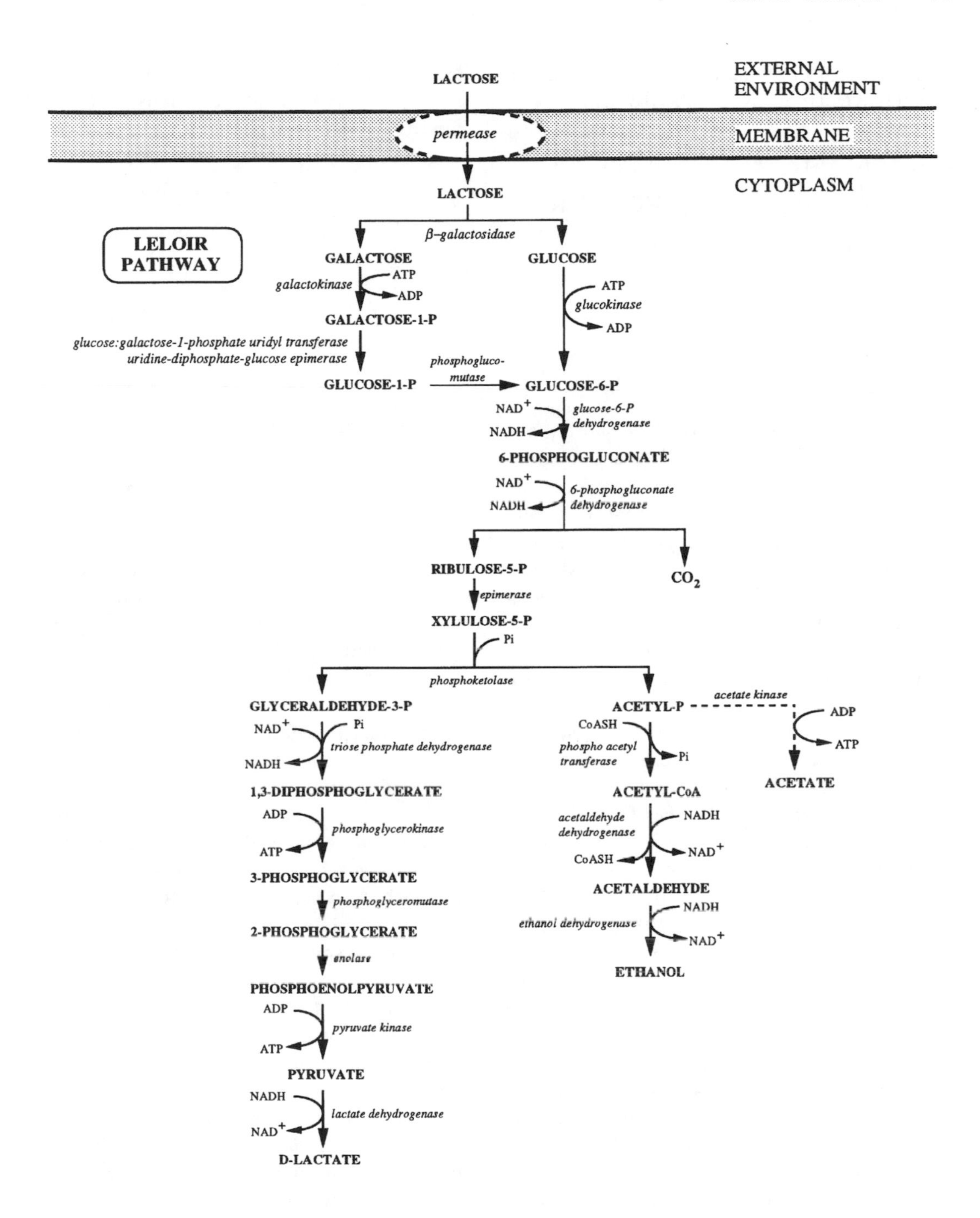

Figure 5–9 Phosphoketolase pathway of lactose metabolism in lactic acid bacteria.

Lactic acid contains an asymmetric carbon and hence can exist as D and L isomers:

```
     COOH                 COOH
      |                    |
  H—C—OH              HO—C—H
      |                    |
     CH3                  CH3

D-lactic acid        L-lactic acid
```

The isomer of lactate produced is useful in the identification of the various genera and species of starter LAB. For example, *Leuconostoc* and *Lb. delbrueckii* produce only the D isomer, *Lactococcus* and *Sc. thermophilus* produce only the L form, and *Lb. helveticus* produces a mixture of the D and L isomers, owing to the presence of two lactate dehydrogenases in the cell, one of which is specific for the L and the other for the D isomer.

Before lactose can be fermented, it must be transported into the cell. Starter LAB use two different systems to transport lactose, the permease and the phosphotransferase (PTS), both of which require energy. Energy for the permease system is derived from ATP, and lactose is transported without being transformed. In the PTS system, the energy is derived from phosphoenolpyruvate (PEP) in a complex series of reactions involving two enzymes, EI and EIII, and a heat-stable protein, HPr. During transport, the high-energy phosphate in PEP is eventually transferred to lactose to form lactose-phosphate rather than lactose (Figure 5–10). Lactococci use the PTS system to transport lactose, while all other starter LAB use the permease system.

The initial enzyme involved in the metabolism of lactose depends on the transport system used. In the PTS system, lactose-P is formed during transport and is hydrolyzed by phospho-β-galactosidase (Pβgal) to glucose and galactose-6-P, while in the permease transport system, lactose is transported intact and is hydrolyzed by β-galactosidase (βgal) to glucose and galactose (Table 5–6).

In the lactococci, the subsequent fermentation of glucose is via the glycolytic pathway, and galactose-6-P is metabolized through several tagatose derivatives to glyceraldehyde-3-P and dihydroxy acetone phosphate (Figure 5–8). Tagatose is a stereoisomer of fructose:

```
       CH2OH              CH2OH
        |                  |
        C=O                C=O
        |                  |
   HO—C—H             HO—C—H
        |                  |
       HC—OH          HO—CH
        |                  |
       HC—OH              HC—OH
        |                  |
       CH2OH              CH2OH

   D-fructose          D-tagatose
```

Fermentation of glucose and galactose by thermophilic starter LAB also occurs via glycolysis but some of these bacteria, including *Sc. thermophilus, Lb. delbrueckii* subsp. *bulgaricus*, and some strains of *Lb. delbrueckii* subsp. *lactis*, ferment only the glucose moiety of lactose and excrete galactose in proportion to the amount of lactose transported. Initially, this was thought to be due to the lack of galactokinase, the first enzyme of the Leloir pathway, but more recent studies have suggested that galactose is excreted as an exchange molecule in the transport of lactose into the cell, in which process a single transporter, lactose permease, simultaneously transports lactose into and galactose out of the cell. *Leuconostoc* spp. ferment the glucose and galactose by the phosphoketolase (PK) pathway; the galactose is first transformed to glucose-1-P via the Leloir pathway (Figure 5–9).

The products of both pathways are quite different. In glycolysis, 1 mole of lactose is transformed to 4 moles of lactic acid (or 2 in the case of *Sc. thermophilus, Lb. delbrueckii* subsp. *bulgaricus*, and the strains of *Lb. delbrueckii* subsp. *lactis*, which excrete galactose), whereas in the PK pathway it is transformed to 2 moles each of lactate, ethanol, and CO_2.

The reason for the production of lactate in both pathways and of ethanol in the PK pathway is the need to reoxidize the NADH and NADPH produced in the earlier steps of the pathway to

Table 5–6 Salient Features of Lactose Metabolism in Starter Organisms

Organism	*Transport*[a]	*Pathway*[b]	*Cleavage Enzyme*	*Products (mol/mol lactose)*	*Isomer of Lactate*
Lactococcus	PEP-PTS	GLY	Pβgal	4 lactate	L
Leuconostoc	Permease	PK	βgal	2 lactate + ethanol + 2 CO_2	D
Sc. thermophilus	Permease	GLY	βgal	2 lactate[c]	L
Lb. delbrueckii	Permease	GLY	βgal	2 lactate[c]	D
Lb. helveticus	Permease	GLY	βgal	4 lactate	DL

[a]PTS = phosphotransferase system.
[b]GLY = glycolysis; PK = phosphoketolase.
[c]These species metabolize only the glucose moiety of lactose.

allow fermentation to continue. The purpose of the fermentations is to produce sufficient ATP to sustain growth. Production of ATP by fermentation is much less efficient than the production of ATP by respiration (e.g., in glycolysis, 4 moles of ATP are produced per mole of lactose fermented, compared with a possible 76 moles by respiration). Therefore, to produce the same amount of ATP by fermentation as by respiration, a large amount of sugar must be fermented, and consequently large amounts of lactic acid are produced by LAB.

The growth of some strains of *Lactococcus* on galactose or low levels of glucose leads to the production of other compounds, besides lactate, from pyruvate, such as ethanol, acetate, and acetaldehyde (Figure 5–11). In these bacteria, lactic dehydrogenase (LDH) and pyruvate-formate lyase (PFL) are allosteric enzymes. LDH requires the presence of fructose-1,6-bisphosphate for activity, whereas pyruvate-formate lyase (PF) activity is inhibited by triose-phosphates. Normally, both effectors are present at high concentrations in the cell, favoring LDH activity and lactate production. Growth at low sugar concentrations results in lower intracellular concentrations of the effectors, and therefore LDH activity is reduced and PFL activity increased, which allows pyruvate to be channelled to ethanol and acetate rather than lactate. Normally, during anaerobic growth, PFL activity is favored over PDH activity in the initial formation of acetyl CoA.

The end-products of lactose fermentation are mainly acidic and will, unless excreted, acidify the cell cytoplasm. LAB have two mechanisms for excreting lactate and protons. One involves the transmembrane reversible F_oF_1-ATPase and is responsible for the secretion of protons. The second involves the simultaneous secretion of lactate anions and protons in symport with each other. This mechanism occurs especially when the external concentration of lactate is low and the internal concentration is high. Energy can also be derived from this reaction through the creation of a proton-motive force.

5.4.4 Citrate

Cit^+ *Lactococcus* and *Leuconostoc* spp. present in mesophilic starters metabolize citrate to acetate, CO_2, diacetyl, acetoin, and 2,3-butanediol by the pathway shown in Figure 5–12. Citrate is not used as an energy source but is co-metabolized with lactose or some other sugar. The organisms involved are Cit^+ lactococci, *Ln. mesenteroides* subsp. *cremoris,* and *Ln. lactis*. Citrate is not metabolized by thermo-

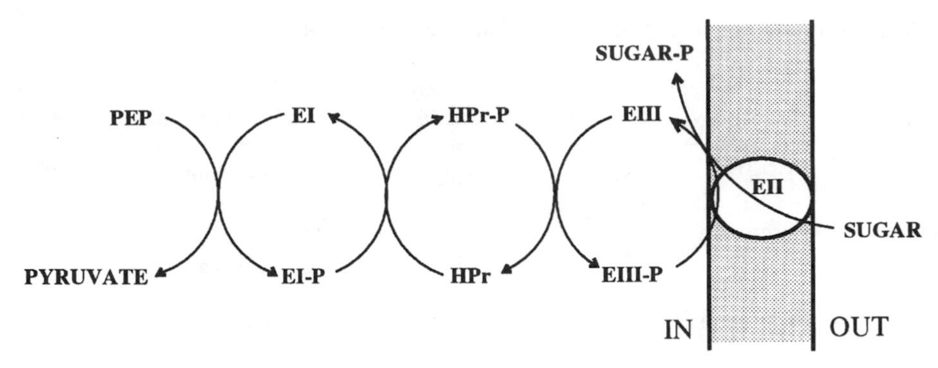

Figure 5–10 Phosphoenolpyruvate-phosphotransferase system for transport of sugar in *Lactococcus lactis.*

philic starters. The CO_2 produced is responsible for the small eyes in Edam and Gouda cheese, while diacetyl and acetate are important contributors to the flavor of many fermented products, including Quarg, Fromage frais, and Cottage cheese. Diacetyl is produced in only small amounts (< 10 mg/ml or 0.11 mM in milk), and acetoin production is generally 10–50 times greater than that of diacetyl. One mole of acetate is produced from each mole of citrate used, but recent studies suggest that about 1.2 moles of acetate are produced per mole of citrate used. This is probably due to the production of small amounts of acetate from sugar metabolism. There is very little information on the concentration of 2,3-butanediol produced by starters.

There is still controversy on how diacetyl is actually produced. One of the reasons for this is that the putative enzyme, diacetyl synthase, has never been categorically found in LAB. Another, and probably more important, reason is that acetolactate (AL) is very unstable and readily autodecarboxylates nonoxidatively to acetoin or oxidatively to diacetyl. AL is so unstable that it is available commercially only as a double ester to protect it from autodecarboxylation; it is normally hydrolyzed with 2 equivalents of NaOH just before use. Acetoin is produced by AL decarboxylase activity, but it is generally believed that diacetyl is produced chemically rather than enzymatically from AL. Despite this, it is difficult to see how diacetyl can be produced oxidatively from AL by starter cultures, which essentially grow anaerobically and produce very low E_h values (≈ -250 mV). Acetolactate is not normally found as an end-product unless the strain lacks acetolactate decarboxylase activity.

Growth and metabolite production by a D and a DL mesophilic mixed-strain culture in milk at 21°C are shown in Figure 5–13. The cell numbers and the production of end-products are plotted semi-logarithmically because bacteria grow and produce end-products exponentially. Arithmetic plots are quite different. The graphs show examples, and it is important to remember that individual cultures will exhibit some variation. Production of the various metabolites is in the following order: lactate > acetate > acetoin >> diacetyl. Citrate utilization is generally slower in L than in D or DL cultures because of the slower growth of the Cit^+ *Leuconostoc* (compared with Cit^+ *Lactococcus*). Production of diacetyl and acetoin ceases as soon as all the citrate has been used, after which the levels of acetoin and diacetyl may decrease due to the activity of acetoin dehydrogenase. The same enzyme is probably responsible for the reduction of diacetyl to acetoin and of acetoin to 2,3-butanediol. Therefore, to retain the maximum amount of diacetyl in unripened cheese, the product should be cooled as soon as possible after citrate utilization is complete.

Little, if any, acetolactate (AL) accumulates in cultures, because most Cit^+ lactococci contain an active AL decarboxylase (ALD). However, the Cit^+ lactococci in the D culture in Figure 5–13 contain an inactive ALD and hence accumulate AL. Nucleotide sequences of the *ald* gene in this

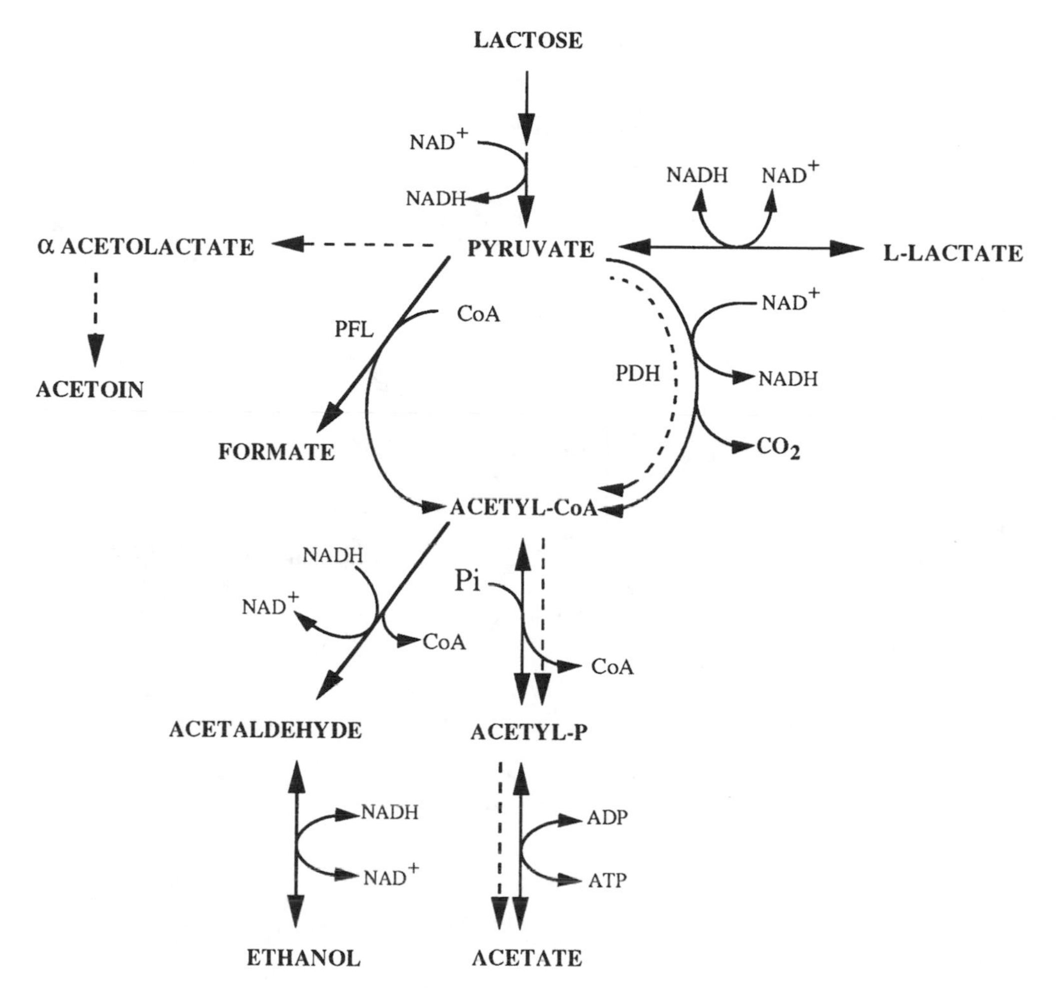

Figure 5–11 Pathways of pyruvate metabolism in lactic acid bacteria.

strain and in a strain containing ALD activity showed only one substitution (cytosine for thymine at position 659), which resulted in a change of histidine to tyrosine in a motif conserved in all ALDs. It is thought that this change is sufficient to lead to the synthesis of an inactive enzyme (Goupil, Corthier, Ehrlich, & Renault, 1996). Autodecarboxylation of AL to acetoin (mainly) and diacetyl probably occurs throughout growth, but is observed only as soon as AL production ceases (when all the citrate is used). The level of diacetyl produced from autodecarboxylation can be increased by aeration at acid pH, and this property is used to produce diacetyl in the manufacture of lactic butter by this particular culture.

Cit^- lactococci are the dominant bacteria in mesophilic mixed-strain cultures. The levels of the Cit^+ lactococci and leuconostocs in these cultures vary but generally comprise less than 10% of the total number of bacteria present. It is obvious from Figure 5–13 that lactic acid production follows the increase in cell numbers very closely. This explains why the amount of lactic acid produced is a good indicator of the growth of LAB. As lactic acid is produced, the pH decreases, so pH is also a good indicator of growth. However, the relationship between the increase in acid production (or cell numbers) and pH is not linear.

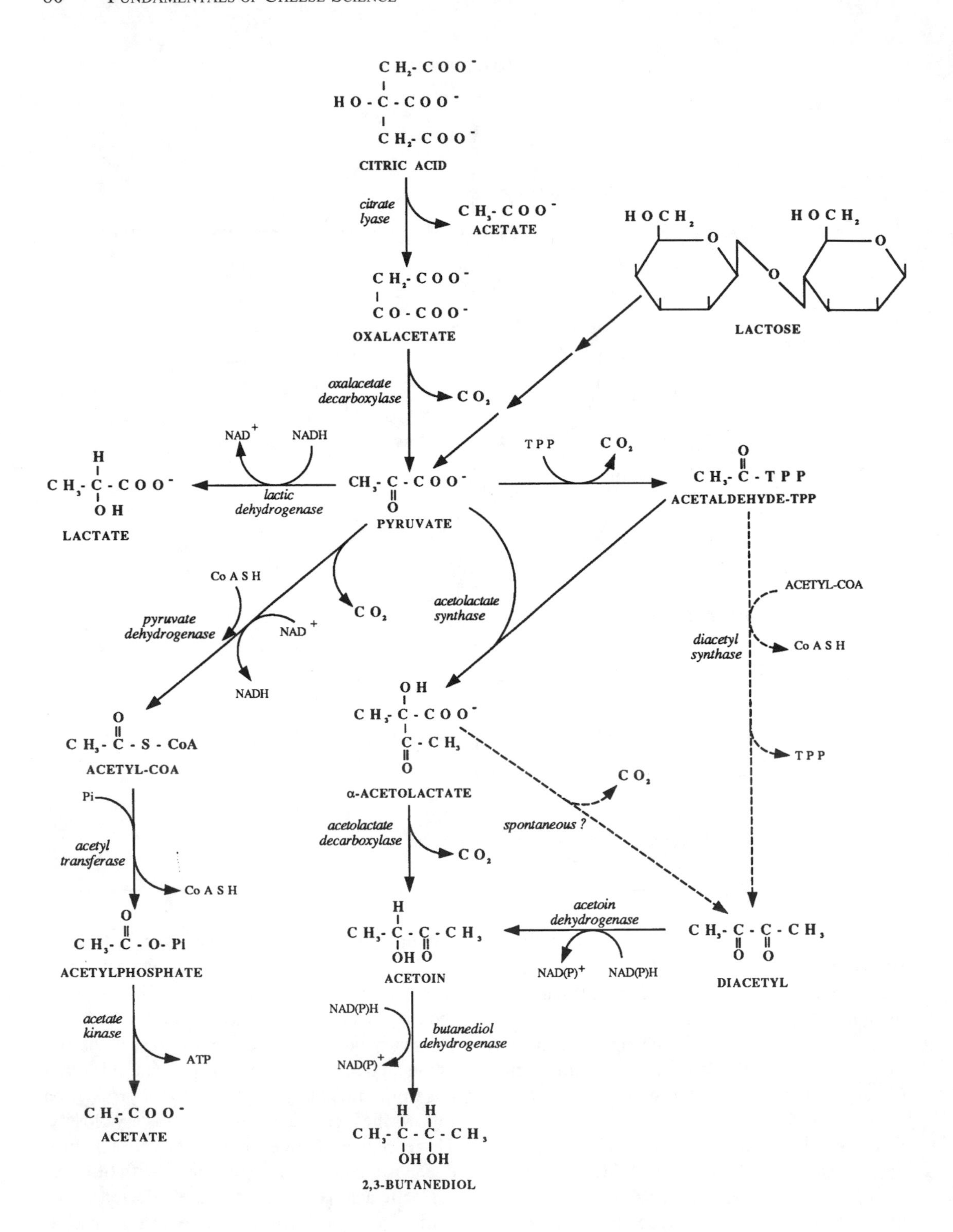

Figure 5–12 Citric acid metabolism in lactic acid bacteria.

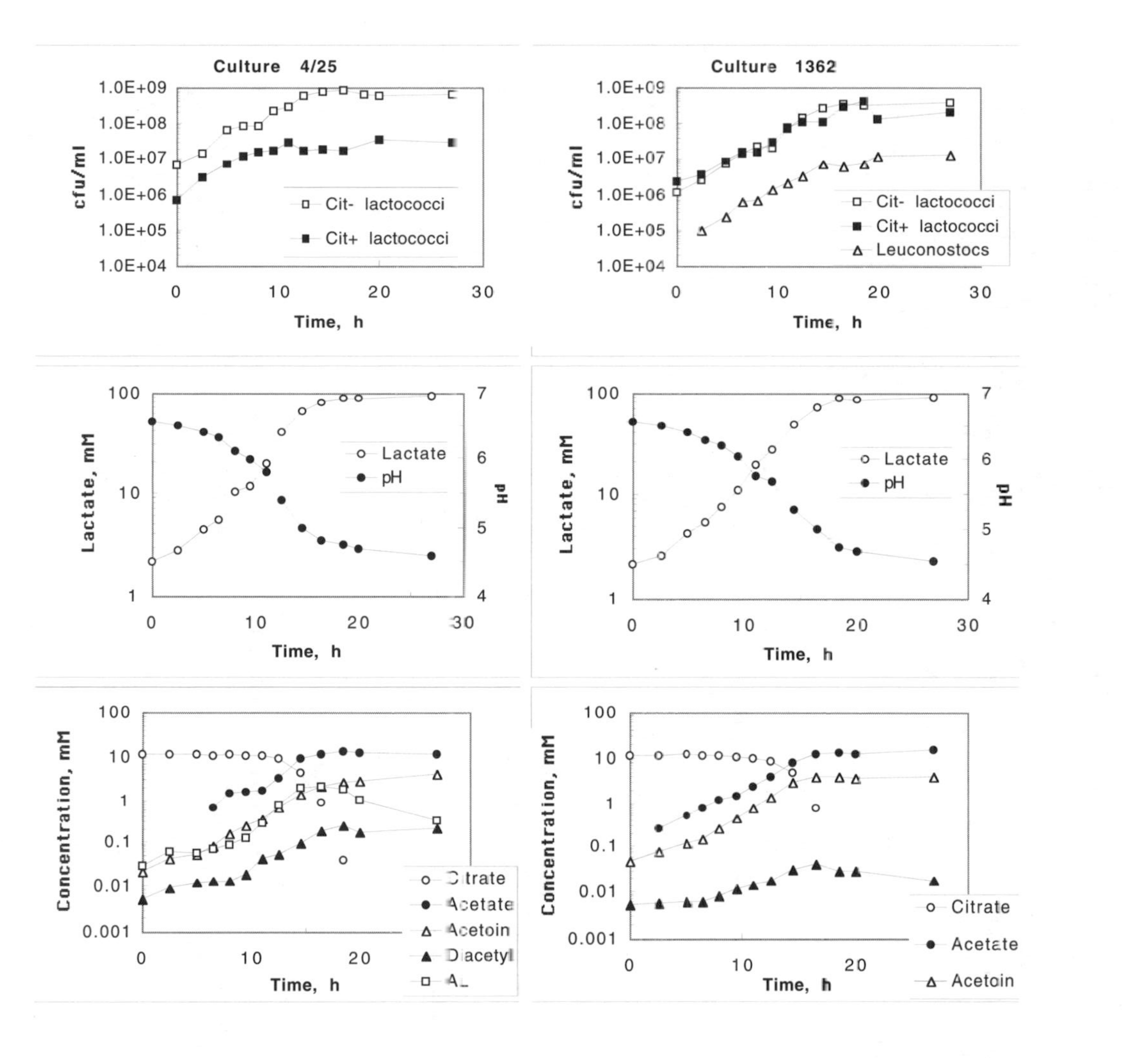

Figure 5–13 Growth and metabolite production by a D (4/25), a DL (1362), and an L (Fr8) culture in sterile 10% (w/v) reconstituted skim milk at 21°C. The DL and L cultures are commercial cheese cultures; the D culture is used in the production of lactic butter.

Nevertheless, pH is easy to measure and is the most widely used indicator of starter growth in the cheese industry.

Pure cultures of Cit+ *Lactococcus* and *Leuconostoc* differ in the products produced during co-metabolism of citrate and lactose. The former produce lactate, acetoin, and CO_2 in a manner similar to that of mixed-strain mesophilic cultures. In contrast, Cit+ *Leuconostoc* spp. produce no acetoin or diacetyl. Instead, they produce increased amounts of lactate and acetate. Pyruvate is an intermediate in the metabolism of both lactose and citrate. In leuconostocs, the pyruvate produced from the metabolism of both substrates is converted to lactic acid. This relieves the cells from forming ethanol to regenerate the pyridine nucleotides. Instead, the acetyl-P produced from lactose is used to produce acetate and ATP:

$$\text{Acetyl-P} + \text{ADP} \rightarrow \text{Acetate} + \text{ATP}$$

This extra production of ATP also results in much faster growth of the organism. However, leuconostocs will produce diacetyl and acetoin from citrate in the absence of an energy source, and proportionately more of the citrate is converted to these compounds as the pH decreases. The question then may be asked, how do leuconostocs produce diacetyl and acetoin in mixed cultures? The answer is not clear, but it is probably due to the fact that leuconostocs cannot take up much lactose at pH values below 5.5.

5.4.5 Acetaldehyde

Acetaldehyde is produced by both mesophilic and thermophilic cultures and is an important component of the flavor of yogurt. The amount produced varies but can reach 30 mg/ml. It is generally considered to be a product of carbohydrate (pyruvate) metabolism (Figure 5–12), but it can also be produced from threonine via threonine aldolase activity:

$$\text{Threonine} \rightarrow \text{glycine} + \text{acetaldehyde}$$

This is the mechanism by which acetaldehyde is produced by lactococci. The only strain of *Lactococcus* that has a requirement for glycine also lacks threonine aldolase, implying that the physiological function of this reaction is the provision of glycine for growth. It is not clear how acetaldehyde is produced by thermophilic cultures.

Both types of bacteria present in yogurt cultures produce acetaldehyde, but *Sc. thermophilus* produces greater amounts than *Lb. delbrueckii* subsp. *bulgaricus*, and both types produce more when grown together than when grown individually, due to symbiosis between the two.

In some fermented dairy products, particularly those prepared with mesophilic cultures, the ratio of diacetyl to acetaldehyde is important. The optimum ratio is 4:1, but when the ratio falls to 3:1, a "green" off-flavor defect, reminiscent of yogurt, develops. One of the functions of *Leuconostoc* spp. in mixed cultures is to reduce the acetaldehyde produced by *Lactococcus* to ethanol, which, at the concentrations found in cultures, has no effect on their flavor.

5.5 PLASMIDS

Plasmids are extrachromosomal pieces of DNA that are much smaller than the chromosome; their molecular mass ranges from about 2 to 100 kDa. Several of the commercially important properties of starters—including the proteinase production, transport of citrate, several of the enzymes involved in transport and metabolism of lactose, exopolysaccharide production, bacteriocin production (Section 5.8), and resistance to phage (Section 5.7.5)—are commonly encoded on plasmids. The lactose plasmid encodes the enzymes of the tagatose pathway (galactose-6-phosphate isomerase, tagatose-6-phosphate kinase, and tagatose-1,6-bisphosphate aldolase), Pβgal, and enzyme I and enzyme III of the PTS system. Exopolysaccharide production by starter LAB is responsible for thickening several Scandinavian fermented milk products (e.g., Taette and Skyr).

Plasmids are easily lost upon subculturing, and once this occurs, the properties encoded by the genes on that plasmid are also lost. For this reason, subculturing of starters should be limited. Instead, several aliquots of the starter can be frozen. When required, an aliquot is thawed, and, after two or three subcultures, it is discarded and replaced by another frozen aliquot.

Cells that lose the proteinase plasmid become proteinase negative (Prt^-) and consequently grow poorly in milk. Those that lose the lactose plasmid are unable to metabolize lactose and therefore cannot grow in milk, whereas those that lose the citrate transport plasmid are unable to metabolize citrate, even though they contain the necessary enzymes. In some strains of starters, the proteinase and lactose genes are encoded on the same plasmid, but in other strains two separate plasmids are involved. Prt^- strains are often isolated from mesophilic mixed-strain starters. In these cultures, Prt^- strains rely on Prt^+ strains to produce the amino acids and peptides required for growth. Conjugative plasmids (i.e., plasmids that can mediate their own transfer to other cells) played a major role in our understanding of the genetics and metabolism of LAB.

5.6 INHIBITION OF ACID PRODUCTION

Slow acid development during cheesemaking is an important cause of poor quality cheese. There are four main causes of slow acid production: natural inhibitors, the presence of antibiotics in the milk, bacteriophage, and bacteriocins. Of these, bacteriophage is the most important.

Milk contains natural inhibitors, called *lactenins*, that inhibit the growth of some strains of starter bacteria. The lactenins have been identified as immunoglobulins and lactoperoxidase (LP). The immunoglobulins cause susceptible starter bacteria to aggregate. This causes localized acid production and precipitation, and in severe cases the aggregates settle on the bottom of the cheese vat. The starters still continue to grow, but localized acid production is so great that they eventually inhibit themselves. Immunoglobulins are inactivated (denatured) by pasteurization.

LP requires H_2O_2 and thiocyanate (SCN^-) for activity. SCN^- is normally present in milk, and the concentration is higher in milk from cows fed *Brassica* plants (cabbage and kale), while the H_2O_2 can be produced by the starter bacteria during growth or through xanthine oxidase or glucose oxidase activity (see Chapter 4). The three components are required together to inhibit the growth of starters. The actual inhibitor has not been identified but is thought to be $OSCN^-$. LP is quite heat resistant and is not inactivated by HTST pasteurization, but it is inactivated by heating to 80°C for a few seconds (so-called flash pasteurization). Inhibition of starters by lactenins is unusual in modern cheesemaking because the strains used have been selected so as not to be affected to any great extent by lactenins.

Antibiotic residues occur in milk because of their use to control mastitis in dairy cows. An effective way of curing mastitis is to infuse the cow's udder with antibiotics, especially penicillin or its derivatives. This results in contamination of the milk with antibiotics. The concentration of antibiotics in the milk decreases with each milking, and generally all the antibiotics will be excreted within 72 hr, depending on the preparation used. Milk from cows treated with antibiotics should be withheld for the length of time prescribed for the antibiotic preparation. The sensitivity of starter cultures to various antibiotics used in mastitis treatment is shown in Table 5–7.

Thermophilic cultures are much more sensitive to penicillin and more resistant to streptomycin than mesophilic cultures. In the past, antibiotic residues were a major cause of slow acid production in cheese manufacture, but nowadays, with the availability of simple and sensitive tests for the detection of antibiotic residues in milk and with better education of farmers, problems due to antibiotic residues in milk are rare.

5.7 BACTERIOPHAGE

The major cause of slow acid production in cheese plants today is bacteriophage (phage). This can significantly upset manufacturing schedules and, in extreme cases, result in complete failure of acid production or "dead vats." Phage for *Lactococcus* were first reported in New Zealand in 1935 and since then they have been described for all starter LAB.

Phage are viruses that can multiply only within a bacterial cell. They are ubiquitous in

Table 5–7 Concentration of Some Antibiotics That Cause 50% Inhibition of Growth of Starter Bacteria in Milk

		Antibiotic (μg/ml ± SD)			
Organism	*No. of Strains*	Penicillin	Cloxacillin	Tetracycline	Streptomycin
Lc. lactis subsp. *cremoris*	4	0.11 ± 0.028	1.69 ± 0.38	0.14 ± 0.02	0.67 ± 0.15
Lc. lactis subsp. *lactis*	4	0.12 ± 0.025	2.16 ± 0.41	0.15 ± 0.05	0.53 ± 0.18
Sc. thermophilus	3	0.01 ± 0.002	0.42 ± 0.07	0.19 ± 0.06	10.5 ± 0.29
Lb. delbrueckii subsp. *bulgaricus*	2	0.03 ± 0.006	0.29 ± 0.04	0.37 ± 0.04	3.0 ± 2.0
Lb. delbrueckii subsp. *lactis*	1	0.024	0.24	0.60	2.29

nature and are so small that they can be "seen" only with the electron microscope. They have a head, which contains the DNA, and a tail, which is composed of protein. A photomicrograph of a phage for *Lc. lactis* is shown in Figure 5–14. Morphologically, three types of phage for lactococci can be distinguished: small isometric-headed (spherical-headed) phage (the most common), prolate-headed (oblong-headed) phage, and large isometric-headed phage. The tail varies in length from 20 to 500 nm. Other features, such as collars between the head and the tail, base plates at the end of the tail, and fibers on the base plates, may also be present. Phage multiplication occurs in one or two ways, called the *lytic* and *lysogenic cycles*. Lytic and lysogenic phages are sometimes called *virulent* and *temperate*, respectively.

5.7.1 The Lytic Cycle

In the lytic cycle (Figure 5–15), the first step involves adsorption of the phage onto special attachment sites, called phage receptors, on the cell surface of the host. This step requires Ca^{2+}, and prevention of this step is the basis for the use of phosphate and citrate as chelators in phage-inhibitory media. Phage adsorb to the cell through their tails. Once a phage has attached to the receptors, it injects its DNA into the host cell. Immediately, phage DNA and phage proteins are produced rather than host cell DNA and proteins. The phage DNA is packaged in a concentrated form in the phage head, and when phage synthesis is completed, the cell lyses, releasing new phage particles, which start the process again. Lysis is caused by a lytic enzyme, called lysin, which is encoded in the phage DNA and hydrolyzes the cell wall of the host cell.

The growth of virulent phage in a sensitive host is characterized by both a *latent period* and a *burst size*, which are determined in one-step growth experiments (Figure 5–16). For such experiments, phage and whole cells are mixed so that the ratio of cells to phage (the multiplicity of infection) is low, and the number of phage is monitored periodically during incubation. At the beginning, the number of phage remains low, since new phage are being synthesized inside the cell. This is called the latent period and spans the time from the initial adsorption of the phage to the host cell until the detection of phage progeny after cell lysis. The suddenly increased number of phage, called the *burst size*, is caused by lysis of the host cells by phage lysin. For lactococcal phage, the latent period varies from 10 to 140 min and the burst size varies from about 10 to 300 phage. Compared with starters, phage multiplication is very rapid. Assuming a latent period of 1 hr and a burst size of 150, 1 phage will result

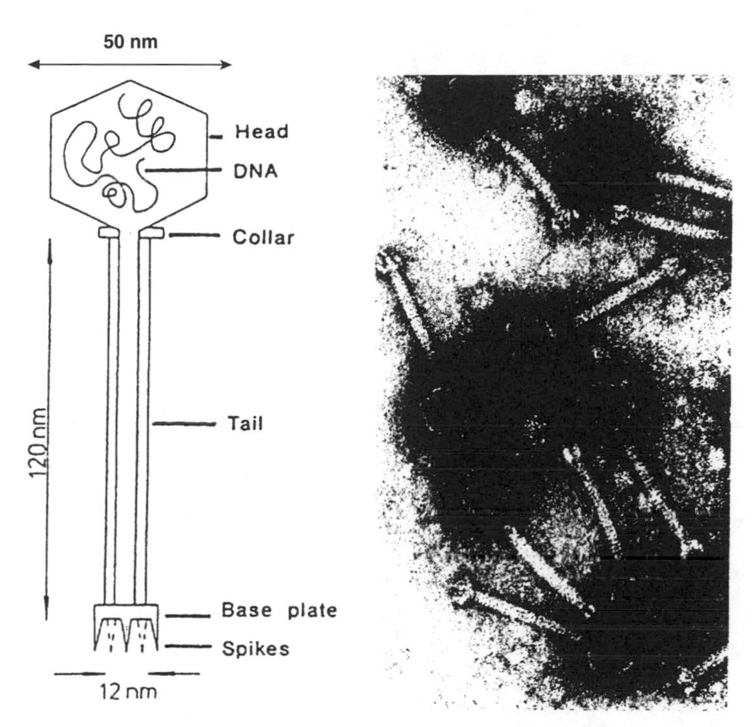

Figure 5–14 Schematic drawing (left) and corresponding electron micrograph (right) of a bacteriophage of *Lactococcus lactis*. The important structural components of the phage particle are shown in the drawing.

in the production of 22,500 phage (150 × 150) in a little over 2 hr. In 3 hr, the number of phage will increase to 3.4 × 10^6. In 3 hr, a *Lactococcus* cell will multiply three times, producing only 8 cells. Thus, the phage will vastly outnumber the bacterial cells very quickly. This clearly indicates the serious problems that occur following contamination with phage. Phage multiplication during cheesemaking is shown in Figure 5–17. The initial number of phage was 100/ml (i.e., log 2), and within a few hours multiplication to 10^7/ml (equivalent to log 7) had occurred.

5.7.2 The Lysogenic Cycle

The second phage multiplication process is called the *lysogenic cycle*. Adsorption and injection of DNA occur as in the lytic cycle, but, instead of phage multiplication, the phage DNA is inserted into the bacterial chromosome and multiplies with the chromosome. Under these conditions, the phage is called a *temperate phage* or a *prophage*, and the cells are considered to be lysogenized. Most strains of LAB are lysogenic, and in this state the host cell is immune to attack by its own phage. Generally, the prophage also immunizes the host cell to closely related strains of phage. This is called *superimmunity*.

In certain circumstances, some temperate phage can be induced, become lytic, and multiply. The host cells in which these phage multiply are called indicator strains. The conditions that cause induction in commercial practice are unclear, but in the laboratory UV light and the antibiotic mitomycin C are used. In many bacteria, lysogenic (temperate) phage are considered to be the source of phage, but this has not been shown for starter LAB, except for the strain of *Lb. casei* used in the production of the Japanese fermented milk Yakult.

5.7.3 Pseudolysogeny

Many mixed-strain starters are permanently infected with a low number of virulent phage. These are called "own" phage to distinguish

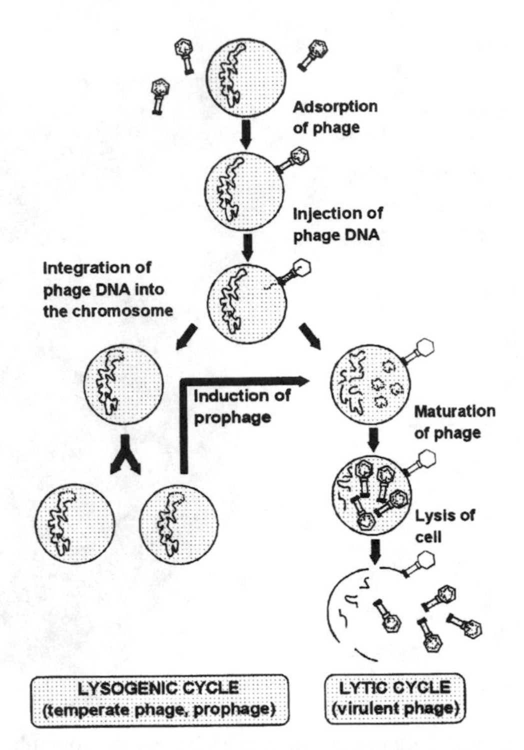

Figure 5–15 Schematic drawing of the propagation of virulent (right branch) and temperate (left branch) bacteriophage in a host cell. The cell is shown with its chromosomal DNA. The phage DNA is indicated by dotted lines.

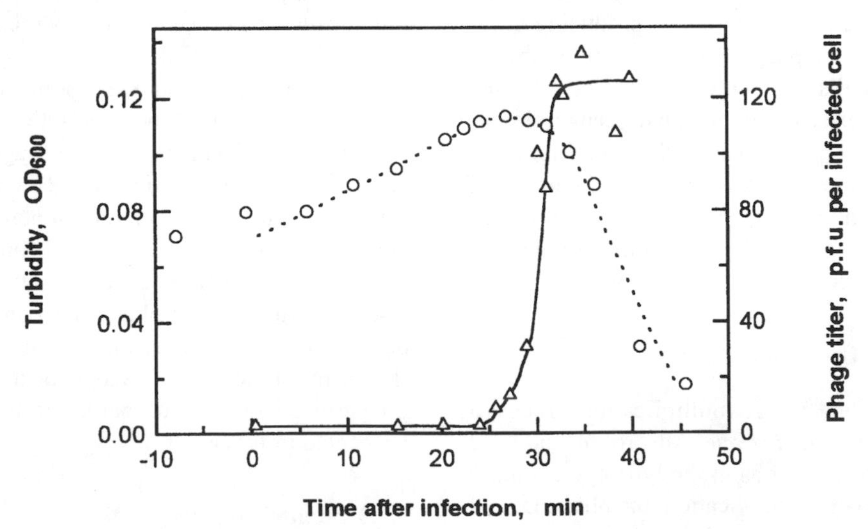

Figure 5–16 Results of a one-step growth experiment on *Lactococcus lactis* infected with a lytic phage. The release of phage (Δ) began 25 min (latent period) after infection (burst size = 124). The increase of free phage (as plaque-forming units [pfu]) was accompanied by a decrease in culture turbidity (O) due to cell lysis.

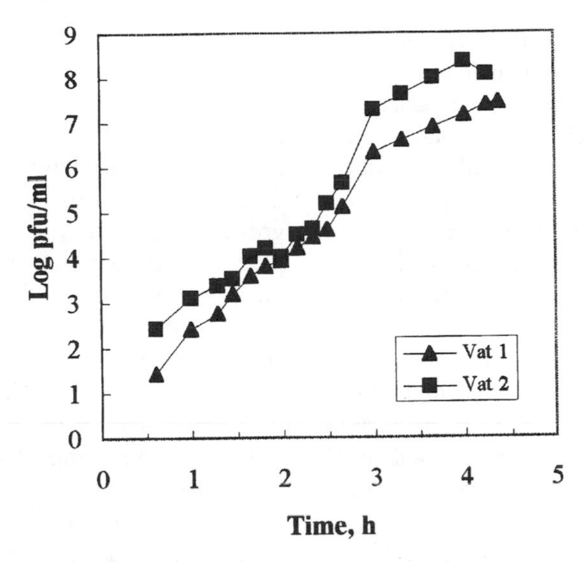

Figure 5–17 Phage multiplication (plaque-forming units [pfu]) during Cheddar cheese manufacture in two vats.

them from lytic or "disturbing" phage and they multiply on any phage-sensitive strains present in the culture. Normal growth of the mixed culture is unaffected by own phage because of the presence of large numbers of acid-producing, phage-insensitive cells. This phenomenon is called pseudolysogeny, and the phage insensitivity of the whole culture is stable as long as no infection with disturbing phage occurs.

5.7.4 Classification of Phage

Phage cannot multiply outside of their hosts, and therefore classical bacterial identification methods cannot be used to identify them. Other techniques have been developed, and some of these have already been mentioned (e.g., phage morphology). Phage protein composition, host range, serology, and DNA homology are also used. Host range is of considerable practical significance, since starter cultures attacked by the same phage should not be used in rotations. Host ranges can be broad (one phage attacks several strains) or narrow (the phage attacks only one or two strains). DNA homology has resulted in lactococcal phage being divided into 12 "species," 3 of which, P36 (isometric), P335 (isometric), and C2 (prolate), are commonly found in cheese plants.

5.7.5 Phage-Resistance Mechanisms

Several phage-resistance mechanisms, including inhibition of phage adsorption, restriction-modification mechanisms, and abortive infection mechanisms, are found in LAB. All of these are commonly encoded on plasmids.

In adsorption inhibition, the receptor sites for the phage on the cell surface are masked, so the phage cannot attach to the cell and hence are unable to multiply. In *Lc. lactis* subsp. *cremoris* SK110, adsorption inhibition has been shown to be plasmid encoded, and the masking agent has been identified as a galactose-containing lipoteichoic acid.

Restriction-modification involves two enzymes, one of which, the restriction enzyme, hydrolyzes the phage DNA. The other (modification) enzyme modifies the host cell DNA, usually by methylation of some of the nucleic acid bases, in such a way that the restriction enzyme cannot hydrolyze it. This mechanism is operative only after adsorption and injection of phage DNA.

Abortive infection (abi) is the term used for phage-resistance mechanisms that do not involve either inhibition of adsorption or restriction-modification systems. Generally, a total loss of plaque formation (see Section 5.7.6) or a reduction in plaque size occurs, as a result of a reduction in both the latent period and burst size, although in some cases only one of these is affected.

Many phage-resistance plasmids are conjugative, and this fact has been used to improve the phage resistance of phage-sensitive commercial cultures. The technique is relatively simple (Figure 5–18): Lac$^-$ mutants of the strain harboring the phage-resistance plasmid are isolated (lac$^-$ phage^r) and mixed with lac$^+$ phage^s recipients. Lac$^+$ phage^r transconjugants are selected in the presence of excess virulent phage on lactose agar, which contains a dye to indicate acid production. The lac$^-$ phage^r cells do not grow very well on this medium, and they are destroyed by the phage present in the agar. The lac$^+$ phage^r transconjugants are then isolated and checked for their ability to produce acid and for the presence of the phage-resistance plasmid. This is a totally natural process that does not involve genetic engineering techniques, and it has been used in the production of phage-resistant strains for commercial use. A new strategy involving the sequential use of the same strain of starter containing different phage defense mechanisms has also been suggested (Sing & Klaenhammer, 1993).

5.7.6 Detection of Phage

Phage are much smaller than their hosts and can be easily separated from them by filtration through a 0.45 μm filter. The host cells are retained by the filter while the phage particles are small enough to pass through into the filtrate. Phage are easily detected. A small volume (e.g., 0.1 ml) of a filter-sterilized (host-free) sample of material suspected of containing phage is added to 10 ml of milk that has been inoculated with a susceptible host. After incubation at the optimum temperature of the host for 6–10 hr, the pH is measured. A difference of more than 0.3 units between the pH of host grown in the absence or presence of the material suspected of containing phage indicates the presence of phage (Figure 5–19). The decrease in pH can also be visualized by adding a suitable pH indicator (bromocresol purple) to the milk. In broths, measurement of the optical density at 600 nm is used. The optical density increases for a little while as the host grows but then decreases as the cells lyse. These procedures give no indication of the number of phage present.

The number of phage in a sample can be counted relatively easily. The material suspected of containing phage is filter sterilized, and 10-fold dilutions of the filtrate are made. Then 1 ml of each dilution is mixed with 0.1 ml of the host culture (containing 10^8 cells), 0.1 ml of 0.185 M Ca^{2+}, and 2.5 ml of a suitable molten medium containing 0.7% agar tempered to 45°C. (This medium is referred to as "sloppy agar," and its function is to allow the phage to diffuse and form fairly large plaques.) The mixture is then poured immediately onto a prehardened plate of the same medium containing the normal amount of agar. After incubation at the optimum temperature of the host for several hours, clear zones, called plaques, are seen in the background lawn of bacterial growth if phage are present (Figure 5–20). Each plaque is considered to have arisen from one phage, and counting the number of plaques and multiplying by the dilution factor gives the number of plaque-forming units (pfu) per ml.

5.7.7 Source of Phage

To control phage, it is important to identify their source. However, the ultimate source of phage for LAB is still unclear. In most bacteria, lysogenic phage are considered to be the source of lytic phage. Many mixed cultures contain lysogenic phage, but no DNA homology has been found between these phage and lytic phage that attack these cultures. Whey contains a large number of phage, and aerosols produced during separation of fat from whey are probably the primary source of phage in cheese plants. Raw milk is also considered to be an important source of phage, but only relatively small numbers of lactococcal phage have been isolated from it.

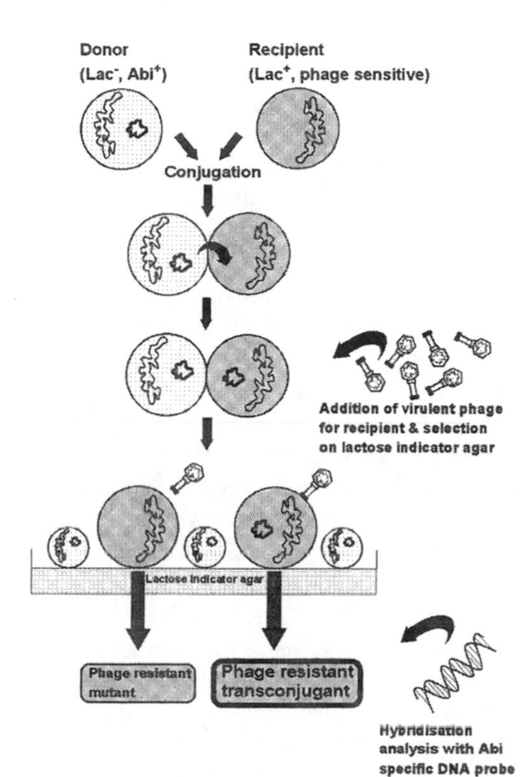

Figure 5–18 Schematic representation of a genetic strategy for the construction of phage-insensitive starter cultures by conjugation. A phage-resistance plasmid with an abortive infection determinant is transferred from a donor lacking the ability to metabolize lactose (Lac^-) to a recipient culture that can metabolize lactose (Lac^+). Phage-resistant transconjugants are selected on lactose indicator agar in the presence of the virulent phage. Under these conditions, the Lac^- donor cells do not grow well. Finally, a hybridization experiment is carried out to distinguish true transconjugants from phage-resistant mutants.

5.7.8 Control of Phage

Starters are often produced in large volumes; a 500 L tank of starter contains around 5×10^{14} cells. One phage getting into such a tank could have serious consequences. Therefore, the most important factor in producing good quality cheese is to ensure that the bulk starter is free of phage. This point cannot be overstressed, but it is often overlooked in commercial practice. Daily determination of phage levels in starters and cheese whey should be an integral part of any well-designed quality control scheme in a modern cheese factory. The important factors in controlling phage are

- use of a limited number of cultures
- aseptic inoculation
- use of phage-inhibitory media heat-treated to 85°C for 30 min
- rotation of phage-unrelated strains
- physical separation of starter and cheese production areas (there should be no direct access between the starter unit and cheese production)
- addition of starter and rennet together
- chlorination of vats between fills
- use of closed vats
- physical separation of the cheese production area from the separator used to recover fat from whey
- heat treatment (> 90°C for 45 min) of starter containing phage before discarding it

Many of these factors are common sense but a few need further elaboration. The greater the number of strains used in a plant, the greater is the likelihood of a phage attack. Therefore, defined strain cultures should be used instead of mixed cultures. Phage are quite resistant to heat. Some phage withstand heating at 75°C for a few minutes, and hence the medium used for growing the starter must be heat-treated at a high temperature (e.g., 85°C for 30 min) to inactivate any phage that might be present. The medium should be heated in the tank in which the culture is to be grown. Some cheese manufacturers heat-treat the milk in a pasteurizer and then fill the starter tank with the heated milk. This practice is not recommended because of the danger that phage might be present in an otherwise clean starter tank. One phage in a tank can be sufficient to cause a reduction in the activity of a culture. The heat treatment has other effects beside the inactivation of phage: it improves the nutritional value of the milk as a culture medium because it inactivates the natural inhibitors and causes slight hydrolysis of the proteins. Bulk cultures contami-

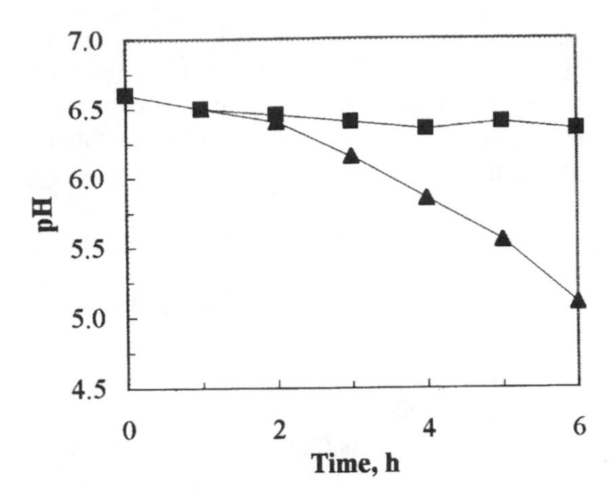

Figure 5–19 Effect of phage on acid production by *Lactococcus lactis* in 10% (w/v) reconstituted skim milk at 30°C; control, ▲; infected with phage, ■.

nated with phage should be heat-treated (> 90°C for 45 min) to prevent their spread.

In the past, milk was the medium used for growth, but nowadays phage-inhibitory media are used (see Section 5.9). Inoculation should be done as aseptically as possible; an aseptic inoculation device has been developed and is commonly used in the Netherlands for starter production (Figure 5–21).

During cooling, the air entering the starter tank should be filtered through a high-efficiency particulate air filter to prevent phage in the air from entering the tank. Transfer of airborne phage into the starter tank during incubation may be eliminated by a slight positive air pressure in the tank and starter room.

Defined cultures are also very useful in controlling phage. Such cultures contain only a small number of phage-unrelated strains and help to reduce the numbers of different phage present in factories. The same strains are used each day and wheys from bulk starter and cheesemaking are checked daily for phage against all the strains in use. Bulk starter whey should always be clear of phage. If a phage infection does occur in the bulk starter culture, the culture should not be used, even if the phage level is low, as phage will multiply more rapidly than bacteria when subcultured in the cheese milk. Instead, the bulk culture should be replaced by a bacteriophage-insensitive mutant (BIM) or another phage-unrelated strain. The bulk culture should be heat-treated at higher than 85°C for 30 min before discarding it to prevent the spread of phage throughout the factory. One can tolerate the presence of phage in the cheese whey and, indeed, it is probably inevitable, but if the level increases to more than 10^6 pfu per ml of whey, then the infected strain should be removed and replaced with a phage-unrelated culture or a BIM.

BIMs are relatively easily isolated. The phage-sensitive strain is grown with the phage in milk until the milk coagulates. Coagulation is due to the growth of a small number of phage-insensitive strains that are present in almost every strain of starter. The coagulated culture is then plated on a suitable medium and several colonies examined for their ability to coagulate milk rapidly, for their resistance to the particular phage, and for their ability to grow in milk without causing off-flavors. It is sometimes not possible to isolate BIMs from cultures; presumably, such cultures do not contain phage insensitive cells.

Addition of rennet as soon as possible after the starter has been added to the milk in the cheese vat also helps, because the coagulum will physically separate phage-infected cells from noninfected cells, and the phage are unable to penetrate the curd to locate noninfected cells.

As little as 1 ml of residual whey in a cheese vat can be a potent source of lytic phage. Therefore, cheese vats should be cleaned routinely and chlorinated between fills. Chlorination is a very effective phagicide. The final step in cleaning equipment, such as vats and filling lines, should include a chlorination component. Exposure to 100 µg of available or "active" chlorine per ml for 10 min is usually sufficient to inactivate all phage on equipment surfaces. The residual chlorine should not be rinsed from the equipment to prevent further contamination with phage from hoses or water. Any residual chlorine on the equipment is immediately inactivated when it comes in contact with

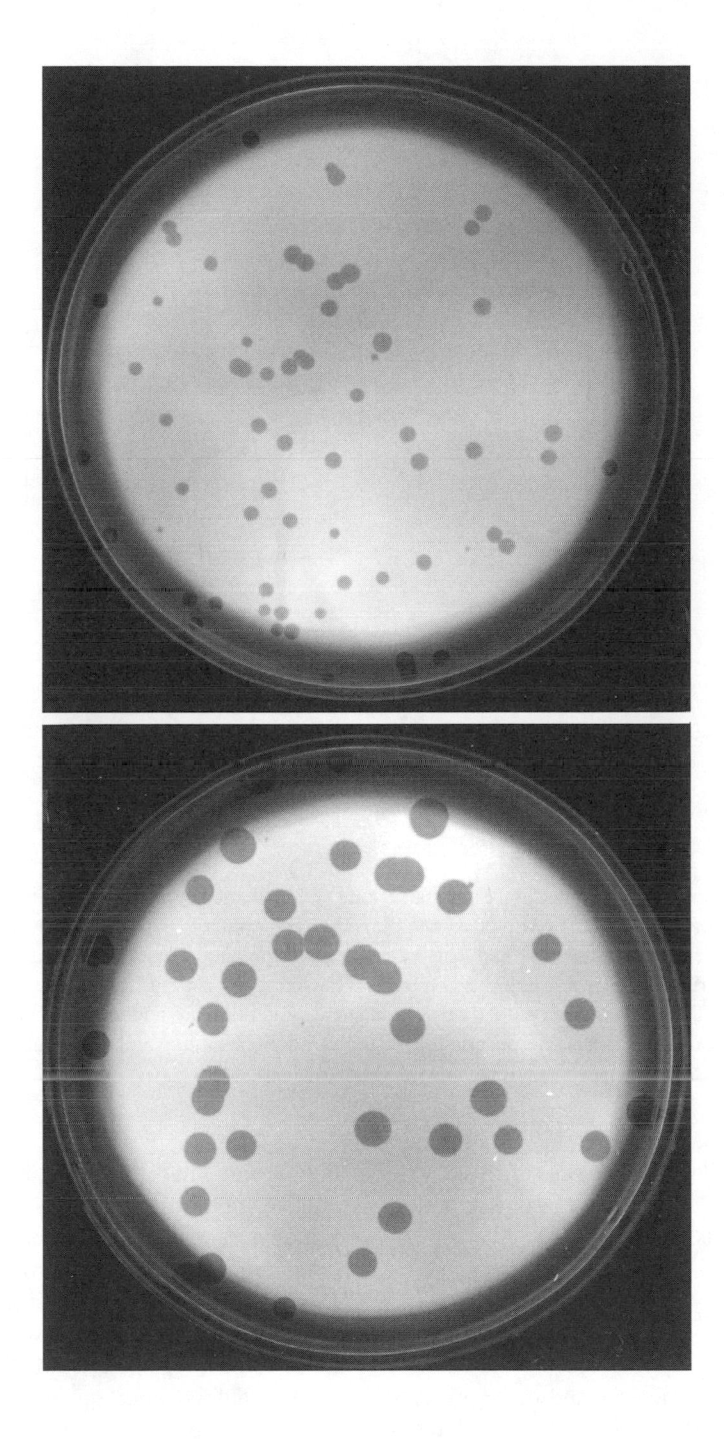

Figure 5–20 Agar plates with a lawn of *Lactococcus lactis* cells showing clear zones (plaques) due to phage infections. Each plaque is the result of the infection of a single cell with a single phage. The progeny phage multiply subsequently on neighboring cells.

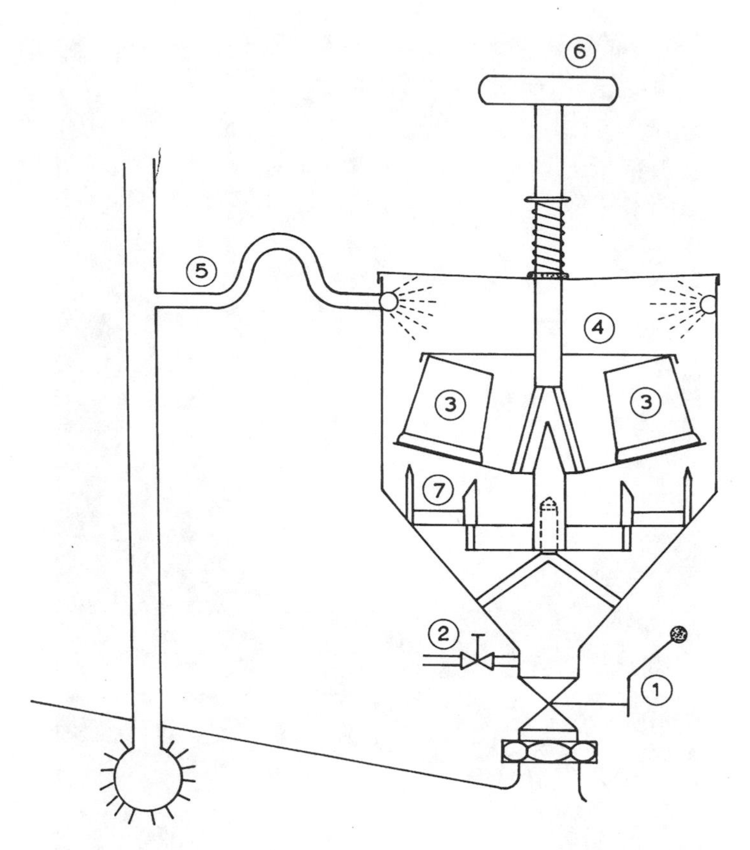

Figure 5–21 An aseptic inoculation device for starter tanks. The device is opened using the screw (6). Cartons (3) containing the frozen inoculum are inverted and placed on the holder (4). The device is assembled and filled with chlorinated water through the pipe (5). This helps to thaw the culture, and as soon as thawing begins, the chlorinated water is run to waste through the valve (2). The screw (6) is then turned, causing the prongs (7) to puncture the foil lid. The cartons empty and their contents (the inoculum) are added to the starter tank through the valve (1). Any residual chlorine has no effect on the starter, as it will be diluted and inactivated on contact with the fluid in which the starter cells are suspended.

organic material, like milk. The use of closed vats is also recommended to prevent contamination from whey aerosols.

5.8 BACTERIOCINS

Many bacteria produce proteins that inhibit the growth of other bacteria, and these proteins are called *bacteriocins*. Generally, they have a narrow host range, inhibiting only closely related bacteria, but bacteriocins with broad host ranges are not unknown. Bacteriocins produced by Gram-positive bacteria usually do not inhibit Gram-negative bacteria and vice versa. Usually, they are of small molecular mass, but bacteriocins of high molecular mass also occur. Their proteinaceous nature and their relatively narrow host range distinguish them from antibiotics. During the past decade, food safety has become a major issue, and a concerted effort has been made to identify bacteriocins that inhibit pathogens and food spoilage organisms. LAB are ideal for this purpose, because they are generally regarded as safe organisms. Bacteriocins that inhibit *Listeria* are particularly useful, since *L. monocytogenes* in soft cheeses has been incriminated as a cause of human listeriosis (see Chapter 20). For use in foods, the bacteriocin should

- be heat stable
- be acid stable
- be resistant to potential proteinases present in the food
- be active over a prolonged period
- be active at the pH of food (4.5 to 7.0)
- have a bactericidal rather than a bacteriostatic mode of action
- have a broad host range, inhibiting several pathogens and spoilage microorganisms

In assaying bacteriocins produced by LAB, it is important to distinguish between bacteriocins *per se* and other potential inhibitors that can be produced by these bacteria, including H_2O_2, lactate, and acetate. This can be done by determining the effect of catalase on the activity and by neutralizing culture supernatants before assaying them.

Bacteriocin production by LAB is common. The best known is nisin, which is produced by some strains of *Lc. lactis* subsp. *lactis*. Nisin has a molecular weight of 3,353 Da, contains 34 amino acids, and normally occurs as dimers and tetramers. It is soluble only at low pH, which reduces its potential use significantly. Its heat stability depends very much on pH; for example, it is stable to autoclaving at 115°C at pH 2 but loses 40% of its activity at pH 5. Nisin has a broad spectrum of activity, inhibiting *Bacillus*, *Clostridium*, *Staphylococcus*, *Listeria*, and *Streptococcus*. It is particularly active against spores of *Cl. tyrobutyricum*, which is the cause of late-gas production in hard natural and processed cheese. In the case of spores, nisin acts by preventing spore germination, but in preventing the growth of bacteria, it acts by creating pores in the cell membrane of sensitive cells, destroying the proton-motive force and permitting the release of cytoplasmic components from the cell. Nisin can be used as a replacement for NO_3 in foods. Nisin is synthesized as a 57 amino acid peptide that is posttranslationally modified to a 34 amino acid peptide containing unusual amino acids: dehydroalanine, dehydrobutyrine, lanthionine (Ala-S-Ala), and β-methylanthionine (Ala-S-Aba [aminobutyric acid]). Nisin is called a lantibiotic because it contains lanthionine and β-methylanthionine. Besides nisin, LAB produce other lantibiotics, such as lactocin 481, which is produced by *Lc. lactis* subsp. *lactis* CNRZ 481, and lactocin S, which is produced by *Lactobacillus sake* L45. Other starter bacteria, including *Lb. helveticus* and *Ln. mesenteroides*, have been shown to produce broad-spectrum bacteriocins, called *helveticin J* and *mesentericin*, respectively. Helveticin is temperature sensitive and is a large, complex molecule, which limits its usefulness in foods.

There are several forms of nisin arising from amino acid substitutions; for example, nisin Z differs from nisin A in having asparagine instead of histidine at position 27. Because of its greater solubility, nisin Z also has greater inhibitory activity than nisin A.

Bacteriocins may also be complexed with other macromolecules, including lipids and polysaccharides. They are encoded either on the chromosome or on plasmids. Bacteriocin-producing bacteria must also have a gene that encodes immunity to it.

The bacteriocins produced by LAB can be divided into 3 groups:

Class I: lantibiotics
Class II: small heat-stable nonlantibiotics
- IIa: single-peptide bacteriocins (e.g., pediocin pAH)
- IIb: two-peptide bacteriocins (e.g., lactococin G and plantaricin E)
- IIc: sec-dependent secreted bacteriocins

Class III: large heat-labile proteins (e.g., helveticin J and caseicin 80)

A fourth class containing complex bacteriocins composed of proteins, lipids, and carbohydrates (e.g., plantaricin S and lactocin 27) has been proposed, but the evidence suggests that these complex molecules are artifacts caused by interaction between cell constituents or the growth medium and the bacteriocin. Class IIa bacteriocins have been isolated from species of several genera, including *Pediococcus* (after which they are named), *Leuconostoc*, *Lactobacillus*, and *Enterococcus*, and they share considerable amino acid sequence homology (38-55%).

It is difficult to compare the reported host ranges of bacteriocin producers because different methods and, more importantly, different strains have been used. Most bacteriocins produced by LAB have a bactericidal mode of action, due to the destruction of the proton motive force involved in transport, which also results in the release of intracellular compounds. Lactocin 27, produced by *Lb. helveticus*, is an exception and is bacteriostatic. Bacteriocin-producing strains, if present in mixed cultures, will reduce the number of strains in these cultures upon subculturing, eventually leading to a mixture of perhaps 1 or 2 strains. Such cultures will also be much more prone to attack by phage than similar mixed cultures that do not contain bacteriocin producers.

5.9 PRODUCTION OF STARTERS IN CHEESE PLANTS

Until perhaps 30 years ago, milk was the medium used for starter production in cheese factories. The milk was selected from cows not suffering from mastitis and so was free of antibiotics. Spray-dried skim milk powder replaced cheese milk when it became more readily available; the solids concentration used was 10–12% (w/v). Today, phage-inhibitory media have replaced skim milk powder for the production of bulk cultures. These are generally carefully formulated proprietary media that contain milk or whey solids, phosphate and/or citrate, and yeast extract. The function of the citrate and phosphate is to chelate Ca^{2+}, which are essential for phage multiplication, and the function of the yeast extract is to stimulate growth of the starter.

The medium is generally heated at 85°C or higher for 30 min in the tank in which the starter is to be grown, and it is cooled to the incubation temperature of about 42°C, in the case of thermophilic cultures, or 21°C, in the case of mesophilic cultures. It is then inoculated with about 1% (v/v) of the culture, and after incubation for 8–10 hr, in the case of thermophilic cultures, or overnight (16 hr), in the case of mesophilic cultures, the culture is fully grown.

In the past, the inoculum for bulk cultures was built up progressively from a small volume of mother culture to larger volumes over several days (Figure 5–22). Minimal subculturing of starters is desirable to prevent loss of plasmids and to maintain the balance between strains in mixed cultures. Nowadays, inocula for bulk cultures are generally obtained from specialized laboratories in which the starters are grown under optimum conditions (e.g., pH 6.3 and 28°C, in the case of lactococci) in an appropriate medium, the composition of which is proprietary. After growth, the cells are harvested by ultrafiltration or centrifugation and frozen in liquid N or freeze-dried in sufficient volumes to inoculate 300, 500, or 1,000 L of medium directly. Such cultures generally contain about 5×10^9 cfu/g, or roughly 5 times more than a normal milk-grown culture. Cryoprotectants (e.g., glycerol, sucrose, or monosodium glutamate) are often added to protect the cells during freezing or freeze-drying. Where pure cultures (single strains) are used, it is much more economical to start from a small volume and build up the necessary inoculum over a few days. Superconcentrated cultures are also commercially available to inoculate the cheese milk in the vat directly. These are often called DVS (direct-vat-set) or DVI (direct-vat-inoculation) cultures. Such cultures are expensive and are often kept as standby cultures for use in a crisis, such as a phage outbreak.

Thermophilic cultures are usually grown at their optimum temperature (≈ 42°C), but mesophilic cultures are grown at 21°C, which is about 9°C below their optimum temperature. The reason for the lower temperature of incubation in the case of mesophilic cultures is that if the milk is inoculated at, say, 3 PM, the cultures are fully grown 16 hr later, at 7 AM the following morning, which is often the starting time for cheesemaking in small cheese plants.

Fully grown cultures of *Lactococcus* and *Sc. thermophilus* generally reduce the pH of milk from its initial value of 6.6 to 4.6 and produce about 90 mM (0.8%, w/v) lactic acid, whereas many thermophilic lactobacilli can produce up to 200 mM (≈ 2%, w/v) lactic acid and reduce

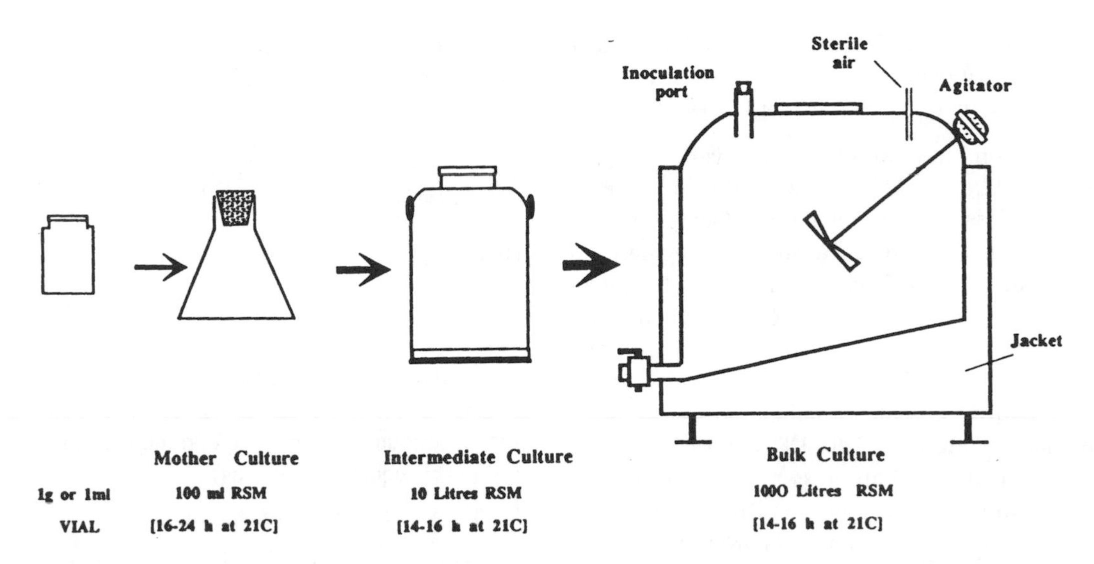

Figure 5–22 A possible protocol for scaling up intermediate cultures for bulk production of *Lactococcus* starters in cheese factories.

the pH to around 3.0. The actual amount of acid produced depends on the buffering capacity of the medium. Such cultures contain about 10^9 cfu/ml. However, in cheese factories, the lactobacilli are usually grown only to a pH of around 4.0 (equivalent to 1%, w/v, lactic acid). Cultures are then generally cooled to either 4°C or 10°C, in the case of mesophilic and thermophilic cultures, respectively, and checked for activity (i.e., their ability to produce lactic acid). Activity is normally assessed by measuring acid production or the decrease in the pH of the culture grown under standardized conditions, such as in 10% (w/v) reconstituted skim milk after 6 hr at 30°C for mesophiles or 5 hr at 40°C for thermophiles using a standardized inoculum, generally 1% (v/v). Sometimes incubation is carried out over the cheese temperature profile, to simulate their potential activity in cheese. For day-to-day comparisons between cultures, it is important to standardize the inoculum and the incubation conditions very carefully. Under the inoculation and incubation conditions just described, *Lc. lactis* subsp. *lactis*, *Sc. thermophilus*, *Lb. helveticus,* and *Lb. delbrueckii* subsp. *lactis* will reduce the pH from 6.6 to less than 5.3 (equivalent to ≈ 0.5%, w/v, lactic acid). Both pH and the amount of acid produced (i.e., the titratable acidity) are used to monitor growth. pH is much easier and faster to measure, and automated equipment able to measure the pH of up to 24 samples continuously is commercially available. When cooled to 4°C, mesophilic cultures will retain activity for 2–3 days, and thermophilic cultures for up to 12 days.

The pH of the medium used in the production of bulk cultures is often controlled during growth. Control can be external or internal. External control involves the use of a pH control unit and a neutralizer (e.g., NH_4OH), and internal control involves the use of an insoluble buffer that dissolves as lactic acid is produced and maintains the pH above 5.3. pH control increases the number of cells per unit volume and therefore reduces the volume of starter required for cheesemaking. Neutralization of the pH of the bulk culture to 6.5 after growth and further incubation for a few hours will also result in an increase in the number of cells in the culture.

The exact amount of bulk starter made each year throughout the world is difficult to calculate exactly, but the following three assumptions allow us to estimate it:

1. Annual cheese production is about 15 million tonnes (i.e., 15×10^9 kg).

2. Ten liters of milk are required to make 1 kg of cheese.
3. An inoculation rate of 1.0% (v/v) is used.

The second assumption is only valid for hard cheeses like Cheddar. Soft and semi-soft cheeses require less milk (in the case of Quarg, about 5 liters per kg); while Parmigiano-Reggiano, a very hard Italian cheese, requires about 12 liters per kg). This assumption also does not take into account the differences in solids levels of cow, goat, sheep, and buffalo milk. The third assumption also only applies to some cheeses. An inoculation rate of 1.5 to 1.8% (v/v) is used in Cheddar and perhaps less than 0.3% (v/v) in Emmental. These considerations mean that about 15×10^{10} liters of milk are used to make cheese annually which requires production of 15×10^{8} liters of starter. Assuming that starters contain about 10^9 cells per ml, this amounts to production of 15×10^{20} starter cells per year.

5.10 MEASUREMENT OF GENERATION TIMES

If bacterial growth is balanced, the following differential equation describes the increase in cell mass with time:

$$dx/dt = kx$$

where x is the cell mass in dry weight/g, t is time and k is the growth rate constant. Since lactic acid production is proportional to cell mass,

$$d(\text{lactic acid})/dt = k\,(\text{lactic acid})$$

Integrating we get

$$\ln(\text{lactic acid}_t) - \ln(\text{lactic acid}_0) = k\,(t - t_0)$$

where lactic acid_0 is the amount of lactic acid present at time_0, lactic acid_t is the amount of lactic acid present at time t, and k is the growth rate. Converting to $\log_{10}$ we get

$$\log_{10}(\text{lactic acid}_t) - \log_{10}(\text{lactic acid}_0) = k(t - t_0)/2.303$$

Rearranging we get

$$\log_{10}(\text{lactic acid}_t) = k(t - t_0)/2.303 + \log_{10}(\text{lactic acid}_0)$$

This equation is that of a straight line ($y = mx + c$), where $m = k/2.303$ or $k = m \times 2.303$.

The generation time (g) is the time required for the amount of lactic acid to double. Substituting in this equation gives us

$$\log_{10} 2 = k\,(g)/2.303 + \log_{10} 1$$

Rearranging we get

$$g = \log_{10} 2 \times 2.303/k + \log_{10} 1$$

And by substituting for k we get

$$g = \log 2/m + 0 = 0.301/m$$

The slope, m, is easily calculated by regression analysis of the data. It is important to correct each value for the inherent acidity of the *uninoculated* milk. The data in Figure 5–2 are plotted correctly (i.e., semi-logarithmically), but the generation time cannot be calculated accurately because the acidity of the uninoculated milk was not reported. However, the data for the cultures shown in Figure 5–13 were corrected for the amount of lactic acid in the uninoculated milk. Based on lactic acid production at 21°C, these cultures had a generation time of around 2.2 hr; at 30°C, the generation time would be 1 hr. The data need to be transformed to $\log_{10}$ before the regression equation is calculated.

REFERENCES

Garvie, E.I., & Farrow, J.A.E. (1982). *Streptococcus lactis* subsp. *cremoris* (Orla-Jensen) comb. nov. and *Streptococcus lactis* subsp. *diacetilactis* (Matuszewski et al.) nom. rev. comb. nov. *International Journal of Systematic Bacteriology, 32*, 453–455.

Goupil, N., Corthier, G., Ehrlich, S.D., & Renault, P. (1996). Imbalance of leucine flux in *Lactococcus lactis* and its use for the isolation of diacetyl overproducing strains. *Applied and Environmental Microbiology, 62*, 2636–2640.

Kunji, E.R.S., Mierau, I., Hagting, A., Poolman, B., & Konings, W.N. (1996). The proteolytic systems of lactic acid bacteria. *Antonie van Leewenhoek, 70*, 187–221.

Lodics, T., & Steenson, L. (1990). Characterisation of bacteriophages and bacteria indigenous to a mixed-strain cheese starter. *Journal of Dairy Science, 73*, 2685–2696.

Nomura, M., Kimoto, H., Someya, Y., & Suzuki, I. (1999). Novel characteristic for distinguishing *Lactococcus lactis* subsp. *lactis* from subsp. *cremoris. International Journal of Systematic Bacteriology, 49*, 163–166.

Schleifer, K.H., & Kilpper-Balz, R. (1987). Molecular and chemotaxonomic approaches to the classification of streptococci, enterococci and lactococci: A review. *Systematic and Applied Microbiology, 10*, 1–9.

Schleifer, K.H., Kraus, J., Dvorak, C., Kilpper-Balz, R., Collins, M.D., & Fischer, W. (1985). Transfer of *Streptococcus lactis* and related streptococci to the genus *Lactococcus* gen. nov. *Systematic and Applied Microbiology, 6*, 183–195.

Sherman, J.M. (1937). The streptococci. *Bacteriology Reviews, 1*, 3–97.

Sing, W.D., & Klaenhammer, T.R. (1993). A strategy for rotation of different bacteriophage defences in a lactococcal single-strain starter culture system. *Applied and Environmental Microbiology, 59*, 365–372.

Taillez, P., Tremblay, J., Ehrlich, S.D., & Chopin, A. (1998). Molecular diversity and relationships within *Lactococcus lactis* as revealed by randomly amplified polymorphic DNA (RAPD). *Systematic and Applied Microbiology, 21*, 530–538.

SUGGESTED READINGS

Board, R.G., Jones, D., & Jarvis, B. (Eds.). (1995). Microbial fermentations: Beverages, foods and feeds [Special issue]. *Journal of Applied Bacteriology, 79*, 1S–139S.

Chapman, H.R., & Sharpe, M.E. (1990). Microbiology of cheese. In R.K. Robinson (Ed.), *Dairy microbiology* (vol. 2). London: Elsevier Applied Science Publishers.

Cocaign-Bousquet, M., Garriguess, C., Loubiere, P., & Lindley, N.D. (1996). Physiology of pyruvate metabolism in *Lactococcus lactis. Antonie van Leewenhoek, 70*, 253–267.

Cogan, T.M., & Accolas, J.P. (1996). *Dairy starter cultures.* New York: VCH.

Cogan, T.M., & Hill, C. (1993). Cheese starter cultures. In P.F. Fox (Ed.), *Cheese: Physics, chemistry and microbiology* (2d ed., Vol 1). London: Chapman & Hall.

The Dairy *Leuconostoc* [Symposium]. (1994). *Journal of Dairy Science, 77*, 2704–2737.

Daly, C., Fitzgerald, G.F., & Davis, R. (1996). Biotechnology of lactic acid bacteria with special reference to bacteriophage. *Antonie van Leeuwenhoek, 70*, 99–110.

De Vuyst, L., & Vandamme, E.J. (1994). *Bacteriocins of lactic acid bacteria.* Glasgow: Blackie Academic and Professional.

Fifth Symposium on Lactic Acid Bacteria: Genetics, Metabolism and Applications (1996). *Antonie van Leeuwenhoek, 12*, 1–271.

Fourth Symposium on Lactic Acid Bacteria: Genetics, Metabolism and Applications. (1993). *FEMS Microbiology Reviews, 12*, 1–272.

Gasson, M.J., & de Vos, W.M. (1994). *Genetics and biotechnology of lactic acid bacteria.* London: Chapman & Hall.

Hoover, D.G., & Steenson, L.R. (1993). *Bacteriocins of lactic acid bacteria.* New York: Academic Press.

Hugenholtz, J. (1993). Citrate metabolism in lactic acid bacteria. *FEMS Microbiology Reviews, 12*, 165–178.

Lawrence, R.C., & Heap, H.A. (1986). *The New Zealand starter system* (Bulletin No. 199). Brussels: International Dairy Federation.

Lodics, T., & Steenson, L. (1993). Phage-host interactions in commercial mixed-strain cultures: Practical significance, A review. *Journal of Dairy Science, 76*, 2380–2391.

Nes, I.F., Diep, D.B., Håverstein, L.S., Bruberg, M.R., Eijsink, V., & Holo, H. (1996). Biosynthesis of bactericins in lactic acid bacteria. *Antonie van Leeuwenhoek, 70*, 113–128.

Pritchard, G.G., & Coolbear, T. (1993). The physiology and biochemistry of the proteolytic system in lactic acid bacteria. *FEMS Microbiology Reviews, 12*, 179–206.

Second Symposium on Lactic Acid Bacteria: Genetics, Metabolism and Applications. (1987). *FEMS Microbiology Reviews, 46*, 201–379.

Stadhouders, J. (1986). The control of starter activity. *Netherlands Milk and Dairy Journal, 40*, 155–173.

Stiles, M.E. (1996). Biopreservation by lactic acid bacteria. *Antonie van Leeuwenhoek, 70*, 331–335.

Third Symposium on Lactic Acid Bacteria: Genetics, Metabolism and Applications. (1990). *FEMS Microbiology Reviews, 87*, 1–188.

CHAPTER 6

Enzymatic Coagulation of Milk

As discussed in Chapter 2, the milk for most cheese varieties is coagulated through the action of selected proteinases, called rennets. The rennet-induced coagulation of milk is in fact a two-stage process (Figure 6–1). The primary phase involves the specific enzymatic modification of the casein micelles to produce paracasein micelles that aggregate in the presence of Ca^{2+} at temperatures above about 20°C. Aggregation of the rennet-altered micelles is referred to as the secondary phase of coagulation. The primary phase of rennet action is well characterized, but the secondary phase is less clear. The subject has been reviewed by Dalgleish (1992, 1993); Fox (1984); Fox and McSweeney (1997); Fox and Mulvihill (1990); and Fox, O'Connor, McSweeney, Guinee, and O'Brien (1996).

6.1 THE PRIMARY PHASE OF RENNET COAGULATION

As discussed in Chapter 3, the caseins exist as micelles stabilized by a surface layer of κ-casein. Following the isolation of κ-casein in 1956, it was shown that this protein is the micelle-stabilizing protein and that its stabilizing properties are destroyed on renneting. Shortly afterwards, it was shown that κ-casein is the only protein hydrolyzed during the rennet coagulation of milk and that it is hydrolyzed specifically at the bond Phe_{105}-Met_{106} (Figure 6–2). The N-terminal part of the molecule, κ-CN f1-105, referred to as para-κ-casein, remains attached to the casein micelle, whereas the C-terminal part, referred to as the (caseino) macropeptide (CMP; or glycomacropeptide, since it contains the carbohydrate moieties of κ-casein) is lost in the surrounding aqueous medium. It has been recognized since the end of the 19th century that small peptides are produced upon renneting. As discussed in Chapter 3, there are about 10 forms of κ-casein that differ in sugar content; hence, 10 CMPs are produced. All the CMPs are soluble in 2% trichloroacetic acid (TCA) but only the glycosylated forms are soluble at higher concentrations of TCA. Thus, TCA-soluble N, or more specifically TCA-soluble sugars (e.g., N-acetyl neuramic acid), can be used to monitor the primary phase of rennet coagulation (Figure 6–3).

The unique sensitivity of the Phe-Met bond of κ-casein has aroused interest. The dipeptide H.Phe-Met.OH is not hydrolyzed, nor are tri- or tetrapeptides containing a Phe-Met bond. However, this bond is hydrolyzed in the pentapeptide H.Ser-Leu-Phe-Met-Ala-OMe, and reversing the positions of serine and leucine, to give the correct sequence of κ-casein, increases the susceptibility of the Phe-Met bond to chymosin. Both the length of the peptide and the sequence around the Phe-Met bond are important determinants of enzyme-substrate interaction. $Serine_{104}$ appears to be particularly important, and its replacement by Ala or even L-Ser in the above pentapeptide renders the Phe-Met bond very resistant to hydrolysis by chymosin. Extension of

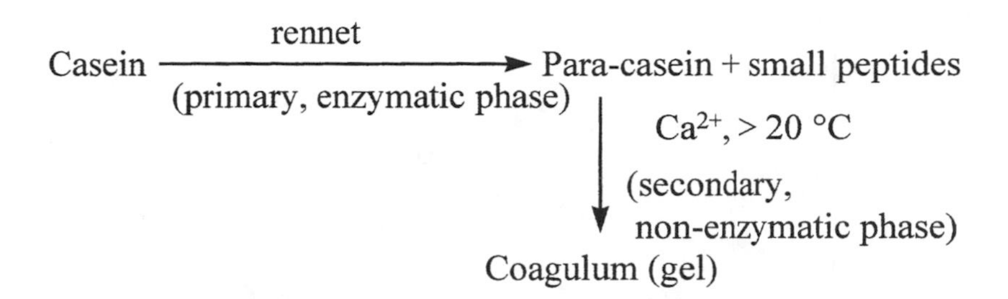

Figure 6–1 Summary of the rennet coagulation of milk. The primary phase involves enzymatic hydrolysis of κ-casein, while the secondary stage involves aggregation of the rennet-altered (paracasein) micelle into a three-dimensional gel network or coagulum.

the pentapeptide H.Ser.Phe.Met.Ala.Ile.OH (i.e., κ-CN f104-108) from the N- and/or C-terminal to reproduce the sequence of κ-casein around the chymosin-susceptible bond increases the efficiency with which the Phe-Met bond is hydrolyzed by chymosin (Table 6–1). The sequence κ-CN f98–111 includes all the residues necessary to render the Phe-Met bond as susceptible to hydrolysis by chymosin at pH 4.7 as it is in intact κ-casein; it is hydrolyzed around 66,000 times faster than the parent pentapeptide (κ-CN f104-108), with a k_{cat}/K_M of about 2 M^{-1} s^{-1}, which is similar to that for intact κ-casein. κ-Casein and the peptide κ-CN f98-111 are also readily hydrolyzed at pH 6.6, but smaller peptides are not.

1
Pyro **Glu-Glu-Gln-Asn-Gln-Glu-Gln-Pro-Ile-Arg-Cys-Glu-Lys-Asp-Glu-Arg-Phe-Phe-Ser-Asp-**

21
Lys-Ile-Ala-Lys-Tyr-Ile-Pro-Ile-Gln-Tyr-Val-Leu-Ser-Arg-Tyr-Pro-Ser-Tyr-Gly-Leu-

41
Asn-Tyr-Tyr-Gln-Gln-Lys-Pro-Val-Ala-Leu-Ile-Asn-Asn-Gln-Phe-Leu-Pro-Tyr-Pro-Tyr-

61
Tyr-Ala-Lys-Pro-Ala-Ala-Val-Arg-Ser-Pro-Ala-Gln-Ile-Leu-Gln-Trp-Gln-Val-Leu-Ser-

81
Asn-Thr-Val-Pro-Ala-Lys-Ser-Cys-Gln-Ala-Gln-Pro-Thr-Thr-Met-Ala-Arg-His-Pro-His-

101 105↓106
Pro-His-Leu-Ser-Phe-Met-Ala-Ile-Pro-Pro-Lys-Lys-Asn-Gln-Asp-Lys-Thr-Glu-Ile-Pro-

121 **Ile (Variant B)**
Thr-Ile-Asn-Thr-Ile-Ala-Ser-Gly-Glu-Pro-*Thr*- Ser-*Thr*-Pro-*Thr*- -Glu-Ala-Val-Glu-
***Thr* (Variant A)**

141 **Ala (Variant B)**
Ser-*Thr*-Val-Ala-Thr-Leu-Glu- -*SerP* - Pro-Glu-Val-Ile-Glu-Ser-Pro-Pro-Glu-Ile-Asn-
Asp (Variant A)

161 169
Thr-Val-Gln-Val-Thr-Ser-Thr-Ala-Val.OH

Figure 6–2 Amino acid sequence of κ-casein, showing the principal chymosin cleavage site (↓); oligosaccharides are attached at some or all of the threonine residues shown in italics.

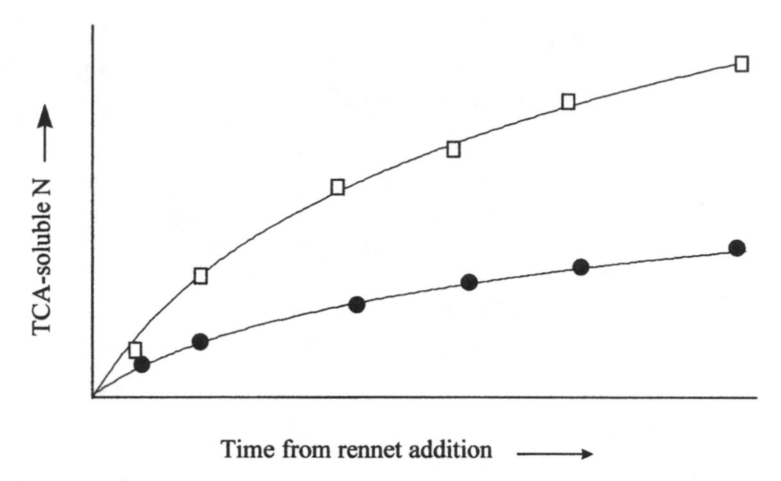

Figure 6–3 Release of nitrogen soluble in 2% (❑) or 12% (●) TCA by rennet from casein in milk.

The Phe and Met residues in the chymosin-sensitive bond of κ-casein are not intrinsically essential for chymosin action. There are numerous Phe and a substantial number of Met residues in all milk proteins. In porcine and human κ-caseins, the chymosin-sensitive bond is Phe-Ile, while in rat and mouse κ-caseins, it is Phe-Leu; yet, these proteins are readily hydrolyzed by calf chymosin, although more slowly than bovine κ-casein. In contrast, porcine milk is coagulated more effectively than bovine milk by porcine chymosin, indicating that unidentified subtle structural features influence chymosin action. Peptides in which Phe is replaced by Phe (NO_2) or cyclohexylamine are also hydrolyzed by chymosin, although less effectively than those with a Phe-Met bond. Oxidation of Met_{106} reduces k_{cat}/K_M roughly tenfold, but substitution of Ile for Met increases it about threefold.

A genetically engineered mutant of κ-casein in which Met_{106} was replaced by Phe_{106} (i.e., the chymosin-sensitive bond was changed from Phe_{105}-Met_{106} to Phe_{105}-Phe_{106}) was hydrolyzed 1.8 times faster by chymosin than natural κ-casein. These findings suggest that the sequence around the Phe-Met bond, rather than the residues in the bond itself, contains the important determinants of hydrolysis by chymosin. The particularly important residues are Ser_{104}, the hydrophobic residues Leu_{103} and Ile_{108}, at least one of the three histidines (residues 98, 100, or 102, as indicated by the inhibitory effect of photo-oxidation), and Lys_{111}. Studies on chemically or enzymatically modified peptide analogues of κ-CN f98-112 indicated the relative importance of residues in the sequences of 98–102 and 111–112. It has been suggested that the sequence Leu_{103} to Ile_{108} of κ-casein, which probably exists as an extended β-structure, fits into the active site cleft of acid proteinases. The hydrophobic residues Leu_{103}, Phe_{105}, Met_{106}, and Ile_{108} are directed toward hydrophobic pockets along the active site cleft, while the hydroxyl group of Ser_{104} forms part of a hydrogen bond with some counterpart in the enzyme. It has been proposed that the sequences 98–102 and 109–111 form β-turns around the edges of the active site cleft of the enzyme. This conformation is stabilized by Pro residues at positions 99, 101, 109, and 110. The three His residues at positions 98, 100, 102, and Lys_{111} are probably involved in electrostatic bonding between enzyme and substrate; none appears to have a predominant role. Lys_{112} appears not to be important in enzyme-substrate binding as long as Lys_{111} is present.

The significance of electrostatic interactions in chymosin-substrate complex formation is in-

Table 6–1 Kinetic Parameters for Hydrolysis of κ-Casein Peptides by Chymosin at pH 4.7

Peptide	Sequence	k_{cat} (s^{-1})	K_M (mM)	k_{cat}/K_M ($s^{-1}mM^{-1}$)
S.F.M.A.I.	104–108	0.33	8.50	0.038
S.F.M.A.I.P.	104–109	1.05	9.20	0.114
S.F.M.A.I.P.P.	104–110	1.57	6.80	0.231
S.F.M.A.I.P.P.K.	104–111	0.75	3.20	0.239
L.S.F.M.A.I.	103–108	18.3	0.85	21.6
L.S.F.M.A.I.P.	103–109	38.1	0.69	55.1
L.S.F.M.A.I.P.P.	103–110	43.3	0.41	105.1
L.S.F.M.A.I.P.P.K.	103–111	33.6	0.43	78.3
L.S.F.M.A.I.P.P.K.K.	103–112	30.2	0.46	65.3
H.L.S.F.M.A.I.	102–108	16.0	0.52	30.8
P.H.L.S.F.M.A.I	101–108	33.5	0.34	100.2
H.P.H.P.H.L.S.F.M.A.I.P.P.K.	98–111	66.2	0.026	2509
	98–111[a]	46.2[a]	0.029[a]	1621[a]
κ–Casein[b]		2–20	0.001–0.005	200–2,000
L.S.F.(NO_2)Nle A.L.OMe		12.0	0.95	12.7

[a] pH 6.6.
[b] pH 4.6.

dicated by the effect of added NaCl on the rennet coagulation time (RCT) of milk. Addition of NaCl up to 3 mM reduces RCT but higher concentrations have an inhibitory effect. It is claimed that the effect of NaCl is on the primary, enzymatic phase rather than on the aggregation of rennet-altered micelles. Increasing ionic strength (0.01–0.11) reduced the rate of hydrolysis of κ-CN f His_{98}-$Lys_{111/112}$ in a model system; the effect became more marked as the reaction pH was increased, but it was independent of ion type.

As well as serving to elucidate the importance of certain residues in the hydrolysis of κ-casein by chymosin, small peptides that mimic or are identical to the sequence of κ-casein around the Phe-Met bond are very useful substrates for determining the activity of rennets in absolute units, that is, independent of variations in the nonenzymatic phase of coagulation of different milks. Standard methods for such quantification have been developed and chromogenic peptide substrates are available commercially. Since the specific activity of different rennets on these peptides varies, methods for quantifying the proportions of acid proteinases in commercial rennets have been proposed.

6.2 RENNET

Several proteinases will coagulate milk under suitable conditions but most are too proteolytic relative to their milk-clotting activity (MCA). Consequently, they hydrolyze the caseins in the coagulum too quickly, causing a reduced cheese yield. (MCA is the inverse of RCT, i.e., MCA = 1/RCT.) Excessive proteolysis or incorrect specificity may also lead to defects in the flavor (especially bitterness) and texture of the cheese. Although plant proteinases appear to have been used as rennets since prehistoric times, gastric proteinases from calves, kids, or lambs have been traditionally used as rennets, with very few exceptions.

Animal rennets are prepared by extracting the dried (usually) or salted gastric tissue (referred to as *vells*) with 10% NaCl and activating and standardizing the extract. Standard calf rennet contains about 60–70 RU/ml and is preserved by making the extract up to 20% NaCl and adding sodium benzoate or sodium propionate. A rennet unit (RU) is the amount of rennet activity that will coagulate 10 ml of milk (usually low-heat skim milk powder reconstituted in 0.01% $CaCl_2$ and perhaps adjusted to pH 6.5) in 100 s. Chymosin (an aspartyl acid proteinase, i.e., a proteinase with two aspartic acid residues at the active site and with a pH optimum of 2–4) represents more than 90% of the MCA of good-quality veal rennet, the remaining activity being due to pepsin. As the animal ages, especially when fed solid food, the secretion of chymosin declines and that of pepsin increases.

Like many other animal proteinases, chymosin is secreted as its zymogen, prochymosin, which is autocatalytically activated on acidification to pH 2–4 by removal of a 44-residue peptide from the N-terminal of the zymogen (see Foltmann, 1993).

Chymosin is well characterized at the molecular level (see Chitpinityol & Crabbe, 1998; Foltmann, 1993). The enzyme, which was crystallized in the 1960s, is a single-chain polypeptide that contains about 323 amino acid residues and has a molecular mass of 35,600 Da. Its primary structure has been established, and a considerable amount of information is available on its secondary and tertiary structures. The molecule exists as two domains separated by an active site cleft in which the two catalytically active aspartyl groups (Asp_{32} and Asp_{215}) are located.

Calf rennet contains three chymosin isoenzymes, principally A and B, with lesser amounts of C. Chymosins A and B are produced from the corresponding zymogens, prochymosins A and B, but chymosin C appears to be a degradation product of chymosin A that lacks three residues, Asp_{244}–Phe_{246}. The specific activity of chymosin A, B, and C is 120, 100, and 50 RU/mg, respectively. Chymosins A and B differ by a single amino acid substitution, Asp and Gly, respectively, at position 244 and have an optimum pH at 4.2 and 3.7, respectively.

The properties of different rennets are discussed in Section 6.8.

6.3 FACTORS THAT AFFECT THE HYDROLYSIS OF κ-CASEIN AND THE PRIMARY PHASE OF RENNET COAGULATION

The hydrolysis of κ-casein is influenced by many factors, some of which are discussed below. While many factors influence both the primary and secondary stages, the effects on each are discussed separately.

- The *pH* optimum for chymosin and bovine pepsin on small synthetic peptides is about 4.7 but is 5.3–5.5 on κ-CN f His_{98}-$Lys_{111/112}$. Chymosin hydrolyzes insulin, acid-denatured hemoglobin and Na-caseinate optimally at pH 4.0, 3.5, and 3.5, respectively. The pH optimum for the first stage of rennet action in milk is about 6.0 with 4°C or 30°C.
- The influence of *ionic strength* on the primary phase of rennet coagulation is discussed in Section 6.1.
- The optimum *temperature* for the coagulation of milk by calf rennet at pH 6.6 is around 45°C. Presumably, the optimum for the hydrolysis of κ-casein is around this value. The temperature coefficient (Q_{10}) for the hydrolysis of κ-casein in solutions of Na-caseinate is about 1.8; the activation energy, E_a, is about 40,000 $J{\bullet}mol^{-1}$; and the activation entropy, ΔS, is about –90 $J{\bullet}K^{-1}{\bullet}mol^{-1}$. Generally similar values have been reported for the hydrolysis of isolated κ-casein by chymosin.
- *Heat treatment* of milk at temperatures above 65°C adversely affects its rennet coagulability. If the heat treatment is very severe (> 90°C for 10 min), the milk fails to coagulate upon renneting. Although changes in salts equilibria are contributory factors, the principal causative factor is intermolecular disulphide bond formation between κ-casein and β-lactoglobulin and/or α-lactalbumin. Both the primary and especially the secondary phases of rennet action are inhibited in heated milk, as reflected by

the marked decrease in the curd-firming rate and in the strength of the resulting gel. The adverse effects of heating can be reversed by acidification to pH values in the region 6.6–6.0 before or after heating or by the addition of $CaCl_2$ (which causes a reduction in pH). The secondary, rather than the primary, phase of rennet action probably benefits from these treatments.

6.4 THE SECONDARY (NONENZYMATIC) PHASE OF COAGULATION AND GEL ASSEMBLY

Hydrolysis of κ-casein by chymosin or similar enzymes during the primary phase of rennet action releases the highly charged, hydrophilic C-terminal segment of κ-casein (macropeptide), as a result of which the zeta potential of the casein micelles is reduced from –10/–20 to –5/–7 mV and the protruding peptides (hairs) are removed from their surfaces, thus destroying the principal micelle-stabilizing factors (electrostatic and steric) and their colloidal stability. When roughly 85% of the total κ-casein has been hydrolyzed, the stability of the micelles is reduced to such an extent that when they collide, they remain in contact and eventually build into a three-dimensional network, referred to as a coagulum or gel (Figure 6–4). Gel formation is accompanied by sharp increases in viscosity and elastic shear modulus, G′, which is a measure of gel firmness (Figure 6–5; see Section 6.6.2). Reducing the pH or increasing the temperature from the normal values (≈ 6.6 and ≈ 31°C, respectively) permits coagulation at a lower level of κ-casein hydrolysis. Although the precise reactions involved in aggregation are not known, the kinetics of aggregation have been described.

The assembly of rennet-altered micelles into a gel has been studied using various forms of viscometry, electron microscopy, and light scattering. Viscosity measurements show that the viscosity of renneted milk remains constant or decreases slightly during a period equivalent to roughly 60% of the visually observed RCT (Figures 6–4 and 6–5). It has been suggested that the decrease in viscosity is due to a decrease in the voluminosity of the casein micelles following release of the macropeptides, which form a "hairy layer" about 10 nm thick (Figure 6–4). The decrease in micelle size has been confirmed by quasi-elastic light scattering.

It is generally agreed that following the initial lag period, the viscosity increases exponentially up to the onset of visual coagulation or gelation, that is, 100% RCT (Figure 6–5). The gelation process, generally referred to as the secondary phase of rennet coagulation, involves initially the formation of chains and clumps of micelles and leads eventually to the formation of a network of partly fused micelles. During the first 60% of the visually observed RCT, the micelles exist as individual particles; the primary enzymatic reaction is about 85% complete at 60% of the visual RCT. Between 60% and 80% of the RCT, the rennet-altered micelles begin to aggregate steadily, with no sudden change in the type or extent of aggregation. Small chain-like aggregates, rather than clumps, form initially (Figure 6–6). At 100% of the RCT, most of the micelles have aggregated into short chains, which then begin to aggregate to form a network. Initially, most micelles are linked by bridges (655 nm long and 40 nm wide) and do not touch. An unexpectedly large proportion of the surface of the participating micelles is involved in the bridging, indicating a large amount of material, the origin of which is unknown. Although the micelles themselves appear the most probable source, if they are the source, micellar rearrangement would be necessary. No change is observed in the size, surface structure, or general appearance of the micelles up to 60% of the RCT (i.e., lag phase). Thus, if a micellar rearrangement is a prerequisite for aggregation, it must occur during the latter half of the RCT. Linkage of the rennet-altered micelles probably occurs at definite surface sites.

Aggregation of the rennet-altered micelles can be described by the von Smoluchowski theory for diffusion-controlled aggregation of hydrophobic colloids when allowance is made

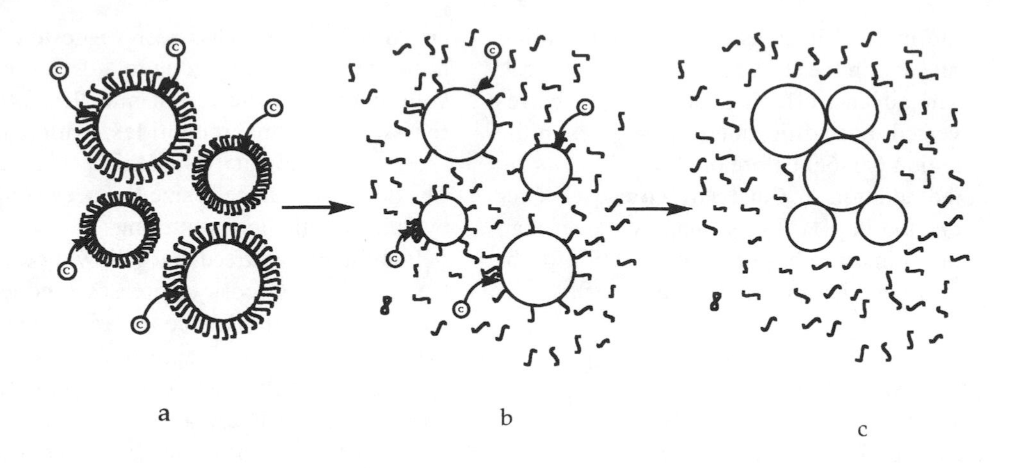

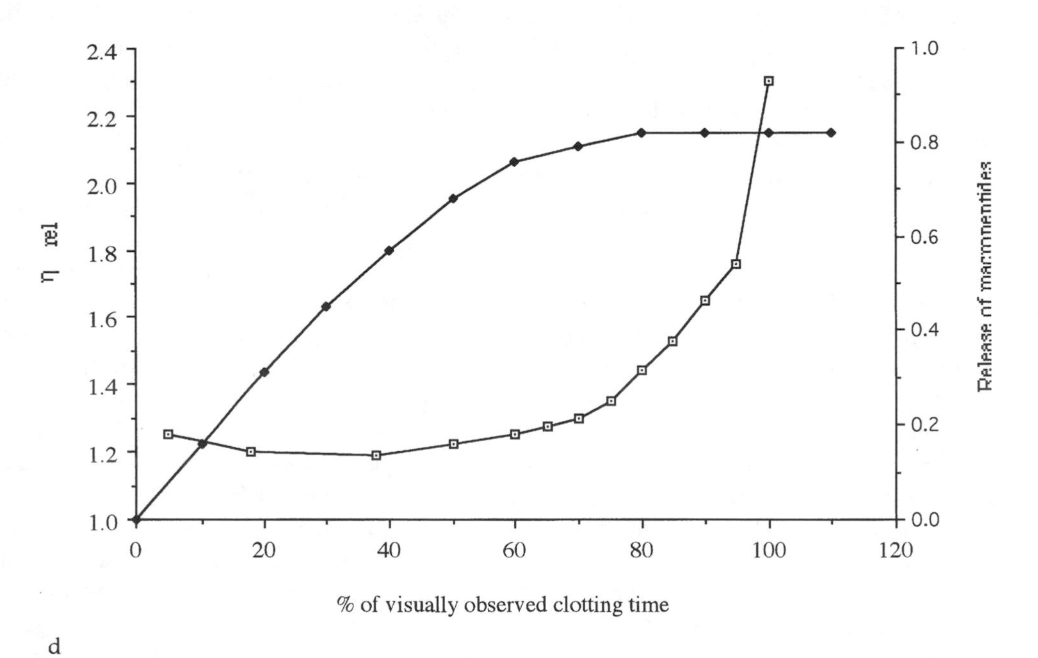

d

Figure 6–4 Schematic representation of the rennet coagulation of milk. (a) Casein micelles with intact κ-casein layer being attacked by the chymosin (C); (b) micelles partially denuded of κ-casein; (c) extensively denuded micelles in the process of aggregation; (d) release of macropeptides (◆) and changes in relative viscosity (◘) during the course of rennet coagulation.

for the need to produce, by enzymatic hydrolysis, a sufficient concentration of particles capable of aggregating (i.e., casein micelles in which > 97% of the κ-casein has been hydrolyzed). The diffusion of the particles is rate limiting and is determined by the random fruitful collision of particles (rennet-altered micelles). The rate of aggregation is not consistent with a branching process model, since the micellar functionality is 1.8 whereas an average function-

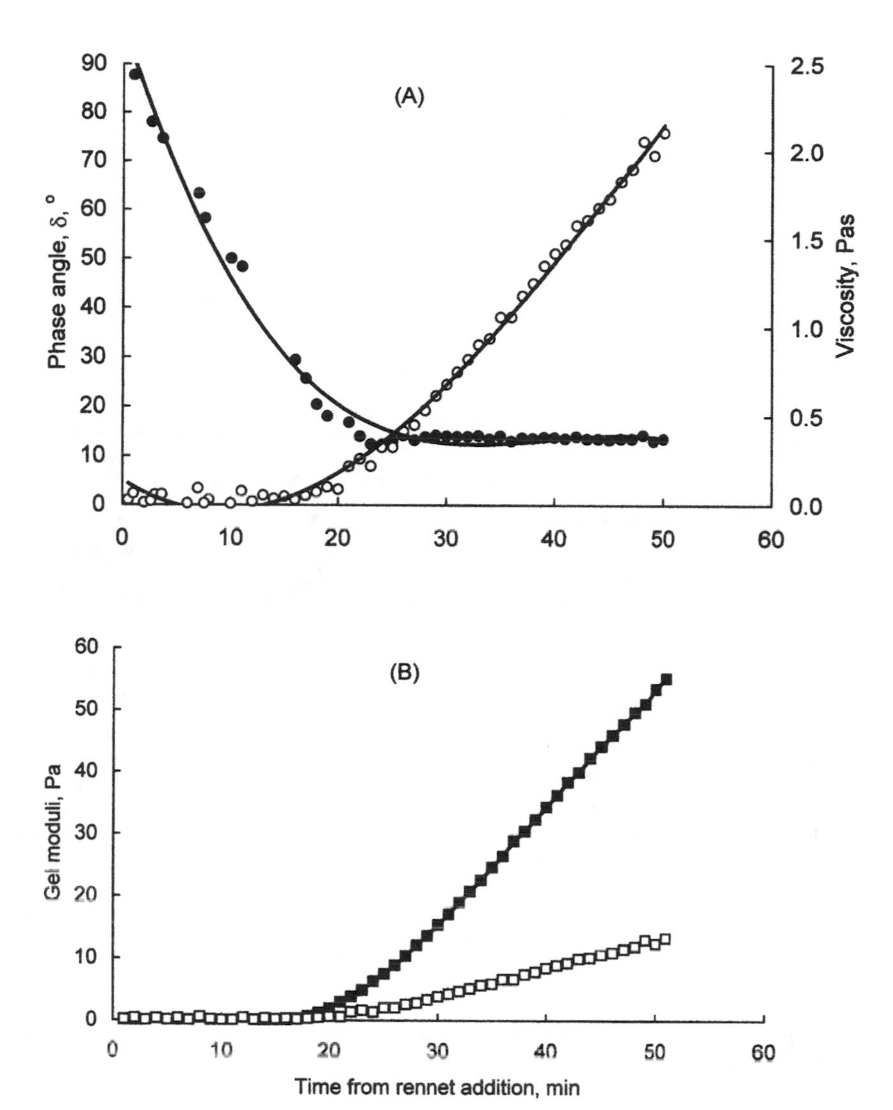

Figure 6–5 Rheological changes in milk during rennet coagulation under quiescent conditions. (A) Phase angle (●) and viscosity (○); (B) elastic modulus (■) and loss modulus (□).

ality greater than 2 is required for network formation.

According to Dalgleish (1980), the overall rennet coagulation of milk can be described by combining three factors:

1. proteolysis of κ-casein, which may be described by Michaelis-Menten kinetics
2. the requirement that around 97% of the κ-casein on a micelle be hydrolyzed before it can participate in aggregation
3. aggregation of paracasein micelles via a von Smoluchowski process

The overall clotting time t_c is the sum of the enzymatic phase and the aggregation phase:

$$t_c = t_{\text{prot}} + t_{\text{agg}} \cong \frac{K_m}{V_{\max}} \ln\left(\frac{1}{1-\alpha_c}\right) + \frac{\alpha_c}{V_{\max}} S_0 + \frac{1}{2k_s C_0}\left(\frac{M_{\text{erit}}}{M_0} - 1\right)$$

where: K_m and $V_{\max}$ are the Michaelis-Menten

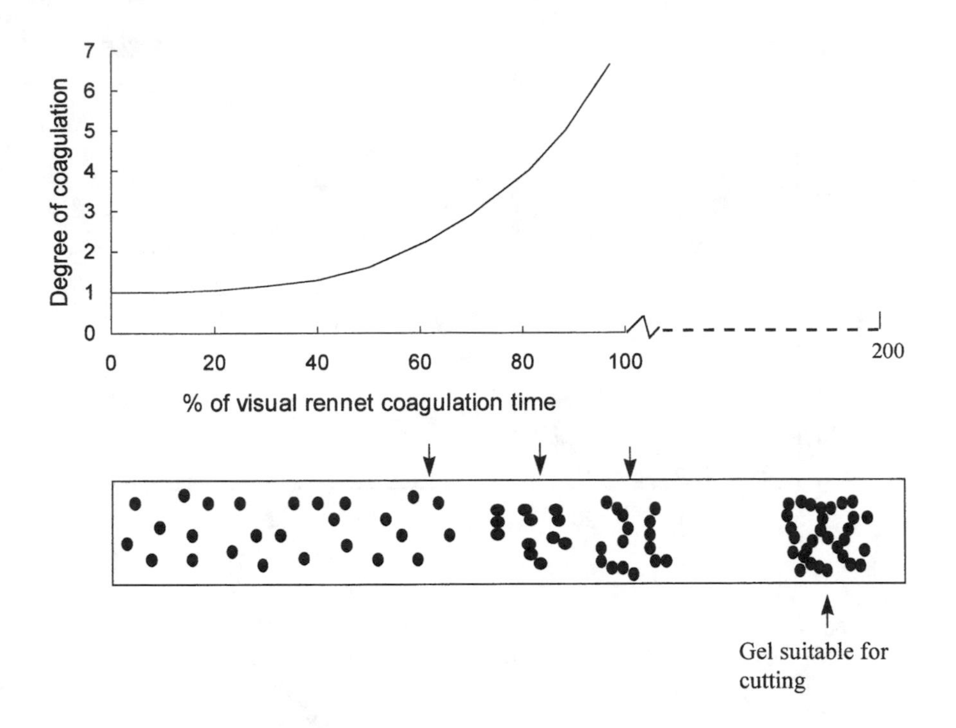

Figure 6–6 Schematic representation of the progress of micelle aggregation during the rennet coagulation of milk. Aggregation of rennet-altered micelles results in the formation aggregates that fuse to form chain-like structures, and these eventually overlap and cross-link to form a three-dimension casein network or gel after a time greater than the visual rennet coagulation time.

---

parameters, α_c is the extent of κ-casein hydrolysis, S_o is the initial concentration of κ-casein, k_s is the rate constant for aggregation, C_o is the concentration of aggregating material, M_{crit} is the weight average molecular weight at t_c (≈10 micellar units), and M_o is the weight average molecular weight at t_o.

Darling and van Hooydonk (1981) proposed an alternative model for rennet coagulation, again by combining Michaelis-Menten enzyme kinetics with von Smoluchowski aggregation kinetics. The stability factor in von Smoluchowski's theory is considered as a variable determined by the concentration of unhydrolyzed surface κ-casein. The coagulation time, t_c, is given by

$$t_c = \frac{1}{V} = \left[S_0 + \frac{1}{C_m}(\exp(-C_m \cdot S_0) - 1 \right] + \frac{W_0 \exp(-C_m \cdot S_0)}{k_s} \left[\frac{1}{n_c} - \frac{1}{n_0} \right]$$

where V is the velocity of enzymatic hydrolysis of κ-casein, S_o is the initial concentration of κ-casein, C_m is a constant relating the stability of the casein micelle to κ-casein concentration, W_o is the initial stability factor for casein micelles, n_o is the initial concentration of casein micelles, and n_c is the concentration of casein aggregates at the observed clotting time t_c. It is claimed that this theoretical model explains the experimentally observed influence of protein concentration, enzyme concentration, and temperature on RCT and the occurrence of a lag phase equal to 60% of RCT.

With milk of normal concentration, about 90% of the micelles are incorporated into the curd at 100% of the visual RCT, but only about 50% are incorporated in a fourfold concentrate when the same level of rennet is used. The micelles that are "free" at or after the RCT may react differently from those free prior to RCT. That is, before visual coagulation, all micelles are freely dispersed in the serum and can aggregate randomly, but once a gel matrix has started to form, free micelles may react either with the gel matrix or with other free micelles. Therefore, a gel assembly may be regarded as a two-stage process, and the properties of the final curd may be affected considerably by the amount of casein free during the RCT. Since this amount is particularly high in concentrated milks, it may explain the coarser structure of curd made from these milks.

Measurements made using the Instron Universal Testing machine showed that the rate of firming of a renneted milk gel as a function of time has two maxima (Storry & Ford, 1982): under the experimental conditions used, firming was first observed about 2.5 min after the visual assessment of clotting, and the firming rate increased to a maximum 12–15 min after clotting. The rate decreased over the next 10–15 min to about 80% of the maximum value, after which it either remained constant or increased slightly for a further 15 min and thereafter decreased steadily. However, this two-stage gel firming process has not been observed when dynamic rheometry, a very sensitive technique, is used.

Based on viscometric data, Tuszynski (1971) suggested that gel assembly is a two-stage process consisting of what he called "flocculation" and "gelation," but the meaning of these terms is not clear. Turbidity experiments also suggest a two-stage gel assembly process (Surkov, Klimovskii, & Krayushkin, 1982), although again the terminology is not very clear:

$$E + S \underset{k_{-1}}{\overset{k_1}{\rightleftharpoons}} ES \xrightarrow{k_2} E + P \xrightarrow{k_c} P^* \xrightarrow{k_s} P_n$$

where E is the enzyme, S is the substrate, P is the reaction product, P^* is the paracasein micelle with transformed quaternary structure, and P_n is the gelled micelle aggregate. The first two steps are the Michaelis-Menten model for the primary, enzymatic phase and are essentially as proposed by Payens, Wiersma, and Brinkhuis, 1977):

$$E + S \underset{k_{-1}}{\overset{k_1}{\rightleftharpoons}} ES \xrightarrow{k_2} E + P_1 + M$$

where P_1 is para-κ-casein and M is a macropeptide. Payens et al. (1977) suggested that the second, nonenzymatic phase may be represented by

$$iP_1 \overset{k_s}{\rightleftharpoons} Pi$$

where i is any number of the aggregating particle P_1.

Surkov et al. (1982) suggested that the enzyme-altered micelles (paracasein micelle P) undergo a cooperative transition in quaternary structure to yield clot-forming particles (P^*) with a rate constant k_c and with E_a = 191 kJ mol^{-1} and $Q_{10°C}$ = 16. These values are close to those reported by Tuszynski (1971).

The sites involved in the aggregation process are not known. Following reduction of the micellar zeta potential by proteolysis of κ-casein, linkage of particles is facilitated. Interparticle linkage could be via calcium bridges and/or hydrophobic interactions (which the marked temperature-dependence of the secondary phase would indicate). Changes in the surface hydrophobicity of casein micelles during renneting has been demonstrated through changes in the binding of the fluorescent marker 8-anilino naphthalene-1-sulfonate (Iametti, Giangiacomo, Messina, & Bonomi, 1993; Peri, Pagliarini, Iametti, & Bonomi, 1990). The hydrophobic amino terminal segment (residues 14–24) of

α_{s1}-casein appears to be important in the establishment of a rennet curd structure. It has been suggested that the matrix of young cheese curd consists of a network of α_{s1}-casein molecules linked together via hydrophobic patches that extends throughout the cheese structure. The softening of texture during the early stages of ripening is considered to be due to breaking of the network upon hydrolysis of the Phe_{23}-Phe_{24} bond of α_{s1}-casein. Modification of histidyl, lysyl, and arginyl residues in κ-casein inhibits the secondary phase of rennet coagulation, suggesting that a positively-charged cluster on para-κ-casein interacts electrostatically with unidentified negative sites. In native micelles, this positive site may be masked or covered by the macropeptide segment of κ-casein but becomes exposed and reactive when this peptide is released.

Following the RCT, the network appearance becomes gradually more apparent. The strands of the network, which are more or less in parallel, are roughly 5 micelles thick and 10 micelle diameters apart. The bridges between the micelles contract slowly, forcing the micelles into contact and eventually causing them to fuse partially. The fate of the bridging material upon micelle fusion has not been explained, but it may be that its disappearance is responsible for the second maximum in the firming rate time curve and for the reported flocculation and gelation stages in the gel assembly process.

Normally, the rate of an enzymatic reaction increases linearly with enzyme concentration, within certain limits. In the case of rennet coagulation, RCT is inversely related to enzyme concentration, as expressed by the formula

$$Et_c = k$$

where E = enzyme concentration and t_c is the RCT.

This equation, which assumes that visually observed coagulation is dependent only on the enzymatic process, has been modified to take account of the duration of the secondary, nonenzymatic phase:

$$E(t_c - x) = k$$

where x is time required for the coagulation of the enzymatically altered casein micelles and $t_c - x$ is the time required for the enzymatic stage. Rearrangement of this equation results in a more convenient form:

$$t_c = k(1/E) + x$$

which is valid within a certain range of rennet concentrations and under certain conditions of temperature and pH. A very good linear relationship exists between clotting time and the reciprocal of enzyme concentration (Figure 6–7).

The coagulation equations developed by Dalgleish (1980) and by Darling and van Hooydonk (1981) might be regarded as greatly refined versions of these simpler equations and reduce to them on first approximations. Rennet clotting time (t_c) has also been expressed (Payens and Wiersma, 1977) by the equation:

$$t_c \approx \sqrt{\frac{2}{k_s V_{max}}}$$

where k_s, the diffusion-controlled flocculation rate constant according to von Smoluchowski's theory (nonenzymatic phase), is proportional to the concentration of reactive (coagulable) particles (proteolyzed micelles) and hence to enzyme concentration; V_{max} is the maximum velocity in Michaelis-Menten kinetics (enzymatic phase) and is proportional to enzyme concentration; and c is a constant that depends on the experimental conditions.

6.5 FACTORS THAT AFFECT THE NONENZYMATIC PHASE OF RENNET COAGULATION

The coagulation of renneted micelles is very temperature dependent ($Q_{10} \approx 16$), and bovine milk does not coagulate at less than around 18°C unless the Ca^{2+} concentration is increased. The marked difference between the temperature dependence of the enzymatic and nonenzymatic

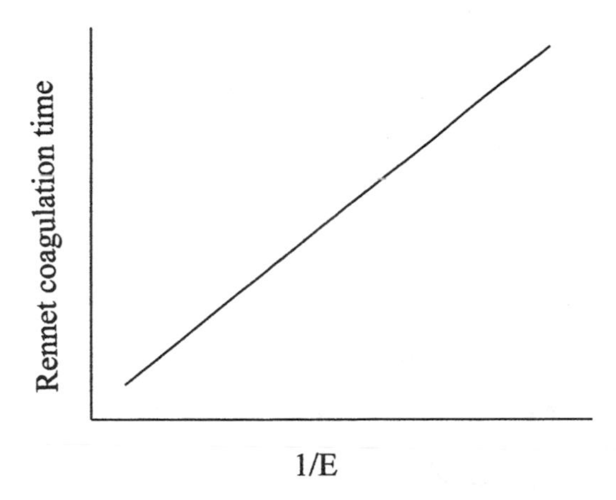

Figure 6–7 Relationship between enzyme concentration (*E*) and rennet coagulation time.

phases of rennet coagulation has been exploited in studies on the effects of various factors on the rennet coagulation of milk—studies done as part of attempts to develop a system for the continuous coagulation of milk for cheese or rennet casein manufacture and to discover possible applications of immobilized rennets. The very high temperature dependence of rennet coagulation suggests that hydrophobic interactions are important.

Coagulation of rennet-altered micelles depends on a critical concentration of Ca^{2+}, which may act by cross-linking rennet-altered micelles, possibly via serine phosphate residues or simply by charge neutralization. Colloidal calcium phosphate is also essential for coagulation but can be replaced by increased Ca^{2+}. Partial enzymatic dephosphorylation of casein, which reduces micellar charge, reduces coagulability; interaction of casein micelles with various cationic species predisposes them to coagulation by rennet and may even coagulate unrenneted micelles. Chemical modification of histidine, lysine, or arginine residues inhibits coagulation, presumably by reducing micellar positive charge.

The apparent importance of micellar charge in the coagulation of rennet-altered micelles suggests that pH should have a major influence on the secondary phase of coagulation. Reduction in pH in the range 6.6–6.0 is accompanied by increases in the rates of the enzymic and coagulation reactions, reductions in gelation time (time for gel onset) and the degree of κ-casein hydrolysis necessary for the onset of gelation (e.g., from ≈ 97% to ≈ 80% of total κ-casein), and increases in the curd-firming rate and firmness after a given renneting time. Although it is claimed that pH has essentially no effect on the coagulation process, the rate of firming of the resultant gel is significantly increased upon reducing the pH (Figure 6–8).

The rate of firming of renneted milk gels is influenced by the type of rennet, especially under unfavorable conditions, such as high pH or low Ca^{2+}. Perhaps differences in the firming rate reflect the effect of pH on rennet activity or perhaps some general proteolysis by rennet substitutes.

Heat treatment of milk under conditions that denature β-lactoglobulin and promote its interaction with κ-casein via sulfydryl-disulfide interaction adversely affects all aspects of rennet coagulation but especially the buildup of a gel network (Figure 6–9) (van Hooydonk, deKoster, & Boerrigter, 1987). Presumably, the attachment of denatured β-lactoglobulin to the surface of the casein micelles (as is evident from electron micrographs of casein micelles) prevents their aggregation in a form capable of building up a gel network.

6.6 MEASUREMENT OF RENNET COAGULATION PROPERTIES

The rennet gelation of milk under quiescent conditions involves the conversion of milk from a colloidal dispersion of stable micelles to a network (gel) of aggregated paracasein micelles, which forms a continuous phase, entrapping moisture and fat globules in its pores. The gel becomes more elastic and firm with time (i.e., upon aging). The transformation is accompanied by a number of physicochemical changes, including hydrolysis of κ-casein, with a concomitant increase in the concentration of the gly-

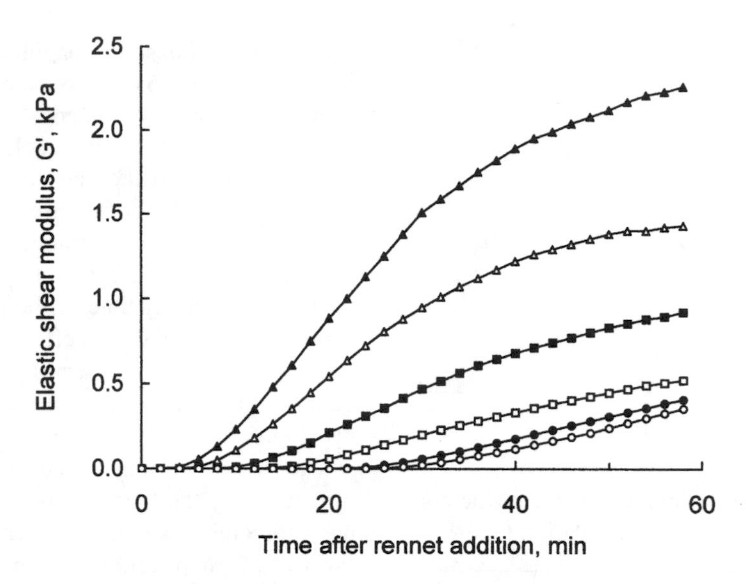

Figure 6–8 Development of elastic shear modulus (G′, equivalent to curd firmness) in rennet-treated, high-protein (≈180 g/kg) milk retentate obtained from skim milk heated to 100°C for 120 s and renneted at pH 6.67 (○), 6.55 (●), 6.45 (□), 6.3 (■), 6.15 (Δ), and 6.0 (▲).

comacropeptide; aggregation of the sensitized paracasein micelles; increases in viscosity and elasticity; and a decrease in the ratio of the viscous to elastic character of the milk. Such changes may also alter some of the physical properties of the milk, such as light reflectance and thermal conductivity.

Numerous methods, the principles of which are based on detection of one or more of the above changes, have been developed to measure the rennet coagulation characteristics of milk or the activity of rennets. Owing to the commercial importance of gel formation in milk as a means of recovering milk fat and casein in the form of cheese curd, most methods measure gel formation (also referred to as curd formation or rennet coagulability)—that is, the combined first and second stages—but some specifically monitor the hydrolysis of κ-casein. Various terms or descriptors, some of which are used interchangeably, are employed to describe the rennet coagulation of milk. These are defined below:

- *Aggregation*. The joining of particles, such as micelles, by various types of electrostatic or hydrophobic bonds. The aggregates are visible by electron microscopy.
- *Coagulation* or *flocculation*. The collision and joining of aggregates, especially under nonquiescent conditions, to form flocs visible to the naked eye.
- *Gelation*. The aggregation of particles (e.g., micelles or aggregates of micelles) to form particulate strands whose particles undergo limited touching and that eventually form a gel network.
- *Elasticity*. The ability of the gel to recover, instantaneously, its original shape and dimensions after removal of an applied stress. Viscoelastic materials, such as a renneted milk gel, are elastic at relatively small strains (e.g., 0.025, which is much less than fracture strain). In this region of strain, known as the linear viscoelastic stress-strain region, the strain is directly propor-

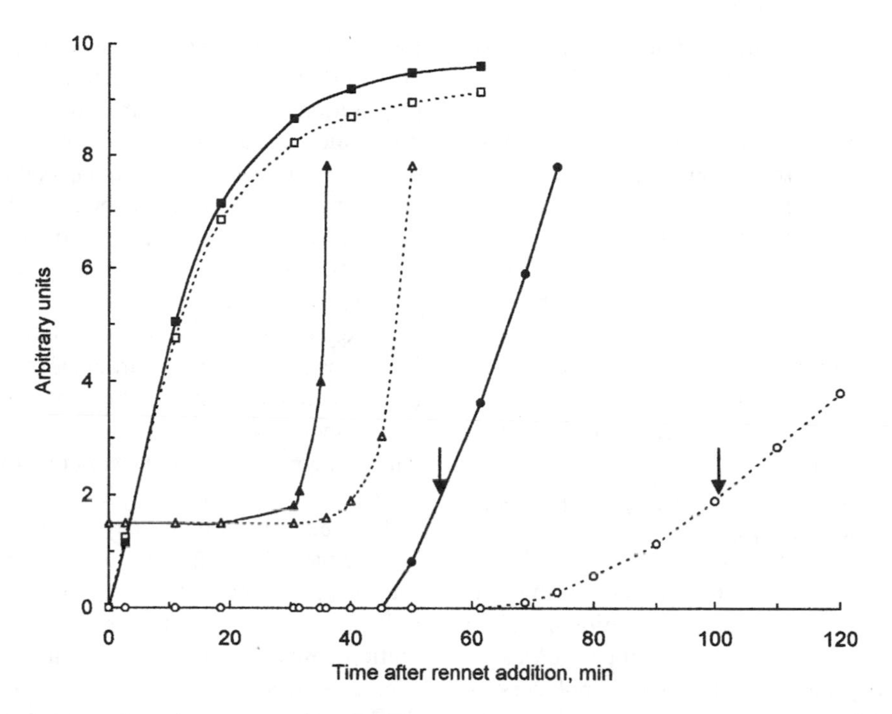

Figure 6–9 Release of casein macropeptides (■, □) and changes in the viscosity (▲, Δ) and curd firmness (●, ○) in skim milk unheated (closed symbols) or heated at 95°C for 15 s (open symbols). Arrows indicate the time at which cutting is initiated during cheese manufacture.

tional to the applied stress, and the material (e.g., a section of a gel strand that bears the applied stress and is strained) recovers its original dimensions immediately upon removal of the stress.

- *Viscosity*. The physical property of a gel given by the ratio between the stress and strain rate.
- *Curd firmness*, *curd strength*, or *curd tension*. The stress required to cause a given strain or deformation. (*Curd tension* is a term frequently used to express the firmness of formed gels.)

Types of measurement used to evaluate gel-forming characteristics include

- measurement of flocculation time under nonquiescent conditions (e.g., rennet coagulation test)
- dynamic measurement of the viscous drag of gelling milk by suspending a pendulum in the milk and determining the tilt of the pendulum over time using instruments such as the Thromboelastrograph, Formagraph, and Gelometer
- dynamic measurement of the ability of gelling milk to transmit a pressure by using, for example, a hydraulically operated oscillating diaphragm apparatus
- measurement of the apparent viscosity of gelled milk after a given time at a fixed shear rate (e.g., by using various types of rotational viscometers) or, alternatively, measurement of the firmness of the gel using various types of penetrometer
- dynamic measurement of parameters such as viscosity, elastic shear modulus (G′), loss modulus (G″), and phase angle (δ) by applying a low-amplitude oscillating strain or

stress to the milk sample using, for example, controlled strain or controlled stress rheometers
- dynamic measurement of some physical properties of the gelling milk in the cheese vat using special probes, including thermal conductivity (using a hot wire probe) and reflectance of near infrared light (using a near infrared diffuse reflectance probe)

Some of the more commonly used laboratory and online methods for monitoring the rennet-induced gelation of milk are described below.

6.6.1 Measurement of the Primary Phase of Rennet Coagulation

The primary phase of rennet action may be monitored by measuring the formation of either product, that is, para-κ-casein or the CMP. Para-κ-casein may be measured by SDS-polyacrylamide gel electrophoresis (PAGE), which is slow and cumbersome, or by ion-exchange high-performance liquid chromatography (HPLC). The CMP is soluble in TCA (2–12% depending on its carbohydrate content) and may be quantified by the Kjeldahl method or more specifically by determining the concentration of N-acetylneuraminic acid (Figure 6–3) or by RP-HPLC. The activity of rennets can be determined easily using chromogenic peptide substrates, a number of which are available. The latter method is generally used as a research tool to study rennet characteristics and/or the kinetics of the primary phase of rennet coagulation.

6.6.2 Methods for Assessing Coagulation, Gel Formation, and/or Curd Tension

Measurement of Rennet Coagulation Time

The simplest laboratory method for measuring the overall rennet coagulation process is to monitor the time between the addition of a measured amount of diluted rennet to a sample of milk in a temperature-controlled water-bath at, for example, 30°C and the onset of visual coagulation. If the coagulating activity of a rennet preparation is to be determined, a "reference" milk, such as low-heat milk powder reconstituted in 0.01% $CaCl_2$ and perhaps adjusted to a certain pH (e.g., 6.5), should be used. A standard method has been published (International Dairy Federation [IDF], 1992), and a reference milk powder may be obtained from Institut National de la Recherche Agronomique, Poligny, France. If the coagulability of a particular milk is to be determined, the pH may or may not be adjusted to a standard value (e.g., 6.55) to reflect that which is typical at setting (rennet addition) during cheese manufacture.

The coagulation point may be determined by placing the milk sample in a bottle or tube that is rotated in a water-bath (Figure 6–10); the fluid milk forms a film on the inside of the rotating bottle or tube, but flocs of protein form in the film upon coagulation. The rennet coagulation time (RCT) provides a very good index of the gelation potential of milk; a low RCT usually indicates potentially good gel formation and high gel strength after a given renneting time. The method is simple and enables the accurate measurement of several samples simultaneously. It has been used to accumulate much of the extensive information reported in the scientific literature on the effects of various processing parameters on the rennet coagulability of milk. However, in contrast to cheese manufacture, where milk is renneted under quiescent conditions to ensure gel formation, this method determines the time for coagulation (i.e., aggregation and flocculation) of the paracasein under agitation.

Nondynamic Assessment of Viscosity Curd Firmness Tests

The apparent viscosity and firmness of the coagulum may be measured using various types of viscometers and penetrometers, respectively. However, use of these instruments permits measurement at only a single point in time, which is a serious limitation in kinetic studies, and it also

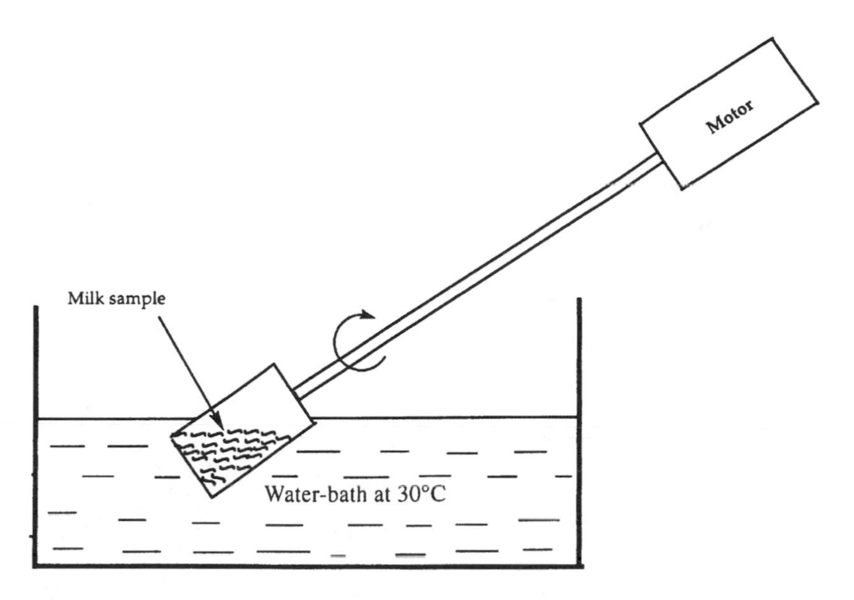

Figure 6–10 Schematic of apparatus for visual determination of the rennet coagulation time of milk.

requires meticulous test conditions, since curd strength increases with time after renneting.

Dynamic Curd Firmness Tests

Various instruments, involving different principles, have been developed to monitor changes continuously throughout the gelation process. These are discussed below.

Formagraph. The most popular of the dynamic measuring instruments, although it is not widely used, is the Formagraph (e.g., Type 1170, Foss Electric, Denmark), a diagram of which is shown in Figure 6–11. The apparatus consists of

- an electrically heated metal block
- a sample rack with cavities (usually 10) into which sample cuvettes fit
- a set of pendulums, each having an arm with an attached mirror
- an optical source that beams a light ray onto the mirror attached to each pendulum
- a chart recorder with photosensitive paper onto which the reflected light from each mirror is directed

Samples of milk to be analyzed are placed in the cuvettes and tempered to the desired temperature (typically 31°C) in the heating block. Rennet is then added, and the cuvettes are replaced in the instrument so that a loop-shaped pendulum is suspended in each sample. The metal block is moved back and forth, creating a "drag" on the pendulum in the milk. A flashing light is directed onto the mirror on the arm of each pendulum and reflected onto photosensitive paper, creating a mark. While the milk is fluid, the viscosity is low and the drag on the pendulum is slight. It scarcely moves from its vertically suspended "zero-time" position, and hence a single straight line appears on the paper. As the milk coagulates, its viscosity increases and the pendulum is dragged from its zero-time position, resulting in bifurcation of the trace. The rate at and extent to which the arms of the trace diverge are indicators of the gel-forming characteristics of the milk. A typical trace (see Figure 6–11) may be used to calculate the following parameters:

- The rennet coagulation time (r) is the time (in minutes) from rennet addition to the on-

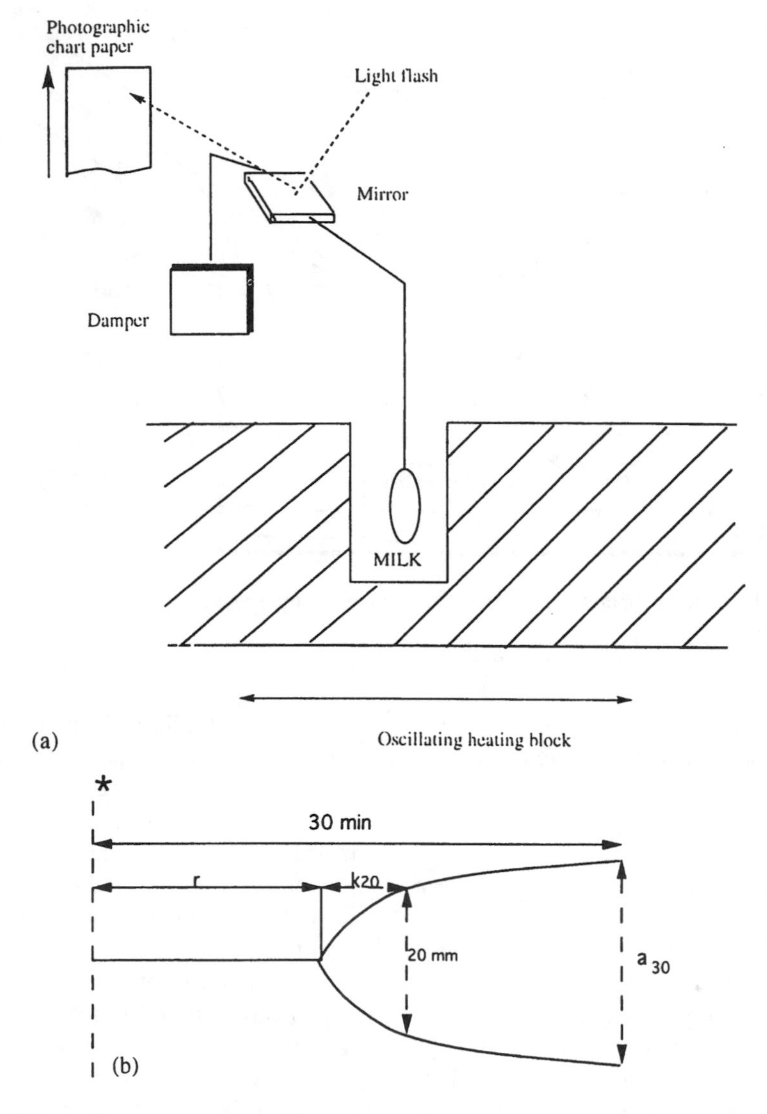

Figure 6–11 (a) Schematic representation of the Formagraph apparatus for determining the rennet coagulation of milk. (b) Typical formagram. The asterisk represents the point of rennet addition, r is the rennet coagulation time, k_{20} is the time required from coagulation for the arms of the formagram to bifurcate by 20 mm, and a_{30} is the extent of bifurcation 30 min after rennet addition (the approximate time at which the coagulum is cut in cheesemaking).

set of gelation (i.e., point where the trace begins to fork).

- k_{20}, essentially the inverse of the curd-firming rate, is the time from the onset of gelation until a firmness corresponding to a trace width of 20 mm is obtained.
- a_t, which represents the curd firmness at time t after rennet addition, is the trace width (in millimeters) at time t.

Good gel-forming properties are characterized by a relatively rapid coagulation time (low r

value), high-curd firming rate (low k_{20} value), and a high-curd firmness or strength after a given renneting time (high a_{30} value). Typical values for these parameters for a pasteurized midlactation milk (3.3%, w/w, protein) renneted under normal conditions (rennet dosage, ≈ 16 RU/L; pH, 6.55; temperature, 31°C) are $r = 5.5$ min, $k_{20} = 11$ min, and $a_{30} = 48.5$ mm.

While the latter parameters have no precise rheological significance, the Formagraph method offers many advantages over the RCT test:

- The method simulates gel formation during cheesemaking.
- The results of the assay are less subjective, being independent of operator judgment.
- The test parameters provide more information on the changes in curd strength over time.

However, production of the instrument has been discontinued.

Hydraulically oscillating diaphragm. In a hydraulically operated oscillating diaphragm apparatus, a sample of milk is placed between two diaphragms (Figure 6–12) and rennet is added. One diaphragm (the transmitting diaphragm) is made to vibrate through the cyclical application of hydraulic pressure. When the milk is liquid, the effect of the vibration is dissipated rapidly and does not affect the receiving diaphragm. When a gel is formed, the vibrations emitted by the transmitting diaphragm reach the receiving diaphragm, causing it to vibrate. These vibrations are detected and quantified by a suitable

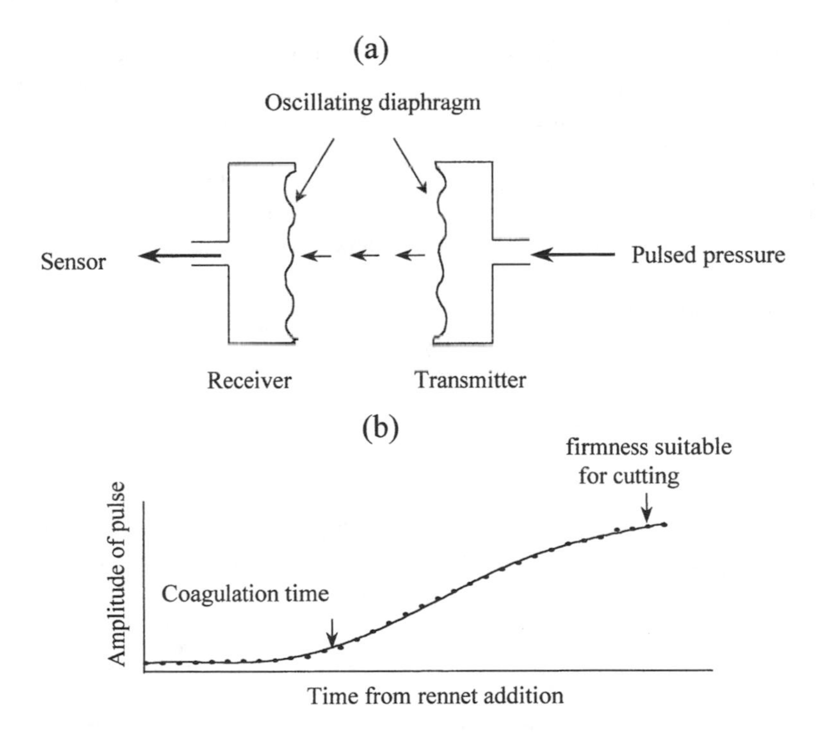

Figure 6–12 Schematic representation of a pressure transmission apparatus for measuring the rennet coagulation time of milk and the strength of the resulting gel.

sensing device. An output generally similar to that of the Formagraph is obtained, from which the coagulation time and a measure of gel strength can be determined.

Low-Amplitude Stress or Strain Rheometry. Since the 1980s, several controlled-strain rheometers (e.g., Bohlin VOR, Bohlin Rheologi, Sweden; Rotovisco RV 100/CV 100, Haake Bucchler Instruments, United States) and controlled-stress rheometers (e.g., Bohlin CS, Bohlin Rheologi, Sweden; Cari-med CSL^2, TA Instruments, United States; Rheometric Scientific SR5, Rheometric Scientific Inc, United States) have been used increasingly as research tools for the continuous measurement of the viscoelastic properties of renneted milks as a function of time from rennet addition. A rheometer (Figure 6–13) essentially consists of:

- a DC motor
- a gear box
- an electromagnetic clutch
- a direct drive position servo actuator, which, combined with the clutch, enables low-amplitude angular deflection (and hence strain) to be transferred from the motor via the gear box to the stalk of the outer cylinder of the sample cell
- a sample cell, which can be of different geometries but typically consists of two coaxial cylinders—an outer cup (e.g., internal diameter, 27.5 mm), into which the milk sample is placed, and an inner bob (e.g., internal diameter, 25 mm); the strain on the sample results in a stress that, when released, creates a strain on or angular displacement of the bob)
- a temperature sensor, which assists in accurate temperature control within the heating/cooling chamber surrounding the outer cylinder of the sample cell and hence within the sample
- a frictionless air bearing, which ensures that the strain or torque created on the bob by the sample is transferred to a torsion bar
- a torque-measuring transducer, which measures the torque, digitizes it, and relays it to the software of the interfacing computer

Dynamic measurements are performed by applying a low-amplitude oscillating shear stress (σ) or shear strain (γ), depending on the type of rheometer, to the milk sample via oscillations of the outer cylinder. The value of σ or γ is maintained sufficiently low in order to stay within the linear viscoelastic limits of the sample (i.e., the region where σ and γ are directly proportional). Hence, the terms *low-amplitude stress oscillation* and *low-amplitude strain oscillation.* (*Amplitude* refers to the maximum displacement of any point on the oscillating cup [and hence in the milk sample or on the inner bob] from its mean [or "zero"] position.) Under these conditions, the gel strands of the gelling milk are strained to a fixed displacement (within their elastic limit) and recover instantaneously when the stress is removed. The stress required to achieve a fixed strain (e.g., displacement of a gel strand at a given position) increases as the gel strands become more elastic and firm; hence, measurement of stress energy provides a measure of the gel strength.

When a controlled-strain rheometer is used, the sample of renneted milk is subjected to an harmonic, low-amplitude shear strain, (γ) of angular frequency ω:

$$\gamma = \gamma_0 \cos \omega t$$

where: γ_0 is the shear strain amplitude, ω is the angular frequency (i.e., $2\pi\upsilon$), and $\cos \omega t$ is a term of the simple harmonic function. The shear strain results in an oscillating shear stress (σ) on the milk that is of the same angular frequency but is out of phase by the angle δ:

$$\sigma = \sigma_0 \cos (\omega t + \delta)$$

where: σ_0 is the stress amplitude and σ is the phase angle between the shear stress and shear strain oscillations, the magnitude of which de-

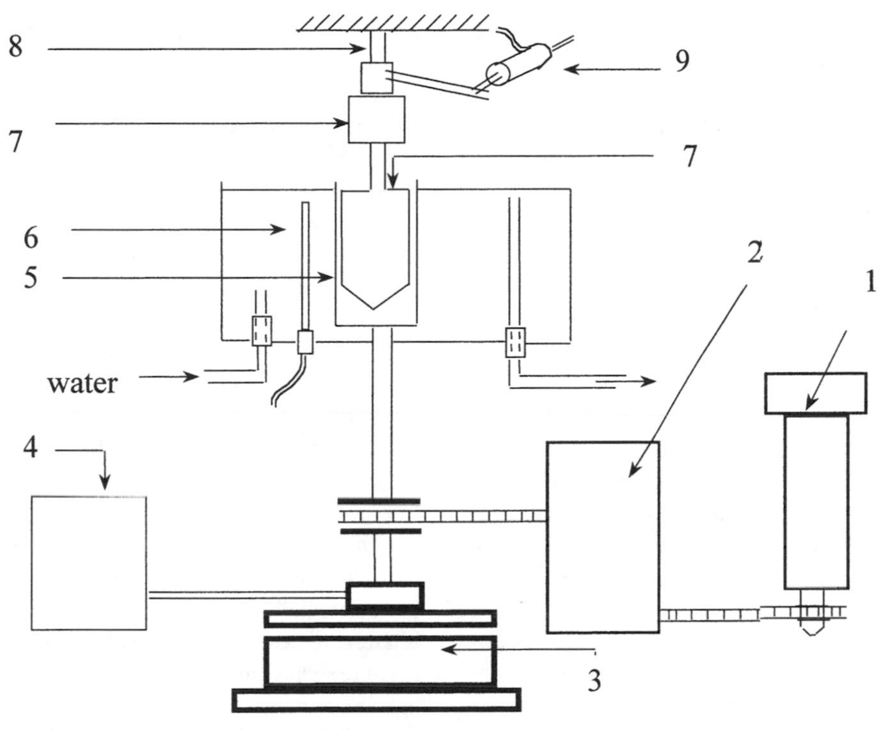

Figure 6–13 Schematic representation of the main parts of a controlled-strain rheometer: DC motor (1), gear box (2), electromagnetic clutch (3), direct drive position servo actuator (4), sample cell (5), temperature sensor (6), frictionless air bearing (7), torsion bar (8), and torque-measuring transducer (9).

pends on the viscoelasticity (ratio of viscous to elastic properties) of the gelling system. The phase angle ranges from 0° for an elastic solid to 90° for a Newtonian liquid and has intermediate values for viscoelastic materials (Figure 6–6). The rheological parameters σ, G', and G'' are computed continually from the measurement of stress energy over time. The first, σ, is of course the phase angle. G' is the storage or elastic shear modulus, which represents elastically stored stress energy and thus gel elasticity or firmness. It is given by the equation:

$$G' = \sigma_0 / \gamma_0) \cos \delta$$

G'', the viscous or loss modulus, which represents energy dissipated in flow, is given by the equation:

$$G'' = \sigma_0 / \gamma_0) \cos \delta$$

Typical changes in the above rheological parameters from the time of rennet addition are presented in Figure 6–5. The onset of gelation is marked by sharp increases in G' and G'' and a decrease in δ, whose abrupt decline from about 80° in milk to about 10° marks the transition from a viscoelastic material that is largely viscous (i.e., milk) to a gel that is largely elastic in character. Structural elements that impart elasticity include the relatively weak, continuous paracasein gel (formed from overlapping strands of aggregated paracasein micelles), while those that contribute to viscosity include aggregates and/or short dangling strands not yet connected to the main network. Following the onset of gelation, the gel undergoes a progressive increase in firmness over time. However, the rate of curd firming increases initially to a maximum and then decreases (as reflected by the changes in slope of the G'/time curve). The increase in curd firmness ensues from the following changes:

- the inclusion of free aggregates or dangling strands into the network
- the increase in contact surface area and number of attractions between neighboring micelles in the gel (Walstra & van Vliet, 1986), which result in thicker strands with higher stress-bearing capacity and thus more elasticity

G'' and δ are useful parameters for monitoring the viscoelastic changes in the gel during aging but are not directly related to gel strength and are thus not discussed further. In contrast, G' is a direct measure of curd firmness and is thus of significance in cheese manufacture. Various objective rennet coagulation parameters pertinent in cheesemaking may be derived from the G' time curve, upon modelling, as described below (Guinee, O'Callaghan, Mulholland, Pudja, & O'Brien, 1996):

- gel time, defined as the time at which G' reaches a threshold value, G_g, arbitrarily set at 0.2 Pa
- the firmness after a fixed renneting time (e.g., 30 or 60 min, G'_{30} or G'_{60})
- maximum curd-firming rate, defined as the maximum slope, S_{max}, of the $G't$ graph
- set-to-cut time (SCT), which is the time between rennet addition and curd cutting at a suitable firmness (e.g., 40Pa, SCT40Pa)

Low-strain or -stress rheometry gives parameters that are rheologically precise and accurately quantify the dynamic rheological changes that occur during renneting without altering the process of gel formation. Hence, it accurately reflects the changes in curd firmness that occur upon renneting milk in the cheese vat under quiescent conditions. However, limitations compared to other instruments, such as the Formagraph, include the high level of operator skill required for accurate measurement and the fact that only one sample can be analyzed at any one time. Moreover, instruments currently available cannot be used for online measurement of gel formation in the cheese vat.

Online Sensors for Predicting Curd Firmness and Cutting Time during Cheese Manufacture. Following the onset of gelation, there is a progressive increase in curd firmness, and the gel eventually attains an optimum firmness (e.g., 40 Pa after 40–50 min, depending on milk composition and renneting conditions), which, for a given vat design, allows it to withstand the mechanical action of the cutting knives in the cheese vat without shattering. Curd shattering corresponds to the fracturing of the individual curd particles (e.g., by the cheese knife and by impact with other curd particles and/or the knife and walls of the cheese vat), especially if the coagulum is soft at cutting. Shattering results in an excessively large curd particle surface area, through which fat is lost into the cheese whey, and it is conducive to the formation of very small curd particles (i.e., curd fines < 1 mm), which are also lost in the whey. Thus, cutting at the optimum firmness and the rate of curd firming are crucial for obtaining the correct particle size, minimizing the losses of fat and fines in the whey, and maximizing cheese yield (Figure 6–14).

The firmness of the coagulum after a given set-to-cut time is influenced by many factors, such as the concentrations of fat and casein in the milk, the stage of lactation and the diet of the cow, the milk pH, the starter activity (which influences pH at set and at cut time), and the rennet:casein ratio. Variation in the firmness at cutting can, in turn, lead to variations in cheese composition (especially moisture), yield, and quality. Hence, standardizing and optimizing the firmness at cutting is essential for consistently ensuring high cheese yields and good-quality cheese.

Formerly, the time at which the coagulum was cut was usually determined by the cheese maker, who subjectively assessed firmness by various means, such as by cutting a small portion with a

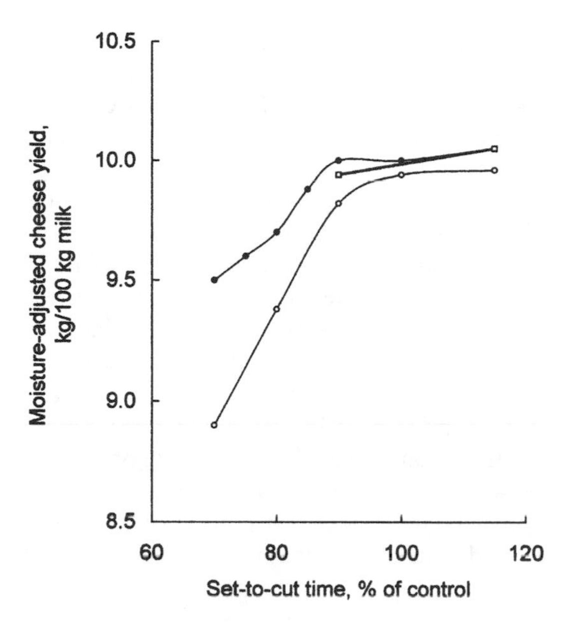

Figure 6–14 Effect of set-to-cut time (and hence firmness at cutting) and healing time on moisture-adjusted Cheddar cheese yield. Healing times were 0 min (○), 5 min (●), and 10 min (□).

hand knife and observing the "cleanness" of the cut and the clarity of the exuding whey. However, in large modern factories, conditions are not conducive to testing gel firmness in cheese vats from separate milk silos or from an individual day's milk, because of the large scale of operations (frequently more than 10^6 L of milk are processed per day) and the use of preprogrammed vats with limited operator access. Hence, most of the cheesemaking operations are performed on the basis of a preset time schedule rather than on the basis of objective criteria, such as gel firmness at cutting or pH at whey drainage. The criteria for determining gel cut times are probably based on one or more of the following:

- data from previous production years that suggest that milk at different periods of the year requires different set-to-cut times (e.g., due to differences in milk composition)
- data from laboratory analysis from a previous day's production showing the level of fat and fines in cheese whey
- recovery of fat and casein in cheese

The above methods are not sufficiently precise to allow cutting at a constant gel firmness in every vat. The limitations of these methods have led to the development of in-vat gel-firmness sensors, which dynamically monitor milk coagulation and, when incorporated into an integrated system, activate the curd-knives to cut the gel when it has attained the desired firmness (strength). The mechanisms employed to date in designing in-vat sensors to monitor the development of curd firmness over time include monitoring related changes in

- convective heat transfer from a probe (a "hot wire") to the surrounding milk, as in hot wire probe sensors (Bellon, Quiblier, Durier, & Noel, 1988; Hori, 1985; LeFevre & Richardson, 1990)
- turbidity (McMahon, Brown, & Ernstrom, 1984)
- diffuse reflectance of visible (e.g., λ, 660 nm), near infrared (e.g., λ, 820 nm), or infrared light (e.g., λ, 950 nm), as in various fiber-optic probes such as the CoAguLite fiber-optic sensor and the Omron E3XA (Payne, 1995)
- near-infrared light transmissions as in various infrared probes such as TxPro and Gelograph NT (O'Callaghan, O'Donnell, & Payne, in press)
- absorption and attenuation of ultrasound waves, or pulses, of different frequencies (e.g., > 0.5 MHz) passed through the milk (Benguigui, Emerery, Durand, & Busnel, 1994; Gunasekaran & Ay, 1994).

The hot wire and fiber-optic probes are manufactured commercially and are being used in cheese factories as online curd-firmness sensors; these are discussed briefly below. Sensors based on measuring changes in tubidity or ultrasound attenuation have not yet, to the authors' knowledge, been successfully developed for use in cheese plants.

The Hot Wire Probe. The hot wire probe has achieved the furthest development toward commercial application of all curd firmness sensors, with most of the research occurring at the Snow Brand Dairy Products Company (Japan) and the National Institute for Agronomic Research (France). Stoelting Inc. (Kiel, Wisconsin) has commercialized the Optiset II hot wire sensor (from Snow Brand), which is now operating in several cheesemaking plants in the United States.

The principle of measurement is based on changes in heat transfer from a hot wire to the milk. A thin platinum wire probe is immersed in the milk. A constant current is passed through the wire, generating heat, which is dissipated readily, by convection currents near the wire, while the milk is liquid. As the milk coagulates, its viscosity increases, and generated heat is no longer readily dissipated. The temperature of the wire increases, causing an increase in its resistance. The resistance and temperature of the wire are dynamically measured by monitoring changes in voltage across the wire, giving a continuous output signal.

A typical output signal is shown in Figure 6–15. The peak in the first derivative of the output signal corresponds to the onset of gelation. The instrument does not detect a gel cutting time; the increase in gel firmness beyond the onset of gelation (i.e., the gel point) is not readily detected by the hot wire, as it has a relatively small effect on heat dissipation (compared with the transition from a liquid to a gel). However, empirical equations have been developed to relate the gel point, as detected by the hot wire, to cut times (at a particular firmness, e.g., 40 Pa) as determined by low-amplitude strain oscillation rheometry, the Formagraph, or other laboratory methods.

Diffuse Reflectance Fiber-Optic Probe. An infrared diffuse reflectance probe, designed at the University of Kentucky, was installed in two cheese plants in the United States in 1993. The principle of measurement is based on changes in the light-scattering properties of milk. Infrared light is emitted from an LED and transmitted through one branch of a bifurcated cable containing optical fibers to the tip of a probe in contact with the renneted milk (Figures 6–16 through 6–18). Light reflected by both the fat globules and casein micelles is detected by the optical fibers in the other branch of the bifurcated cable and transmitted to a photodetector. As the milk coagulates, more light is reflected (due to the aggregation of the paracasein micelles) and transmitted to the photodetector, the output signal from which is directly proportional to the amount of light received. As in the case of the hot wire probe, the peak in the first derivative of the output signal corresponds to the onset of gelation, which is then related to the cut time at a given firmness as determined by laboratory instruments.

6.7 FACTORS THAT AFFECT RENNET COAGULATION

The strength of the resulting gel (curd tension) is as important as the coagulation time, if not more so, especially from the point of view of cheese yield. The gel assembly process is quite slow (see Figure 6–6), and in the case of most cheese varieties a period roughly equal to the RCT is allowed from the onset of visual coagulation for the gel to become sufficiently firm prior to cutting. If the gel is too soft when cut, fat and casein losses in the whey will be high (see Bynum & Olson, 1982, for a description of the influence of curd firmness on cheese yield and for references on this subject). In general, there is an inverse relationship between RCT and curd tension, which means any factor that reduces RCT increases curd tension and vice versa. The effects of various compositional and environmental factors on the primary and secondary phases of rennet coagulation are summarized in Table 6–2.

6.7.1 Milk Protein Level

The coagulation time of milk decreases markedly with protein (and thus casein) content, in

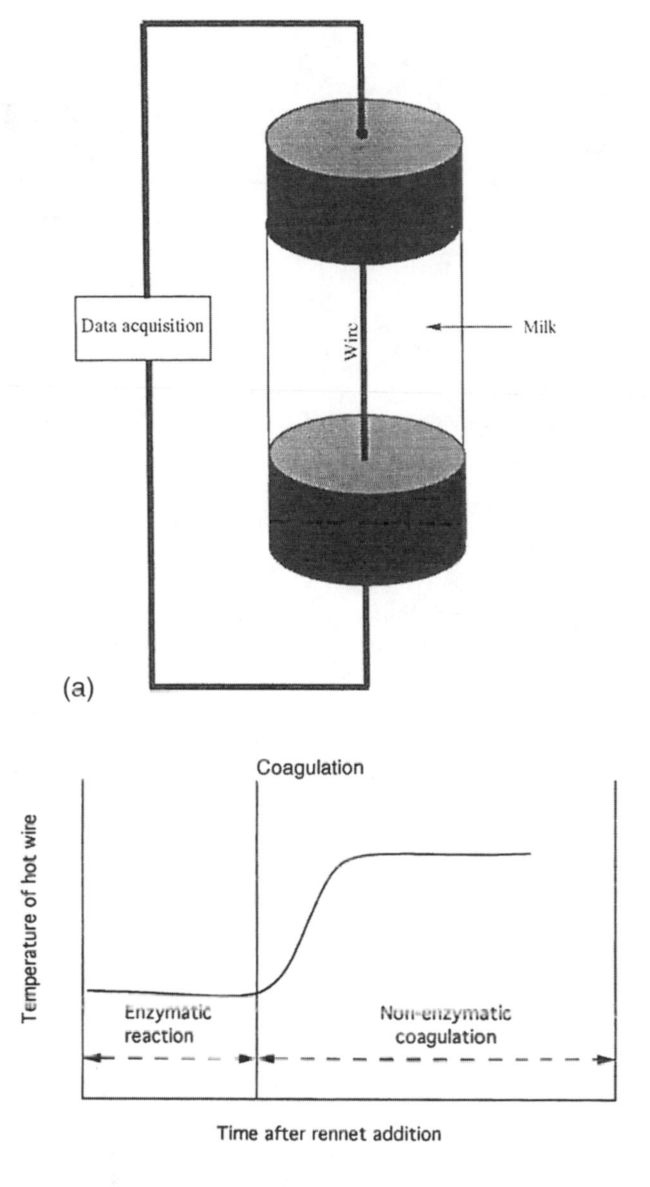

Figure 6–15 (a) Hot wire sensor for objectively measuring the rennet coagulation of milk. (b) Changes in the temperature of the hot wire during the course of the rennet coagulation of milk.

the range 2.0–3.0% (w/w), when rennet is added on a volume basis (Figures 6–19 and 6–20). Further increases in milk protein level (i.e., > 3.0%, w/w) result in a slight increase in gelation time, an effect attributable to the decreasing rennet:casein ratio, which necessitates an increase in the time required to generate sufficient hydrolysis of κ-casein to induce aggregation of paracasein micelles. At a constant rennet:casein ratio, the RCT decreases with increasing casein

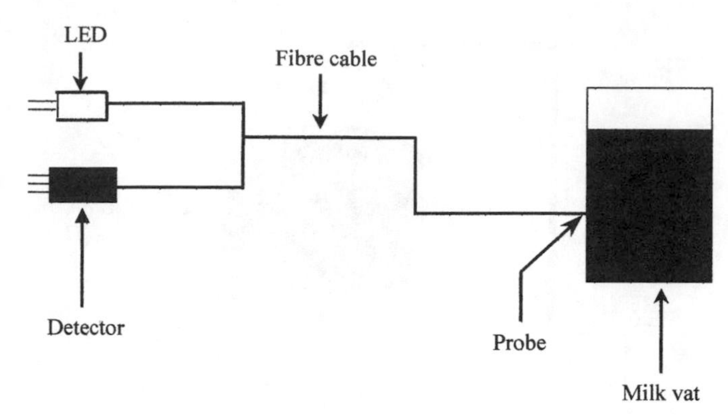

Figure 6–16 Schematic representation of the fiber-optic sensor developed to measure the diffuse reflectance of coagulating milk. LED = light emitting diode.

concentration, such as obtained by ultrafiltration, and vice versa. From a practical viewpoint, a minimum protein level of 2.5–3.0% (w/w) is necessary for gel formation in cheese manufacture (i.e., within 40–60 min). The maximum curd-firming rate (S_{max}) and curd firmness (G') increase more than proportionally with protein level (Figure 6–18), with a power law dependence of the latter parameters and protein concentration (i.e., $S_{max} \propto P^{n1}$ and $G' \propto P^{n2}$, where n1 and n2 > 1.0, typically ≈ 2.0 [Guinee et al., 1996]). Hence, small variations in the protein content of milk, as can occur throughout the cheesemaking season, exert a relatively large effect on the coagulation properties of rennet. The positive effects of the higher milk protein content on the rennet coagulation properties probably ensue from the higher level of gel-forming

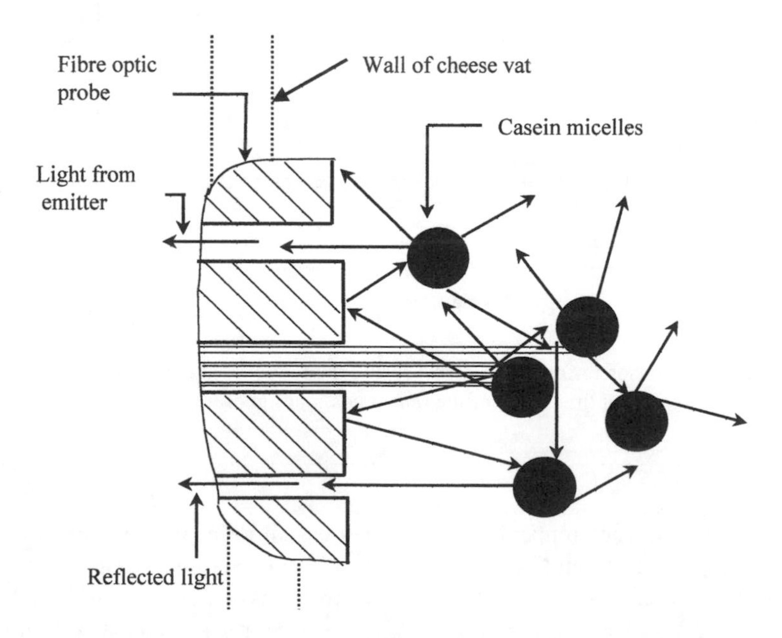

Figure 6–17 Schematic representation of the diffuse reflectance probe of the fiber-optic sensor.

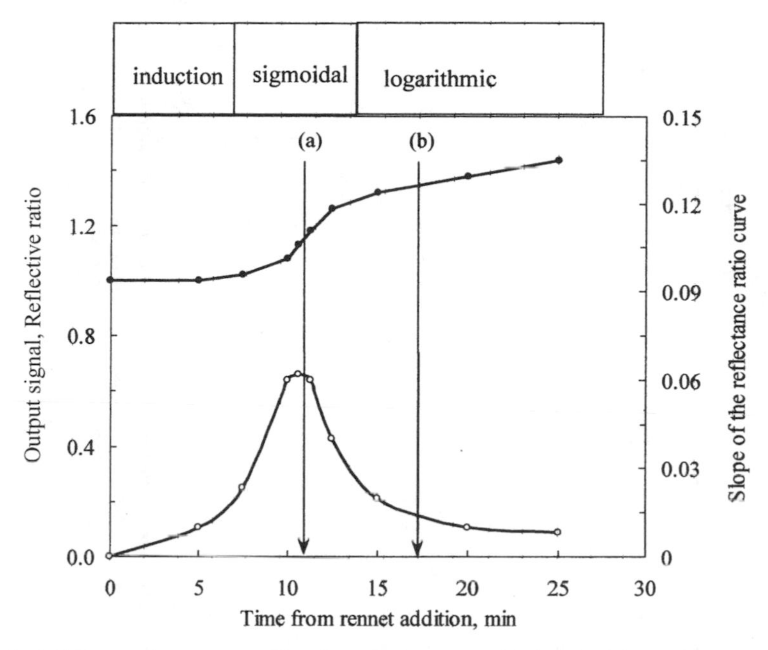

Figure 6–18 Typical diffuse reflectance profile of rennet-treated milk showing the various stages of rennet coagulation process: induction (rennet hydrolysis of κ-casein), sigmoidal (aggregation of paracasein micelles and gel formation), and logarithmic (continued fusion of paracasein micelles and curd firming). The reflectance slope is obtained from the first derivative of the reflectance ratio; the maximum slope (arrow a) corresponds to the time of the maximum aggregation rate; arrow b indicates the time at which the gel has become sufficiently firm for cutting.

protein, which increases the proximity of casein micelles and thus augments the rate of casein aggregation.

One of the economic attractions in using ultrafiltration-concentrated milk in cheesemaking is the savings that accrue from using less rennet. Cheese made from ultrafiltration-concentrated milk ripens more slowly than normal, due partly to slower proteolysis, for which there may be a number of causes, including the lower ratio of rennet to casein.

6.7.2 Milk Fat Level

Increasing fat content in the range 0.1–10% (w/w) while maintaining the protein level constant (e.g., at 3.3%, w/w) enhances the rennet coagulation properties, as reflected by decreases in coagulation time and set-to-cut time and higher values for S_{max} and G' (Figure 6–20).

However, the positive effects are much lower than those obtained by increasing protein content in the same range. Indeed, in a milk in which the level of fat plus protein is maintained constant, increasing the fat level results in significant decreases in S_{max} and G'. In commercial cheese manufacture, where standardization of milk protein to a fixed level (e.g., by ultrafiltration of skim milk) is not normally practiced, S_{max} and G' increase progressively upon adding cream to a fat level of about 4% (w/w) and decrease rapidly thereafter. The decrease is due to the dilution effect on the protein, which eventually offsets the benefits of increasing the fat content. From physical and structural considerations, the effect of increasing the fat level in a milk where the absolute level of gel-forming protein is constant is probably twofold:

Table 6–2 Effects of Some Compositional and Processing Factors on Various Aspects of the Rennet Coagulation of Milk

Factor	First Phase	Second Phase	Overall				
			RCT	*GT*	*Curd Firming Rate*	*Curd Firmness after a Fixed Renneting Time*	*Set-to-Cut Time*
Increasing the casein level when rennet is added on a volume basis	–	+++	ND	⇓	⇑	⇑	⇓
Increasing the fat content of milk							
when casein level is constant	ND	+	ND	⇓	⇑	⇑	⇓
when fat plus casein levels are constant	ND	–	ND	⇑	⇓	⇓	⇑
Pasteurization temperature:							
60°C × 15 s	++	+++	⇓	ND	ND	ND	ND
≥ 72°C × 15 s	– –	– –	⇑	⇑	⇓	⇓	⇑
Milk homogenization	ND	ND	ND	⇓	NE	⇑	⇓
Added $CaCl_2$							
0.2–10 mM	NE	+	⇓	⇓	⇑	⇑	⇓
> 10 mM	ND	–	ND		⇓	⇓	⇑
Gelation temperature (4→35°C)	+	+++	⇓	⇓	⇑	⇑	⇓
Decreasing gelation pH (6.6→6.0)	+++	+	⇓	⇓	⇑	⇑	⇓
Rennet concentration	+++	ND	⇓	⇓	⇑	⇑	⇓

Key: RCT = rennet coagulation time as determined by the rennet coagulation time assay; GT = gelation time as determined by dynamic methods such as the Formagraph method or low-amplitude strain oscillation rheometry; NE = no effect; ND = no data available; – = slight negative effect; – – = moderate negative effect; + = slight positive effect; ++ = moderate positive effect; +++ = large positive effect; ⇑ = magnitude of the rennet coagulation parameter increases; ⇓ = magnitude of rennet coagulation parameter decreases.

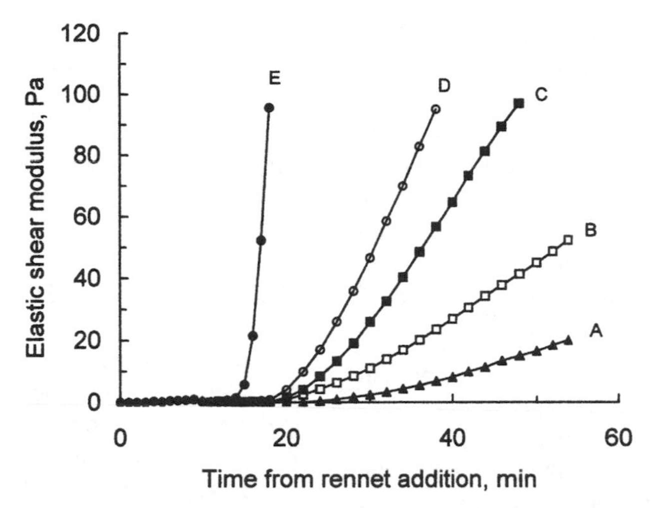

Figure 6–19 Effect of milk protein level (30 [A], 35 [B], 45 [C], 69 [D], or 82 [E] g/kg) on the elastic modulus of rennet-coagulated milk. Milks B–E were prepared by ultrafiltration of milk A. Coagulation parameters that may be derived from the curve are gelation time (point at which G' begins to increase, curd-firming rate (slope of G' time curve in the linear region), and curd firmness (the value of G' at a given time from rennet addition).

1. The concomitant increase in viscosity with fat content probably restricts the movement of gel strands and thereby contributes to a higher gel rigidity.
2. Simultaneously, the increasing number of fat globules causes the gel strands to become more elongated to surround and occlude the obstructing fat globules; this results in thinner and weaker gel strands.

6.7.3 Pasteurization Temperature

Preheating milk up to about 65°C has a beneficial effect on rennet coagulation, owing to heat-induced precipitation of calcium phosphate and a concomitant decrease in pH. These changes occur also at higher temperatures, but their beneficial effects on rennet coagulation are offset and eventually overridden by the combined effects of

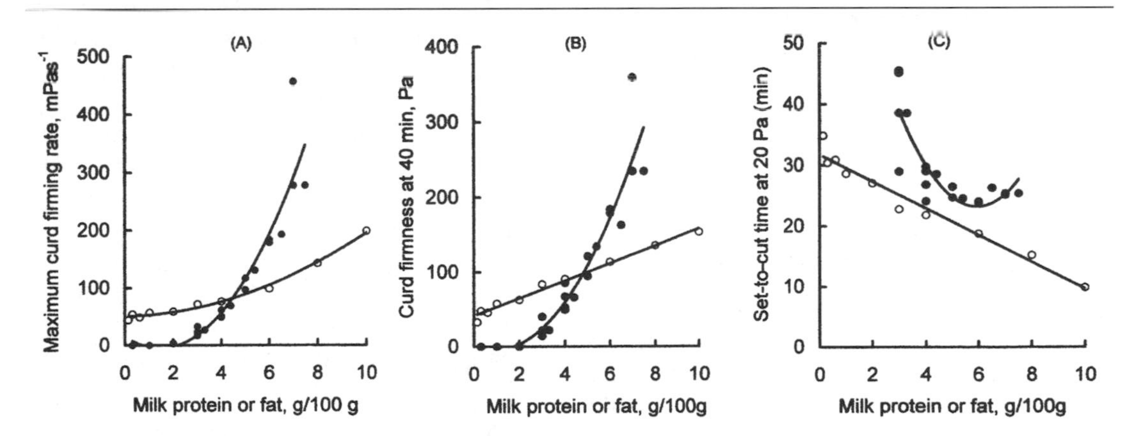

Figure 6–20 Effect of increasing level of protein (●) in skim milk or fat (○) in milks containing 33 g/kg protein on rennet coagulation properties: maximum curd-firming rate (A), curd firmness at 40 min after rennet addition (B), and set-to-cut time at a firmness of 20 Pa (C).

- whey protein denaturation and the interaction of denatured β-lactoglobulin with micellar κ-casein
- the deposition of heat-induced, insoluble calcium phosphate and the consequent reduction, upon subsequent cooling, in the concentration of native micellar calcium phosphate, which is important for cross-linking paracasein micelles, and hence aggregation, during gel formation

The complexation of denatured whey protein with κ-casein adversely affects both the enzymatic and nonenzymatic phases of rennet coagulation but especially the latter. Increasing the extent of whey protein denaturation (as a percentage of total) to a level greater than 15% by high heat treatment (e.g., > 80°C × 15 s) impairs the rennet coagulation characteristics to such an extent that the milk is unsuitable for commercial cheese manufacture (Figure 6–21). Very severely heated milk (e.g., 90°C × 10 min, 80–90% total whey protein denatured) is not coagulable by rennet.

If heated milk is cooled, the RCT increases further (Figure 6–22), a phenomenon referred to as *rennet hysteresis*. The effect can be explained as follows: the adverse influence of the interaction of β-lg with κ-CN on rennet coagulation is offset to some extent by the beneficial effect of heat-precipitated calcium phosphate and reduced pH. However, heat-induced changes in calcium phosphate are at least partially reversible upon cooling, and hence the full adverse effects of the protein interaction become fully apparent upon cooling. In practice, milk should be pasteurized immediately before cheesemaking and should not be cold-stored before use.

6.7.4 Cooling and Cold Storage of Milk

Cooling and cold storage of milk (raw or heated) have adverse effects on the cheesemaking properties of milk. Apart from the growth of psychrotrophs, two undesirable changes occur:

1. Some indigenous colloidal calcium phosphate dissolves, with a concomitant increase in pH.
2. Some proteins, especially β-casein, dissociate from the micelles.

These changes are reversed by HTST pasteurization or by heating at a lower temperature, such as 31°C, for a longer period.

6.7.5 Milk Homogenization

Homogenization of milk is practiced in the manufacture of some cheese varieties in which lipolysis is important for flavor development, such as Blue cheese. The objective is to increase the accessibility of the fat to fungal lipases and thus to increase the formation of fatty acids and their derivatives (e.g., methyl ketones). Moreover, homogenization is a central part of the manufacturing process for cheeses made from recombined milks. Homogenization reduces fat globule size and increases the interfacial area of the fat surface by a factor of 5–6. Simultaneously, the fat globules become coated with a protein layer consisting of casein micelles, micelle subunits, and whey proteins. Hence, the newly formed fat globules behave as pseudoprotein particles and are able to become part of the gel network. Numerous studies have been undertaken to evaluate the effect of homogenization under different temperatures, pressures, and/or milk fat levels. While some discrepancies exist between the results of these studies, the main trends indicate that homogenization lowers the gelation time slightly, has no effect on the curd-firming rate, and causes a slight increase in G'. However, the higher moisture content of cheese made from homogenized milk, compared with that made from nonhomogenized milk, suggests that homogenization may alter the rate of casein aggregation during the later stages of cheese manufacture (i.e., after cutting).

6.7.6 Renneting (Set) Temperature

The principal effect of set temperature is on the secondary, nonenzymatic phase of coagulation, which does not occur at temperatures be-

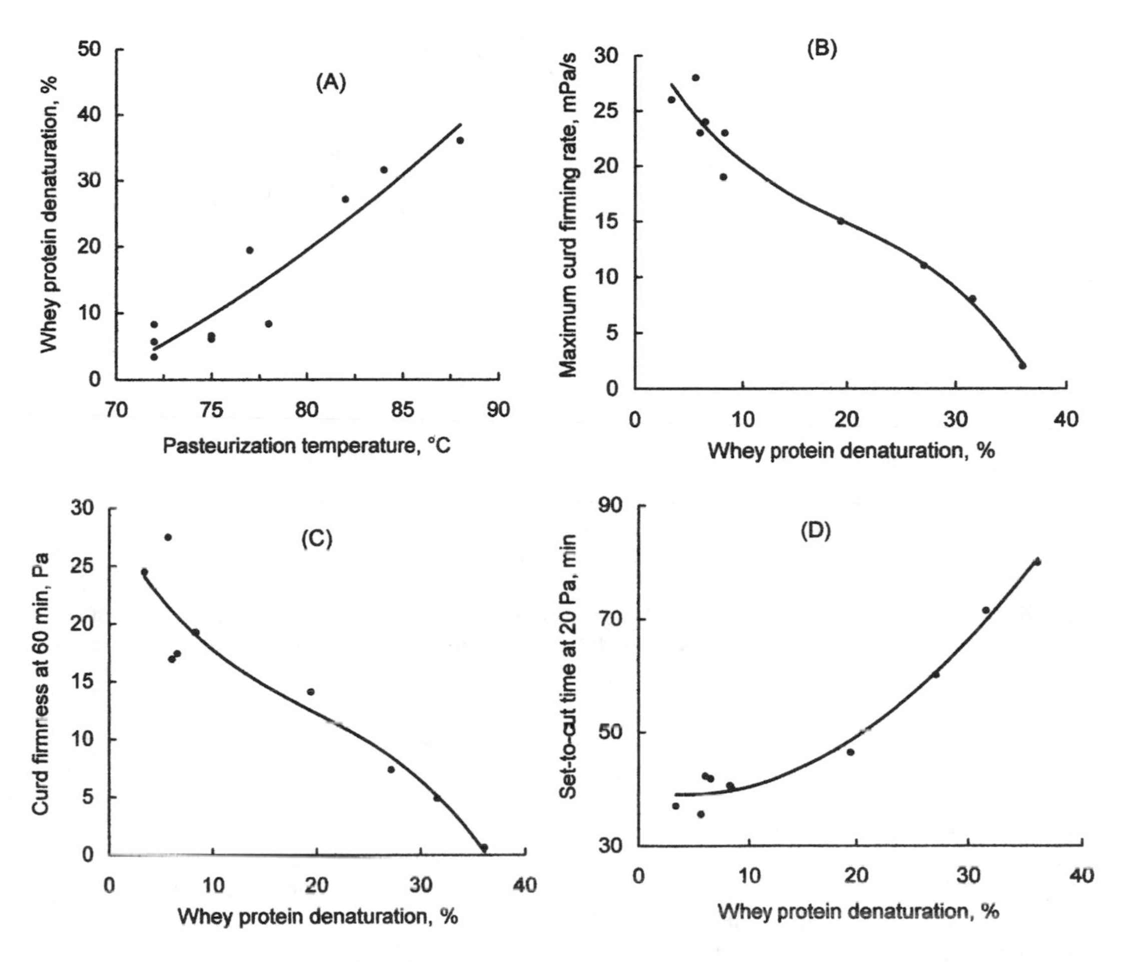

Figure 6–21 Effect of pasteurization temperature (for 15 s) on the level of whey protein denaturation (A), and effect of the level of whey protein denaturation on maximum firming rate (B), curd firmness 60 min after rennet addition (C), and set-to-curd time at a firmness of 20 Pa (D).

low around 18°C. Above this temperature, the coagulation time decreases to a broad minimum at 40–45°C and then increases again as the enzyme becomes denatured. In cheesemaking, rennet coagulation normally occurs at around 31°C, well below the optimum temperature. The lower temperature is necessary to optimize the growth of mesophilic starter bacteria, which have an optimum growth temperature of about 27–28°C and will not grow, nor perhaps even survive, above 40°C. In addition, the structure of the coagulum is improved at the lower temperature, which is therefore used even for cheeses made using thermophilic cultures.

6.7.7 pH

Due to the effect of pH on the activity of the enzyme, the rennet coagulation time increases with increasing pH, especially above pH 6.4 (Figure 6–23). The sensitivity to pH depends on the rennet used. Porcine pepsin is particularly sensitive, while the microbial rennets are relatively insensitive. Owing to the pH dependence of the rennet coagulation of milk, factors that

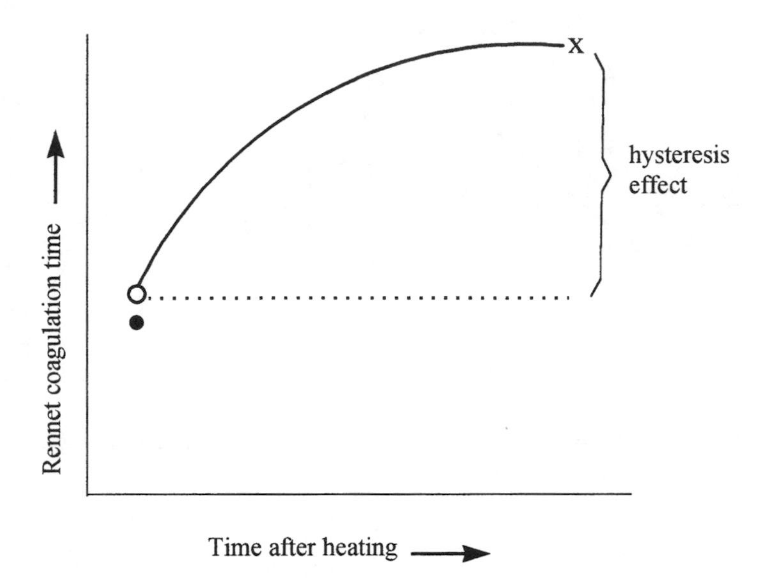

Figure 6–22 Schematic representation of the hysteresis effect on the rennet coagulation time (RCT) of heated milk. The symbols represent the RCT of raw milk (●), milk immediately after pasteurization (○), and milk 6 hr after pasteurization (X).

might affect the pH of milk (e.g., amount and form of starter added, addition of $CaCl_2$, ripening of milk, pH adjustment by addition of acid or acidogen, mastitis, and stage of lactation) will affect rennet coagulability. Curd firmness increases markedly with decreasing pH to a maximum at pH 5.9–6.0. The decrease in curd tension at lower pH values may be due in part to the solubilization of colloidal calcium phosphate as the pH is reduced. The pH of milk increases markedly in response to mastitic infection and may exceed 7.0 (i.e., it approaches the pH of blood, which is around 7.4). Mastitic milk has a longer RCT and lower curd tension than milk from healthy cows, probably owing to a combination of factors, such as high pH, low casein content, and the generally high somatic cell count (e.g., > 10^6 cells/ml) and associated proteolytic activity (which causes extensive hydrolysis of α_{s1}- and β-caseins). The pH of first colostrum is around 6; the pH increases to the normal value (6.7) within about 1 week and then remains relatively constant for the main part of lactation, before increasing substantially (to pH 7 or even higher) at the end of lactation.

6.7.8 Added $CaCl_2$

The addition of $CaCl_2$ to milk, which is common practice, promotes rennet coagulation via three beneficial changes:

1. an increase in $[Ca^{2+}]$
2. an increase in the concentration of colloidal calcium phosphate
3. a concomitant decrease in pH (the addition of $CaCl_2$ to 0.02%, i.e., 1.8 mM Ca, reduces the pH by ≈ 0.05–0.1 units, depending on protein level)

Hence, the addition of $CaCl_2$ (to 0.02 g/L, i.e., ≈ 2 mM Ca) enhances the rennet coagulation properties as reflected by a reduction in gel time and by increases in the curd-firming rate and curd firmness (Figure 6–23). However, at addi-

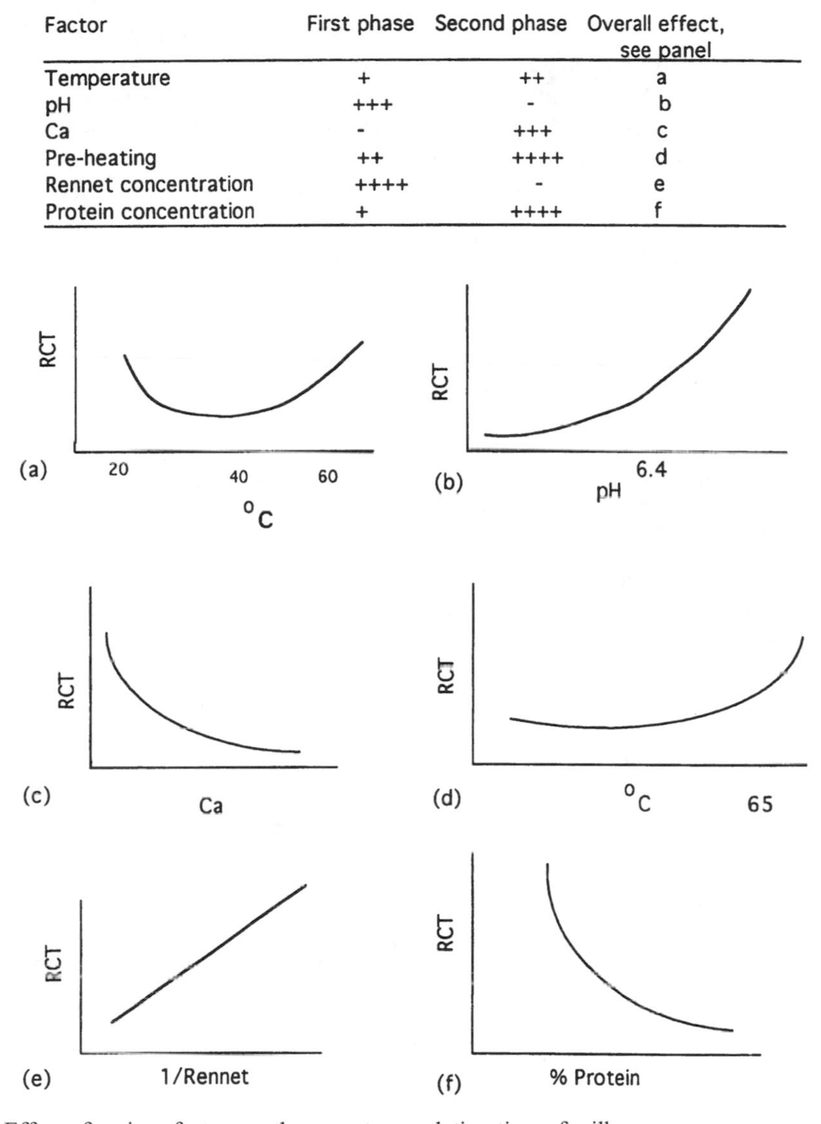

Factor	First phase	Second phase	Overall effect, see panel
Temperature	+	++	a
pH	+++	-	b
Ca	-	+++	c
Pre-heating	++	++++	d
Rennet concentration	++++	-	e
Protein concentration	+	++++	f

Figure 6–23 Effect of various factors on the rennet coagulation time of milk.

tion levels above 0.02 g $CaCl_2$/L, the curd-firming rate and curd firmness plateau and decrease again at levels greater than or equal to 0.1 g/L (i.e., ≥ 9 mM Ca). The decrease in curd tension at the higher Ca levels may be due to the effect of the interaction of the excess Ca with the negatively charged carboxyl groups on the para-casein; this would have the result of increasing the positive charge on the casein, making it less prone to aggregation. As expected, the addition of calcium chelators (e.g., EDTA and sodium phosphates) reduces gel firmness. The addition of NaCl or KCl increases gel firmness up to 100 mM but markedly decreases it at higher concentrations, possibly via displacement of micellar Ca.

6.7.9 Rennet Concentration

Obviously, the rate of the enzymatic phase of rennet coagulation is directly related to the amount of rennet used; there is a good linear relationship between enzyme concentration and milk-clotting activity (MCA) (Figure 6–23). However, there are discrepancies as to the effect of rennet level on the curd-firming rate and curd firmness, with some studies showing increases in the latter parameters and others no effect or slight decreases, depending on the stage of lactation.

In cheesemaking, the amount of rennet added is sufficient to coagulate the milk in 30–40 min (200–220 ml of standard calf rennet [≈ 60 RU/ml] per 1,000 L of milk). This level of rennet is traditional and is presumably based on experience. From a strictly coagulation viewpoint, more or less rennet could probably be used without adverse effects (other than the change in RCT). However, the amount of rennet retained in the curd is proportional to the amount of rennet added to the milk (at least for calf rennet), and this has a major effect on the rate of proteolysis during ripening. The retention of gastric rennets (e.g., calf chymosin and bovine pepsin) increases with decreases in pH at gel cutting and at whey drainage. On the other hand, the retention of *R. miehei* (Rennilase) and *R. pusillus* (Emporase) proteinases is not influenced by pH at cutting or at whey drainage.

There are numerous reports that curd tension is strongly influenced by the type of rennet used: calf chymosin gives a more rapid increase in curd tension than microbial rennets, although the substrate on which the rennets were standardized for clotting activity is of some significance. The fact that rennets standardized to equal clotting activity give different rates of curd firming and behave differently in response to compositional factors, such as Ca^{2+}, suggests possible differences in the extent and/or specificity of proteolysis during the primary, enzymatic phase of rennet coagulation. As far as is known, the primary phase of coagulation by all the principal coagulants involves cleavage of the Phe_{105}-Met_{106} bond except the enzyme from *C. parasitica*, which hydrolyzes Ser_{104}-Phe_{105}. Possibly, other bonds are also hydrolyzed by microbial rennets, although this is not obvious from gel electrophoretic studies. Further studies on this point are required.

The amount of rennet used seems to be optional, but the strategy of increasing concomitantly both the level of rennet used and starter cell numbers does not seem to have been investigated as a possible means of accelerating cheese ripening.

6.7.10 Other Factors

The rennet coagulation properties of milk may be influenced by stage of lactation and diet, which cause changes in milk composition (i.e., casein, fat, mineral, and pH levels); degree of casein hydrolysis (e.g., as influenced by plasmin and other proteinases); and the health of the cow. These effects tend to be more marked in countries, such as Ireland, New Zealand, and Australia, where milk is largely collected from spring-calving herds fed predominantly on pasture. Late lactation milk, especially when the lactose level is below 4.1% (w/w), is frequently associated with long coagulation times and low curd firmness. These defects may be alleviated by drying off cows at milk yields of above 8 L/d, improving the quality of the feed, blending late lactation milk with early lactation milk, and standardizing the cheesemaking process (e.g., pH at set and the rennet:casein ratio).

6.8 RENNET SUBSTITUTES

Owing to the increasing world production of cheese (roughly 2–3% per annum over the past 30 years) and the reduced supply of calf vells (due to a decrease in calf numbers and a tendency to slaughter calves at an older age), the supply of calf rennet has been inadequate for many years. This has led to an increase in the price of veal rennet and to a search for rennet substitutes. Despite the availability of numerous potentially useful milk coagulants, only six rennet substitutes (all aspartyl proteinases) have been found to be more or less acceptable for cheese production: bovine, porcine, and chicken pepsins and the acid proteinases from *Rhizomucor miehei*, *R. pusillus*, and *Cryphonectria parasitica*. (*Rhizomucor* and *Cryphonectria* were previously known as *Mucor* and *Endothia*, respectively.)

In addition to fulfilling the criteria laid down by legislative agencies regarding purity, safety, and absence of antibiotics (IDF, 1990), rennet substitutes must possess the following characteristics (Guinee & Wilkinson, 1992):

- A high MCA:proteolytic activity ratio. A high ratio, as for example with calf rennet, prevents excessive nonspecific proteolysis during manufacture and hence protects against a weak gel structure, high losses of protein and fat in the whey, and reduced yields of cheese solids. Moreover, it avoids excessive proteolysis during maturation and thus ensures the correct balance of peptides of different molecular weights and hence desirable flavor, body, and functional characteristics in the ripened cheese, making it suitable for certain applications (e.g., processed cheese products and cheese powder). Excessive proteolysis, especially of β-casein, is associated with the development of a bitter flavor.
- An MCA that is not very pH dependent in the region 6.5–6.9. A sharp decrease in MCA combined with increasing pH may lead to slow gelation and a low curd tension at cutting, especially if the milk pH at setting is high (e.g., 6.7–6.8, as may occur in late lactation) or when the casein concentration is low (e.g., < 2.4%, w/w). These conditions are conducive to low recovery of fat and reduced cheese yield and can occur in large factories, where the duration of milk ripening is short (especially with the use of direct vat starters) and production steps (including cutting) are generally carried out according to a fixed time schedule. The addition of $CaCl_2$ or acidulants (e.g., gluconic acid-δ-lactone) may overcome the latter problems.
- Thermostability comparable to that of calf rennet at the pH values and temperatures used during cheesemaking. This can markedly influence the level of residual rennet in high-cook cheeses such as Emmental, Romano, Provolone, and low-moisture Mozzarella and hence the level of proteolysis, texture, and functionality of the cheese during maturation (see Chapters 11 and 19).
- Low thermostability of rennets during whey processing. A heat-stable rennet in the whey (≈ 90% of that added to the cheese milk) may lead to coagulation of formulated milks, which normally include whey (e.g., infant formulae and calf milk replacers) upon reconstruction.
- The ability to impart desired flavor, body, and texture characteristics to the finished cheese.

Chicken pepsin is the least suitable of the commercial rennet substitutes and was used widely only in Israel, where it has now been replaced by microbial chymosin. Owing to its low MCA:proteolytic activity ratio, chicken pepsin promotes extensive degradation of both α_{s1}- and β-caseins in Cheddar cheese, leading to the development of flavor defects (e.g., bitterness) and textural defects (soft body and greasiness) during maturation. Bovine pepsin is probably the most satisfactory. Good quality veal rennet contains about 10% bovine pepsin, and many commercial "calf rennets" contain about 50%. Its proteolytic specificity is similar to that of calf chymosin, and it gives generally satisfactory results with respect to cheese yield and quality. The activity of porcine pepsin is very sensitive to pH greater than 6.6, and it may be denatured extensively during cheesemaking, impairing proteolysis during cheese ripening. A 50:50 mixture of porcine pepsin and calf rennet gave generally acceptable results, but porcine pepsin has been withdrawn from most markets.

Although the proteolytic specificity of the three commonly used fungal rennets is considerably different from that of calf chymosin, they have given generally satisfactory results when used in the manufacture of most cheese varieties. However, the proteolytic activity of all the rennet substitutes is higher than that of calf chymosin, resulting in higher levels of protein in the cheese whey and lower cheese yields (Figure 6–24). Prior to the introduction of genetically engineered chymosin, microbial rennets were used widely in the United States but not in most European countries, Australia, or New Zealand. The extensive literature on rennet substitutes has been reviewed (see Fox & McSweeney, 1997, for references).

Like chymosin, all commercially successful rennet substitutes are acid (aspartyl) proteinases. The molecular and catalytic properties of the principal rennet substitutes are generally similar

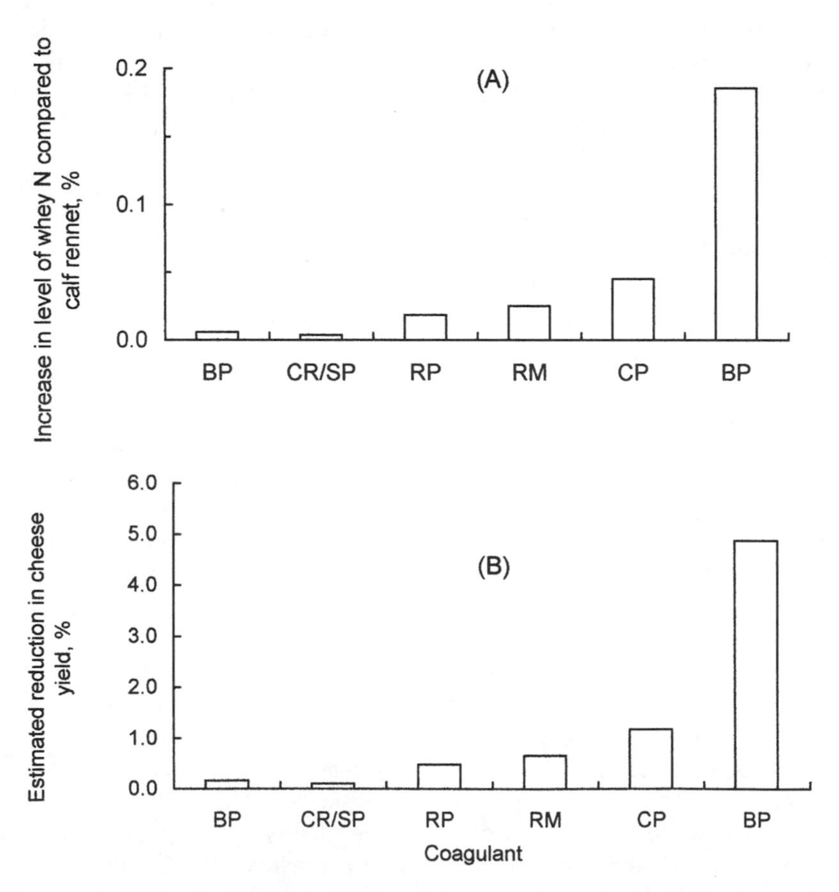

Figure 6–24 Increase in the level of N in whey, expressed as percentage protein (A), and estimated decrease in moisture-adjusted (370 g/kg) yield of Cheddar cheese (B), compared with calf rennet (94% chymosin and 6% bovine pepsin) when using different coagulants. BP, bovine pepsin (91% bovine pepsin + 9% calf chymosin); CR/SP, 50:50 blend of calf rennet and swine pepsin; RP, *Rhizomucor pusillus*; RM, *R. miehei;* CP, *Cryphonectria parasitica*; BP, *Bacillus polymyxa*.

to those of chymosin (see Chitpinityol & Crabbe, 1998; Foltmann, 1993). Acid proteinases have a relatively narrow specificity, with a preference for peptide bonds to which a bulky hydrophobic residue supplies the carboxyl group. Their narrow specificity is significant for the success of these enzymes in cheese manufacture. The fact that the pH of cheese is far removed from their optima (≈ 2 for porcine pepsin) is probably also significant. However, not all acid proteinases are suitable as rennets, because they are too active even under the prevailing relatively unfavorable conditions in milk and cheese. The specificity of porcine and bovine pepsins on α_{s1}- and β-caseins is quite similar to that of chymosin, but the specificity of the fungal rennet substitutes is quite different (see Chapter 11). Like chymosin, the Phe_{105}-Met_{106} bond of κ-casein is also preferentially hydrolyzed by pepsins and the acid proteinases of *Rhizomucor miehei* and *R. pusillus*, but the acid proteinase of *Cryphonectria parasitica* preferentially cleaves the Ser_{104}-Phe_{105} bond. However, unlike chymosin, the *Rhizomucor* and *Cryphonectria parasitica* proteinases also cleave several other bonds in κ-casein.

The MCA of commercial rennets (calf rennet, *R. miehei*, *R. pusillus*, and *C. parasitica*) increases with temperature in the range 28–36°C. The MCA of porcine pepsin, calf rennet, and bovine pepsin at pH 6.6 increases with temperature up to 44°C, 45°C and 52°C, respectively. The fungal enzymes (*R. miehei*, *R. pusillus*, and *C. parasitica*) lose activity at 47°C, 57°C, and 57°C, respectively. The MCA of the pepsins, especially porcine pepsin, is more pH dependent than that of chymosin, while that of the fungal rennets is less sensitive in the pH region 6.2–6.8 (Figure 6–25). The coagulation of milk by *C. parasitica* proteinase is also less sensitive to added Ca^{2+} than coagulation by calf rennet, but coagulation by *Rhizomucor* proteinases is more sensitive. For a given MCA, the rate of gel firming depends on the rennet used; this aspect of milk coagulation should be independent of rennet type and may indicate nonspecific proteolysis by the fungal enzymes.

The thermal stability of rennets differs considerably (Figure 6–26). Thermal stability is important when the whey is to be used in food processing. The early fungal rennets were considerably more thermostable than chymosin or pepsins, but the present products have been modified (by oxidation of methionine residues in the molecule) and have thermal stability similar to that of chymosin. The thermal stability of *C. parasitica* proteinase is less than that of chymosin at pH 6.6. The thermal stability of all rennets increases markedly with decreasing pH (Figure 6–26) (Thunell, Duersch, & Ernstrom, 1979).

Although they are relatively cheap, rennets represent the largest single industrial application of enzymes, with a world market of about 25×10^6 L of standard rennet per annum. Therefore, rennets have attracted the attention of industrial enzymologists and biotechnologists. The gene for prochymosin has been cloned in *E. coli*, *Saccharomyces cerevisiae*, *Kluyveromyces marxianus* var. *lactis*, *Aspergillus nidulans*, *A. niger*, and *Tricoderma reesei* (see Foltmann, 1993, and Pitts et al., 1992, for references). The enzymatic properties of the recombinant enzymes are indistinguishable from those of calf chymosin, although they may contain only one of the isoenzymes, A or B. The cheesemaking properties of recombinant chymosins have been assessed on many cheese varieties, always with very satisfactory results (see review by Fox & Stepaniak, 1993). Recombinant chymosins have been approved for commercial use in many, but not all, countries. Three recombinant chymosins are now marketed commercially: Maxiren, secreted by *K. marxianus* var. *lactis* and produced by Gist Brocades (the Netherlands); Chymogen, secreted by *A. niger* and produced by Hansen's (Denmark); and Chymax secreted by *E. coli*, and developed by Pfizer (United States). The genes for Maxiren and Chymogen were isolated from calf abomasum, while that used for Chymax was synthesized. Microbial chymosins have taken market share from both calf rennet and especially fungal rennets and now represent about 35% of the total market.

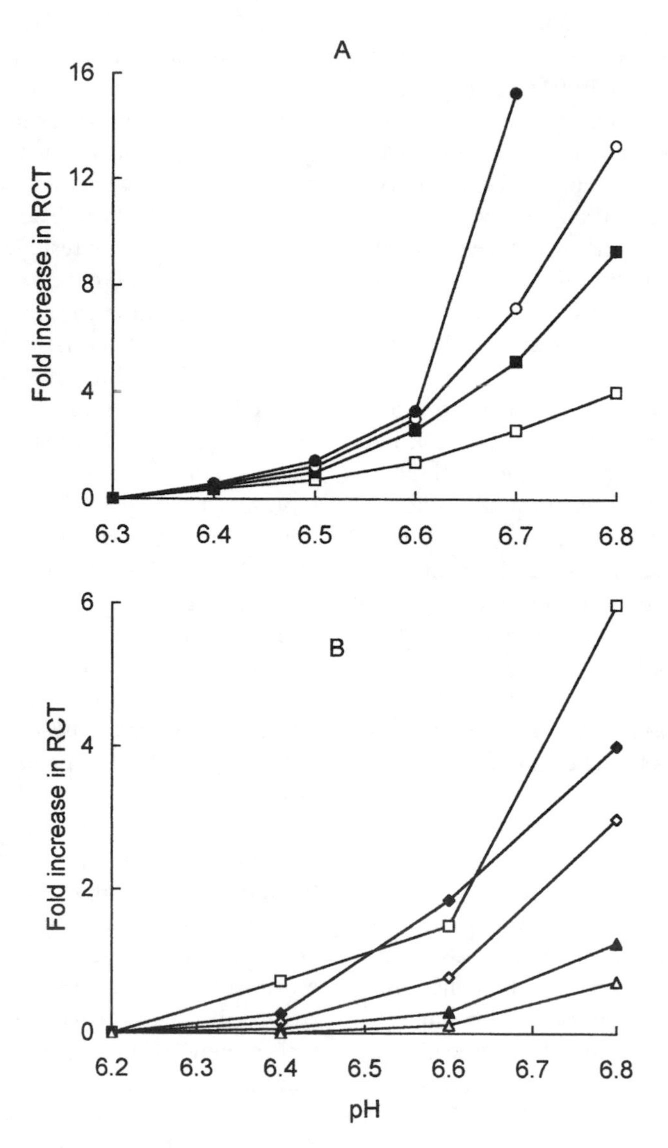

Figure 6–25 Effect of pH on the rennet coagulation time (RCT) of milk using (A) calf chymosin (❑), bovine pepsin (■), ovine pepsin (○), and porcine pepsin (●); (B) calf rennet (❑) and *Rhizomucor miehei* (◆), *R. pusillus* (◇), *Cryphonectria parasitica* (▲), and *Bacillus polymyxa* (Δ) proteinases.

The recombinant chymosins currently available are identical, or nearly so, to calf chymosin, but there are several published studies on engineered chymosins (Fox & McSweeney, 1997). At present, attention is focused on elucidating the relationship between enzyme structure and function, but this work may lead to rennets with improved MCA or modified general proteolytic activity (i.e., on α_{s1}- and/or β-casein). The natural function of chymosin is to coagulate milk in the stomach of the neonate. It was not intended for cheesemaking, and the wild-type enzyme

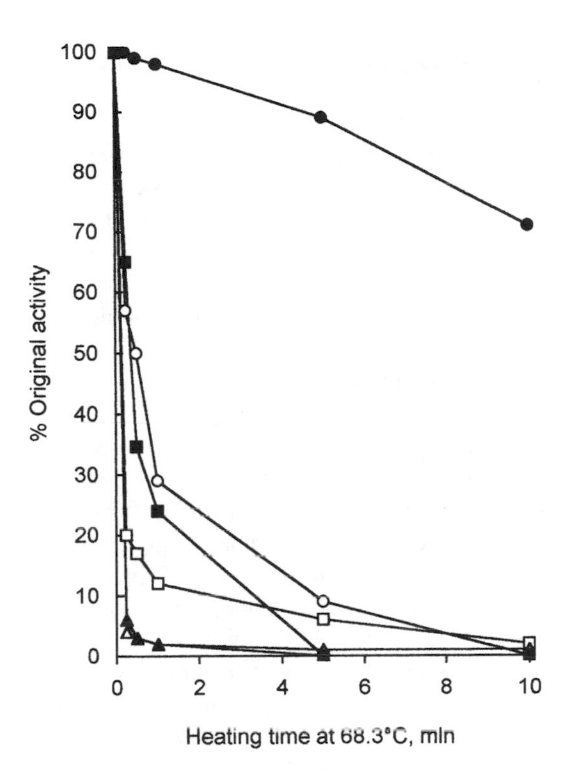

Figure 6–26 Effect of heating time at 68.3°C on the residual activity of various coagulants in whey at pH 5.2: calf rennet (□), bovine pepsin (■), porcine pepsin (▲), *Rhizomucor miehei* (●), *R. pusillus* (○), and *Cryphonectria parasitica* (Δ) proteinases.

probably is not the most efficient or effective proteinase for catalyzing proteolysis in cheese during ripening. Therefore, it may be possible to modify chymosin so as to accelerate its action on specific bonds of casein during ripening and/or to reduce its activity on others, hydrolysis of which may lead to undesirable consequences, such as bitterness. To date, the pH optimum, thermal stability, k_{cat}, and K_M of synthetic peptides have been modified through genetic engineering. We are not aware of any cheesemaking studies using engineered chymosins, and approval has not been obtained for their use.

The gene for *R. miehei* proteinase has been cloned in and expressed by *A. oryzae* (Novo Nordisk A/S, Denmark). It is claimed that this new rennet (Marzyme GM) is free of other proteinase or peptidase activities that are present in fungal rennets and may reduce cheese yield. Excellent cheesemaking results with Marzyme GM have been reported. Cloning of the gene for *R. miehei* proteinase has created the possibility for site-directed mutagenesis of the enzyme.

6.9 IMMOBILIZED RENNETS

Most of the rennet added to cheese milk is lost in the whey (more than 90% in the case of Cheddar). The loss of rennet represents an economic loss and creates potential problems for whey processors. Both problems could be solved through the use of immobilized rennets. A further incentive for immobilizing rennets is the possibility of producing cheese curd continuously by using a cold renneting technique (i.e., renneting at around 10°C, which allows the primary phase but not the secondary phase to occur), which should facilitate process control. The feasibility of continuous coagulation using cold renneting principles has been demonstrated, but the technique has not been commercially successful to date. As discussed in Chapter 11, the chymosin (or rennet substitute) retained in cheese curd plays a major role in cheese ripening. Consequently, if an immobilized rennet was used to coagulate milk, it would be necessary to add some chymosin (or similar proteinase) to the curd, and uniform incorporation of any such enzyme would be problematic, as has been demonstrated by the use of exogenous proteinases to accelerate cheese ripening (see Chapter 15).

In modern cheesemaking, most operations are continuous or nearly so. The actual coagulation step is the only major batch operation remaining, although the use of small "batches" of milk, as in the Alpma process for Camembert, makes coagulation, in effect, a continuous process. However, in large modern Cheddar and Gouda cheese factories, very large vats are used (20,000–30,000 L).

There is interest in the manufacture of rennet-free curd for studies on the contribution of enzymes from different sources to cheese ripening. A number of approaches have been used to produce rennet-free curd (see Fox, Law, McSweeney, & Wallace, 1993), but an effective completely immobilized rennet would be very useful.

Several investigators have immobilized different rennets on a range of supports and have claimed that these can coagulate milk. However, it appears that in such studies some enzyme leached from the support and that this solubilized enzyme was responsible for coagulation. An irreversibly immobilized rennet was unable to coagulate milk although it could hydrolyze nonmicellar casein. Presumably, the κ-casein on the surface of casein micelles is unable to enter the active site cleft of the immobilized enzyme owing to steric factors.

Even if immobilized rennets could coagulate milk, they may not be cost competitive (rennets are relatively cheap) and would be difficult to use in factory situations. The strategy envisaged for their use involves the passage of cold milk (e.g., 10°C) through a column of immobilized enzyme where the enzymatic phase of renneting would occur without coagulation (owing to the low temperature). The rennet-altered micelles would then be coagulated by heating the milk exiting the column to about 30°C. Heating would have to be conducted under quiescent conditions to ensure the formation of a good gel and to minimize losses of fat and protein; quiescent heating may be difficult on an industrial scale (e.g., many cheese factories process 10^6 L of milk per day). Hygiene and phage-related problems may present serious difficulties since cleaning the column by standard regimes would inactivate the enzyme. Plugging of the column and loss of activity have been problems even on a laboratory scale, and power cuts long enough to lead to an increase in temperature would be disastrous, as the column reactors would become plugged with cheese curd that would be difficult or impossible to remove. In short, the prospects for the use of immobilized rennets on a commercial scale are not bright, and such rennets are not currently being used.

REFERENCES

Bellon, J.L., Quiblier, J.P., Durier, C., & Noel, Y. (1988). Un noveau capteur industrial de mesure du temps de coagulation du lait: Le coagulomètre. *Technique Latière and Marketing, 1031*, 29–32. Cited from *Dairy Science Abstracts, 50*, 775.

Benguigui, L., Emerery, J., Durand, D., & Busnel, J. (1994). Ultrasonic study of milk clotting. *Le Lait, 74*, 197–206.

Bynum, D.G., & Olson, N.F. (1982). Influence of curd firmness at cutting on Cheddar cheese yield and recovery of milk constituents. *Journal of Dairy Science, 65*, 2290–2291.

Chitpinityol, S., & Crabbe, M.J.C. (1998). Chymosin and aspartic proteinases. *Food Chemistry, 61*, 395–418.

Dalgleish, D.G. (1980). Effect of milk concentration on the rennet coagulation time. *Journal of Dairy Research, 47*, 231–235.

Dalgleish, D.G. (1992). The enzymatic coagulation of milk. In P.F. Fox (Ed.), *Advanced dairy chemistry: Vol. 1. Proteins.* London: Elsevier Applied Science Publishers.

Dalgleish, D.G. (1993). The enzymatic coagulation of milk, In P.F. Fox (ed.), *Cheese: Chemistry, physics and microbiology* (2d ed., Vol. 1). London: Chapman & Hall.

Darling, D.F., & van Hooydonk, A.C.M. (1981). Derivation of a mathematical model for the mechanism of casein micelle coagulation by rennet. *Journal of Dairy Research, 48*, 189–200.

Foltmann, B. (1993). General and molecular aspects of rennets. In P.F. Fox (Ed.), *Cheese: Chemistry, physics and microbiology* (2d ed., Vol. 1). London: Chapman & Hall.

Fox, P.F. (1984). Proteolysis and protein-protein interactions in cheese manufacture. In B.J.F. Hudson (Ed.), *Developments in food proteins* (Vol. 3). London: Elsevier Applied Science Publishers.

Fox, P.F., Law, J., McSweeney, P.L.H., & Wallace, J. (1993). Biochemistry of cheese ripening, In P.F. Fox (Ed.), *Cheese: Chemistry, physics and microbiology* (2d ed., Vol. 1). London: Chapman & Hall.

Fox, P.F., & McSweeney, P.L.H. (1997). Rennets: Their role in milk coagulation and cheese ripening. In B.A. Law

(Ed.), *Microbiology and biochemistry of cheese and fermented milk* (2d ed.). London: Chapman & Hall.

Fox, P.F., & Mulvihill, D.M. (1990). Casein. In P. Harris (Ed.), *Food gels.* London: Elsevier Applied Science Publishers.

Fox, P.F., O'Connor, T.P., McSweeney, P.L.H., Guinee, T.P., & O'Brien, N. (1996). Cheese: Physical, biochemical and nutritional aspects. *Advances in Food Science and Nutrition, 39,* 163–328.

Fox, P.F., & Stepaniak, L. (1993). Enzymes in cheese technology. *International Dairy Journal, 3,* 509–530.

Guinee, T.P., O'Callaghan, D.J., Mulholland, E.O., Pudja, P.D., & O'Brien, N. (1996). Rennet coagulation properties of retentates obtained by ultrafiltration of skim milks heated to different temperatures. *International Dairy Journal, 6,* 581–596.

Guinee, T.P., & Wilkinson, M.G. (1992). Rennet coagulation and coagulants in cheese manufacture. *Journal of the Society of Dairy Technology, 45,* 94–104.

Gunasekaran, S., & Ay, C. (1994). Evaluating milk coagulation with ultrasonics. *Food Technology, 48,* 774–780.

Hori, T. (1985). Objective measurement of the process of curd formation during rennet treatment of milks by the hot wire method. *Journal of Food Science, 50,* 911–917.

Iametti, S., Giangiacomo, R., Messina, G., & Bonomi, F. (1993). Influence of processing on the molecular modifications of milk proteins in the course of enzymic coagulation. *Journal of Dairy Research, 60,* 151–159.

International Dairy Federation. (1990). *Use of enzymes in cheesemaking* (Bulletin No. 247). Brussels: Author.

International Dairy Federation. (1992). *Bovine rennets. Determination of total milk-clotting activity* (Provisional Standard 157). Brussels: Author.

LeFevre, M.J., & Richardson, G.H. (1990). Monitoring cheese manufacture using a hot wire probe. (Abstract). *Journal of Dairy Science, 73* (Suppl. 1), 74.

McMahon, D.J., Brown, R.J., & Ernstrom, C.A. (1984). Enzymic coagulation of milk. *Journal of Dairy Science, 67,* 745–748.

O'Callaghan, D.J., O'Donnell, C.P., & Payne, F.A. (in press). Effect of protein content of milk on the storage and loss moduli in renneting milk gels. *Journal of Food Process Engineering.*

Payens, T.A.J., & Wiersma, A. (1977). On enzymic clotting processes: V. Rate equations for the case of arbitrary rate of production on the clotting species. *Biophysical Chemistry, 11,* 137–146.

Payens, T.A.J., Wiersma, A., & Brinkhuis, J. (1977). On enzymic clotting processes: I. Kinetics of enzyme-triggered coagulation reactions. *Biophysical Chemistry, 6,* 253–261.

Payne, F.A. (1995). Automatic control of coagulum cutting in cheese manufacture. *Applied Engineering in Agriculture, 11,* 691–697.

Peri, C., Pagliarini, E., Iametti, S., & Bonomi, F. (1990). A study of surface hydrophobicity of milk proteins during enzymic coagulation and curd hardening. *Journal of Dairy Research, 57,* 101–108.

Pitts, J.E., Dhanaraj, V., Dealwics, C.G., Mantafounis, D., Nugent, P., Orprayoon, P., Cooper, J.B., Newman, M., & Blundel, T.L. (1992). Multidisciplinary cycles for protein engineering: Site-directed mutagenesis and X-ray structural studies on aspartic proteinases. *Scandinavian Journal of Clinical and Laboratory Investigation, 52* (Suppl. 210), 39–50.

Storry, J.E., & Ford, G.D. (1982). Development of coagulum firmness in renneted milk: A two-phase process. *Journal of Dairy Research, 49,* 343–346.

Surkov, B.A., Klimovskii, I.I., & Krayushkin, V.A. (1982). Turbidimetric study of kinetics and mechanism of milk clotting by rennet. *Milchwissenschaft, 37,* 393–395.

Thunell, R.K., Duersch, J.W., & Ernstrom, C.A. (1979). Thermal inactivation of residual milk clotting enzymes in whey. *Journal of Dairy Science, 62,* 373–377.

Tuszynski, W.B. (1971). A kinetic model of the clotting of casein by rennet. *Journal of Dairy Research, 38,* 115–125.

van Hooydonk, A.C.M., de Koster, P.G., & Boerrigter, I.J. (1987). The renneting properties of heated milk. *Netherlands Milk and Dairy Journal, 41,* 3–18.

Walstra, P., & van Vliet, T. (1986). The physical chemistry of curd making. *Netherlands Milk and Dairy Journal, 40,* 241–259.

CHAPTER 7

Postcoagulation Treatment of Renneted Milk Gel

7.1 INTRODUCTION

The rennet coagulation process is essentially similar for all cheese varieties and the structure of the coagulum (gel) is also similar. The gel is subjected to a series of treatments (see Chapter 2), the principal object of which is to remove whey from the gel and effectively concentrate the casein and fat to the degree characteristic of the variety. The principal treatments are described in this chapter. A summary on the manufacturing protocol for each of a number of cheeses is given in Chapter 17.

Rennet- or acid-coagulated milk gels are quite stable if left undisturbed, but when they are cut or broken or subjected to external pressure, the paracasein matrix contracts, expressing the aqueous phase of the gel (known as whey). This process, known as *syneresis*, enables the cheese maker to control the moisture content of the cheese; hence the activity of microorganisms and enzymes in the cheese, and hence the biochemistry of ripening and the stability and quality of the finished cheese. The higher the moisture content of cheese, the faster it will mature but the less stable it will be. High-moisture cheeses have a much greater propensity to develop off-flavors than low-moisture varieties. Although the starter and adventitious microflora of cheese have a major impact on the biochemistry of cheese ripening, they do so only in as far as the composition of the cheese curd permits. Syneresis is under the control of the cheese maker, and, via syneresis, so is the composition and quality of the cheese.

Many of the treatments to which rennet-coagulated milk gels are subjected may be classified generically as dehydration. Cheese manufacture essentially involves concentrating the fat and casein of milk approximately tenfold, and removing lactose, whey proteins, and soluble salts in the whey. Although there are certain common features, the factors that promote and regulate syneresis (dehydration) in a cheese variety or family of varieties are specific to that variety or family. In the case of Cheddar- and Swiss-type cheeses, dehydration is accomplished mainly in the cheese vat by fine cutting the coagulum, extensive "cooking" of the curds-whey mixture (to ≈ 40°C for Cheddar-type and ≈ 55°C for Swiss-type cheeses) and vigorous agitation during cooking. For the softer (high-moisture) varieties, the gel may be scooped directly into the molds without cutting or cooking, and whey explosion occurs mainly in the molds as the pH decreases. Curds for some varieties (e.g., Cheddar and Swiss) are subjected to considerable pressure in the molds to aid whey removal, while curds for the softer varieties are pressed only under their own weight.

Most of the published studies on syneresis have been concerned mainly with the factors that affect it during the early stages of dehydration in Cheddar- and Dutch-type cheeses, that is, mainly during cooking, but it is assumed that basically the same mechanisms operate in all varieties throughout the dehydration process.

Despite its accepted importance in the control of cheese moisture, the mechanism of syneresis of renneted milk gels is not well understood. There is a considerable amount of empirical information on factors that influence syneresis but the actual mechanism of syneresis has received very little study. Poor methodology is mainly responsible for the lack of information; the number of principles exploited in methods used to measure syneresis attests to their unsuitability. Some authors have attempted to simulate cheese manufacture, such as by stirring, observing a cooking profile, even adding starter, but the accuracy and precision of many of the methods are poor. Many of the methods have been used only by the original investigator.

The literature on the syneresis of milk gels has been reviewed by Green and Grandison (1993) and Walstra (1993) and is summarized here.

7.2 METHODS FOR MEASURING SYNERESIS

A variety of methods have been employed to quantify syneresis. These include

- measuring the volume of whey expressed from curd pieces under standard conditions, following cutting of the gel
- measuring changes in the moisture content, volume, or density of curd pieces over time
- using tracers or markers to indirectly measure whey volume
- measuring changes in the electrical resistance of the curd

Techniques for measuring the volume of expressed whey are simple and straightforward to execute, but complete recovery of whey is difficult, and syneresis continues during any separation process. Similar constraints apply to methods that depend on the volume or composition of the curd, and, in addition, the actual analytical step may be difficult while avoiding continuing syneresis. Methods based on the use of a tracer or marker involve adding a small volume of a solution of some tracer (e.g., a dye) to the system at the start of syneresis; as the volume of free whey increases, the concentration of tracer in the solution decreases. The principal problems to be avoided are diffusion of the tracer into or its adsorption onto the curd particles. The opposite strategy has also been used—the placing of a small amount of clarified whey on top of the cut gel. Whey expressed from the curd is turbid owing to the presence of fat globules, and therefore the turbidity of the free whey increases as syneresis progresses. As the moisture content of curd decreases, its electrical conductivity decreases. As with many other methods, clean separation of curd particles from the whey without concomitant changes is difficult to achieve.

However, data from studies using these methods and from actual cheesemaking experiments have helped clarify the influence of several factors on syneresis, at least in general terms.

7.3 INFLUENCE OF COMPOSITIONAL FACTORS ON SYNERESIS

The syneresis of renneted milk gels is influenced by milk composition, which in turn is affected by the feed, stage of lactation, and health of the animals from which the milk is obtained. Fat tends to reduce syneresis and increase the water-holding capacity of cheese curd, and increasing the fat content of cheese milk increases cheese yield by about 1.2 times the weight of the additional fat. However, syneresis tends to be directly related to casein concentration, which is to say that good syneresis occurs at high casein levels. Since the fat and casein levels in milk tend to change in parallel, they have offsetting effects on syneresis. Concentration of milk suppresses syneresis, possibly because of its effect on gel strength, although the rigidity modulus (see Chapter 6) at the time of cutting appears to have little effect on syneresis.

The rate of syneresis is directly related to the acidity and therefore is inversely related to pH; it is optimal at the isoelectric point (i.e., pH 4.6–4.7). The addition of $CaCl_2$ to milk promotes syneresis, but the effect appears to be less than might be expected and may be negative at certain pH values and at high calcium concentrations, especially if the gel is held for a long period before cutting. The adverse effect of a high concentration of calcium has been attributed to interaction of Ca^{2+} with the aspartate and glutamate groups of proteins, leading to an increased net positive charge, swelling of the protein, and suppression of syneresis. It is likely that a firmer gel, such as would be obtained on longer holding, would also be more resistant to syneresis. The influence of colloidal calcium phosphate on syneresis does not appear to have been investigated. Addition of low levels of NaCl increases the rate of syneresis but higher levels retard it.

7.4 INFLUENCE OF PROCESSING VARIABLES ON SYNERESIS

The extent of syneresis, and hence the moisture content of cheese, is influenced by various factors associated with cheesemaking procedures. Many of these are exploited by cheese makers to control cheese composition and thus its flavor and texture. The principal factors are described below.

7.4.1 Size of the Curd Particles

Everything else being equal, the smaller the curd pieces, the faster the rate and the greater the extent of syneresis, reflecting the greater surface area available for loss of whey. For some high-moisture cheeses, the coagulum may not be cut but scooped, unbroken, into cheese molds. For Cheddar and Dutch-type cheeses the coagulum is cut into cubes of about 1 cm size using knives with vertical or horizontal wires or bars (Figure 7–1). Traditionally, the coagulum for many Swiss or Italian varieties was cut with a harp or a spino, respectively (Figure 7–1), which is used in a swirling action around the hemispherical or conically shaped vats used traditionally for these varieties. In the large modern vats used for Cheddar, Dutch, and other varieties, the cutting knives are fixed in the vats and serve to cut the coagulum and agitate the curd-whey mixture during cooking (Figure 7–2).

7.4.2 Cook Temperature

Heating the curds-whey mixture (a process referred to as scalding or cooking) promotes syneresis (Figure 7–3). The cook temperature is characteristic of the variety. For example, the cook temperature is 31°C for high-moisture varieties such as Camembert (in effect, no cooking), 36°C for Gouda and Edam, 38–40°C for Cheddar, and 52–55°C for Emmental and Parmesan. The cook temperature must match the thermal stability of the starter. Acid production by some *Lactococcus* strains is stopped around 35°C, but other strains withstand cooking at 40–42°C, and cooking cheese curds to a temperature that inhibits the culture may have a negative effect on syneresis owing to the reduced rate of acidification. A cook temperature up to 55°C may be used when a thermophilic starter is used. Such starters survive but do not grow at 55°C, and hence syneresis depends on temperature rather than on pH. In fact, temperature and pH are complementary: syneresis of low-acid curds (e.g., Emmental) depends mainly on temperature, while in high-acid curds (e.g., Camembert) temperature is of little consequence.

For most varieties, cooking is done by circulating hot water or steam through the jacket of the cheese vat (steam is preferable because it can be shut off more readily than hot water, facilitating better control of temperature). Before the availability of hot water or steam for cooking curds and whey, cooking was done over an open fire and would have been difficult to control precisely (open-fire cooking is still practiced in artisanal farmhouse cheesemaking). Illustrations of old cheese factories suggest that jacketed vats (and cooking by hot water or steam) were used from the start of industrialized cheesemaking. For Dutch-type cheeses and a number of other varieties, cooking is done by removing part of the whey (30–40%) and replacing it with warm water to give a blend of the desired temperature. This method was probably developed for

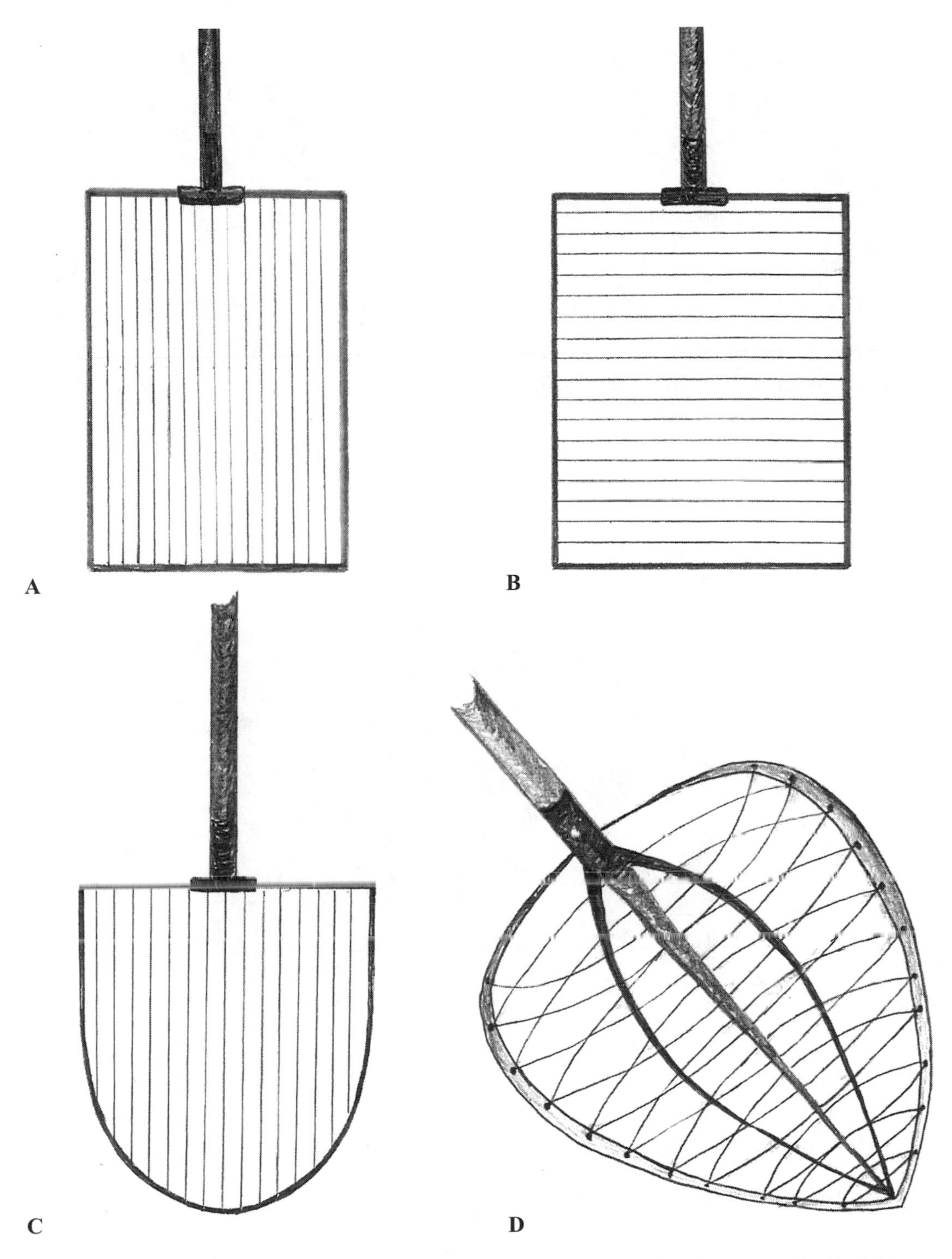

Figure 7–1 Examples of tools used to cut renneted milk gels: (A) vertical knife, (B) horizontal knife, (C) harp, and (D) spino.

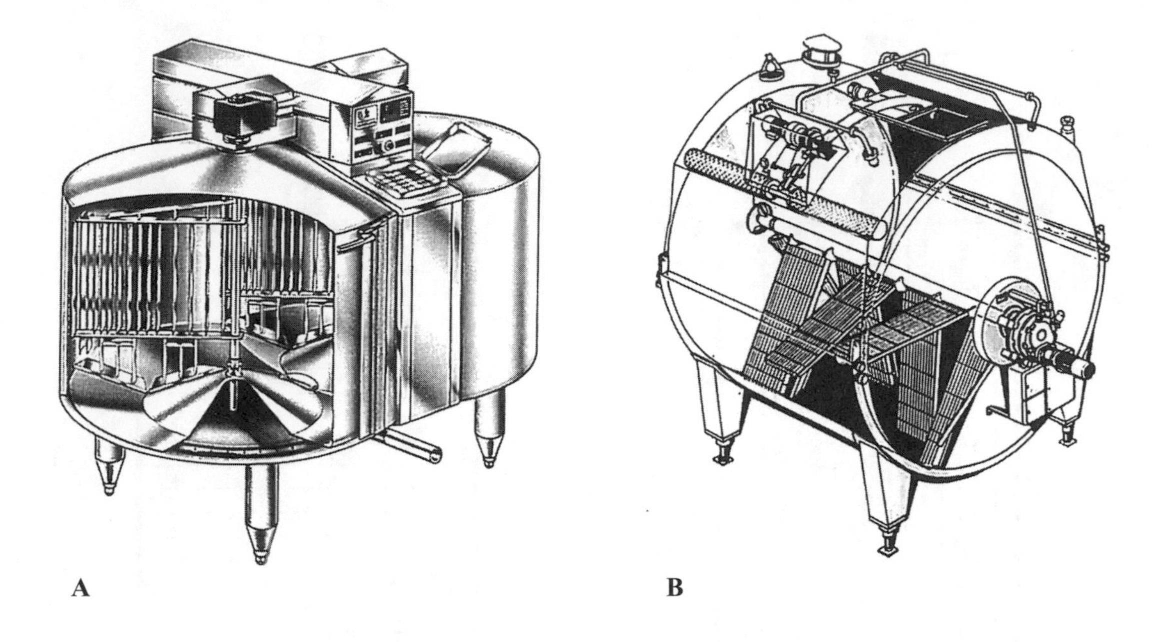

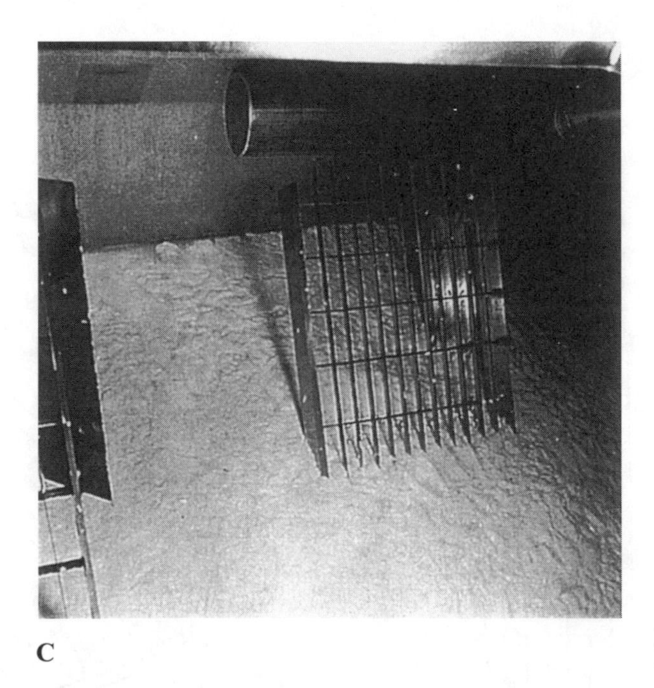

Figure 7–2 Illustrations of cheese vats showing the curd knife–stirrer blades. The curd knife–stirrer blades are shown arranged vertically in a Double-O Multicurd vat (A); arranged horizontally in a staggered mode along a central horizontal shaft in a cylindrical OST vat (B); and located toward the end of the cutting cycle in a horizontal cheese vat (C). The blades are tapered and move in the direction of the sharp edge (knife) when cutting and in the reverse direction (blunt edge, stirrer) when stirring.

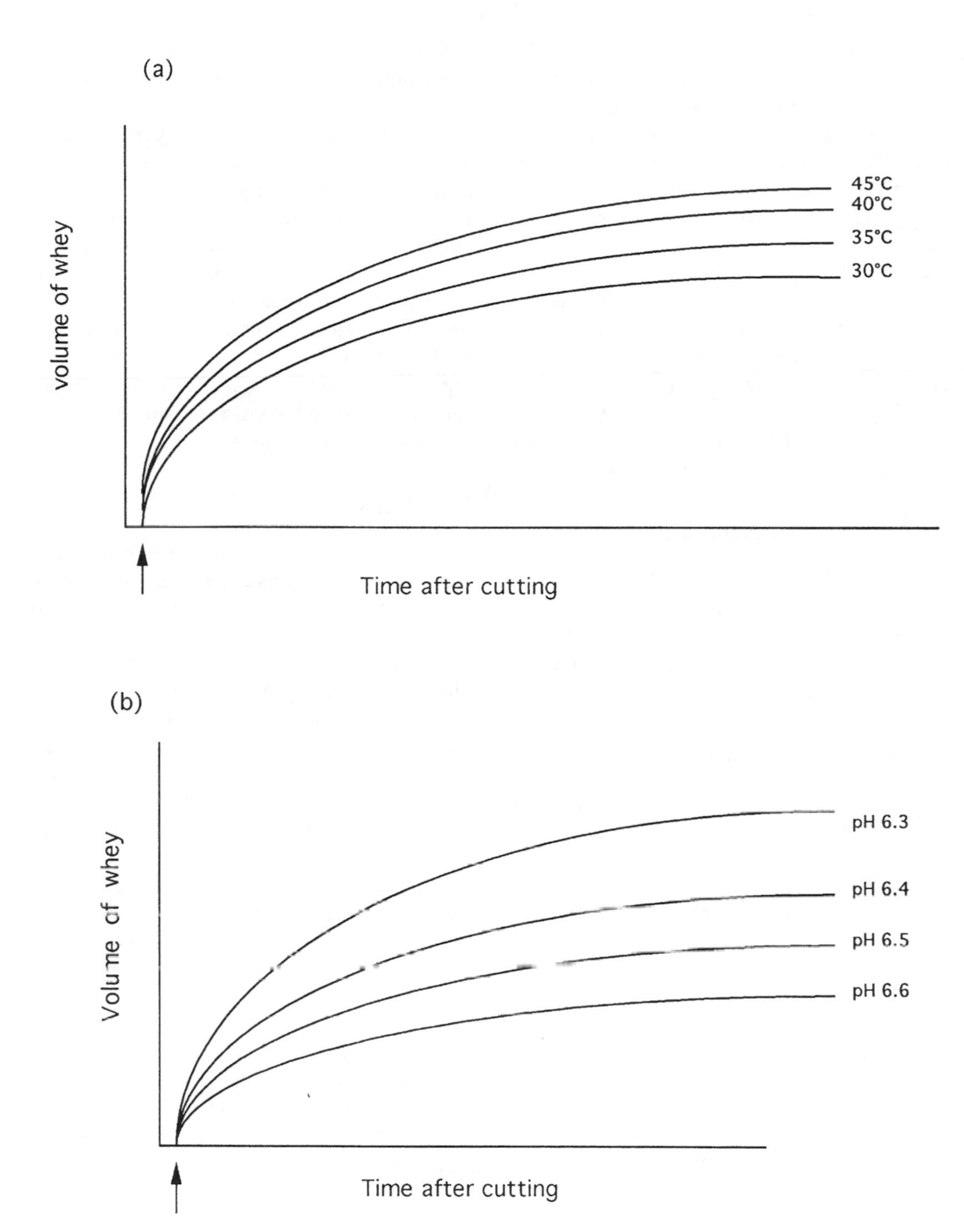

Figure 7–3 Effect of temperature (a) and pH (b) on the rate and extent of syneresis in cut or broken renneted milk gels.

cheesemaking on farms, which would have lacked the means to circulate hot water or steam through a jacketed vat. It was probably used for many other varieties, but it is now mainly restricted to Dutch-type cheeses, where its main function is to reduce the lactose content of the cheese curd and thereby control the pH of the cheese.

The rate of cooking is characteristic of the variety (see Chapter 17). If the rate of cooking is too fast, especially during the early stages, excessive dehydration will occur at the curd surface, leading to the formation of a skin (case hardening), which will retard syneresis and the removal of whey from the interior of the curd pieces and result in a high-moisture content.

7.4.3 Rate of Acid Development

The lower the pH, the faster the rate and the greater the extent of syneresis (see Figure 7–3). Presumably, this relationship reflects the reduced negative charge on the casein molecules as the isoelectric point is approached.

7.4.4 Stirring of the Curd-Whey Mixture

During cooking, the curd-whey mixture is stirred, which serves a number of functions:

- It facilitates cooking.
- It prevents the curd pieces from matting (which would have a strong negative effect on syneresis).
- It promotes syneresis via collisions between curd pieces and between curd pieces and the vat wall.

Everything else being equal, syneresis is directly proportional to the intensity of stirring. Initially, the curd is very soft, and gentle stirring should be used. A period of 5–10 min is allowed for the cut surfaces of the curd pieces to "heal," and vigorous agitation of the curd during this period will cause extensive losses of fat and protein into the whey and a decrease in cheese yield (see Chapter 9).

For some varieties, the curd is held in the whey until a certain pH is reached (e.g., 6.2 for Cheddar), after which the curds and whey are separated, usually using metal screens. For other varieties (e.g., Gouda), the curds and whey are separated after holding at the desired cook temperature for a defined period. For yet other varieties (e.g., Parmesan), the curds and some whey are scooped from the vat using a cheesecloth and transferred to perforated molds, where much of the whey drainage occurs. If whey separation occurs in the cheese vat, stirring of the curd during draining (referred to as dry stirring) is a useful way of promoting syneresis, but this method is not applicable to most varieties.

7.4.5 Pressing

After removal of the whey, the curds mat to form a continuous mass. Treatment of this mass of curd is characteristic of the variety and may involve inverting the mass of curd in the molds, turning and piling blocks of curd in the vat (traditional "cheddaring"), and in many cases pressing (see Chapter 17). Syneresis occurs during these operations but is not easily controlled. With the exception of Cheddar-type cheeses, acidification occurs mainly after molding, and this promotes considerable syneresis of the curds. For Cheddar-type cheeses, acidification occurs mainly during "cheddaring" in the vats, and relatively little syneresis occurs after molding.

7.4.6 Salting

As discussed in Chapter 8, all cheeses are salted at the end of manufacture. Salting causes the loss of moisture from the curd (≈ 2 kg H_2O are lost per kg of salt absorbed). However, salting should not be relied upon as a means of controlling cheese moisture.

7.4.7 Milk Quality and Pretreatment

Heating milk under conditions that cause whey protein denaturation and interaction with casein micelles reduces the tendency of renneted milk gels to synerese. Homogenization of whole milk has a similar effect. It is reported that the growth of psychrotrophs in milk reduces its syneretic properties (i.e., leads to a high moisture content). Increased levels of rennet slightly increase the rate of syneresis, while plasmin activity in milk is reported to reduce syneresis.

7.5 KINETICS AND MECHANISM OF SYNERESIS

Data from various investigations indicate that syneresis is initially a first-order reaction, that is, the rate of syneresis depends on the amount of whey remaining in the curd. Within the temperature range 16–45°C at constant pH, curd volume, V_t, as a function of time can be expressed by the formula:

$$2.3 \log V_t = 2.3 \log V_0 - tk(T - T_0)$$

where t is time (min), T is temperature, and k and T_0 are constants having values of 0.225×10^{-3} and 16 under the experimental conditions.

It is generally assumed that syneresis is due to protein-protein interactions and may be regarded as a continuation of the gel assembly process during rennet coagulation. The inhibitory effects of high concentrations of salts ($CaCl_2$, NaCl, KCl) on syneresis suggest that ionic attractions are involved. Urea promotes syneresis, suggesting that hydrogen bonds are not involved. The effectiveness of pH in promoting syneresis is probably due to a reduction of overall charge as the isoelectric point is approached. Studies of artificial milk systems implicate the ε-NH_2 group of the lysine residues in β-casein in syneresis. The apparent importance of ε-NH_2 groups in the second phase of rennet coagulation was discussed in Chapter 6. Lysozyme, which reacts with casein micelles, reducing their charge and rennet coagulation time, also accelerates syneresis when added to milk.

Some authors have concluded that curd-firming and syneresis are different aspects of the same phenomenon, but opinions are not unanimous. As discussed in Chapter 6, electron microscopy studies have shown that the aggregation of casein micelles to form a gel is followed by increasingly closer contact between the micelles, leading to fusion. The syneretic pressure in an uncut gel is very small (≈ 1 Pa), but when the coagulum is cut, the whey leaks out. Syneresis is initially a first-order reaction, because the pressure depends on the amount of whey in the curd. Holding curd in whey retards syneresis owing to back pressure of the surrounding whey, while removing whey promotes syneresis. When the curd is reduced to roughly 70% of its initial volume, syneresis becomes dependent on factors other than the volume of residual whey in the curd. It has been proposed that hydrophobic and ionic interactions within the casein network are probably responsible for the advanced stages of syneresis. This view is in accord with the promotion of syneresis by reduced pH and low levels of $CaCl_2$, which reduce micellar charge and increase hydrophobicity, and by increased temperature, which increases hydrophobic interactions.

The foregoing discussion of syneresis pertains especially to the syneresis occurring in the cheese vat, that is, mainly to hard and semi-hard varieties of cheese. Syneresis continues after hooping (molding), and syneresis during this period represents the major part of syneresis in soft varieties. Presumably, the mechanism of syneresis in the molds is the same as in the cheese vat, although the range of treatments that can be applied at this stage is rather restricted. External pressure is applied to the curds for many cheese varieties after molding and makes a significant contribution to whey removal. In general, the drier the cheese curd at hooping, the higher the pressure applied, which is probably a reflection of the greater difficulty in ensuring fusion of low-moisture curds.

7.6 TEXTURED CHEESE

The development of a recognizably fibrous texture is part of the manufacturing procedure for a small number of cheese varieties, and this texture was traditionally regarded as an essential organoleptic characteristic of these cheeses. Texturized cheeses belong to two classes: Cheddar and some closely related varieties, in which a fibrous texture is developed prior to pressing, and pasta filata types, such as Mozzarella, Kashkaval, and Provolone, in which texturization is accomplished by heating, stretching, and kneading the curd.

In traditional Cheddar manufacture, the drained curds are piled along the sides of the vat, with the result that matting (fusion) of curd particles occurs. To enable faster turnover of the cheese vats, it became common practice in the 1960s to transfer the curd-whey mixture after cooking to cheaper cheddaring "sinks," where whey drainage and cheddaring occurred. The piles of curd are cut into blocks (30 × 10 cm), which are inverted frequently and piled over a period of about 2 hr. This operation, known as cheddaring, was considered by many researchers and cheese makers as the most characteristic part of the Cheddar cheese manufacturing process. During cheddaring, the curd flows under its own weight, leading to fusion and deformation of curd particles, which was believed to be responsible for the "chicken breast meat" structure of fresh Cheddar curd and for the characteristic texture of mature Cheddar cheese. Cheddaring promotes a number of physicochemical conditions that are conducive to curd flow and texturization:

- Solubilization of micellar calcium, which is bound to the casein and acts as a cementing agent between the casein micelles/submicelles
- A decrease in the concentration of micellar Ca, resulting in an increase in the ratio of soluble to casein-bound Ca (soluble Ca as a percentage of total Ca in the curd increases from ≈ 5% to 40% as the pH decreases from 6.15 to 5.2)
- An increase in paracasein hydration, which increases with decreasing pH in the range 6.6–5.15
- An increase in the viscous character of the curd

The increase in casein hydration with decreasing pH is a probably a consequence of the increase in the ratio of soluble to micellar Ca. It has been found in model casein systems that casein hydration is inversely related to the concentration of casein-bound Ca (Sood, Sidhu, & Dewan, 1980).

As a consequence of the decrease in casein-bound Ca and the increase in casein hydration, the viscoelastic casein matrix, with occluded liquid fat and moisture phases, flows if unrestricted, especially when piled and pressed under its own weight. The flow of curd gives the desired planar orientation of the strands of the paracasein network (Figure 7–4). The physicochemical changes in curd during cheddaring are summarized in Figure 7–5. Note that there is little scientific support for the necessity of cheddaring. On the contrary, there is strong evidence that cheddaring is of no consequence to Cheddar cheese quality and serves only to allow the desired degree of acid development and syneresis to occur.

Various forms of restricted flow under different degrees of external pressure result in Cheddar cheese with a lower moisture content than curd cheddared in the traditional manner. Differences in the extent of curd deformation caused by modified cheddaring processes diminish during milling, salting, and pressing and have little effect on the flavor and textural characteristics of the final cheese. The development of a fibrous texture results in loss of micelle structure, but this change in structure is not essential, as the amount of deformation is very small and is probably altered by the subsequent and more extensive deformation during pressing.

In modern practice, most Cheddar cheese curd is manufactured in continuous, mechanical cheddaring systems in which little flow occurs in comparison with traditional methods. Indeed, matting is prevented in the manufacture of some Cheddar-type cheeses, such as stirred-curd Cheddar. The textural quality of Cheddar cheese

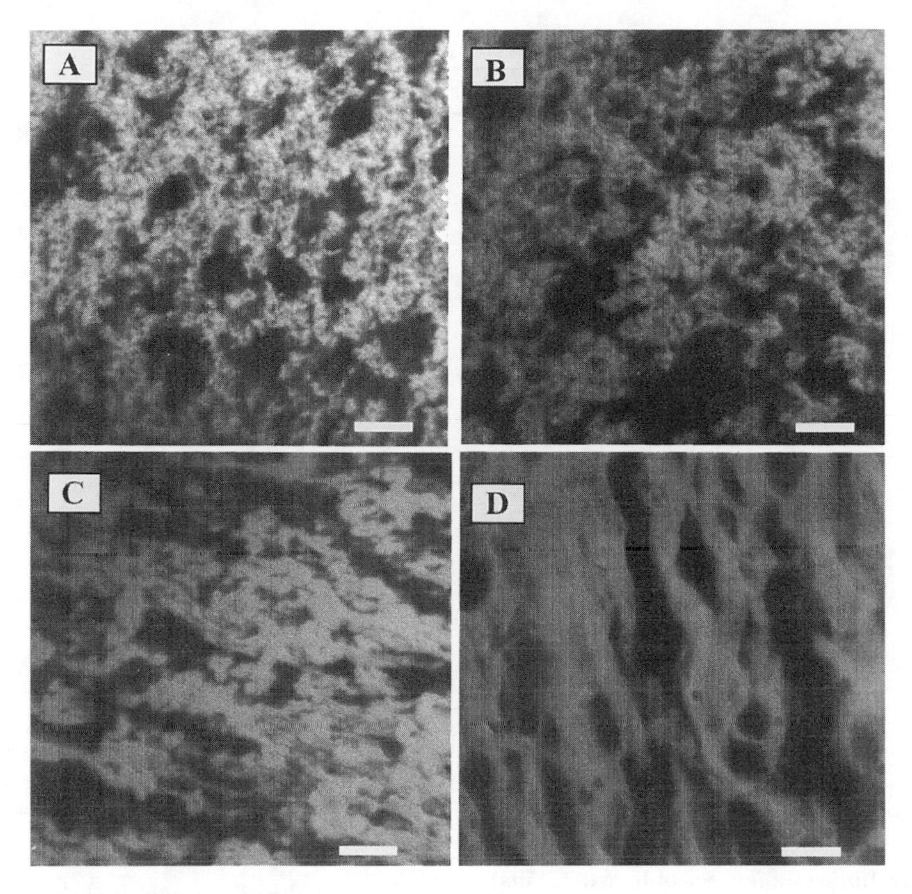

Figure 7–4 Confocal laser scanning micrographs of Mozzarella cheese curd at various stages of manufacture: after cutting (A), at whey drainage (B), after cheddaring (C), and after plasticization (D). The white-gray areas represent the paracasein matrix, and the black areas represent the occluded fat and moisture phases. Bar = 25 μm.

produced by these systems is acceptable, indicating that flow during manufacture is not essential.

Presumably, the various interactions, ionic and/or hydrophobic, that are considered to be responsible for syneresis continue during the cheddaring process, but there appear to have been no studies on this aspect of cheesemaking.

In the manufacture of Mozzarella and other pasta-filata (stretched-curd) cheeses, the cheddared curd is shredded, heated to around 58–60°C by kneading in hot water (≈ 78°C), and stretched in equipment designed to promote extension of the hot molten curd (Figure 7–6). The process by which the curd is converted into a plastic molten mass is referred to as *plasticization* and was originally developed in hot climates as a means of pasteurizing and hence extending the shelf-life of curd of poor microbiological status. Successful plasticization of the curd requires that the viscoelastic paracasein matrix undergoes limited flow and stretches into hot molten sheets without breaking. Plasticization is accompanied by microstructural changes in the cheddared curd, including further linearization of the paracasein matrix into fibers and coalescence of fat into elongated pools trapped between and showing the same orientation as the protein fibers (Figure 7–4).

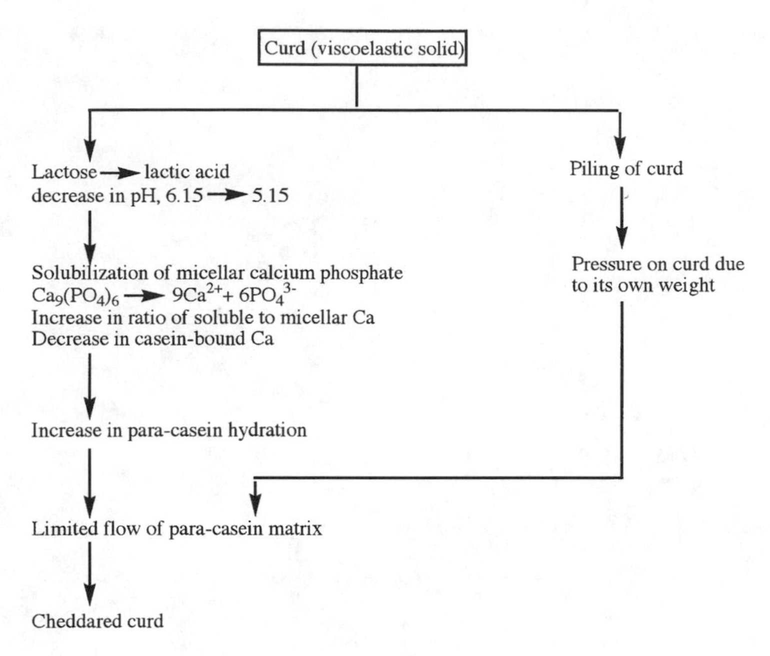

Figure 7–5 Physicochemical changes in cheese curd during cheddaring.

The physicochemical changes responsible for plasticization of the curd have not been fully elucidated. However, based on the behavior of curds of different composition and pH when subjected to texturization (Guinee, unpublished results), the microstructural changes that accompany placticization (e.g., Figure 7–4), and the viscoelastic changes in curd when heated to temperatures similar to those during the plasticization process (Guinee et al., 1998), it may be speculated that successful texturization is a consequence of

- an adequate degree of casein hydration in the cheddared curd, which is controlled by its pH, its total calcium content, and the ratio of soluble to micellar Ca
- heat-induced coalescence of free fat (formed as a consequence of shearing of the fat globule membrane), which lubricates the flow of the paracasein matrix
- extension and shear stresses applied to the curd, which assist in the displacement of contiguous planes of the paracasein matrix (see Chapter 13)

The relationship between paracasein hydration and pH may be explained by the dominance of two opposing factors over the pH range 6.0–5.0:

1. neutralization of negative charges, which leads to contraction of the paracasein and thereby limits hydration and impedes the flow of the paracasein matrix
2. solubilization of micellar calcium, which is conducive to casein hydration and promotes the flow of the paracasein matrix

At pH values in the range 6.0–5.2, solubilization of micellar calcium appears to be dominant, as decreasing pH results in an increase in hydration of paracasein. In contrast, the reduction in pH is the dominant factor at pH 5.2–4.6, as decreasing pH results in a marked decrease in paracasein hydration. The total calcium content of the curd, which is controlled mainly by the pH of the milk at setting and that of the curd at whey drainage, determines the curd pH at which plasticization is possible. In the normal manufacture of Mozzarella, the milk is typically set at pH 6.55, the whey is drained at pH 6.15, and the ideal pH for plasti-

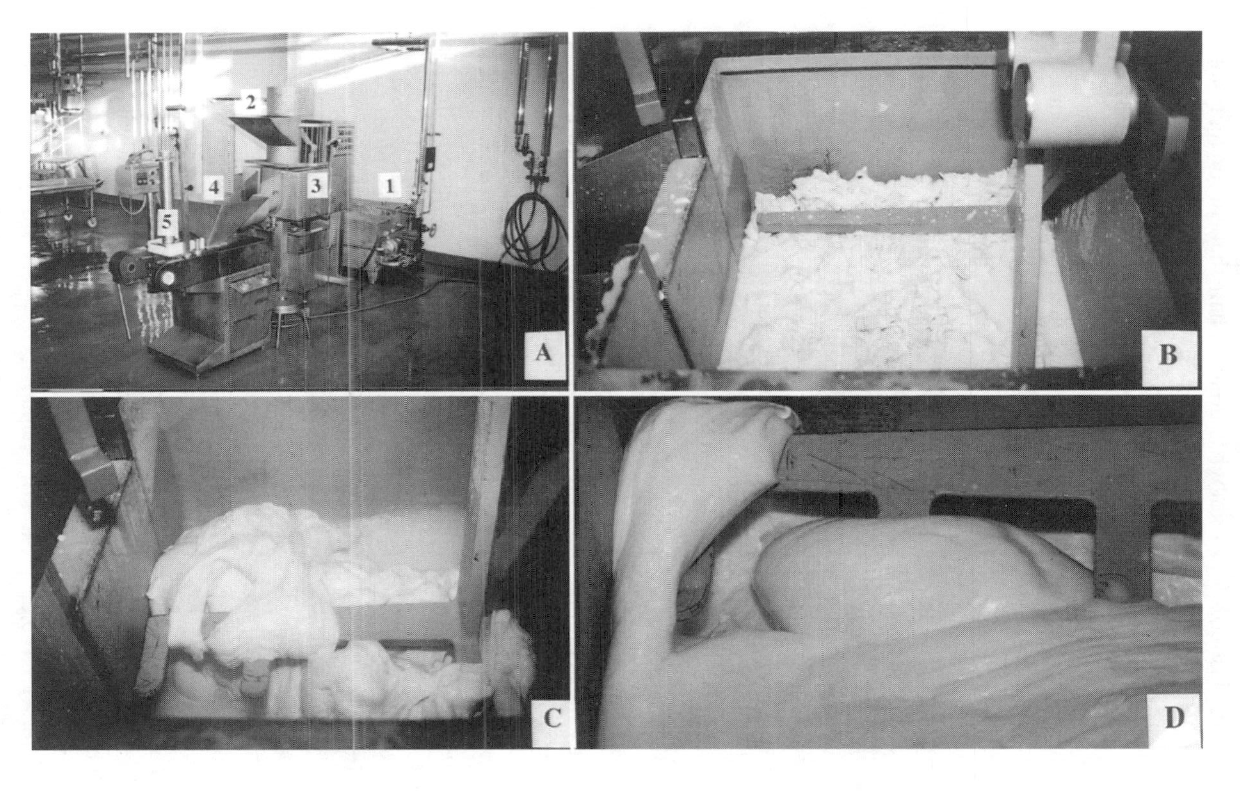

Figure 7–6 Plasticization of low-moisture Mozzarella. (A) Kneading-plasticizing equipment (from Costruzioni Meccaniche e Tecnologia S.p.a., Perveragno, Cuneo, Italy) consisting of a hot water heating unit (1); a cheese-shredding unit (2); a plasticization chamber, where the curd is kneaded and stretched in hot water by toothed arms that oscillate backwards and forwards in opposite directions (3); and an auger (4), which conveys the plasticized cheese to the molding unit (5). (B) Initial stages of plasticization; shredded curds have not yet fused. (C) Midstage of plasticization; shredded curds have begun to fuse but plasticization is not yet complete, as shown by the presence of lumps in the curd mass. (D) Fully plasticized molten curd mass, which exhibits a long consistency and an oily surface sheen.

cizing cheddared curd is about 5.15. At this pH, the concentration of calcium in the curd (≈ 27 mg/g protein) and the proportion of soluble calcium (≈ 40% of total) ensure that the paracasein is sufficiently hydrated to enable successful plasticization. At increasingly higher curd pH, the curd becomes progressively less smooth after plasticization, reflecting the decrease in paracasein hydration resulting from the reduced ratio of soluble calcium to micellar calcium. Similarly, in processed cheese manufacture, heating of cheese is accompanied by aggregation of the protein and exudation of moisture and free fat unless emulsifying salts (e.g., sodium orthophosphates) are added to chelate the micellar Ca (see Chapter 18). At a curd pH greater than 5.4, the curd fails to plasticize correctly. Instead, a nonplastic mass with a rough, dull, short, lumpy consistency is obtained. However, successful plasticization may be achieved at a higher curd pH (e.g., 5.6–5.8) if the Ca level of the curd is sufficiently low (e.g., < 18 mg/g protein), as in the case of directly acidified Mozzarella.

In the manufacture of directly acidified Mozzarella, acidification is achieved by the addition of food-grade organic acids rather than the conversion of lactose to lactic acid by starter culture. The milk pH is typically adjusted to about 5.6 prior to rennet addition, and no further change in pH occurs during curd manufacture, which is otherwise similar to that for conventional Mozzarella made using a starter culture. Following whey drainage, the curd, typically with a pH of around 5.6, plasticizes readily upon heating and stretching. The ability of curd made by direct acidification to plasticize at a higher than normal pH can be explained on the basis of the interactive effects of total curd calcium and soluble Ca: micellar Ca ratio (which changes with pH) on paracasein hydration. While soluble Ca as a percentage of total Ca decreases from around 40% to 20% as the pH is increased from 5.15 to 5.6, the total concentration of calcium is lower, and hence the level of micellar Ca is probably similar to that obtained at pH 5.15 in conventional cheese manufacture. Consequently, as there is an inverse relationship between casein-bound Ca and casein-bound moisture, the degree of paracasein hydration obtained in directly acidified Mozzarella curd at pH 5.6 is similar to or somewhat greater than that in conventionally produced Mozzarella curd at pH 5.3. Indeed, comparative studies have shown that the water-binding capacity of directly acidified Mozzarella cheese curd (pH 5.6) is higher than that of conventionally produced Mozzarella curd (pH 5.2) during the first 3 weeks of aging (Kindstedt & Guo, 1997).

7.7 MOLDING AND PRESSING OF CHEESE CURD

At some stage in the manufacturing process (e.g., just after coagulation for Camembert, after cooking for Emmental, or after acidification for Cheddar), the curds are transferred to molds of the cheese's characteristic shape and size. The principal purpose of molding is to allow the curd to form a continuous mass; matting of high-moisture curds occurs readily under their own weight but pressing is required for low-moisture cheese. It is important that the curds are warm during pressing, especially for low-moisture cheeses.

Various pressing systems have been developed, ranging from the very simple to the continuous-pressing systems. In modern Cheddar cheese factories, the salted curds are formed and pressed under their own weight and under a slight vacuum in towers (Wincanton towers) for about 30 min. Upon exiting the tower, the column of curd is cut into 20 kg blocks by a guillotine, and the blocks are placed in plastic bags sealed under vacuum. For examples of cheese-pressing systems, the reader is referred to Kosikowski and Mistry (1997) or Robinson and Wilbey (1998) or to any text on cheese technology referenced in the "Suggested Readings" section of Chapter 1.

Cheeses are made up in characteristic shapes and sizes (see Chapter 17). At first glance, it might appear that the shape and size of a cheese is cosmetic. While this may be so in many cases, size and shape are very significant in some varieties. For example, surface-ripened cheeses (mold or smear) are made up in small, low cylinders to allow ripening from the surface toward the center. If such a cheese were made large, the surface would become overripe while the center remained unripe. For cheeses with large eyes (e.g., Emmental), a large size is required, as otherwise the leakage of CO_2 through the surface would be excessive and the pressure of gas within the cheese would not build up to the level needed to form eyes.

7.8 PACKAGING

Like other sectors of the food industry, indeed, of industry in general, packaging has become a major feature of cheese production, distribution, and retailing. Kosikowski and Mistry (1997) include a useful chapter on various aspects of the packaging of cheese and fermented milks. Kadoya (1990) provides a more general discussion of food packaging, including a chapter on cheese and fermented milks. The science and technology of packaging are specialized subjects that will not be discussed here.

The objectives of cheese packaging, as of food packaging generally, are as follows:

- To protect the cheese against physical, chemical, or microbial contamination. Mold growth is of particular concern. Since molds are aerobic, their growth can be prevented by covering the cheese with wax or plasticote or vacuum-packing it in plastic film with low permeability to oxygen and free of pin holes. By preventing contamination, packaging serves a public health function as well as reduces losses due to spoilage.
- To reduce loss of moisture from the surface and therefore increase economic return. To achieve this, the packaging material should have low permeability to moisture.
- To prevent physical deformation of the cheese, especially soft cheeses, and thus facilitate stacking during ripening, transport, and retailing.
- To allow for product labelling and brand identification, which in turn provides an opportunity for advertising and the provision of nutritional information.

After salting (see Chapter 8), those cheeses on which the growth of molds (surface or internal) or of a surface smear is encouraged are transferred to a room at a controlled temperature (≈ 15°C) and humidity (90–95% equilibrium relative humidity). Even at this high humidity, some loss of moisture from the surface occurs, but the loss is insufficient to create a rind (low-moisture surface layer). After adequate growth of the mold or smear has occurred, such cheeses may be wrapped in foil or grease-proof paper to avoid further loss of moisture.

Traditionally, the development of a rind was encouraged on internally bacterial-ripened cheese by controlled drying of the surface. If properly formed, the rind effectively sealed off the interior of the cheese, preventing excessive loss of moisture and the growth of microorganisms on the surface. To further stabilize the surface of such cheeses, they were rubbed with oil (e.g., butter oil or olive oil) or coated with paraffin wax. Sometimes wax of a particular color was used (e.g., red for Edam, black for extra-mature Manchego and Cheddar). The color of the wax was characteristic of the variety or of its maturity and was recognized by the consumer as an index of variety or quality.

Today, many internally bacterial-ripened cheeses are packaged in plastic bags of low gas permeability or coated with film-forming plastic material. A variety of plastic packaging materials are used for cheese, including cellophane, cellophane-polyethylene, polyvinyl chloride, polyvinylidene chloride, polystyrene, polypropylene, ethylene vinyl acetate, co-extruded polyolifin, metal foils, and paper.

Gases such as CO_2 and H_2S are produced in many cheeses during ripening. CO_2 will cause bulging of the package, while H_2S has an obnoxious aroma that will render the cheese unacceptable. To avoid such problems, the package should be permeable to these gases.

Packaging is particularly important for soft cheeses, such as Cottage cheese, Quarg, Cream cheese, and processed cheeses. Metal foils are widely used for consumer or catering packages of processed cheese. Much processed cheese is commercialized as individual slices wrapped in plastic material. High-moisture fresh cheeses are commercialized in plastic tubs; plastic-, wax-, or foil-lined cardboard containers; and plastic packages.

Metal cans or glass jars may be used to package natural and processed cheese to offer a novelty presentation feature or, in the case of cans, to provide extra physical protection during distribution and storage.

As with other foods, the packaging of cheese has led to the development of specialized packaging equipment, much of which is highly automated and computerized.

REFERENCES

Green, M.L., & Grandison, A.S. (1993). Secondary (non-enzymatic) phase of rennet coagulation and post-coagulation phenomena. In P.F. Fox (Ed.), *Cheese: Chemistry, physics and microbiology* (2d ed., Vol. 1). London: Chapman & Hall.

Guinee, T.P., Auty, M.A.E., Harrington, D., Corcoran, M.O., Mullins, C., & Mulholland, E.O. (1998). Characteristics of different cheeses used in pizza pie (Abstract). *Australian Journal of Dairy Technology, 53*, 109.

Kadoya, T. (1990). *Food packaging*. San Diego: Academic Press.

Kindstedt, P.S., & Guo, M.R. (1997). Chemically-acidified pizza cheese production and functionality. In T.M. Cogan, P.F. Fox, & R.P. Ross (Eds.), *Proceedings of the Fifth Cheese Symposium*. Dublin: Teagasc.

Kosikowski, F.V., & Mistry, V.V. (1997). *Cheese and fermented milk foods* (3d ed.). Westport, CT: F.V. Kosikowski, LLC.

Robinson, R.K., & Wilbey, R.A. (1998). *Cheesemaking practice* (3d ed.). Gaithersburg, MD: Aspen Publishers.

Sood, S.M., Sidhu, K.S., & Dewan, R.K. (1980). Voluminosity and hydration of casein micelles from abnormal milks. *New Zealand Journal of Dairy Science and Technology, 15*, 29–35.

Walstra, P. (1993). The syneresis of curd. In P.F. Fox (Ed.), *Cheese: Chemistry, physics and microbiology* (2d ed., Vol. 1). London: Chapman & Hall.

Chapter 8

Salting of Cheese Curd

8.1 INTRODUCTION

The use of salt (NaCl) as a food preservative dates from prehistoric times and, together with fermentation and dehydration by exposure to low-humidity air, is one of the classical methods of food preservation. Until the development in the 19th century of modern methods such as pasteurization or sterilization, chilling or freezing, and "hot air" drying, salting was probably the most widely used method for the long-term preservation of many foods. Salt was a highly valued item of trade and was exchanged for goods and services. One can readily envisage how fermentation and "natural" dehydration could have been discovered by accident, but the use of salt as a preservative required direct intervention and was a very significant discovery at an early stage of human civilization. The three classical methods of food preservation, fermentation, salting, and dehydration, along with refrigeration, are used to preserve cheese and/or control its maturation.

The preservative action of NaCl is due to its effect on the water activity (a_w) of the medium:

$$a_w = p/p_0$$

where p and p_0 are the vapor pressure of the water in a system and of pure water, respectively. If the system is at equilibrium with its gaseous atmosphere, then $a_w = ERH/100$, where ERH is the equilibrium relative humidity.

Due to the presence of various solutes in foods, the vapor pressure of water in a food system is always less than that of pure water (i.e., $a_w < 1.0$). The relationship between water activity and the moisture content of food is shown (Figure 8–1). Three zones are usually evident:

- Zone I represents monolayer water that is tightly bound to polar groups in the food, such as the -OH group of carbohydrates and the $-NH_3^+$ and $-COO^-$ groups of proteins.
- Zone II consists of multilayer water in addition to the monolayer water.
- Zone III contains bulk phase water in addition to monolayer and multilayer water.

General discussions on the general concept of water activity in relation to foods are provided by Fennema (1996), Rockland and Beuchat (1987), and Rockland and Stewart (1981). More specific aspects of water activity in relation to dairy products are discussed by Kinsella and Fox (1986) and Roos (1997).

The water activity of food depends on its moisture content and the concentration of low molecular mass solutes. The water activity of young cheese is determined almost entirely by the concentration of NaCl in the aqueous phase:

$$a_w = 1 - 0.033\,[NaCl_m] = 1 - 0.00565\,[NaCl]$$

where $[NaCl_m]$ is the molality of NaCl (i.e., moles of NaCl per liter of H_2O) and [NaCl] is the concentration of NaCl as g/100 g cheese moisture (Marcos, 1993).

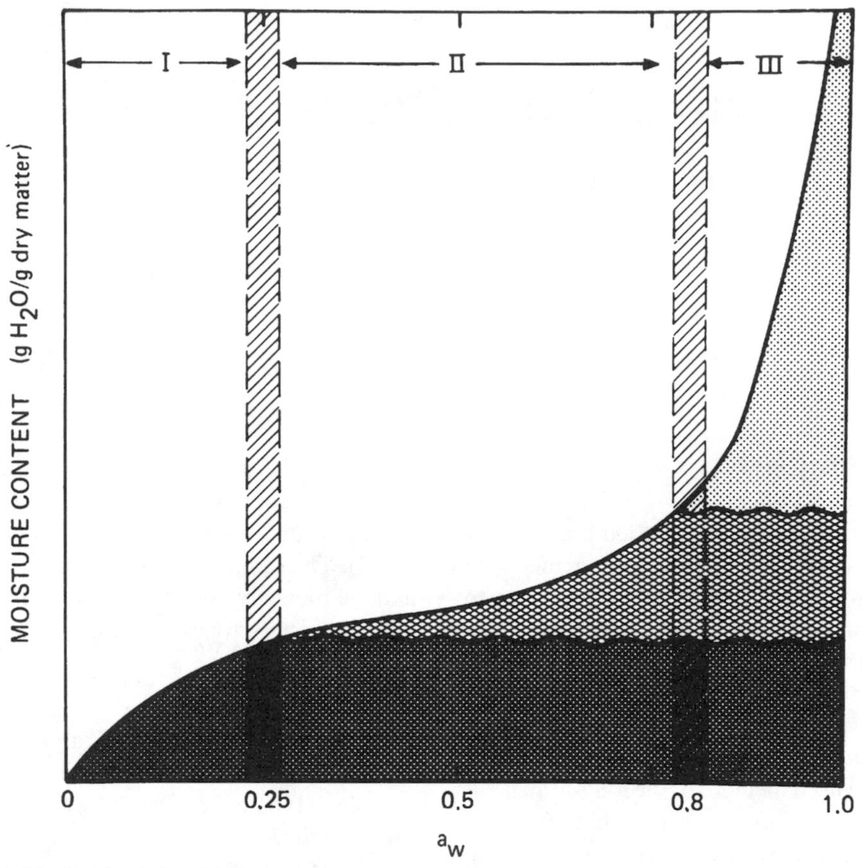

Figure 8–1 Idealized relationship between the water activity (a_w) of the food and its water content.

The salt content of cheese varies from about 1.0% for Emmental to about 5% for Domiati (Table 8–1). Typical values for the a_w of some cheese varieties are shown in Table 8–2. Other compounds besides NaCl including lactic and other acids, amino acids, very small peptides, and calcium phosphate (from milk), contribute to the depression of a_w, especially in extra mature cheeses.

Salt increases the osmotic pressure of the aqueous phase of foods, causing dehydration of bacterial cells and killing them or at least preventing their growth. It will be apparent from Figure 8–2 that the water activity of most cheese varieties is not low enough to prevent the growth of yeasts and molds, but together with a low pH it is quite effective in controlling bacterial growth. Enzyme activity is also affected strongly by water activity (Figure 8–2).

Measurement of the salt content of cheese is an important quality control step in cheese production. As described above, the water activity of cheese can be calculated from its composition, but it can also be determined experimentally (see Marcos, 1993, and Chapter 23).

The concentration and distribution of salt in cheese have a major influence on various aspects of cheese quality. Among the principal effects of salt are the following:

- Salt inhibits or retards the growth and activity of microorganisms, including patho-

genic and food-poisoning microorganisms (Table 8–2), and hence increases the safety of cheese (see Chapter 20).
- It inhibits the activity of various enzymes in cheese.
- It affects the syneresis of cheese curd, resulting in whey expulsion and thus in a reduction in the moisture of cheese, which also influences the activity of microorganisms and enzymes.
- It causes changes in cheese proteins that influence cheese texture, protein solubility, and probably protein conformation.
- It affects cheese flavor directly and indirectly via its influence on microorganisms and enzymes in cheese.
- High levels of salt in cheese may have undesirable nutritional effects (see Chapter 21).

Various aspects of the significance of salt in cheese are discussed comprehensively by Guinee and Fox (1993) and are summarized below.

8.2 SALTING OF CHEESE CURD

Cheese is salted by one of four methods. The method used is characteristic of the variety:

1. Dry salt is added to and mixed with small pieces of fresh cheese curd prepared by milling or breaking larger blocks of cheese (as in Cheddar-type cheeses) prior to molding and pressing. This method is commonly referred to as dry salting.
2. Dry salt or a salt slurry is rubbed on the surface of the molded cheese. This method is used for some Blue cheeses.
3. The molded cheese is immersed in a concentrated NaCl brine (15–23%). Edam, Gouda, Emmental, and Camembert are salted in this manner.
4. A combination of two of these methods is used for quite a few varieties. For example, milled (broken) Mozzarella curd may be partially dry-salted before stretching and molding, then subjected to brining or the surface application of dry salt.

8.2.1 Dry-Salting of Cheese

Dry-salting is used mainly for British varieties of cheese, such as Cheddar, Cheshire, and Stilton, and American Cottage cheese. Traditionally, the salt was added to the milled curd and mixed manually, but the process is now done mechanically, permitting better control of the level of salt added and its distribution. When small curd pieces are salted, the concentration of NaCl rapidly reaches a level throughout the pieces that is sufficiently high to retard the growth of the starter culture, and hence the pH of the curd at salting is close to the final desired value in the freshly made cheese (e.g., 5.4 vs. 5.1 for Cheddar curd). However, although growth of the starter ceases shortly after salting, metabolism of lactose continues, and the concentration of lactic acid typically increases, for example, from 0.7% to 1.5% during the first 24 hr post salting. The decrease in pH during this period is much smaller (e.g., from 5.4 to 5.2–5.1) than would be expected from the amount of lactic acid produced, owing to the fact that the buffering capacity of cheese curd has a maximum at about pH 5.2 and to the fact that pH is logarithmic.

It should be possible to achieve very precise and uniform control of salt concentration in dry-salted milled cheese curd. When dry-salting is properly executed, it is possible to get to within ± 0.1% of the desired salt concentration. However, if salt distribution is poor initially, uniform distribution of salt throughout the cheese is not attained during the life of the cheese. This is because each piece of curd behaves as a mini-cheese in which salt equilibrium is attained within about 24 hr. Thereafter, there may be several centers of high or low salt concentration throughout the cheese and hence a very low driving force for the attainment of overall equilibrium throughout the cheese.

When dry-salting milled curd, it is easy to calculate the correct amount of salt that should be added. When a continuous mechanized mixing

Table 8–1 Typical Composition of Major Cheeses

Cheese (%)	*Total Fat (%)*	*Total Solids (%)*	*Protein (%)*	*Salt (%)*	*Ash (%)*	*pH (%)*
Blue	29.0	58.0	21.0	4.5	6.0	6.5
Brick	30.0	60.0	22.5	1.9	4.4	6.4
Bulgarian White	32.3	68.0	22.0	3.5	5.3	5.0
Camembert	23.0	47.5	18.5	2.5	3.8	6.9
Cheddar	32.0	63.0	25.0	1.5	4.1	5.5
Edam	24.0	57.0	26.1	2.0	3.0	5.7
Emmental	30.5	64.5	27.5	1.2	3.5	5.6
Gouda	28.5	59.0	26.5	2.0	3.0	5.8
Grana (Parmesan)	25.0	69.0	36.0	2.6	5.4	5.4
Gruyère	30.0	66.5	30.0	1.1	4.1	5.7
Limburger	28.0	55.0	22.0	2.0	4.8	6.8
Muenster	29.0	57.0	23.0	1.8	4.4	6.2
Provolone	27.0	57.5	25.0	3.0	4.0	5.4
Romano	24.0	77.0	35.0	5.5	10.5	5.4
Roquefort	31.0	60.0	21.5	3.5	6.0	6.4
Domiati[a]	25.0	45.0	12.0	4.8	–	4.6
Feta	26.0	47.0	16.7	3.0	–	4.5

[a]made from buffalo milk

system is used, localized variations will occur unless adequate metering and mixing systems are used. An effective system involves the use of an oscillating boom that releases salt in response to a signal indicating either the depth or weight of the bed of curd beneath the boom. However, the temperature and humidity conditions of storage and transportation of salt to the oscillatory boom are critically important for uniform salt deposition and distribution. Adequate mixing of the salted curd chips can be achieved using rotating pegged rollers or a rotating drum. Although the latter should give more uniform mixing, the former is more common in commercial cheese production.

Even when properly measured and mixed, variations in salt concentration still occur, due to variability in the uptake of salt added to the curd.

Table 8–2 Water Activity (a_w) of Some Cheese Varieties

a_w	*Cheese*
1.00	Cheese curd, Whey cheese
0.99	Beaumont, Cottage cheese, Fresh cheese, Quarg
0.98	Belle des Champs, Münster, Pyrénées, Processed, Taleggio
0.97	Brie, Camembert, Emmental, Fontina, Limburger, St. Paulin, Serra da Estrêla
0.96	Appenzeller, Chaumes, Edam, Fontal, Havarti, Mimolette, Norvegia, Samsø, Tilsit
0.95	Bleu de Bresse, Cheddar, Gorgonzola, Gouda, Gruyère, Manchego
0.94	Idiazábal, Majorero, Mozzarella, Norzola, Raclette, Romano, Sbrinz, Stilton
0.93	Danablu, Edelpilzkäse, Normanna, Torta del Casar
0.92	Castellano, Parmesan, Roncal, Zamorano
0.91	Provolone, Roquefort
0.90	Cabrales, Gamalost, Gudbrandsdalsost, Primost

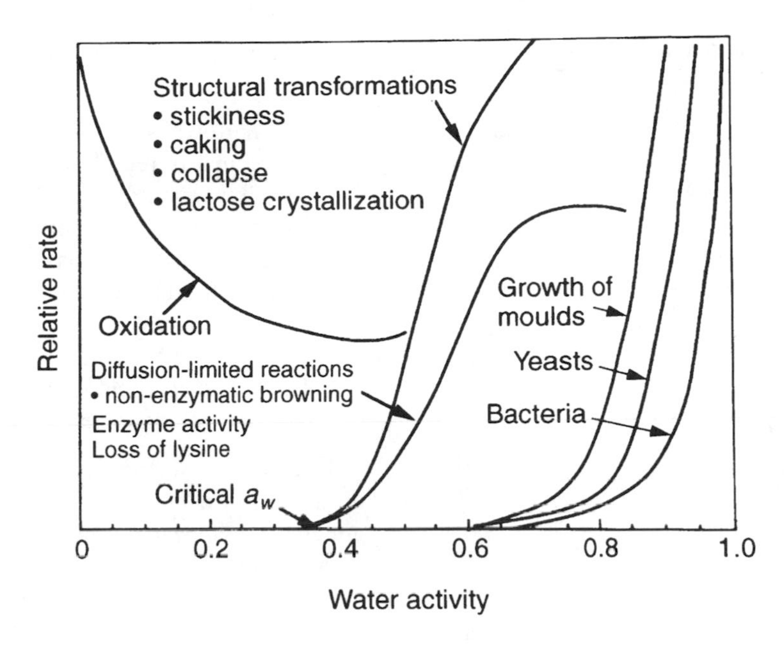

Figure 8–2 Generalized deterioration reaction rates in food systems as a function of water activity (a_w) at room temperature.

The principal factors that affect the uptake of salt by Cheddar cheese curd are summarized in Figure 8–3.

- There is nonlinear relationship between salt uptake and the amount of salt added (Figure 8–4).
- Salt is lost in whey expressed from the curd due to syneresis caused by salt. The volume of whey expressed from the curd increases almost linearly with the amount of salt added, and approximately 2 kg of H_2O are lost per kg of salt absorbed.
- The amount of salt lost in the whey increases slightly with increasing temperature at salting.
- Increasing the duration of the salt and curd mixing has little effect on the volume of whey released but reduces the amount of salt lost and hence increases the salt and salt-in-moisture content of the cheese.
- Salt losses are reduced substantially—and hence the salt and salt-in-moisture percentages are increased—by extending the prepressing holding period.
- Upon salting, about 0.25 kg of fat are lost per 100 kg cheese. Loss of fat increases markedly with increasing temperature.
- The level of salt lost increases with increasing moisture content of the cheese curd.
- Higher losses of salt occur from high-acidity than from low-acidity curd.
- The size of the salt crystals has little effect on salt retention. Salt retention increases as the size of the curd particles is reduced.
- Increasing the depth of the bed of salted curd during holding prior to molding reduces the level of moisture, salt, and salt-in-moisture in the cheese.

8.2.2 Brine-Salting of Cheese

When cheese is placed in brine, there is a net movement of Na^+ and Cl^- from the brine into the cheese as a consequence of the difference in os-

Table 8–3 Minimum Water Activity (a_w) for the Growth of Pathogenic Bacteria in Foods

Pathogen	*Minimum* a_w
Aeromonas hydrophila	0.970
Bacillus cereus	0.930
Campylobacter jejuni	0.990
Clostridium botulinum A	0.940
Clostridium botulinum B	0.940
Clostridium botulinum E	0.965
Clostridium botulinum G	0.965
Clostridium perfringens	0.945
Escherichia coli	0.935
Listeria monocytogenes	0.920
Salmonella spp.	0.940
Shigella spp.	0.960
Staphylococcus aureus (anaerobic)	0.910
Staphylococcus aureus (aerobic)	0.860
Vibrio parahaemolyticus	0.936
Yersinia enterocolitica	0.960

motic pressure between the moisture phase of the cheese and the brine. Water in the cheese diffuses out through the cheese matrix to establish osmotic equilibrium. Cheese can be viewed as a spongelike matrix consisting of strands of fused paracasein micelles. The properties of the interstitial fluid are generally not appreciably different from those of corresponding solutions. Hence, the diffusion of salt from the brine through the moisture phase of cheese would be expected not to differ significantly from that of salt molecules through pure water (e.g., where pure water and the salt solution are separated by a semipermeable membrane). However, model experiments on the brining of cheese designed to obey Fick's laws for unidimensional flow have shown that the rate of diffusion of salt in cheese moisture is much lower than that in pure water. The diffusion coefficient (D) of salt in cheese moisture at 12°C is roughly 0.1–0.2 cm²/day, compared with 1.0 cm²/day for salt in pure water. The difference is due to a number of factors associated with the cheese that retard and/or impede the movement of salt. The factors include these:

- The narrowness of the pores of the paracasein matrix retard the movement of Na^+ and Cl^- passing through them.
- Fat globules and casein particles in the cheese obstruct the passage of salt molecules/ions. This circuitous route taken by the ions increases the effective distance over which they must move.
- The higher apparent viscosity of the cheese moisture compared to that of pure water.

Thus, there is a strong salt concentration gradient within a cheese after salting, but this gradually disappears during storage, and if ripening is long enough, equilibrium will be established throughout the cheese (Figure 8–5).

The absorption and diffusion of salt in cheese curd are affected by several compositional and environmental factors (Figure 8–6). These are discussed next.

8.2.3 Concentration Gradient

Although the diffusion coefficient for NaCl in cheese is essentially independent of brine concentration, the rate of NaCl uptake increases at a decreasing rate with increasing brine concentration (Figure 8–7).

8.2.4 Salting Time

The quantity of salt absorbed increases with salting time, but the rate of salt absorption decreases with time, owing to a decrease in the NaCl concentration gradient between the cheese moisture and the brine. The quantity of NaCl taken up by a cheese is proportional to the square root of brining time, t. The theoretical relationship for the quantity of salt absorbed through a flat surface as a function of brining time is:

$$M_t = 2(C - C_o)(D^* t/\pi)^{1/2} \, V_w$$

where M_t is the quantity of salt absorbed over time (g NaCl/cm²), C is the concentration of NaCl in the brine (g NaCl/ml), C_o is the original salt con-

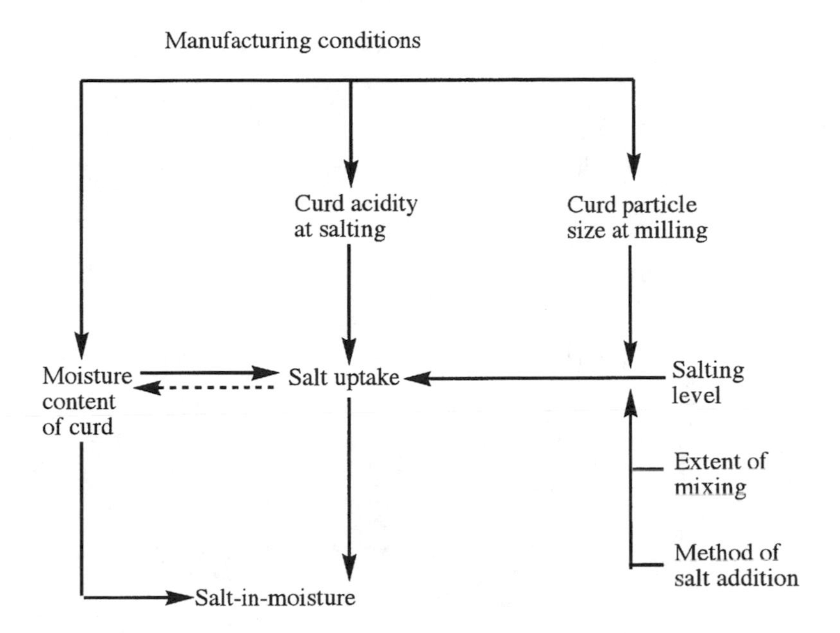

Figure 8–3 Principal factors that affect the uptake of salt by Cheddar curd.

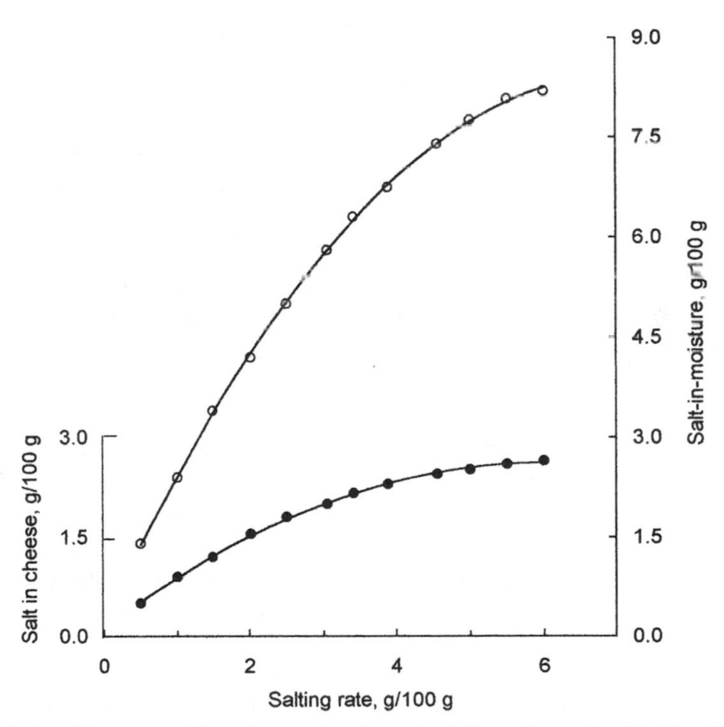

Figure 8–4 The relationship between the salt content (●) and salt-in-moisture level (○) of batches of curd that were from the same vat but salted at different levels.

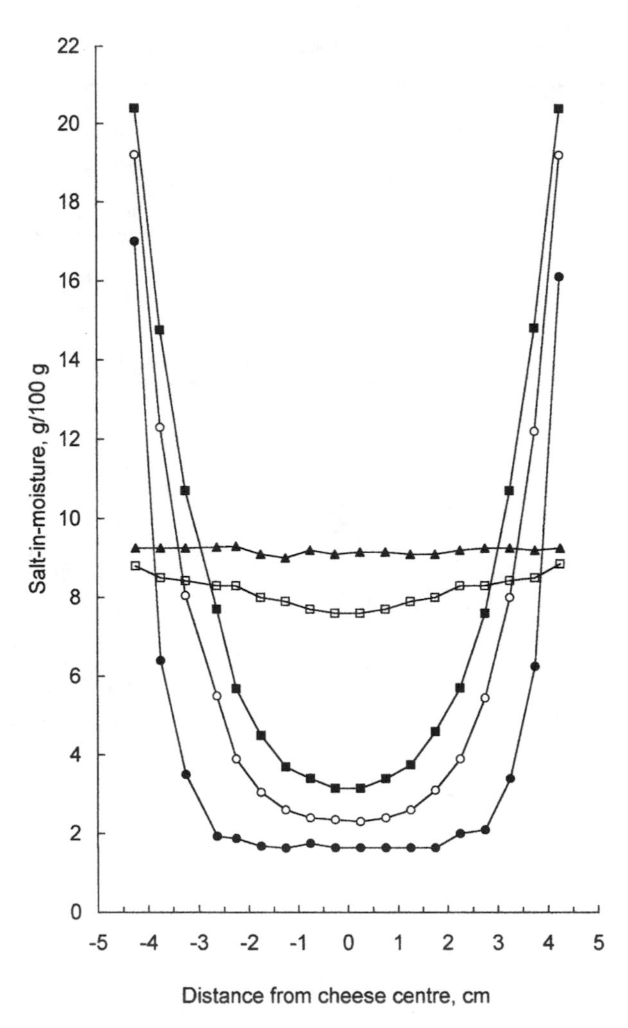

Figure 8–5 The mean salt-in-moisture level throughout cylindrical Romano-type cheese salted in 19.5% NaCl brine at 23°C for 1 day (●), 3 days (○), or 5 days (■) or salted for 5 days and stored wrapped at 10°C for 30 days (□) and 83 days (▲).

centration in the cheese (g/ml cheese moisture), D^* is the pseudodiffusion coefficient of NaCl in cheese moisture (cm^2/d), t is the duration of salting period (days), and V_w is the average water content throughout the cheese at time t (g/g).

8.2.5 Cheese Size and Geometry

The rate of salt absorption increases with an increasing surface area:volume ratio for the cheese. Thus, for cheeses of the same shape and relative dimensions, the mean salt content will decrease with increasing size after brining for equal intervals. The quantity of salt absorbed from the brine by a cheese depends on its shape. The quantity of NaCl absorbed per cm^2 of cheese surface is greater for a flat slab than for a sphere (and the relative reduction increases with the degree of curvature and duration of brining) and for a rectangular-shaped cheese (three effective directions of salt penetration) than for a cylindrical cheese (two effective directions).

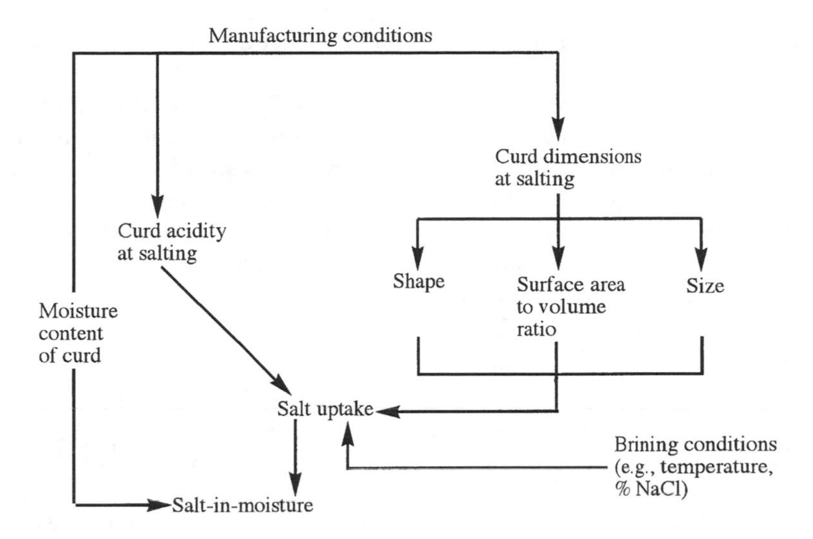

Figure 8–6 Principal factors that affect salt uptake by brine-salted cheeses.

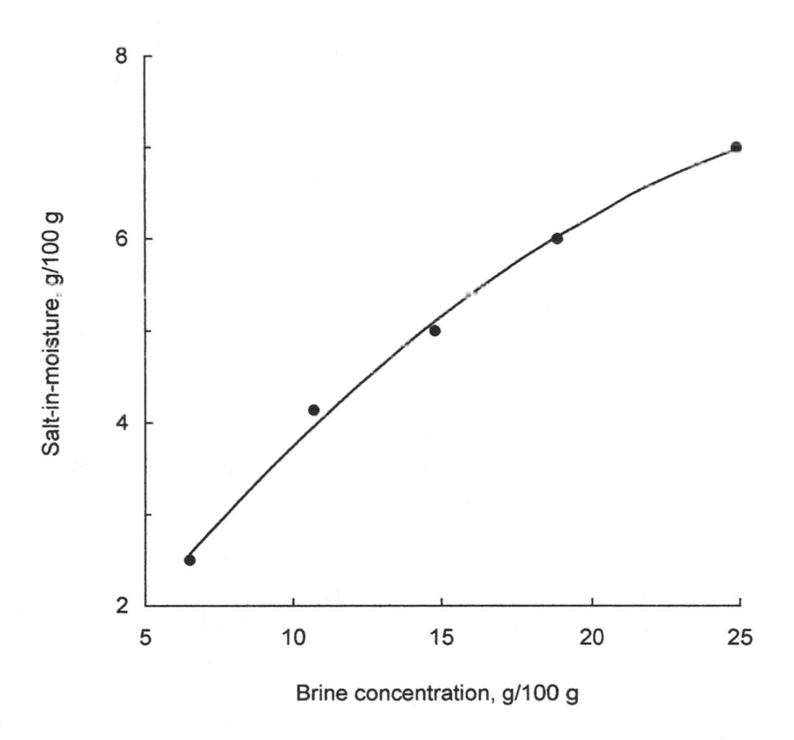

Figure 8–7 Salt level in cheese slices (7 cm diameter, 0.5 cm thick) salted in brines of different concentration for 200 min at 20°C.

8.2.6 Temperature of Curd and Brine

Model brining experiments with small cheeses (curd chips) have shown that salt uptake by curd tempered at any temperature in the range 27–43°C increases with increasing brining temperature in the same range. However, curd tempered to 32°C absorbs salt less readily than curd tempered to a lower or higher temperature. This effect has been attributed to a layer of exuded fat on the surface of curd particles at 32°C that impedes salt uptake; less fat is exuded at temperatures below 32°C, while at higher temperatures exuded fat is liquid and disperses in the brine. Increasing the brining temperature increases the mobility of NaCl and the amount absorbed, partly because of an increase in true diffusion and partly because of an increase in the effective width of the pores in the protein matrix as nonsolvent water decreases with increasing temperature. The reverse situation generally occurs when curd chips are dry-salted (as in Cheddar), where an increase in curd temperature in the range 24–41°C is paralleled by a decrease in the salt and salt-in-moisture in the final cheese. This effect is due to the greater expulsion of whey from the curd as the temperature is increased, which in turn causes more salt to be lost in the whey, making less available for absorption by the curd. In contrast, in brine-salting the cheese is immersed in brine, which is usually agitated; hence the salt available for absorption is not limiting.

8.2.7 Curd pH

Curd salted at a low acidity retains more salt than more acidic curd. The opposite might be expected, since low acidity curd normally contains more moisture than acidic curd and hence will undergo greater syneresis, resulting in a higher loss of salt. However, everything else being equal, the rate of salt uptake and diffusion are higher in high-moisture curd. In addition, the protein in high pH curd may be more salt-soluble than that in more acid curd, which might improve salt retention.

8.2.8 Moisture Content of Cheese Curd

The diffusion coefficient and the quantity of salt absorbed upon brine-salting increases as the moisture content of the curd increases (Figure 8–8), which has been attributed to an increase in the relative pore size in the protein matrix, resulting in less retardation of the diffusing salt molecules/ions.

The reverse situation occurs upon dry-salting milled Cheddar curd. That is, as the initial moisture content increases, the level of salt and salt-in-moisture in the cheese decreases for a fixed salting level. This effect has been attributed to greater losses of whey and salt from the high-moisture curds. Thus, while the diffusion of salt within each curd particle increases with moisture content, less salt is available.

8.3 EFFECT OF SALT ON CHEESE COMPOSITION

For any particular variety, there is an inverse relationship between the levels of moisture and salt in cheese, everything else being equal. In the case of dry-salted cheese, this reflects the syneresis of the curd upon salting, the extent of which is directly related to the amount of salt added to the curd. The relative fluxes of NaCl and H_2O in unidimensional brine-salted cheese are related as $-\Delta W_x \approx p\Delta S_x$, where ΔW_x and ΔS_x are the changes (from the unsalted cheese) in g H_2O and g NaCl, respectively, per 100 g nonsalt cheese solids in planes of the cheese x cm from the cheese-brine interface. The experimental value of p is about 2 (Figure 8–9). That is, the weight of water lost is about twice that of salt taken up, but the value varies from 1.5 to 2.34 (or from < 1 to 3.75 in another study), depending on the distance from the surface. Changes in the texture and appearance of the cheese can be seen as the salt front moves through the cheese.

Uptake of salt during brining is sometimes accompanied by an increase in moisture content in the vicinity of the cheese-brine interface, especially in calcium-free weak brines (< 10%, w/v,

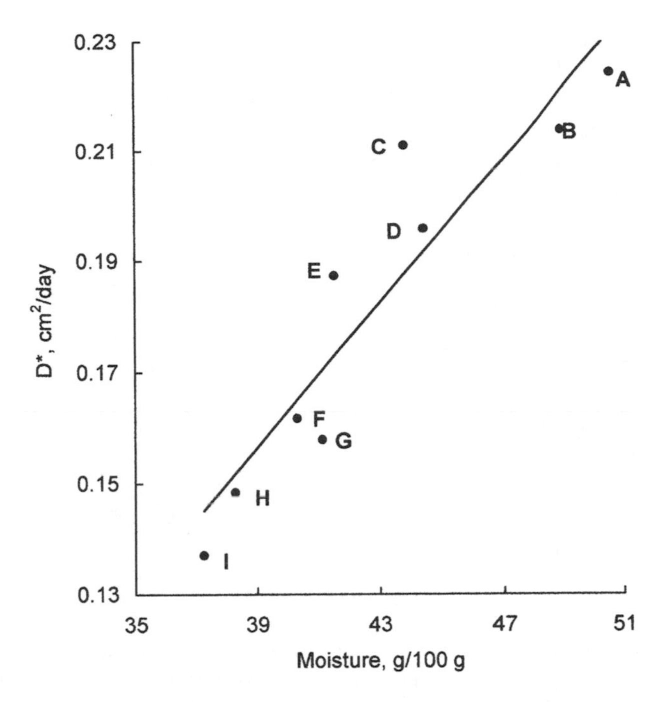

Figure 8–8 Dependence of the pseudodiffusion coefficient of salt in cheese moisture (D*) on the initial moisture content of cheese salted in ≈ 20% NaCl brine at 15–16°C. Blue cheese (A, B), Gouda (C, D), Romano (E), Jarlsberg (F), Emmental (G, I), unsalted milled Cheddar (H).

NaCl) (Figure 8–10). Such an effect is responsible for the "soft-rind" defect and swelling in cheese, and it is attributed to "salting-in" (solubilization) of the cheese protein in dilute NaCl solutions.

Higher salt concentrations are associated with increased levels of fat and protein due to the loss of water from the cheese. There is a significant inverse relationship between fat and moisture levels in mature Cheddar cheese.

The concentrations of lactose and lactic acid in and the pH of cheese depend on the continued activity of the starter and hence on salt content.

8.4 EFFECT OF NaCl ON THE MICROBIOLOGY OF CHEESE

The concentration of salt-in-cheese moisture has a major effect on the growth of microorganisms in and on the cheese (see Chapter 10). Probably the most extreme example of the use of salt to control bacterial growth is the Domiati-type cheese, where 8–15% NaCl is added to the cheese milk to inhibit bacterial growth and maintain milk quality. In the manufacture of most cheese varieties, salt is added after curd formation.

The growth of *Lactococcus* strains used as starters is stimulated by low levels of NaCl but is strongly inhibited at levels above 5% NaCl. In dry-salted cheeses, the concentration of NaCl rapidly reaches an inhibitory level throughout the cheese. In Cheddar-type cheese, starter growth ceases shortly after salting, but metabolism of lactose continues unless the level of salt-in-moisture exceeds about 5% (Figures 8–11 and 8–12). Strains of *Lc. lactis* ssp. *lactis* are more salt tolerant than *Lc. lactis* ssp. *cremoris*: the former grow in the presence of 4% NaCl

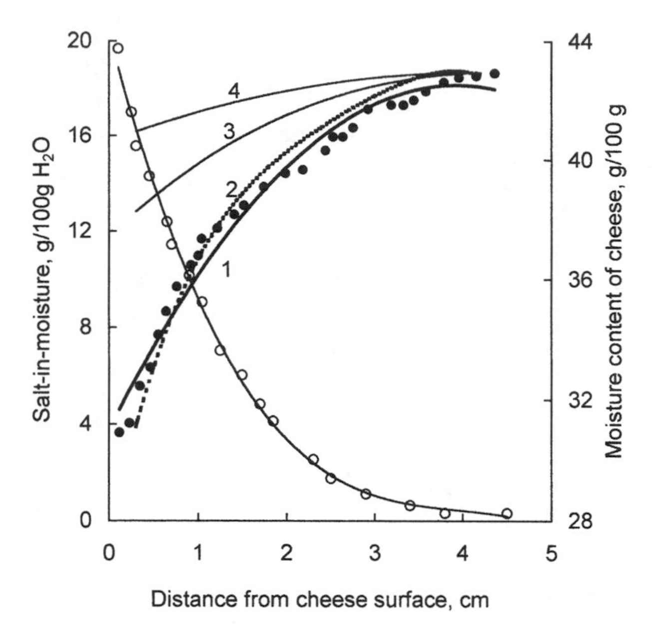

Figure 8–9 Experimental (●) and theoretical (1–4) moisture levels and experimental salt-in-moisture level (○) in a full-fat Gouda cheese (pH 5.64) after brining for about 8 days as a function of distance from the cheese surface in contact with the brine (20.5 g NaCl/100 g H_2O; temperature, 12.6°C). Theoretical moisture levels were calculated using the relationship $\Delta W_x = p\Delta S_x$, where W and S are g H_2O and g NaCl, respectively, per 100 g nonsalt cheese solids, x is the distance (cm) from the cheese surface in contact with the brine, and p is a coefficient denoted as the flux ratio. The theoretical moisture levels were calculated for $p = 2.5$ (1); p varying from 1.7 at the salt front (i.e., maximum distance to which salt had penetrated) to 2.9 at the cheese surface (2); p = 1 (3); and p = 0 (4).

whereas the latter grow in the presence of 2% but not 4% NaCl. However, there is considerable interstrain variation in both subspecies. A low level of salt-in-moisture may lead to high numbers of starter cells in Cheddar cheese, which may lead to bitterness.

If starter activity is inhibited after manufacture owing to an excessively high level of salt-in-moisture, residual lactose will be metabolized relatively late in ripening, when the number of nonstarter lactic acid bacteria (NSLAB) is high. In modern Cheddar cheese, the number of NSLAB is low initially (< 100 cfu/g) and they grow at a rate that is largely determined by the rate of cooling of the pressed curd and the ripening temperature. NSLAB vary in their ability to grow in the presence of NaCl. Most strains can grow in the presence of 6% but not 8% NaCl, and those strains that can grow in the presence of 8% NaCl are inhibited by 10% NaCl. Normally, the number of NSLAB is too low to rapidly metabolize the residual lactose, unless it persists for several weeks (see Chapter 10).

In brine-salted or dry surface–salted cheese, NaCl diffuses slowly from the surface to the center, and even in small cheeses salt probably does not reach an inhibitory concentration in the interior until starter growth has ceased, owing to depletion of lactose or perhaps low pH. Thus, although thermophilic starters, *Streptococcus thermophilus* and *Lactobacillus* spp., are more sensitive to NaCl than *Lactococcus* spp., this is probably not significant in cheese acidification, since cheeses made using these starters are usually brine-salted.

Although data on the salt sensitivity of *Propionibacterium* spp. are variable, most studies show that they are quite salt sensitive and cannot

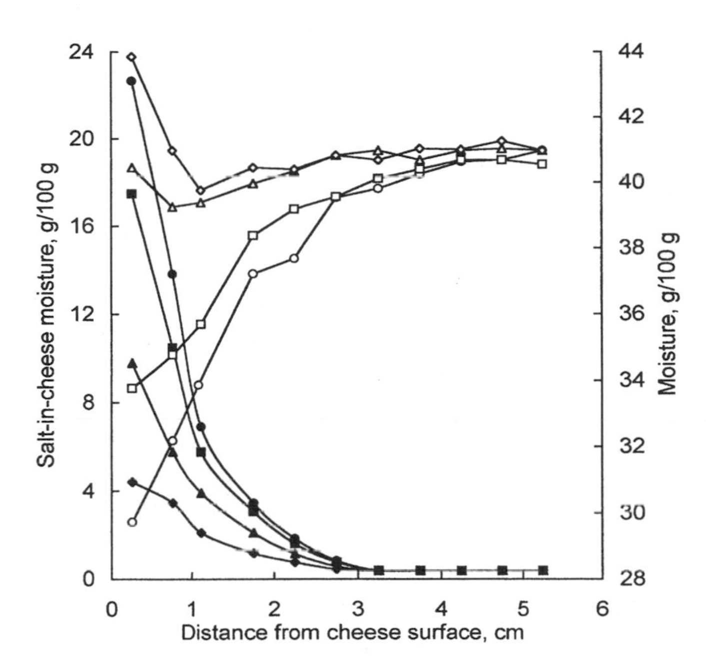

Figure 8–10 Moisture content (open symbols) and salt-in-moisture concentration (black symbols) in Gouda cheese as a function of distance from the salting surface after brine-salting for 4 days at 20°C in 5 (◇, ◆), 12 (Δ, ▲), 20 (❑, ■), and 24.8 (○, ●) percent NaCl solution (without calcium).

grow in the presence of 3% NaCl. The salt-in-moisture in Emmental is only around 2%.

Blue cheeses, at 3–5% NaCl, are among the most heavily salted varieties. Germination of *P. roqueforti* spores is stimulated by 1% NaCl but inhibited by 3–6% NaCl, depending on the strain. Germinated spores can grow in the presence of up to 10% NaCl. It is fairly common commercial practice to add 1% NaCl directly to Blue cheese curd, perhaps to stimulate spore germination, although it also serves to give the cheese a more open structure, which facilitates mold growth. Since most Blue cheeses are dry surface–salted, a salt gradient from the surface to the center exists for a considerable period after manufacture, and the salt-in-moisture percentage in the outer layer of the cheese may be high enough to inhibit spore germination during a critical period, resulting in a mold-free zone on the outside of the cheese.

Growth of *P. camemberti* is also stimulated by low levels of NaCl. Mold growth on Camembert cheese is poor and patchy at levels below 0.8% NaCl.

Since smear-ripened cheeses are brine-salted, a salt gradient from the surface to the center exists initially. However, most of these cheeses are relatively small and have a relatively high moisture content. Therefore, salt equilibrates throughout the cheese relatively quickly. These cheeses are also rubbed with brine occasionally to distribute the microorganisms evenly over the surface. The surface microflora of these cheeses is very complex, but the principal microorganisms are yeasts and coryneform bacteria, both of which are quite salt tolerant (see Chapter 10).

8.5 INFLUENCE OF NaCl ON ENZYMES IN CHEESE

8.5.1 Coagulant

With the exception of high-cooked cheeses, such as Emmental and Parmesan, in which the

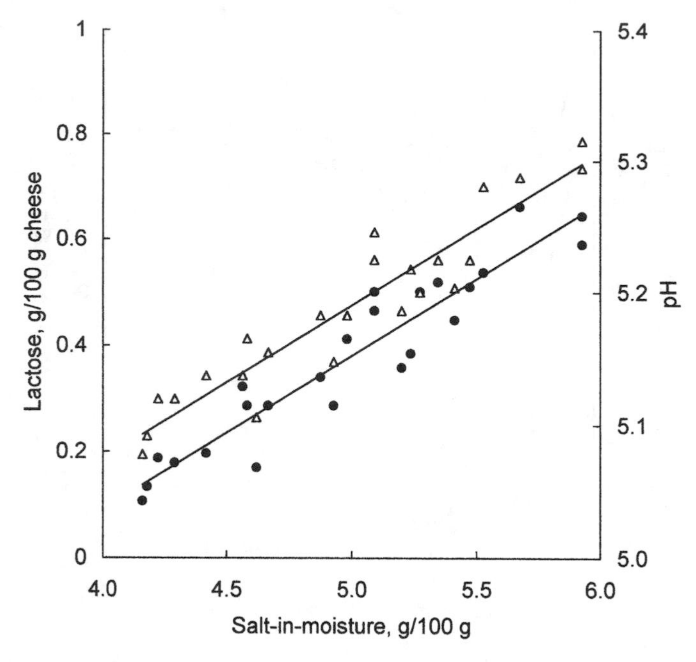

Figure 8–11 Effect of salt-in-moisture concentration on the concentration of lactose (Δ) and pH (●) within a single block of Cheddar cheese analyzed 14 days after manufacture.

rennet is denatured extensively during cooking, primary proteolysis is catalyzed mainly by the residual coagulant (see Chapter 11). Although chymosin, pepsins, and *Rhizomucor* proteinases readily hydrolyze β-casein in solution, α_{s1}-casein is the principal substrate in cheese. β-Casein is less susceptible to hydrolysis by the coagulant, probably mainly because of hydrophobic interactions between adjacent C-terminal regions, which contain the primary chymosin-susceptible bonds (see Chapter 11); these interactions are intensified at high ionic strength. The concentration of salt in cheese has a large effect on the rate of proteolysis (see Chapter 11).

8.5.2 Milk Proteinases

The principal indigenous proteinase in milk, plasmin, contributes to proteolysis in all cheese varieties that have been studied, as indicated by the formation of γ-caseins. It is a major contributor to proteolysis in high-cooked cheeses, owing to partial or complete inactivation of the coagulant. Plasmin is associated with the casein micelles in milk and is incorporated into cheese curd. The activity of plasmin in cheese is stimulated by low levels of NaCl, up to a maximum at 2%, but it is inhibited by higher concentrations, although some activity remains at 8% NaCl. The influence of NaCl on the activity of the indigenous acid milk proteinase (cathepsin D) has not been investigated.

8.5.3 Microbial Enzymes

The effect of NaCl on the stability and activity of microbial enzymes, especially in the cheese environment, has received little attention and appears to warrant research.

8.6 EFFECT OF SALT ON CHEESE QUALITY

Considering that salt has a major influence on the microbiology, enzymology, pH, and mois-

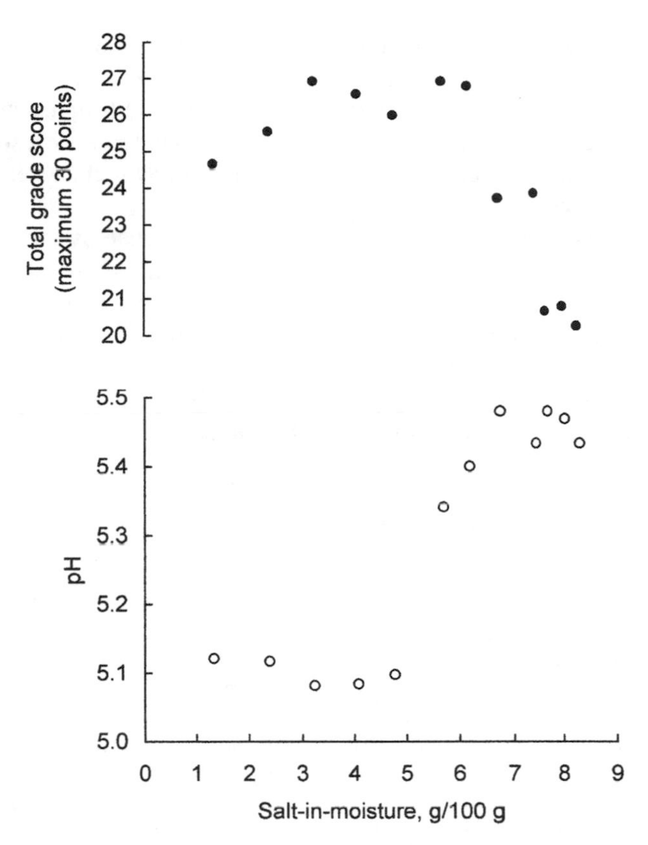

Figure 8–12 Effect of salt-in-moisture levels on pH (○) at 8 weeks and the total grade score (maximum 30) (●) for Cheddar cheese made from curd from the same vat but salted at different levels.

ture content of cheese, it is not surprising that the concentration of salt in cheese has a major effect on its quality (Figure 8–12). Ripening is retarded at high salt concentrations, whereas defects such as bitterness are common at low concentrations. The optimum concentration for Cheddar is about 5% salt-in-moisture. Although the impact of NaCl concentration on cheese quality is well recognized, its effects at the molecular level are not known. It is likely that high concentrations of NaCl retard ripening through a general inhibitory effect on several enzymes in cheese. High concentrations of NaCl (e.g., > 8% salt-in-moisture in Cheddar) probably inhibit the growth of NSLAB, but concentrations in the range normally encountered in Cheddar appear to have little or no effect. Flavor defects encountered at low salt concentrations probably arise from excessive or unbalanced enzyme activity. For example, bitterness can occur in Dutch-type cheeses if excessive proteolysis of β-casein by chymosin occurs, which releases bitter C-terminal peptides (e.g., β-CN f193–209).

NaCl makes a direct positive contribution to cheese flavor, as most consumers appreciate a salty taste in foods. Salt-free cheese has a rather insipid, watery taste; 0.8% NaCl is sufficient to overcome this defect.

8.7 NUTRITIONAL ASPECTS OF NaCl IN CHEESE

High intake of salt in the diet is undesirable, since it increases hypertension and the risk of osteoporosis via increased excretion of calcium. Sodium rather than chloride is the responsible agent (see Chapter 21). Even in countries with a high consumption of cheese, cheese contributes only about 5% of total sodium intake. Nevertheless, there has been considerable interest in the production of reduced-sodium cheeses. Approaches used include

- reducing the level of salt added (the degree of reduction is limited owing to the development of off-flavors in low-salt cheese)
- replacing some of the NaCl by KCl, $MgCl_2$, or $CaCl_2$ (about 50% of the NaCl may be replaced by KCl without undesirable consequences, but higher levels of replacement lead to bitterness caused by KCl)
- use of flavor enhancers to mask defects

Processed cheeses and cheese products contain much higher levels of Na than natural cheeses owing to the addition of sodium-rich "emulsifying" salts (see Chapter 18). There would appear to be greater opportunities to reduce the concentration of Na in these products than in natural cheeses.

REFERENCES

Fennema, O.R. (Ed.). (1996). *Food chemistry* (3d ed.). New York: Marcel Dekker.

Guinee, T.P., & Fox, P.F. (1993). Salt in cheese: Physical, chemical and biological aspects. In P.F. Fox (Ed.), *Cheese: Chemistry, physics and microbiology* (2d ed., Vol. 1). London: Chapman & Hall.

Kinsella, J.E., & Fox, P.F. (1986). Water sorption by proteins: Milk and whey proteins. *CRC Critical Reviews in Food Science and Nutrition*, *24*, 91–139.

Marcos, A. (1993). Water activity in cheese. In P.F. Fox (Ed.), *Cheese: Chemistry, physics and microbiology* (2d ed., Vol. 1). London: Chapman & Hall.

Rockland, L.B., & Beuchat, L.R. (Eds.). (1987). *Water activity: Theory and applications to food*. New York: Marcel Dekker.

Rockland, L.B., & Stewart, G.F. (1981). *Water activity: Influences on food quality*. New York: Academic Press.

Roos, Y. (1997). Water activity in milk products. In P.F. Fox (Ed.), *Advanced dairy chemistry: Vol. 3. Lactose, water, salts and vitamins*. London: Chapman & Hall.

CHAPTER 9

Cheese Yield

9.1 INTRODUCTION

World production of cheese is 15 million tonnes per annum, with an estimated value of $55 billion. Approximately 7% of world cheese is traded on the global market, the major suppliers being the European Union (≈ 50%), New Zealand (~ 16%), and Australia (≈ 11%). The yield of cheese and its control are of great economic importance, determining the profit of cheese plants and the price of milk accruing to farmers. Owing to its economic importance, cheese yield and the factors that affect it have been investigated extensively, and several comprehensive reviews on the subject have been published (International Dairy Federation [IDF] 1991, 1994; Lucey & Kelly, 1994).

9.2 DEFINITION OF CHEESE YIELD

The definition of cheese yield is important for two main applications:

1. measuring the efficiency of and determining the economic viability of a cheesemaking operation
2. measuring the results of experiments, which is essential for evaluating the potential usefulness of a particular process or change in technology

Yield may be expressed in many ways, as discussed below. The format used is generally determined by the needs of the particular situation. Cheese yield may be expressed simply as the quantity of cheese of a given dry matter produced from a given quantity of milk with a defined protein and fat content (kg/100 kg milk). Actual cheese yield (Y_a) is often loosely expressed as the "kilogram of cheese per 100 kilograms of milk" or "percent yield." An alternate, frequently used way of stating cheese yield is "the number of liters of milk (of a given composition) required to manufacture one tonne of cheese," which in the case of Cheddar cheese is around 10,000 liters. Such a definition of cheese yield is suitable only when one variety of cheese is being manufactured to a relatively constant composition from milk with a relatively constant composition. However, the composition of milk and hence the yield of cheese from a given quantity of milk can vary depending on several factors, including species (e.g., cow, goat, or sheep), breed, stage of lactation, plane of nutrition, lactation number, and animal health. Moreover, the composition of cheese milk may be altered by technological interventions, including the following:

- Standardization. Standardization is a process whereby the casein:fat (CF) ratio is adjusted to produce cheese of the required composition. As milk is generally standardized to a different CF ratio for each cheese variety, it is necessary to state the milk composition when expressing the yield of different varieties from a given quantity of milk. Hence, the yield of Cheddar may be expressed as kg/100 kg of milk with a fat content of 33 g/kg and a protein content of, for example, 31.7 g/kg.

- Low concentration factor ultrafiltration (LCF-UF) or fortification with reconstituted extra-low heat skim milk powder. Ultrafiltration and fortification are undertaken to increase and maintain the level of casein at a fixed value (e.g., 38 g/kg) throughout the year. LCF-UF, which is widely practiced, is a very effective method for preparing milk of a more uniform composition and hence producing cheese of more consistent quality, especially in regions where large variations in milk composition occur throughout the manufacturing season.
- Pasteurization practices. For a given level of protein in the standardized milk, the effective protein level (i.e., the level of gel-forming protein that is potentially recoverable in cheese curd) may be increased by changing the heat treatment of the milk. At normal pasteurization temperature (72°C × 15 s), a low level of whey protein is denatured (≈ 5% of total), depending on the levels and proportions of different whey proteins present. It is generally assumed that the denatured whey proteins complex with κ-casein and are retained in the cheese curd (see Chapter 6). As the severity of the heat treatment is increased, the extent of whey protein denaturation and hence the effective milk protein level increase and contribute to increased cheese yield.

Likewise, the composition of cheese differs with variety. Hence, a more precise definition of cheese yield is "kilograms of actual cheese type (e.g., Cheddar) per 100 kilograms of milk containing specified levels of fat and protein (or preferably casein)." When comparing actual cheese yield from milks of different composition using the latter definition, yield may be defined as "kilograms of cheese type (e.g., Cheddar) per 100 kilograms of milk adjusted to a standard concentration of casein (or protein) plus fat" (abbreviated YaCFAM).

The composition of most cheese varieties must fall within certain specifications prescribed by national or international standards of identity (e.g., for Cheddar cheese in the United States, fat ≥ 330 g/kg; moisture ≤ 390 g/kg). However, intravarietal differences in composition are usual. The moisture content of Cheddar cheese, for instance, typically varies from around 350 to 380 g/kg. Table cheese and most cheese supplied to the food service and catering industries are sold on the basis of total weight. Hence, increasing the moisture content to maximize cheese yield is desirable, provided that the composition of the cheese is within legal specifications and that quality is not impaired. Generally, for most cheese varieties, there are defined bands for the various compositional parameters (moisture, moisture in nonfat substances, fat in dry matter, pH, and Ca) that give optimum cheese quality (see Chapter 14). In some applications, cheese is incorporated, along with water (and other optional ingredients), as a formulation ingredient in the preparation of another food, such as pasteurized processed cheese products and cheese powders. In these applications, a low moisture content (e.g., 330–340 g/kg for Cheddar) is often preferred and specified, as it leads to a significant reduction in transport costs, especially when the cheese is transported over long distances. The ratio of moisture to cheese solids in the final product is easily regulated by adding water during the formulation of these products. Hence, for these applications, buyers prefer to purchase cheese on a dry weight basis and cheese yield is best expressed as moisture-adjusted cheese yield, that is, kilograms of moisture-adjusted cheese type (e.g., 380 g/kg moisture Cheddar) per 100 kilograms of milk adjusted for casein (e.g., 25 g/kg) and fat (e.g., 33.3 g/kg fat), or as dry matter cheese yield, that is, kilograms of cheese solids per 100 kilograms of milk adjusted for casein (e.g., 25 g/kg) and fat (e.g., 33.3 g/kg fat).

The percentage recovery of a particular component (e.g., milk fat) in cheese affects cheese yield and determines the efficiency of the cheesemaking operation. Moreover, information on the recovery of fat and casein is useful, as it may help trace the cause(s) of high fat losses, such as inadequate curd firmness at cutting and blunt curd knives. Hence, the amounts of milk fat, protein, and/or casein recovered are used as indirect measures of yield. Alternatively, the levels of fat and curd fines in the whey, which represent unrecovered milk solids, may also be used as an index of cheesemaking efficiency and provide indirect information on cheese yield. Curd fines (curd dust) are fragments of curd broken off the curd particles during cutting and/or the initial phases of stirring.

From the foregoing, it is clear that cheese yield may be expressed in different ways, depending on milk composition and the use of the final cheese. The definition of yield for a particular cheese plant should be chosen to ensure maximum profitability. However, there may be differences between cheese yield and profitability, depending on plant type and cheese variety. Maximization of cheese yield is profitable only if the cost of implementing new procedures is reasonable and savings are significant. Hence, increasing fat recovery improves cheese yield but scarcely affects the profitability of cheesemaking, as whey cream, recovered by separation of whey, has almost the same value as sweet cream. Similarly, improving the recovery of fines from whey may improve cheese yield, but the capital cost of new equipment may be too high to improve profitability, at least in the short term. Formulae for the determination of cheese yield and recoveries are described in the next section.

Table 9–1 Typical Mass Balance for a Full-Fat Cheddar Cheese

	Weight (*kg*)
Inputs	
Pasteurized milk	454.8
Starter	6.37
Rennet solution	1.00
Salt	1.44
Fat in cheese milk + starter	16.41
Protein in cheese milk + starter	16.46
Total weight of inputs	463.61
Weight of fat + protein	32.87
Outputs	
Cheese	46.98
Bulk whey[a]	409.86
White whey[b]	5.97
Fat in cheese	14.55
Fat in bulk whey	1.71
Fat in white whey	0.14
Protein in cheese	12.46
Protein in bulk whey	3.92
Protein in white whey	0.06
Total weight of outputs	462.81
Weight of fat + protein	32.84

[a] Bulk whey is whey removed at whey drainage and during cheddaring.

[b] White whey is whey expressed during salting and pressing.

9.3 MEASUREMENT OF CHEESE YIELD AND EFFICIENCY

The determination of actual cheese yield requires measurement of the weight of all inputs and outputs. A typical mass balance for the experimental production of Cheddar is presented in Table 9–1.

In pilot-scale cheesemaking experiments, an accurate mass balance is easily achievable because of the batch nature and small scale, which facilitate weighing of all materials. The fitting of load cells to pilot-scale cheese vats further simplifies and increases the accuracy of mass balances in experimental cheesemaking. In contrast, achieving a mass balance (especially for individual vats) in commercial cheesemaking is more difficult because of the continuous nature of the different operations (e.g., overlapping of curds from two or more vats on the drainage belt or cheddaring tower of a Cheddar Master system). Hence, at the industrial level, a mass balance tends to be performed on a day's production. Quantities of milk, starter, and whey are usually measured using on-line flow meters.

The actual cheese yield may be then calculated using Equation 9.1:

[Equation 9.1]

$$\text{Actual yield } (Y_a) = 100\left(\frac{\text{weight of cheese}}{\text{weight of milk} + \text{starter culture} + \text{salt}}\right)$$

The units of actual yield are usually kg/100 kg. An alternate term is *percent yield.*

While each cheese variety has a maximum permitted moisture content (e.g., 390 g/kg for Cheddar), variations in moisture content for a given variety are common. Comparison of the actual yield of a given variety from milk of a given composition may thus reflect differences in both moisture content and/or recovery of milk constituents. Hence, comparison of actual yields, when moisture levels are different, may conceal inefficiencies in the recovery of milk components such as fat or protein. However, it is sometimes desired to compare the yield of two or more batches of a given variety of cheese but with a different moisture content and from milk of a given composition. In this situation, adjusting the moisture content of the different batches to a reference or desired value eliminates the effects of variations in moisture content on yield. The resulting yield expression is termed "moisture-adjusted cheese yield" (MACY):

[Equation 9.2]

$$\text{MACY (kg / 100 kg)} = (\text{actual yield})\left(\frac{100 - \text{actual cheese moisture content}}{100 - \text{reference cheese moisture content}}\right)$$

Many factors affect the yield of a particular variety of cheese, including the milk composition, the moisture content of the cheese, the cheesemaking process, and the type of plant equipment, as discussed in Section 9.5. The latter two factors influence cheese yield, since they affect the recovery of milk fat and protein in the cheese. The actual yield is also influenced by the variety of cheese being manufactured, which determines the moisture content, and the number of operations where fat and protein may be lost. In the manufacture of Cheddar cheese, the salted curd is vacuum packed in polyethylene liners following pressing, and little or no further loss of moisture, fat, or protein occurs during storage. In contrast, brine-salted cheeses lose moisture during salting. The weight of moisture lost is about twice that of salt absorbed. Moreover, other constituents, e.g., soluble nitrogenous material, Ca and fat, may be lost to a greater or lesser degree, depending on the temperature and the duration of brining. For Gouda cheese, there is an approximate net weight loss of about 3% during salting. Moreover, moisture may be lost from brine-salted cheeses during storage, depending on shape and size, the type of packaging material (e.g., natural rind, plasticoat, or wax), storage humidity, and temperature. In wheel-shaped Romano-type cheese (≈ 2.8 kg) with a natural rind, the moisture content decreases by about 50 g/kg during storage (7°C, 85% relative humidity [RH]) for 120 days.

The recovery of milk components (fat and protein or casein) may also be determined when their concentrations in the inputs (milk and starter) and outputs (cheese and whey) are known. This enables the recovery or loss of fat, casein, nonfat solids, and/or protein to be calculated using Equations 9.3 and 9.4.

[Equation 9.3]

$$\text{Percentage of fat recovered in cheese} = 100\left[\frac{(\text{weight of cheese})(\text{fat content of cheese})}{(\text{weight of milk})(\text{fat content of milk} + \text{starter})}\right]$$

[Equation 9.4]

$$\text{Percentage of fat lost in whey} = \left[\frac{(\text{weight of whey})(\text{fat content of whey})}{(\text{weight of milk} + \text{starter})(\text{fat content of milk} + \text{starter})}\right]$$

The recovery of fat and protein for a particular variety of cheese is influenced by many factors, as described in Section 9.5. Fat recovery is very dependent on the cheese variety being manufactured, which determines the number and types of operations where fat may be lost. In industrial Cheddar cheese manufacture, about 8.5% of total fat is lost in the whey, of which about 76%, 17%, 5%, and 2% are lost in the whey from the cheese vat, cheddaring tower, salting belt, and block former, respectively. Similar levels of fat loss have been reported during the commercial manufacture of Gouda and Emmental. In batch pilot-scale manufacture of low-moisture Mozzarella cheese curd, a higher percentage milk fat is lost than for Cheddar (≈ 20 vs. ≈ 8.5%) due to high losses during kneading and stretching of the curd in hot water (about 80°C); ~ 60% of the total fat lost during the manufacture of low-moisture Mozzarella cheese curd occurs in the stretch water. Typically, 25–27% of total protein is lost during cheese manufacture, about 71%, 14%, and 14% of which is whey protein, casein, and nonprotein nitrogen (expressed as protein), respectively. Loss of casein (≈ 5% of total casein) ensues mainly from the glycomacropepeptide, which is soluble in whey following its release from the κ-casein by rennet.

It may be of interest to compare the Y_a or MACY obtained on different days or different years. Such comparison enables a cheese company to monitor its efficiency over time. However, when the composition of milk changes, especially the levels of fat and casein, during the cheesemaking season, then comparison of Y_a or MACY for a particular variety at different times has little value as an indicator of the efficiency of cheesemaking. The differences in yield due to different recoveries (e.g., of fat) could be completely masked by those that occur as a result of differences in milk composition. In this situation, if the composition of the cheesemilk (protein and fat) and the cheese (protein, fat, and salt) at the different manufacturing times is known, the yield of cheese at the different times may be meaningfully compared by adjusting the protein and fat content of the milk to reference values, as in Equation 9.5. The resulting yield expression is termed the moisture-adjusted cheese yield/100 kg milk adjusted for protein and fat (MACYPFAM):

[Equation 9.5]

$$\text{MACYPFAM (kg / 100 kg)} = \text{MACY}\left[\frac{100}{\left(\dfrac{100 \times (\text{actual content of protein and fat in milk} + \text{starter})}{\text{content of protein and fat in reference milk} + \text{starter}}\right)}\right]$$

In Equation 9.5, it is assumed that casein, as a percentage of total protein, does not change over time. However, if casein concentration changes over time, substitution of casein for protein in Equation 9.5 allows comparison of milks with different levels of casein.

9.4 PREDICTION OF CHEESE YIELD

Predictive yield formulae are used to estimate the yield of a particular variety of cheese from milk of a given composition. Prediction of cheese yield is useful in that it allows a cheese plant to

- measure its efficiency by comparing actual and predicted yields
- plan production (i.e., capacity and technology) in the event of an anticipated increase in milk supply
- plan product mix and milk-pricing schemes

Predictive yield formulae for a particular variety are compiled on the basis of information obtained from

- cheese yield experiments in which yield and component recovery are related to milk composition
- theoretical consideration of the cheesemaking process's influence on the partition of the various components of milk (e.g., loss of glycomacropeptide, milk salts, and fat) between the curd and whey

Predictive yield formulae and their application to different types of cheese, especially Cheddar, have been reviewed recently (see IDF, 1994). Predictive formulae are of the following general type: $Y = aF + bC$ or $Y = aF + bC + k$, where y = yield; F and C are the fat content and casein contents of the milk (with added starter culture), respectively; k is a constant, the magnitude of which depends on the loss of casein and the levels of nonfat, noncasein solids in the cheese; and a and b are coefficients, the magnitude of which depends on the contributions of fat and casein to yield. Probably the simplest and most widely applied formula for predicting the yield of different cheese varieties is that of van Slyke, which was developed for Cheddar cheese in 1936. The van Slyke yield formulae for actual and moisture-adjusted cheese yields are

[Equation 9.6]

$$Y_a = \frac{\left[Fx\ (\%FR\ /\ 100)\ - C - a\right] \times b}{1 - \left(\dfrac{\text{actual moisture}}{100}\right)}$$

[Equation 9.7]

$$Y_{\text{MACY}} = \frac{\left[Fx\ (\%FR\ /\ 100)\ - C - a\right] \times b}{1 - \left(\dfrac{\text{reference moisture}}{100}\right)}$$

where, F and C are the fat content and casein content of the cheesemilk (with added starter culture), $\%FR$ is the fat recovery, a is the coefficient for casein loss, and b is the coefficient to account for cheese solids nonfat, nonprotein (SNFC). The values for $\%FR/100$, a, and b for Cheddar cheese, as predicted by van Slyke, are 0.93, 0.1, and 1.09, respectively. These formulae can be rewritten in the format $Y = aF + bC$, where the values of coefficients a and b are 1.66 and 1.78, respectively, for Cheddar cheese containing 390 g moisture/kg. The van Slyke formula has been modified for other cheese types, based on results from cheesemaking experiments. For low-moisture Mozzarella, reported values of $\%FR/100$, a, and b are typically 0.86, 0.36, and 1.09, respectively. The lower values of $\%FR/100$ and a for low-moisture Mozzarella compared with Cheddar reflect the higher losses of fat and casein in the hot water used to heat the curd during stretching. For commercial Finnish Edam and Emmental cheeses, the mean values of $\%FR/100$ and the coefficient a were 88.7 and 88.1, and 0.13 and 0.15, respectively. There is considerable intravarietal variation in the reported values of $\%FR/100$ and the coefficients a and b. For example, the $\%FR/100$ for Cheddar cheese has been found to range from about 83 to 93. Discrepancies between studies undoubtedly reflect differences in milk composition, milk quality and storage conditions, milk heat treatment, cheesemaking conditions, and cheesemaking technology (see Section 9.5).

Variations in the coefficients also occur between cheese plants due to the above factors, which cause interplant differences in efficiency. Moreover, deviations between the actual and predicted yields may vary between plants. Hence, the application of a generic predictive cheese yield formula for a given cheese variety may not accurately predict cheese yield in all plants. As an alternative to generic predictive formulae, plant-specific formulae may be developed for each factory. Plant-specific yield formulae may be developed by the statistical analysis of historical data (collected weekly or monthly) on milk composition, milk quality (e.g., somatic cell count), fat and protein recovery, and cheese moisture. Plant-specific formulae tend to give very accurate predictions of cheese yield because they reflect, more than generic prediction formulae, the composition and quality of the milk, and the actual cheesemaking conditions used in the plant. The formulae are useful in that they compare the actual yield with that expected from a given weight of milk of a given composition using a given process. If actual yield is less than the predicted yield, remedial action is taken to redress process inefficiencies. However, a close relationship between actual and predicted yields does not indicate that a cheese plant is operating at maximum efficiency. Hence, a plant-specific yield formula should be updated regularly (e.g., annually) to reflect improvements in milk quality and cheesemaking technology, factors that influence cheesemaking efficiency (see Section 9.5).

9.5 FACTORS THAT AFFECT CHEESE YIELD

The principal factors that influence cheese yield are discussed below.

9.5.1 Milk Composition

The single most important factor affecting cheese yield is the composition of the milk, particularly the concentrations of fat and casein, which together represent around 94% of the dry matter of Cheddar cheese. Yield increases linearly as the concentrations of fat and casein are increased (Figures 9–1 and 9–2). In the range 4–33 g fat/kg, the yield of Cheddar cheese typically increases by 1.16 kg cheese/kg milk fat. Similarly, cheese yield increases with casein level in the range 20–30 g/kg, typically by 2.39 kg/kg casein for Cheddar.

Formulae of the general type $Y = aF + bC$, which relate cheese yield to the concentrations of milk fat (F) and casein (C), have been developed for prediction of cheese yield (see Section 9.4). These formulae were derived from experimental cheesemaking data and consideration of many factors, including the chemistry of the conversion of milk to cheese curd, milk composition, cheese composition, retention of fat and casein, and partition of components such as milk salts and lactose between the whey and the cheese curd.

The values of the coefficients a and b have been found to range from about 1.47–1.6 and 1.44–1.9, respectively, for Cheddar cheese. A recent pilot-scale (500 L) study (M.A. Fenelon and T.P. Guinee, unpublished results) showed that the actual (Ya) and moisture-adjusted (MACY) yields of Cheddar cheese were accurately described by Equation 9.8 (Figure 9–3):

[Equation 9.8]

$$Y_{\text{cheddar}} = 1.56F + 1.71C$$

Undoubtedly, the values of the coefficients depend on the composition of the milk, the make procedure, and the equipment design, which influence cheese composition and/or the retention of fat and casein. While formulae of the type $Y = aF + bC$ are empirical, they indicate that in general casein contributes significantly more to Cheddar cheese yield than fat. The separate effects of increasing levels of casein and fat in milk on cheese yield are illustrated in Figures 9–1 and 9–2. The greater contribution of casein is expected, as it forms the continuous paracasein spongelike network that occludes the fat and moisture (serum) phases. In contrast, fat on its own has little water-holding capacity. Occluded moisture contributes directly to cheese yield and indirectly owing to the presence of dissolved solids, including whey proteins, κ-casein glycomacropeptide, lactate, and soluble milk salts. In

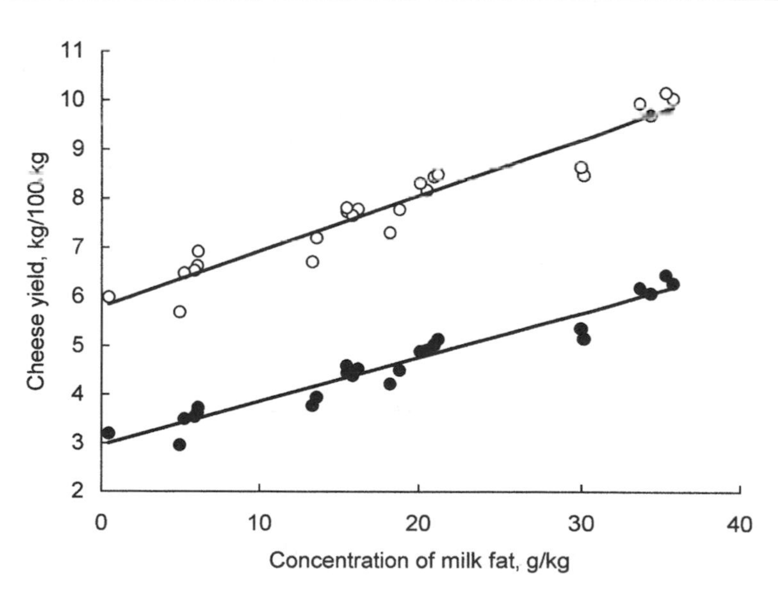

Figure 9–1 Effect of milk fat concentration on actual (○) and dry matter (●) cheese yield at a constant milk casein level of 25.5 g/kg.

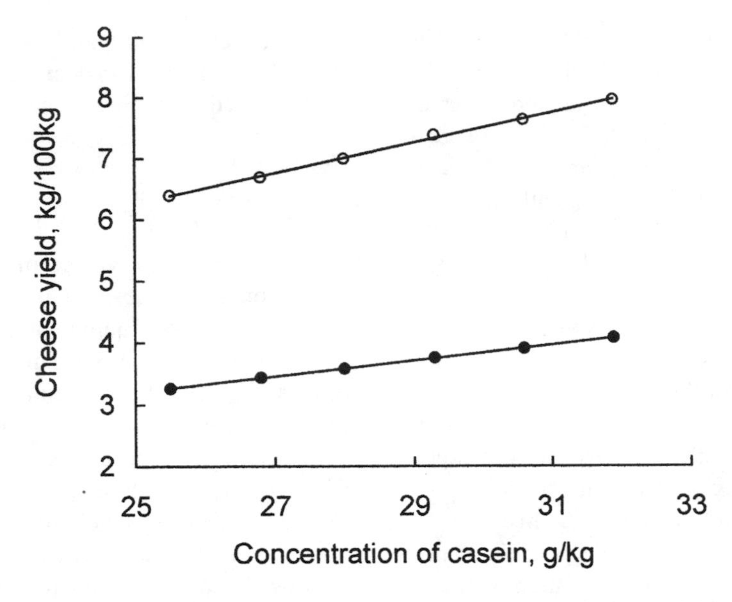

Figure 9–2 Effect of casein concentration on the actual (○) and dry matter (●) yield of skim milk cheese.

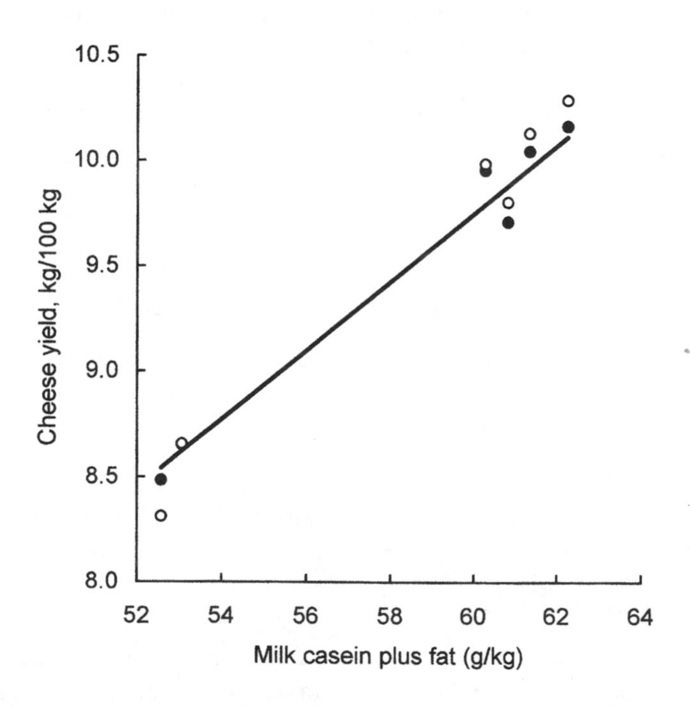

Figure 9–3 Comparison of actual (●) and moisture-adjusted (to 380g/kg; ○) Cheddar cheese yield, with the yield predicted using the formula $Y = 1.56F + 1.71C$, where F and C are the concentrations of casein and fat, respectively, in the pasteurized cheese milk. The regression line for predicted yield and the means from six replicate trials are shown.

milk, micellar Ca and PO_4, which are associated with the casein micelles, are present at concentrations of around 21 and 16 mM, respectively, and around 90% Ca and 98% PO_4 are retained during the manufacture of Cheddar curd. The combined contribution of dissolved solids and micellar Ca and PO_4 to the yield of Cheddar cheese ranges from around 9% for full-fat Cheddar (330 g/kg) to around 14% for low-fat Cheddar (60 g/kg) (Fenelon and Guinee, unpublished results).

Fat generally contributes more than its own weight to Cheddar-type cheese (yield increases by about 1.16 kg/kg milk fat). This greater than pro-rata increase is due to the increase in the level of moisture in nonfat substance as the fat content of the cheese increases. Fat is occluded in the pores of the paracasein network of the cheese and impedes syneresis. The occluded fat globules physically limit aggregation of the surrounding paracasein network and therefore reduce the degree of matrix contraction and moisture expulsion. Hence, as the fat content of the curd is increased, it becomes more difficult to expel moisture, and the moisture:protein ratio increases. However, if the moisture in nonfat substance is maintained constant (e.g., by process modifications such as reduction of curd particle size and slight elevation of the scald temperature), fat contributes less than its own weight to cheese yield (≈ 0.9 kg/ kg), owing to the fact that about 8–10% of the milk fat is normally lost in the whey.

The contribution of fat and casein to the yield of a particular cheese type, such as Cheddar or Cream cheese, is critically dependent on the casein:fat ratio to which the milk is standardized. Other factors that affect the retention of fat and casein, including equipment design and operation and make procedure, also have an influence. For any cheese variety, increasing the casein:fat ratio (e.g., by reducing the level of fat) results in a higher moisture content (and dissolved solids) and, apart from acid-curd cheeses such as Quarg, in higher levels of ash. Conversely, reducing the casein:fat ratio (e.g., by maintaining the casein level constant and increasing the fat content) increases the level of fat and decreases the level of moisture in the cheese. Similarly, the contributions of fat and casein to the yield of different varieties of cheese depend on the ratio of protein (mainly casein) to fat in the cheese, which is controlled by standardization to the desired casein:fat ratio in the milk. Thus, only the casein is important in determining the yield of skim milk cheeses, such as Quarg and Cottage cheese, whereas fat is much more important than casein in the yield of Cream cheese.

Many factors contribute to variations in both the concentration and state (e.g., level of casein hydrolysis and free fat) of casein and fat in milk. These variations are, in turn, associated with variations in the rennet coagulation properties, cheese composition, cheese yield, and quality.

Species and Breed

The species of animal has a major influence on the concentrations of fat and casein and the concentrations of different caseins (Table 9–2).

Table 9–2 Mean Gross Composition and Estimated Yield of Cheddar Cheese from Standardized Milks (Casein:Fat = 0.75) from Different Species

Species	*Fat_{rm} (g/100 g)*	*Fat_{sm} (g/100 g)*	*Casein (g/100 g)*	*Lactose (g/100 g)*	*Ash (g/100 g)*	*Cheese Yield (kg/100 kg)*
Sheep	7.2	5.20	3.9	4.8	0.9	14.78
Water buffalo	7.4	4.26	3.2	4.8	0.8	12.11
Cow	3.9	3.46	2.6	4.6	0.7	9.86
Goat	4.5	3.46	2.6	4.3	0.8	9.84

Key: Fat_{rm} = fat in raw milk; fat_{sm} = fat in milk standardized to a casein:fat ratio of 0.725.

The estimated yield of cheese from bovine milk is lower than that from sheep or water buffalo milk and similar to that from goat milk.

The breed of cow has a marked influence on milk composition and its cheese-yielding capacity, with breeds having higher protein and fat levels also having higher cheese yields. Jersey cow milk, which contains 52.2 g/kg fat and 92.7 g/kg nonfat solids, yields 13.48 kg Cheddar cheese/100 kg milk compared to 10.1 kg/100 kg for Friesian milk, which contains 38.0 g/kg fat and 89.0 g/kg nonfat solids. However, the average weight of milk solids (fat and protein) per lactation is higher for Friesian than for Jersey cows, indicating a higher cheese-yielding potential of the latter over the course of a lactation.

Low Concentration Factor Ultrafiltration (LCF-UF)

Low concentration factor (1.5–2.0 ×) UF is widely practiced in the manufacture of rennet-curd cheeses as a means of standardizing the protein level in milk and thereby obtaining more consistent cheese yield and quality. Variations in gel strength at cutting, buffering capacity, and the rennet:casein ratio are minimized. However, when using conventional cheesemaking vats, concentration is limited to a maximum concentration factor of about 1.5, or 40–55 g protein/kg. Increasing the protein level results in faster curd-firming rates and higher curd firmness after a given renneting time (Figure 9–4). Owing to the rapid curd-firming rate, it becomes increasingly difficult to cut the coagulum cleanly (without tearing) before the end of cutting, especially if the cut program is relatively long (e.g., > 10 min). Reflecting the tearing of the coagulum and consequent shattering of curd particles, fat losses in the whey are markedly higher than those predicted on the basis of volume reduction (due to UF) for milks with a protein concentration greater than 45 g/kg (Figure 9–5), and the yield decreases. The poorer fat-retaining ability of the coagulum from high-protein milk (> 45 g/kg) may also be due in part to a coarser, more porous protein network.

Reduction of the set temperature normalizes the rate of aggregation of rennet-altered micelles, gel-firming rate, and set-to-cut time for high-protein milks. Hence, at 27°C, the gel-firming rate and set-to-cut time of milk containing 45 g/L protein are similar to those for milk with a protein content of 33 g/L renneted at around 31°C. Under these conditions, increasing the concentration of protein in the milk in the range 30–45 g/kg protein has no effect on the

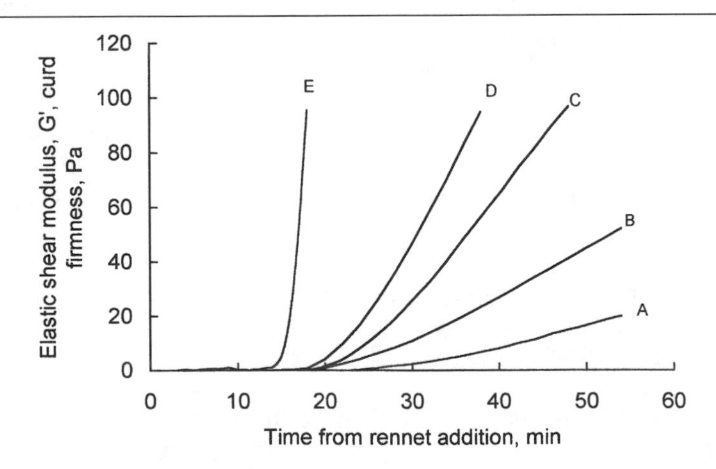

Figure 9–4 Effect of milk protein level (30 g/kg [A], 35 g/kg [B], 45 g/kg [C], 69 g/kg [D], and 82 g/kg [E]) on the rennet coagulation properties of milk. Milks B–E were prepared by ultrafiltration of milk A. The rennet coagulation properties obtained from the curve are gelation time (point at which G′ begins to increase), curd-firming rate (slope of G′/time curve in the linear region), and curd firmness (value of G′ at a given time after rennet addition).

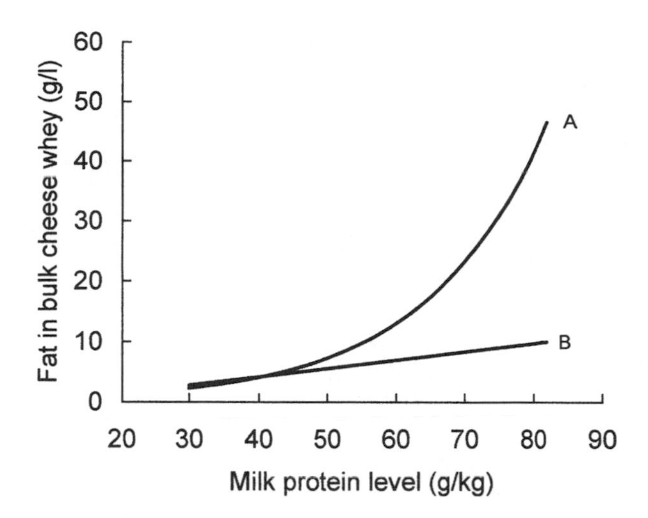

Figure 9–5 Effect of milk protein concentration, varied by ultrafiltration, on actual (A) and predicted (B) fat levels in bulk Cheddar cheese whey. Predicted values were calculated by adjusting the value for the control (30 g protein/kg) for the volume reduction due to ultrafiltration.

percentage recovery of fat and protein. Moreover, the moisture-adjusted Cheddar cheese yield increases with milk protein concentration at a rate similar to that predicted by the van Slyke cheese yield equation (Equation 9.7) on the basis of milk fat and casein (Figure 9–6). However, because of the inverse relationship between the concentration of protein in the milk and the moisture content of the cheese, the actual cheese yield increases less than predicted on the basis of casein and fat (Equation 9.6). The moisture content of cheese made from high-protein milk (> 30 g protein/kg) may be increased to that of the control milk (30 g protein/kg) by manipulating the cheesemaking process, such as by cutting at a higher gel firmness, increasing the cut size, lowering the scald temperature, and/or increasing the rate of cooking. With such alterations, the increases in actual and predicted cheese yields with milk protein level in the range 30–45 g/kg are similar.

9.5.2 Somatic Cell Count and Mastitis

The influence of somatic cells and mastitis on the composition of milk and its suitability for cheese manufacture has been studied extensively. There are three main types of somatic cells: lymphocytes (L), phagocytes, and mammary gland epithelial cells (E) (Burvenich, Guidry, & Paape, 1995). Lymphocytes function in humoral and cell-mediated immunity while phagocytes, of which there are two types, polymorphonuclear leucocytes (PMN) and macrophages (M_ϕ), ingest and kill pathogenic microorganisms that invade the mammary gland. Low numbers of somatic cells (e.g., < 100,000/ml) are present in normal milk from healthy animals during mid lactation, with M_ϕ, L, PMN, and E cells typically in a ratio of roughly 2.1:1.0:0.4:0.2. Somatic cells are released from the blood to combat udder infection and thereby prevent or reduce inflammation (mastitis). Factors that contribute to increases in the somatic cell count (SCC) of bulk manufacturing milk include subclinical mastitis, advanced stage of lactation, lactation number, stress, and poor nutrition. During clinical mastitis, there is a rapid increase in SCC, due primarily to PMN. Depending on the type and extent of bacterial infection, milk from infected (mastitic) quarters of the udder may have a SCC of 200–5,000 × 10^3/ml. However, the milk from animals suffering from clinical mastitis is excluded from the commercial milk supply. Such milk frequently forms clots, which are a mixture of somatic cells and precipitated

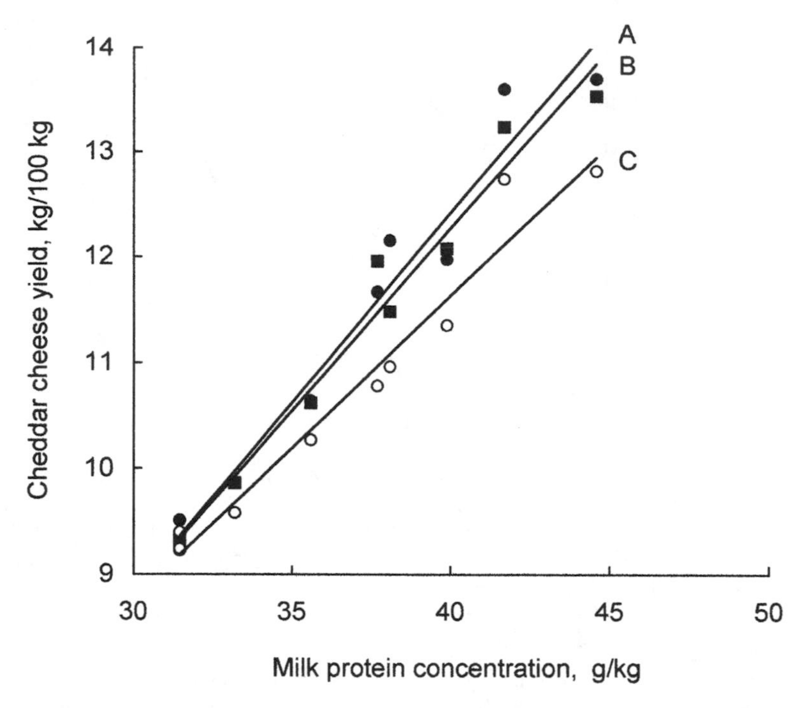

Figure 9–6 Effect of milk protein level on the yield of full-fat Cheddar cheese: moisture-adjusted (to 380 g/kg) yield (●, A), moisture-adjusted (to 380 g/kg) yield predicted using the van Slyke cheese yield formula (■, B; Equation 9.7), and actual yield (○, C).

milk proteins, within the udder. In severe mastitis, these clots block the drainage ductules and ducts in the mammary gland, thereby preventing milk drainage. The initial stage of mastitic infection is subclinical, with inflammation so slight that it is not detectable by visual examination. Hence, the milk from cows suffering from subclinical mastitis becomes part of bulk herd milk and bulk manufacturing milk unless individual cows are tested routinely at farm level for subclinical mastitis (e.g., SCC), which is seldom the case. While bulking dilutes such milk, subclinical mastitis may contribute to an increased SCC in the milk and thereby reduce its suitability for cheese manufacture.

An increased SCC in milk is associated with marked changes in both the concentration and the state (e.g., degree of hydrolysis) of milk constituents that lead to a deterioration in rennet coagulation properties, curd syneresis, and reduced cheese yield. Increasing SCC in the range 10^5–10^6/ml results in progressive increases in the concentration of whey proteins, especially Ig, and decreases in the levels of casein and fat (Figure 9–7). Elevated SCC results in a marked decrease in β-casein as a proportion of total casein and concomitant increases in γ-caseins, proteose peptones, and the ratio of soluble to micellar casein. These changes ensue from hydrolysis of β- and α_{s2} caseins by the elevated activity of plasmin (and probably other proteinases) in the milk that parallels increased SCC; α_{s1}- and κ-caseins are hydrolyzed more slowly by plasmin than β- and α_{s2}-caseins. These proteinases may come directly from the blood (e.g., plasmin and its zymogen plasminogen) or from the somatic cells. The types of casein-derived peptides produced by the two types of proteinases differ somewhat.

Increasing SCC in the range 10^5 to 6×10^5/ml results in an increase in rennet coagulation time and a decrease in curd-firming rate (reciprocal of k20; see Section 6.6.2) and curd firmness (Politis & Ng-Kwai-Hang, 1988b). Typical

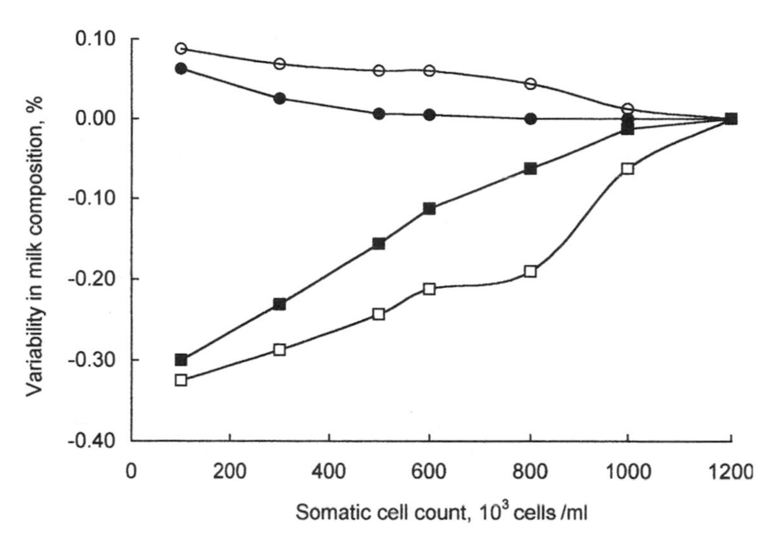

Figure 9–7 Variability in the composition of milks from individual cows with somatic cell count. Casein (❍), fat (●), total protein (■), and whey protein (❑).

trends in rennet coagulation properties with SCC are shown in Figure 9–8. Losses of fat and protein during Cheddar cheese manufacture increase, more or less linearly, by about 0.7% and 2.5%, respectively, with SSC in the range 10^5–10^6/ml (Politis & Ng-Kwai-Hang, 1988a). An increase in SCC from 10^5 to 6×10^5/ml resulted in an 11% decrease in moisture-adjusted (370 g/kg) Cheddar cheese yield (Figure 9–9). It is noteworthy that there was a relatively large decrease in yield (≈ about 0.4 kg/100 kg milk) on increasing the SCC from 10^5 to 2×10^5, a range that would be considered relatively low for good quality bulk milk. Other studies have reported similar trends. For example, Auldist et al. (1996) found that an increase in SCC from above 3×10^5 to above 5×10^5 in late lactation (220 d) resulted in a 9.3% decrease in moisture-adjusted (to 355 g/kg) yield of Cheddar and decreases in the recovery of fat (from 90.1% to 86.6%) and protein (from 78.3% to 74.4%). A decrease of 4.3% has been reported in the yield of uncreamed Cottage cheese upon increasing the SCC from 8.3×10^4 to 8.7×10^5 cells/ml.

The negative impact of SCC on cheese yield and recovery is due largely to the increased proteolysis of α_s-caseins (α_{s1}- + α_{s2}-) and β-caseins to products that are soluble in the serum and are not recoverable in the cheese. Moreover, the lower effective concentration of gel-forming protein results in a slower curd-firming rate and hence a lower degree of casein-casein interaction in the gel following cutting (at a given firmness) and during early stirring. A gel with the latter characteristics exhibits

- greater susceptibility to shattering during cutting and the early stages of stirring, resulting in higher losses of curd fines and milk fat
- an impaired syneretic capacity, with a consequent increase in moisture level

A high SCC may also inhibit the activity of some strains of lactococci during cheese manufacture, an effect expected to further impair the curd-firming rate and reduce firmness at cutting. In commercial practice, the gel is generally cut, not on the basis of firmness, but rather on the basis of a preset renneting time, based on curd firmness within the acceptable range for normal milk. In large modern factories, conditions are not conducive to assessing the firmness of gels in individual vats because of the large scale of operations (more than a million liters per day) and the use of preprogrammed vats with limited operator access. In such factories,

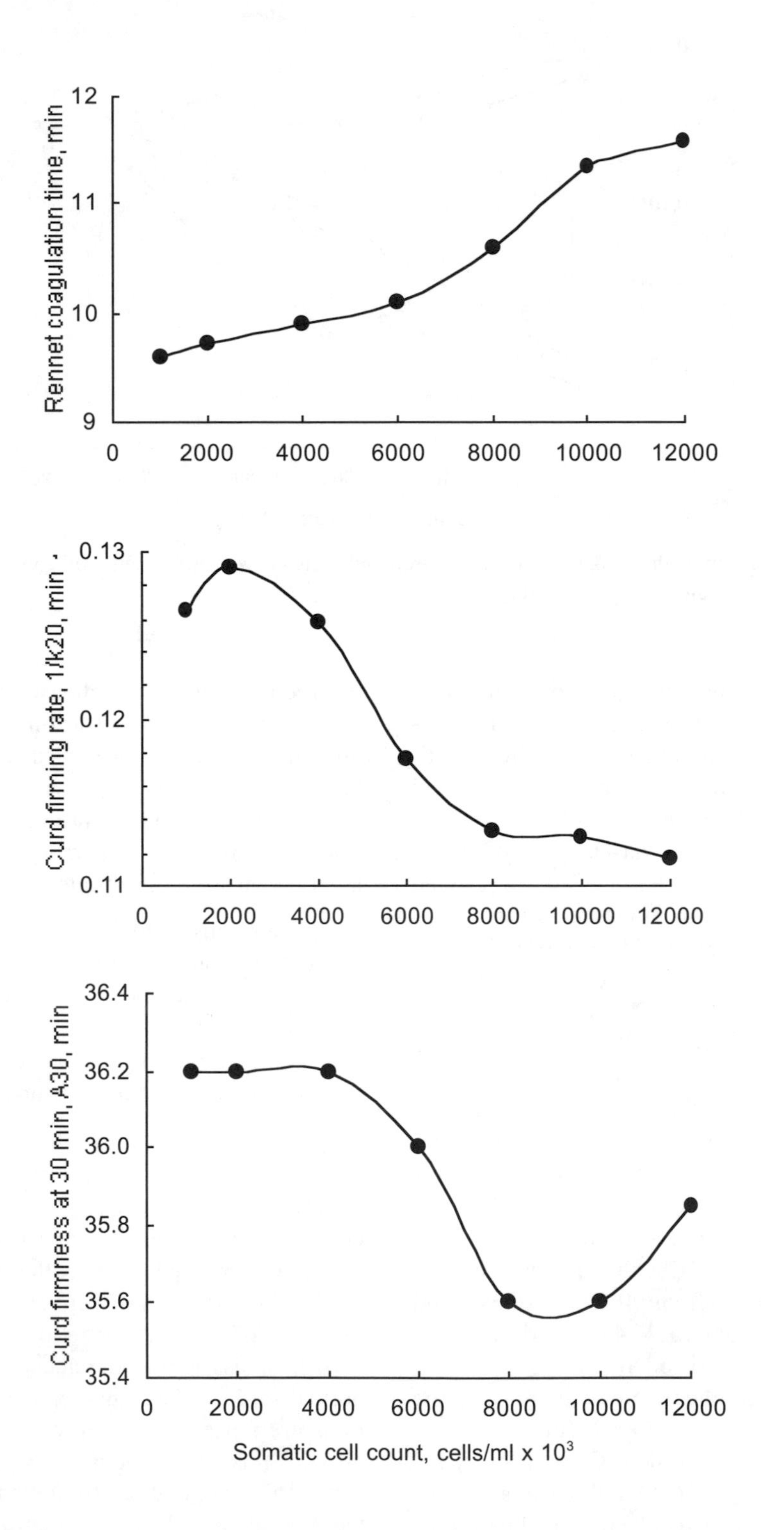

Figure 9–8 Effect of somatic cell count on the rennet coagulation properties of milks from individual cows.

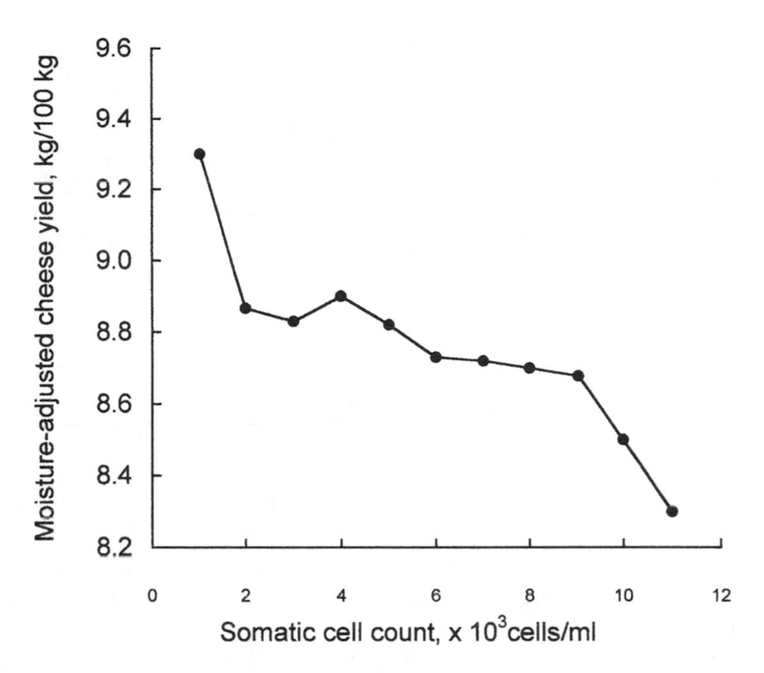

Figure 9–9 Influence of somatic cell count on the moisture-adjusted (to 370 g/kg) Cheddar cheese yield from milks from individual cows.

the effects of increased SCC may be accentuated, as the slower than normal curd-firming rate is conducive to lower than optimum firmness at cutting. Another factor that contributes to poor yield from high SCC milks is the increased susceptibility of the milk fat globule membrane to damage, which increases the susceptibility of the fat to lipolysis by indigenous milk lipoprotein lipase and lipases from psychrotrophic bacteria.

In conclusion, high SCC is detrimental to cheese yield and the profitability of cheesemaking. It is estimated that the monetary loss resulting from a 2% decrease in cheese yield due to increasing the SCC from 10^5 to 5×10^5 cells/ml is about $6,000 per day in a Cheddar cheese plant processing 1 million liters per day, based on a value of $3 per kilogram of curd. The SCC in milk is being lowered through the use of good on-farm practices, such as reducing the percentage of animals with subclinical mastitis, and is being driven by new regulations for lower SCC. For example, in the European Union, the permitted maximum of SCC in milk for the manufacture of dairy products bearing the health mark was reduced in 1998 from 5×10^5 cells/ml to 4×10^5 cells/ml (EU directive 92/46/EEC).

9.5.3 Stage of Lactation and Season

Marked changes occur in the composition of milk throughout the year, especially when milk is produced mainly from spring-calving herds fed predominantly on pasture, as in Ireland, New Zealand, and parts of Australia. These changes in milk composition are due principally to the physiologically induced changes in the biosynthetic performance of the mammary gland as influenced by stage of lactation and diet (see Chapter 3). Lactation-related changes in milk composition may be defined as those that occur during the period of milk production, between parturition and drying off, mainly as a result of physiological changes in the mammary gland of healthy cows fed on a standard good quality diet. On the other hand, seasonal changes in the milk from individual cows may be defined as those arising due to lactation and the superimposed effects of other factors, such as variation in diet, illness (e.g., mastitis), environmental factors, and climate. However, in practice, individual-cow milks are bulked at the farm level, and the ensuing herd milks are further bulked during the transport to and assimilation at the manufacturing plant.

Seasonal changes in milk composition, which are most pronounced at the extremes of lactation (Figure 9–10), result in variations in rennet coagulation properties, cheese composition, recovery of fat and casein, cheese yield, and quality. Seasonal studies in Ireland during the early 1980s showed that between March and November protein increased from around 30 to 42 g/kg, casein from 22 to around 30 g/kg, and proteose peptone from around 16 to 53 mg/100 g. The estimated yield of Cheddar cheese varied from about 8.5 to 11.5 kg/100 kg milk. Similar studies in Scotland reported smaller seasonal variations in milk composition and cheese yield: protein ranged from around 30.9 to 35.6 g/kg, casein from 24.1 to 27.4 g/kg, fat from 35.6 to 41.3 g/kg, and Cheddar cheese yield from 9.3 to 10.5 kg/100 kg. Yield studies in the United States have shown even smaller variations in the composition of commercial milk. For example, in New York, the range for casein was 23–25 g/kg; for milk fat, 34–38 g/kg; for casein number, around 76.5 to 81.7; and for actual Cheddar cheese yield, around 9.1 to 9.76 kg/100 kg. Marked seasonal variations in the actual yield of Emmental cheese have been observed in Germany (≈ about 8.94–9.45 kg/100 kg) and Finland (≈ about 10.8–12.0 kg/100 kg). Seasonal variations in milk composition are paralleled by variations in the recovery of fat and/or casein, reflecting variations in curd-firming rates and the susceptibility of the resulting curds to shattering during cutting and stirring (see Section 9.5.12).

In addition to changes in gross composition, the relative concentrations of the individual caseins as a percentage of total casein show seasonal changes, especially in factory milk obtained largely from spring-calving herds (Figure 9–10). These changes probably occur as a result of an increase in plasmin activity in milk over the course of lactation (Figure 9–11). β-Casein and α_{s2}-casein are readily cleaved by plasmin, whereas κ-casein and α_{s1}-casein are more resistant. The authors are not aware of any studies reporting the effects of variations in the propor-

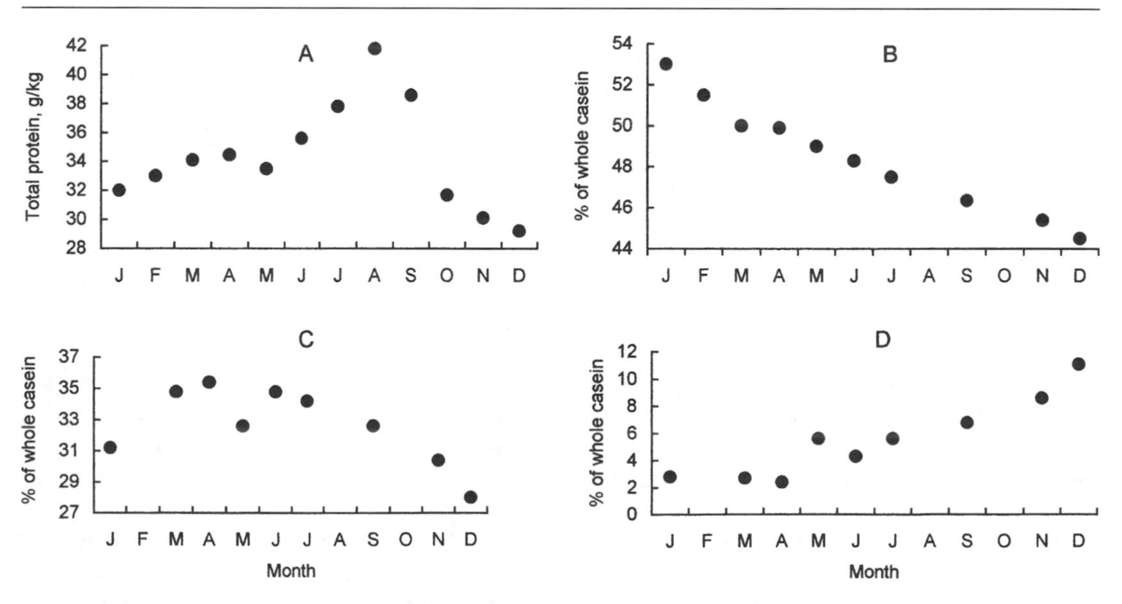

Figure 9–10 Seasonal changes in the level of total protein in winter-spring–calving herd milk (A) and in the proportions of α_s-casein (B), β-casein (C), and γ-casein (D) in Irish bulk creamery milk, mainly from winter-spring–calving herds (B–D).

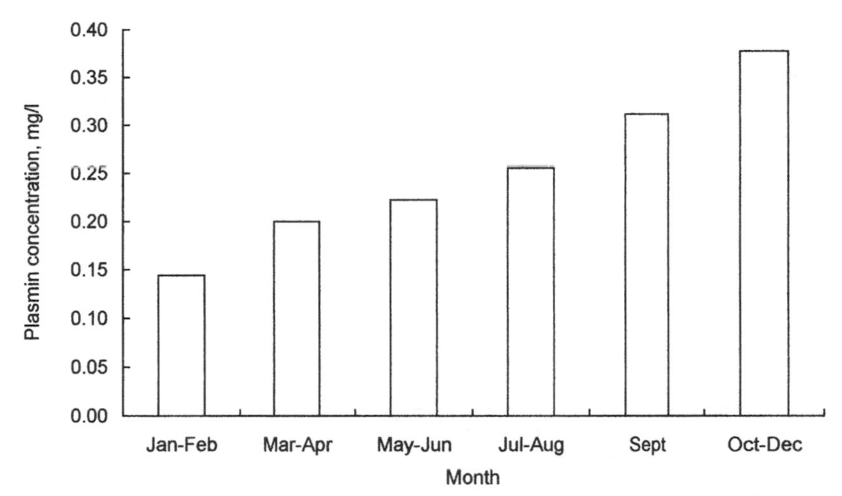

Figure 9–11 Effect of stage of lactation on the concentration of plasmin in milk from individual cows.

tions of different caseins on cheese yield. However, considering the different properties of caseins (see Chapter 3), it is likely that variations in the ratio of the caseins would influence rennet coagulation properties, recovery of components, and cheese composition and yield.

Numerous reports have shown that the rennet coagulation and curd-syneretic properties of late lactation milk are inferior to those of mid-lactation milk. Consequently, cheese made from late lactation milk tends to have a high moisture content and to be of inferior quality. All other factors being equal, the gel-firming rate is positively correlated with the milk casein level (see Chapter 6). Considering its high casein content, late lactation milk would be expected to have good cheesemaking properties and to give high yields. However, the reverse is observed, possibly owing to a high pH, changes in the proportions of the caseins (Figure 9–10), proteolysis by plasmin, and perhaps unidentified factors. Cheese made from late lactation milk tends to have a high moisture content, reflecting poor syneretic properties, which have not been fully explained. The high moisture content of cheese from late lactation milk contributes to a higher actual cheese yield.

Seasonal changes in milk composition can be minimized and its suitability for cheese manufacture increased through good husbandry and milk-handling practices, such as

- maintenance of a high-energy diet (e.g., dry matter intake of ≥ 17 kg/cow/day, with a dry matter digestibility of 820 g/kg organic matter) by good grassland management practices and concentrate supplementation of herbage-based diets, especially when grass is in short supply
- elimination of extreme late lactation milk by drying off cows at a suitable stage (e.g., at a milk yield ≥ 8 kg/cow/day)
- reducing stress and infection

9.5.4 Genetic Polymorphism of Milk Proteins

All the major proteins in milk (i.e., α_{s1}-, α_{s2}-, β-, and κ-caseins, β-lactoglobulin, and α-lactalbumin, exhibit genetic polymorphism (see Chapter 3).

The genetic variants that have been investigated most thoroughly for their effects on the rennet coagulation and cheesemaking characteristics of milk are those of κ-casein and β-lactoglobulin. Compared to the AA variants, the BB genotypes of β-lactoglobulin and κ-casein are generally associated with higher concentrations of casein and superior rennet coagulation properties, as reflected by higher gel-firming rates

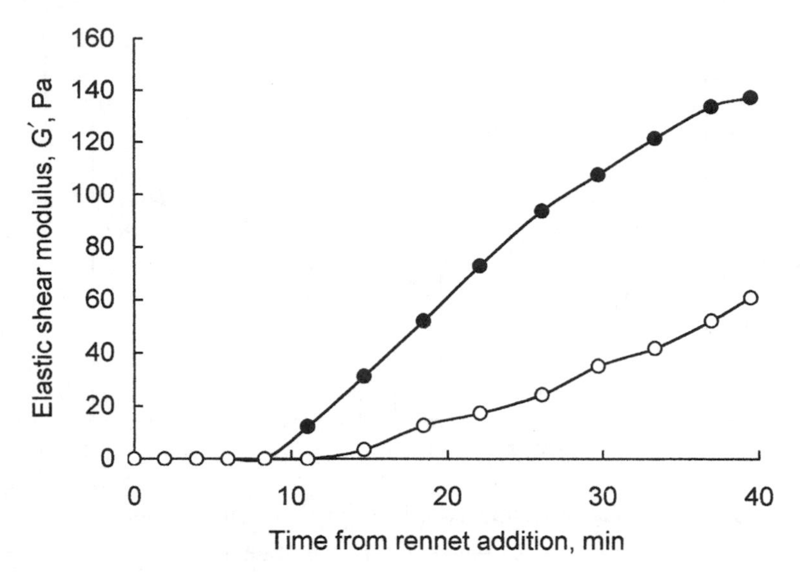

Figure 9–12 The development of elastic shear modulus (curd firmness) of rennet-treated milk for different κ-casein variants. κ-casein AA, 34.4 g protein/kg (○); κ-casein BB, 35.3 g protein/kg (●).

and gel firmness after a given renneting time (Figure 9–12). The BB variants of κ-casein and β-lactoglobulin are also associated with superior cheesemaking properties, as reflected by a higher recovery of fat, lower levels of curd fines in cheese whey, and higher actual and moisture-adjusted cheese yields for a range of varieties, including Cheddar, Sveciaost, Parmigiano-Reggiano, Edam, Gouda, and low-moisture Mozzarella. Reported increases in moisture-adjusted yield with the κ-casein BB variant range from about 3% to 8%, depending on milk composition and cheese type. The superior rennet coagulation and cheese-yielding characteristics of κ-casein BB, compared with the AA variant, probably result from the higher level of κ-casein as a percentage of total casein, its smaller micelles, and its lower negative charge. These properties are conducive to a higher degree of casein aggregation and a more compact arrangement of the sensitized paracasein micelles, which in turn favors more intermicellar bonds during gel formation. Model rennet coagulation studies have shown that, for a given casein concentration, the curd-firming rate of renneted micelles is inversely proportional to the cube of the micelle diameter. The generally higher level of casein accompanying κ-casein BB, compared with the AA variant, also contributes to its superior rennet coagulation and cheese-yielding properties. Milk containing κ-casein AB generally exhibits rennet coagulation and cheese-yielding characteristics intermediate between those of κ-casein AA and BB.

Compared with α_{s1}-CN BB, α_{s1}-CN BC casein is associated with higher levels of α_{s1}-casein, total casein, and total protein and higher estimated yields of Parmesan cheese. However, owing to the higher yields of milk, milk fat, and milk protein in milks containing α_{s1}-CN BB casein over a complete lactation, cows producing this variant give higher cheese yields during a lactation than those producing α_{s1}-CN BC casein.

In conclusion, the BB variants of κ-casein and β-lactoglobulin enhance cheese yield and have no adverse effects on cheese quality or on functionality in the case of low-moisture Mozzarella cheese (Walsh et al., 1998).

9.5.5 Cold Storage of Milk

In modern farm and milk collection, milk is cooled rapidly to below 8°C following milking, and milk collection from the farm often occurs every second or third day. Moreover, cold milk is hauled over long distances and is often cold-stored at the cheese plant for 1–3 days, depending on time of year and the manufacturing schedules. Consequently, milk can be cold-stored for 2–5 days prior to processing. During this period, the cold milk is subjected to varying degrees of shear from pumping, flow in pipelines, and agitation.

Cold storage and shearing result in a number of physicochemical changes that may alter the cheesemaking properties of milk, including

- solubilization of micellar caseins, especially β-casein, and colloidal calcium phosphate, leading to an increase in serum casein
- increased susceptibility of serum casein to hydrolysis by proteinases from psychrotrophic bacteria or somatic cells and/or plasmin and the concomitant increase in nonprotein N (Figure 9–13)
- damage to the milk fat globule membrane and hydrolysis of free fat by lipases from psychrotrophic bacteria and/or milk, resulting in a decrease in the level of fat (Figure 9–13)

Cold storage impairs rennet coagulation properties, reduces the recovery of protein and fat, and reduces cheese yield (Figure 9–13 and Table 9–3). However, there is disagreement between reported studies as to the magnitude of the effect of cold storage (cold aging). Discrepancies between reports may be attributed to variations in experimental conditions, such as prehandling and temperature history of milk prior to experimentation, milk pH, somatic cell count, bacterial count, and species/strains of psychrotrophic bacteria in milk, storage temperature and time, and cheesemaking conditions. It is generally agreed that, at a level of less than 10^6 cfu/ml, psychrotrophs have little effect on the cheesemaking properties of milk.

On storage at 4°C for 48 hr, the increase in the rennet coagulation time (RCT) of milk from individual cows ranged from 10% to 200% and that of bulk milk from 9% to 60% (Fox, 1969). A direct relationship was found between the increase in the RCT of cold-aged milk and the initial RCT before aging. The large variation in the RCT of cold-aged milk from individual cows is probably a consequence of differences in composition, microbiological status, and somatic cell count. The chemical changes (i.e., the increases in serum casein, in the ratio of soluble Ca to micellar Ca, and in RCT associated with cold storage) are almost complete after 24 hr and are largely reversed by pasteurization (72°C × 15 s) or milder

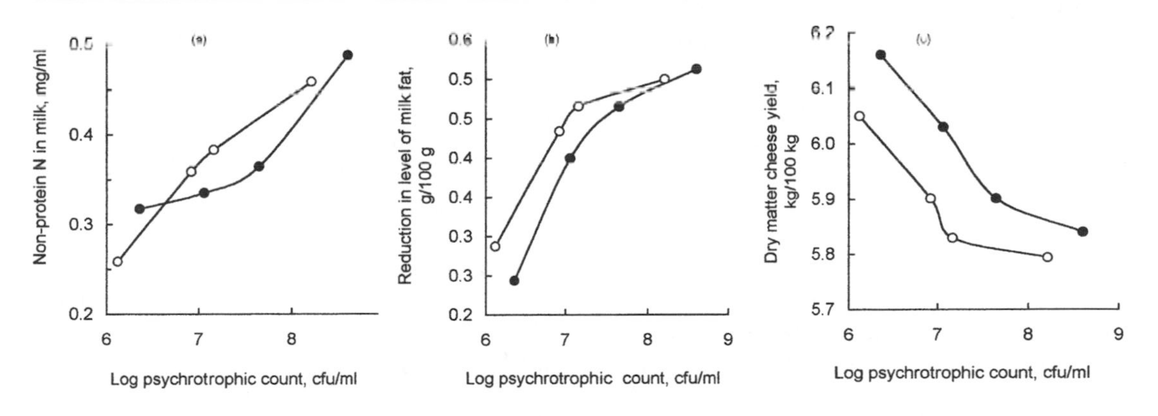

Figure 9–13 Effect of type and population of psychrotrophic bacteria on milk composition (a, b) and the dry matter yield of experimental cheese curd (c). Pasteurized milk was inoculated at various levels (10^0–10^6 cfu/ml) with *Bacillus* (○) or *Pseudomonas* (●) species isolated from raw milk (stored at 7°C for 3 days) and incubated for 6 days (*Pseudomonas*) or 10 days (*Bacillus*) before cheese manufacturing. Data are the means of eight trials with each microorganism.

Table 9–3 Effect of Cold-Storing Milk at 3°C on the Recovery of Milk Fat and Solids and on Actual Cheese Yield

	Recovery (%)		
Storage Time (Days)	*Fat (%)*	*Solids (%)*	*Yield (kg/100 kg)*
0	89.2	57.0	9.7
3	87.7	49.2	9.1
5	86.0	48.3	9.0

heat treatment (e.g., 50°C × 300 s).

The increase in RCT—and consequently the longer time required for curd formation during cheese manufacture—that generally accompanies cold storage beyond 24 hr appears to be due, at least in part, to enzymatic degradation of casein. Proteolysis reduces the concentration of gel-forming casein to an extent dependent on the proteolytic activity in the milk. Peptides, which are soluble in the serum phase (as nonprotein N), do not coagulate upon renneting and are largely lost in the whey. Moreover, the reduced casein level results in slow curd formation and a soft coagulum at cutting (see Chapter 6), a situation conducive to curd shattering, high losses of fat in the whey, and reduced cheese yields. In commercial practice, the coagulum is usually cut at a fixed time after rennet addition rather than at a given firmness.

In the European Union, the permitted maximum total bacteria count (TBC) in milk for the manufacture of dairy products was reduced from 4×10^5 cfu/ml to 1×10^5 cfu/ml in 1998 (EU directive 92/46/EEC). Improved dairy husbandry practices, combined with the more stringent standards for TBC and SCC, should reduce the level of storage-related proteolytis and lipolysis in milk. The fact that the chemical changes that occur during cold storage are reversed by pasteurization suggests that, with modern milk production practices, cold storage of milk for several days probably has little influence on its cheesemaking properties.

9.5.6 Thermization of Milk

Prolonged holding of milk prior to processing can occur in the factory when milk is in oversupply, such as during the spring flush, or on the farm when daily production is low, such as during winter. Extended cold storage may lead to the development of high psychrotrophic populations that produce proteolytic and lipolytic enzymes resistant to pasteurization. These enzymes can reduce cheese yield (as discussed in Section 9.5.5) and adversely affect product quality. Thermization of milk at a subpasteurization temperature (e.g., 57–68°C for 10–15 s) on reception at cheese factories is widely practiced. Thermization reduces the number of psychrotrophs in milk to an extent dependent on the heat treatment. Studies on Cottage cheese showed significantly higher actual and moisture-adjusted (to 820 g/kg) yields from milk heated at 74°C × 10 s prior to storage than from the unpreheated control (16.85 vs. 16.0 kg/100 kg), when milk was cold stored at 3°C for 7 days. Hence, it has been suggested that when milk is stored for a long period on farms, on-farm thermization may prove advantageous for cheese yield.

9.5.7 Pasteurization of Milk and Incorporation of Whey Proteins

Pasteurization of milk (72°C × 15 s) results in a low level of denaturation of whey proteins (≤ 5% of total), which complex with κ-casein and are retained in the cheese curd, where they contribute to a yield increase of about 0.1–0.4%. However, most of the native whey proteins (≈ 94–97%, depending on the cheese moisture level), which account for 20% of the total milk protein, are lost in the whey. Unlike casein, native whey proteins are stable to rennet treatment and acidification to pH 4.6 and thus remain soluble in whey during the manufacture of rennet- and acid-curd cheeses. Theoretically, if all whey proteins were retained without adversely affecting cheese moisture or quality, a yield increase of approximately 12% (10.7 vs. 9.54 kg/100 kg) would be achievable for Cheddar cheese with around 380 g/kg moisture. Indeed, since an increase in the severity of milk heat treatment is paralleled by an increase in cheese moisture (Figure 9–14), the increase in actual yield would be even higher (e.g., ≈15% following pasteurizing at 88°C × 15 s). However, inclusion of high levels of whey protein (> 35% of total whey protein), either in denatured or native form, can adversely affect rennet coagulation properties (see Chapter 6), cheese rheology (e.g., reduced elasticity and firmness), functionality (e.g., reduced flowability), and the overall quality of most rennet-curd cheeses (see IDF, 1991). Owing to an effect on cheese rheology, high levels of denatured whey proteins in cheese milk may be exploited as a means of improving the texture (reducing the firmness and elasticity) of low-fat cheeses, which tend to be too firm and rubbery (chewy), compared with their full-fat equivalents. Inclusion of high levels of whey proteins in some fresh acid-curd cheeses (e.g., Quarg, Cream cheese), while altering the textural properties, generally does not impair eating quality. However, even here functionality may be altered. In acid-heat coagulated cheese types (e.g., Ricotta, Paneer, some types of Queso blanco), the incorporation of a high level of denaturated whey protein is a feature of the manufacturing process and may impart certain desirable attributes, such as flow resistance and lack of elasticity. Whey proteins may be incorporated into cheese in a number of ways:

- Whey proteins may be denatured in situ by high heat treatment (HHT) of the cheese milk (e.g., ≈ 65% of total whey proteins at 100°C × 120 s).
- High concentration factor ultrafiltration (HCF-UF) can be used, with or without HHT of the milk before UF or HHT of the rententate after UF, as in the production of pre-cheese.
- Partially denatured whey protein concentrates, prepared by high heat treatment and/or acidification of cheese whey, can be added to the cheese curd. Commercial whey protein preparations of this type (e.g., Dairy Lo and Simplesse 100) have been used to improve the textural characteristics of reduced-fat cheese.

High Heat Treatment of Cheese Milk

In situ denaturation of whey proteins by HHT is widely used in the manufacture of fresh acid-curd cheeses, such as Quarg, Fromage frais, and Cream cheese, with typical heat treatments ranging from 72°C × 15 s to 95°C × 120–300 s. The extent of whey protein denaturation and the yield of Quarg (18% moisture) from milk subjected to these treatments are about 3% and 18.6 kg/100 kg and about 70% and 21.3 kg/100 kg, respectively. Increasing the level of whey protein denaturation from 0% to 50% by HHT of milk increases protein recovery and actual and moisture-adjusted cheese yield for a range of hard and semi-hard rennet curd cheeses (Table 9–4). The recovery of fat also increases as the level of whey protein denaturation increases (at least to ≈ 21% of total) provided that the cheesemaking process is altered to restore the gel-forming properties of the HHT milk to those of the unheated milk, for example, by extending the set-to-cut time when the pH of milk is normal, by adding $CaCl_2$, and/or by lowering the set pH by the addition of food grade acid or acidogen.

The effects of HHT on the rennet coagulation properties of milk are discussed in Chapter 6. These effects are attributable mainly to the increase in the concentration of the gel-forming protein and the reduction in the porosity of the more finely structured gel produced from the HHT milk (see Chapter 16), which increases the water- and fat-holding capacities of the gel. The complex between κ-casein and denatured β-lactoglobulin results in the formation of filamentous appendages that protrude from the surface of the casein micelles. These appendages render the κ-casein less susceptible to hydrolyis by rennet and probably limit the degree of aggregation of the paracasein micelles. The rate of aggregation is also impaired as a consequence of the reduction in the level of micellar calcium phosphate, which is the intramicellar "cementing agent." The more finely structured gel (compared to that from pasteurized milk) has impaired syneretic properties, and consequently the moisture content (Figure 9–14) and actual cheese yield are increased. However, rennet-induced gels from HHT milks with a level of whey protein denaturation (WPD) above 30% tend to be very fragile and prone to shattering upon cutting and stirring, even when the milk has been acidified to pH 6.0 prior to setting. Hence, high losses of fat (up to 15%) and curd fines (up to 800 mg/kg) in the whey are possible unless the gel is treated very gently. The susceptibility of curd particles from HHT milk to fracture persists for a longer time than normal into the stirring period, because of the lower tendency to synerese and the slower rate of firming of the curd particles.

Addition of Denatured Whey Protein to Cheese Milk

An alternative to in situ denaturation of whey protein in milk is the addition of denatured whey protein (recovered from whey) to the cheese milk. Potential advantages of this method are as follows:

- The curd-forming properties of the casein are not (or only slightly) impaired.
- The cheese yield is increased but the texture is not altered to the same extent as in curd from HHT milk with an equivalent level of denatured whey protein.

Potential disadvantages of adding denatured whey proteins include these:

- The addition of a milk protein fraction (i.e., whey protein) may not comply with the le-

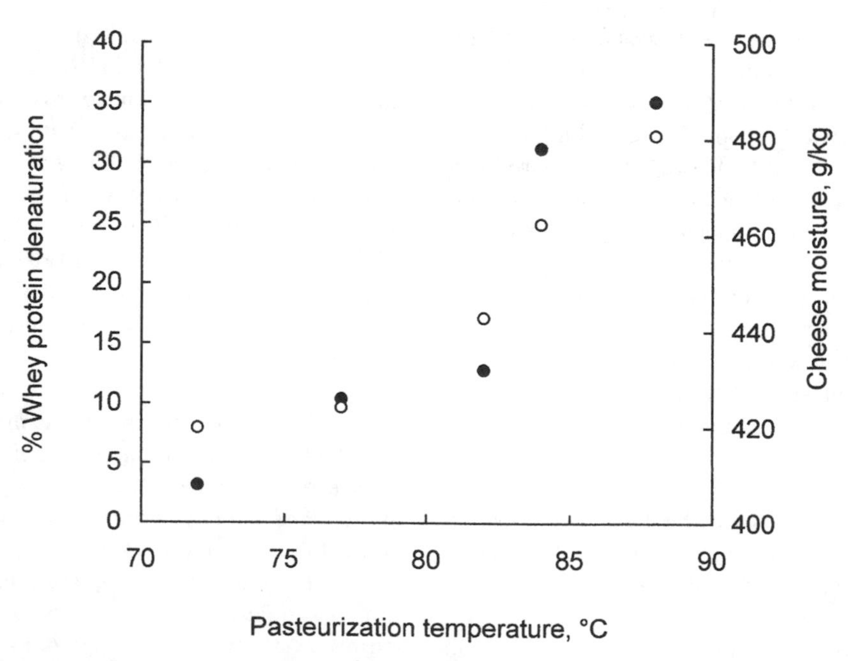

Figure 9–14 Effect of milk pasteurization temperature (for 15 s) on whey protein denaturation (●) and the moisture content of half-fat Cheddar cheese (○).

Table 9–4 Effect of Whey Protein Incorporation by in situ Denaturation at High Temperature on the Recovery of Components and Yield of Different Cheese Types

				Recovery				Percentage Increase in Yield Relative to Control[a]			
Heat Treatment (°C × s)	*WPD (%)*	*Cheese Type*	*Cheese Moisture (g/kg)*	*Fat (%)*	*Protein (%)*	*Dry Matter (%)*	*SNF (%)*	*Actual*	*Moisture Adjusted*	*Dry Matter*	*Reference*[b]
C: 0 x 0 (raw)	0		363	92.4	74.5	NP	NP	–	–	–	1
T: 63 x 1800	5	Cheddar	371	93.5	75.1	NP	NP	0.78	2.0	2.1	
C: 80 x 2	3		515	97.4	76.0	NP	32.5	–	–	–	
T1: 110 x 2	21		516	98.1	77.4	NP	33.2	1.6	NP	NP	
T2: 130 x 2	32	Mozzarella	496	96.3	79.5	NP	34.3	3.4	NP	NP	2
C: 72 x 17	3?		NP	NP	NP	NP	NP	–	–	–	
T: 97 x 15	32	Cheshire	NP	NP (0.7)[c]	NP (6.7)	NP (4.5)	–	NP	4.5	NP	3
C: 72 x 16	3?		371	94.3	77.8	NP	NP	–	–	–	
T: 110 x 60	50	Cheddar	387	95.8	86.2	NP	NP	11.9	4.7	NP	4[d]

Key: SNF = solids-not-fat; WPD = whey protein denaturation; C = control; T = treatment; NP = data not presented; ? = data not presented but value estimated.

[a] The percentage increase in cheese yield was calculated by expressing the difference between the control and the treatment as a percentage of the control value.

[b] References: 1. K.Y. Lau, D.M. Barbano, & R.D. Rasmussen, Influence of pasteurization on fat and nitrogen recoveries and Cheddar cheese yield, *Journal of Dairy Science, 73* (1990), 561–570. 2. H.W. Shafer & N.F. Olson, Characteristics of Mozzarella cheese made by direct acidification from ultra-high-temperature processed milk, *Journal of Dairy Science, 58* (1974), 494–501. 3. R.J. Marshall, Increasing cheese yield by high heat treatment of milk, *Journal of Dairy Research, 53* (1986), 313–322. 4. J.M. Banks, G. Stewart, D.D. Muir, & I.G. West, Increasing the yield of Cheddar cheese by the acidification of milk containing heat denatured whey protein, *Milchwissenschaft, 42* (1987), 212–215.

[c] Values in parentheses for recoveries denote presented data for the percentage increases in fat, protein, and dry matter.

[d] In this experiment, the milk was heated at 110°C x 60 s, cooled to 30°C, acidified to pH 5.8, set and cut after a 15 min set-to-cut time. Cutting the gel from the treated milk after a time similar to that for the control (45 min) resulted in a reduced fat recovery (8.4%).

gal standard of identity of the cheese in all countries.
- It may be too expensive.

Several processes have been developed for the recovery of protein from sweet whey, most of which involve flocculation, by heat denaturation and/or acification, and recovery by centrifugation. The Centriwhey process yields concentrates containing around 120 g solids/L and the Lactal process yields concentrates containing around 180 solids g/L. These concentrates are added at the desired level to the cheese milk, which is then treated as in normal cheese manufacture. Commercial partially denatured whey protein may also be prepared by concentrating sweet whey by ultrafiltration and heat-treating and shearing the retentate to give a controlled level of denaturation. The heat-treated retentate may be supplied as a viscous, wet slurry or spray dried.

Variable results have been obtained with the addition of partially denatured whey protein concentrate (PDWPC, prepared by the Centriwhey, Lactal, or ultrafiltration processes) to milk for the manufacture of hard and semi-hard cheeses, such as Cheddar and Gouda. There is general agreement that the addition of PDWPC increases the moisture content, actual yield, and moisture-adjusted yield, with the extent of the increases depending on the amount of PDWPC added and the degree of whey protein denaturation of the concentrate. However, the addition of PDWPC generally leads to defective body characteristics (i.e., greasy, soft) and flavor characteristics (i.e., unclean, astringent) in Gouda and Cheddar cheese.

Commercial PDWPCs (e.g., Simplesse 100 or Dairy Lo) have been used as fat mimetics to improve the texture of reduced-fat cheeses, which tend to be more firm and elastic than their full-fat equivalents. These materials are added at levels of around 10–20 g/kg to increase the level of milk protein by about 3.5 g/kg. Their use in half-fat Cheddar cheese gives a higher level of cheese moisture (e.g., 450 vs. 427 g/kg) and higher actual cheese yields (e.g., 7.7 vs. 7.2 kg/100 kg) and moisture-adjusted cheese yields (Fenelon & Guinee, 1997; IDF, 1994). These materials also improve the texture of reduced-fat Cheddar (by reducing fracture stress, fracture strain, and firmness) and have little influence on flavor or aroma.

High Concentration Factor Ultrafiltration

High concentration factor ultrafiltration (HCF-UF) has been widely used as an alternative to centrifugation, especially in Europe since the 1980s, for concentrating the gelled milk in the commercial production of fresh acid-curd cheeses, such as Quarg and Cream cheese. This method, which concentrates the milk sixfold or more, allows complete recovery of whey proteins and gives higher yields of cheese of very acceptable quality. The use of HCF-UF instead of centrifugation for the manufacture of fresh acid-curd cheeses has necessitated slight changes in the manufacturing procedure to give comparable sensory quality (see Chapter 16). Functionality is sometimes a quality attribute in Cream cheese (e.g., desired degree of flowability in cordon bleu poultry). While no published information is available on this aspect of Cream cheese, it is envisaged that Cream cheese produced from HCF-UF retentate would be resistant to flow upon cooking owing to the thermal gelation of the whey proteins at the high cook temperature (e.g., 90–100°C). In contrast, the production of Cream cheese with customized flowability is easily achieved by manipulating the manufacturing procedure when using centrifugation to concentrate the curd.

HCF-UF has also been used for the commercial manufacture of rennet-curd cheeses, including cast Feta in Denmark and Cheddar in Australia made using the Siro-Curd process. Manufacture involves HCF-UF of the milk to produce a retentate or liquid pre-cheese with a high dry matter content (e.g., 400–500 g/kg), which has a composition close to that of the finished cheese. Rennet and starter cultures are added to the pre-cheese, which is then treated in the normal manner, except that little (e.g., Cheddar) or no (cast Feta) expulsion of whey occurs. Since the upper limit for concentration by UF is about 7:1 for whole milk, it is not possible to achieve the dry matter level required for hard cheeses, such as Cheddar or Gouda. Hence, further whey expulsion from the pre-cheese occurs after coagulation and cutting on conventional equipment (e.g., drainage belts, finishing vats) or on specialized equipment (e.g., Siro-Curd).

The attraction of HCF-UF for the manufacture of rennet-curd cheese is the increased yield owing to the high level of retention of whey proteins and glycomacropeptides. The exact degree of retention depends on

- the heat treatment of the milk prior to renneting, which determines the extent of whey protein denaturation and hence whey protein solubility
- the level of whey expulsion from the pre-cheese following coagulation and cutting (native whey proteins and glycomacropeptides are soluble and are lost in the whey)

When complete recovery of the native whey proteins and glycomacropeptides is achievable (as in cast Feta), the estimated saving in skim milk is about 9%, which makes the process economically viable (see IDF, 1994).

Extensive research has been undertaken on the characteristics of different hard cheeses produced at pilot or laboratory scale from HCF-UF retentate. The sensory characteristics of HCF-UF cheeses tend to differ from those of traditional cheese to a degree dependent on the level of whey protein included, the extent of whey protein denaturation, and other manufacturing conditions that influence the composition and rate of proteolysis and/or lipolysis during maturation.

9.5.8 Homogenization and Microfluidization

Homogenization of milk reduces fat globule size and increases the surface area of the fat by a factor of 5 to 6. The fat globules become coated with a protein layer consisting of casein micelles, submicelles, and to a lesser extent whey proteins. Hence, the newly formed fat globules behave as pseudocasein particles and exhibit the ability to become part of the gel network formed upon renneting or acidification. Homogenization of milk or cream is not widely practiced in the manufacture of rennet-curd cheeses, as it tends to produce cheese with a high moisture content and altered texture (e.g., lower elasticity and firmness), altered flavor (e.g., hydrolytic rancidity), and altered functionality (e.g., reduced flow) to a degree dependent on milk composition and homogenization temperature and pressure. The reduction in firmness has been exploited as a means of improving the texture of reduced-fat cheeses, which tend to be excessively firm and elastic as a consequence of their relatively high protein:fat ratio. Cheese produced from homogenized milk is whiter than that from unhomogenized milk, an attribute that may be desirable (e.g., in the case of Blue cheese or Mozzarella) or undesirable (e.g., in the case of Swiss-type cheese). The main applications of homogenization in cheese manufacture are these:

- Cheeses made from recombined milk are formed by homogenizing oils (butter oil and/or vegetable oils) in aqueous dispersions of milk protein (e.g., reconstituted or reformed skim milks). This method is used in countries where the demand for cheesemaking exceeds the local supply of fresh milk.
- In the case of fresh acid-curd cheeses (e.g., Cream cheese), especially when the fat content of the milk is high (e.g., 10%, w/w), homogenization contributes to product homogeneity, as it retards creaming of the fat globules during the relatively long gelation period (≈ 12 hr) and improves product texture by increasing the level of effective protein, as the casein-coated fat globules become part of the gel rather than being occluded within the gel, as with native fat globules in milk.
- In Blue cheese manufacture, the casein-based fat globule membrane allows access for lipases from the mold to the fat and thereby enhances the formation of free fatty acids, the main substrates for the production of methyl ketones, which are very important for flavor.

The effects of homogenization on composition, quality, and yield have been investigated for many rennet-curd cheeses (Jana & Upadhyay, 1992). Homogenization of milk or cream increases the yield of practically all varieties, the effect being more pronounced with

- increasing homogenization pressure (in the range 5–25 MPa)
- homogenization of standardized cheese milk rather than homogenization of the cream used for standardization

The reported yield increase, which ranges from 2.8% to 6.3% for Cheddar, is attributed to the increased moisture content (e.g., 20–30 g/kg for Cheddar, depending on homogenization and cheesemaking conditions) and the lower losses of fat in whey (e.g., 2–6% milk fat) and/or in stretch water in the case of pasta-filata cheeses. However, the efficiency of recovery of fat from the whey from homogenized milk is only about 30–40% of that from nonhomogenized milk using centrifugal separators.

Microfluidization is a relatively new technology that has been applied in the health care industry since the late 1980s for the reduction of mean particle size in the manufacture of products such as antibiotic dispersions, parenteral emulsions, and diagnostics. Microfluidization differs from homogenization in the types of forces applied to the fluid and also in the size distribution and mean diameter of the resulting particles.

It is generally accepted that, given the application of equivalent pressures to the milk, microfluidization results in a lower mean fat globule diameter (e.g., 0.03–0.3 μm vs. 0.5–1.0 μm) and a narrower size distribution than homogenization. Moreover, the fat globule membrane in microfluidized milk has a high proportion of fragmented casein micelles and little or no whey protein compared with that in homogenized milk. Microfluidization of milk (at 7 MPa) or cream (at 14 or 69 MPa) results in a higher moisture content in Cheddar, higher retention of milk fat, and higher actual and moisture-adjusted yields (Table 9–5).

9.5.9 Type of Starter Culture and Growth Medium

Casein is a major contributor to cheese yield, directly and indirectly, as it forms the matrix that occludes the fat and moisture; the latter contains dissolved substances, including native whey proteins, nonprotein nitrogen (NPN), lactate, and soluble salts. All other conditions being equal, the recovery of casein from cheese milk is theoretically higher when cheese is made by direct acidification through the addition of acid and/or acidogen (e.g., gluconic acid-δ-lactone) rather than by a starter culture. Starter cultures hydrolyze casein to varying degrees, depending on their proteolytic activity, during preparation of the bulk culture and during curd manufacture to small peptides that are ingested by the cells for growth (see Chapter 5). Model studies, using skim milk acidified by starter culture or direct acidification, have shown that starter cultures produce significantly higher losses of casein (0.7–6.6%) than direct acidification, depending on the proteolytic activity of the culture. Yet starter cultures are generally used in preference to direct acidification as a means of acidification, principally because of their contribution to cheese flavor development (see Chapters 11 and 12). In addition, the starter culture ferments residual lactose in cheese curd to lactic acid, thereby removing it as a carbon source for the growth of nonstarter LAB (NSLAB). However, direct acidification is sometimes used as the principal or sole means of acidification in the manufacture of some cheeses, for example, where flavor resulting from starter activity is not a major quality attribute (e.g., low-moisture Mozzarella, Cream cheese, Queso blanco, Ricotta, and Paneer) or where flavor can be provided by alternative means, such as flavored dressing (e.g., creamed Cottage cheese).

Proteinase-negative single-strain starters generally give higher dry matter yields of Cheddar cheese than the corresponding proteinase-positive starters, with the yield increase ranging from 1.4% to 2.4%, depending on the starter strain. Similarly, in model acidified skim milk systems, the recovery of casein with proteinase-

Table 9–5 Effects of Microfluidization on the Yield of Cheddar Cheese

	Milk		*Cream*	
	0 MPa	*7 MPa*	*14 MPa*	*69 MPa*
Cheese composition				
Moisture (g/kg)	350	376	384	393
Fat (g/kg)	344	340	334	330
Bulk whey composition				
Fat (g/kg)	5.2	2.3	1.7	1.3
Curd fines (g/kg)	0.9	1.0	1.1	1.2
Cheese yield				
Actual (kg/100 kg)	9.4	10.2	10.4	10.6
Moisture-adjusted (kg/100 kg)	9.7	10.1	10.2	10.2

negative starter is significantly higher (3.1%) than with the corresponding proteinase-positive strain and only marginally lower (–0.69%) than that obtained using direct acidification. Proteinase-negative strains, which rely on the indigenous amino acids and small peptides in the milk for growth, reproduce very slowly and therefore reduce the pH too slowly for cheese manufacture. Moreover, their use may lead to slow proteolysis and flavor development during maturation. Hence, proteinase-negative strains are generally not used alone but rather in blends with proteinase-positive strains. Such blends are commonly used as cheese cultures.

There is disagreement in the literature on the relative effects of direct-vat starter versus bulk starter on the recovery of milk components and cheese yield. Discrepancies between results may arise from several sources, including

- the method of calculating yield (i.e., whether starter solids are included in the yield calculation)
- the starter medium used in the preparation of the bulk starter (i.e., whether skim milk based or whey based)
- the temperature-time treatment of the bulk starter medium prior to inoculation
- milk composition, pH at set, firmness at cutting, and proteolytic activity of the starter strains in direct-vat and bulk starters
- cheese moisture

Differences in these factors can influence the level of whey protein denaturation, level of recoverable solids, degree of casein hydrolysis during manufacture, and susceptibility of the coagulum to shattering at cutting. Approximately 40% of the bulk starter solids (casein plus denatured whey proteins) are retained in Cheddar cheese when the starter medium is 10% reconstituted low-heat skim milk powder heated at 85–90°C for 1 hr (Banks, Tamime, & Muir, 1985). Some authors have reported increases in actual yield (e.g., ≈ 2%) and recovery of milk solids (e.g., 49.2% vs. 48.5%) in Cheddar cheese when direct-vat starter rather than bulk starter was used (Salji & Kroger, 1981). In contrast, a Scottish study showed that the use of bulk starter grown in 10% reconstituted skim milk powder resulted in higher actual (1.1%) and moisture-adjusted (1.05%) cheese yields and higher retention of milk solids (i.e., 53.1% vs. 52.6%) compared to DVS (Banks et al., 1985). Moreover, further studies by the same group (Banks & Muir, 1985) showed that, compared with a commercial casein-free starter medium, the use of reconstituted skim milk powder as starter medium resulted in higher actual yield (1.8%) and recovery of milk solids (1.7%). The greater recovery of solids was attributed to the retention of casein and denatured whey protein from the starter, which are coagulable following acidification to pH 4.6 in the bulk starter and reneutralization to around pH 6.5, when the starter is added to the milk. Obviously, further studies in which factors such as starter strain(s), milk composition, set pH, and gel firmness at cutting are standarized are required to clarify the comparative effects of bulk and direct-vat starters on Cheddar cheese yield.

Significant increases in Cottage cheese yield have been reported when whey-based media with external pH control (i.e., pH constantly maintained at the initial value, such as 6.6, by addition of base) were used instead of conventional skim milk–based media without pH control. In the authors' experience, it is important that when a direct-vat starter is used, the pH at set should be standardized to that normally obtained with a bulk starter (e.g., to 6.55) by using $CaCl_2$ or allowing sufficient ripening time, especially when the level of casein in milk is low. Otherwise, gel firmness after a specified set-to-cut time may be low, resulting in shattering of the curds and reduction in cheese yield due to high losses of fat and curd fines. The addition of direct-vat starter has little or no immediate effect on milk pH, whereas the addition of bulk starter gives an immediate decrease in pH of around 0.1. Moreover, during the ripening or vat filling, the decrease in pH is greater when bulk starter is used.

9.5.10 Addition of $CaCl_2$

The addition of $CaCl_2$ at a level of about 0.02 g/L (i.e., ≈ 2 mM Ca) to milk is common commercial practice, especially when using late lactation milk. Addition of $CaCl_2$ generally improves the rennet coagulation properties, an effect attributable to the reduction in pH and the increase in the concentration of Ca^{2+} (see Chapter 6). While the effect of $CaCl_2$ addition on rennet coagulation properties has been studied extensively, comparatively few studies have considered its effect on cheese yield. An investigation on the commercial manufacture of Swiss-type cheese showed that the addition of $CaCl_2$ (0.1 g/L) produced insignificant increases in the mean recovery of milk fat (85.3% vs. 84.7%) and nonfat milk solids (33.85% vs. 33.75%) and a significant increase in the mean cheese yield (0.038 kg/100 kg) (Wolfschoon-Pombo, 1997). The proportion of large curd particles (5.5–7.5 mm) was increased, and the proportion of small particles (< 3.5 mm) was reduced. These trends suggest that the positive effects of $CaCl_2$ on the recovery of fat and protein and cheese yield probably ensue from an enhanced degree of casein aggregation, which reduces the susceptibility of the curd to fracturing during cutting and the initial phase of stirring (see Section 9.5.12).

9.5.11 Rennet Type

Ideally, rennets should hydrolyze only the Phe_{105}-Met_{106} bond of κ-casein during milk coagulation, with further cleavage of caseins occurring only after complete removal of whey. In this situation, the recovery of casein is maximized and cheese yield increased. The various rennets used in cheesemaking differ in their milk clotting:proteolytic activity ratio and thus hydrolyze casein to a greater or lesser degree during cheese manufacture, depending on the length of time the curd is in contact with the whey and the curd pH at whey drainage. Some breakdown products of casein are soluble in whey and are removed and lost in the whey at whey drainage. Calf chymosin is the least proteolytic of the gastric proteinases, the proteolytic activity of which decreases in the following general order: chicken pepsin > porcine pepsin > ovine pepsin > bovine pepsin > calf rennet (chymosin) ≈ fermentation-produced chymosin. Microbial rennets are also more proteolytic than calf chymosin, with proteolytic activity being in the following order: *Cryphonectria parasitica* proteinase >> *Rhizomucor miehei* > *R. pusillus* > calf chymosin.

However, the relative proteolytic activity of different rennets is not always reflected in cheese yield. Indeed, many discrepancies exist between reported results on the effects of coagulant on cheese yield. In a laboratory-scale Canadian cheesemaking study (Emmons & Beckett, 1990), the increase in nonprotein N level (expressed as whey protein) in Cheddar cheese whey, compared with calf rennet, ranged from 0.006% for bovine pepsin to 0.19% (w/w) for *Bacillus polymyxa* proteinase (Figure 9–15).

The corresponding estimated reduction in moisture-adjusted cheese yield ranged from 0.16% to 4.5%, respectively. In a pilot-scale study involving 9 to 12 replicate trials, Barbano and Rasmussen (1994) found that the losses of fat and protein in whey obtained with calf rennet or recombinant chymosin were similar and lower than those with *R. miehei*, *R. pusillus*, or bovine rennet (78% bovine pepsin, 22% chymosin). The moisture-adjusted Cheddar yield was highest for calf rennet and recombinant chymosin and lowest for *R. pusillus* (Figure 9–16). A similar study by Ustanol and Hicks (1990) showed that the above coagulants had no significant effect on N or fat levels in whey or on dry matter cheese yield. However, compared with the other rennets, *C. parasitica* proteinase produced significantly higher levels of N and fat in the whey and a lower dry matter cheese yield. The decrease in dry matter yield of Cheddar cheese that resulted from using a *C. parasitica* coagulant was eliminated upon the addition of $CaCl_2$ at a level of 0.02%. The discrepancy between studies may be attributed in part to differences in curd pH at whey drainage (e.g., 5.85 vs. 6.1). At the lower pH, all coagulants are more proteolytic, especially those with a low ratio of milk clotting to proteolytic activity, and therefore produce more soluble peptides, which are lost in the whey. Hence, cheese yield decreases as the pH at whey drainage decreases.

Generally, no significant differences have been reported between the commercially available recombinant chymosins and calf rennet in relation to fat and N levels in the whey and moisture-adjusted cheese yield.

In conclusion, the extent of casein hydrolysis during the manufacture of cheese curd is lowest with calf rennet and recombinant chymosins, intermediate with bovine pepsin and *Rhizomucor* rennets, and highest with *C. parasitica* and *Bacillus polymyxa* proteinases. Whether these differences in proteolytic activity impact significantly on cheese yield probably depends largely on the pH at whey drainage. Rennets with a high level of proteolytic activity (compared with calf rennet) probably have little effect on yield when the pH at whey drainage is high (e.g., ≥ 6.15), as in the case of Cheddar, Gouda, and Emmental, but reduce yield when the pH at drainage is below 6.0, as in the case of Blue cheese and Camembert. The thermostability of the different rennets at the cook temperature for a given variety probably also determines how differences in proteolytic activity impact yield.

9.5.12 Firmness at Cutting

Cutting the gel is a central part of cheese manufacture, being the first step in the dehydration process by which the colloidal constituents of milk (fat, casein, and micellar salts) are concentrated to form cheese curd. The effect of firmness at cutting on cheese yield has been a subject of extensive research in recent years owing to the increasing competitiveness of cheese manufacture.

During gel formation, firmness increases progressively from the onset of gelation as a consequence of ongoing aggregation of paracasein micelles (see Chapter 6). Eventually, the gel reaches a firmness that allows it to withstand mechanical cutting by the knives in the cheese vat without shattering. Traditionally, in commercial cheese manufacture, and still in most reported experimental studies, the curd particles are allowed to sit quiescently in the whey after cutting. During this period, referred to as healing, syneresis proceeds rapidly and the curd particles heal, that is, become firmer and develop a surface film, which is essentially an outer layer with a higher casein:fat ratio than the interior. The combined effects of the film and the cushioning effect of the expressed whey limit the damage inflicted on the curd particles by impact with the agitators and vat surfaces and by the velocity gradients during the initial phases of stirring. Hence, healing reduces the tendency of curd particles to shatter (i.e., fracture along their weakest points into smaller particles with jagged edges). The surface film becomes progressively stronger as a consequence of the dehydrating effects of heat, acid, and stirring (which creates pressure gradients over the surface, forcing new aggregation sites in the interior of the curd particle), and it seals the fat and casein within the curd particles. The skin develops into curd gran-

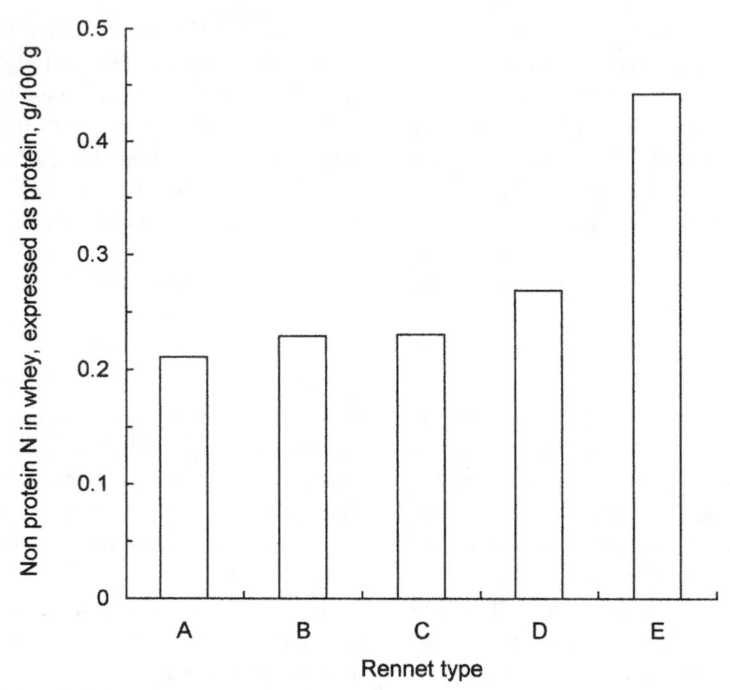

Figure 9–15 Effect of different rennets on the level of nonprotein nitrogen, expressed as percentage protein, in bulk Cheddar cheese wheys. Calf rennet (A), *R. miehei* (B), *R. pusillus* (C), *C. parasitica* (D), and *B. polymyxa* (E).

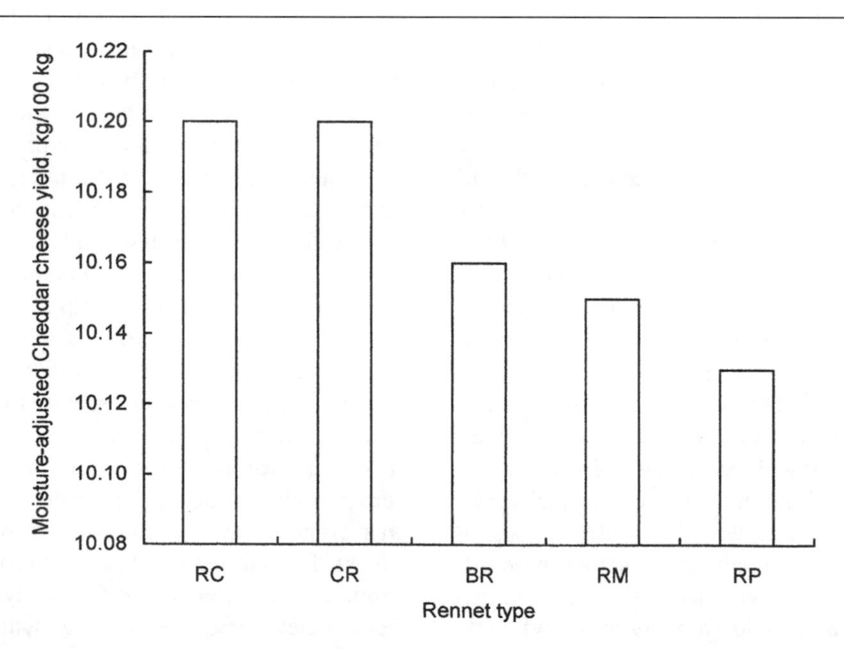

Figure 9–16 Effect of coagulant type on moisture-adjusted Cheddar cheese yield. Coagulants: RC, recombinant chymosin (Chymax); CR, calf rennet (94% chymosin); BR, bovine rennet (78% bovine pepsin, 22% chymosin); RM, *Rhizomucor miehei* (Morcurd plus); RP, *Rhizomucor pusillus* (EMPORASE sf 100).

ule junctions in the molded cheese curd, which are readily recognizable upon microstructural analysis of the cheese.

In large modern factories, the curds are not given a defined period for healing. Instead, the curds may heal during the cutting program, to a greater or lesser degree depending on the program, which determines the number and duration of alternate cutting and rest cycles. In practice, cutting cycles for Cheddar may range from around 10 to 20 min, depending on knife speed during successive cutting cycles.

Shattering is undesirable, as it results in an increase in the surface area through which fat globules can escape from the surface of the curd particle, along with the outflow of whey immediately after cutting and during the early stages of stirring. Moreover, curd shattering results in the formation of curd fines (curd particles less than 1 mm), which may be lost from the curd mass to a greater or lesser degree depending on downstream curd-handling equipment (ex-vat).

Consideration of the microstructure of the renneted milk gel and its interaction with the cheese knives suggests that cutting when the gel is too soft or too firm increases the propensity to shattering. When the gel is too soft, the gel structure is insufficiently developed and fractures even under the small strains applied upon gentle cutting. In an overfirm gel, the degree of casein aggregation is relatively high, and the gel tends to be brittle (low fracture strain) and susceptible to breakage. Observations at the practical level largely substantiate this analysis. In large modern cheese plants, the gel is generally cut after a specified set-to-cut time (e.g., 40 min) to conform to factory schedules. However, many factors that affect gel firmness do not remain constant throughout the cheesemaking season. Hence, the firmness at cutting can vary, resulting in variations in cheese yield. Several factors influence the firmness of the rennet-induced milk gel after specified times (see Chapter 6), including milk composition, stage of lactation, somatic cell count, milk heat treatment, culture type, and pH. Hence, much attention has been focused on quantifying the effect of cutting the gel after different set times or at different firmness (when curd firmness sensors were available) on cheesemaking efficiency.

Increasing the set-to-cut time from 80% to 200% of the control optimum value (a value determined subjectively by the cheesemaker) while maintaining a constant heal time results in higher levels of cheese moisture and moisture-adjusted Cheddar cheese yield and greater retention of milk fat and solids (Figure 9–17). These trends have been attributed to the more complete development of the gel structure at cutting, which enhances the retention of milk solids. However, there is an interactive effect between firmness at cutting and heal time (Figure 9–18). The positive effect of healing was small at set-to-cut times greater than 85% of the control but became increasingly greater at set-to-cut times less than 85% of the control. As the healing time is reduced, the yield-enhancing effects of increasing firmness at cutting become markedly greater, especially when the curd is underset at cutting, that is, at set-to-cut times less than 95% of the control (Figure 9–18). Factors that contribute to undersetting for a given set-to-cut time include increases in SCC and milk pH and a decrease in milk protein level. Bynum and Olson (1982) reported that the effects of increasing set-to-cut time on fat recovery and Cheddar cheese yield depended on vat size. Increasing curd firmness had no effect on fat recovery or moisture-adjusted Cheddar yield when small experimental vats (460 L) were used and had a positive effect (i.e., resulted in higher yield and higher recovery of fat and casein) when larger vats (2,400 L) were used. The observed differences between large and small vats may be attributed to differences in stirrer design and to how rapidly clumps of curd particles are broken up during the initial phase of stirring. In large vats, clumps of curd particles tend to disintegrate more slowly, which has the effect of increasing healing time. Other studies (Banks & Muir, 1984) have shown that varying set-to-cut times (to produce underset and overset gels at cutting) had no significant effect on moisture-adjusted Cheddar cheese yield.

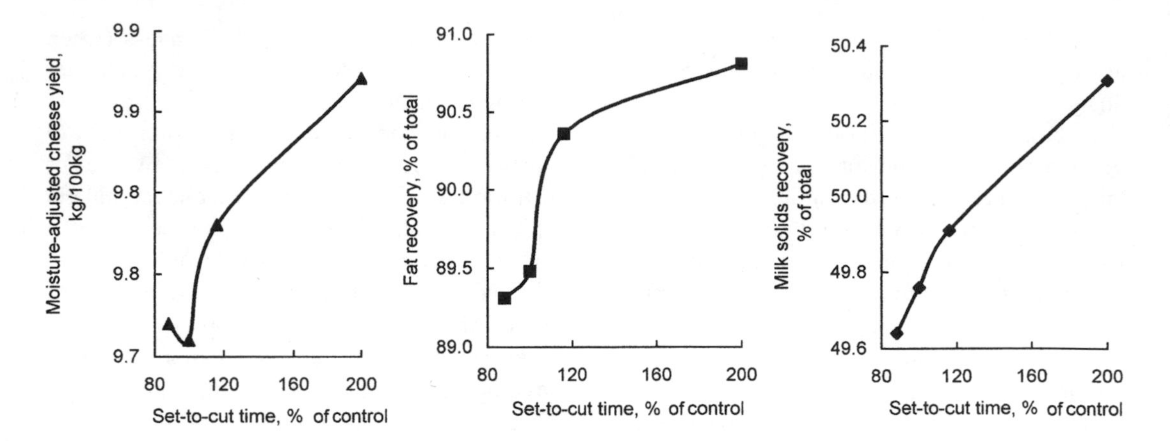

Figure 9–17 Effect of firmness at cutting (varied by changing the set-to-cut time) on the efficiency of Cheddar cheese manufacture.

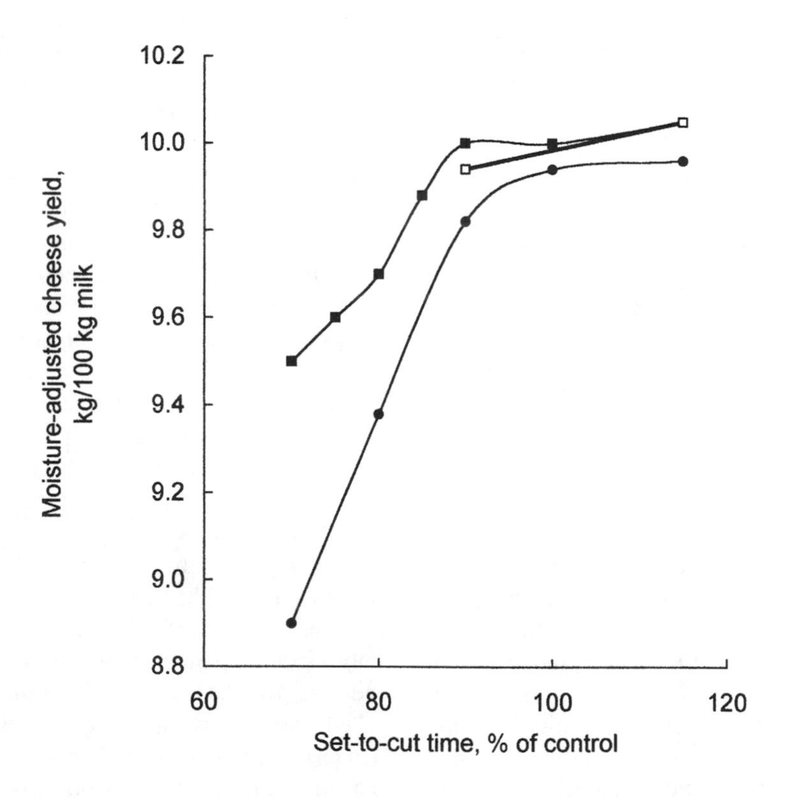

Figure 9–18 Effect of set-to-cut time and healing time on moisture-adjusted Cheddar cheese yield. Healing time was 0 min (●), 5 min (■), and 10 min (❑) min.

9.5.13 Particle Size

The curd particle size distribution (CPS_d), during the initial phase of stirring affects yield efficiency, as it determines the surface area through which fat escapes into the whey. However, it is not possible to measure the CPS_d at this stage of cheese manufacture, as the curd particles are still very fragile and would fracture during the sieving process involved in the determination. Moreover, such a measurement is not very relevant, as the curd particles fracture during the initial stages of stirring to a greater or lesser degree, depending on their size, the speed of stirring, and the vat design (Johnston, Dunlop, & Lawson, 1991). In practice, curd particle size is measured just prior to whey drainage, when the curd particles are resilient enough to resist fracture during assay.

Johnston et al. (1991) investigated the effects of variations in the speed and duration of cutting in 20,000 L Damrow (variable speed motor) vats on the efficiency of commercial Cheddar manufacture. Cut programs with a short duration of cutting at slow speeds produced a low $\%CPS_d$ and relatively high fat losses in the whey where $\%CPS_d$ was defined as the percentage of total curd particles with a size < 7.5 mm. These cutting conditions resulted in large curd particles (i.e., high $\%CPS_d$), which shattered quickly during subsequent stirring, thereby increasing the surface area for loss of fat. A maximum $\%CPS_d$ and a minimum fat level were reached after the knives in the Damrow vat had completed 37–40 rotations (Figure 9–19). Hence, the time required to obtain the maximum $\%CPS_d$ decreased as knife speed increased (e.g., ≈ 18 min at 2 rpm and ≈ 8 min at 5 rpm). Fat losses in whey decreased and hence cheese yield increased as the number of revolutions completed by the knives reached 37–40 (Figure 9–19). A further increase in the number of knife revolutions had little effect on fat losses. This study indicates that for a particular vat and knife design, the $\%CPS_d$ and hence fat losses in whey are influenced by a combination of the speed and duration of cutting and the subsequent speed of stirring prior to cooking. For a given vat design, proper maintenance of knives (i.e., edge and knife angle) is essential to enable clean cutting and thereby reduce the risk of tearing the curds and causing high fat losses in cheese whey.

9.5.14 Design and Operation of the Cheese Vat

The design and operation of the cheese vat have a large influence on cheesemaking efficiency, cheese composition, and cheese quality. In all mechanized vats, the coagulum moves to a greater or lesser degree as soon as the knives are switched on, with a firm coagulum moving faster than a soft gel. Cheese vats of different design and mode of operation that enable the knives to cut the moving curd efficiently have been developed. Design features that allow the knives to "catch up" with the coagulum moving before it include

- side-mounted baffles that push the curd onto the knives
- continuously variable-speed knife drives capable of speeds up to 12 rpm
- intermittent cutting cycles, followed by rest periods, which permit the knives to cut through the settling curd layers

A survey (Phelan, 1981) of Irish commercial Cheddar-manufacturing facilities showed that the fat content of whey varied markedly with the type of vat used and season of the year (Figure 9–20). The increased level of fat in the whey as the cheesemaking season progressed was attributed to the increase in the level of milk fat and not to variations in the percentage of milk fat recovered, which remained relatively constant throughout the season. In a series of Dutch studies (during the period 1976–1987), 6 six different vat types were compared. The level of fat in whey ranged from 5.7% to 7.2% of the total and curd fines from 97 to 179 mg/kg whey during the production of Gouda cheese. A further Irish study (see IDF, 1994) showed that Tebel Ost IV vats and APV OCT vats gave similar mean fat levels in the whey (0.30%, w/w), levels significantly lower than that for a W-vat (0.47%, w/w). These studies indicate that under normal operat-

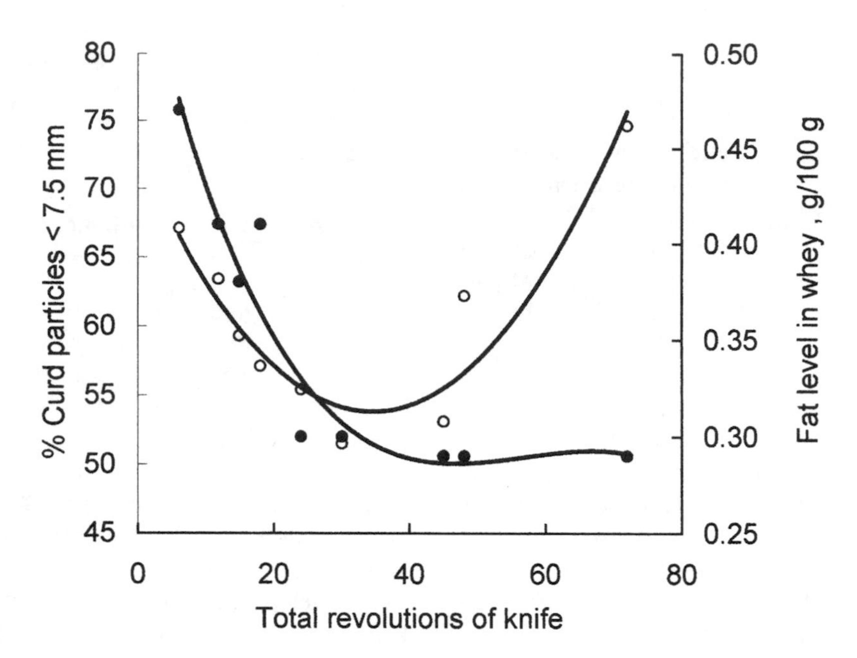

Figure 9–19 Effect of number of cheese knife revolutions (rpm × min) on the amount of curd particles smaller than 7.5 mm as percentage of total (○) and on the fat level in cheese whey (●).

ing conditions some vat designs are inherently better than others in maximizing the retention of fat in cheese curd, probably as a consequence of a more favorable curd particle size distribution (CPS_d). However, for any vat design, maximum efficiency requires in-plant studies to optimize the interactive effects of coagulum firmness, cutting program, and stirring speed so as to achieve the best CPS_d and fat retention results.

9.5.15 Curd-Handling Systems

Most of the losses during cheese manufacture occur in the cheese vat. For example, about 6.5% of milk fat and about 4–5% of the casein (owing to loss of the glycomacropeptides) are lost during commercial Cheddar manufacture (Table 9–6). The losses that occur ex-vat are comparatively small (e.g., ≈ 0% of the casein and 2.0% of the total milk fat.) Nonetheless these losses are important determinants of cheesemaking efficiency. Following removal of most of the whey at drainage—on drainage belts with overhead stirrers, which agitate the curd—the curd is subjected to variety-specific conveying and handling processes, such as stirring, cheddaring, milling, salting, and prepressing. During these operations, moisture and fat are lost to varying degrees, thereby affecting actual cheese yield. Milling of cheddared curd exposes fresh surfaces from which fat is lost to an increasing extent in the whey (so-called salt whey) with elevation of temperature, reduction in chip size, and severity of mechanical squeezing either in the mill or on worm conveyors. It is also conceivable that the level of fat in whey from block formers increases with the level of vacuum. Owing to the general nonavailability and high cost of downstream pilot-scale curd-handling systems, little published information is available on the comparative effects of different systems (e.g., CheddarMaster vs. Alfamatic or strainer vat vs. Casomatic) under different operating conditions on cheese yield efficiency.

9.6 CONCLUSION

Cheese is a very important trade item in the dairy industry (≈ 35% of total milk production). In the production of commodity cheeses, such as

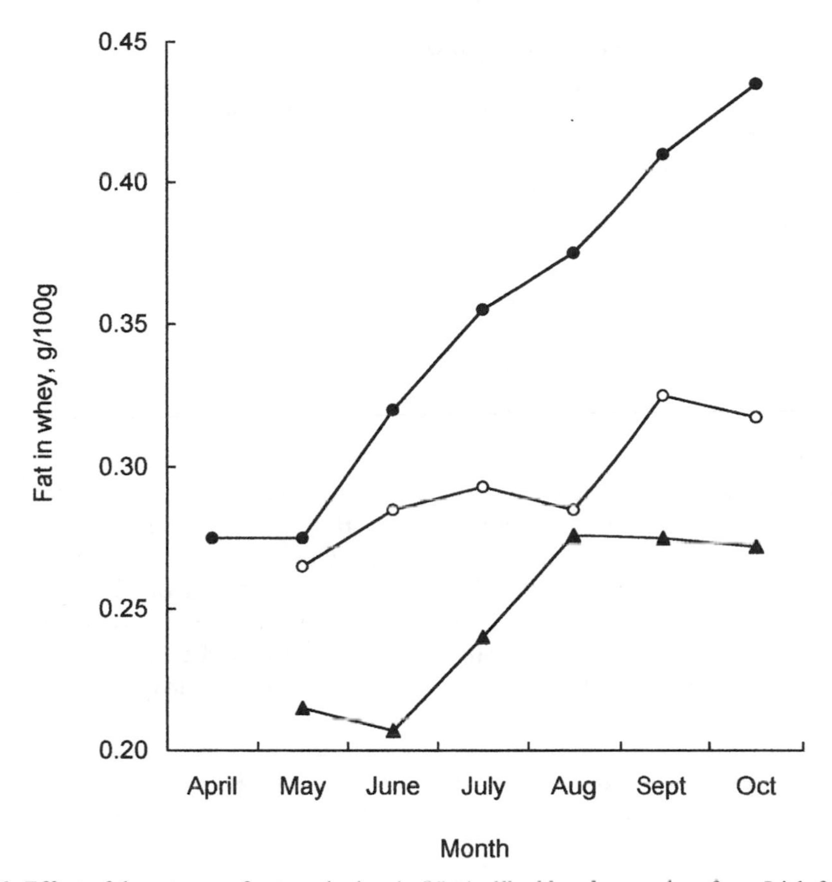

Figure 9–20 Effect of three types of vat on the level of fat in Cheddar cheese whey from Irish factories.

Cheddar and Gouda, increasing the scale of production and cheesemaking efficiency are essential for reducing production costs and ensuring competitiveness and survival in the marketplace. Hence, there is much interest in cheese yield at the commercial level, as it determines the profits that accrue to a cheese plant and the price it can afford to pay for milk. The plant's profits are also determined by its cost-effectiveness in recovery from by-streams (e.g., cheese whey and stretch water) and the production of whey byproducts. These aspects are discussed in Chapter 22.

Cheese yield is influenced by many factors, including the composition and quality of the raw milk, milk-handling and -storage practices, milk pretreatments (e.g., the pasteurization temperature), the cheesemaking process (i.e., the make procedure, equipment, and technology), and cheese composition (e.g., moisture). Maximization of cheese yield requires a comprehensive knowledge of milk composition, the factors that influence it, gelation, and the influence of the cheesemaking process on the gel. Measurement of cheesemaking efficiency is essential so that inefficiencies can be redressed. Indices of cheesemaking efficiency include cheese yield and/or recovery of components, especially casein and fat. Comparison of actual and predicted yields allows a cheese plant to monitor its efficiency over time. However, agreement between actual and predicted yields does not imply that the yield is at a maximum. Moreover, the use of plant-specific yield prediction formulae does not allow comparison of yields from an in-

Table 9–6 Mass Balance of Fat during Cheddar Cheese Manufacture

Constituent	*Weight (kg)*	*Fat (kg)*	*Fat (% Total Fat)*
Milk	10,000	335	100
Cheese	927	306.5	91.5
Whey			
Drain whey	8,750	21.9	6.50
White whey			
Cheddaring tower	227	4.6	1.37
Salting belt	68	1.4	0.42
Block former	28	0.6	0.18

dividual plant with those from other plants or published reports. In modern cheesemaking, full yield potential is not yet achievable, as fat and protein recovery is still less than 100%. Undoubtedly, in the future a closer realization of maximum yield will ensue from continued improvements in a number of areas, including

- milk quality
- seasonal variations in the composition of cheese milk (minimized through improved milk production practices and/or standardization to a consistent casein level using LCF-UF)
- standardization of casein and fat in milk (using online methods)
- starter cultures (development of cultures that cause less hydrolysis of casein during manufacture)
- firmness at cutting (through use of in-vat curd firmness sensors)
- equipment and process design

REFERENCES

Auldist, M.J., Coats, S., Sutherland, B.J., Mayes, J., McDowell, G.H., & Rogers, G.L. (1996). Effects of somatic cell count and stage of lactation on raw milk composition and the yield and quality of Cheddar cheese. *Journal of Dairy Research, 63,* 269–280.

Banks, J.M., & Muir, D.D. (1984). Coagulum strength and cheese yield. *Dairy Industries International, 49*(9), 17–21, 36.

Banks, J.M., & Muir, D.D. (1985). Incorporation of the protein from starter growth medium in curd during manufacture of Cheddar cheese. *Milchwissenschaft, 40,* 209–212.

Banks, J.M., Tamime, A.Y., & Muir, D.D. (1985). The efficiency of recovery of solids from bulk starter in Cheddar cheesemaking. *Dairy Industries International, 50*(1), 11–13, 21.

Barbano, D.M., & Rasmussen, R.R. (1994). Cheese yield performance of various coagulants. In *Cheese yield and factors affecting its control* [Proceedings of IDF Seminar, Cork, Ireland, 1993]. Brussels: International Dairy Federation.

Burvenich, C., Guidry, A.J., & Paape, M.J. (1995). Natural defence mechanisms of the lactating and dry mammary gland. In A. Saran & S. Soback (Eds.), *Proceedings of the Third IDF International Mastitis Seminar.* Haifa, Israel: M. Lachmann.

Bynum, D.G., & Olson, N.F. (1982). Influence of cut firmness on Cheddar cheese yield and recovery of milk constituents. *Journal of Dairy Science, 65,* 2281–2290.

Emmons, D.B., & Beckett, D.C. (1990). Milk clotting enzymes: 1. Proteolysis during cheesemaking in relation to estimated losses of yield. *Journal of Dairy Science, 73,* 8–16.

Fenelon, M.A., & Guinee, T.P. (1997). The compositional, textural and maturation characteristics of reduced-fat Cheddar made from milk containing added Dairy-Lo™. *Milchwissenschaft, 52,* 385–389.

Fox, P.F. (1969). Effect of cold-ageing on the rennet coagulation time of milk. *Irish Journal of Agricultural Research, 8,* 175–182.

International Dairy Federation. (1991). *Factors affecting the yield of cheese* (Special Issue No. 9301). Brussels: Author.

International Dairy Federation. (1994). *Cheese yield and factors affecting its control* [Proceedings of IDF Seminar, Cork, Ireland, 1993]. Brussels: Author.

Jana, A.H., & Upadhyay, K.G. (1992). Homogenisation of milk for cheesemaking [A review]. *Australian Journal of Dairy Technology*, *47*, 72–79.

Johnston, K.A., Dunlop, F.P., & Lawson, M.F. (1991). Effects of speed and duration of cutting in mechanised Cheddar cheesemaking on curd particle size and yield. *Journal of Dairy Research*, *58*, 345–354.

Lucey, J., & Kelly, J. (1994). Cheese yield. *Journal of the Society of Dairy Technology*, *47*, 1–14.

Phelan, J.A. (1981). Standardisation of milk for cheesemaking at factory level. *Journal of the Society of Dairy Technology*, *34*, 152–156.

Politis, I., & Ng-Kwai-Hang, K.F. (1988a). Association between somatic cell count of milk and cheese-yielding capacity. *Journal of Dairy Science*, *71*, 1720–1727.

Politis, I., & Ng-Kwai-Hang, K.F. (1988b). Effects of somatic cell counts and milk composition on the coagulating properties of milk. *Journal of Dairy Science*, *71*, 1740–1746.

Salji, J.P., & Kroger, M. (1981). Effect of using frozen concentrated direct-to-the-vat culture on the yield and quality of Cheddar cheese. *Journal of Food Science*, *48*, 920–924.

Ustanol, Z., & Hicks, C.L. (1990). Effect of milk clotting enzymes on cheese yield. *Journal of Dairy Science*, *73*, 8–16.

Walsh, C.D., Guinee, T.P., Reville, W.D., Harrington, D., Murphy, J.J., O'Kennedy, B.T., & Fitzgerald, R.J. (1998). Influence of k-casein genetic variant on rennet gel microstructure, Cheddar cheesemaking properties and casein micelle size. *International Dairy Journal*, *8*, 707–714.

Wolfschoon-Pombo, A.F. (1997). Influence of calcium chloride addition to milk on the cheese yield. *International Dairy Journal*, *7*, 249–254.

CHAPTER 10

Microbiology of Cheese Ripening

10.1 GENERAL FEATURES

The initial objective of cheesemaking was to conserve the principal constituents of milk, and hence any changes that occurred during storage were unintentional. Cheese curd contains a diversity of microorganisms and enzymes, and therefore biological, biochemical, and probably chemical changes can be expected to occur unless the preservative factors are sufficient to prevent them. In most cases, the cheese environment is not sufficiently severe to prevent the activity of enzymes originating from the milk, the coagulant, and the starter and nonstarter microorganisms. The action of these enzymes and of the secondary microflora induces changes in flavor and texture, a process referred to as ripening or maturation. The changes that occur during ripening are responsible for the characteristic flavor (taste and aroma), texture, and in most cases appearance (e.g., the formation of eyes, growth of molds) of the individual varieties. The changes range from the very limited (e.g., Mozzarella) to the very extensive (e.g., Blue cheeses). The duration of ripening ranges from about 3 weeks (e.g., Mozzarella) to 2 or more years (e.g., Parmesan and extra mature Cheddar). The rate of ripening is directly related to the moisture content of the cheese and inversely related to its salt content.

Although some high-moisture, rennet-coagulated curd is consumed fresh (e.g., the Spanish Burgos cheese), most rennet-coagulated cheeses are ripened to at least some extent. In contrast, all acid-coagulated cheeses are consumed fresh. Why acid-coagulated cheeses are not ripened is not clear. As discussed in Chapter 16, acid-coagulated curds do not synerese as well as rennet-coagulated curds, and consequently they have a high moisture content and would ripen and deteriorate very rapidly; however, moisture content could probably be reduced, if desired. In order to reduce the moisture content of acid-coagulated curds, a high cook temperature is used (e.g., 55°C for Cottage and ≈ 60°C for thermo Quarg). As a result, significant numbers of the starter bacteria are killed and their enzymes are extensively denatured. The heat/acid-coagulated cheeses are subjected to a high temperature and are consumed fresh. Hence, only rennet-coagulated cheeses are ripened.

The quality of cheese is determined mainly by its flavor and texture, and thus considerable effort has been devoted to elucidating the principal microbiological and biochemical changes that occur during ripening. The appearance of many, perhaps most, varieties changes during ripening. These changes include the formation of holes, called *eyes*, in Swiss-type and to a lesser extent in Dutch-type cheeses, the growth of mold on the surface (e.g., Brie and Camembert) or interior (Blue varieties), or the growth of microorganisms on the surface (smear-ripened cheeses). Since the changes in appearance are visually perceptible, they are the criteria by which the consumer initially judges cheese quality and hence are of major significance. These changes are not just cosmetic—they are visual evidence that the flavor and texture are satisfactory. The absence of eyes in Swiss cheese indicates that the propionic acid fermentation has not occurred satisfactorily and that therefore the flavor is unlikely to be satisfactory. The absence of mold in mold-ripened cheese clearly indicates unacceptable quality, and of course the growth of molds on nonmold varieties indicates spoilage.

Traditionally, the surface of cheese was exposed to the atmosphere, and hence loss of moisture occurred. The loss of moisture is especially critical for varieties in which the growth of microorganisms on the surface is a key feature of ripening, and thus such varieties are ripened in high-humidity environments, traditionally in caves with naturally high humidity and frequently now in environments with artificially controlled humidity.

Loss of some moisture through evaporation from the surface is not critical for internally bacterial-ripened cheeses, and if controlled, it leads to the formation of a rind (a low-moisture surface layer), which effectively seals the cheese, restricting continued loss of water and preventing the growth of microorganisms on the surface (owing to the low water activity in the rind). In rinded cheeses, there is a moisture gradient within the cheese, which affects ripening.

As discussed in Chapter 8, initially there is a salt gradient in surface-salted cheese, but equilibrium is established gradually throughout the cheese.

10.2 MICROBIAL ACTIVITY DURING RIPENING

The factors controlling the growth of microorganisms in cheese include water activity, concentration of salt, oxidation-reduction potential, pH, NO_3^-, ripening temperature, and the presence or absence of bacteriocins (produced by some starters). Individually, the effect of these factors may not be very great, but their joint impact as so-called hurdles is the real controlling factor. Other compounds produced during curd manufacture and ripening (e.g., H_2O_2 and fatty acids) also inhibit microbial growth, but the concentrations of these compounds produced by the starters in cheese are not sufficiently high to have a significant effect on microbial growth.

10.2.1 Water Activity

All microorganisms require water for growth, but it is the availability of the water rather than the total amount present that is the important factor. Water availability is expressed in terms of water activity (a_w), which is defined as the ratio of the vapor pressure over the cheese (p) to the vapor presence of pure water (p_0) at that temperature:

$$a_w = \frac{p}{p_0}$$

The value of a_w ranges from 0 to 1.0.

Cheese, unless vacuum packed, loses moisture by evaporation during ripening. The proteins in cheese are hydrated, and this "bound" water is not available for bacterial growth. The hydrolysis of proteins to peptides and amino acids and of lipids to glycerol and fatty acids during ripening reduces the availability of water, since one molecule of water is added at each bond hydrolyzed. In addition, the salt and organic acids (lactate, acetate, and propionate) are dissolved in the moisture of the cheese and reduce the vapor pressure. Each of these factors reduces the a_w of cheese during ripening.

Yeast grow at a lower a_w than bacteria, and molds at still lower values. Most bacteria require a minimum a_w of around 0.92 for growth. The limit for most yeast is around 0.83, but osmophilic yeast grow at a_w values below 0.60, while molds have a lower a_w limit of around 0.75. Growth of microorganisms at low a_w is characterized by a long lag phase, a slow rate of growth (i.e., a long generation time), and a reduction in the maximum number of cells produced. Each of these factors helps to limit the number of cells produced. Lactic acid bacteria (LAB) generally have higher minimum a_w values than other bacteria. The minimum a_w for *Lc. lactis, Sc. thermophilus*, *Lb. helveticus*, and *P. freudenreichii* is about 0.93, 0.98, 0.96, and 0.96, respectively. The influence of a_w on the growth of some other microorganisms associated with cheese is shown in Table 10–1. *Penicillium camemberti* is the mold responsible for the white coating on Camembert and Brie cheese, and *Brevibacterium linens* and *Debaryomyces hansenii* are important microorganisms on the surface of smear-ripened cheeses. *P. camemberti, B. linens*, and *D. hansenii* can grow slowly in the presence of 10%, 12%, and 15% NaCl, respectively. *Staphylococcus aureus* and micrococci can grow quite well in the presence of 6.5% NaCl, which is equivalent to an a_w value of 0.96. Compared with other fungi, *Geotrichum candidum* is very sensitive to a_w whereas *B. linens* is quite resistant. Propionibacteria are also particularly sensitive to a_w. Facultative anaerobes have different minimum a_w values depending on whether the organisms are growing aerobically or anaerobically. For example, in the presence of O_2, *S. aureus* has a minimum a_w of 0.86, but in the absence of O_2, the minimum is 0.91.

Evaporation of water from the cheese surface during ripening also contributes to the reduction of the a_w of cheese (examples for Emmental and Gruyère are shown in Figure 10–1). The faster rate of decrease in the a_w of Gruyère may be due to the surface salting of Gruyère during ripening. In addition, the a_w of cheese can vary throughout its mass (Figure 10–2). Variations are much greater in large cheeses, like Emmental (50–60 kg), than in small cheeses, like Appenzeller (6–8 kg). This is due to several factors, including the temperature gradient in the cheese during the early stages of the fermentation, the loss of moisture during ripening, the NaCl gradient in the cheese, and microbial activity on the rind. These factors must be taken into account in determining the significance of a_w, especially in large cheeses. Typical a_w values for cheese are listed in Table 10–2. As a comparison, the a_w of milk is 0.995. Since the a_w of cheese decreases during ripening, some of these values must be interpreted with care; however, they are useful as a guide. Except for the soft cheeses like Brie and Camembert, most of these values are close to the minima for starter growth cited above.

10.2.2 Salt

The use of NaCl to prevent microbial spoilage of food is probably as old as food production itself. The concentration required depends on the nature of the food, its pH, and its moisture content, but generally less than 10% is sufficient. The major inhibitory factor is probably the reduction in a_w that occurs when salt (or any solute) is dissolved in water. The relationship between salt concentration and a_w is shown in Figure 10–3 and is almost, but not quite, linear. The linear equation is

$$a_w = -0.0007x + 1.0042$$

and describes the relationship very well, since the r^2 value is 0.997, and x = concentration of NaCl (g/kg). It is generally considered that an a_w value of less than 0.92 is necessary to prevent bacterial growth; this is equivalent to a salt concentration of 12.4%. In cheese, the salt concentration varies from 0.7% to 7%. Therefore, other factors are involved in preventing bacterial growth in cheese, such as pH and temperature. The ions themselves are also important (e.g., Na^+ is a much more effective inhibitor than K^+). In calculating the inhibitory effect of salt in cheese, the concentration of salt dissolved in the water of the cheese rather than the actual concentration of salt is the important parameter. For example, in a Cheddar cheese with 38g moisture/100g and 1.9g salt/100g, the salt-in-water percentage is 5%. Generally, the salt-in-water percentage in Cheddar cheese varies from 4% to 6%.

Cheese is either dry-salted (e.g., Cheddar) or brine-salted (most cheeses). In brine-salted cheeses, the salt concentration is influenced directly by the size of the cheese, the concentration of salt in the brine, the temperature of the brine, and the length of time the cheese is immersed in the brine (see Chapter 8). Data for the effect of

Table 10–1 Influence of Water Activity (a_w) on the Growth of Different Microorganisms in Nutrient Broth, pH 6.6 after 10 Days at 25°C (Results are Expressed as Percentage of Maximum Development)

	NaCl Concentration (g/100 ml)				
	0	*5*	*10*	*15*	*20*
			a_w		
	0.992	*0.975*	*0.947*	*0.916*	*0.880*
Molds					
Mucor mucedo 54_o	100.0	47.4	11.6	–	–
Penicillium candidum 53ll	100.0	80.9	36.4	4.1	1.1
Cladosporium herbarum 53b	82.6	100.0	62.4	13.9	3.5
Scopulariopsis fusca[a] 53L1	100.0	78.4	76.9	65.4	14.6
Yeast					
Rhodotorula spp. 44a	100.0	69.5	21.8	1.0	–
Debaryomyces spp.*54k*	100.0	49.7	30.2	10.5	8.2
Geotrichum candidum 53aa	100.0	46.9	–	–	–
Trichsporon 57k	100.0	62.6	–	–	–
Bacteria					
Micrococcus saprophyticus[a] 55a	100.0	96.3	67.2	19.1	–
Micrococcus saprophyticus[a] 56b	84.7	100.0	61.2	16.7	–
Micrococcus lactis 57h	100.0	45.9	–	–	–
Brevibacterium linens 58a	100.0	44.1	29.9	13.9	4.1
Brevibacterium linens BL107	100.0	67.0	30.0	15.6	3.2
Arthrobacter citrans KR3	100.0	19.4	7.0	–	–
Escherichia coli Strain 54i	100.0	23.4	–	–	–
Escherichia coli Strain SL	100.0	19.9	–	–	–

Note: – indicates no growth

[a] These are the names given in the original reference; *Micrococcus saprophyticus* is likely to be *Staphylococcus saprophyticus.*

brining time on the salt concentration and the a_w of Camembert cheese are shown in Figure 10–4. The brine normally used contains about 20% NaCl, has a pH of about 5.2 (adjusted with lactic acid), and a Ca content of 0.2% (adjusted with $CaCl_2$). The pH and Ca concentration simulate the levels in cheese and help to prevent the efflux of lactate and Ca from the cheese.

10.2.3 Oxidation-Reduction Potential

Oxidation-reduction potential (E_h) is a measure of the ability of chemical or biochemical systems to oxidize (lose electrons) or reduce (gain electrons). E_h is generally measured using a platinum electrode coupled with a calomel reference electrode and is expressed in mV. It can also be estimated using indicator dyes that change color at different redox potentials. A positive value indicates an oxidized state and a negative value indicates a reduced state.

The E_h of milk is about +150 mV whereas that of cheese is about –250 mV. The exact mechanism by which the E_h of cheese is reduced is not clear but is almost certainly related to the fermentation of lactose to lactic acid by the starter

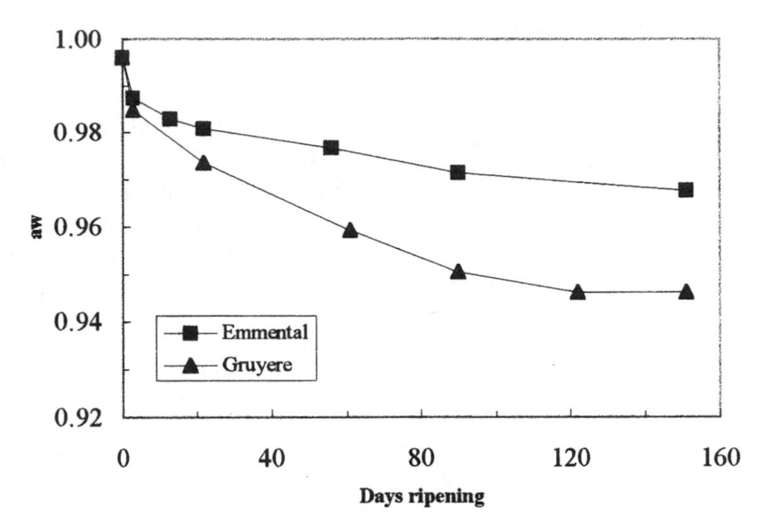

Figure 10–1 Decrease in the a_w of Emmental and Gruyère cheese during ripening. The a_w at time zero (0.995) corresponds to that of milk.

during growth and is probably related to the reduction of the small amounts of O_2 in the milk to H_2O (or to H_2O_2 and then to H_2O). Because of these reactions, cheese is essentially an anaerobic system, in which only facultatively or obligately anaerobic microorganisms can grow. Obligate aerobes, like *Pseudomonas* spp., *Brevibacterium* spp., and *Micrococcus* spp., will not grow within the cheese, even when other conditions for growth are favorable. E_h is therefore important in determining the types of microorganisms that grow in cheese. The bacteria that develop on the surface of cheese are mainly obligate aerobes and are unable to grow within the anaerobic cheese environment.

10.2.4 pH and Organic Acids

Most bacteria require a neutral pH value for optimum growth and grow poorly at pH values below 5.0. The pH of cheese curd after manufacture generally lies within the range 4.5–5.3, so pH is also a significant factor in controlling bacterial growth in cheese. LAB, especially lactobacilli, generally have pH optima below 7, and *Lactobacillus* spp. can grow at pH 4.0. Most yeast and molds can grow at pH values below 3.0, although their optima range is from 5 to 7. *B. linens*, an important organism in smear-ripened cheese, cannot grow below pH 6.0. *Micrococcus* spp., which are commonly found on the surface of soft cheeses, cannot grow at pH 5 and only slowly at pH 5.5.

The efficacy of organic acids as inhibitors of microbial growth is thought to depend on the amount of undissociated acid present and therefore on the dissociation constant (pK_a) and pH. The pK_as of propionic, acetic, and lactic acids, the principal acids found in cheese, are 4.87, 4.75, and 3.08, respectively, so at the same concentration lactic acid is the least and propionic acid the most effective inhibitor. Propionic acid is very effective at repressing the growth of molds. However, the concentration of the acid is also important, and lactate is invariably present at much greater concentrations in cheese and cheese curd than either of the other two acids. The pH of mold- and smear-ripened cheeses characteristically increases during ripening, particularly on the surface, due to the growth of yeast and molds. Sometimes, it is thought that the difference between pH 5.2, the pH of a well-made cheese, and pH 5.4, the pH of a poorly made cheese, is not very great. However, this is

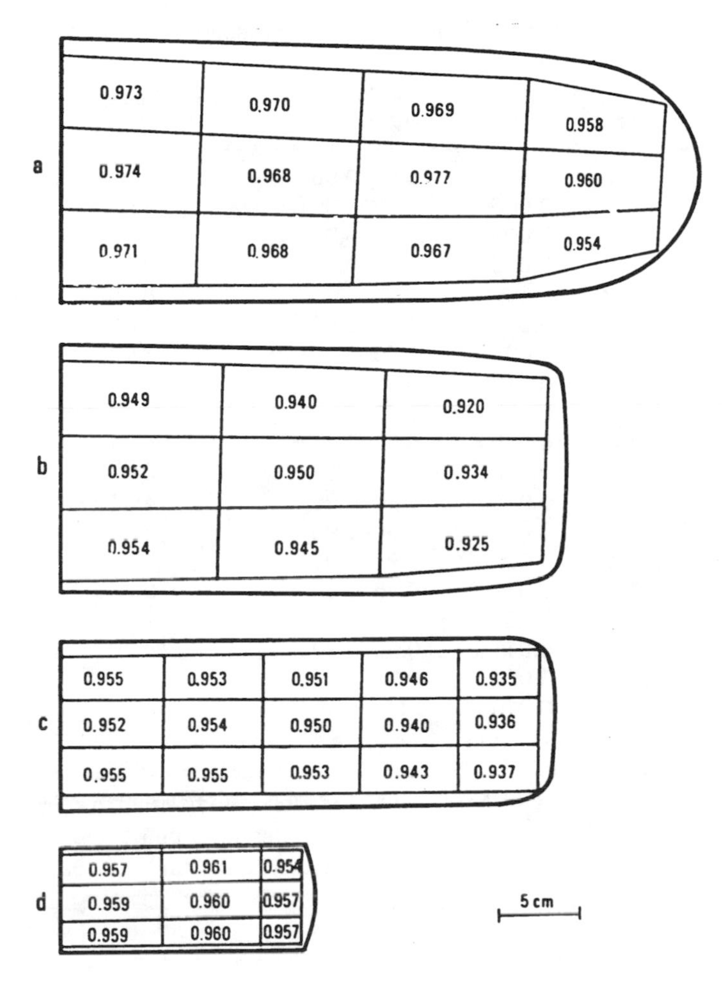

Figure 10–2 Typical variations in the a_w of slices, from the center to the surface, of (a) Emmental, (b) Sbrinz, (c) Gruyère, and (d) Appenzeller cheese. The cheeses were about 5 months old, and the a_w of the rinds was (a) 0.90–0.95, (b) 0.80–0.90, and (c, d) 0.92–0.98.

not so. pH is a log scale, and a difference of 1 pH unit is equivalent to a tenfold difference in the H^+ concentration. The difference in $[H^+]$ between 5.2 and 5.4 is twofold.

10.2.5 Nitrate

NO_3^-, as KNO_3 (saltpeter) or $NaNO_3$, is added to the milk (20g/100L) for some cheeses, especially Dutch-type cheeses like Gouda and Edam, to prevent the production of early and late gas by coliform and *Clostridium tyrobutyricum*, respectively. Much of the NO_3^- is lost in the whey. The maximum amount of NO_3^- permitted in cheese is 50 mg/kg, calculated as $NaNO_3$. The real inhibitor is NO_2^-, which is formed from NO_3^- by the indigenous xanthine oxidase present in the milk or curd. How NO_2^- acts in preventing microbial growth is not clear. NO_2^- can also react with aromatic amino acids in cheese to produce nitrosamines, many of which are carcinogenic (see Chapter 21).

Table 10–2 Typical Water Activity (a_w) Values for Various Cheeses

Type	*Typical a_w**	*SD*	*Range of a_w in Rind*	*Typical Moisture (%)*	*Typical Salt (%)*
Appenzeller	0.960	0.011	0.97–0.98		
Brie	0.980	0.006	0.98–0.99	48.4	1.91
Camembert	0.982	0.008	0.98–0.99	51.8	2.5
Cheddar	0.950	0.010	0.94–0.95	36.8	1.5
Cottage cheese	0.988	0.006		82.5	1.0
Edam	0.960	0.008	0.92–0.94	41.5	2.0
Emmental†	0.972	0.007	0.90–0.95	37.2	1.2
Fontal	0.962	0.010	0.93–0.96		
Gorgonzola	0.970	0.017	0.97–0.99		3.5
Gouda	0.950	0.009	0.94–0.95	41.4	2.0
Gruyère†	0.948	0.012	0.92–0.98	34.5	1.06
Limburger	0.974	0.015	0.96–0.98	48.4	2.74
Munster	0.977	0.011	0.96–0.98	41.8	1.8
St. Paulin	0.968	0.007	0.96–0.97		
Parmesan	0.917	0.012	0.85–0.88	29.2	2.67
Quarg	0.990	0.005		79.0	0.70
Sbrinz†	0.940	0.011	0.80–0.90	42.9	1.90
Tilsiter	0.962	0.014	0.92–0.96		2.63
Processed cheese	0.975	0.010			

* Measured at 25°C.

†Values for Emmental, Gruyère, and Sbrinz were measured after ripening for 4–5, 6–7, and 10–11 months, respectively. The other values were determined in commercial samples of unknown age.

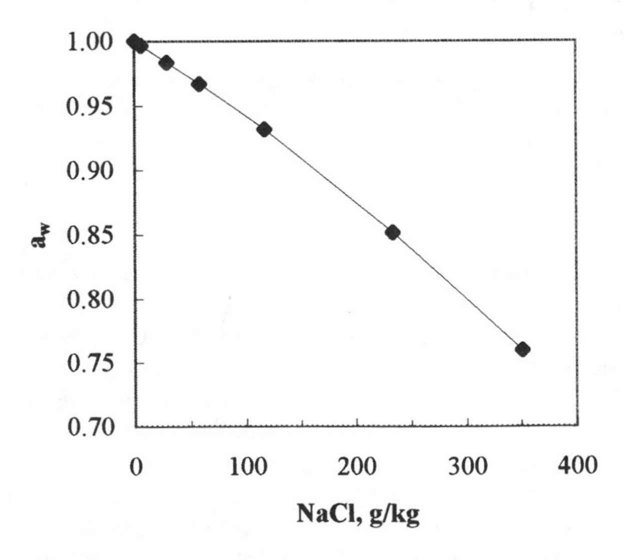

Figure 10–3 Effect of salt concentration on the a_w of brine.

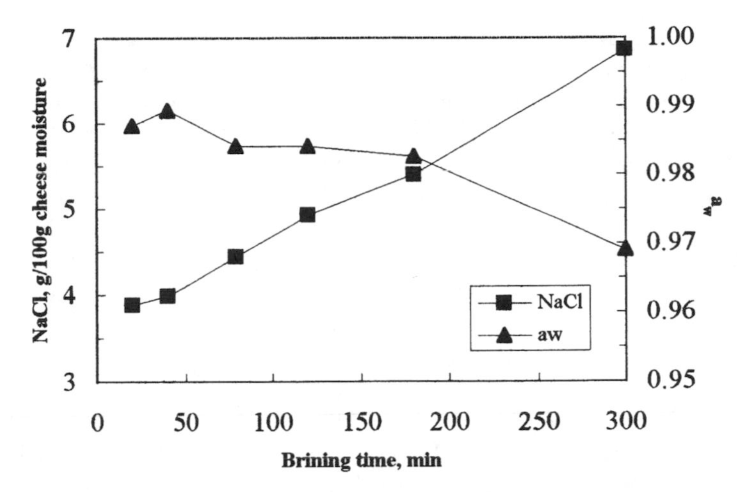

Figure 10–4 Influence of the duration of brining at 14°C in 20% NaCl brine on the a_w of Camembert cheese. The NaCl concentration and a_w levels were determined 15 days after manufacture.

10.2.6 Temperature

Generally, the optimum temperature for the growth of bacteria is around 35°C for mesophiles and around 55°C for thermophiles. Thermophilic starters have an optimum temperature of around 42°C. Psychrophilic bacteria have an optimum temperature below 20°C, but true psychrophiles are not found in cheese. At temperatures below the optimum, growth is retarded. The temperature at which cheese is ripened is dictated by two opposing requirements—on the one hand, the need to control the growth of potential spoilage and pathogenic bacteria and, on the other, the need to promote the ripening reactions and the growth of the secondary microflora (in the case of soft and Swiss-type cheeses). Higher temperatures promote faster ripening by the starter and nonstarter microorganisms but also allow the growth of spoilage and pathogenic bacteria. Generally, Cheddar cheese is ripened at 6–8°C while Camembert and other mold and bacterial smear-ripened cheeses are ripened at 10–15°C. Emmental cheese is ripened initially for 2–3 weeks at a low temperature (≈ 12°C), after which the temperature is increased to 20–24°C for 2–4 weeks to promote the growth of propionic acid bacteria and the fermentation of lactate to propionate, acetate, and CO_2. The temperature is then reduced again to around 4°C. For soft cheeses, the humidity of the environment is also controlled to prevent excessive evaporation of moisture from the cheese surface.

Traditional Emmental cheese is made from raw milk, and because of the relatively high temperature of ripening for this cheese, great attention must be paid to the microbial quality of the raw milk. The need for such attention is mitigated to some extent by the higher cooking temperature (≈ 52°C) and longer cooking time (60–90 min) to which Emmental curd is subjected during manufacture, compared with other hard cheeses like Cheddar, for which the maximum cooking temperature is around 38°C.

Increasing the temperature of ripening is probably the simplest and most cost-effective way of accelerating the ripening of cheese (see Chapter 15), but it also increases the rate of growth of other bacteria that may be present.

10.3 GROWTH OF STARTER BACTERIA IN CHEESE

The initial number of starter bacteria in cheese milk ranges from about 10^5 to 10^7 cfu/ml and depends on the level of inoculation. Subsequent growth of the starter and syneresis (contraction) of the curd during manufacture results in starter counts of about 10^9 cfu/g in almost all cheeses within 1 day of manufacture. During ripening, starter organisms dominate the microflora of cheeses but most die off and lyse relatively rapidly. This is shown in Figure 10–5 for 5 strains of *Lactococcus* in Cheddar cheese and in Figure 10–6 for *Sc. thermophilus* and *Lb. helveticus* in Comté cheese. In the case of Cheddar cheese, the rate of death depends on the strain, and in the case of Comté cheese, the rate of death of *Sc. thermophilus* is faster than that of the thermophilic lactobacilli *Lb. helveticus* and *Lb. delbrueckii* subsp. *lactis*. Many artisanal cheeses, especially Spanish varieties, are made without the deliberate addition of a starter. In these cheeses, lactococci also make up the major part of the microflora and, except for La Serena, also show significant rates of death during ripening (Figure 10–7). The reason for the slow rate of death of lactococci in La Serena cheese may be due to its relatively low salt concentration during the early weeks of ripening.

Once the starter counts begin to decrease, lysis usually occurs and intracellular enzymes, particularly peptidases, are released, which, together with chymosin and the starter proteinase, hydrolyze the caseins to peptides and amino acids, which are the precursors of the flavor compounds in cheese (see Chapters 11 and 12). Starters vary in their ability to lyse: some strains lyse relatively quickly while others hardly lyse at all. Lysis is caused by an intracellular muramidase which hydrolyzes the cell wall peptidoglycan. This enzyme is under stringent regulation; otherwise the cells would not grow. Generally, strains of *Lc. lactis* subsp. *cremoris* lyse faster than strains of *Lc. lactis* subsp. *lactis*, which may partly explain why the former are thought to produce a better-flavored cheese than the latter. Lysis is influenced by several factors, including the level of salt and the presence of prophage, which is thought to be induced by cooking. The presence of small numbers of lytic phage may also have a role in lysis. Cheese

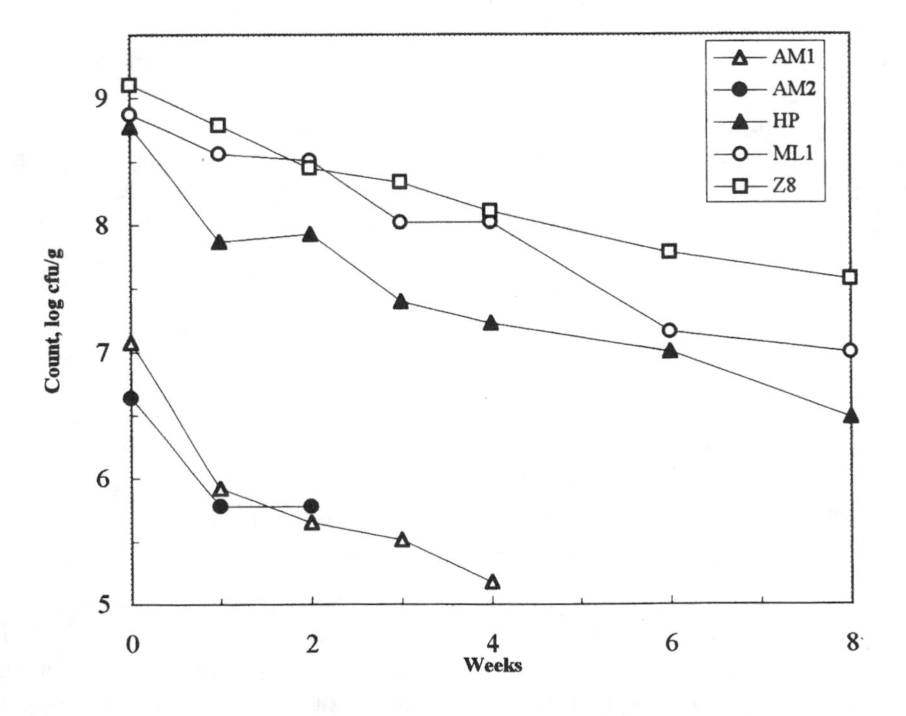

Figure 10–5 Numbers of different strains of *Lactococcus* in Cheddar cheese during ripening.

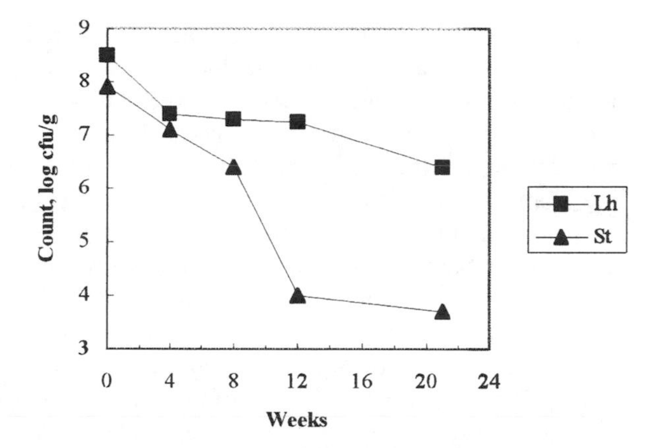

Figure 10–6 Numbers of *Streptococcus thermophilus* (St) and *Lactobacillus helveticus* (Lh) in Comté cheese during ripening.

made with a fast-lysing starter will ripen more rapidly than those made using slow-lysing cultures.

10.4 GROWTH OF NONSTARTER LACTIC ACID BACTERIA IN CHEESE

Most, if not all, cheeses, whether made from raw or pasteurized milk, contain adventitious nonstarter LAB. These bacteria are mainly facultatively heterofermentative lactobacilli (group II), such as *Lb. casei*, *Lb. pararcasei*, *Lb. plantarum*, and *Lb. curvatus*, but *Pediococcus* spp. and obligately heterofermentative *Lactobacillus* spp. (group III), such as *Lb. brevis* and *Lb. fermentum*, are also found occasionally. The species of group II and III lactobacilli found in cheese are referred to as *mesophilic* to distin-

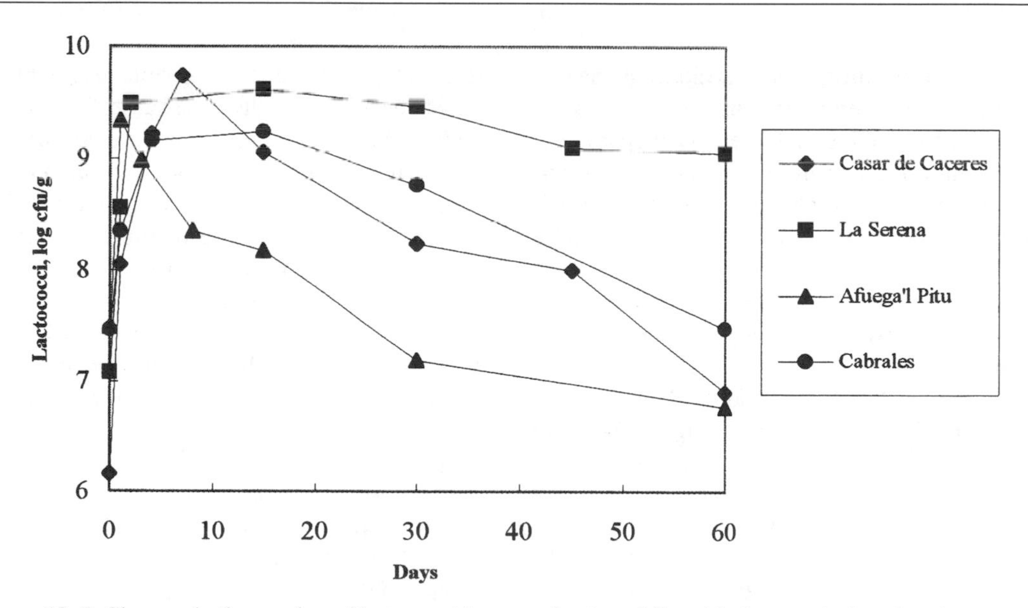

Figure 10–7 Changes in the number of lactococci in several artisanal Spanish cheeses during ripening.

guish them from the thermophilic lactobacilli used as starters. The sources of these bacteria are the raw milk and the factory environment. Small numbers of some lactobacilli survive pasteurization and the cooking temperature (52°C) used for hard cheeses, like Emmental, which is traditionally made from raw milk. All of these bacteria are salt and acid tolerant and are facultative anaerobes, and therefore they grow quite well in cheese. Many of the nonstarter lactobacilli and pediococci found in cheese can grow in the presence of 10% NaCl. They need a fermentable carbohydrate for energy production, but the energy source used by them in cheese is unclear, since at the time of exponential growth of nonstarter LAB (NSLAB) no lactose is present. Possible substrates include citrate and/or amino acids. In model systems, mesophilic lactobacilli can utilize the sugar of the glycoproteins of the milk fat globule membrane as an energy source.

In contrast to starter cells, the initial number of NSLAB in cheese varies considerably, from about 100 cfu/g in Cheddar cheese to 10^6 cfu/g in Casar de Cáceres (Figure 10–8). Within the first few weeks of ripening, however, they grow relatively quickly to high numbers ($\approx 10^8$ cfu/g) in all cheese at a rate that depends primarily on the ripening temperature. Their generation time in Cheddar cheese ripened at 6°C is 8.5 days. NSLAB grow much more rapidly in Casar de Cáceres and La Serena cheese than in Comté or Cheddar cheese (Figure 10–8). The difference is due to the higher moisture content of the first two cheeses, which is around 50% at day 4 but subsequently decreases to about 35% and 45% after 45 days ripening for Casar de Cáceres and La Serena, respectively; the moisture content of Comté and Cheddar cheese is around 38% from the beginning of ripening. The faster growth rates in Casar de Cáceres and La Serena cheese may also reflect the higher ripening temperature used for these cheeses. In addition, Cheddar is the only one of the cheeses in Figure 10–8 that is made from pasteurized milk, which probably explains the low initial number of NSLAB in this cheese. The higher rate of growth of NSLAB in Comté cheese, compared with Cheddar, is likely to be due to the higher ripening temperature of Comté (3 weeks at 14°C, followed by 9 weeks at 18°C, at which time the temperature is reduced to 7°C); Cheddar is kept at 6°C throughout ripening. A higher temperature is used in the ripening of Comté to promote the growth of propionic acid bacteria. In raw milk cheese, the number of NSLAB in the curd is higher, they grow faster, and the population is more heterogeneous than for pasteurized milk cheeses.

Despite extensive study, the role of NSLAB in the development of cheese flavor is not clear. In contrast to starter cells, mesophilic lactobacilli die off very slowly in hard cheese (Figure 10–8). They appear to lack a cell envelope–bound proteinase, and since they die off only slowly in cheese, their intracellular enzymes are probably not released into the cheese matrix. Nevertheless, cells of NSLAB are viable and, at the high cell densities found in cheese, exhibit considerable metabolic activity. NSLAB do contribute to ripening, but the significance of this contribution is open to question. At least in Cheddar cheese, they transform L lactate to D lactate; racemases are not involved. It is likely that L lactate is oxidized to pyruvate, which is then reduced to D lactate. A racemic mixture of both isomers is formed eventually. Some NSLAB can also oxidize lactate to acetate and CO_2 on the cheese surface in the presence of O_2, sharpening the taste of the cut surfaces of the cheese, especially if the cut surfaces remain uncovered for several hours. Pediococci are much more active than lactobacilli in forming acetate from lactate.

Cheddar cheese is one of the few varieties without a deliberately added secondary microflora, but there is considerable interest in inoculating milk for Cheddar with selected mesophilic lactobacilli with the objective of accelerating ripening and/or intensifying its flavor (see Chapter 15).

In virtually all artisanal cheeses, *Leuconostoc* and *Enterococcus* spp. are also found in large numbers and contribute to flavor development. Further information on enterococci is given in Chapter 20.

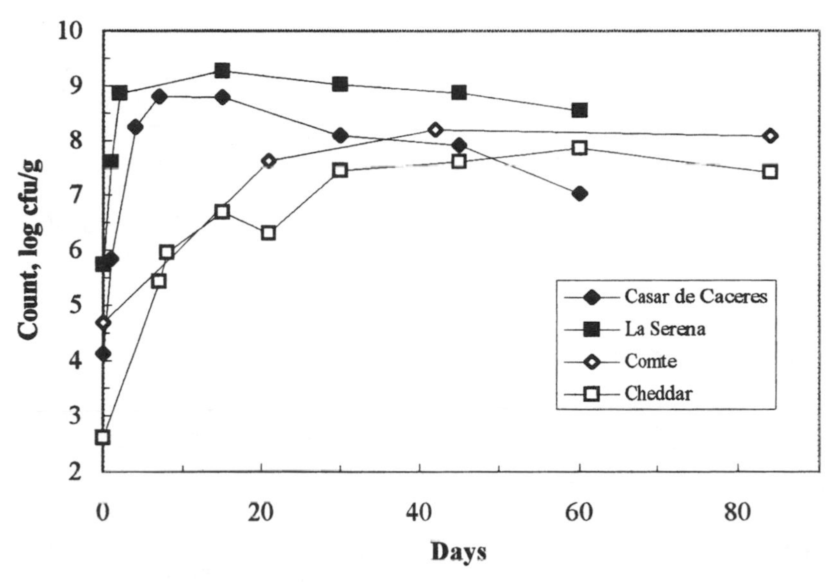

Figure 10–8 Growth of mesophilic (mainly facultatively heterofermentative) lactobacilli in Casar de Cáceres, La Serena, Comté, and Cheddar cheese during ripening.

10.5 OTHER MICROORGANISMS IN RIPENING CHEESE

Many cheese varieties contain a secondary, non-LAB microflora, the function of which is to produce some specific characteristic change in the cheese, such as surface growth in the case of bacterial-ripened (smear-ripened) and mold-ripened cheeses and the production of CO_2, propionate, and acetate in the case of some Swiss varieties (e.g., Emmental and Comté). CO_2 is responsible for eye formation in these cheeses. In all of these cheeses, flavor development is dominated by the metabolic activity of the secondary flora.

Several microorganisms are involved, including bacteria (*Arthrobacter, Brevibacterium, Brachybacterium, Corynebacterium, Microbacterium, Propionibacterium*, and *Micrococcus* spp.), yeasts (*Kluyveromyces marxianus* and *Debaryomyces hansenii*), and molds (*Geotrichum candidum*, *Penicillium camemberti*, and *P. roqueforti*). They are all involved in ripening and, except for *Propionibacterium* spp. and *P. roqueforti*, develop only on the cheese surface. The surface microflora has two important functions in ripening:

1. production of enzymes
2. deacidification of the cheese surface

The enzymes include lipases, proteinases, and peptidases. The lipases and proteinases hydrolyze the fat and protein to fatty acids and peptides, respectively, while the peptidases hydrolyze the smaller peptides to amino acids. Both fatty acids and amino acids are the precursors of many of the flavor compounds in mold- and smear-ripened cheese (see Chapters 11 and 12).

During the first few days of ripening smear- and mold-ripened cheese, yeast and molds grow on the cheese surface and deacidify it by oxidizing the lactate to H_2O and CO_2. In turn, this causes the pH of the surface to increase from an initial value of about 4.8 to 5.8 or higher. The surface bacteria grow poorly, if at all, at the low initial pH, and the increase in pH promotes their growth considerably.

The presence of molds and yeasts on the cheese surface is to be expected, since cheese has a relatively low pH (both types of microorganism can grow at pH values below 3), a readily fermentable substrate (lactate) and a relatively low a_w. Traditionally, cheese became contaminated by these microorganisms from the environment, and their growth was promoted by high relative humidity and/or high temperature (12–24°C) in the ripening rooms or caves. Wooden shelves are extremely porous and are a likely source of contamination. Generally, the bacteria and molds are added deliberately to the milk or cheese, but the yeasts are adventitious contaminants of the cheese surface. In those cheeses in which they are found, yeasts grow to 10^6/cm^2 of surface within a few days of ripening, after which they generally decrease and stabilize at 10^4 to 10^5/cm^2. Several methods are used to inoculate the cheese with molds, including addition to the curd before molding, addition to the brine, dusting spores on the surface of the curd, and smearing with an aqueous suspension of mold spores. In modern practice, the milk for mold-ripened varieties is inoculated with a pure culture of *P. roqueforti*, in the case of Blue cheeses, or *P. camemberti*, in the case of Camembert and Brie, at the same time as the starters. The curd for Blue cheese is subsequently pierced to allow limited entry of O_2 to promote growth of *P. roqueforti*. Surface- or smear-ripened cheeses, like Tilsit, Munster, and Limburger, are dipped, sprayed, or brushed with aqueous suspensions of *G. candidum* and *B. linens* as soon as the cheeses are removed from the brine. Both mold- and bacterial-ripened cheese are then ripened at 10–15°C at a high relative humidity to prevent the loss of moisture from the cheese surface. Traditionally, natural contamination of the milk was relied upon as the source of propionic acid bacteria in the case of Emmental and Comté cheeses, but nowadays these bacteria generally are added deliberately to the milk with the starter cultures.

For a long time, *B. linens* was thought to be the most important bacterium growing on the surface of smear-ripened cheeses. Recent evidence (Table 10–3) shows that other bacteria, particularly *Arthrobacter globiformis*, *A. nicotianae*, *Corynebacterium ammoniagenes*, *C. variabilis*, *Microbacterium imperiale*, and *Rhodococcus facians* are also important. It should be noted that many of the bacteria have not been identified, implying that the surface microflora is very complicated. In addition, *Micrococcus* spp. are found on the surface of Roquefort and Comté cheese and probably most other cheese, but information on the species involved is very limited. *Brachybacterium alimentarium* and *Br. tyrofermentans* are also found on Comté and Beaufort cheese. All of the smear bacteria are salt-tolerant (the surface layer of surface-ripened cheese can contain up to 15% NaCl), aerobic, or facultatively anaerobic microorganisms and hence grow easily at the high salt level in the surface layer of these brine-salted cheeses. *B. linens* does not grow at pH values below 6.0. This is also probably true of the other bacteria found on the surface of cheese.

Arthrobacter, Brevibacterium, Brachybacterium, Corynebacterium, and *Microbacterium* are generically called coryneform bacteria. All of them are Gram-positive, catalase-positive, non-spore-forming, and generally nonmotile rods. A major feature of their growth is that exponential-phase cells are pleomorphic, showing the presence of irregularly shaped rods, including wedge, club, V, and curved shapes. In addition, *Arthrobacter*, *Brevibacterium*, and *Brachybacterium* spp. go through a marked rod-coccus cycle during growth, with rod forms dominating the exponential phase of growth (1–2 days) and coccal forms dominating the stationary phase (5–7 days). Except for *Corynebacterium* and *Brachybacterium*, metabolism of sugars by coryneforms, if it occurs, is respiratory, although the evidence for acid production from glucose by *Micrococcus* spp. is conflicting. Cell-wall composition (particularly the amino acids [lysine, ornithine, or 2,6-diaminopimelic acid] and sugars [arabinose or galactose] found in the peptidoglycan), the presence and type of menaquinones, the presence or absence of mycolic acids, and whether they go through a rod-coccus cycle during growth are important criteria in identifying coryneform bacteria. Despite this, their taxonomy is very confusing. All belong,

Table 10–3 Species of Bacteria Found in Smear-Ripened Cheeses

Species	*Limburger, Romadour, Weinkase, and Harzer Cheese*[a] *(6 Plants)*	*Tilsit Cheese Mainly*[b] *(15 Plants)*
Arthrobacter citreus	1	19
Arthrobacter globiformis		102[c]
Arthrobacter nicotianae	14	10
Brevibacterium fermentans	3	
Brevibacterium imperiale		8
Brevibacterium linens	25	77[d]
Brevibacterium fuscum		2
Brevibacterium oxydans	1	3[e]
Brevibacterium helvolum		1
Corynebacterium ammoniagenes	36	53
Corynebacterium betae		4
Corynebacterium insidiosum		8
Corynebacterium variabilis	12	14
Curtobacterium poinsettiae		12
Microbacterium imperiale	5	8
Rhodococcus fascians	15	
Total number identified	112	321
Total number not identified	36	73

[a] N. Valdes-Stauber, S. Scherer, & H. Seiler, Identification of yeasts and coryneform bacteria from the surface microflora of brick cheeses, *International Journal of Food Microbiology, 34* (1997), 115–119.

[b] F. Eliskases-Lechner and W. Ginzinger, The bacterial flora of surface-ripened cheeses with special regard to corneforms, *Lait, 75* (1995), 571–584.

[c] Seventy-two from 1 plant.

[d] From 11 of 15 plants.

[e] Only from 1 plant.

however, to the actinomycete branch of the Gram-positive bacteria. There is now overwhelming biochemical and genetic evidence that *Corynebacterium sensu stricto*, which contain mycolic acids, are quite unrelated to the coryneform bacteria (*Arthrobacter, Brevibacterium*, and *Microbacterium*), which lack mycolic acids.

A brief discription of the above genera and of other microorganisms found in cheese is given below.

10.5.1 Arthrobacter

These are Gram-positive, catalase-positive, nonmotile, obligately aerobic rods that go through a marked rod-coccus cycle during growth. Their peptidoglycan contains lysine. They have nonexacting nutritional requirements; generally, biotin is the only vitamin required. Their habitat is soil, and they do not withstand HTST pasteurization of milk.

10.5.2 Brachybacterium

These are Gram-positive, facultatively anaerobic short rods that exhibit a rod-coccus growth cycle. Their optimum temperature is around 30°C. They contain *meso*-diaminopimelic acid and glucose, galactose, and rhamnose, but not mycolic acids, in their cell walls. Five species are recognized and two of these, *Br.*

alimentarium and *Br. tyrofermentans*, have been isolated from Comté and Beaufort cheese, respectively (Schubert et al., 1996). Their nutritional characteristics do not appear to have been studied, but *Br. alimentarium* and *Br. tyrofermentans* can grow in the presence of 14% and 16% NaCl, respectively.

10.5.3 Brevibacterium

These are Gram-positive, catalase-positive, nonmotile, obligately aerobic rods that go through a marked rod-coccus cycle during growth. Their cell walls contain *meso*-diaminopimelic acid, and they metabolize sugars by respiration. There are four species: *B. linens, B. casei, B. iodinum*, and *B. epidermidis*. The first two species have been isolated from cheese, the third from milk, and the fourth from skin. *B. linens* produces yellow, orange, red, or brown colonies, while those of *B. iodinum* are purple, due to the production of a phenazine derivative. The other two species produce gray-white colonies. Brevibacteria grow poorly, if at all, at 5°C, have an optimum temperature of 20–25°C, and grow in the presence of a high concentration of NaCl. *B. linens* and *B. iodinum* grow in the presence of 8–10% NaCl, and *B. casei* and *B. epidermidis* in the presence of 15% NaCl. Their nutrition has not been studied in depth, but most strains of *B. linens* require amino acids and vitamins for growth. Their metabolism is respiratory, and they do not produce acid from glucose. They are easily confused with *Arthrobacter* spp. *B. linens* metabolizes methionine to methional, which is thought to be responsible from the characteristic "dirty sock" odor of smear-ripened cheeses. *B. linens* can be determined specifically through the production of a stable pink color within 2 min following treatment of a small amount of a colony with a drop of 5 M KOH or 5 M NaOH or a salmon pink color within 1 min following treatment with glacial acetic acid. Brevibacteria are acid sensitive and will not grow at a pH value below 6.0. Their major habitats are dairy products, especially cheese, activated sludge, and human skin.

10.5.4 Corynebacterium

These are Gram-positive, catalase-positive, nonmotile, facultatively anaerobic slightly curved rods with tapered ends; club-shaped forms may be found also. Currently, 16 species of *Corynebacterium* are recognized. A rod-coccus cycle does not occur. Metachromatic granules, which stain deeply with methylene blue, are formed. *Meso*-diaminopimelic acid and short-chain mycolic acids (22–36 C atoms) are found in their cell walls. They are nutritionally exacting, requiring several vitamins, amino acids, purines, and pyrimidines for growth. Two bacteria, *Microbacterium flavum* and *Caseobacter polymorphus,* which were isolated from cheese, have been reclassified as *C. flavescens* and *C. variabilis*, respectively. *C. variabilis* was isolated from the surface of Dutch smear-ripened cheese. This organism produces gray-white, slightly pink, or slightly red colonies.

10.5.5 Microbacterium

These are small, Gram-positive, nonmotile or motile rods that do not go through a rod-coccus cycle. However, in older cultures (3–7 days), the rods are short and a proportion may be coccoid. Currently, there are 13 species, only one of which, *M. lacticum*, has been found in milk. Their optimum temperature is 30°C. Colonies vary in color from gray-white to pale green or yellow. Their cell wall peptidoglycan contains lysine. Generally, their metabolism is respiratory, but acid is produced from glucose and some other sugars in peptone-containing media. Most strains require biotin, pantothenic acid, and thiamine for growth. The main species found in milk is *M. lacticum*, which is thermoduric and survives heating at 63°C for 30 min. The organism is not found in aseptically drawn milk, and there is strong evidence that the major source of contamination of milk with this organism is improperly cleaned dairy equipment.

10.5.6 Rhodococcus

These are Gram-positive, catalase-positive, aerobic rods that usually produce an extensive mycelium, which may fragment into rods and cocci in older cultures. Most grow well on normal laboratory media at 30°C, producing orange, pink, red, or brown colonies. Their cell walls contain *meso*-diamnopimelic acid, arabinose, galactose, and mycolic acids having 32–66 carbon atoms and up to 4 double bonds. They are closely related to *Corynebacterium* and are found extensively in soil and dung.

10.5.7 Propionibacterium

These are Gram-positive, nonmotile, pleomorphic rods, which may be coccoid, bifid, or sometimes branched in shape. They may occur singly, in pairs, in short chains, or in clumps with "Chinese lettering" arrangements. Colonies vary in color and can be white, gray, pink, red, yellow, or orange. Despite the fact that these organisms are catalase positive, they are essentially anaerobic or microaerophilic bacteria. The genus *Propionibacterium* is divided into two groups, the classical group and the acnes group. The classical propionibacteria are found mostly in dairy products, particularly cheese, although they are also found in silage and olive fermentations. The acnes group are found on human skin. The classical group is divided into four species: *P. freudenreichii,* which is the most common species; *P. jensenii*; *P. thoenii*; and *P. acidipropionici*. The peptidoglycan of *P. freudenreichii* contains *meso*-diaminopimilic acid, while the L isomer is found in the other three species. *P. freudenreichii* was considered to exist as two subspecies, *P. freudenreichii* subsp. *freudenreichii* and *P. freudenreichii* subsp. *shermanii.* Genetic studies have shown that both subspecies are identical; the only phenotypic difference is that *P. freudenreichii* subsp. *freudenreichii* is able to ferment lactose while the other subspecies cannot. *P. freudenreichii* and *P. jensenii* produce cream-colored colonies, while *P. thoenii* and *P. acidipropionici* produce red-brown and cream to orange-yellow colonies, respectively. *P. thoenii* is β-hemolytic. Propionic acid bacteria have relatively simple nutritional requirements, although they generally require pantothenic acid, biotin, or thiamine for growth, and many of them can use NO_3^- as the sole source of N.

10.5.8 Pediococcus

These are Gram-positive, catalase-negative cocci that occur in tetrads. Tetrad formation is due to cell division in two directions in a single plane and is characteristic of this genus. Currently, eight species are recognized. They are homofermentative, producing either DL or L lactate from sugars, and most strains can grow in the presence of 6.5% NaCl. They generally have complex nutritional requirements. Most pediococci do not ferment lactose and therefore grow poorly in milk. Those strains that can metabolize lactose may lack a proteinase to hydrolyze the milk protein to the amino acids and peptides required for growth in milk.

Some pediococci can grow at pH 8.5 and 4.2, and some metabolize citrate to acetate and formate rather than diacetyl and acetoin. Pediococci are found occasionally as a minor part of the NSLAB flora in some hard cheeses, but their influence on the production of cheese flavor is not clear.

10.5.9 Micrococcus

Micrococci are Gram-positive, catalase-positive, strictly aerobic, nonmotile cocci (0.2–2.0 mm in diameter) that occur in pairs, clusters, or tetrads. Division occurs in several planes, resulting in formation of regular and irregular clusters. Their natural habitat is skin, and currently 17 species are recognized. All grow in the presence of 5% NaCl and many in the presence of 10–15% NaCl. Many species produce yellow, orange, or red colonies. Their nutritional requirements are variable. *M. luteus*, the type species, produces yellow colonies and grows on glutamate as the sole source of C and N in the presence of thiamin and/or biotin. Some species can utilize ammonium phosphate as a N source, but many species have complex nutritional requirements. They are commonly found on the surface of smear-ripened cheese but the species found are not clear. They dominate the surface of Comté and Blue cheese.

10.5.10 Staphylococcus

These are Gram-positive, catalase-positive, facultatively anaerobic, nonmotile cocci (0.5 to 1.5 μm in diameter), which characteristically divide in more than one plane to form clusters. They can also occur in pairs and tetrads. Currently, 19 species of staphylococci are recognized and many produce yellow or orange colonies. They are facultative anaerobes and grow better aerobically than anaerobically. Most strains grow in the presence of 10% NaCl and between 10°C and 40°C. Acid is produced anaerobically from several sugars, including glucose and lactose. They are fastidious, requiring 5–12 amino acids and several B vitamins for growth. *S. aureus* causes mastitis in cows and boils and carbuncles in humans and is considered to be a pathogen. Many strains of *S. aureus* produce a heat-stable enterotoxin that causes food poisoning. Growth to about 10^6 cfu/g in food is necessary to produce sufficient toxin (0.1–1.0 mg/kg) to cause food intoxication. Enterotoxins are difficult to measure. Coagulase is accepted as the indicator of pathogenicity in staphylococci, and *S. aureus*, *S. intermedius*, and *S. hyicus* produce it. *S. intermedius* has been found in the nasal passages of horses, dogs, mink, and foxes, and *S. hyicus* has been found on the skin of pigs and less frequently on the skin and in the milk of cows. Major habitats of staphylococci include the nasal membranes, skin, and the gastrointestinal and genital tracts of warm-blooded animals.

Traditionally, *Micrococcus* and *Staphylococcus* have been placed in the family *Micrococcaceae,* indicating that they are closely related. However, phylogenetic studies show that they are quite distant from each other; *Staphylococcus* spp. belong to the *Clostridium* branch of the Gram-positive bacteria, and *Micrococcus* belong to the actinomycete branch. This is also reflected in the guanine plus cytosine levels, with staphylococci containing 30–39% and micrococci 63–73%. Phylogenetically, *Micrococcus* is closely related to *Arthrobacter* and may be a degenerate form of this genus.

It is relatively easy to distinguish micrococci from staplylococci. The simplest way is to check for acid production from glucose under aerobic and anaerobic conditions. Staphylococci produce acid from glucose aerobically and anaerobically, whereas micrococci either do not produce acid or produce it only aerobically. In addition, micrococci are resistant to lysostaphin, a cell-wall degrading enzyme, and are sensitive to erythromycin (0.04 mg/ml) whereas staphylococci have the opposite reactions.

10.5.11 Yeasts and Molds

Yeasts and molds are generally not nutritionally demanding and are larger and grow more slowly than bacteria. Therefore, they do not compete with bacteria in environments in which bacteria grow, for example, at pH values around 7. However, they grow quite well at pH values of 2 to 4, where bacteria either do not grow or grow only very poorly. The low pH of freshly made cheese is therefore partially selective for their growth. Yeast and molds are eukaryotes—that is, they contain a clearly identifiable nucleus—and most of them also contain chitin, a β-1,4 polymer of N-acetylglucosamine, which is responsible for their rigid structure.

Colonies of yeast generally have a soft consistency, while those of molds are hard and large and often exhibit several different colors. In addition, they look quite different under the microscope: yeast are generally round and pear-shaped whereas molds show a mycelial network of filamentous hyphae. Some fungi are dimorphic, producing hyphae under one set of circumstances and yeastlike cells under another. The human pathogen *Candida albicans* is the best example of dimorphism; it grows like a yeast in body fluids but develops hyphae to invade tissue.

Both yeast and molds are classified as fungi and are divided into three major groups: Ascomycetes, Zygomycetes, and Deuteromycetes. Classification of fungi is very complex, and only a few important criteria are mentioned here. These include whether cells in the mycelium are septate (possess cross-walls) or nonseptate (lack cross-walls), the type of spores and how they are produced, and whether reproduction is sexual or asexual. Ascomycetes and Zygomycetes are septate, and Deuteromycetes are nonseptate. The spores produced by Ascomycetes are formed in a sac called the ascus (for this reason these spores are called ascospores) and are involved in sexual reproduction. The spores produced by Deuteromycetes and Zygomycetes are called conidia (see below) and sporangiophores, respectively, and are not involved in sexual reproduction. Yeast generally multiply by budding: a protuberance is formed on the wall of the cell and eventually breaks off to form a new cell, in which further budding occurs. Sometimes, several buds are produced by the same cell and remain attached to it. Some yeast (*Schizo saccharomyces* spp.) multiply by binary fission. Sexual reproduction is given the generic name *teleomorph*, and asexual reproduction is called *anamorph*. The same fungus has often been given different names, depending on the type of reproduction. Some examples of this are shown in Table 10–4. Taxonomically, the teleomorphic name is normally used, but there are exceptions. For instance, the anamorphic name *Geotrichum candidum* is more commonly used than the teleomorphic name *Galactomyces candidum*.

The species of yeast found on the surface of different cheeses show considerable diversity (Table 10–4). The dominant species in all cheeses, except Romadour from one plant, is *D. hansenii. Kluvyeromyces* spp. are also dominant in the French cheeses (Roquefort, Camembert, and St. Nectaire) and in the Spanish cheese (Cabrales) but appears to be absent from the German and Austrian cheeses (Weinkase, Limburg, Romadour, and Tilsit). *Saccharomyces cerevisiae* and *Pichia* spp. are also important in Camembert and Cabrales cheese. All these yeasts are members of the Ascomycetes group. *S. cerevisiae* is also involved in wine, beer, and bread making. Very few yeast are capable of fermenting lactose, but *Kluvyeromyces lactis* is an exception. This may be one reason for its dominance in some surface-ripened cheese. Whether variation occurs within the same cheese has not been studied to any great extent. The evidence in Table 10–4 suggests that it does occur, at least in Limburg. Both cheeses examined contained *D. hansenii* and *Geotrichum candidum* in significant numbers, but in addition *Torulaspora delbrueckii* was found in one cheese and *Yarrowia lipolytica* in the other.

The most important molds in cheese are *P. camemberti, P. roqueforti*, and *G. candidum*; all are members of the Deuteromyces group. *P. camemberti* is responsible for the white growth on the surface of Camembert and Brie, and *P. roqueforti* is responsible for the blue veins in Roquefort and other Blue cheeses. It is generally thought that *G. candidum* is present on the surface of most mold- and bacterial-ripened cheese. The results in Table 10–4 suggest that it is found only in Weinkase, Romadour, Limburg, and Tilsit. Scanning electron micrographs of Camembert cheese show the presence of *G. candidum*, and it is likely that the reason it was not reported to be present in the other cheeses in Table 10–4 is that the various workers involved considered it to be a mold rather than a yeast.

Microscopic observation is very important in classifying fungi because their various structures can be seen clearly. Both *P. camemberti* and *P. roqueforti* reproduce asexually from conidia (spores) extruded from a flask-shaped cell called a *phialide*, which is borne on the conidiophore or spore-bearing hyphae (Figure 10–9). The multiplication of *G. candidum* is quite different. The hyphae grow to a considerable extent, then stop, and septa are formed transversely, separating the hyphae into short compartments that eventually fragment into separate conidia, which start the reproductive process again. Many molds produce toxins that are carcinogenic (e.g., the aflatoxins produced by *Aspergillus flavus*). However, the strains involved in cheese do not produce toxins. The physiological conditions for the production of

Table 10–4 Species of Yeast Found in Different Cheeses

		Weinkase[1]		*Romadour*[1]			*Limburg*[1]						
Teleomorph	*Anamorph*	*Factory A*	*Factory B*	*Factory C*	*Factory D*	*Factory A*	*Factory A*	*Factory C*	*Tilsit*[2]	*Roquefort*[3]	*Cabrales*[4]	*Camembert*[5]	*St. Nectaire*[6]
Candida catenulata			2	3				12	2				
Candida intermedia			2		10								
Candida mogii												6	
Candida rugosa											4		
Candida saitoana										11			
Candida versatilis													
Cryptococcus flavus													
Debaryomyces hansenii	*Candida famata*	86	95	69		55	64	85	79	16	30	6	86
Dipodascus capitatus	*Geotrichum capitatum*										15		
Galactomyces geotrichum	*Geotrichum candidum*	4	1	6	3	21	17	2	5				
Kluyveromyces lactis	*Candida sphaerica*									35	7	52	10
Kluyveromyces marxianus	*Candida kefyr*									6		9	1
Leucosporidium scottii	*Candida scottii*												
Pichia anomala	*Candida pelliculosa*									15			
Pichia fermentans	*Candida lambica*										16		
Pichia kluyveri											6		
Pichia membranaefaciens	*Candida valida*									7	21		
Rhodotorula spp.										5			
Saccharomyces cerevisiae	*Candida robusta*										1	3	
Saccharomyces unisporus											3		
Torulaspora delbrueckii	*Candida colliculosa*					24					1		
Trichosporon beigelii		3		22									
Yarrowia lipolytica	*Candida lipolytica*				87		19		7	5			
Zygosaccharomyces rouxii											1	8	

Note: Results from reference 1 are as a percentage of the surface yeast microflora; other results are as a percentage of the number of strains isolated and/or identified. Factories A and C produced more than one variety of cheese.

[1] N. Valdes-Stauber, S. Scherer, & H. Seiler, Identification of yeasts and coryneform bacteria from the surface microflora of brick cheeses, *International Journal of Food Microbiology, 34* (1997), 115–129.

[2] E. Eliskases-Lechner & W. Ginzinger, The yeast flora of surface-ripened cheese, *Milchwissenschaft, 50* (1995), 458–462.

[3] J.J. Devoyod & D. Sponem, La flore microbiennee du fromage de Roquefort. 6: Les levures, *Lait, 50* (1970), 524–543.

[4] M. Nunez, M. Medina, P. Gaya, & C. Dias-Amado, Les levures et les moissures dans le fromage bleu de Cabrales, *Lait, 61* (1981), 62–79.

[5] C. Baroiller & J. Schmidt, Contribution a l'etude de l'origine des levures du fromage Camembert, *Lait, 70* (1980), 67–84.

[6] J. Vergeade, J. Guiraud, J.P. Larpent, & P. Galzy, Etude de la flore de levure du Saint-Nectaire, *Lait, 56* (1976), 275–285.

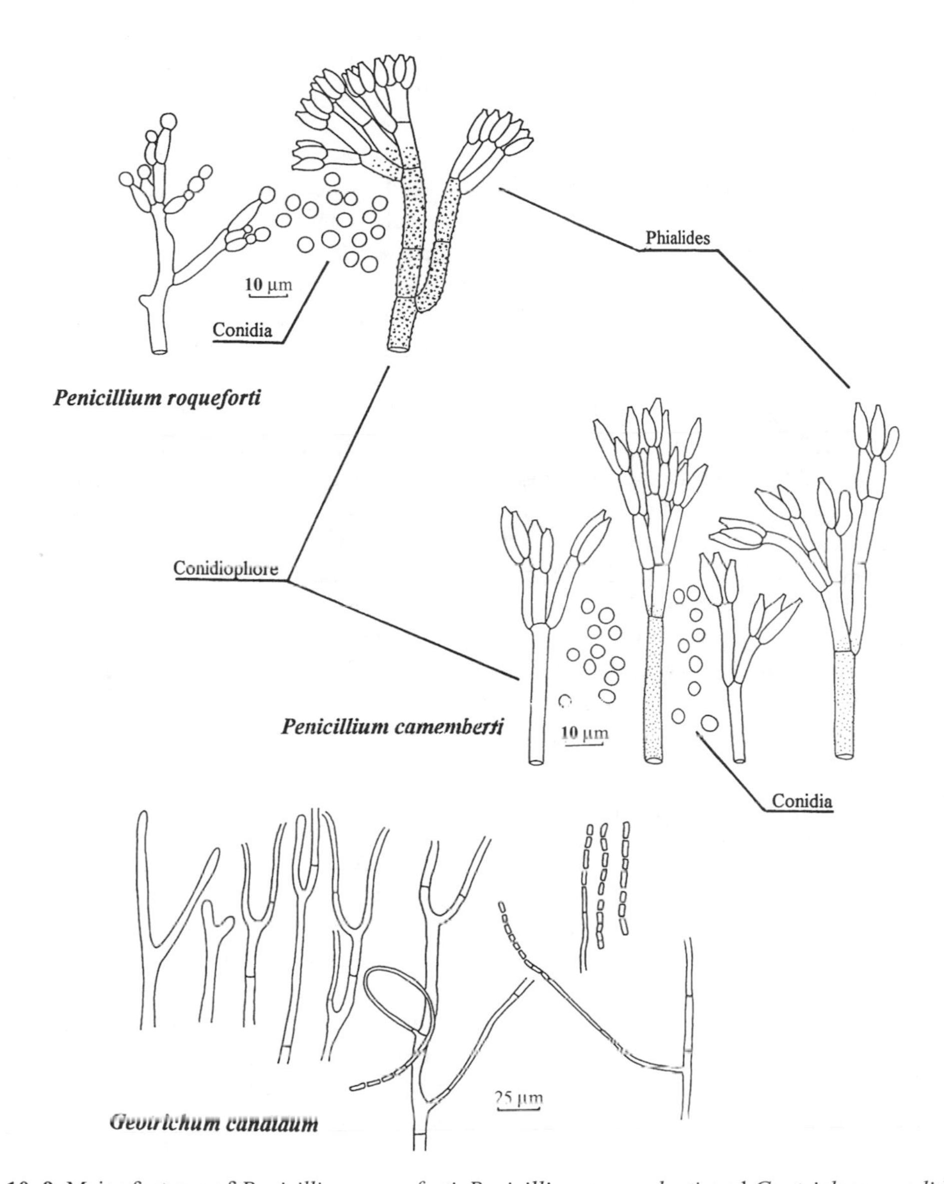

Figure 10–9 Major features of *Penicillium roqueforti*, *Penicillium camemberti*, and *Geotrichum candidum*.

toxins by microorganisms are generally much narrower than those for growth.

G. candidum has characteristics of both a yeast and a mold, and in the past it was often called a yeastlike fungus. It was initially called *Oidium lactis*, later called *Oospora lactis*, and then given its present name. It is commonly known as the dairy mold. Its natural habitat is soil, where it is involved in the decay of organic matter.

Most fungi grow quite well at the pH of cheese, and most of those found in cheese are also quite tolerant to salt. For example, the growth of *P. camemberti* is largely unaffected by 10% NaCl (Table 10–1), and some strains of *P. roqueforti* can tolerate 20% NaCl. An exception is *G. candidum*, which is quite sensitive to salt. A slight reduction in its growth occurs in the presence of 1% NaCl, and it is completely inhibited at about 6%. Therefore, too much brining will prevent its growth on the cheese surface. Perhaps its intolerance to salt explains why it is sometimes deliberately added in the manufacture of some surface-ripened cheeses, the hope being that some cells will grow.

Generally, yeast are facultative anaerobes whereas molds are considered to be obligate aerobes. However, *P. roqueforti* can grow in the presence of limited levels of O_2, as demonstrated by its growth throughout the mass of Blue cheese. Yeast and molds are generally heat sensitive and are killed by pasteurization.

Occasionally, yeast have been incriminated in the spoilage of cheese, either through the production of gas (CO_2) or the development of off-flavors. Unripened cheeses (e.g., Cottage cheese and Quarg), which can contain a high level of lactose, are particularly prone to spoilage by yeast. Reviews of the role of yeast in dairy products include those of Fleet (1990) and Jakobsen and Narvhus (1996).

10.6 EXAMPLES OF MICROBIAL GROWTH IN CHEESE

Cheese is a very complex microbiological ecosystem in which molds, yeasts, and bacteria coexist and multiply. Some examples are given below.

10.6.1 Cheddar

This is a hard, dry-salted cheese made with a mesophilic starter, which grows very rapidly in the cheese from an initial level of about 10^7/g to 10^8 or 10^9/g at salting (≈ 5.5 hr after inoculation). Mesophilic starters generally die out relatively rapidly during the first few weeks of ripening (Figure 10–5), but the rate is strain dependent and probably reflects the ability of the strain to withstand the cooking temperature of the cheese and its ability to lyse. Phage may also be involved in reducing the number of cells.

Normally, the fermentation of lactose by the starter LAB in cheese is rapid and is complete within 1 day. However, in dry-salted cheeses, like Cheddar, a relatively large amount of lactose (≈ 10g/kg of cheese) is present in the curd after overnight pressing. This is due to inhibition of the metabolism of the starter cultures by the salt and the relatively low pH. The S/M in Cheddar cheese determines the subsequent rate of lactose fermentation by the starter. High S/M levels reduce the rate and low levels increase it (Figure 10–10). For example, at an S/M of 4.1 the fermentation is virtually complete in 7 or 8 days, while at a 6% level it takes more than 50 days. These values are only indicative and vary depending on the sensitivity of the particular culture to salt. *Lc. lactis* subsp. *cremoris* is much more sensitive to salt than *Lc. lactis* subsp. *lactis* strains. The former cannot grow in the presence of 4% salt whereas the latter can. Salt can also uncouple acid production from growth.

NSLAB, particularly facultative heterofermenters like *Lb. paracasei* and *Lb. casei*, are facultative anaerobes and are also acid and salt tolerant. They can grow at pH 4.5 and in the presence of more than 6% salt. In fact, many of them can grow in the presence of 8 or 10% salt, so salt is of little consequence in preventing their growth in cheese. They grow relatively rapidly in Cheddar cheese during ripening from low initial numbers (≈10^2/g) to final numbers of 10^7 to 10^8/g. Generation times of 8.5 days have been reported for Cheddar cheese ripened at 6°C. Such high numbers of lactobacilli must have some role in flavor development in the cheese, but that role is unclear. They do transform L-lactate to D-lactate, eventually forming a racemic mixture, but this transformation has no effect on the flavor of the cheese. However, Ca D-lactate is very insoluble and can form small crystals throughout the cheese. An example of the growth of lactobacilli and the racemization of the L-lactate in Cheddar cheese during ripening is shown in Figure 10–11.

10.6.2 Emmental and Comté

Both of these cheeses are made with thermophilic cultures, and little acid is produced in the vat during their manufacture. During the initial hours in the press, most of the lactic acid is pro-

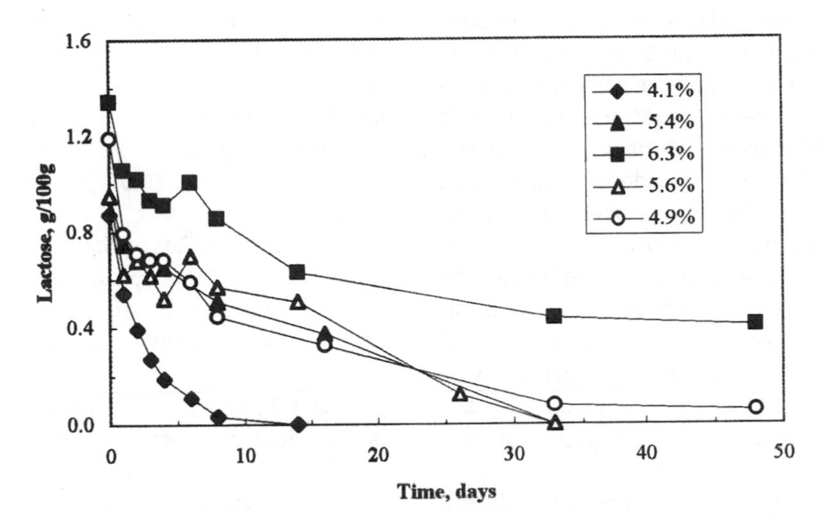

Figure 10–10 Effect of salt-in-moisture (S/M) on lactose metabolism in Cheddar cheese made with *Lc. lactis* ssp. *cremoris* C13 and 266 and ripened at 12 ° C.

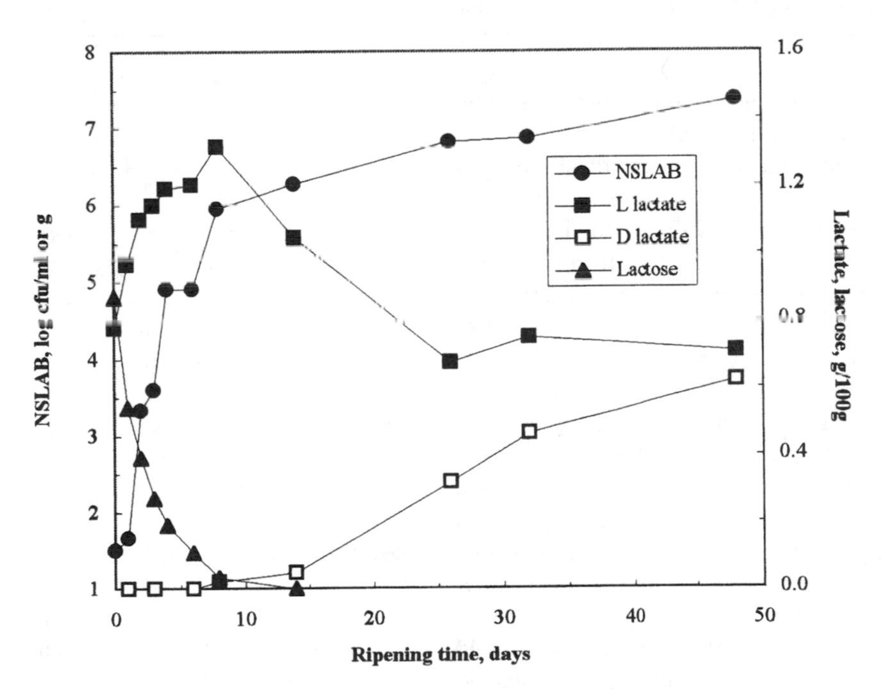

Figure 10–11 Relationship between nonstarter lactic acid bacteria (NSLAB; ●), metabolism of lactose (▲), and production of L-lactate (■) and D-lactate (❑) in Cheddar cheese (4.1% salt-in-moisture) during ripening at 12°C.

duced by *Sc. thermophilus*, but, as the temperature and pH decrease, *Lb. helveticus* begins to grow, reaching maximum numbers 12–20 hr after the addition of starter. Counts of both *Sc. thermophilus* and *Lb. helveticus* in Comté cheese, and presumably Emmental also, are higher at the periphery than at the center (Figure 10–12). In both cheeses, growth of the starter is limited by the high cooking temperature (52–54°C), but growth begins again as soon as the temperature decreases. The temperature falls more rapidly at the periphery of the cheese than in the center, and hence greater bacterial growth (and acid production) occurs at the periphery.

Sc. thermophilus metabolizes only the glucose moiety of lactose and excretes galactose, which, along with residual lactose, is metabolized by *Lb. helveticus*. All the lactose is fermented during the first 10 or 12 hr of manufacture. The L isomer of lactate is produced by both *Sc. thermophilus* and *Lb. helveticus*, while the D isomer is produced only by the latter organism. After several weeks ripening at a low temperature (4–14°C), the cheese is placed in a "warm room" at 18–24°C, during which the propionic acid bacteria grow and transform the lactate to propionate, acetate, and CO_2, which is responsible for eye formation. The eyes in Emmental cheese are much larger than in Comté cheese because Emmental is ripened at 22°C and Comté at 18°C. In traditionally made Comté and Emmental, the propionic acid bacteria are natural contaminants of the raw milk, but, in the industrial production of Emmental, they are normally added deliberately to the milk to give initial counts of about 10^3 to 10^5/ml.

The pathway of lactate fermentation by propionic acid bacteria is complicated (Figure 10–13) and involves two separate cycles, one in which propionate is produced and the other in which acetate is produced. The lactate is first oxidized to pyruvate, two moles of which are converted to propionate and one mole to acetate. ATP is only

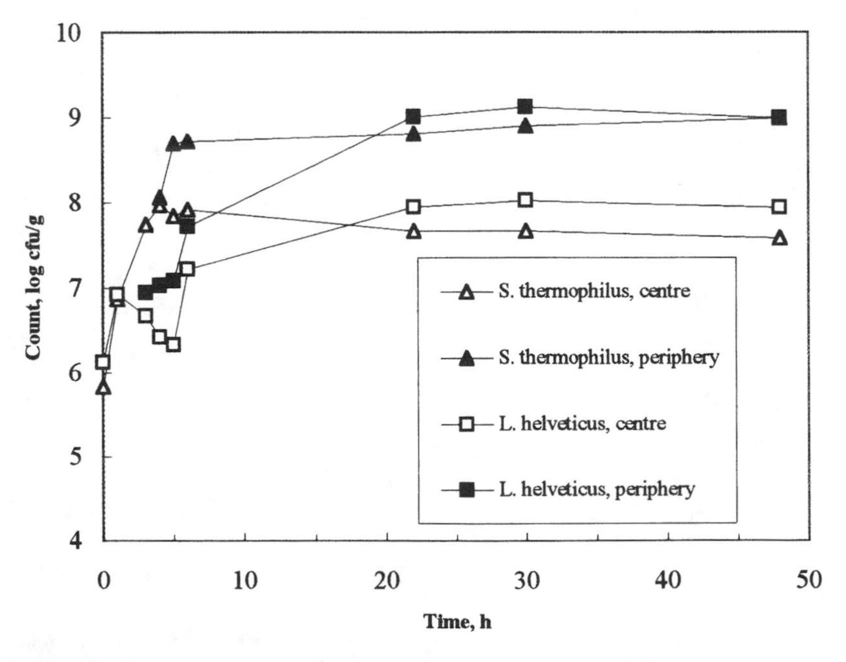

Figure 10–12 Growth of *Streptococcus thermophilus* (Δ, ▲) and *Lactobacillus helveticus* (❑, ■) at the center (open symbols) and periphery (closed symbols) of Gruyerè cheese during manufacture.

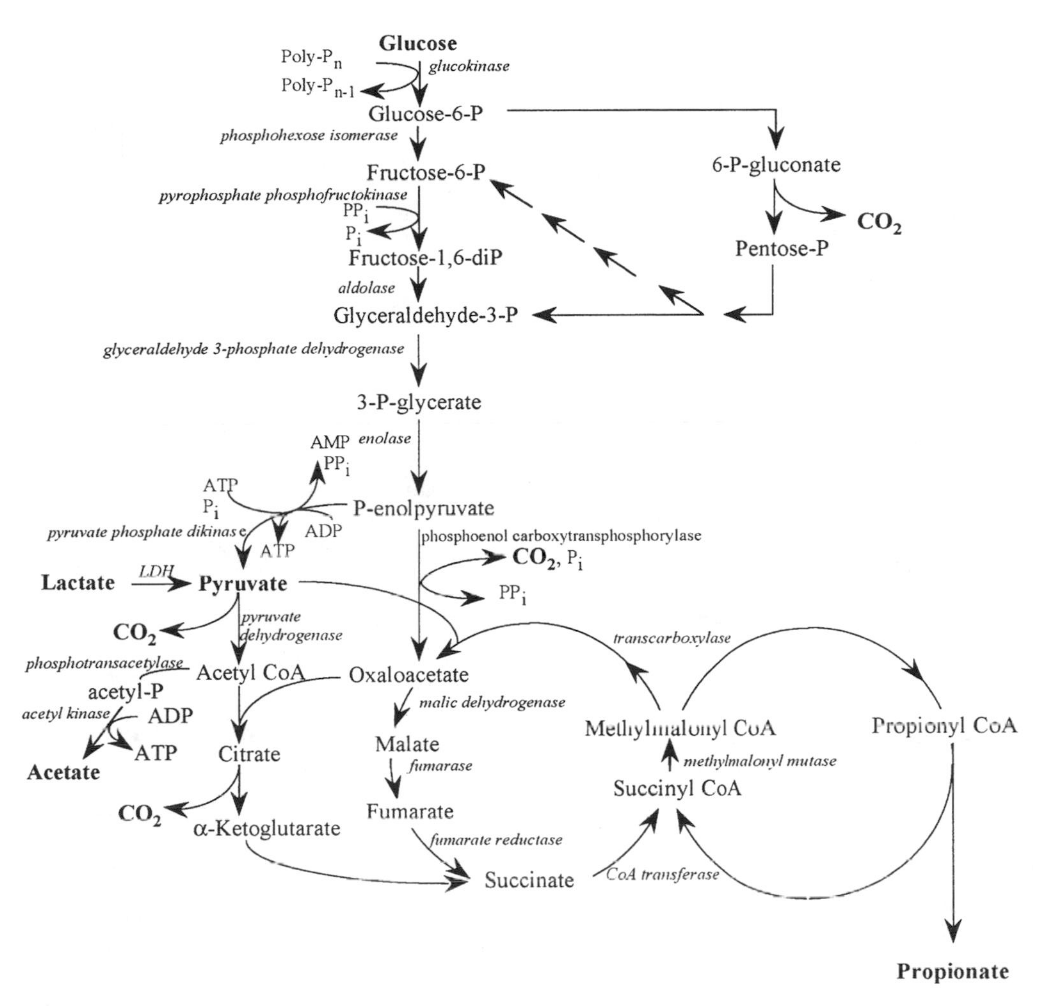

Figure 10–13 Cycles involved in the propionic acid fermentation (LDH, lactose dehydrogenase; Poly-Pn, polyphosphate; PPi, pyrophosphate). For reasons of clarity, only the pyrophosphate-dependent conversion of fructose-6-P to fructose-1,6-diP is shown, and the generation of ATP by the electron transfer system is omitted. All the reactions are directed toward propionate production, even though the reactions are reversible.

generated in the production of acetate. The overall stoichiometry is

$$3 \text{ Lactate} \rightarrow 2 \text{ Propionate} + 1 \text{ Acetate} + 1\ CO_2 + 1\ H_2O$$

Transcarboxylase is the key enzyme in the production of propionate and requires biotin for activity. Generally, propionic acid bacteria are able to metabolize both isomers of lactate, but, in a mixture of the two, they preferentially metabolize the L rather than the D isomer. The complex interrelationships between lactose and lactate utilization and production of propionate and acetate by the propionic acid bacteria in Emmental cheese are shown in Figure 10–14.

Comté cheese is also covered by an orange smear, called the *morge*, composed mainly of corynebacteria, micrococci, and yeast. About 10^{10} microorganisms/cm^2 are present on the surface of ripened cheese, and it has been calculated that the total number of bacteria in the smear of Comté cheese is equal to the total number in the cheese mass, but, except for *Brachybacterium*, the species involved do not appear to have been identified.

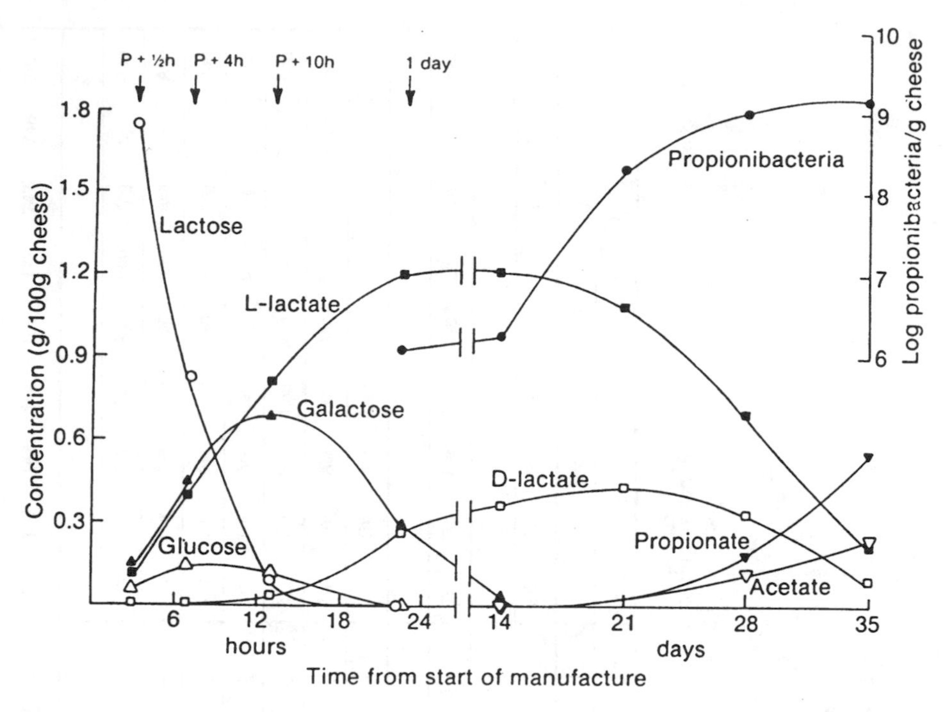

Figure 10–14 Relationship between lactose and lactate metabolism, growth of propionibacteria, and production of propionate and acetate in Swiss-type cheese.

10.6.3 Camembert

This is a mold-ripened cheese with a relatively high moisture content (≈ 50%) and short ripening time. Spores of *P. camemberti* are either added to the milk with the mesophilic starter or are inoculated directly on the surface of the cheese after salting. This organism is an obligate aerobe and grows only on the cheese surface. Adventitious acid-tolerant yeast and *G. candidum* also grow on the surface of Camembert cheese within a few days of manufacture, while *P. camemberti* is generally not visible on the surface until about the 6th day of ripening. Scanning electron micrographs of the development of these on the surface of the cheese have been published (Rousseau, 1984). The yeast and molds oxidize lactate to CO_2 and H_2O, causing a significant increase in the pH of the cheese, particularly on the surface. Significant proteolysis also occurs on the surface, and production of NH_3 through deamination of the resulting amino acids may also be involved in increasing the pH of the cheese. Some microbiological changes that occur on the surface and in the interior of Camembert cheese are shown in Figure 10–15. Numbers of *Lactococcus* reach about 10^9/g at the beginning of ripening and remain at this level both in the interior and on the surface throughout ripening. In contrast, growth of yeast and *Micrococcus* is much greater on the surface than in the interior of the cheese. *S. cerevisiae* is an important yeast in Camembert cheese (Table 10–4), and yeast counts on the surface can reach 10^8/g within the first 2 weeks of ripening, after which they began to decrease very slowly.

10.6.4 Cabrales

Cabrales is a Spanish Blue cheese made from raw milk without the deliberate addition of starters or molds. Adventitious LAB in the milk are responsible for acid production during manufacture and ripening. The coagulum is cut 2 hr after

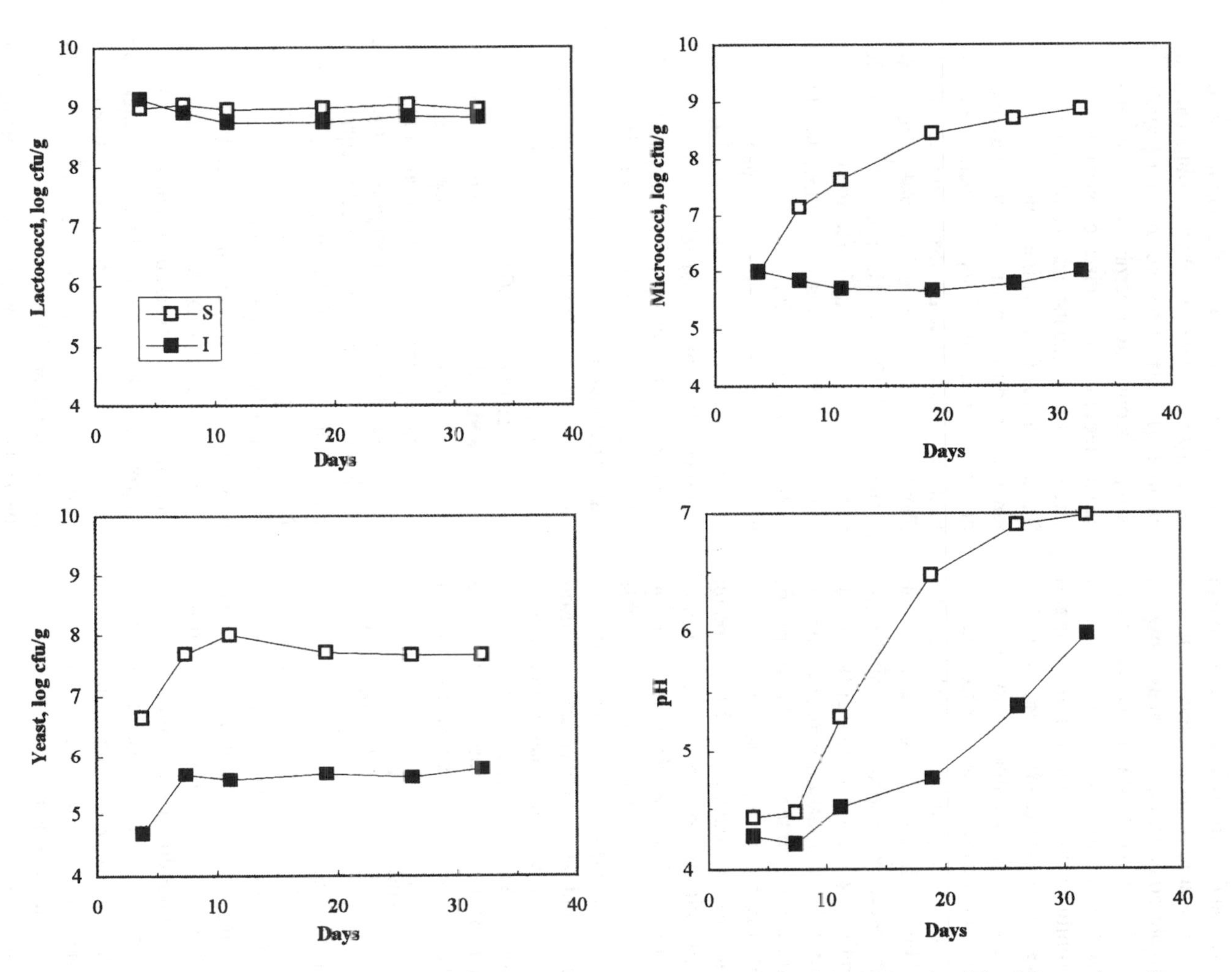

Figure 10–15 Growth of lactococci, micrococci, and yeast and changes in the pH on the surface (□) and in the interior (■) of Camembert cheese.

addition of rennet and is then scooped into molds, which are held at 16–18°C for 48 hr to allow whey drainage to occur. Then the curd is removed from the mold, covered with coarse salt, held for a further 48 hr at 16–18°C, ripened at 10–12°C for 10–15 days, and transferred to caves for further ripening at 9–10°C and at 90–95% relative humidity.

The microbiological changes that occur in this cheese during ripening are shown in Figure 10–16. In each graph, the first and second points refer to the counts in the milk and the curd at 1–2 days, respectively. Growth of lactococci is relatively rapid during the first few days, after which their number decreases (more rapidly on the surface than in the interior of the cheese). The number of mesophilic lactobacilli remains more or less constant at 10^6/g on the surface but increases to 10^8/g in the interior, after which it decreases. *Micrococcus* spp. show the opposite trend, being more abundant on the surface than in the interior, and yeast show significant growth on the surface of the cheese early in ripening, after which they decrease. Coliforms grow during cheesemaking, but their number decreases rapidly over the next 2 weeks. This is probably due to the very rapid decrease in pH, which reaches about 5.0 in 48 hr due to the growth of the lactococci, after which it increases to 6.5 in the interior and to 7 on the surface due to metabolism of lactate to CO_2 by the yeast and molds.

10.6.5 Tilsit

Tilsit, a smear-ripened cheese made with a mesophilic starter, is particularly popular in Germany, Austria, and Switzerland. Sometimes commercial cultures containing some or all of the following microorganisms are used to smear the cheese: *B. linens, G. candidum, Candida utilis, D. hansenii*, and *K. lactis*. In Austria, only *B. linens* is deliberately inoculated onto the surface of the cheese. The yeast are natural contaminants of the cheese and arise from the milk, air, equipment, brine, and smear water. In Germany, the young cheese is smeared with so-called old smear from well-ripened cheese. This can be problematic, because such smear can be contaminated with pathogens, particularly *Listeria monocytogenes* (see Chapter 20), which serve as an inoculum for the new cheese. The function of the yeast is to metabolize the lactate on the surface, which results in an increase in pH to a point where the bacteria, particularly *B. linens* and corynebacteria, can grow. The microbiological changes in Tilsit cheese are shown in Figure 10–17.

The pH increases steadily from about 5.5 to 7.5 during the first 2 weeks of ripening, after which it remains more or less constant. Simultaneously, the number of yeast and salt-tolerant bacteria (coryne forms) also increase, with the bacteria increasing more rapidly than the yeast. In Figure 10–17, bacterial numbers are plotted per cm^2; counts/g of surface would be 100 times greater. Despite the fact that deliberate addition of *B. linens* is used to inoculate the surface, there is a high diversity in the different species of bacteria and yeast on the surface of Tilsit cheese (Tables 10–3 and 10–4). *D. hansenii* and *Y. lipolytica* are the dominant yeasts, and *A. globiformis*, *A. citreus*, *B. ammoniagenes*, and *B. linens* are the dominant bacteria.

10.7 MICROBIAL SPOILAGE OF CHEESE

The most common microbial defects of cheese are early and late gas, both of which are relatively uncommon in cheese today, mainly because of better hygiene and better quality control in cheese plants.

Early gas generally occurs within 1 or 2 days after manufacture. It is characterized by the appearance of many small holes and is caused by coliform bacteria and/or yeast. The gas produced by coliform is mainly H_2, which is produced by formic hydrogenylase activity from formate, a product of lactose metabolism. It is more problematic in soft and semi-soft cheese than in hard cheese because of higher a_w in the former. An effective way of controlling early gas is to add KNO_3 or $NaNO_3$ at low levels (0.2%) to the milk. NO_3 does not prevent the growth of coliform but acts as an alternative electron acceptor, allowing complete oxidation of lactose to CO_2 and H_2O

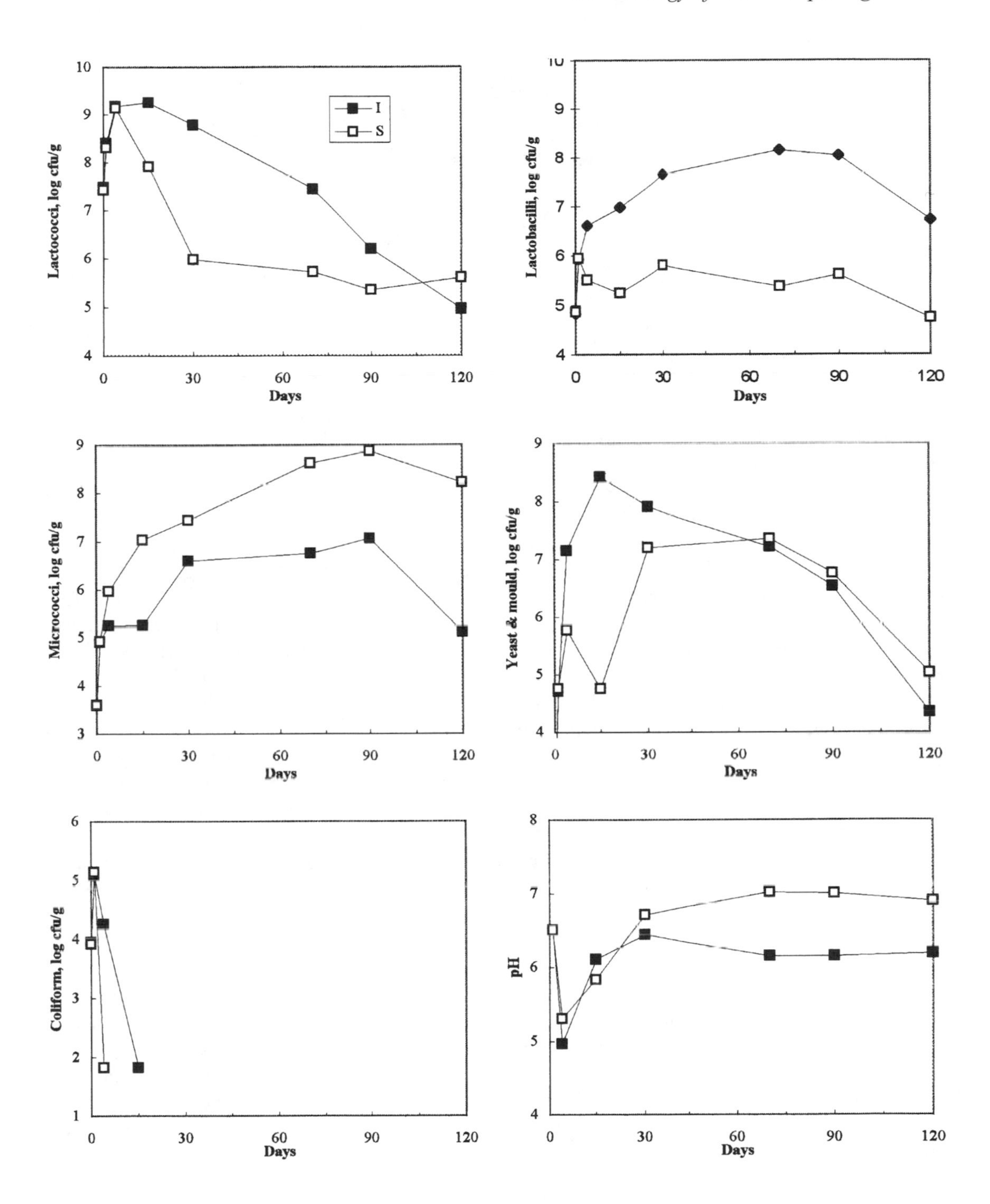

Figure 10–16 Growth of different organisms on and changes in the pH of Cabrales cheese during ripening. Surface (❑), interior (■).

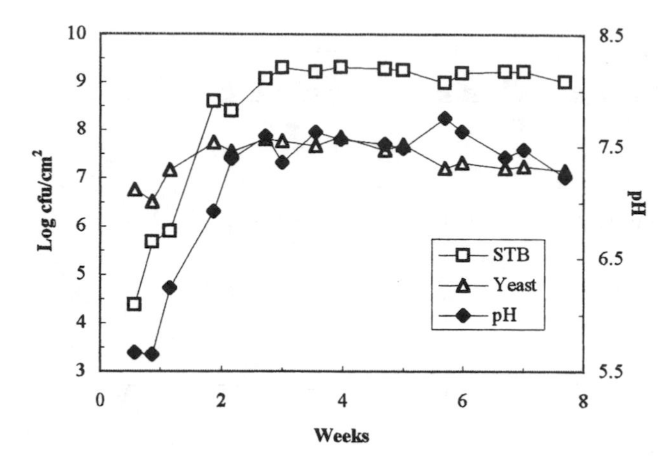

Figure 10–17 Growth of yeast and salt-tolerant bacteria (STB) (on plate count agar containing 8% salt) and changes in the pH on the surface of Tilsit cheese during ripening.

rather than fermentation to formate, thus effectively reducing the production of H_2 from formate. Early gas production by yeast is due to the production of CO_2 from lactose or lactate.

Late gas formation, or late blowing, does not occur until late in ripening. It is due to fermentation of lactate to butyrate and the production of copious amounts of H_2 by *Clostridium tyrobutyricum* and *Cl. butyricum*. Consequently, large holes are generally produced. The butyrate is responsible for off-flavor development in the cheese. Late gas can be particularly prevalent in Swiss-type cheese, where clostridia can grow with the propionic acid bacteria during the "hot room" ripening period. Silage is a potent source of these bacteria, and for this reason it is forbidden in Switzerland to feed it to cows whose milk is intended for cheesemaking. In addition, many thermophilic cultures are thought to stimulate the growth of clostridia through the production of peptides and amino acids. Late gas production can be controlled by bactofugation of the milk (see Chapter 4), but this often results in inferior quality cheese.

The bacteriocin nisin, which is produced by some strains of *Lc. lactis* subsp. *lactis,* is effective in controlling the growth of clostridia and is used for this purpose in processed cheese. However, it is not suitable for use in natural cheese, because many starters are sensitive to it. Increasing the level of salt, lowering the pH of the cheese rapidly through the use of an active starter, adding NO_3^-, or adding lysozyme can also be effective in preventing late gas production. Lysozyme, which is found in milk, saliva, tears, and other body fluids, hydrolyzes the cell walls of sensitive bacteria (e.g., *Cl. tyrobutyricum*), causing them to lyse. Commonly used in Italy, it is added to the milk with the starter at a level of 25 mg/L. It is generally considered to have no effect on the growth of starters, although some strains in Italian natural whey cultures are inhibited by it.

NO_3^- is an effective inhibitor of clostridia though not coliform. It was initially thought that NO_2^- combined with some of the enzymes involved in respiration, particularly those containing an -SH group. However, this cannot be true if NO_3^- is effective against an obligate anaerobe such as *Cl. tyrobutyricum*. It is possible that NO_3^- interacts with the Fe in the ferredoxin protein involved in oxidation-reduction reactions in clostridia, but this hypothesis has not been proven.

Other microorganisms have occasionally been implicated as spoilage organisms. Citrate-metabolizing lactobacilli have been incriminated as the cause of open texture in Cheddar cheese due to the production of CO_2 from citrate. The optimum pH for uptake of citrate ranges from 4 to 5, and significant metabolism of citrate occurs in the absence of an energy source at pH 5.2, the pH of many semi-hard and hard cheeses. *Ec. malodoratus*, which, as its name implies, causes the production of bad flavors, has been found in Gouda cheese. The surface of cheese, especially when it is moist (e.g., unwrapped soft or semi-soft cheese), is an ideal environment for the growth of molds and yeast. These cause little damage to the cheese but are unsightly. They can be washed off the cheese surface with a dilute brine solution.

Yeasts and molds are occasionally incriminated as spoilage organisms in cheese. Sorbic acid is allowed in Italy as a preservative for hard, fresh, and processed cheese. Sorbate-resistant molds, *Paecilomyces variotti*, and the yeast *D. hansenii* from Crescenza and Provolone cheese are able to grow in the presence of 3 mg of sorbic acid/g and are also able to transform sorbic acid to *trans*-1,3-pentadiene, which has a taste and odor like kerosene (Sensidini, Rondinini, Peressini, Maifreni, & Bartolomeazzi, 1994).

Cladosporium cladosporioides, *Penicillium commune*, *C. herbarum*, *P. glabrum*, and a *Phoma* species are responsible for the "thread mold" defect of Cheddar cheese (Hocking & Faedo, 1992). This defect occurs as a black, dark brown, or dark green spot or thread in the folds, creases, and gusset ends of the plastic bags used to wrap Cheddar cheese during ripening. It can occur on the cheese surface but is more often associated with free whey drawn from the fresh cheese block during vacuum packaging. These molds are obviously able to grow in the presence of low levels of O_2.

Growth of *P. commune*, which is closely related to *P. camemberti*, can result in discoloration of cheese surfaces and the production of off-flavors (Lund, Filtenberg, & Frisvad, 1995).

REFERENCES

Fleet, G.H. (1990). Yeasts in dairy products. *Journal of Applied Bacteriology*, *68*, 199–211.

Hocking, A.D., & Faedo, M. (1992). Furgi causing thread mould spoilage of vacuum packaged Cheddar cheese during maturation. *International Journal of Food Microbiology*, *16*, 123–130.

Jakobsen, H., & Narvhus, J. (1996). Yeasts and their possible beneficial and negative effects on the quality of dairy products. *International Dairy Journal*, *6*, 755–768.

Lund, F., Filtenberg, O., & Frisvad, J.C. (1995). Associated mycoflora of cheese. *Food Microbiology*, *12*, 173–180.

Rousseau, M. (1984). Study of the surface flora of traditional Camembert cheese by scanning electron microscopy. *Milchwissenschaft*, *39*, 129–135.

Schubert, K., Ludwig, W., Springer, N., Kroppenstedt, R.M., Accolas, J.P., & Fiedler, F. (1996). Two coryneform bacteria isolated from the surface of French Gruyère or Beaufort cheeses are new species of the genus *Brachybacterium*: *Brachybacterium alimentarium* sp. nov and *Brachybacterium tyrofermentair* sp. nov. *International Journal of Systematic Bacteriology*, *46*, 81–87.

Sensidini, A., Rondinini, G., Peressini, D., Maifreni, M., & Bartolomeazzi, R. (1994). Presence of an off-flavour associated with the use of sorbates in cheese and margarine. *Italian Journal of Food Science*, *6*, 237–242.

SUGGESTED READINGS

Deacon, J.W. (1997). *Modern mycology* (3d ed.). Oxford: Blackwell Science.

Eck, A. (1986). *Cheesemaking, science and technology* (English ed.). Paris: Lavoisier.

Kurtzman, C.P., & Fell, J.W. (1998). *The yeasts: A taxonomic study* (4th ed.). Amsterdam: Elsevier Science.

Pill, J.I., & Hocking, A.D. (1997). *Fungi and food spoilage* (2d ed.). London: Blackie Academic and Professional.

CHAPTER 11

Biochemistry of Cheese Ripening

11.1 INTRODUCTION

As discussed in Chapter 10, the original objective of cheese production was to conserve the principal constituents of milk—the lipids and caseins. However, although well-made cheese is a hostile environment for microbial growth and possesses several preservative hurdles, these hurdles are not sufficient to prevent the growth of certain microorganisms (see Chapter 10) and the activity of those enzymes from various sources that may be present. These microorganisms and enzymes catalyze a complex series of biochemical reactions, which, if unbalanced, cause off-flavors and textural defects but, if properly controlled and balanced, lead to the desirable and characteristic flavors and textures of the numerous cheese varieties. Although the biochemistry of cheese ripening is not yet fully characterized, a considerable body of information is now available. The objective of this chapter is to describe the principal biochemical reactions that occur in cheese during ripening.

Cheese curd is a relatively simple mixture of casein (and very little whey proteins except in cheese made from milk concentrated by ultrafiltration), lipids, a little lactose (≈ 1% at pressing), lactic acid (≈ 1%), citric acid (≈ 0.2%), NaCl (0.7 to ≈ 6%), and water. Not surprisingly, the primary features of ripening involve the two principal organic constituents, proteins and lipids. However, the metabolism of lactose and citrate, although they are present at low concentrations, is important in all varieties and critical in some. Most of the primary reactions are well characterized. Many of the products of the primary reactions undergo further modifications, which are not fully understood but which are probably responsible for the characteristic flavor of cheese.

11.2 RIPENING AGENTS IN CHEESE

The ripening of cheese is catalyzed by the metabolic activity of living organisms and enzymes from these organisms or from other sources:

- *Coagulant.* The coagulant usually contributes chymosin or other suitable proteinase (see Chapter 6), but rennet paste used in some Italian varieties contributes both proteinase and lipase. In high-cooked cheeses (e.g., Emmental and Parmesan), enzymes in the coagulant are extensively or completely denatured.
- *Milk.* As discussed in Chapter 3, milk contains about 60 indigenous enzymes, at least some of which are significant in cheese ripening, including proteinases, especially plasmin; lipase; acid phosphatase; and xanthine oxidase. Most of the indigenous enzymes are quite heat stable and fully or partially survive pasteurization. Furthermore,

they are either associated with the casein micelles or present in the fat globule membrane and are therefore incorporated into the cheese curd. Enzymes present in the serum phase are largely lost in the whey and thus are of little importance in cheese ripening.

- *Starter culture*. Starter cultures were discussed in Chapter 5. Live starter cells probably contribute little to cheese ripening but they possess a diversity of enzymes (see Chapter 10 and Section 11.7.4), which are located mainly intracellularly and are released upon cell death and lysis. These starter enzymes are major contributors to ripening.
- *Secondary microflora*. Many cheese varieties contain a secondary (nonstarter) microflora (see Chapter 10), the function of which is not acid production but rather some specific secondary function. In many cases, flavor development is dominated by the metabolic activity of the secondary culture. The microorganisms involved include propionic acid bacteria, coryneform bacteria, yeasts, and molds. In addition to these, the growth of which is characteristic and encouraged, cheeses contain adventitious nonstarter lactic acid bacteria (NSLAB) that originate from the milk or the environment. Owing to the selective nature of the interior of cheese (see Chapter 10), this adventitious microflora is composed mainly of mesophilic lactobacilli and, to a lesser extent, pediococci.
- *Exogenous enzymes*. With the objective of accelerating ripening, cheese makers have experimented with adding exogenous enzymes, usually proteinases and perhaps peptidases and lipases, to the cheese curd (see Chapter 15).

11.3 CONTRIBUTION OF INDIVIDUAL AGENTS TO RIPENING

The role of the individual ripening agents in cheese has been studied using model cheese systems in which the action of one or more of the ripening agents is eliminated (see Fox, Law, McSweeney, & Wallace, 1993). Pioneering studies on this subject used milk obtained from selected cows by aseptic milking techniques. However, in our experience, bulk herd milk with a total bacteria count (TBC) less than 10^3 cfu/ml can be obtained from a healthy commercial herd using good but not special milking practices.

Heat treatment of aseptically drawn milk is necessary to further reduce bacterial counts. Batch pasteurization (68°C × 5 min or 63°C × 30 min), HTST pasteurization (≈ 72 or 77°C × 15 s), or ultra-high temperature treatment have been used. A heat treatment of 83°C × 15 s or 72°C × 58 s is necessary to ensure an 8-log reduction in the bacterial population, which is deemed necessary to produce cheese with a nonstarter count below 10 cfu/kg cheese for milk with an initial TBC of 10^3 cfu/ml. HTST pasteurization (72°C × 15 s) is sufficient for milk with an initial count of 10 cfu/ml.

To avoid contamination from the environment, cheese is manufactured under aseptic conditions, which can be achieved using enclosed cheese vats, a sterile room with a filtered air supply, or a laminar air-flow unit (the latter is probably the simplest of these techniques). Using aseptic conditions, it is relatively easy to produce curd free of NSLAB, but in our experience NSLAB always grow in such cheese, sometimes only after a long lag period (e.g., 100 days). A cocktail of antibiotics (penicillin, streptomycin, and nisin) extends the lag period and reduces the final number of NSLAB. The growth of NSLAB is strongly retarded, essentially prevented, by ripening at about 1°C, although the whole ripening process is also retarded.

The acidifying role of starter can be simulated closely using an acidogen, usually gluconic acid-δ-lactone (GDL), although the rate of acidification is faster than occurs in biologically acidified cheese. Incremental additions of lactic acid and GDL give the best results, but precise control of pH is difficult.

To study the role of the coagulant in cheese ripening, it is necessary to inactivate the rennet

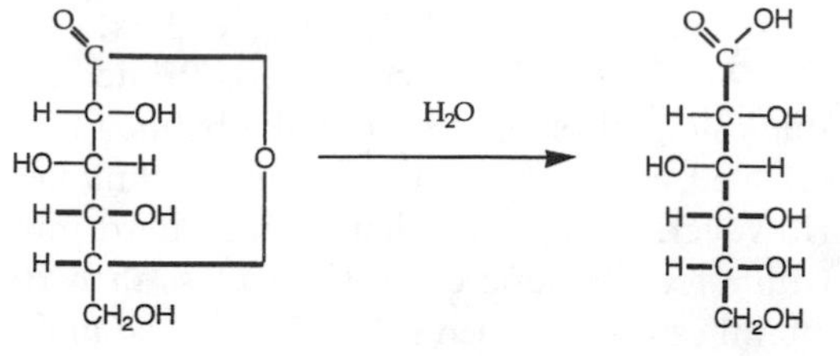

$$\text{Gluconic acid-}\delta\text{-lactone} \xrightarrow{H_2O} \text{Gluconic acid}$$

Gluconic acid-δ-lactone

Gluconic acid

after coagulation, for which four techniques have been developed. One approach involves separating the first and second stages of rennet action. Milk depleted of Ca^{2+} and Mg^{2+} by treatment with an ion-exchange resin is renneted (but does not coagulate), heated (72°C × 20 s) to inactivate the rennet, and cooled to below 15°C; $CaCl_2$ is added. The renneted milk is then heated dielectrically to induce coagulation. Porcine pepsin may be inactivated after coagulation of the milk by adjusting the pH of the curd-whey mixture to 7.0. Piglet gastric proteinase, which hydrolyzes bovine κ-casein but has little effect on α_{s1}- or β-casein, has been used to prepare rennet-free curd in small-scale cheesemaking trials. Chymosin and all commercial rennet substitutes are aspartyl proteinases and are inhibited by pepstatin. The effectiveness of adding pepstatin to Cheddar cheese curd at salting has been demonstrated on a small scale.

6-Aminohexanoic acid (AHA), a noncompetitive inhibitor of plasmin, has been used to study the significance of plasmin in cheese ripening. It is necessary to use a high concentration of AHA, which affects curd syneresis and the moisture content of the cheese. Also, since AHA contains N, the background level of soluble N is increased greatly. Plasmin is inhibited by several proteins, including soybean trypsin inhibitor, which may be suitable for the inhibition of plasmin in cheese. It is also specifically inhibited by dichloroisocoumarin, but neither it nor the inhibitory proteins have been investigated in cheesemaking. Plasmin activity is increased by high cooking temperatures, probably owing to inactivation of the indigenous inhibitors of plasmin or of plasminogen activators. The high heat stability of plasmin suggests that it may be possible to develop a model system that is based on aseptic curd in which the rennet is denatured by a suitable cook temperature and acidified by GDL and that can be used to assess plasmin activity alone.

Some or all of these techniques, in various combinations, have been used to study the contribution of various agents to cheese ripening, especially proteolysis and flavor development.

The biochemistry of cheese ripening will be considered in three principal sections based on the principal biochemical events: glycolysis, lipolysis, and proteolysis.

11.4 GLYCOLYSIS AND RELATED EVENTS

The primary glycolytic event, the conversion of lactose to lactate, is normally mediated by the starter culture during curd preparation or the early stages of ripening. In cases where glycolysis has not been completed by the starter, NSLAB may contribute. The metabolism of lactose by LAB was discussed in Chapter 5.

Approximately 96% of the lactose in milk is removed in the whey as lactose or lactate. However, fresh curd contains a considerable amount of lactose, the fermentation of which has a significant effect on cheese quality. Obviously, the concentration of lactose in fresh curd depends on its moisture content, the extent of fermentation prior to molding, and whether the curd is washed with water or not. Cheddar curd, which is extensively drained and has reached a pH of about 5.4 at milling, contains 0.8–1.0% lactose. In the manufacture of Dutch-type cheese, part of the whey is removed and replaced by water, but the curd is subjected to less syneresis and the pH is high (≈ 6.2–6.3) at molding. Hence, Gouda cheese curd contains around 3.0% lactose at molding. Although Emmental cheese curd is cooked to a high temperature (52–55°C) and hence undergoes extensive syneresis in the vat, it is transferred with the whey to molds at around pH 6.4 and contains around 2% lactose at molding.

Compared with other varieties, the residual lactose in Cheddar is fermented relatively slowly at a rate and to an extent dependent on the percentage salt-in-moisture (S/M) in the curd (see Figure 8–11). At low S/M concentrations and low populations of NSLAB, residual lactose is converted mainly to L-lactate by the starter. At a high population of NSLAB (e.g., at a high storage temperature), a considerable amount of D-lactate is formed, partly by fermentation of residual lactose and partly by isomerization of L- to D-lactate. At high S/M levels (e.g., 6%) and low NSLAB populations, the concentration of lactose decreases slowly and changes in lactate are slight. The quality of Cheddar cheese is strongly influenced by the fermentation of residual lactose: the pH decreases after salting at S/M levels below 5%, owing primarily to the continued action of the starter, but at higher levels of S/M, starter activity decreases abruptly, as indicated by a high level of residual lactose and a high pH, accompanied by a sharp decrease in cheese quality (see Figure 8–12).

Dutch-type cheese contains about 3.0% lactose at pressing, but the amount of lactose decreases to undetectable levels within about 12 hr.

Typically, Emmental cheese curd contains roughly 1.7% lactose 30 min after molding. Neither *Streptococcus thermophilus* nor starter *Lactobacillus* spp. grow in Emmental curd during cooking owing to the high cook temperature (52–55°C). The curds are transferred to molds while still hot, but as they cool in the molds, *Sc. thermophilus* begins to grow and metabolize lactose. Only the glucose moiety of lactose is metabolized by *Sc. thermophilus*, and consequently galactose accumulates to a maximum of around 0.7% at about 10 hr, when the galactose-positive lactobacilli begin to multiply. These metabolize galactose and residual lactose to a mixture of D- and L-lactate, which reach around 0.35% and 1.2%, respectively, at 14 days, when all the sugars have been metabolized (see Figure 10–13).

11.4.1 Effect of Lactose Concentration on Cheese Quality

Although the concentration of lactose in milk decreases with advancing lactation (see Figure 3–1), its concentration in bulked factory supplies from cows on a staggered calving pattern is essentially constant. However, when synchronized calving is practiced, as in New Zealand, Ireland, and Australia, substantial seasonal changes occur in the concentration of lactose in milk and consequently in fresh cheese curd. Variations in the concentration of lactose in cheese curd probably affect the final pH of the cheese, which, in turn, affects cheese texture, enzyme activity, and perhaps the nonstarter microflora. Cheese flavor is likely to vary owing to variations in the concentration of lactic and acetic acids and to variations in the metabolic activity of the cheese microflora.

The concentration of lactose in cheese curd is influenced by some features of the manufacturing process. As far as is known, the concentration of lactose in cheese curd is not increased intentionally for any variety of rennet-coagulated cheese. However, curd made from milk concentrated to a high factor by ultrafiltration (i.e., precheese) contains a high level of lactose owing to the lack of syneresis, and lactose may have to be reduced to an appropriate level by diafiltration. If the concentration of lactose in cheese curd is too high, the concentration of D-Ca-lactate will exceed its solubility and crystallize on the surface of the cheese. The concentration of lactose in the curd of several varieties, including Gouda and Edam, is reduced by replacing part of the whey by warm water. This process, which was probably introduced as a simple method for cooking the curds on farms lacking jacketed cheese vats, effectively controls the pH of the cheese. In these cheeses, the level of wash water added is based on the concentrations of lactose and casein in the milk. This washing protocol minimizes variations in the pH of cheese curd express that might otherwise occur due to seasonal variations in the lactate:casein ratio. Curds are washed in the washed-curd variants of Cheddar cheese and perhaps in the production of low-fat cheese to increase its moisture content.

The effect of variations in the concentration of lactose in cheese curd on the quality of the mature cheese has received minimal attention. In an attempt to vary the concentration of lactose in Cheddar cheese curd, Huffman and Kristoffersen (1984) added lactose to the curd-whey mixture after cutting the coagulum, but owing to the strong outflow of whey from the curd at that stage, due to syneresis, the achieved increase in the concentration of lactose within the curd was quite small. Waldron and Fox (unpublished study) reduced the lactose content of Cheddar cheese curd by replacing 35–45% of the whey shortly after cutting the coagulum by an equal volume of warm water. The curd contained 0.25% lactose compared with about 1% in the control. A curd containing only 0.03% lactose was obtained by repeating the whey removal-replacement treatment. The lactose level was also reduced by washing the curd with water prior to salting, but this was less effective than whey replacement, possibly because little syneresis occurred at this late stage of curd production. The lactose content of other batches of curd was increased by using lactose-supplemented milk (6.4% or 8.4% lactose) for curd manufacture. Overall, the concentration of lactose in the 1-day-old cheese ranged from 0.03% to 2.5%.

Changes in the concentration of lactose in and the pH of the cheese during ripening are shown in Figure 11–1. The lactose in both types of washed-curd cheese was completely metabolized within about 2 weeks, but it persisted in the high-lactose cheeses throughout ripening. Not surprisingly, the pH of the cheeses was inversely proportional to the concentration of lactose in the curd. The pH of high-lactose cheeses continued to decrease (to ≈ 4.8) throughout ripening, whereas in the washed-curd cheeses the pH increased once the lactose had been exhausted. This increase in pH is common in many varieties of cheese, probably due to proteolysis and the production of NH_3 from amino acids. Flavor development was substantially faster in the high-lactose cheese than in the washed-curd cheeses, although it was considered to be rather harsh, perhaps due to the low pH. The flavor of the low-lactose cheeses was clean and mild. The rate of growth and the final number of NSLAB were not affected by the concentration of lactose, suggesting that NSLAB do not depend on lactose as a growth substrate. (Factors affecting the growth of NSLAB in cheese are discussed in Chapter 10 and by Fox, McSweeney, & Lynch, 1998.)

The results of this study suggest that the concentration of lactose in cheese curd has a substantial effect on the quality of Cheddar and probably other cheeses. Replacing some of the whey by water or washing the cheese curd might be considered when a mild, clean flavor is desired. Normal variations in the lactose content of milk from mixed-calving herds are probably not significant but may have a substantial effect when synchronized calving is practiced. Under such circumstances, the pH of the cheese may vary in an undesirable manner. The use of low concentration factor ultrafiltration (LCF-UF) milk for cheese manufacture is not expected to influence cheese pH, as the lactose content of cheese is increased only very slightly (by ≈ 0.25% when milk is concentrated 1.5-fold). The problem is more serious when high concentration factor ultrafiltration retentate is used and diafiltration of the cheese milk or washing of the curd would appear to be desirable or essential. Cheese made from milk with a high content of fat and casein may have a reduced lactose content.

The presence of residual lactose or its component monosaccharides in cheese may lead to Maillard (nonenzymatic) browning, especially if the cheese is heated (e.g., as a food ingredient). Lactose per se is unlikely to be a problem except when very young cheese curd is used in processed cheese. Browning is most likely to be problematic in cheeses made with thermophilic cultures. As discussed in Chapters 5 and 10, *Sc. thermophilus* is unable to metabolize the galactose moiety of lactose, which it excretes. If a galactose-positive strain of *Lactobacillus* is used, the galactose will be metabolized to L- or DL-lac-

A

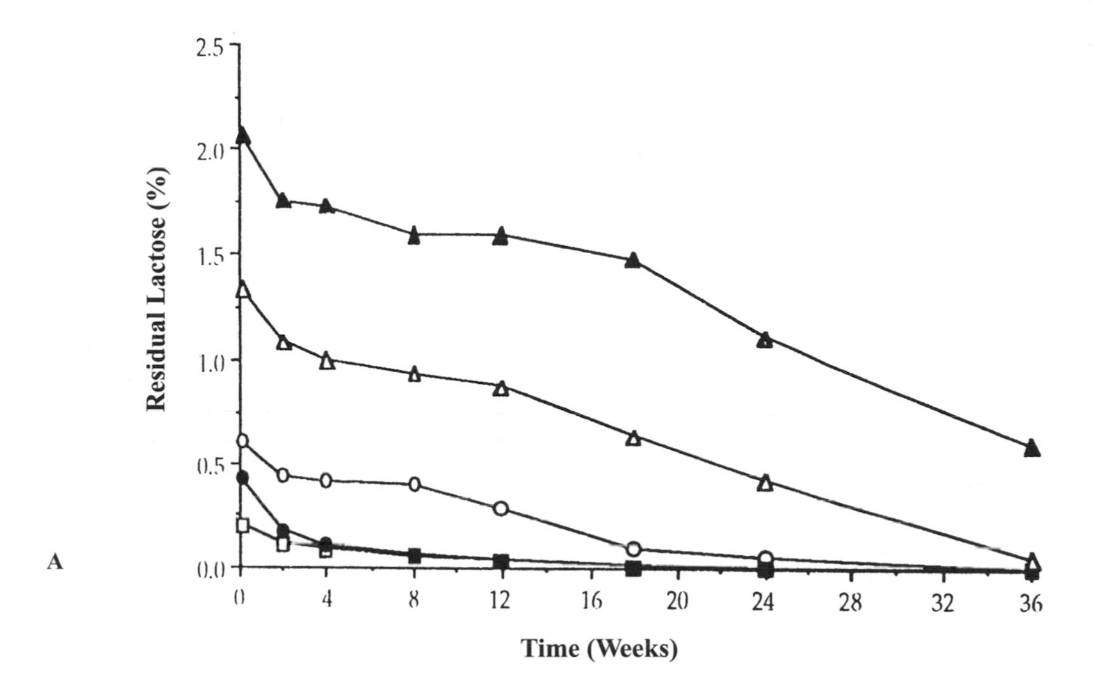

B

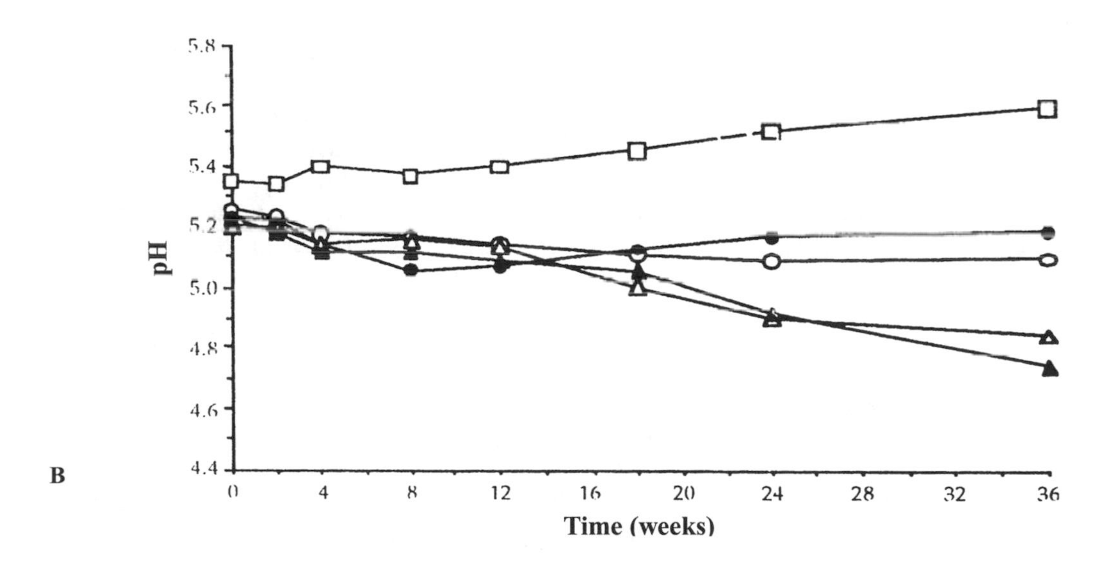

Figure 11–1 Changes in the concentration of lactose (A) and in the pH (B) of cheese made from curd with modified levels of lactose during ripening. Control cheese, (○); 35% whey-replaced cheese, (❑); washed-curd cheese, (●); and lactose-enriched cheese 6.4%, Δ or 8.4%, (▲).

tate, but most strains of *Lactobacillus delbrueckii* and *Lb. lactis* are galactose-negative, and therefore galactose accumulates. The residual galactose may cause undesirable browning in many cheeses but is a particularly serious problem in Mozzarella, which is subjected to considerable heating during the cooking of pizza. Browning may also be problematic in cheeses that are typically grated, such as Parmesan and Grana, which have a low-moisture content and approach a water activity level that is optimal for Maillard browning (≈ 0.6). Since grating cheeses are extensively ripened, other carbonyls (e.g., diacetyl and glyoxals), which are very active in Maillard browning, may contribute to browning.

11.4.2 Modification and Catabolism of Lactate

The fate of lactic acid during cheese ripening has some significance in all varieties and is of major consequence in some types. Lactic acid has a direct effect on the taste of cheese, especially young cheese, which lacks other flavor compounds. Obviously, lactic acid affects the pH of cheese and consequently its texture (see Chapter 13). pH affects the solubility of $Ca_3(PO_4)_2$ and hence also indirectly affects cheese texture. Perhaps most importantly, lactic acid is an important substrate for microbial growth in many cheese varieties, and its catabolism has major effects on cheese flavor (see Fox, Lucey, & Cogan, 1990).

Typical concentrations of lactate in Camembert, Swiss, Romano, and Cheddar are 1.0%, 1.4%, 1.0%, and 1.5%, respectively. The fate of lactic acid in cheese depends on the variety. Initially, Cheddar contains only L(+) lactic acid, but as the cheese matures the concentration of D-lactate increases, and eventually a racemic mixture is formed (see Figure 10–11). D-Lactate could be formed from residual lactose by lactobacilli or by racemization of L-lactate by NSLAB, including pediococci. Except in cases where the post-milling activity of the starter is suppressed (e.g., by S/M > 6%), racemization is probably the principal mechanism. Racemization of L-lactate appears to occur in several cheese varieties (Thomas & Crow, 1983), and a racemic mixture will be formed if the duration of ripening is long enough (Table 11–1). Racemization is not significant from the flavor viewpoint, but calcium D-lactate, which is less soluble than L-lactate, may crystallize in cheese, especially on the surface, causing undesirable white specks.

Lactate in cheese may be oxidized to acetate. Pediococci produce 1 mole of acetate and 1 mole of CO_2 and consume 1 mole of O_2 per mole of lactate utilized. Lactate is not oxidized until all sugars have been exhausted. The oxidation of lactate to acetate in cheese depends on the NSLAB population and on the availability of O_2, which is determined by the size of the block and the oxygen permeability of the packaging material (Figure 11–2). Acetate, which may also be produced by starter bacteria from lactose or citrate or from amino acids by starter bacteria and lactobacilli, is usually present at fairly high concentrations in Cheddar cheese and is considered to contribute to cheese flavor, although high concentrations may cause off-flavors. Thus, the oxidation of lactate to acetate probably contributes to Cheddar cheese flavor.

In Romano cheese, L-lactate predominates initially, reaching a maximum of around 1.9% at 1 day. The concentration of L-lactate begins to decrease after 10 days, reaching 0.2–0.6% after 150–240 days of ripening. Some of the decrease is due to racemization to D-lactate, which reaches a maximum (up to 0.6% in some cheeses) at around 90 days and then declines somewhat. In some cheeses, acetate reaches very high levels (1.2%) at around 30 days, but decreases thereafter. The agents responsible for the metabolism of acetate have not been identified, but yeasts (*Debaryomyces hansenii*) may be involved. Presumably, the oxidation of lactate to acetate also occurs in other hard and semi-hard cheeses, but studies in this area are lacking.

The metabolism of lactate is very extensive in surface mold–ripened varieties, such as Camembert and Brie. The concentration of lactic acid

Table 11–1 Concentration of L(+)- and D(−)-Lactate in Various Cheeses

		Lactate (g/100 g Cheese)	
Cheese Type	*Age (Weeks)*	*L(+)*	*D(−)*
Cheshire	15	0.75	0.66
	23	0.74	0.73
Colby	16	0.71	0.68
	–	0.57	0.51
Egmont	20	0.62	0.37
Gouda	10	0.84	0.31
	15	0.66	0.55
	–	0.69	0.42
Blue	8	0.65	0.61
	–	0.74	0.43
Camembert	6	0.17	0.02
	8	0.04	0.01
	3	0.57	0.02
Feta	–	0.97	0.88

in these cheeses is around 1.0% at 1 day. The lactic acid is produced exclusively by the mesophilic starter and hence is L-lactate. Secondary organisms quickly colonize and dominate the surface of these cheeses, initially *Geotrichum candidum* and yeasts, followed by *Penicillium camemberti,* and, in traditional manufacture, by *Brevibacterium linens* and other coryneform bacteria (see Chapter 10). *G. candidum* and *P. camemberti* rapidly metabolize lactate to CO_2 and H_2O, causing an increase in pH (Figure 11–3). Deacidification occurs initially at the surface, resulting in a pH gradient from the surface to the center and causing lactate to diffuse outward. When the lactate has been exhausted, *P. camemberti* metabolizes proteins, producing NH_3, which diffuses inward, further increasing the pH. The concentration of calcium phosphate at the surface exceeds its solubility at the increased pH, and it precipitates as a layer of $Ca_3(PO_4)_2$ at the surface, thereby causing a calcium phosphate gradient within the cheese (Figure 11–4). The elevated pH stimulates the action of plasmin, which contributes significantly to proteolysis. Although surface microorganisms secrete very potent proteinases, they diffuse into the cheese to only a very limited extent; however, peptides produced at the surface may diffuse into the cheese. The combined action of increased pH, loss of calcium (necessary for the integrity of the protein network), and proteolysis is necessary for the very extensive softening of the body of Brie and Camembert (Karahadian & Lindsay, 1987; Lenoir, 1984).

The pH of Blue cheese also increases substantially during ripening (Figure 11–5), but in contrast to surface mold–ripened cheeses, the extent of the increase is greater at the center than at the surface. One would expect that catabolism of lactic acid would be responsible for the increase in pH, but the only published data available suggest that Blue cheese contains a high concentration of lactic acid (≈ 1.2%, see Table 11–1). Perhaps the increase in pH is due to the production of NH_3 on the catabolism of amino acids. Among Blue cheeses, the increase in pH appears to be least in Danablue, in which low levels of NH_3 are produced.

Large changes in pH also occur in surface smear–ripened cheeses, especially at the surface

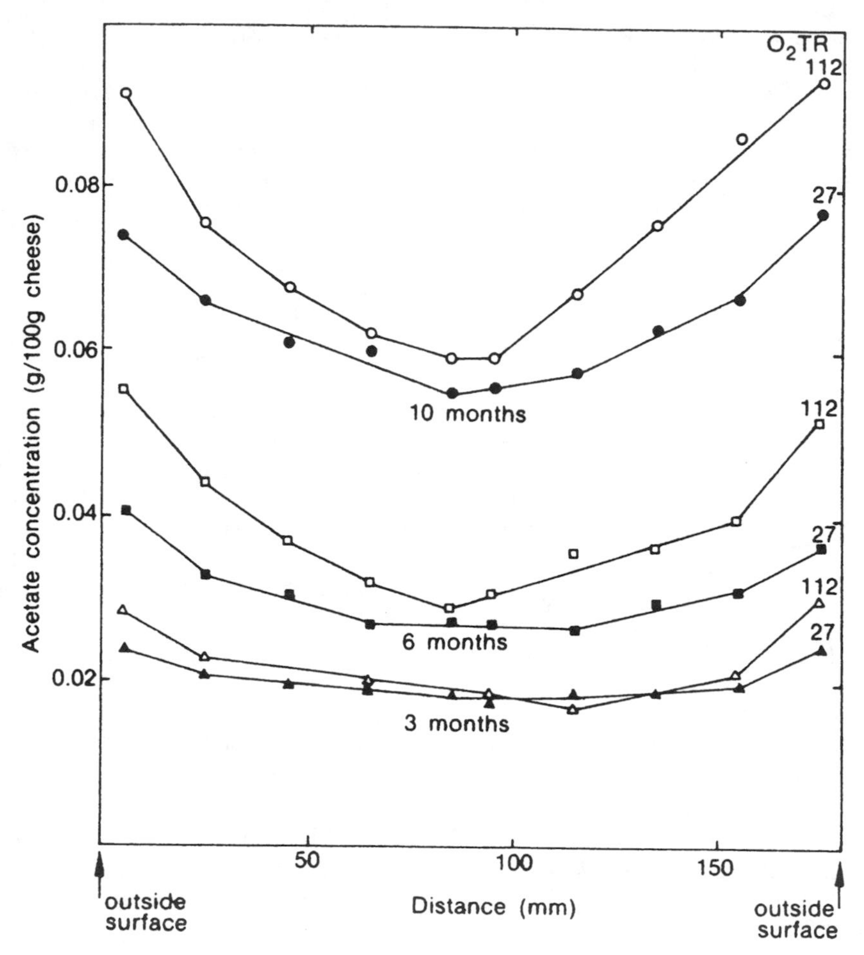

Figure 11–2 Acetate concentration gradients in 20 kg blocks of Cheddar cheese inoculated with *P. pentosaceus* 1220 and ripened at 12°C and 85–90% RH for 3, 6, and 10 months. The cheeses were wrapped in plastic film with an oxygen transmission rate (O_2TR) of 27 ml O_2 m^{-2} 24 hr^{-1} (black symbols) or 112 ml O_2 m^{-2} 24 hr^{-1} (white symbols).

(Figure 11–6). In these cheeses, lactate in the surface layer is catabolized by yeasts, which are the first microorganisms to colonize the surface. The increase in pH at the surface is a critical factor in the ripening of these cheeses, since *Brevibacterium linens*, the characteristic microorganism in the smear, does not grow at a pH below 5.8 (see Chapter 10).

The catabolism of lactic acid is also critical in Swiss-type cheeses, but the causative agents and effects are different from those in surface mold–ripened and smear-ripened cheeses. On transfer to the warm room, *Propionibacterium* sp., the characteristic microorganisms in Swiss-type cheeses, multiply by 2–3 log cycles and metabolize lactate, preferentially the L-isomer, to propionate, acetate, and CO_2 (Figures 10–13 and 10–14):

$$\underset{\text{Lactic acid}}{3\,CH_3\text{-}\overset{\overset{\displaystyle H}{|}}{\underset{\underset{\displaystyle OH}{|}}{C}}\text{-}COOH} \longrightarrow \underset{\text{Propionic acid}}{2\,CH_3CH_2COOH} + \underset{\text{Acetic acid}}{CH_3COOH} + CO_2 + H_2O$$

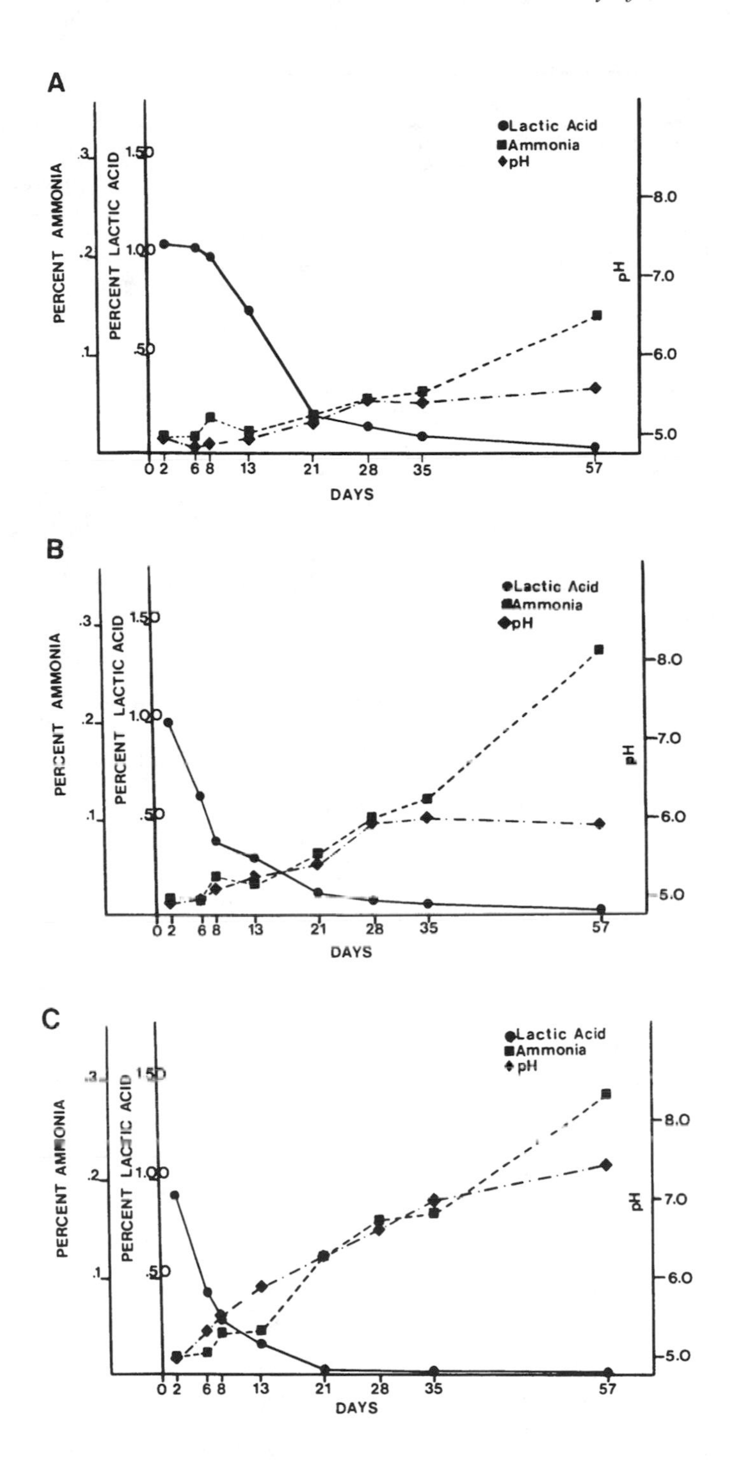

Figure 11–3 Rate of lactic acid metabolism (●), ammonia production (■), and pH changes (◆) in Brie cheese sampled at the center (A), surface (B), and corner (C) during ripening.

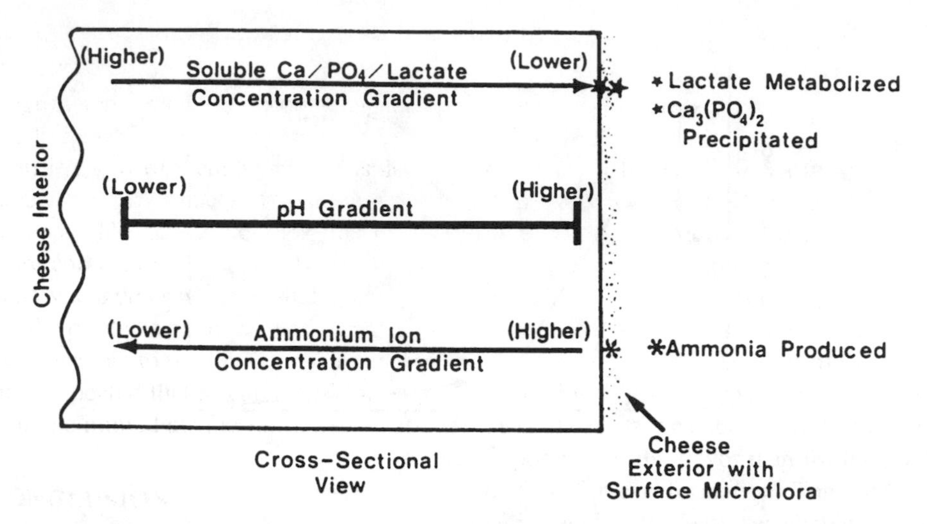

Figure 11–4 Schematic representation of the gradients of calcium, phosphate, lactic acid, pH, and ammonia in ripening of Camembert cheese.

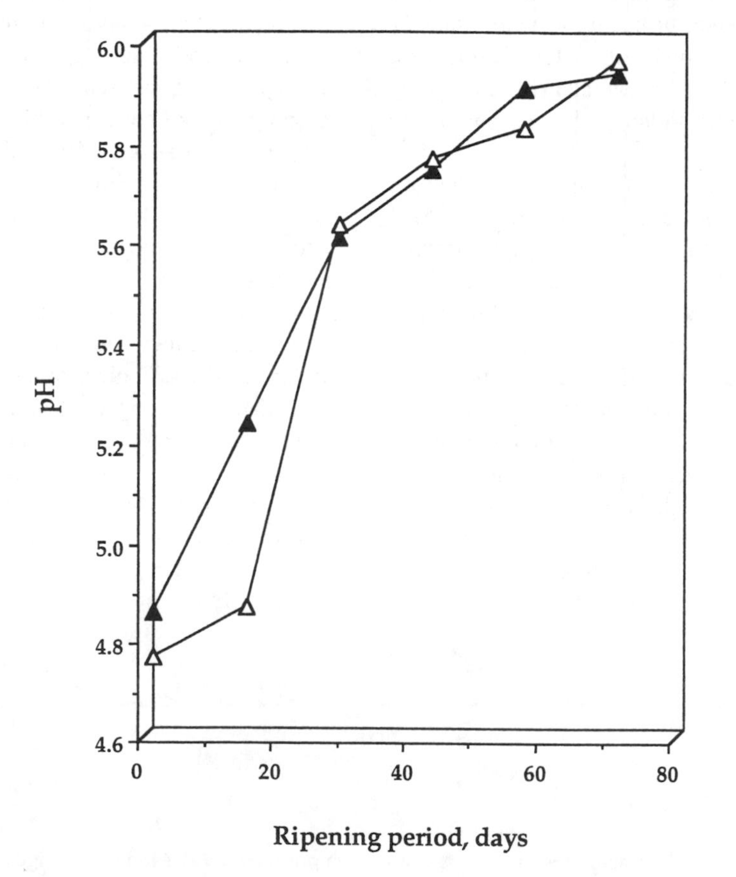

Figure 11–5 Changes in the pH of two batches (▲, Δ) of an Irish Blue cheese during ripening.

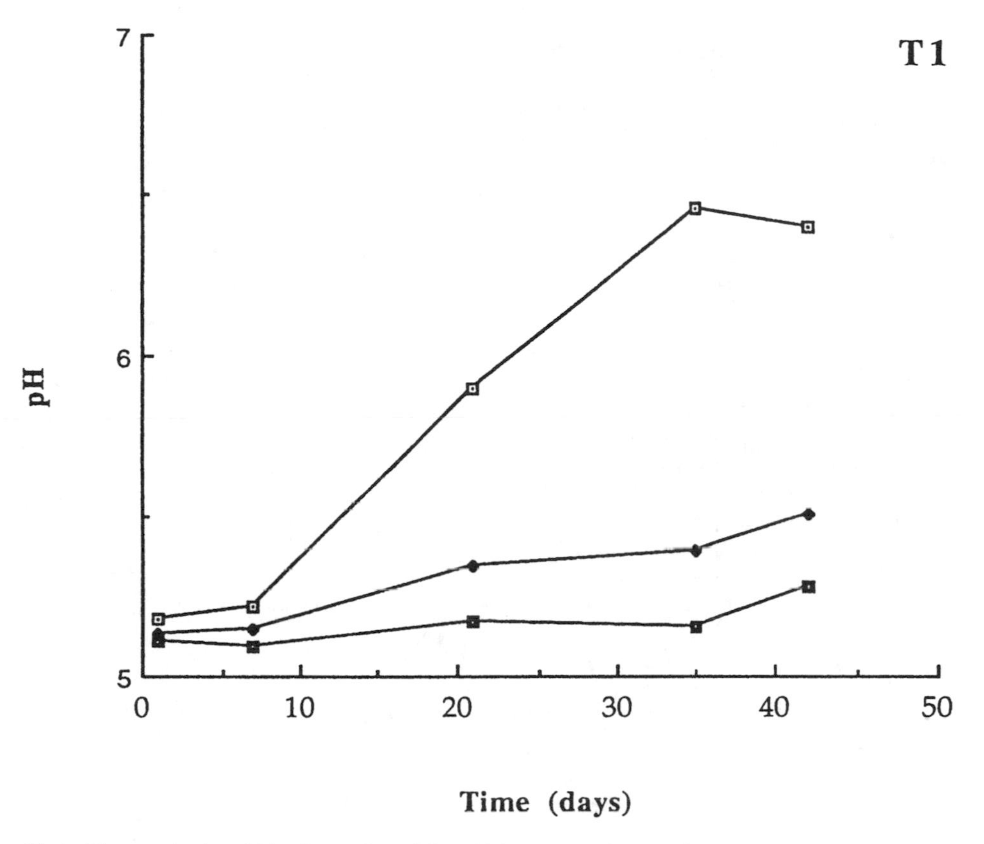

Figure 11–6 Changes in the pH in the surface (◘), middle (◆), and core (□) layers of Taleggio cheese during ripening.

The CO_2 generated is responsible for eye development, a characteristic feature of these varieties. Most of the CO_2 produced diffuses through the curd and is lost, but if the growth of *Propionibacterium* sp. is adequate, sufficient CO_2 is produced to induce good eye formation. The acetic acid and especially the propionic acid produced in this fermentation contribute to the flavor of Swiss-type cheeses.

A common defect in many cheeses arises from the metabolism of lactate (or glucose) by *Clostridium* spp. to butyrate, H_2, and CO_2 (Figure 11–7). This reaction leads to late gas blowing and off-flavors in many cheese varieties unless precautions are taken, such as good hygiene, addition of $NaNO_3$ or lysozyme, bactofugation, or microfiltration.

The significance of the primary fermentation of lactose to L lactate in cheese manufacture is well recognized (see Chapters 5, 6, and 10). The foregoing discussion indicates that the metabolism of lactose and lactate in cheese during ripening is well understood. Quantitatively, these changes are among the principal metabolic events in most cheese varieties. In comparison with other biochemical changes during cheese ripening, however, the conversion of lactose to lactate may have relatively little direct effect on the flavor of mature cheese. Nonetheless, since it determines the pH of cheese, it is of major significance in regulating the various biochemical reactions that occur in cheese during ripening. The isomerization of lactate probably has little impact on cheese flavor, but its conversion to

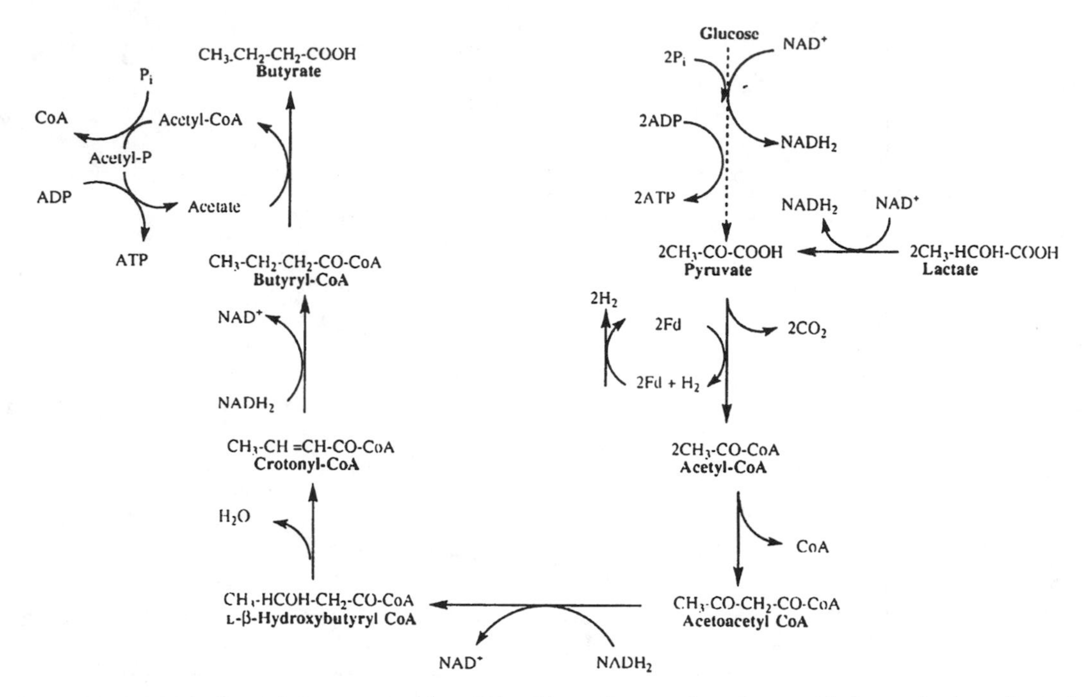

Figure 11–7 Metabolism of glucose or lactic acid by *Clostridium tyrobutyricum*, with the production of butyric acid, CO_2, and H_2.

propionate and/or acetate is probably significant, and when it occurs, the metabolism of lactate to butyrate has a major adverse effect on cheese quality.

11.5 CITRATE METABOLISM

The relatively low concentration of citrate in milk (≈ 8 mM) belies the importance of its metabolism in some cheeses made using mesophilic cultures (for review, see Cogan & Hill, 1993). Most of the starters used in cheese production—*Lc. lactis* subsp. *lactis, Lc. lactis* subsp. *cremoris, Lactobacillus* ssp. and *Sc. thermophilus*—do not metabolize citrate, but a minor component of mixed-strain mesophilic starters, such as those used for Dutch-type cheeses, contain strains of *Lc. lactis* subsp. *lactis* and *Leuconostoc* spp., which metabolize citrate to diacetyl in the presence of a fermentable sugar during manufacture and early ripening (see Figure 5–12). The CO_2 produced is responsible for the small eyes characteristic of Dutch-type cheeses.

Diacetyl is a very significant compound for the aroma and flavor of unripened cheeses, including Cottage cheese and Quarg, and many fermented milks. It contributes to the flavor of Dutch-type cheeses and possibly also Cheddar. Acetate may also contribute to the flavor of these cheeses.

Approximately 90% of the citrate in milk is soluble and is lost in the whey. However, the concentration of citrate in the aqueous phase of cheese is roughly 3 times that in whey, reflecting the concentration of colloidal citrate. Cheddar cheese contains 0.2–0.5% (w/w) citrate, which decreases to 0.1% at 6 months through the metabolic activity of some mesophilic lactobacilli, many of which catabolize citrate to ethanol, acetate, and formate (see Figure 5–11) late in the ripening when numbers of NSLAB have increased sufficiently.

11.6 LIPOLYSIS AND RELATED EVENTS

Pure triglycerides elicit an oily sensation in the mouth but are devoid of flavor *stricto senso*. However, lipids have a major effect on the flavor and texture of foods, including cheese. The influence of lipids on cheese texture is discussed in Chapter 13.

Lipids contribute to cheese flavor in three ways:

1. They are a source of fatty acids, especially short-chain fatty acids, which have strong and characteristic flavors. Fatty acids are produced through the action of lipases in a process referred to as *lipolysis*. In some varieties, the fatty acids may be converted to other sapid and aromatic compounds, especially methyl ketones and lactones.
2. Fatty acids, especially polyunsaturated fatty acids, undergo oxidation, leading to the formation of various unsaturated aldehydes that are strongly flavored and cause a flavor defect referred to as *oxidative rancidity*. Lipid oxidation appears to be very limited in cheese, probably owing to its low redox potential (–250 mV).
3. Lipids function as solvents for sapid and aromatic compounds produced not only from lipids but also from proteins and lactose. Lipids may also absorb from the environment compounds that cause off-flavors.

Of the various possible contributions of lipids to cheese flavor, lipolysis and modification of the resultant fatty acids are the most significant.

The degree of lipolysis in cheese varies widely between varieties, from about 6 mEq free fatty acids in Gouda to 45 mEq/100 g fat in Danish Blue (Gripon, 1987, 1993). Lipases in cheese originate from milk, rennet preparation (paste), starter, adjunct starter, or nonstarter bacteria. Lipolysis in internally bacteria-ripened varieties, such as Gouda, Cheddar, and Swiss, is generally low but is extensive in mold-ripened and some Italian varieties. In general, in those varieties in which extensive lipolysis occurs, lipases originate from the coagulant (rennet paste, which contains pre-gastric esterase, as used in some Italian varieties) or from the adjunct culture (*Penicillium* spp., which produce a number of lipases [Gripon, 1987, 1993] in mold-ripened varieties).

11.6.1 Lipases and Lipolysis

Lipases are hydrolases that hydrolyze esters of carboxylic acids (EC group 3.1.1):

$$R^1{-}C({=}O){-}OR^2 + H_2O \longrightarrow R^1C({=}O){-}OH + R^2OH$$

Lipases have little or no effect on soluble esters, and they prefer to act at the oil-water interface of emulsified esters. Thus, lipases are distinguished by the physical state of the substrate rather than by the type of bond hydrolyzed.

Lipases exhibit various types of specificity:

- They are usually specific for the outer ester bonds of tri- or diglycerides (i.e., sn1 and sn3 positions). Thus, initially they hydrolyze triglycerides to 1,2- and 2,3-diglycerides and later to 2-monoglycerides. The fatty acid at the sn2 position migrates to the vacant sn1 or sn3 position and is then released by lipase (Figure 11–8). Therefore, lipases eventually hydrolyze triglycerides to glycerol and 3 fatty acids. In most cheeses, lipolysis probably does not go beyond the first step.
- Lipases usually exhibit specificity for fatty acids of certain chain length.
- Some lipases exhibit specificity for saturated or unsaturated fatty acids.

Although some lipases may be optimally active at neutral or acid pH values, most have an alkaline pH optimum. Lipases are inactivated by fatty acids and therefore are activated by Ca^{2+}, which precipitates the fatty acids as insoluble

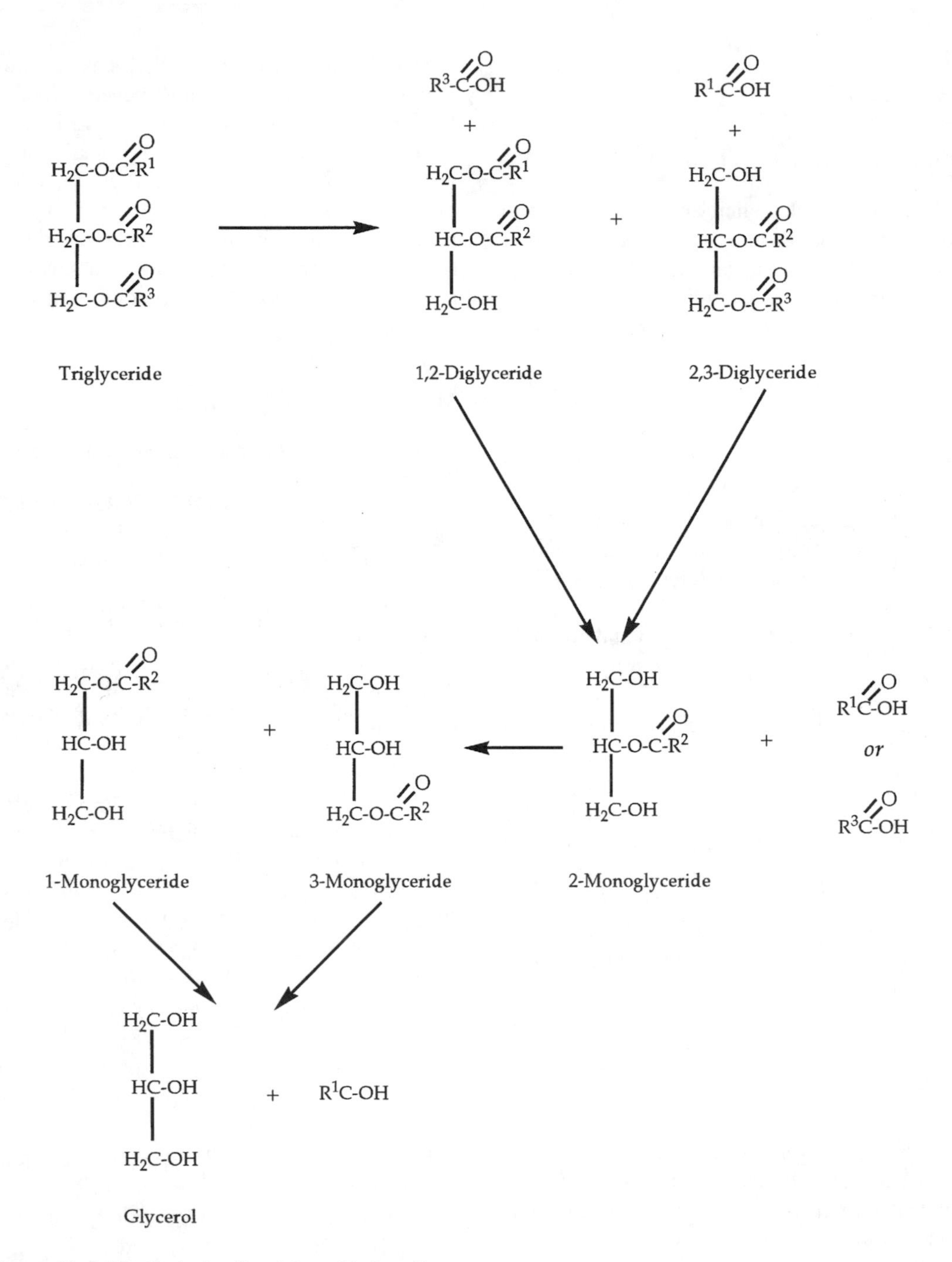

Figure 11–8 Hyydrolysis of a triglyceride by a lipase.

soaps and removes them from the reaction environment. The whey proteins β-lactoglobulin and blood serum albumin bind fatty acids and stimulate lipase activity. Lipases are activated by bile salts, which emulsify the triglyceride substrate. Some lipases, referred to as lipoprotein lipases, are also stimulated by lipoproteins, which promote the adsorption of the enzyme at the oil-water interface. An example is blood serum lipase, which is the indigenous lipase in milk.

11.6.2 Indigenous Milk Lipase

Milk contains an indigenous lipoprotein lipase (LPL) that is well characterized (Olivecrona & Bengtsson-Olivecrona, 1991; Olivecrona, Vilaro, & Bengtsson-Olivecrona, 1992). The enzyme enters milk as a result of leakage through the mammary cell membrane from the blood where it is involved in the metabolism of plasma triglycerides. Bovine milk contains 12 mg lipase/L (10–20 nM). Under optimum conditions, it has a turnover of 3,000 s^{-1} and could theoretically release sufficient fatty acids in 10 s to cause hydrolytic rancidity. However, most of the lipase (> 90%) is associated with the casein micelles, and the fat occurs in globules surrounded by a lipoprotein membrane (the milk fat globule membrane [MFGM]). Thus, the substrate and enzyme are compartmentalized, and lipolysis does not occur unless the MFGM is damaged by agitation, foaming, freezing, or homogenization, for example. Inappropriate milking and milk-handling techniques at the farm and/or factory may cause sufficient damage to the MFGM to permit significant lipolysis and thus off-flavors in cheese and other dairy products.

LPL is rather nonspecific for the type of fatty acid but is specific for the sn1 and sn3 positions of mono-, di-, and triglycerides. Therefore, lipolysis in milk leads to preferential release of short- and medium-chain acids, which in milk triglycerides are esterified predominantly at the sn3 position. Since more than 90% of the LPL in bovine milk is associated with the casein micelles, it is incorporated into cheese curd. LPL is relatively heat labile and is extensively inactivated by HTST pasteurization although heating at the equivalent of 78°C for 10 s is required for complete inactivation. Significantly more lipolysis occurs in raw milk cheese than in pasteurized milk cheese (Figure 11–9). Milk LPL probably contributes to this difference but the NSLAB microflora of raw and pasteurized milk cheeses also differ markedly.

11.6.3 Lipases from Rennet

Good quality rennet extract contains no lipolytic activity. However, rennet paste used in the manufacture of hard Italian varieties (e.g., Romano, Provolone) contains a potent lipase, pregastric esterase (PGE), which is responsible for the extensive lipolysis in and the characteristic "piccante" flavor of such varieties. The literature on PGE was comprehensively reviewed by Nelson, Jensen, and Pitas (1977) and updated by Fox and Stepaniak (1993).

PGE, also called *lingual* or *oral lipase*, is secreted by glands at the base of the tongue. Suckling stimulates the secretion of PGE, which is subsequently washed into the abomasum by milk and siliva. Rennet paste is prepared from the abomasa of calves, kids, or lambs slaughtered after suckling. The abomasa are partially dried and ground into a paste, which is slurried in milk or water before being added to cheese milk. Rennet pastes are considered unhygienic, and their use is not permitted in several countries, including the United States. Instead, partially purified PGEs are used.

Calf, kid, and lamb PGEs have been partially purified from commercial preparations and calf PGE from oral tissue. The enzyme is a glycoprotein with an isoelectric point of 7.0 and a molecular weight of about 49 kDa. The gene for rat lingual lipase has been cloned and sequenced, and the primary structure of the enzyme has been deduced. PGE is highly specific for short-chain acids esterified at the sn3 position and therefore releases high concentrations of highly flavored short- and medium-chain acids from milk fat. The specificity of calf, lamb, and kid PGEs differ slightly, and consequently the flavor characteristics of Italian cheese differ slightly, depending on the source of the PGE used. Most other lipases are unsuitable for the manufacture of Italian cheese because of incorrect specificity, but it has been claimed that certain fungal li-

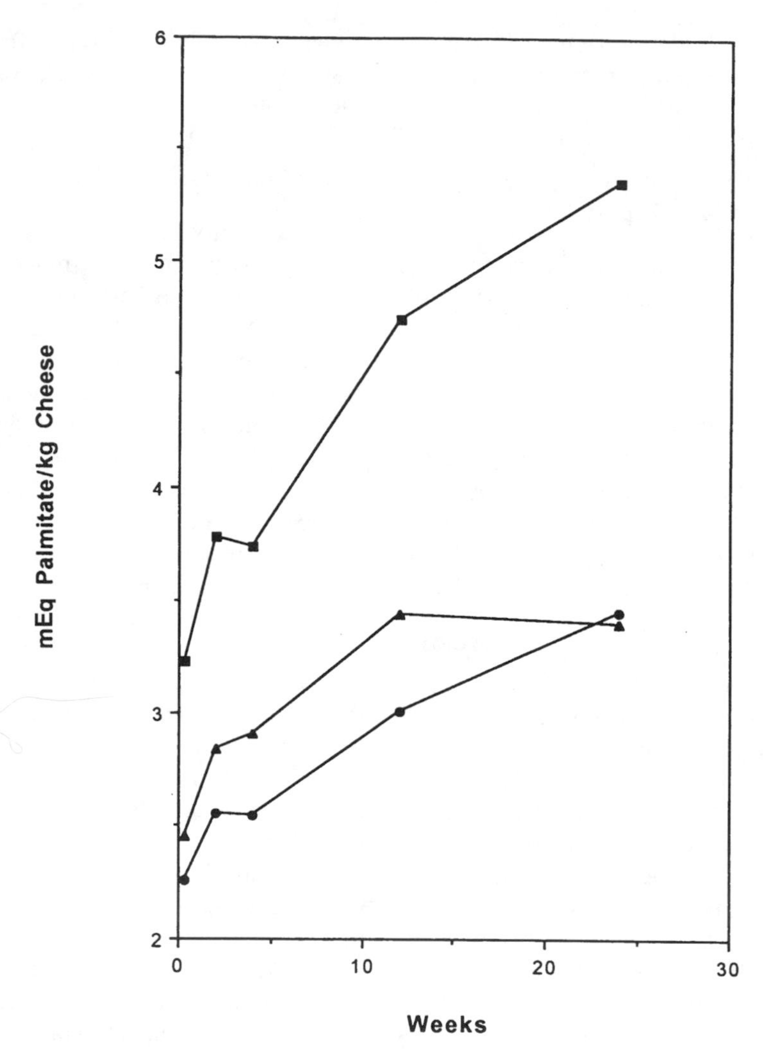

Figure 11–9 Liberation of free fatty acids in Cheddar cheese made from raw milk (■), pasteurized milk (●), and (▲) microfiltered milk.

---

pases may be acceptable alternatives (see Fox & Stepaniak, 1993). The use of PGE to accelerate the ripening of other cheese varieties is discussed in Chapter 15.

11.6.4 Microbial Lipases

Lactococcus spp. and *Lactobacillus* spp. have lower levels of lipolytic activity than other bacteria (e.g., *Pseudomonas*) and molds. However, in the absence of strongly lipolytic agents and when present at high numbers over a long period, as in ripening cheese, lipases and esterases of lactococci and lactobacilli are probably the principal lipolytic agents in Cheddar and Dutch-type cheeses made from pasteurized milk. Aseptic cheeses acidified with GDL instead of starter have low concentrations of free fatty acids that do not increase during ripening. The lipase and esterase activity of LAB appears to be entirely intracellular. Cell-free extracts of various dairy LAB are most active on tributyrin at pH 6–8 and at 37°C. They have little or no activity on triglycerides of long-chain fatty acids (e.g., > C_{10}). There appears to be considerable interstrain variation in esterase and lipase activity, and some strains appear to possess two esterases. Starter bacteria can liberate free fatty acids from mono- and diglycerides produced in milk by other lipases (e.g., milk LPL or lipases from gram-negative bacteria).

The intracellular esterase and lipase of two *Lactococcus* strains have been isolated and characterized (Chich, Marchesseau, & Gripon, 1997; Holland & Coolbear, 1996). At present, little is known about the genetics of these enzymes. Isolation of lipase- and esterase-negative variants of *Lactococcus* would permit the significance of these enzymes in cheese ripening to be assessed.

Both mesophilic and thermophilic lactobacilli possess (mainly) intracellular esterolytic and lipolytic activity (Gobbetti, Fox, & Stepaniak, 1996; Khalid, El-Soda, & Marth, 1990). The esterolytic and lipolytic activity in cell homogenates of a number of lactobacilli was characterized by El-Soda, Abd El-Wahab, Ezzat, Desmazeaud, and Ismail (1986), and an intracellular lipase and an intracellular esterase from *Lb. plantarum* were purified and characterized by Gobbetti, Fox, and Stepaniak (1997) and Gobbetti, Fox, Smacchi, Stepaniak, and Damiani (1997).

Micrococcus, which constitutes part of the nonstarter microflora of cheese, especially the surface microflora, produces lipases that may contribute to lipolysis during ripening (Bhowmik & Marth, 1990a, 1990b). The nonstarter microflora of cheese may also include *Pediococcus* spp., which are weakly esterolytic and lipolytic (Bhowmik & Marth, 1989; Tzanetakis & Litopoulou-Tzanetaki, 1989).

An intracellular lipase of *Propionibacterium shermanii* was partially characterized by Oterholm, Ordal, and Witter (1970); it probably contributes to lipolysis in Swiss varieties. *Brevibacterium linens,* a major component of the surface of smear-ripened cheeses, possesses intracellular lipases and esterases. An intracellular esterase has been purified and characterized (Rattray & Fox, 1997).

Extensive lipolysis occurs in mold-ripened cheese, particularly Blue varieties. In some cases, up to 25% of the total fatty acid may be free (see Gripon, 1987, 1993). However, the impact of free fatty acid on the flavor of Blue mold-ripened cheeses is less than in hard Italian varieties, possibly owing to neutralization as the pH increases during ripening and to the dominant influence of methyl ketones on the flavor of Blue cheese. Lipolysis in mold-ripened varieties is due primarily to the lipases of *Penicillium roqueforti* or *P. camemberti*, which secrete potent, well-characterized extracellular lipases (see Gripon, 1993). *P. camemberti* appears to excrete only one lipase, which is optimally active at around pH 9.0 and 35°C. *P. roqueforti* excretes two lipases, one with a pH optimum at around 8.0, the other at around 6.0. The acid and alkaline lipases exhibit different specificities. *Geotrichum candidum* produces two lipases with different substrate specificities (see Charton, Davies, & McCrae, 1992; Sidebottom et al., 1991).

Psychrotrophs, which usually dominate the microflora of refrigerated milk, are a potentially important source of potent lipases in cheese but are considered not to be very important unless their numbers exceed 10^7 cfu/ml. Many psychrotroph lipases are heat stable and thus may cause rancidity in cheese over the course of a long ripening period. The subject of psychrotroph enzymes in cheese was discussed by Mottar (1989). Unlike psychrotroph proteinases, which are largely water-soluble and are lost in the whey, psychrotroph lipases adsorb onto the fat globules and are therefore concentrated in cheese.

11.6.5 Pattern and Levels of Lipolysis in Selected Cheeses

Lipolysis is considered to be undesirable in most cheese varieties. Cheddar, Gouda, and Swiss-type cheeses containing even a moderate level of free fatty acids would be considered rancid. However, certain cheese varieties are characterized by extensive lipolysis (e.g., Romano, Parmesan, and Blue cheeses). Only small qualitative and quantitative differences in free fatty acids ($C_{2:0}$–$C_{18:3}$) occur between Cheddar cheeses differing widely in flavor. The proportions of free fatty acids ($C_{6:0}$–$C_{18:3}$) in cheese are similar to those in milk fat, indicating that they are released in a nonspecific manner. However, free butyric acid is usually present at a higher concentration than can be explained by its proportion in milk fat, suggesting that it is liberated selectively. Lipolysis in hard Italian varieties is extensive and due primarily to the action of PGE in the rennet paste used in the manufacture of these cheeses. Lipolysis in Blue cheese varieties is extensive owing to the action of lipases from *Penicillium* spp. The free fatty acid levels in a number of cheese varieties are listed in Table 11–2.

11.6.6 Catabolism of Fatty Acids

The taste and aroma of Blue cheese is dominated by saturated n-methyl ketones, a homologous series which, containing an odd number of carbon atoms from C_3 to C_{17}, is present. Concentrations of methyl ketones in Blue cheese fluctuate, presumably due to reduction to secondary alcohols, but heptan-2-one, nonan-2-one, and undecan-2-one dominate.

The metabolism of fatty acids in cheese by *Penicillium* spp. involves four main steps (Figure 11–10; see Kinsella & Hwang, 1976):

1. release of fatty acids by the lipolytic systems (see Sections 11.6.1 to 11.6.5)
2. oxidation of β-ketoacids
3. decarboxylation to methyl ketone with one less carbon atom
4. reduction of methyl ketones to the corresponding secondary alcohol (this step is reversible under aerobic conditions)

The concentration of methyl ketones is related to lipolysis. Methyl ketones can also be formed by the action of the mold on the ketoacids naturally present at low concentrations in milk fat (≈ 1% of total fatty acids). They could also be formed by the oxidation of monounsaturated acids, but evidence for such a pathway is equivocal.

Table 11–2 Typical Concentrations of Free Fatty Acids (FFAs) in Some Cheese Varieties

Variety	*FFA (mg/kg)*
Sapsago	211
Edam	356
Mozzarella	363
Colby	550
Camembert	681
Port Salut	700
Monterey Jack	736
Cheddar	1,028
Gruyère	1,481
Gjetost	1,658
Provolone	2,118
Brick	2,150
Limburger	4,187
Goat's milk cheese	4,558
Parmesan	4,993
Romano	6,754
Blue (US)	32,230
Roquefort	32,453

A number of factors affect the rate of methyl ketone production, including temperature, pH, physiological state of the mold, and the ratio of the concentration of fatty acids to the dry weight of spores. Both resting spores and fungal mycelium are capable of producing methyl ketones. The rate of production of methyl ketones does not depend directly on the concentrations of free fatty acid precursors. Indeed, high concentrations of free fatty acids are toxic to *P. roqueforti*.

Lactones are cyclic esters resulting from the intramolecular esterification of a hydroxyacid through the loss of water to form a ring structure:

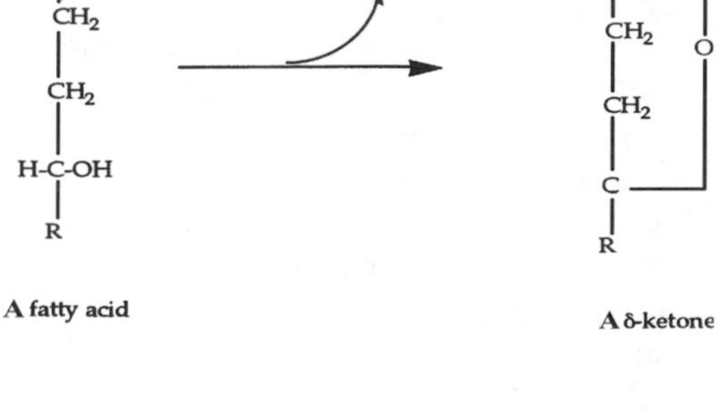

A fatty acid

A δ-ketone

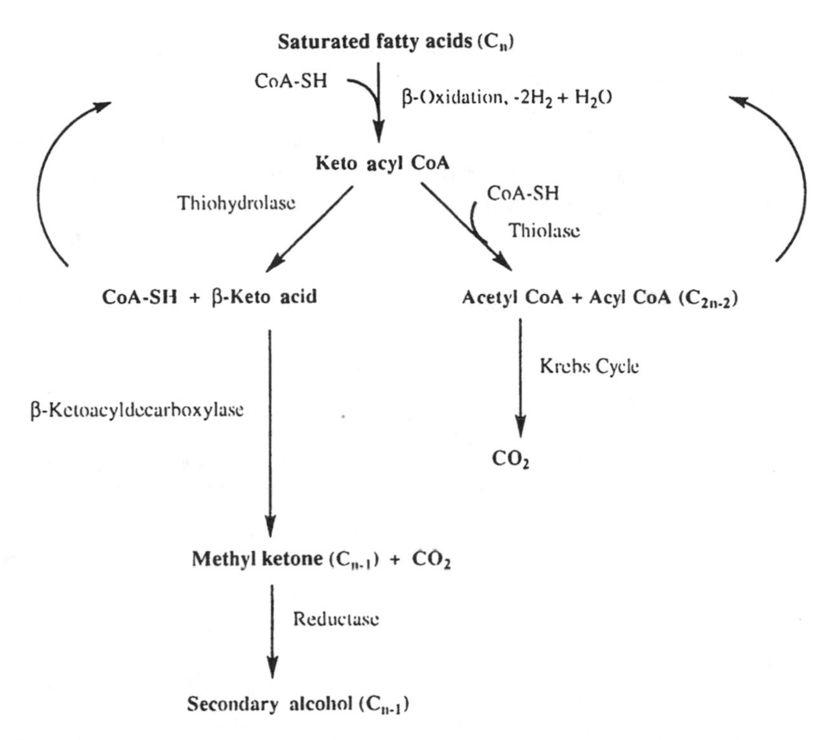

Figure 11–10 β-Oxidation of fatty acids to methyl ketones by *Penicillium roqueforti* and subsequent reduction to secondary alcohols.

α-Lactones and β-lactones are highly reactive and are used, or occur, as intermediates in organic synthesis; γ- or δ-lactones are stable and have been found in cheese. Lactones possess a strong aroma, which, although not specifically cheese-like, may be important in the overall flavor of cheese.

γ-Lactones and δ-lactones in freshly secreted milk probably originate from the corresponding γ- and δ-hydroxyacids following release from triglycerides; they are formed spontaneously following release of the corresponding hydroxyacid. Lactones could also be produced from keto acids released by lipolysis, followed by reduction to hydroxyacids. It has been reported that the mammary gland of ruminants has a δ-oxidation system for fatty acid catabolism, and thus oxidation within the mammary gland may be the primary source of lactone precursors. The potential for lactone production depends on such factors as feed, season, stage of lactation, and breed.

δ-Lactones have very low flavor thresholds. γ-C_{12}, γ-C_{14}, γ-C_{16}, δ-C_{10}, δ-C_{12}, δ-C_{14}, δ-C_{15}, δ-C_{16}, and δ-C_{18} lactones have been identified in Cheddar cheese, and their concentration correlates with age and flavor intensity, suggesting that certain lactones are significant in Cheddar cheese flavor, although this has not been confirmed.

The concentration of lactones in Blue cheese is higher than in Cheddar, probably reflecting the extensive lipolysis that occurs in Blue cheese. The principal lactones in Blue cheese are δ-C_{14} and δ-C_{16}.

11.7 PROTEOLYSIS

11.7.1 Introduction

Of the three primary biochemical events (glycolysis, lipolysis, and proteolysis) that occur in cheese during ripening, proteolysis is the most complex and, in the view of most investigators, the most important. It is primarily responsible for textural changes—in hardness, elasticity, cohesiveness, fracturability, stretchability, meltability, adhesiveness, and emulsifying properties (see Chapter 13)—and makes a major contribution to cheese flavor and the perception of flavor (through release of sapid compounds). Unfortunately, some small peptides are bitter and, if present at sufficient concentrations, will cause bitterness, a common flavor defect in cheese (see Chapter 12).

Proteolysis during maturation is essential in most cheese varieties. The extent of proteolysis varies from very limited (e.g., Mozzarella) to very extensive (e.g., Blue varieties), and the products range in size from large polypeptides only slightly smaller than the intact caseins through a succession of medium and small peptides to free amino acids. Clearly, no one proteolytic agent is responsible for such a wide range of products. Small peptides and amino acids contribute directly to cheese flavor, and the latter may be catabolized to a range of sapid and aromatic compounds that are major contributors to cheese flavor (e.g., amines, acids, carbonyls, and sulfur-containing compounds). Although the catabolism of amino acids is not proteolysis, it is dependent on the formation of amino acids and will be treated in this section.

11.7.2 Assessment of Proteolysis

Proteolysis is routinely monitored in studies on cheese ripening and is a useful index of cheese maturity and quality. Considering the complexity of proteolysis, a variety of methods may be used, depending on the depth of information required. These methods fall into two general classes: specific and nonspecific. The latter include determination of nitrogen soluble in or extractable by one of a number of solvents or precipitants (e.g., water, pH 4.6 buffers, NaCl, ethanol, trichloroacetic acid, phosphotungstic acid, and sulfosalicylic acid) or permeable through ultrafiltration membranes and quantified by any of several methods (e.g., Kjeldahl, biuret, Lowry, Hull, absorbance at 280 nm) or by the formation of reactive α-amino groups quantified by reaction with one of several reagents (e.g., trinitrobenzene sulphonic acid [TNBS], o-phthaldialdehyde [OPA], fluorescamine, Cd-ninhydrin, and Li-ninhydrin). Such methods are valuable for assessing the overall extent of proteolysis and the general contribution of each proteolytic agent. Nonspecific techniques are relatively simple and are valuable for the routine assessment of cheese maturity, since soluble nitrogen correlates well with cheese age and to a lesser extent with quality.

Specific techniques involve the use of chromatography and/or electrophoresis, which resolve individual peptides. They permit monitoring proteolysis of the individual caseins and identification of the peptides formed. Various forms of chromatography have been used to study peptides in cheese, including paper, thin-layer, ion-exchange, gel permeation, and metal chelate techniques as well as, more recently, a variety of high-performance techniques, especially reverse-phase high-performance liquid chromatography. Electrophoresis is a very effective and popular technique for assessing primary proteolysis in cheese, especially alkaline urea-PAGE, but SDS-PAGE and isoelectric focusing are also used. Gel electrophoretograms are not easy to quantify accurately, which is a major limitation of these techniques. In recent years, capillary electrophoresis is being applied increasingly to the analysis of peptides in cheese and has given very satisfactory quantifiable results.

Techniques for assessing proteolysis in cheese during ripening have been the subject of a number of recent reviews, including Fox (1989); Fox, McSweeney, and Singh (1995); Grappin, Rank, and Olson (1985); International Dairy Federation (1991); McSweeney and Fox (1993, 1997); and Rank, Grappin, and Olson (1985). See Chapter 23 for further details.

11.7.3 Proteolytic Agents in Cheese and Their Relative Importance

Cheese contains proteolytic enzymes from rennet, milk, starter lactic acid bacteria (LAB), adventitious nonstarter LAB (NSLAB), and, in most varieties, secondary cultures.

Several studies using the model cheese systems described in Section 11.3, especially studies on Cheddar and Gouda, have shown that enzymes in rennet (chymosin or rennet substitute) are mainly responsible for initial proteolysis and the production of most of the water-soluble or pH 4.6-soluble N (Figure 11–11). However, the production of small peptides and free amino acids is due primarily to the action of enzymes from starter bacteria. γ-Caseins, formed from β-caseins by plasmin, have been found in all cheese varieties that have been studied, indicating plasmin activity, and such activity has been confirmed in cheese supplemented with plasmin or containing a plasmin inhibitor. Nonetheless, plasmin activity is probably not necessary for satisfactory cheese ripening. The rennet enzymes are extensively, probably completely, denatured in high-cooked cheeses, such as Mozzarella, Parmesan, and Emmental, and therefore the contribution of plasmin to primary proteolysis is considerably higher in these varieties than in Cheddar- and Dutch-type cheeses. The pH optimum for plasmin is about 7.5, and hence cheese (≈ pH 5.2) is not a very suitable substrate. However, the pH of many cheeses increases during ripening (e.g., the pH of Camembert increases to ≈ 7), and these therefore are more amenable to plasmin action.

Although NSLAB can dominate the microflora of Cheddar-type cheese during much of its ripening, their influence on proteolysis in cheese is relatively limited and mainly at the level of free amino acid formation.

Some adjunct or secondary cultures are very proteolytic (e.g., *P. roqueforti, P. camemberti*, and *B. linens*). Consequently, these microorganisms make a major contribution to proteolysis in those cheeses in which they are used, especially in Blue cheeses, in which extensive mold growth occurs throughout the cheese.

The extent and specificity of proteolysis in representatives of the principal groups of cheese have been characterized (and are described in Section 11.8). The specificity of the principal proteinases and peptidases on the individual caseins in cheese has been established and can be related to proteolysis in cheese. The specificity of the principal proteolytic enzymes found in cheese is described briefly below, and actual proteolysis in some varieties is discussed in Section 11.8.

11.7.4 Specificity of Cheese-related Proteinases

Coagulant

As discussed in Chapter 6, the principal and essential role of the coagulant in cheesemaking is the specific hydrolysis of κ-casein, as a result of which the colloidal stability of the casein micelles is destroyed and coagulation occurs under suitable conditions of temperature and calcium concentration. Most of the rennet added to cheesemilk is lost in the whey or denatured as a result of cooking the curd-whey mixture to a high temperature. Typical values for the percentage of added rennet retained in the curd range from about 0% for high-cook cheese (e.g., Mozzarella, Parmesan, and Emmental) through 6% for Cheddar to about 20% for high-moisture, low-cook cheeses (e.g., Camembert).

Chymosin (EC 3.4.23.4), the principal proteinase in traditional rennets used for cheesemaking, is an aspartyl proteinase of gastric origin, secreted by young mammals. Its action on the B-chain of insulin indicates that chymosin is specific for hydrophobic and aromatic amino acid residues. Chymosin is weakly proteolytic. Indeed, limited proteolysis is one of the characteristics to be considered when selecting proteinases for use as rennet substitutes.

The primary chymosin cleavage site in the milk protein system is the Phe_{105}-Met_{106} bond in κ-casein, which is many times more susceptible to chymosin than any other bond in milk proteins, and its hydrolysis leads to coagulation of the milk (see Chapter 6). Cleavage of κ-casein at Phe_{105}-Met_{106} yields para-κ-casein (κ-CN f1-105) and glycomacropeptides (GMP; κ-CN f106-169). Most of the glycomacropeptides are lost in the whey, but the para-κ-casein remains attached to the casein micelles and is incorporated into the cheese. α_{s1}-casein, α_{s2}-casein, and β-casein are not hydrolyzed during milk coagu-

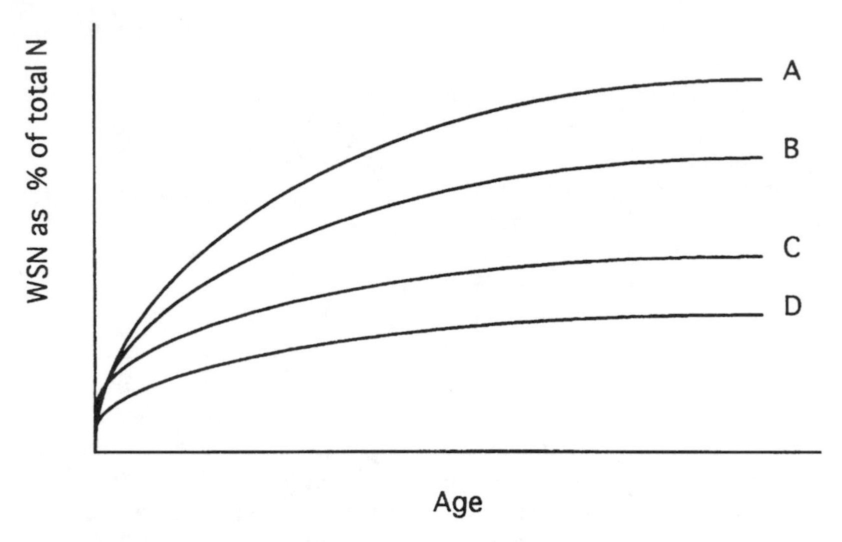

Figure 11–11 Formation of water-soluble nitrogen (WSN) in Cheddar cheese that (A) has controlled microflora (is free of nonstarter bacteria); (B) has controlled microflora and is chemically acidified (starter free); (C) has controlled microflora and is rennet free; and (D) has controlled microflora and is rennet and starter free.

lation but may be hydrolyzed in cheese during ripening.

In solution, chymosin cleaves the Leu_{192}-Tyr_{193} bond of β-casein very rapidly; the larger peptide, β-CN f1-192, is commonly referred to as β-I-casein. At very low ionic strength (e.g., distilled water), this bond is the second most susceptible bond to hydrolysis in the caseins, after the Phe_{105}-Met_{106} bond of κ-casein, with a K_M and k_{cat} of 0.075 mM and 1.54 s^{-1} for the micellar protein and 0.007 mM and 0.56 s^{-1} for the monomeric protein. However, its hydrolysis is strongly inhibited when the ionic strength is increased, even in 50 mM phosphate buffer; it is strongly inhibited by 5% NaCl and completely by 10% NaCl. The bond Ala_{189}-Phe_{190} and bonds in the region of residues Leu_{165} and Leu_{140} of β-casein are hydrolyzed less rapidly. The peptide β-CN f1-189 has the same mobility as β-CN f1-192 on urea-PAGE at pH 9 and is also referred to as β-I-casein. The peptides β-CN f1-165 and β-CN f1-140, referred to as β-II- and β-III-casein, respectively, are readily resolved by urea-PAGE.

β-Casein undergoes very little proteolysis by chymosin in cheese. Undoubtedly, NaCl is an inhibitory factor (Cheddar cheese contains 4–6% salt-in-moisture), but even in salt-free cheese proteolysis is very limited. Perhaps the concentration of milk salts is sufficient to cause inhibition, and protein-protein interactions may also contribute to the low level of proteolysis (the C-terminal region of β-casein is very hydrophobic, and intermolecular hydrophobic interactions may cause the chymosin-susceptible bonds to become inaccessible). The small peptide β-CN f193–209 and ..agments thereof are bitter and hence even limited hydrolysis of β-casein by chymosin may cause bitterness. Because the concentration of NaCl in the interior of most brine-salted cheeses increases slowly due to diffusion from the surface (see Chapter 8), sufficient proteolysis of β-casein by chymosin may occur in Dutch-type and other low-cooked cheese to cause bitterness.

The primary site of chymosin action on α_{s1}-casein is Phe_{23}-Phe_{24}. Cleavage of this bond is believed to be responsible for softening of cheese texture, and the small peptide α_{s1}-CN f1-23 is rapidly hydrolyzed by starter proteinases. The hydrolysis of α_{s1}-casein in solution by chymosin is influenced by pH and ionic strength. In 0.1 M phosphate buffer, pH 6.5, chymosin cleaves α_{s1}-casein at Phe_{23}-Phe_{24}, Phe_{28}-Pro_{29}, Leu_{40}-Ser_{41}, Leu_{149}-Phe_{150}, Phe_{153}-Tyr_{154}, Leu_{156}-Asp_{157}, Tyr_{159}-Pro_{160}, and Trp_{164}-Tyr_{165}. These bonds are also hydrolyzed at pH 5.2 in the presence of 5% NaCl (i.e., conditions similar to those in cheese), and in addition Leu_{11}-Pro_{12}, Phe_{32}-Gly_{33}, Leu_{101}-Lys_{102}, Leu_{142}-Ala_{144}, and Phe_{179}-Ser_{180} are hydrolyzed (McSweeney, Olson, Fox, Healy, & Hojrup, 1993a). The rate at which many of these bonds are hydrolyzed depends on the ionic strength and pH, particularly Leu_{101}-Lys_{102}, which is cleaved far faster at the lower pH. The k_{cat} and K_M for the hydrolysis of Phe_{23}-Phe_{24} bond of α_{s1}-casein by chymosin is 0.7 s^{-1} and 0.37 mM, respectively.

Primary proteolysis (i.e., the formation of large peptides) in low-cooked cheese, including Cheddar, in which the chymosin is not inactivated during cooking, is due mainly to chymosin, and as discussed in Section 11.8, the cleavage sites correspond to those cleaved in α_{s1}-CN in solution at pH 5.2.

α_{s2}-Casein appears to be relatively resistant to proteolysis by chymosin. Cleavage sites are restricted to the hydrophobic regions of the molecule (i.e., residues 90–120 and 160–207). The bonds Phe_{88}-Tyr_{89}, Tyr_{95}-Leu_{96}, Gln_{97}-Tyr_{98}, Tyr_{98}-Leu_{99}, Phe_{163}-Leu_{164}, Phe_{174}-Ala_{175}, and Tyr_{179}-Leu_{180} were reported by McSweeney, Olson, Fox, and Healy (1994) to be the primary cleavage sites. Although para-κ-casein has several potential chymosin cleavage sites, it does not appear to be hydrolyzed either in solution or in cheese.

Good-quality calf (veal) rennet contains about 10% bovine pepsin (EC 3.4.23.1), but values up to 50% have been reported in "calf" rennet. The principal peptides produced from Na-caseinate by bovine pepsin are similar to those produced by chymosin, but the specificity of bovine or porcine pepsins on bovine caseins has not been rigorously determined. The Leu_{109}-Glu_{110} bond of α_{s1}-CN appears to be resistant to chymosin but is relatively susceptible to pepsin. The peptide α_{s1}-CN f110-199 is quite pronounced in electrophoretograms of cheese made with commercial calf rennet but not of cheese made using microbial chymosin, which has the same specificity as purified calf chymosin.

The specificity of the microbial rennet substitutes is quite different from that of chymosin (Figure 11–12); *C. parasitica* proteinase is much more active on β-casein than chymosin. The principal cleavage sites for *R. miehei* proteinase in α_{s1}-casein are Phe_{23}-Phe_{24}, Phe_{24}-Val_{25}, Met_{123}-Lys_{124}, and Tyr_{165}-Tyr_{166}. The principal sites on β-casein are Glu_{31}-Lys_{32}, Val_{58}-Val_{59}, Met_{93}-Gly_{94}, and Phe_{190}-Leu_{191}.

Indigenous Milk Proteinases

Milk contains several indigenous proteinases. Plasmin is the principal indigenous proteinase, but a low level of cathepsin D is also present. Several other proteinases have been reported in milk but are believed to be of little significance in milk and dairy products.

Plasmin. Plasmin (fibrinolysin, EC 3.4.21.7) has been the subject of much study (for reviews see Bastian & Brown, 1996; Grufferty & Fox, 1988). It is a component of blood, where its physiological role is solubilization of fibrin clots. Plasmin is a component of a complex system consisting of the active enzyme, its zymogen (plasminogen), and activators and inhibitors of the enzyme and of its activators (Figure 11–13), all of which are present in milk. The components of the plasmin system enter milk via defective mammocyte membranes and are elevated in late lactation and during mastitic infection. Plasmin, plasminogen, and plasminogen activators are associated with the casein micelles in milk and consequently are incorporated into cheese curd, while the inhibitors of both plasmin and plasminogen activators are in the serum phase and are lost in the whey.

Plasmin is a trypsin-like serine proteinase with a pH optimum at about 7.5 and a high specificity for peptide bonds containing lysine at the N-terminal side. It is active on all caseins but especially on α_{s2}- and β-caseins. Plasmin cleaves β-casein at three primary sites: Lys_{28}-Lys_{29}, Lys_{105}-His_{106}, and Lys_{107}-Glu_{108}, with the forma-

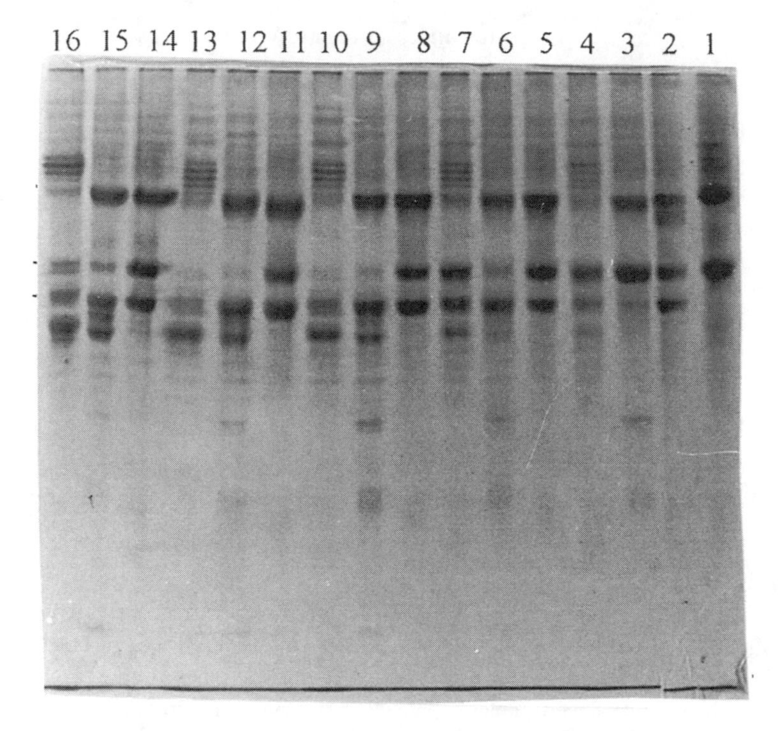

Figure 11–12 Urea-PAGE of Na-caseinate hydrolyzed with different rennets (0.04 RU/ml) at pH 5.2 for 30 min. Lane 1: sodium caseinate; lanes 2, 5, 8, 11, and 14: sodium caseinate containing 0, 1, 2.5, 5, and 10% NaCl hydrolyzed by chymosin; lanes 3, 6, 9, 12, and 15: sodium caseinate containing 0, 1, 2.5, 5, and 10% NaCl hydrolyzed by *R. miehei* proteinase; lanes 4, 7, 10, 13, and 16: sodium caseinate containing 0, 1, 2.5, 5, and 10% NaCl hydrolyzed by *C. parasitica* proteinase.

tion of the polypeptides β-CN f29-209 (γ_1-CN), f106-209 (γ_2-CN), and f108-209 (γ_3-CN); β-CN f1-105 and f1-107 (proteose peptone 5); β-CN f29-105 and f29-107 (proteose peptone 8-slow); and β-CN f1-28 (proteose peptone 8-fast) (see Figure 3–11). Additional cleavage sites include Lys_{29}-Ile_{30}, Lys_{113}-Tyr_{114}, and Arg_{183}-Asp_{184}.

Plasmin cleaves α_{s2}-casein in solution at eight sites—Lys_{21}-Gln_{22}, Lys_{24}-Asn_{25}, Arg_{114}-Asn_{115}, Lys_{149}-Lys_{150}, Lys_{150}-Thr_{151}, Lys_{181}-Thr_{182}, Lys_{188}-Ala_{189}, and Lys_{197}-Thr_{198}—and thereby produces about 14 peptides, 3 of which are potentially bitter.

Although plasmin is less active on α_{s1}-casein than on α_{s2}- and β-caseins, it hydrolyzes α_{s1}-casein in solution at the bonds Arg_{22}-Phe_{23}, Arg_{90}-Tyr_{91}, Lys_{102}-Lys_{103}, Lys_{103}-Tyr_{104}, Lys_{105}-Val_{106}, Lys_{124}-Glu_{125}, and Arg_{151}-Gln_{152} (McSweeney et al., 1993b). The formation of λ-casein, a minor casein component, has been attributed to its action on α_{s1}-casein.

Although κ-casein contains several potential sites, it is very resistant to plasmin, and the products have not been identified. Even though the pH of cheese is quite far removed from the pH optimum of plasmin, the hydrolysis of β-casein in cheese is due mainly to plasmin. Being quite heat stable, plasmin and plasminogen survive HTST pasteurization and cheese cooking, and plasmin is principally responsible for primary proteolysis in high-cooked cheeses, such as Parmesan and Emmental.

Cathepsin D. The indigenous acid proteinase in milk, cathepsin D (EC 3.4.23.5), has received little attention. It is relatively heat labile (completely inactivated by 70°C × 10 min) and has a

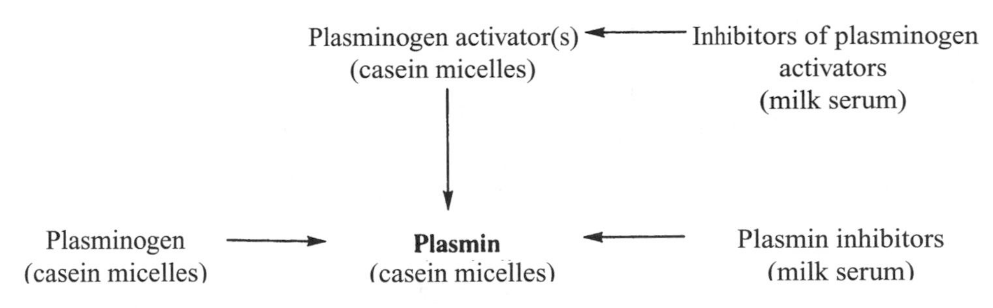

Figure 11–13 Schematic representation of the plasmin system in milk.

pH optimum of 4.0. The specificity of cathepsin D on the caseins has not been determined, although electrophoretograms of hydrolyzates indicate that its specificity is similar to that of chymosin; surprisingly, it is unable to coagulate milk (McSweeney, Fox, & Olson, 1995).

The contribution of cathepsin to cheese ripening is unclear but is very likely to be less than that of chymosin, which is present at a higher concentration and has a similar specificity.

Other Indigenous Milk Proteinases. The presence of other proteolytic enzymes in milk has been reported, including thrombin, a lysine aminopeptidase, and proteinases from leucocytes, but they are considered not to be significant in cheese (see Grufferty & Fox, 1988).

Proteolytic Enzymes from Starter

Although LAB are weakly proteolytic, they do possess a proteinase and a wide range of peptidases, which are principally responsible for the formation of small peptides and amino acids in cheese. The proteolytic system of *Lactococcus* has been studied thoroughly at the molecular, biochemical, and genetic levels. The system of *Lactobacillus* spp. is less well characterized but the systems of both genera are generally similar. *Sc. thermophilus* is less proteolytic than *Lactococcus* or *Lactobacillus* and has been the subject of little research. The extensive literature on the proteolytic systems of LAB has been comprehensively reviewed by Kunji, Mierau, Hagting, Poolman, and Konings (1996); Law and Haandrikman (1997); Monnet, Chapot-Chartier, and Gripon (1993); Tan, Poolman, and Konings (1993); Thomas and Pritchard (1987); and Visser (1993).

The proteinase in LAB is anchored to the cell membrane and protrudes through the cell wall, giving it ready access to extracellular proteins; all the peptidases are intracellular. The oligopeptides produced by the proteinase are actively transported into the cell, where they are hydrolyzed further by the battery of peptidases (see Chapter 5).

Cell envelope–associated proteinases (CEPs) of *Lactococcus* have been classified into three groups: P_I-, P_{III}-, and mixed-type proteinases. P_I-type proteinases degrade β- but not α_{s1}-casein at a significant rate, while P_{III}-type proteinases rapidly degrade both α_{s1}- and β-caseins. The nucleotide sequences of the genes for both P_I- and P_{III}-type proteinases are very similar. Based on their specificity on α_{s1}-CN f1-23 (a peptide produced very rapidly from α_{s1}-CN by chymosin), lactococcal proteinases can be classified into seven groups, a–g (Figure 11–14).

The specificity of the CEPs from several *Lactococcus* strains on α_{s1}-, α_{s2}-, β-, and κ-caseins and short peptide substrates has been established (see Fox & McSweeney, 1996; Fox, O'Connor, McSweeney, Guinee, & O'Brien, 1996; Fox, Singh, & McSweeney, 1994). The lactococcal CEP is involved in the formation of many of the small peptides in cheese, and the peptidases are responsible for the release of amino acids.

Lactococcus spp. contain at least the following intracellular proteolytic enzymes:

Group	Strains	Substr.	Cleavage sites in α_{s1}-casein fragment 1-23 (5, 10, 15, 20) RPKHPIKHQGLPQEVLNENLLRF	Amino acid substitutions at positions relevant for substrate binding								
				131	138	142	144	166	177	747	748	763
	L. lactis											
a	AM1, SK11, US3	α_{s1}, β, κ		Ser	Lys	Ala	Val	Asn	Leu	Arg	Lys	Asn
b	AM2			Thr	Thr	Ala	Leu	Asp	Leu	Arg	Lys	Asn
c	E8	α_{s1}, β, κ		Thr	Thr	Ala	Leu	Asp	Ile	Arg	Lys	Asn
d	NCDO763, UC317	α_{s1}, β, κ		Thr	Thr	Ala	Leu	Asp	Leu	Arg	Lys	His
e	WG2, C13, KH	β, κ		Thr	Thr	Ser	Leu	Asp	Leu	Leu	Thr	Asn
f	Z8, H61, TR, FD27	α_{s1}, β, κ		Thr	Thr	Ala	Leu	Asp	Leu	Leu	Thr	His
g	HP	β, κ		Thr	Thr	Asp	Leu	Asp	Ile	Leu	Thr	His
	Lb. paracasei											
	NCDO151			Thr	Thr	Ala	Leu	Asp	Leu	Gln	Thr	Asn
	Lb. bulgaricus											
	NCDO1489			Ser	Gly	Asp	Ile	Val		Gly	Thr	
	Lb. helveticus											
	L89											

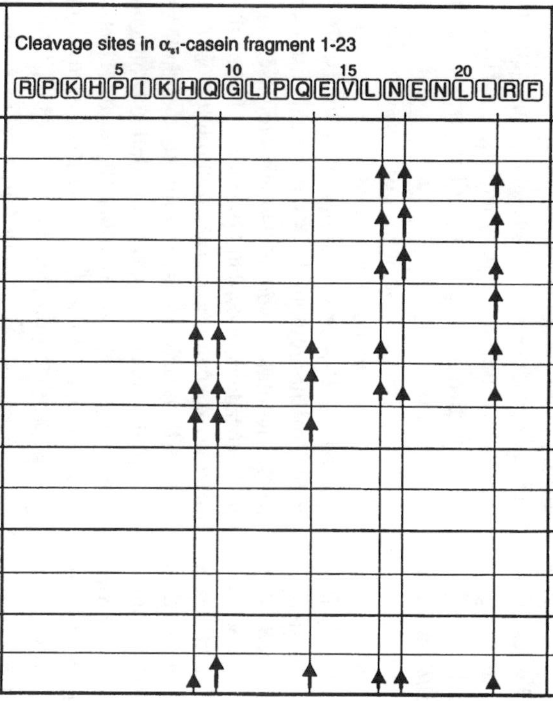

Figure 11–14 Classification of cell envelope–associated proteinases of lactic acid bacteria according to their specificity toward α_{s1}-casein fragment 1-23.

- Proteinases capable of hydrolyzing caseins with a specificity different from that of the CEPs. Their activity in cheese has not been demonstrated.
- Four endopeptidases (PepO_1, PepO_2, PepF_1, PepF_2) that are unable to hydrolyze intact caseins but can hydrolyze internal peptide bonds in large casein-derived peptides (up to 30 amino acid residues). The specificity of these endopeptidases on several casein-derived peptides and on synthetic peptides has been established, but their activity in cheese has not been demonstrated.
- Four aminopeptides (PepN, PepA, PepC, PepP). PepN is a broad specificity metalloenzyme, PepC is a broad specificity thiol enzyme, PepA is a metalloenzyme with high specificity for N-terminal Glu or Asp residues, and PepP is a metalloenzyme that releases the N-terminal residue from peptides with Pro as the penultimate residue.
- An iminopeptidase (PepI) that releases N-terminal Pro.
- A dipeptidyl aminopeptidase (PepX) that has a high but not absolute specificity for peptides with Pro at the penultimate position, releasing X-Pro dipeptides, where X may be one of several residues.
- A pyrrolidone carboxylyl peptidase (PCP) that releases a cyclic pyroglutamic acid from the N-terminal.
- A tripeptidase (PepT).
- A number of dipeptidases, including a general dipeptidase (PepV); PepL, which preferentially hydrolyzes dipeptides and some tripeptides containing an N-terminal Leu; and proline-specific dipeptidases, prolinase (PepR) and prolidase (PepQ), which hydrolyze Pro-X and X-Pro dipeptidases, respectively.

Most of these peptidases have been isolated from at least one strain of *Lactococcus* and characterized (Table 11–3).

The activity and stability of these peptidases in cheese has not been established, but at least some are active, as indicated by the presence of relatively high concentrations of certain peptides and amino acids in cheese. The proteolytic system is capable of hydrolyzing casein completely to free amino acids. The sequential action of the peptidase system is shown schematically in Figure 11–15. This complex proteolytic system is required by LAB for growth to high numbers in milk that contains a low concentration of small peptides and free amino acids.

Thermophilic obligately homofermentative *Lactobacillus* spp. *(Lb. helveticus, Lb. delbrueckii* spp. *bulgaricus, Lb. delbrueckii* ssp. *acidophilus*), alone or paired with *Sc. thermophilus*, are used as starters for high-cooked cheeses. The proteolytic system of the thermophilic lactobacilli is generally similar to that of the lactococci and includes a CEP, the specificity of which has not been determined. These bacteria die and lyse relatively rapidly in cheese (see Chapter 10), releasing intracellular peptidases, which explains the high level of amino acids in cheese made with thermophilic starters.

Sc. thermophilus is weakly proteolytic (no proteinase has yet been isolated), but it possesses substantial peptidase activity. Its contribution to proteolysis in cheese is probably less than that of the thermophilic lactobacilli, but definitive studies are lacking.

Proteolytic System of Nonstarter Microflora

The starter cells, both *Lactococcus* and thermophilic *Lactobacillus*, reach maximum numbers shortly after manufacture and then die off and lyse. In contrast, NSLAB grow from very low initial numbers (< 50 cfu/g) to about 10^7–10^8 cfu/g within about 3 months and dominate the microflora of long-ripened cheeses for most of their ripening period. As discussed in Chapter 10, the interior of cheese is a hostile environment for bacteria. It has a relatively low pH (≈ 5), has a relatively high salt content (2–4%), lacks a fermentable carbohydrate, is anaerobic, and may contain bacteriocins produced by starter bacteria. Hence, cheese is highly selective, and the NSLAB microflora is dominated by

Table 11–3 Peptidases of Lactic Acid Bacteria

Organism	Principal Assay Substrate	Mol. Wt. (kDa)	Optimal Activity		Subunits	Class
			pH	°C		
Oligoendopeptidases (PepO, PepF, LEP, MEP, NOP)						
Lc. lactis ssp. *lactis* CNRZ 267	Peptides	49	–	–	–	–
Lc. lactis ssp. *cremoris* H61	Peptides	98	7–7.5,	40	1	M
Lc. lactis ssp. *cremoris* H61	α_{s1}-CN f1-23	80	6	37	2	M
Lc. lactis ssp. *cremoris* Wg2	Met-enkephalin	70	6–6.5,	30–38	1	M
Lc. lactis ssp. *cremoris* HP	α_{s1}-CN f1-23	180	8–9	42	>2	M
Lc. lactis ssp. *cremoris* C13	α_{s1}-CN f1-23	70	6–7	35	1	N
Lc. lactis ssp. *lactis* MG 1363	α_{s1}-CN f1-23	70	7.5	40	1	M
Lc. lactis ssp. *cremoris* SK11	Bradykinin	70	6.0	–	1	M
Lc. lactis ssp. *lactis* NCDO763	Bradykinin	70	8.0	40	1	M
Lb. delbr. ssp. *bulgaricus* B14	Met-enkephalin	70	7.7	47	1	M
Aminopeptidases						
Aminopeptidase N (general aminopeptidase, AMP, PepN)						
Lc. bv. *diacetylactis* CNRZ 267	Lys-*p*-NA	85	6.5	35	–	M
Lc. lactis ssp. *cremoris* AC1	Lys-*p*-NA	36	7	40	1	M
Lc. lactis ssp. *cremoris* WG_2	Lys-*p*-NA	95	7	40	1	M, –SH
Lb. delbrueckii ssp. *lactis* 1183	Lys-*p*-NA	78–91	6.2–7.2	47.5	1	M
Lb. acidophilus R-26	Lys-*p*-NA	38	–	–	–	M
Lb. delbr. ssp. *bulgaricus* CNRZ 397	Lys-*p*-NA	95	–	–	–	M
Lb. helveticus CNRZ 32	Lys-*p*-NA	97	–	–	–	M
Lb. delbrueckii ssp. *bulgaricus* B14	Lys-*p*-NA	95	7	50	1	M
Lb. helveticus LME-511	Leu-*p*-NA	92	7	37	1	M
Lb. casei ssp. *casei* LLG	Leu-*p*-NA	87	7	39	1	M
Lb. delbr. ssp. *bulgar.* ACA-DC233	Lys-*p*-NA	98	6	40	1	M
Lb. helveticus ITGL1	Lys-*p*-NA	97	6.5	50	1	M
Str. sal. ssp. *thermophilus* CNRZ1199	Lys-*p*-NA	89	6.5	35	1	M
Sc. sal. ssp. *thermophilus* CNRZ302	Lys-*p*-NA	97	7.0	36	1	M
Sc. sal. ssp. *thermophilus* NCDO573	Lys-*p*-NA	96	6.9–7.0	35	1	M, –SH

continues

Aminopeptidase A (glutamyl aminopeptidase, GAP, PepA)						
Lc. lactis ssp. *cremoris* HP	Glu-/Asp-*p*-NA	130	–	50–55	3	M
Lc. lactis ssp. *lactis* NCDO 712	Glu-*p*-NA	245	8	65	6	M
Lc. lactis ssp. *cremoris* HP	Glu-*p*-NA	520	8	50	~10	M
Lc. lactis ssp. *cremoris* AM2	Asp-*p*-NA	240	–	–	6	M
Aminopeptidase C (thiol aminopeptidase, PepC)						
Lc. lactis ssp. *cremoris* AM2	His–β–NA	300	7	40	6	–SH
Lb. delbrueckii ssp. *bulgaricus* B14	Leu-Gly-Gly	220	6.5–7	50	4	–SH
Pyrrolidone carboxylyl peptidase (pyroglutamyl aminopeptidase, PCP)						
Lc. lactis ssp. *cremoris* HP	Pyr-*p*-NA	–	–	–	–	–
Lc. lactis ssp. *cremoris* HP	Pyr-*p*-NA	80	8–8.5	37	2	S
X–Prolyldipeptidyl aminopeptidase (XPDA, PPDA, XAP, PepX)						
Lc. lactis ssp. *cremoris* P8-2-47	X-Pro-*p*-NA	180	7	45–50	2	S
Lc. lactis ssp. *lactis* NCDO 763	Ala-Pro-*p*-NA	190	8.5	40–45	2	S
Lc. lactis ssp. *cremoris* AM2	Gly-Pro-NH-Mec	117	6–9	–	–	S
Lc. lactis ssp. *lactis* H1	X–Pro-*p*-NA	150	6–9	–	–	S
Lb. delbrueckii ssp. *lactis*	X-Pro-*p*-NA	165	7	50–55	2	S
Lb. helveticus CNRZ 32	X-Pro-*p*-NA	72	7	40	1	S
Lb. delbr. ssp. *bulgaricus* CNRZ 397	X-Pro-*p*-NA	82	7	50	–	S
Lb. delbrueckii ssp. *bulgaricus* B14	Ala-Pro-*p*-NA	170–200	6.5	45	2	S
Lb. acidophilus 357	Ala-Pro-*p*-NA	170–200	6.5	45	2	S
Lb. delbr. ssp. *bulgaricus* LBU-147	Gly-Pro-*p*-NA	270	6.5	50	3	S
Lb. delbr. ssp. *lactis* DSM7290	Ala-Pro-*p*-NA	95	7.0	46–50	1	S
Lb. helveticus LHE–511	Gly-Pro-*p*-NA	90	6.5	50	1	S
Lb. casei ssp. *casei* LLG	Gly-Pro-*p*-NA	79	7.0	50	1	S

continues

Table 11–3 continued

Organism	*Principal Assay Substrate*	*Mol. Wt. (kDa)*	*Optimal Activity*		*Subunits*	*Class*
			pH	*°C*		
Proline iminopeptidase (PIP, PepI)						
Pr. freud. ssp. *shermanii 13673*	–	–	–	–	–	–
Lc. lactis ssp. *cremoris* HP	Pro-Gly-Gly	100	8.5	37	2	M
Lb. delbr. ssp. *bulgar. CNRZ 397*	Pro-*p*-NA	100	6–7	40	3	S
Lb. casei ssp. *casei* LLG	Pro-AMC	46	7.5	40	1	–SH
Dipeptidases (DIP, PepV, PepD)						
Lactococcus spp.	Dipeptides	25, 34	7	30		M
Lc. bv. *diacetylactis* CNRZ 267	Leu-Leu	51	7.5	–	1	M
Lc. lactis ssp. *cremoris* H61	Leu-Gly	100	8	–	–	M
Lc. lactis ssp. *cremoris* Wg2	Dipeptides	49	8	50	1	M
Lb. delbr. ssp. *bulgaricus* B14	Dipeptides	51	7	50	1	M
Prolidase (PRD, PepQ)						
Lc. lactis ssp. *cremoris* H61	Leu-Pro	43	6.5–7.5	–	–	M
Lc. lactis ssp. *cremoris* AM2	Leu-Pro	42	7.35–9.0	–	–	M
Tripeptidases (TRP, PepT)						
Lc. bv. *diacetylactis* CNRZ 267	Tripeptides	75	7	35	–	M
Lc. lactis ssp. *cremoris* Wg2	Leu-Leu-Leu	103–105	7.5	55	2	M
Lc. lactis ssp. *cremoris* AM2	Tripeptides	105	8.6	–	2	M
Lc. lactis ssp. *cremoris* IMN-C12	Leu-Leu-Leu	72	5.8	33	3	–SH
Lb. delbrueckii ssp. *bulgaricus* B14	Leu-Gly-Gly	85	6.0	40	>1	M

Key: AMC = aminomethyl coumarin; M = metallo; *p*–NA = *p*–nitroanalide; S = serine; –SH = thiol.

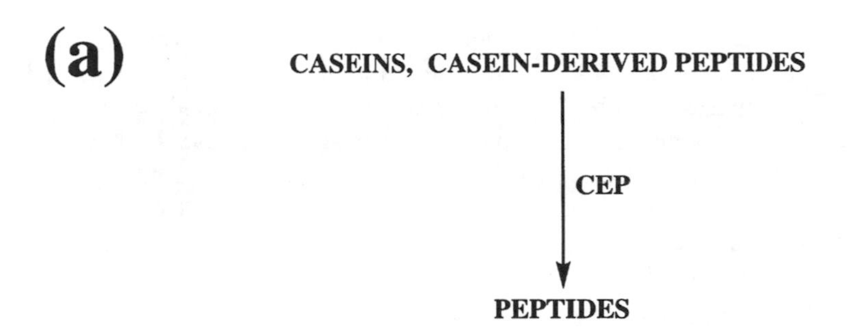

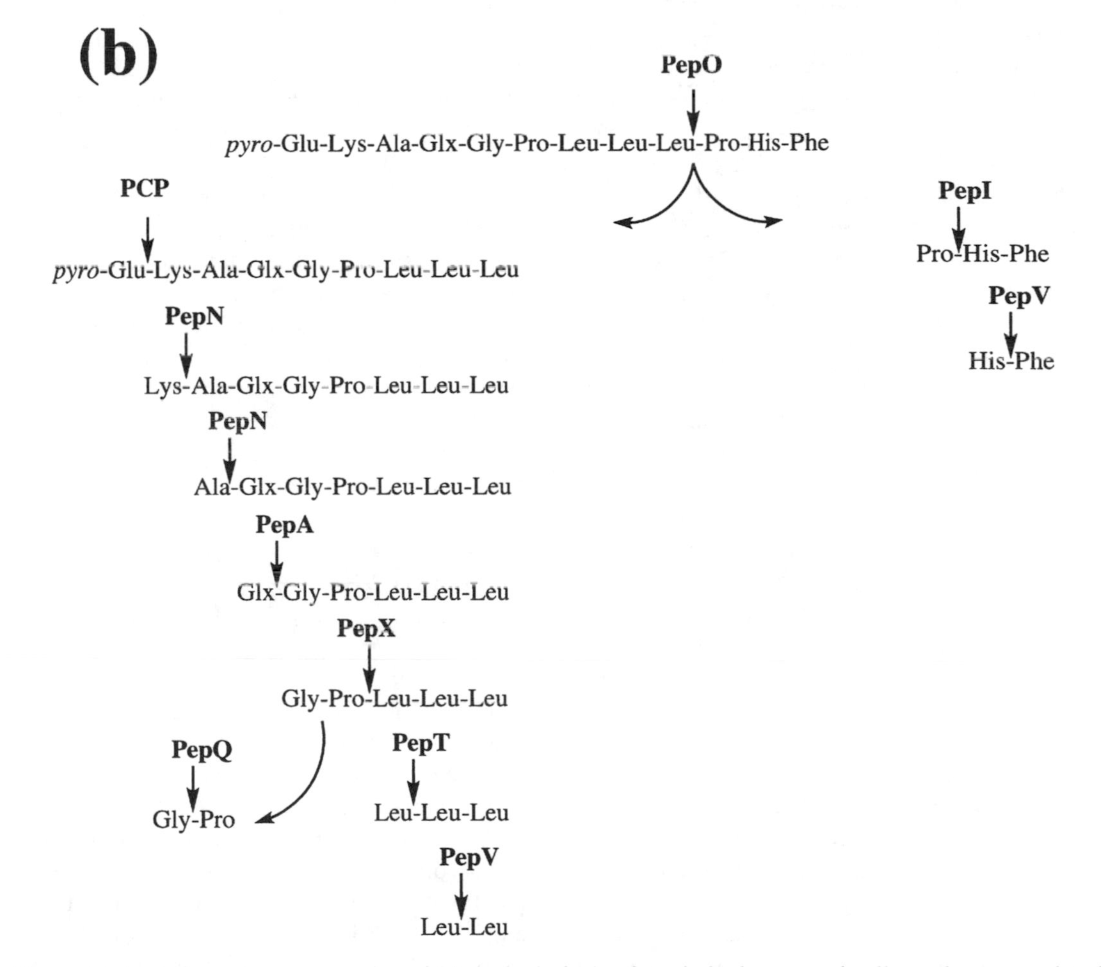

Figure 11–15 Schematic representation of (a) the hydrolysis of casein by lactococcal cell envelope–associated proteinase (CEP) and (b) the degradation of a hypothetical dodecapeptide by the combined action of lactococcal peptidases: oligopeptidase (PepO), various aminopeptidases (PCP, PepN, PepA, PepX), tripeptidase (PepT), prolidase (PepQ), and dipeptidase (PepV).

a few species of mesophilic lactobacilli (*Lb. casei*, *Lb. paracasei*, *Lb. plantarum*, and *Lb. curvatus*). Some authors have reported the presence of pediococci in cheese, but recent studies have failed to find them in significant numbers.

The proteolytic system of these mesophilic lactobacilli is not as well studied as that of the lactococci or thermophilic lactobacilli. Since they do not grow well in milk in the absence of an added source of small peptides or amino acids, they may lack a CEP. Also, the method used to release CEPs from lactococci—washing cells with a calcium-free buffer—fails to release CEPs from mesophilic lactobacilli. These latter bacteria do possess a range of intracellular peptidases, but few of these have been studied.

NSLAB appear to contribute little to primary proteolysis in Cheddar cheese but do contribute to the release of free amino acids (see Fox, McSweeney, & Lynch, 1998). As discussed in Chapter 15, mesophilic lactobacilli have been used as adjunct starters in Cheddar cheese, in which they are reported to modify and perhaps improve flavor.

There are few reports on the proteolytic activity of pediococci (see Fox et al., 1996) and their significance in cheese ripening is unknown.

Proteinases from Secondary Starter

Most cheese varieties have a secondary microflora, the function of which is other than acid production. Originally, this microflora was adventitious, and its development occurred as a result of selective conditions (e.g., pH, humidity, temperature, and a_w). Today, the secondary microflora may be adventitious, but in many cases the milk or curd is inoculated with selected microorganisms. The principal secondary microorganisms are *Penicillium roqueforti* (Blue mold cheese); *P. camemberti* (surface mold cheese, such as Camembert and Brie); *Brevibacterium linens, Arthrobacter*, and other coryneform bacteria (surface smear–ripened cheese); *Propionibacterium freudenreichii* subsp. *shermanii* (Swiss-type cheese); and several species of yeasts (see Chapter 10). Most of these microorganisms are metabolically very active and consequently may dominate the ripening of cheeses in which they occur.

P. roqueforti and *P. camemberti* secrete aspartyl and metalloproteinases, which have been fairly well characterized, including their specificity on α_{s1}- and β-caseins (see Gripon, 1993). Intracellular acid proteinase(s) and exopeptidases (amino and carboxy) are also produced by *P. roqueforti* and *P. camemberti* but have not been well studied (see Gripon, 1993). Tri- and dipeptidases and proline-specific peptidases from *P. roqueforti* and *P. camemberti* do not appear to have been studied.

The proteinase of *Br. linens* ATCC 9174 has been purified and characterized, including its specificity on α_{s1}- and β-casein (see Rattray & Fox, 1998). An extracellular aminopeptidase and an intracellular aminopeptidase have also been purified.

Propionibacterium spp. are weakly proteolytic but strongly peptidolytic. They are particularly active on proline-containing peptides, and consequently Swiss-type cheeses contain high concentrations of proline, which may contribute to the characteristic flavor of these cheeses. Aminopeptidase, iminopeptidase, and X-prolyl dipeptidylaminopeptidase has been isolated from at least one strain of *P. freudenreichii* subsp. *shermanii* (see Fernandez-Espla & Fox, 1997).

11.8 CHARACTERIZATION OF PROTEOLYSIS IN CHEESE

The extent of proteolysis in cheese ranges from very limited (e.g., Mozzarella) to very extensive (e.g., Blue-mold varieties). PAGE shows that the proteolytic pattern, as well as its extent, exhibits marked intervarietal differences; the PAGE patterns of both the water-insoluble and water-soluble fractions are, in fact, quite characteristic of the variety, as shown in Figures 11–16 and 11–17. RP-HPLC of the water-soluble fraction or subfractions thereof also show varietal characteristics (Figures 11–18 and 11–19). Both the PAGE and HPLC patterns vary and become more complex as the cheese matures, and they

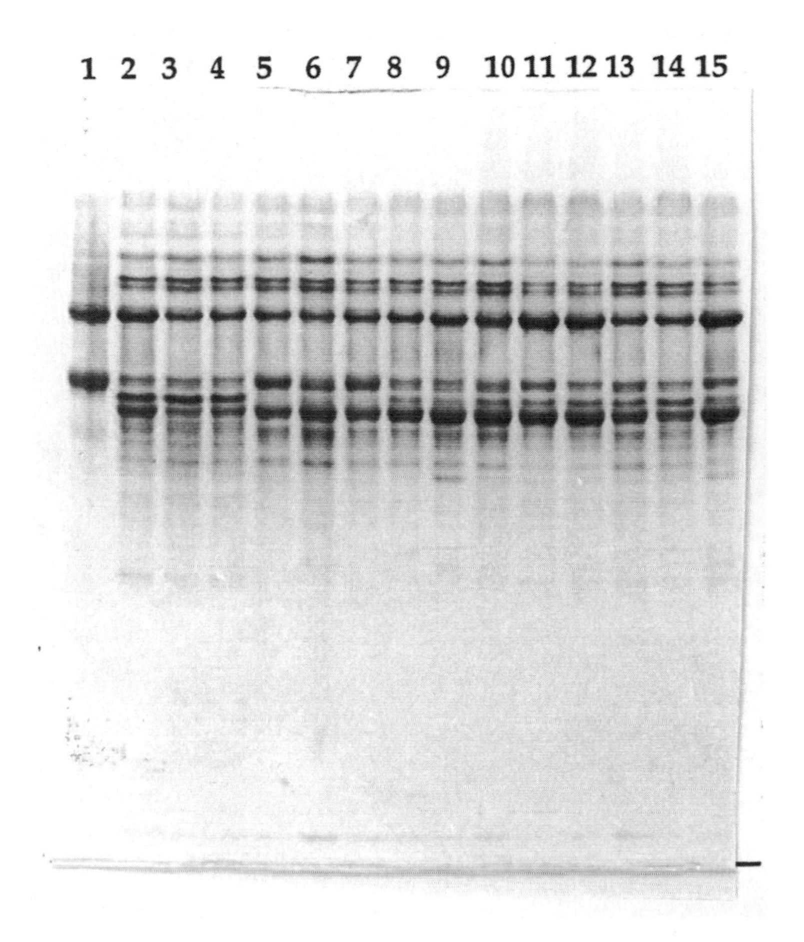

Figure 11–16 Urea-PAGE of the water-insoluble fraction of a selection of cheese varieties: 1, Na-caseinate; 2–4, Cheddar; 5–7, Emmental; 8, Maasdamer; 9, Jarlsberg; 10–12, Edam; 13–15, Gouda.

are in fact very useful indices of cheese maturity and to a lesser extent of cheese quality. Therefore, they have potential as tools in the objective assessment of cheese quality. Urea-PAGE patterns of Cheddar cheeses at various stages of maturity are shown in Figure 11–20.

Complete characterization of proteolysis in cheese requires isolation and identification of the individual peptides. A comprehensive fractionation protocol is shown in Figure 11–21. Many of the water-insoluble and water-soluble peptides in Cheddar cheese have been isolated and identified by amino acid sequencing and mass spectrometry; these are summarized in Figures 11–22 and 11–23. All the principal water-insoluble peptides are produced either from α_{s1}-casein by chymosin or from β-casein by plasmin and represent the C-terminal fragments of these proteins (Figure 11–22). In mature Cheddar (> 6 months old), all of the α_{s1}-casein is hydrolyzed by chymosin at Phe_{23}-Phe_{24}. The peptide α_{s1}-CN f1-23 does not accumulate but is hydrolyzed rapidly at Gln_9-Gly_{10}, Gln_{13}-Glu_{14}, and/or Leu_{16}-Asn_{17} by the lactococcal cell wall proteinase, depending on its specificity (see Figure 11–14). A significant amount of the larger peptide (α_{s1}-CN f24-199) is hydrolyzed at Leu_{101}-Lys_{102}. In 6-month-old Cheddar, about 50% of the β-casein is hydro-

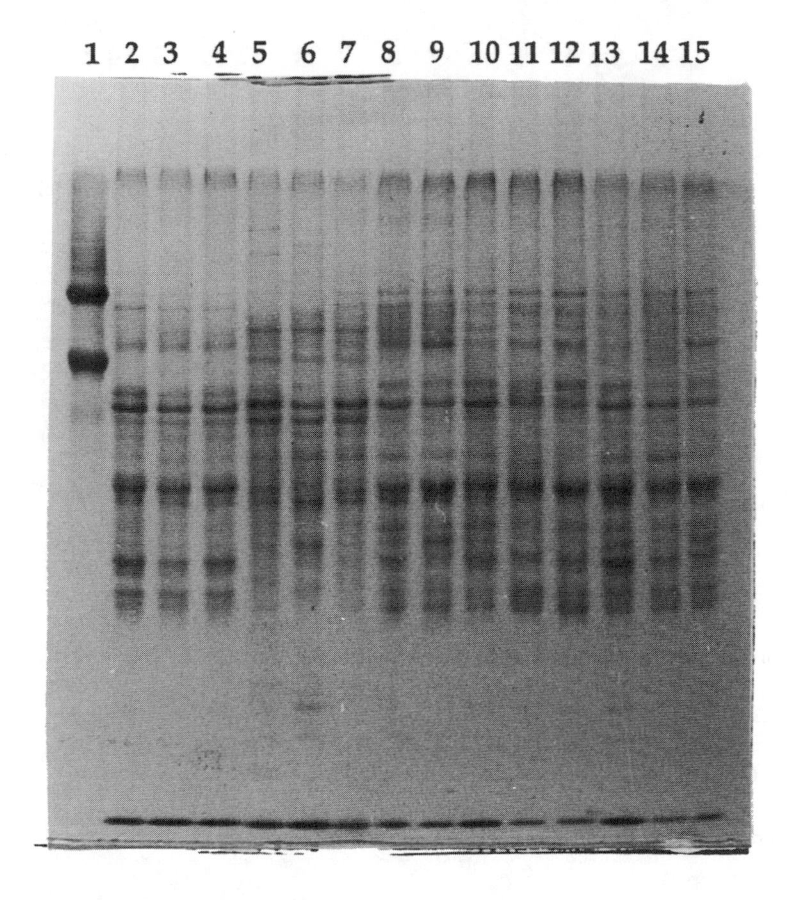

Figure 11–17 Urea-PAGE of the water-soluble fraction of a selection of cheese varieties: 1, Na-caseinate; 2–4, Cheddar; 5–7, Emmental; 8, Maasdamer; 9, Jarlsberg; 1–12, Edam; 13–15, Gouda.

lyzed, mainly by plasmin, to γ-casein (β-CN f29-209, f105-209, and f107-209) and proteose peptones (β-CN f1-28, f1-104, f1-106, f29-104, f29-106). These polypeptides do not appear to be hydrolyzed by chymosin or lactococcal proteinase. Although α_{s2}-casein gradually disappears from PAGE patterns of cheese during ripening, few polypeptides produced from it have been identified. Para-κ-casein is quite resistant to proteolysis, and no peptides produced from it have been identified.

Most of the water-soluble peptides are derived from the N-terminal half of α_{s1}- and β-caseins (Figure 11–23). The N-terminal of many of these peptides corresponds to a chymosin (α_{s1}-CN) or plasmin (β-CN) cleavage site, but some appear to arise from the action of the lactococcal CEP. However, the N-terminal and especially the C-terminal of many peptides do not correspond precisely to the known cleavage sites of chymosin, plasmin, or lactococcal proteinase. The discrepancy in N-terminal suggests the action of bacterial aminopeptidases. Carboxypeptidase activity would explain why the C-terminal of many peptides does not correspond to known proteinase cleavage sites, but this activity has not been reported in *Lactococcus* spp. It must be presumed that another proteinase, perhaps from NSLAB or endopeptidases (PepO or PepF types) from starter and NSLAB, are involved, or perhaps other cleavage sites for lactococcal cell wall proteinase remain to be identified.

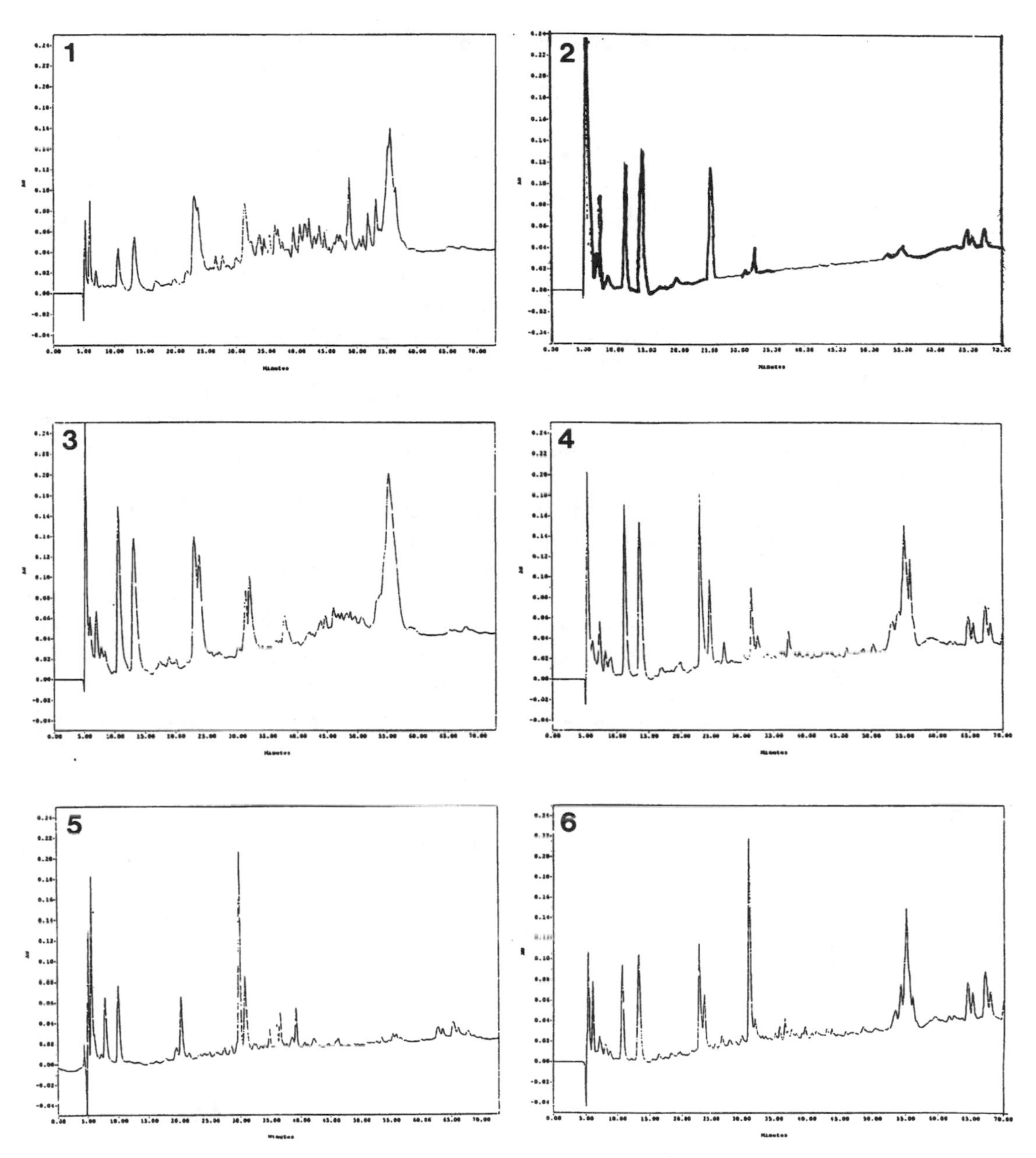

Figure 11–18 RP-HPLC profiles of the 70% ethanol-soluble fractions of Cheddar (1), Parmesan (2), Emmental (3), Leerdammer (4), Edam (5), and Gouda (6) cheese.

The N-terminal sequence of α_{s1}-CN f1-9 and f1-13 is RPKHPIK, which should be susceptible to PepX. The accumulation of these peptides in Cheddar and the apparent absence of peptides with a sequence commencing at Lys_3 of α_{s1}-CN suggest that PepX is not active in cheese.

A number of authors have shown that the very small peptides (< 500 Da) make a significant

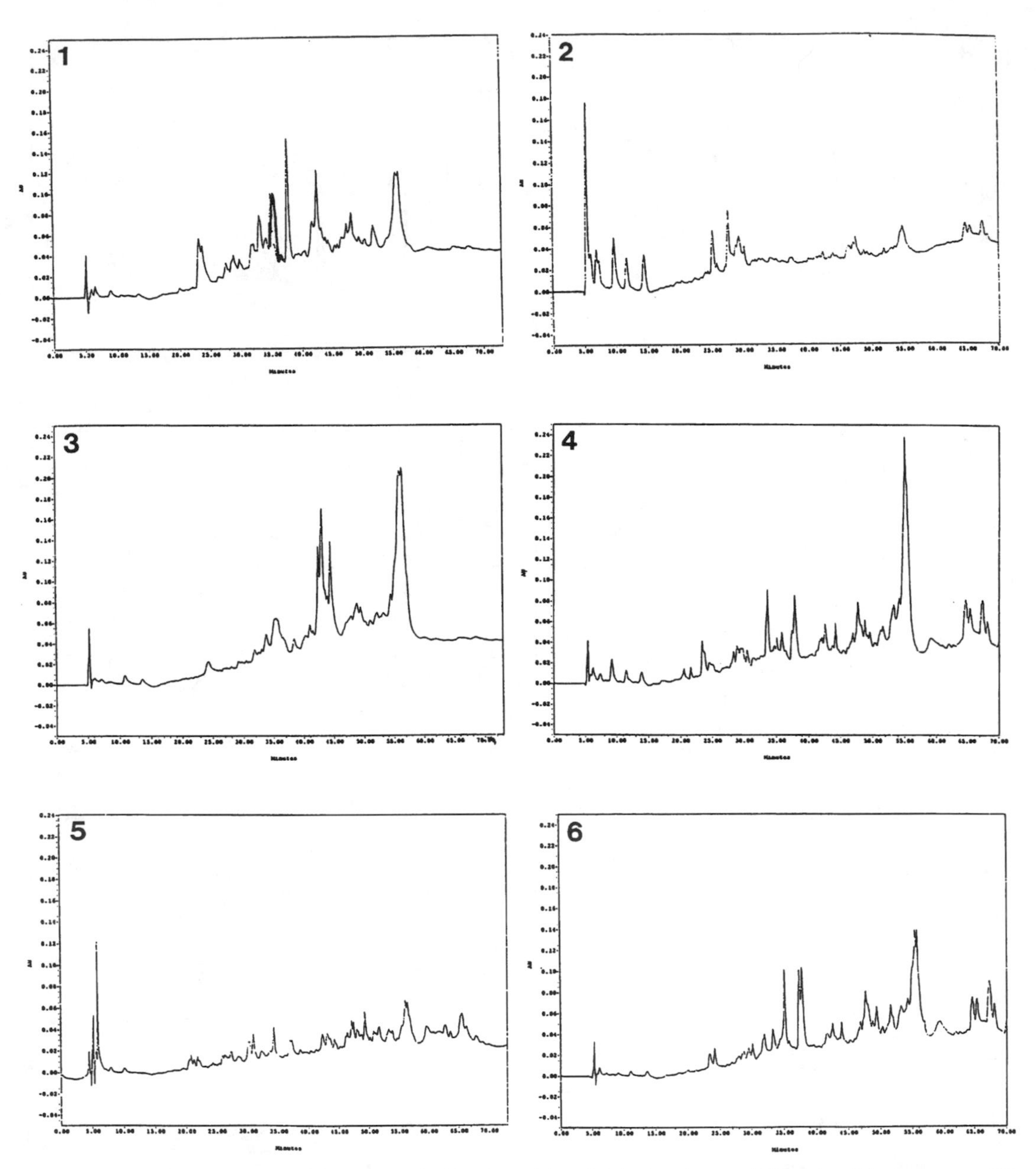

Figure 11–19 RP-HPLC profiles of the 70% ethanol-insoluble fractions of Cheddar (1), Parmesan (2), Emmental (3), Leerdammer (4), Edam (5), and Gouda (6) cheese.

contribution to Cheddar flavor but only a few of these peptides have been identified.

A large number of 12% TCA soluble and insoluble peptides in the water-soluble extract of Parmesan have been identified by fast atom bombardment mass spectrometry (Addeo, Chianese, Sacchi, et al., 1994; Addeo, Chianese, Salzano, et al., 1992). Although Parmesan un-

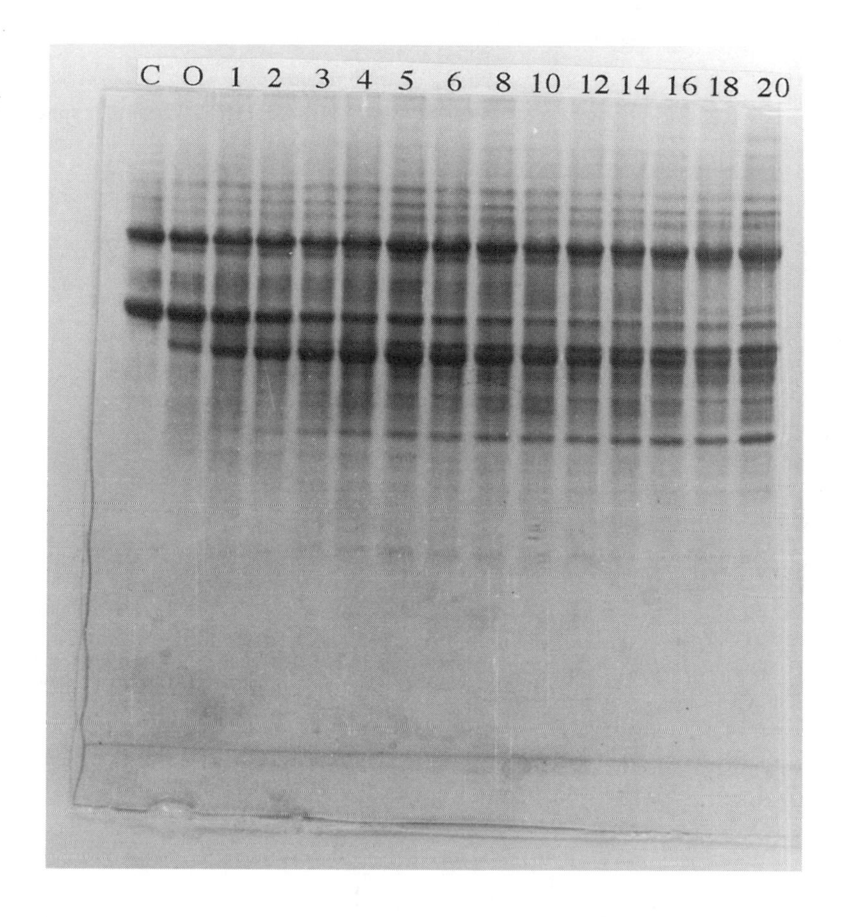

Figure 11–20 Urea-polyacrylamide gel electrophoretograms of Cheddar cheese after ripening for 0, 1, 2, 3, 4, 5, 6, 8, 10, 12, 14, 16, 18, and 20 weeks (lanes 1–14). C, sodium caseinate.

dergoes extensive proteolysis and has a very high concentration of free amino acids, it contains low concentrations of medium-size peptides.

Although very extensive proteolysis occurs in Blue cheeses, and some of the larger peptides detectable by PAGE have been partially identified (see Gripon, 1993), very little work has been done on the small pH 4.6-soluble peptides. Some of the peptides resulting from the cleavage of α_{s1}-CN f1-23 (produced by chymosin) by lactococcal CEP have been identified in Gouda. Proteolysis in Swiss-type cheeses has been studied using PAGE and RP-HPLC, but small peptides have not been isolated and characterized.

Significant concentrations of amino acids, the final products of proteolysis, occur in all cheeses that have been investigated (see Fox & Wallace, 1997). Relative to the level of water-soluble N, Cheddar contains a low concentration of amino acids (see Figure 12–14). The principal amino acids in Cheddar are Glu, Leu, Arg, Lys, Phe, and Ser (Figure 11–24) (see Fox & Wallace, 1997, for a comprehensive compilation of data

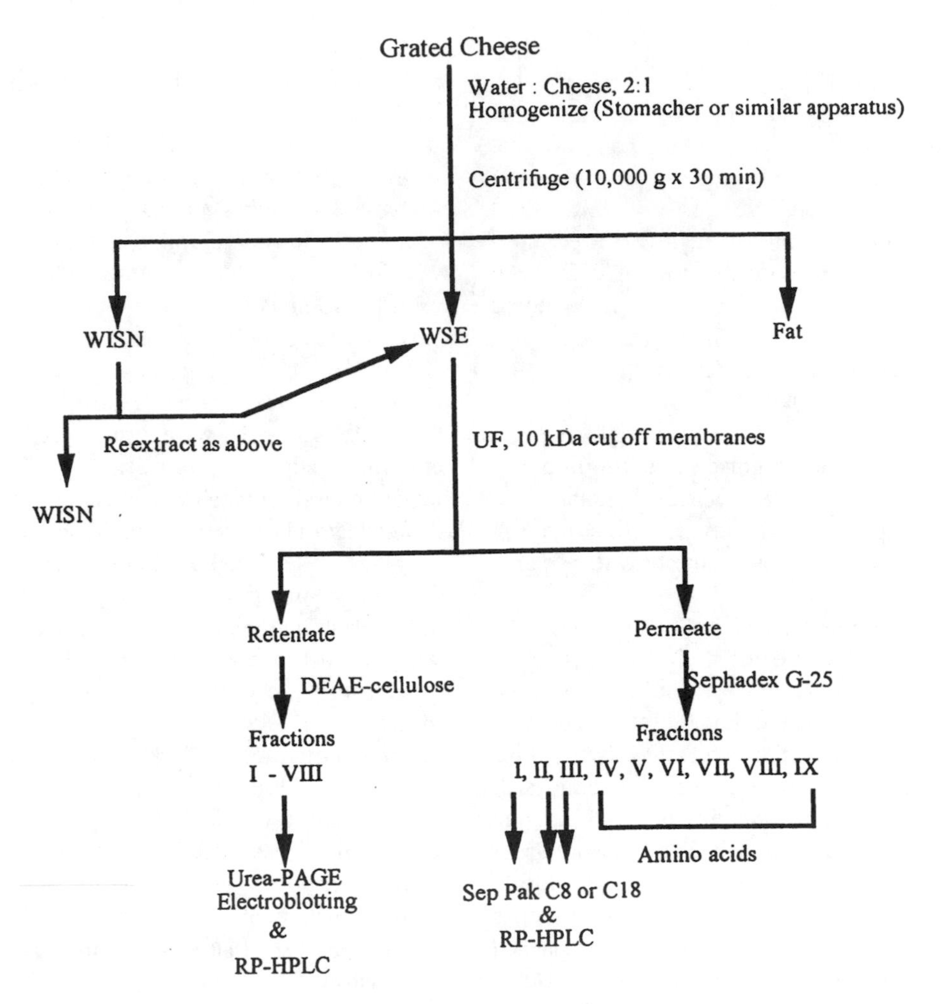

Figure 11–21 Scheme for the fractionation of cheese nitrogen.

for amino acids, in various cheeses). Parmesan contains a very high concentration of amino acids, which appear to make a major contribution to the characteristic flavor of this cheese. The presence of amino acids in cheeses clearly indicates aminopeptidase activity. Since these enzymes are intracellular, their action indicates lactococcal cell lysis. Based on the presumption that amino acids contribute to cheese flavor, a search is now on for fast-lysing lactococcal strains susceptible to heat-, phage-, or bacteriocin-induced lysis (see Chapter 15). Amino acids have characteristic flavors (see Chapter 12), and although none has a cheeselike flavor, it is believed that they contribute to the savory taste of mature cheese.

11.9 CATABOLISM OF AMINO ACIDS AND RELATED EVENTS

Catabolism of free amino acids probably plays some role in all cheese varieties but is particularly significant in mold- and smear-ripened varieties. Catabolism involves decarboxylation, deamination, transamination, desulfuration, and hydrolysis of amino acid side chains leading to the production of a wide array of compounds,

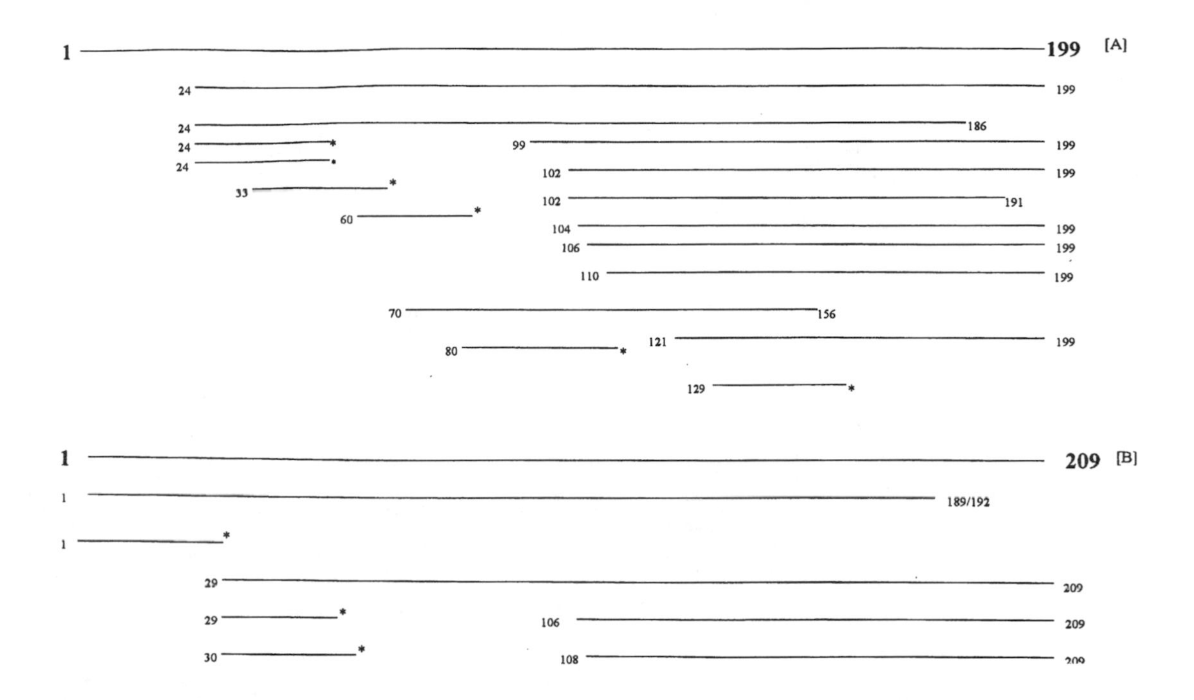

Figure 11–22 Schematic representation of the principal water-insoluble peptides isolated from Cheddar cheese and identified. α_{s1} = casein (A), β = casein (B).

including carboxylic acids, amines, NH_3, CO_2, aldehydes, alcohols, thiols, and other sulfur compounds, phenols, and hydrocarbons. General pathways of amino acid catabolism are summarized in Figure 11–25. The catabolism of amino acids has been reviewed by Fox and Wallace (1997); Hemme, Bouillanne, Métro, and Desmazeaud (1982); and Law (1987).

Decarboxylation involves the conversion of amino acid to the corresponding amine, with the loss of CO_2. The presence of primary amines in cheese can be explained in terms of simple decarboxylation, although the formation of secondary and tertiary amines is more difficult to explain. The principal amine in cheese is tyramine. A number of amines produced in cheese are biologically active (see Chapter 21).

Deamination results in the formation of NH_3 and α-ketoacids. Ammonia is an important constituent of many cheeses, such as Camembert, Gruyère, and Comté. Ammonia can also be formed by oxidative deamination of amines, yielding aldehydes. Transamination results in the formation of other amino acids by the action of transaminases. Aldehydes formed by the above processes can then be oxidized to acids or reduced to the corresponding alcohols.

Amino acid side chains can also be modified in cheese. Hydrolases can release ammonia from Asn and Gln or by the partial hydrolysis of the guanidino group of Arg, forming citrulline or, through degradation, ornithine. Phenol and indole can be produced by the action of C-C lyases on Tyr and Trp.

Volatile sulfur compounds, including hydrogen sulfide (H_2S), dimethylsulfide [$(CH_3)_2$-S], dimethyldisulfide (CH_3-S-S-CH_3), and methanethiol (CH_3SH), are found in most cheeses and can be important flavor constituents. Sulfur-containing compounds are produced mainly from methionine, since Cys is rare in the caseins (it occurs at low levels in only α_{s2}- and κ-caseins, which are not extensively hydrolyzed in cheese). Methanethiol and related compounds are

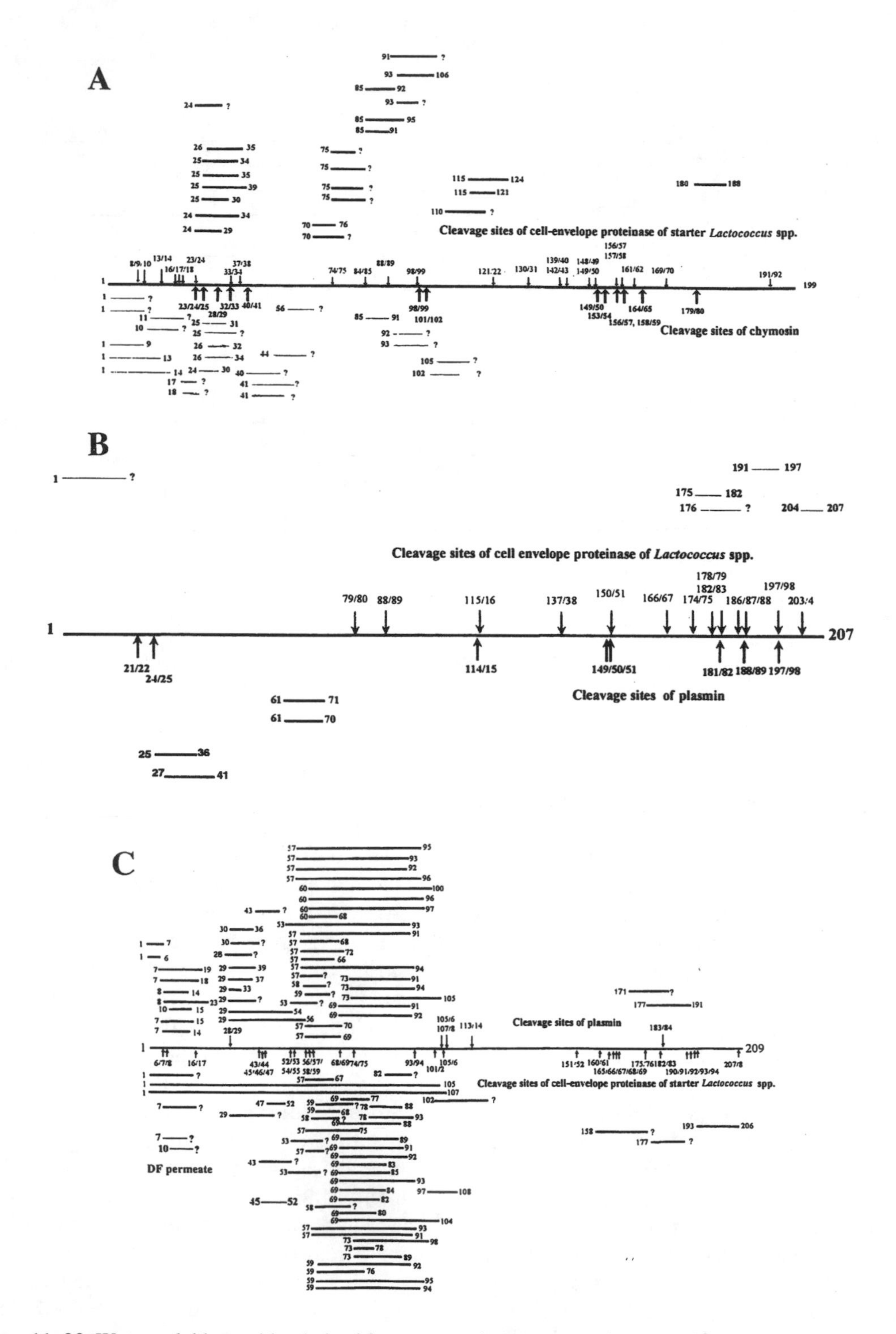

Figure 11–23 Water-soluble peptides derived from α_{s1}-casein (A), α_{s2}-casein (B), and β-casein (C) isolated from Cheddar cheese and identified. The principal chymosin, plasmin, and lactococcal cell–envelope proteinase cleavage sites are indicated by arrows.

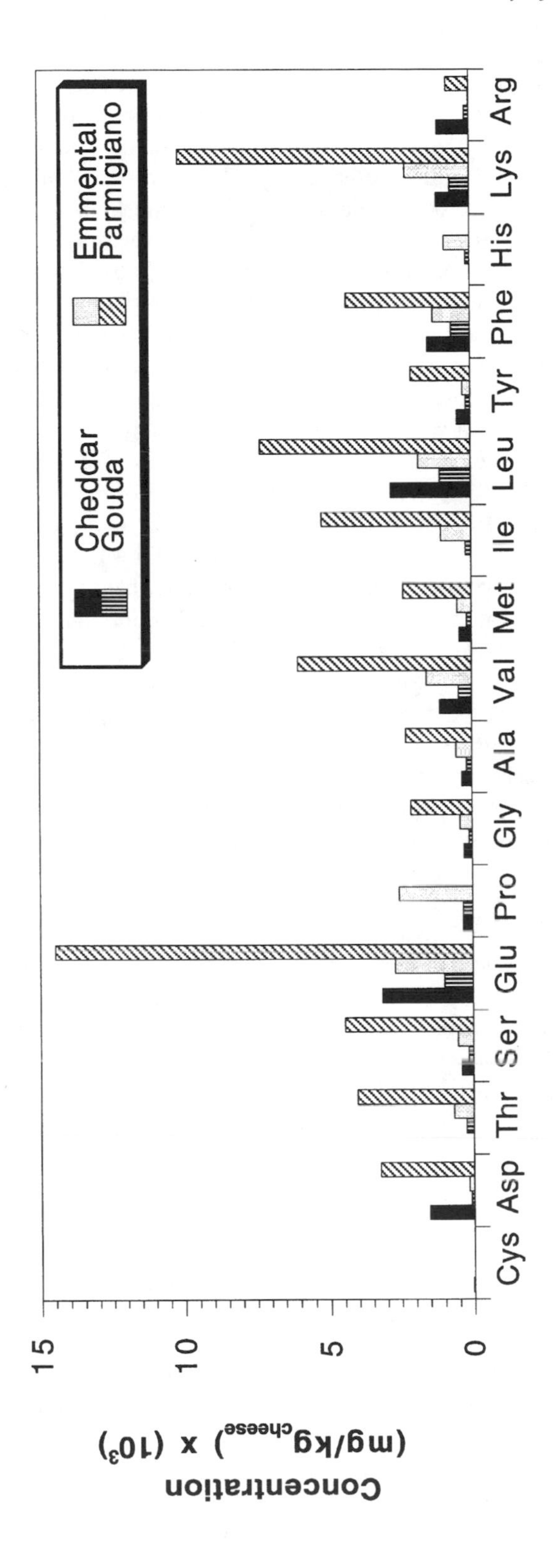

Figure 11–24 Typical concentrations of amino acids in Cheddar, Gouda, Emmental, and Parmigiano-Reggiano.

thought to be particularly important in the flavor of Cheddar cheese.

Although enzymes capable of catalyzing most of the catabolic reactions described above have been identified, few of them have been isolated and characterized from cheese-related microorganisms, especially LAB. Perhaps this is because more complex assay methods are needed to detect these enzymes when present at low levels. Since the products of amino acid catabolism probably contribute to the finer points of cheese flavor, it is expected that research will focus on this area in the immediate future.

11.10 CONCLUSION

Cheese ripening involves a very complex series of biochemical reactions catalyzed by living microorganisms or by enzymes from several sources. The primary events—glycolysis, lipolysis, and proteolysis—have been described rather thoroughly. The fermentation of lactose, mainly to lactic acid, is caused by living microorganisms, principally the starter culture (or adventitious bacteria in the case of artisanal cheeses). The fermentation of lactose has been described thoroughly. Lipolysis is quite limited in most cheese varieties, and in those varieties in which it is important, lipolysis has been well characterized in terms of extent and the enzymes involved. The initial steps in proteolysis and the enzymes responsible have been established for the principal varieties. Secondary proteolysis is characterized to some extent in a few varieties, but proteolysis is so complex and variable, both between and within varieties, that it is probably not possible to characterize it in full detail.

Many of the enzymes responsible for primary ripening have been isolated and characterized. However, the stability and activity of these enzymes in the cheese environment have received little attention, although they appear to warrant research. The primary reactions are probably responsible for changes in cheese texture, such as an increase in pH (due to the catabolism of lactic acid and/or production of NH_3 from deamination of amino acids) or hydrolysis of the protein matrix. With the exception of fatty acids, the products of primary reactions are relatively minor contributors to cheese flavor.

Modifications of the primary products of glycolysis and lipolysis are fairly well characterized. The catabolism of fatty acids to methyl ketones via β-oxidation and decarboxylation is a major contributor to the characteristic flavor of Blue-mold cheeses but is not significant in most varieties. The catabolism of lactic acid has a minor influence on the flavor of most cheeses. An exception is Swiss-type cheeses, but even in these cheeses the catabolism of lactic acid is less important for the flavor than for the production of CO_2 for eye formation. The catabolism of amino acids is the least well characterized aspect of cheese ripening. It is very likely that the products of amino acid catabolism are major contributors to the flavor of many cheese varieties. The ammonia produced in many of these reactions contributes to the pH of cheese during ripening, and this change in pH affects the texture of the cheese and probably affects the stability and activity of many enzymes, which in turn probably influence flavor development. It is very likely that future research on cheese ripening will focus on amino acid catabolism.

Since the biochemistry of cheese ripening is responsible for its flavor, texture, and appearance, elucidation of these biochemical reactions is clearly a prerequisite for controlling and modifying cheese ripening. Such knowledge is essential for the selection of primary and secondary cultures and for their genetic modification. Without this knowledge, selection of starter cultures will be empirical.

The stability and activity of microorganisms and enzymes in cheese depend on its composition. Although the composition of cheese produced in modern factories using modern technology is controlled within quite narrow limits, the quality of the resultant cheese is somewhat variable, even when the cheese is made from pasteurized milk essentially free of indigenous bacteria and using high-quality rennet and starter. This suggests that slight variations in curd composition are important, perhaps owing to their effect on the stability and activity of key enzymes. To date, most studies on these aspects of cheese ripening have been performed on model systems that are well controlled and oversimplified.

Very considerable advances in the biochemistry of cheese ripening have been made during the past 20 years. The general features have been elucidated, but the details remain to be established.

REFERENCES

Addeo, F., Chianese, L., Sacchi, R., Musso, S.P., Ferranti, P., & Molorni, A. (1994). Characterization of the oligopeptides of Parmigiano-Reggiano cheese soluble in 120 g trichloroacetic acid/l. *Journal of Dairy Research, 61*, 365–374.

Addeo, F., Chianese, L., Salzano, A., Sacchi, R., Cappuccio, U., Ferranti, P., & Molorni, A. (1992). Characterization of the 12% trichloroacetic acid insoluble oligopeptides of Parmigiano-Reggiano cheese. *Journal of Dairy Research, 59*, 401–411.

Bastian, E.D., & Brown, R.J. (1996). Plasmin in milk and dairy products [An update]. *International Dairy Journal, 6*, 435–457.

Bhowmik, T., & Marth, E.H. (1989). Esterolytic activities of *Pediococcus* species. *Journal of Dairy Science, 72*, 2869–2872.

Bhowmik, T., & Marth, E.H. (1990a). Esterases of *Micrococcus* species: Identification and partial characterization. *Journal of Dairy Science, 73*, 33–40.

Bhowmik, T., & Marth, E.H. (1990b). Role of *Micrococcus* and *Pediococcus* species in cheese ripening [A review]. *Journal of Dairy Science, 73*, 859–866.

Charton, E., Davies, C., & McCrae, A.R. (1992). Use of specific polyclonal antibodies to detect heterogeneous lipases from *Geotrichum candidum*. *Biochimica Biophysica Acta, 1127*, 191–198.

Chich, J.-F., Marchesseau, K., & Gripon, J.-C. (1997). Intercellular esterase from *Lactococcus lactis* subsp. *lactis* NCDO 763: Purification and characterization. *International Dairy Journal, 7*, 169–174.

Cogan, T.M., & Hill, C. (1993). Cheese starter cultures. In P.F. Fox (Ed.), *Cheese: Chemistry, physics and microbiology* (2d ed., Vol. 1). London: Chapman & Hall.

El-Soda, M., Abd El-Wahab, H., Ezzat, N., Desmazeaud, M.J., & Ismail, A. (1986). The esterolytic and lipolytic activities of the lactobacilli. II. Detection of esterase system of *Lactobacillus helveticus, Lactobacillus bulgaricus, Lactobacillus lactis* and *Lactobacillus acidophilus*. *Lait, 66*, 431–443.

Fernandez-Espla, M.D., & Fox, P.F. (1997). Purification and characterization of X-prolyl dipeptidyl aminopeptidase from *Propionibacterium shermani* NCDO 853. *International Dairy Journal, 7*, 23–29.

Fox, P.F. (1989). Proteolysis during cheese manufacture and ripening. *Journal of Dairy Science, 72*, 1379–1400.

Fox, P.F., Law, J., McSweeney, P.L.H., & Wallace, J. (1993). Biochemistry of cheese ripening. In P.F. Fox (Ed.), *Cheese: Chemistry, physics and microbiology* (2d ed., Vol. 1). London: Chapman & Hall.

Fox, P.F., Lucey, J.A., & Cogan, T.M. (1990). Glycolysis and related reactions during cheese manufacture and ripening. *CRC Critical Reviews in Food Science and Nutrition, 29*, 237–253.

Fox, P.F., & McSweeney, P.L.H. (1996). Proteolysis in cheese during ripening. *Food Reviews International, 12*, 457–509.

Fox, P.F., McSweeney, P.L.H., & Lynch, C.M. (1998). Significance of non-starter lactic acid bacteria in Cheddar cheese. *Australian Journal of Dairy Technology, 53*, 83–89.

Fox, P.F., McSweeney, P.L.H., & Singh, T.K. (1995). Methods for assessing proteolysis in cheese during ripening. In E.L. Malin & M.H. Tunick (Eds.), *Chemistry of structure-function relationships in cheese*. New York: Plenum Press.

Fox, P.F., O'Connor, T.P., McSweeney, P.L.H., Guinee, T.P., & O'Brien, N.M. (1996). Cheese: Physical, biochemical and nutritional aspects. *Advances in Food and Nutrition Research, 39*, 163–328.

Fox, P.F., Singh, T.K., & McSweeney, P.L.H. (1994). Proteolysis in cheese during ripening. In A.T. Andrews & J. Varley (Eds.), *Biochemistry of milk products*. Cambridge: Royal Society of Chemistry.

Fox, P.F., & Stepaniak, L. (1993). Enzymes in cheese technology. *International Dairy Journal, 3*, 609.

Fox, P.F., & Wallace, J.M. (1997). Formation of flavour compounds in cheese. *Advances in Applied Microbiology, 45*, 17–85.

Gobbetti, M., Fox, P.F., Smacchi, E., Stepaniak, L., & Damiani, P. (1997). Purification and characterization of a lipase from *Lactobacillus plantarum* 2739. *Journal of Food Biochemistry, 20*, 227–246.

Gobbetti, M., Fox, P.F., & Stepaniak, L. (1996). Esterolytic and lipolytic activities of mesophilic and thermophilic lactobacilli. *Italian Journal of Food Science, 8*, 127–136.

Gobbetti, M., Fox, P.F., & Stepaniak, L. (1997). Isolation and characterization of a tributyrin esterase from *Lactobacillus plantarum* 2739. *Journal of Dairy Science, 80*, 3099–3106.

Grappin, R., Rank, T.C., & Olson, N.F. (1985). Primary proteolysis of cheese proteins during ripening. *Journal of Dairy Science*, *68*, 531–540.

Gripon, J.-C. (1987). Mould-ripened cheeses. In P.F. Fox (Ed.), *Cheese: Chemistry, physics and microbiology* (Vol. 2). London: Elsevier Science Publishers.

Gripon, J.-C. (1993). Mould-ripened cheeses. In P.F. Fox (Ed.), *Cheese: Chemistry, physics and microbiology* (2d ed., Vol. 2). London: Chapman & Hall.

Grufferty, M.B., & Fox, P.F. (1988). Milk alkaline proteinase [A review]. *Journal of Dairy Research*, *55*, 609–630.

Hemme, D., Bouillanne, C., Métro, F., & Desmazeaud, M.J. (1982). Microbial catabolism of amino acids during cheese ripening. *Science des Aliments*, *2*, 113–123.

Holland, R., & Coolbear, T. (1996). Purification of tributyrin esterase from *Lactococcus lactis* subsp. *cremoris* E8. *Journal of Dairy Research*, *63*, 131–140.

Huffman, L.M., & Kristoffersen, T. (1984). Role of lactose in Cheddar cheese manufacturing and ripening. *New Zealand Journal of Dairy Science and Technology*, *19*, 151–162.

International Dairy Federation. (1991). *Chemical methods for evaluation of proteolysis in cheese maturation* [Bulletin No. 261]. Brussels: Author.

Karahadian, C., & Lindsay, R.C. (1987). Integrated roles of lactate, ammonia, and calcium in texture development of mold surface–ripening cheese. *Journal of Dairy Science*, *70*, 909–918.

Khalid, N.M., El-Soda, M., & Marth, E.H. (1990). Esterase of *Lactobacillus helveticus* and *Lactobacillus delbrueckii* ssp. *bulgaricus*. *Journal of Dairy Science*, *73*, 2711–2719.

Kinsella, J.E., & Hwang, D.H. (1976). Enzymes of *Penicillium roqueforti* involved in the biosynthesis of cheese flavor. *CRC Critical Reviews of Food Science and Nutrition*, *8*, 191–228.

Kunji, E.R.S., Mierau, I., Hagting, A., Poolman, B., & Konings, W.N. (1996). The proteolytic system of lactic acid bacteria. *Antonie van Leeuwenhoek*, *70*, 187–221.

Law, B.A. (1987). Proteolysis in relation to normal and accelerated cheese ripening. In P.F. Fox (Ed.), *Cheese: Chemistry, physics and microbiology* (Vol. 1). London: Elsevier Science Publishers.

Law, J., & Haandrikman, A. (1997). Proteolytic enzymes of lactic acid bacteria. *International Dairy Journal*, *7*, 1–11.

Lenoir, J. (1984). *The surface flora and its role in the ripening of cheese* [Bulletin No. 171]. Brussels: International Dairy Federation.

McSweeney, P.L.H., & Fox, P.F. (1993). Cheese: Methods of chemical analysis. In P.F. Fox (Ed.), *Cheese: chemistry, physics and microbiology* (2d ed., Vol. 1). London: Chapman & Campbell.

McSweeney, P.L.H., & Fox, P.F. (1997). Chemical methods for the characterization of proteolysis in cheese during ripening. *Lait*, *77*, 41–76.

McSweeney, P.L.H., Fox, P.F., & Olson, N.F. (1995). Proteolysis of bovine casein by cathepsin D: Preliminary observations and comparison with chymosin. *International Dairy* Journal, *5*, 321–336.

McSweeney, P.L.H., Olson, N.F., Fox, P.F., & Healy, A. (1994). Proteolysis of bovine α_{s2}-casein by chymosin. *Z. Zeitschrift für Lebensmittel Untersuchung und Forschung 119*, 429–432.

McSweeney, P.L.H., Olson, N.F., Fox, P.F., Healy, A., & Hojrup, P. (1993a). Proteolytic specificity of chymosin on bovine α_{s1}-casein. *Journal of Dairy Research*, *60*, 401–412.

McSweeney, P.L.H., Olson, N.F., Fox, P.F., Healy, A., & Hojrup, P. (1993b). Proteolytic specificity of plasmin on bovine α_{s1}-casein. *Food Biotechnology*, *7*, 143–158.

Monnet, V., Chapot-Chartier, M.P., & Gripon, J.-C. (1993). Les peptidases des lactocoques. *Lait*, *73*, 97–108.

Mottar, J.F. (1989). Effect on the quality of dairy products. In R.C. McKellar (Ed.), *Enzymes of psychrotrophs of raw foods*. Boca Raton, FL: CRC Press.

Nelson, J.H., Jensen, R.G., & Pitas, R.E. (1977). Pregastric esterase and other oral lipases [A review]. *Journal of Dairy Science*, *60*, 327–362.

Olivecrona, T., & Bengtsson-Olivecrona, G. (1991). Indigenous enzymes in milk: Lipase. In P.F. Fox (Ed.), *Food Enzymology* (Vol. 1). London: Elsevier Science Publishers.

Olivecrona, T., Vilaro, S., & Bengtsson-Olivecrona, G. (1992). Indigenous enzymes in milk. II. Lipases in milk. In P.F. Fox (Ed.), *Advanced Dairy Chemistry* (Vol. 1). London: Elsevier Science Publishers.

Oterholm, A., Ordal, Z.J., & Witter, L.D. (1970). Purification and properties of glycerol ester hydrolase (lipase) from *Propionibacterium shermanii*. *Applied Microbiology*, *20*, 16–22.

Rank, T.C., Grappin, R., & Olson, N.F. (1985). Secondary proteolysis of cheese during ripening [A review]. *Journal of Dairy Science*, *68*, 801–805.

Rattray, F.P., & Fox, P.F. (1997). Purification and characterization of an intracellular esterase from *Brevibacterium linens* ATCC 9174. *International Dairy Journal*, *7*, 273–278.

Rattray, F.P., & Fox, P.F. (1998). Recently identified enzymes of *Brevibacterium linens* ATCC 9174 [A review]. *Journal of Food Biochemistry*, *22*, 353–373.

Sidebottom, C.M., Charton, E., Dunn, P.P.J., Mycock, G., Davies, C., Sutton, J.L., MacCrae, A.R., & Slabas, A.R. (1991). *Geotrichum candidum* produces several lipases with markedly different substrate specificities. *European Journal of Biochemistry, 202*, 485–491.

Tan, P.S.T., Poolman, B., & Konings, W.N. (1993). Proteolytic enzymes of *Lactococcus lactis. Journal of Dairy Research, 60*, 269–286.

Thomas, T.D., & Crow, V.L. (1983). Mechanism of D(–)-lactic acid formation in Cheddar cheese. *New Zealand Journal of Dairy Science and Technology, 18*, 131–141.

Thomas, T.D., & Pritchard, G.C. (1987). Proteolytic enzymes in dairy starter cultures. *Federation of European Microbiological Associations Microbiology Review, 46*, 245–268.

Tzanetakis, N., & Litopoulou-Tzanetaki, E. (1989). Biochemical activities of *Pediococcus pentosaceus* isolates of dairy origin. *Journal of Dairy Science, 72*, 859–863.

Visser, S. (1993). Proteolytic enzymes and their relation to cheese ripening and flavor [An overview]. *Journal of Dairy Science, 76*, 329–350.

Chapter 12

Cheese Flavor

12.1 INTRODUCTION

The quality of cheese is determined by its flavor (taste and aroma), texture (hardness, crumbliness, cohesiveness, stretchability, sliceability, etc.), and appearance (color, uniformity, eyes and other fissures, and presence or absence of molds). When a consumer selects cheese, its appearance is the first and perhaps the only criterion of quality applied. For example, an Emmental without eyes, a Gouda with patches of mold, or a Roquefort without mold will not be purchased. Today, cheese offered for sale at reputable outlets is unlikely to be defective in appearance. The relative importance of flavor and texture depends on the variety. For some varieties, such as Mozzarella, which has a very mild flavor, the textural attributes of meltability and stretchability are paramount, whereas flavor is the most important characteristic of Blue cheese varieties. While the nutritional and safety aspects of cheese are essential, most consumers, at least in developed countries, consume cheese mainly for its sensory attributes, which are, therefore, the raison d'être of cheesemaking.

Various aspects of cheese texture are discussed in Chapter 13. Aspects of the flavor of cheese are discussed in this chapter.

Cheese flavor has been the subject of scientific investigation since the beginning of this century. Initially, it was believed that a single compound or small group of compounds might be responsible for cheese flavor, but as data accumulated it became apparent that this was not so and that a large number of sapid and aromatic compounds are present in cheese. This led to the "component balance theory," developed in the 1950s, which proposed that cheese flavor results from the correct balance and concentration of numerous sapid and aromatic compounds. During the intervening 40 years, there has been extensive research on the flavor of several cheese varieties, but complete information is not yet available on the flavor chemistry of any specific variety. The extensive literature on cheese flavor has been reviewed by Adda, Gripon, and Vassal (1982); Aston and Dulley (1982); Bosset and Gauch (1993); Fox, Singh, and McSweeney (1995); Fox and Wallace (1997); Imhof and Bosset (1994); Kristoffersen (1973, 1985); McGugan (1975); McSweeney, Nursten, and Urbach (1997); Olson (1990); Reiter, Fryer, Sharpe, and Lawrence (1966); and Urbach (1993, 1995, 1997). Imhof and Bosset (1994) list 57 reviews on various aspects of cheese flavor.

In contrast to the attempts at chemically defining desirable cheese flavor, efforts to identify the compounds responsible for off-flavors have been moderately successful, because off-flavors arise from a disproportionally high concentration of certain compounds or groups of compounds. For example, bitterness is due mainly to hydrophobic peptides, rancidity to fatty acids, and fruitiness to esters.

Although it is not possible to describe the flavor of cheese in precise chemical terms, very considerable progress has been made regarding the identification of flavor compounds in cheese and the elucidation of the biochemical pathways by which these compounds are produced. It is generally recognized that the aroma of cheese is in the volatile fraction and taste is largely in the aqueous phase; until recently, most researchers focused on the volatile fraction. Intervarietal comparisons should be valuable for identifying key flavor compounds. Although several comparative studies on the volatile compounds have been reported, there have been relatively few comparative studies on the aqueous phase.

One of the major problems encountered in research on cheese flavor is defining what the typical flavor should be. Within any variety, a fairly wide range of flavor and textural characteristics is acceptable. This is particularly true for Cheddar, which makes it especially difficult to chemically define its flavor. In cheese factories, wholesale or retail outlets, and research laboratories, somebody, perhaps a single grader whose views may not be typical, decides what constitutes desirable and undesirable flavor. Systematic attempts to objectively describe the sensory attributes of cheese have been made only recently, such as Hirst, Muir, and Naes (1994); McEwan, Moore, and Colwill (1989); and Muir and Hunter (1992). An international study under the European Union's FLAIR-SENS program (FLAIR Concerted Action No. 2, Cost 902, Relating Instrumental, Sensory and Consumer Data) had a similar objective, especially for cheese varieties with Appellation d'Origine Contrôlée status. Progress in developing an international protocol for sensory profiling of hard cheese is reported by Nielsen et al. (1998). An agreed vocabulary is essential if the results of instrumental studies are to be related to the sensory attributes and quality of cheese. The vocabulary developed by Nielsen et al. (1998) is shown in Exhibit 12–1 and the aroma-taste profiles for 12 hard cheeses using this vocabulary are shown in Figure 12–1.

Exhibit 12–1 Descriptors Selected To Describe Cheese Flavor

Smell

Creamy
Yogurt
Citrus fruit; other fruit; nutty
Grass
Cowshed
Caramel
Sour
Ammonia

Aroma-Taste

Creamy; yogurt
Grass
Citrus fruit; other fruit; nutty
Cowshed
Caramel
Sour
Pungent
Ammonia
Sweet
Salty
Acid
Bitter

Texture

Elastic
Firmness
Crumbliness
Coating
Dryness
Melting/solubility
Grainy

The texture of cheese has a major impact on flavor perception. For example, it has been suggested that the main contribution of proteolysis to cheese flavor is due to its effect on cheese texture, which affects the release of sapid compounds during mastication of the cheese. Therefore, these attributes should be considered together. However, they rarely are, and cheese texture is even less well understood at the molecular level than cheese flavor.

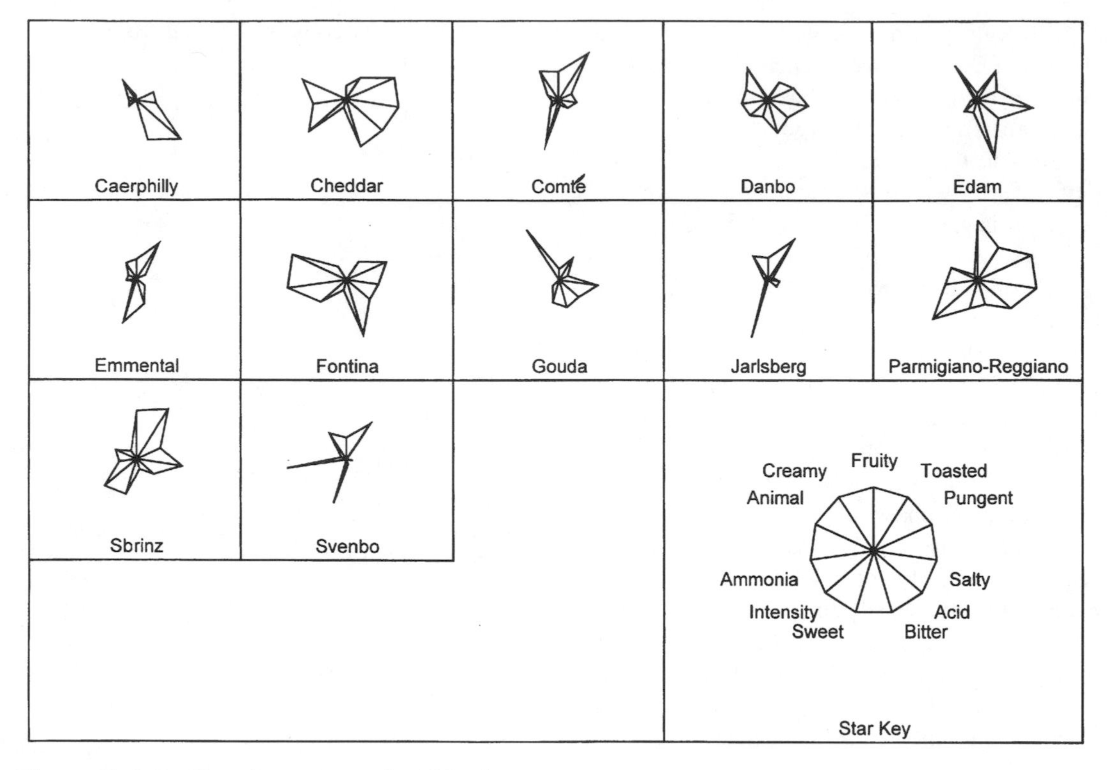

Figure 12–1 Profiles of aroma-taste for 12 hard cheeses.

12.2 ANALYTICAL METHODS

12.2.1 Nonvolatile Compounds

Although studies on cheese flavor date from the beginning of this century, the techniques available prior to the development of gas chromatography (GC) in the 1950s were inadequate to permit significant progress. Early investigators recognized the important contribution of proteolysis and lipolysis to cheese ripening. Studies on proteolysis relied on changes in protein and peptide solubility (e.g., in water, pH 4.6 buffers, TCA, and ethanol). Such techniques, which have been reviewed by Fox (1989), International Dairy Federation (1991), and McSweeney and Fox (1993, 1997) and are described briefly in Chapter 23, are still widely used as indices of cheese maturity but are mediocre as indices of cheese quality. Since the products of proteolysis are nonvolatile, they contribute to cheese taste but not to its aroma.

More specific studies on proteolysis became possible with the development of various types of chromatography (paper, ion exchange, gel permeation, and especially reversed phase high-performance liquid chromatography [RP-HPLC]). Large water-soluble and -insoluble peptides are best characterized by electrophoresis, especially polyacrylamide gel electrophoresis (PAGE), and, recently, capillary electrophoresis. Isoelectric focusing has had limited application in studies on cheese ripening. More than 200 peptides have been isolated from Cheddar cheese and identified by N-terminal sequencing and mass spectrometry (see Chapter 11). Amino acids are usually quantified by ion exchange HPLC with postcolumn derivitization using ninhydrin or by separation of fluorescent amino acid derivatives by RP-HPLC.

The total concentration of free fatty acids is usually determined by extraction or titration methods or spectrophotometrically as Cu-soaps. Early attempts to quantify the concentration of individual short-chain fatty acids involved steam distillation and adsorption chromatography. Complete separation and quantitation of free fatty acids can be achieved by GC (usually as their methyl esters, for which several preparative techniques have been published) or by RP-HPLC (see Chapter 23). Free fatty acids are major contributors to the flavor of some varieties, such as Romano, Feta, and Blue cheeses. In Blue cheeses, up to 25% of the total fatty acids may be free. Short-chain fatty acids are important contributors to cheese aroma, whereas longer chain acids contribute to taste. Excessive concentrations of either cause off-flavors (rancidity), and the critical concentration is quite low in many varieties, such as Cheddar and Gouda.

Several other organic acids, especially lactic acids, are present in cheese and probably contribute to flavor. These acids are routinely analyzed by HPLC. Enzymatic methods are available for several acids, such as D- and L-lactic acids (the enzymatic method is the easiest for distinguishing between the isomers) as well as acetic, pyruvic, and succinic acids (see Chapter 23).

12.2.2 Volatile Compounds

The early GC instruments used packed columns, which gave relatively poor resolution, and thermal conductivity detectors, which lacked sensitivity. Volatiles were stripped by vacuum distillation at, for example, 70°C, and trapped in cold fingers cooled by, for example, liquid nitrogen. Such techniques may cause heat-induced artifacts, and very volatile compounds may be lost. Compounds were identified by comparing their retention times with those of standard compounds. The introduction of flame ionization detectors and capillary columns and the interfacing of GC with mass spectrometry (MS) greatly improved resolution and increased sensitivity. Selective detectors are available for sulfur compounds.

The problem of separating the volatile compounds from cheese remains. Vacuum distillation of whole cheese or the fat fraction using improved apparatus (Figure 12–2) is still widely used. In some cases, steam distillation is used. To avoid the generation of artifacts, the analysis of headspace volatiles has been used frequently. One form of this technique involves removing a plug from a block of cheese and covering the opening of the resulting hole with a septum (Figure 12–3). After a period of time, a sample of gas in the cavity is withdrawn using an airtight syringe and injected into the GC column. The sensitivity of headspace analysis can be improved by trapping headspace volatiles in a Tenax trap, which can be inserted directly into the port of the GC. Alternatively, a sample of grated cheese may be placed in a glass vessel closed with a septum (Figure 12–4). A sample of headspace gas is withdrawn by syringe through the septum. The flask and contents may be heated to increase the release of volatiles, if desired. A more effective apparatus for collecting and concentrating headspace gas is shown in Figure 12–5. Grated cheese is placed in a long tube that is flushed for an extended period with an inert gas (e.g., helium), and the volatiles are collected in a Tenax trap. This approach permits a large volume of headspace gas to be trapped, thereby increasing sensitivity. The tube containing the sample may be heated, if desired.

Some authors have used solvent extraction (e.g., with dichloromethane) or a combination of steam distillation and solvent extraction. Ultrapure solvents are required to avoid artifacts, and the solvent peak may mask peaks due to components of the sample.

A major challenge facing researchers studying cheese flavor is identifying the key compounds responsible for cheese flavor—many of the volatile compounds identified by GC-MS may not contribute to flavor. A common approach adopted to identify key aroma compounds is to sniff the eluate from the GC column, a technique now referred to as gas chromatography/olfactometry (GCO). GCO is hard to quantify, and it is difficult to establish

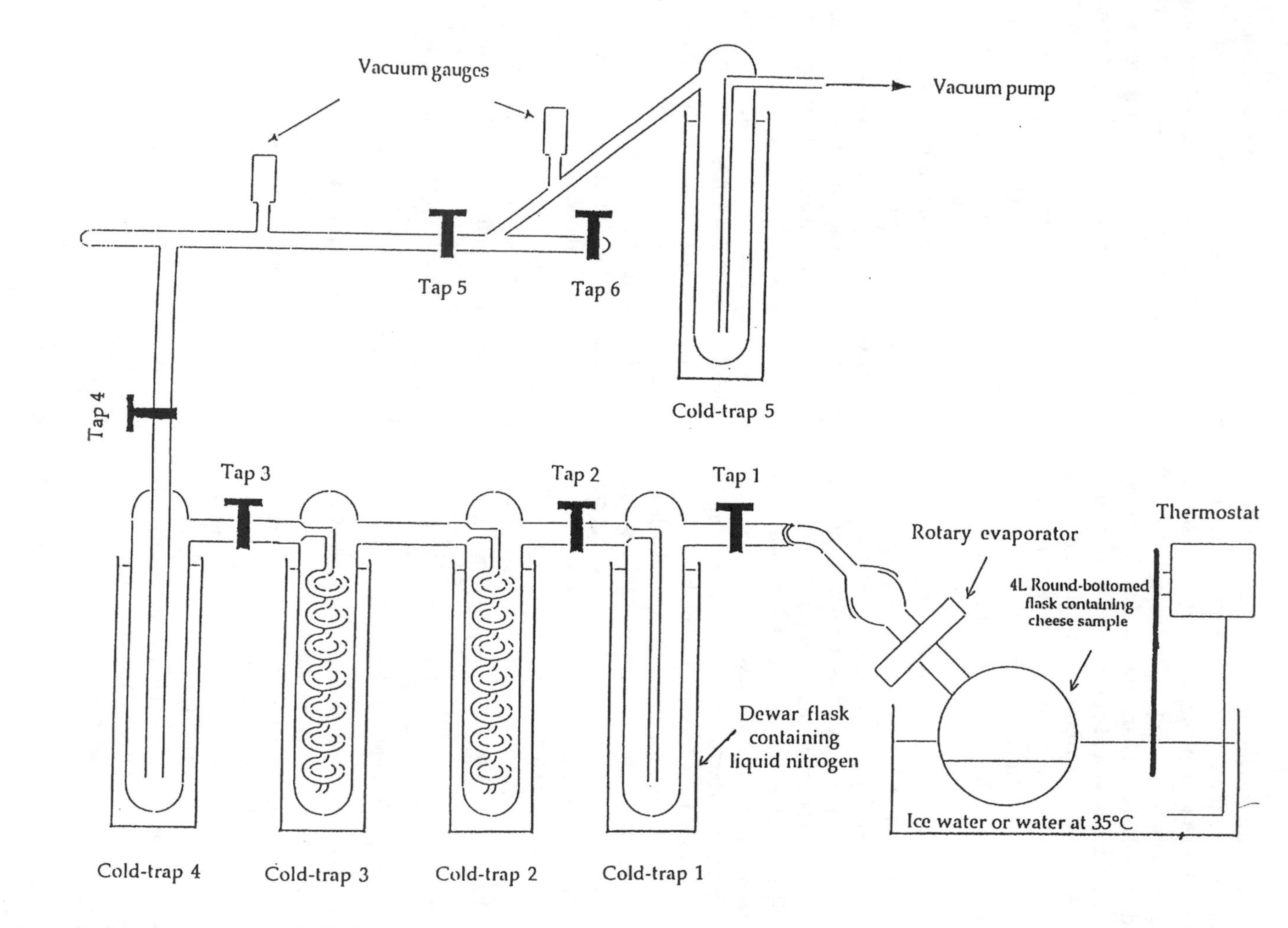

Figure 12–2 Apparatus for primary vacuum distillation of cheese volatiles.

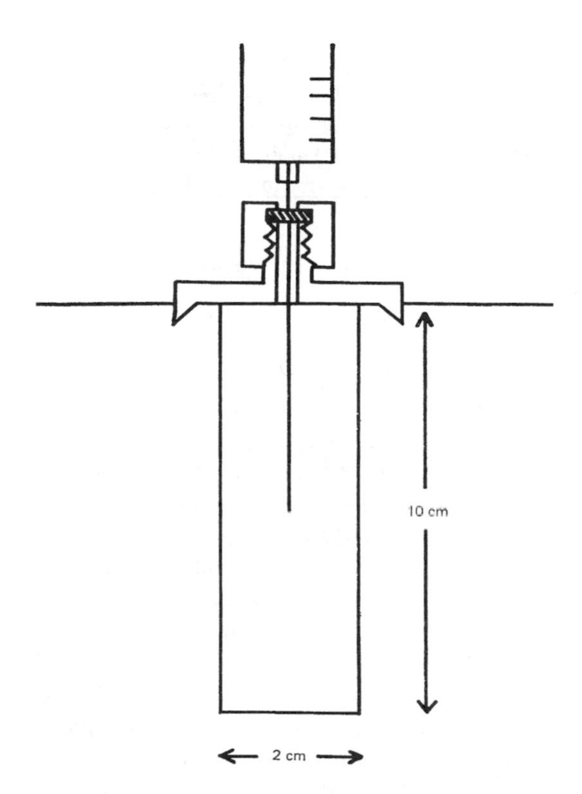

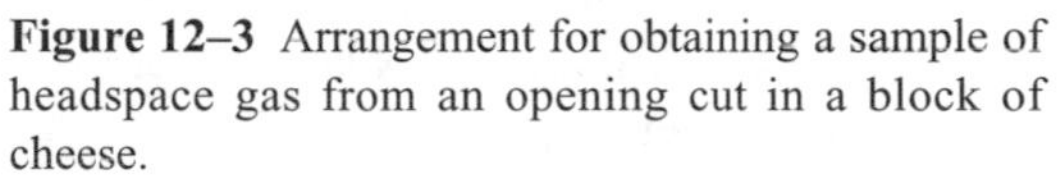

Figure 12–3 Arrangement for obtaining a sample of headspace gas from an opening cut in a block of cheese.

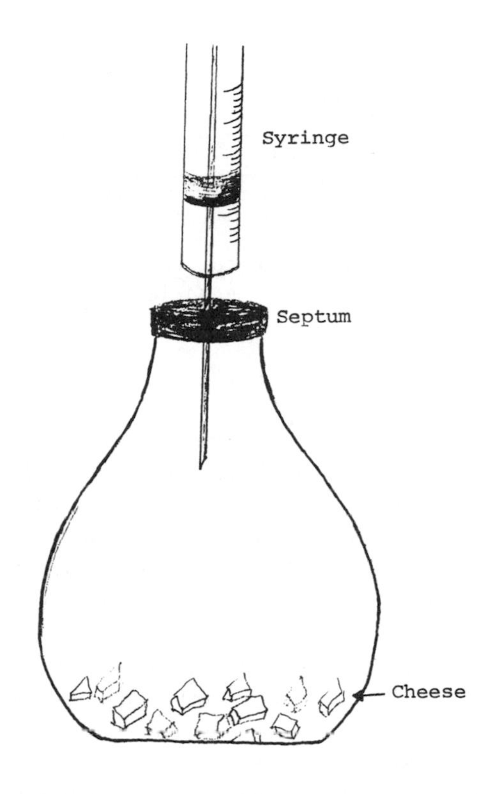

Figure 12–4 Arrangement for obtaining a sample of headspace gas from grated cheese held in a glass vessel.

whether a particular odorous compound is a major contributor to the flavor of the food under investigation. To overcome these problems and to obtain information on the odor activities of the odorants detected during sniffing, two methods have been developed to make GCO semi-quantitative: CHARM-analysis and aroma extract dilution analysis (AEDA). Both methods involve diluting the volatile fraction stepwise with a suitable solvent and evaluating each dilution by GCO. This procedure is performed until no odorant is perceivable in the GC eluate. The highest dilution at which a compound can be smelled is defined as its flavor dilution (FD) factor. The FD factor is the same for CHARM and AEDA, but the former also takes the persistence of smell into account. FD factors may be plotted against retention time (Rt) to give an FD-chromatogram (Figure 12–6). CHARM or AEDA may be applied to the analysis of volatile distillates, solvent extracts, or headspace gases.

Developments in the analysis of volatile flavor compounds have been reviewed by Schieberle (1995).

The human nose is extremely sensitive to aromatic compounds—it is reported that some compounds can be detected at concentrations as low as 0.02 μg/kg. This very high sensitivity, coupled with the very large number of odorous molecules in foods (e.g., > 600 odorous compounds have been detected in coffee), makes it very expensive and very difficult or impossible to detect complex odors by conventional analytical methods. Consequently, traditional sensory

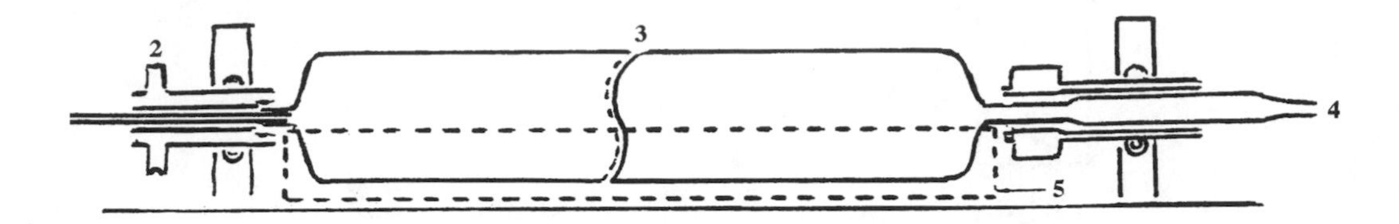

Figure 12–5 Apparatus for the dynamic headspace extraction of solid foods: (1) carrier gas (He) inlet, (2) belt transmission, (3) glass cylinder, (4) Tenax trap, (5) thermostatted bath.

(organoleptic) methods have survived. Recently, attempts have been made to use arrays of sensors that are sensitive to certain compounds and give an electronic response in a manner somewhat analogous to the nose (the device has been referred to as the "electronic nose"). The most commonly used sensors are metal oxides, quartz resonators, lipid layers, phthalocyanines, and conducting polymers. The mechanism of operation of these sensors and the electronic nose will not be discussed here; the reader is referred to Bartlett, Elliott, and Gardner (1997) and Gardner (1996). The electronic nose could be used continuously, for example, as a quality control device, and, unlike GC-MS, it does not require sample preparation, which may generate artifacts. Such instruments have been used to distinguish between different coffees, different brands of sausage, different alcoholic beverages, and different types of meat or fish and to detect spoiled meat or fish. There is interest in the application of the electronic nose to the study of cheese flavor, but the authors are not aware of published data.

12.3 CONTRIBUTION OF THE AQUEOUS PHASE OF CHEESE TO FLAVOR

The ultrafiltration (UF) permeate (obtained using 10 kDa cutoff membranes) of the water-soluble fraction of cheese has a taste essentially similar to that of the original cheese. This suggests that the basic taste of cheese is due to a mixture of water-soluble compounds on which other flavors are superimposed. The UF permeate contains small peptides, amino acids, short-chain fatty acids (C_4–C_{10}), other acids (acetic, succinic, especially lactic, and in some varieties propionic acid), and inorganic salts, especially NaCl. There is a positive, although not a very strong, correlation between the concentration of phosphotungstic acid (PTA)-soluble N (very small peptides and amino acids) and the intensity of cheese flavor (Figure 12–7). On fractionation of the UF permeate of cheese by gel permeation on a column of Sephadex G-25, the fraction containing very small peptides has a savory cheeselike taste. However, no single peptide with a cheeselike taste has been isolated. As discussed in Chapter 11, about 200 peptides have been isolated from cheese, especially Cheddar and Parmesan, but the very small peptides remain to be identified (the smallest peptides identified to date contained 5–6 amino acids). There are probably very many di-, tri-, and tetrapeptides in mature cheese, but these are difficult to isolate and identify. As discussed in Section 12.5.1, some peptides may have a bitter taste. If such peptides accumulate to an excessively high level, they lead to bitterness, a common defect in many cheeses. However, probably all cheeses contain bitter peptides, and when present at certain concentrations and balanced by other sapid compounds, they probably make a positive contribution to cheese flavor.

The exact role of the medium and small peptides in cheese flavor is not clear, although it is likely that they contribute to the background fla-

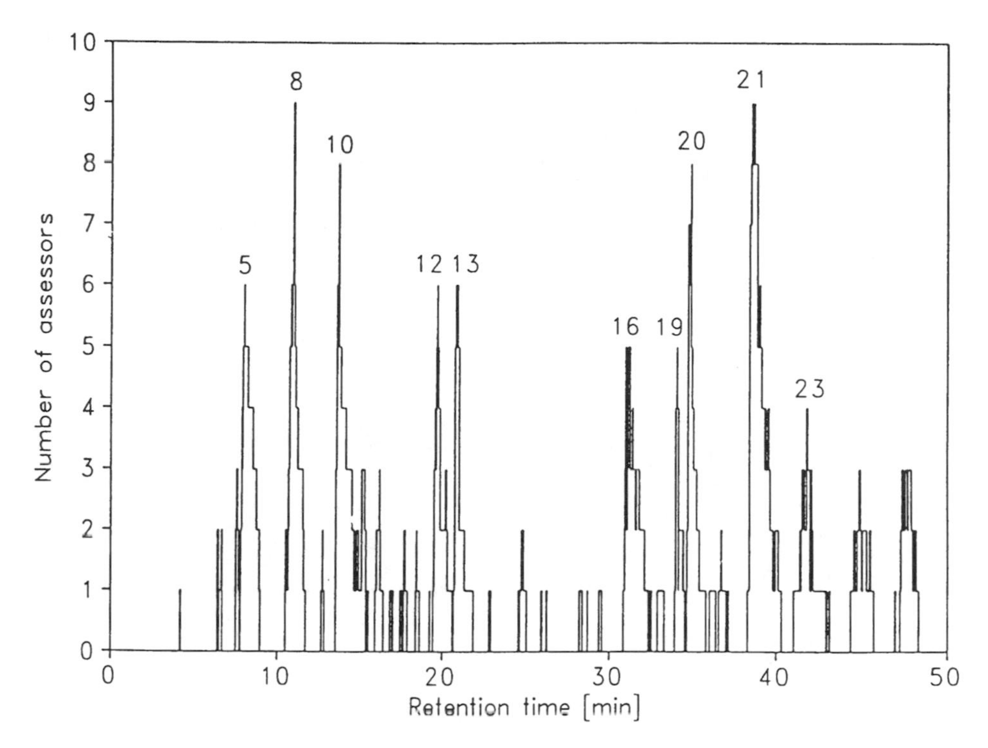

Figure 12–6 Example of a flavor dilution chromatogram.

vor of Cheddar, at least toward a brothy or savory flavor.

γ-Glutamyl peptides, which contain a peptide bond between the γ-carboxylic acid group of L-glutamic acid and another amino acid, have been implicated in cheese flavor. Although no γ-glutamyl bonds occur in the caseins, γ-Glu-Phe, γ-Glu-Leu, and γ-Glu-Tyr (at 9, 20, and 70 mg/kg, respectively) have been isolated from Gruyère de Comté cheese. Presumably, they are formed by γ-glutamyl transferase (GGT, EC 2.3.2.2), an indigenous enzyme in bovine milk. GGT catalyzes the transfer of the γ-glutamyl residue from a γ-glutamyl–containing peptide (e.g., glutathione, which occurs naturally in milk) to an acceptor amino acid:

$$\gamma\text{-glutamyl-peptide} + \text{amino acid} \rightleftharpoons \text{peptide} + \gamma\text{-glutamyl-amino acid}$$

It has also been postulated that the action of certain synthetases, such as γ-Glu-Cys synthetase (EC 6.3.2.2), perhaps in combination with GGT, may be responsible for the formation γ-glutamyl peptides in cheese.

Some γ-glutamyl peptides (e.g., γ-Glu-Glu, γ-Glu-Asp, γ-Glu-Ala, and γ-Glu-Gly) have a sour taste, while the taste of γ-Glu-Phe has been described as umami, slightly sour, salty, or metallic. The concentrations of the 3 γ-glutamyl peptides found in Comté (9–70 mg/kg) are well below the taste threshold of these compounds in solution (200–500 mg/kg). The presence of γ-glutamyl peptides in Comté may be related to the fact that it is a raw milk cheese; GGT activity in milk is attenuated by pasteurization (≈ 80% loss on heating at 75°C × 30 s). To date, there are no reports of the presence of γ-glutamyl peptides in other varieties of cheese. Further work on the

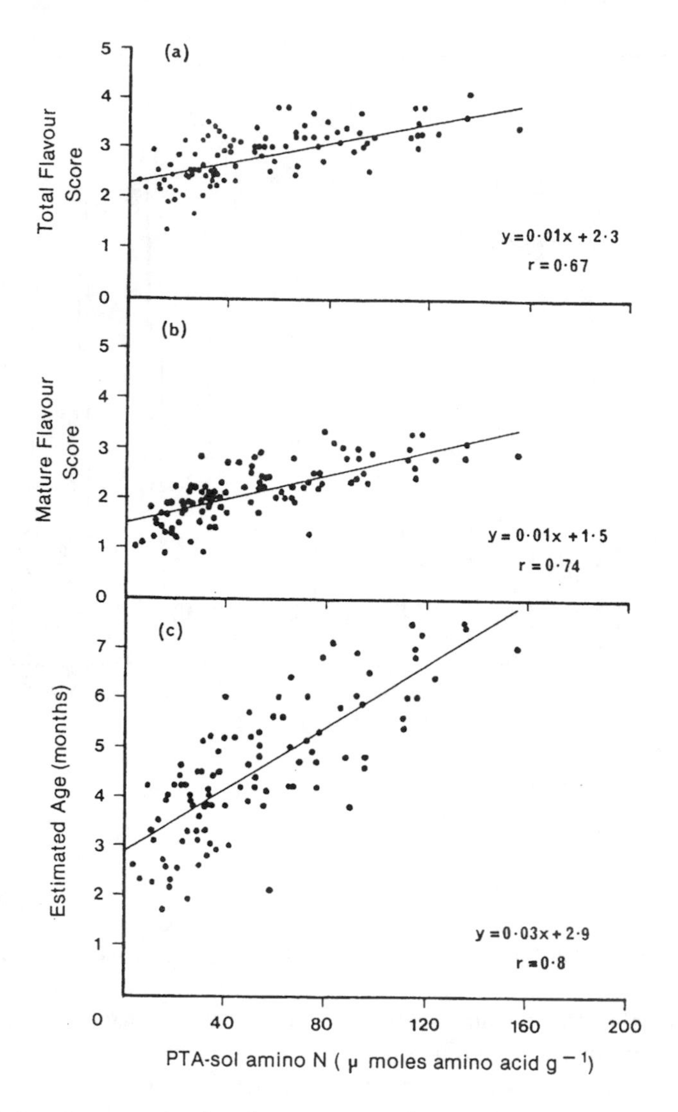

Figure 12–7 Correlations between 5% phosphotungstic acid (PTA)–soluble N (a measure of amino acids and short peptides) and (a) total flavor, (b) mature flavor, and (c) age of Cheddar cheese.

occurrence of these peptides in other cheese varieties appears to be warranted.

Amino acids are produced in cheese by the action of microbial exopeptidases on small and medium-size peptides. The concentration of amino acids in cheese varies with variety and age (see Fox & Wallace, 1997). Some amino acids are bitter (Table 12–1), mainly those with nonpolar or hydrophobic side chains, such as Ile, although Lys (usually charged) and Tyr (polar but normally uncharged) are also considered to be bitter. Pro and Lys are reported to be bittersweet. Arg is bitter, although it has low hydrophobicity. Ala, Gly, Ser, and Thr are sweet, while Glu, His, and Asp are sour. Asp and Glu have the lowest taste thresholds of the amino acids (30 and 50 mg/L, respectively). The concentrations of several amino acids in a water-soluble extract of mature Cheddar cheese are above their flavor thresholds.

The term "sweet flavor" when applied to cheese often implies a lack of acidity but does not imply fruity aroma, which is due principally to esters. Sweetness per se is regarded as a defect in many cheese varieties, although sweet flavor notes are desirable in some varieties (e.g., Swiss). The sweet flavor in Swiss cheese, which is concentrated in the water-soluble, nonvolatile fraction, is not due to sugars (lactose, glucose, and galactose), which are essentially absent from mature Swiss cheese, but to products of proteolysis, especially proline. It has been suggested that the sweet flavor of Swiss cheese results from the interaction of Ca^{2+} and Mg^{2+} with amino acids and small peptides. However, 4-hydroxy-2,5-dimethyl-2(2H)-furanone and 5-ethyl-4-hydroxy-3(2H)-furanone are also considered to contribute to the sweet note in the aroma profile of Emmental cheese.

Thus, it appears that amino acids potentially contribute directly to cheese flavor and quality, although their exact role is not clear. However, products arising from the degradation of amino acids through transamination, deamination, decarboxylation, and other mechanisms (see Chapter 11) are very important to the development of cheese aroma. These compounds include ammonia, amines, acids, carbonyls, sulfur-containing compounds (e.g., H_2S, methional, dimethyl sulfide), and complex products produced via the Maillard and Strecker reactions. Most of these compounds are volatile and are discussed in Section 12.4.

Acid taste is caused by H^+, the taste threshold of which is about 2 mM. The principal acid in cheese is lactic acid. Its concentration (and con-

Table 12–1 Taste Descriptor and Threshold Values of Amino Acids

Amino Acids	*Taste*	*Threshold Value (mg/L)*	*Average Hydrophobicity (cal/mol)*
Histidine	Bitter	200	500
Methionine	Bitter	300	1,300
Valine	Bitter	400	1,500
Arginine	Bitter	500	750
Isoleucine	Bitter	900	2,950
Phenylalanine	Bitter	900	2,500
Tryptophan	Bitter	900	3,400
Leucine	Bitter	1,900	1,800
Tyrosine	Bitter	ND	2,300
Alanine	Sweet	600	500
Glycine	Sweet	1,300	0
Serine	Sweet	1,500	–300
Threonine	Sweet	2,600	400
Lysine	Sweet and bitter	3,000	1,500
Proline	Sweet and bitter	3,000	2,600
Aspartic acid	Sour	30	0
Glutamic acid	Sour	50	0
Asparagine	Sour	1,000	0
Glutamine	Flat	ND	0
Cysteine	–	ND	1,000
Glutamate Na	Umami	300	
Aspartate Na	Umami	1,000	

sequently the pH of cheese) varies considerably with variety as influenced by its initial production by the starter, the extent of loss in the whey, and its metabolism by the secondary microflora of the cheese. Total lactate concentration may not be a good index of cheese acidity, since in certain varieties (e.g., mold-ripened cheeses) the pH increases during ripening, as a result of ammonia released by deamination of amino acids and/or by the metabolism of lactate (see Chapters 10 and 11). The perception of acidity in cheese can be influenced by NaCl concentration. Several other acids have been identified in cheese (principally acetic, propionic, and butyric acids) and presumably contribute to acidity, although they principally affect cheese aroma.

A number of short-chain fatty acids are formed in cheese during ripening. Some (e.g., butanoic acid) are produced principally through lipolysis while others (e.g., propanoic, acetic, and formic acids) result from the action of cheese microflora on lactose, lactate, citrate, and amino acids. Short and medium-chain fatty acids (C_4 to C_{10}) contribute to the acid taste of cheese, although they contribute principally to its aroma. The acid taste of Swiss cheese correlates with the concentration of di- and tripeptides ($r^2 = 0.81$) and amino acids ($r^2 = .80$) but not with cheese pH or the concentration of lactate.

Salt (NaCl) is an important contributor to the taste of the water-soluble, nonvolatile fraction of cheese. Salty taste is stimulated by small inorganic ions. The taste of chlorides of group I elements varies from acid (HCl) through salty (e.g., NaCl) to bitter (e.g., KCl and CsCl). The taste of most high molecular weight salts is bitter rather than salty, and cheeses salted using KCl, $CaCl_2$, or $MgCl_2$ are extremely bitter.

In the context of dairy foods, the compound responsible for salty taste is almost always NaCl. The threshold for NaCl is 30 mM (unspecific taste response) and 40–50 mM for its salty taste. Bovine milk typically contains 25 mM Na^+ and 29 mM Cl^-, so the salty taste of dairy products arises mainly from NaCl added during manufacture. NaCl is particularly important to the flavor and quality of cheese and butter. Salt was originally added to cheese as a preservative, acting to reduce water activity (a_w) (see Chapters 8 and 10). Although its preservative effect is still important, the taste of the final product is now also considered when determining the amount of salt to be added during processing. Reduction in the level of NaCl in cheese (by adding a reduced level or by partial replacement with KCl) for nutritional reasons can lead to bitterness due to excessive or unbalanced proteolysis (see Chapter 11).

Typical values for the concentration of NaCl in cheese range from 0.7% for Quarg to 5.5% for Romano. The majority of cheeses contain 1–3% (w/w) NaCl (see Chapters 8, 17, and 21). Variations in the salty taste of cheese are likely to be related to processing parameters, such as the amount of salt added and its distribution throughout the cheese. Consumers appear to be unable to detect differences in flavor and texture between cheeses containing 1.12% versus 1.44% NaCl, but differences are apparent between cheeses containing 0.75% versus 1.12% NaCl. Low NaCl concentrations facilitate proteolysis and the growth of lactic acid bacteria.

The apparent saltiness of cheese increases with maturity, increasing NaCl concentration, and decreasing pH. Grated cheese is perceived as being more salty than the corresponding intact product but less salty than an aqueous NaCl solution of the same concentration.

12.4 CONTRIBUTION OF VOLATILE COMPOUNDS TO CHEESE FLAVOR

Compounds responsible for cheese aroma are volatile to at least some degree. While some work on the volatile constituents of cheese was done before 1960 (e.g., on short-chain fatty acids and amines), significant progress was not possible until the development of GC. GC was first applied to the study of Cheddar cheese volatiles in 1962. Volatiles were stripped from the sample by vacuum distillation and condensed in cold traps. They were then resolved by

GC on packed columns and located by thermal conductivity detectors. The introduction of flame ionization detectors, capillary columns, and interfacing GC with mass spectrometry (MS) markedly increased sensitivity and greatly extended the number of compounds detected (> 200 for Cheddar).

Based on a survey of the published literature, Maarse and Visscher (1989) listed 213 volatile compounds that had been identified in 40 studies on Cheddar cheese, including 33 hydrocarbons, 24 alcohols, 13 aldehydes, 17 ketones, 42 acids, 30 esters, 12 lactones, 18 amines, 7 sulfur compounds, 5 halogens, 6 nitriles and amides, 4 phenols, 1 ether, and 1 pyran. The concentrations of many of these compounds were reported. The principal volatile compounds identified in Cheddar are listed in Exhibit 12–2.

Thus, a great diversity of potentially sapid and/or aromatic compounds have been identified in one or more varieties. These include small peptides (200 or more) and amino acids and more than 200 volatile compounds (fatty acids, other acids, carbonyls, amines, sulfur compounds, and hydrocarbons). Most cheese varieties contain similar volatile compounds but in different proportions. A comparison of the GC-MS profiles of six cheeses is shown in Figure 12–8. The concentrations and proportion of volatile and nonvolatile flavor compounds are probably responsible for the specific flavor of each cheese variety. However, the flavor of some varieties is dominated by a particular compound or class of compounds, such as short-chain fatty acids for Romano or methyl ketones for Blue cheeses. Although many of the flavor compounds that have been identified in cheese are present at concentrations well below their flavor threshold, they may still modify the overall flavor impact. The compounds considered by Urbach (1997) to be important or key contributors to the flavor of a number of cheese varieties are listed in Table 12–2. Although several hundred volatile and nonvolatile flavor compounds have been identified in Cheddar cheese, Urbach (1997) and many earlier workers consider methanethiol (CH_3SH, derived from methionine) to be the key characteristic flavor compound in Cheddar, although alone it does not have a cheesy aroma. Obviously, the aroma of methanethiol is modified in cheese by the presence of many other compounds.

12.5 OFF-FLAVORS IN CHEESE

Avoiding off-flavors is generally more important than producing a cheese of exceptionally good quality—most consumers will object to cheese with an off-flavor but relatively few will appreciate an excellent cheese. Consequently, cheese produced in large, well-managed factories rarely suffers serious off-flavors. Since off-flavors usually have a specific cause, it is easier to control them than to develop good flavors, which have very complex causes.

Many off-flavors are due to the presence of disproportionately high concentrations of certain compounds that may contribute positively to flavor at lower concentrations. Since specific compounds are, in most cases, responsible for off-flavors, they are generally understood at the molecular level. Some specific off-flavors are considered in the following sections.

12.5.1 Bitterness

Bitterness is probably the principal taste defect in cheese. Although amino acids, amines, amides, substituted amides, long-chain ketones, some monoglycerides, N-acyl amino acids, and diketopiperazines may contribute to bitterness, this defect in cheese usually results from the accumulation of hydrophobic peptides.

Bitterness and Hydrophobicity

Ney (1979) suggested that the mean hydrophobicity (Q) of a peptide, expressed as

$$Q = \frac{\Sigma \Delta f_i}{n}$$

where, Δf_i is side chain hydrophobicity and n is the number of residues, is the principal determi-

Exhibit 12–2 Volatile Compounds Identified in Cheddar Cheese

acetaldehyde	dimethyl sulfide	methyl acetate
acetoin	dimethyl disulfide	2–methylbutanol
acetone	dimethyl trisulfide	3-methylbutanol
acetophenone	δ-dodecalactone	3-methyl-2-butanone
β-angelicalatone	ethanol	3-methylbutyric acid
1,2-butanediol	ethyl acetate	2-nonanone
n-butanol	2-ethyl butanol	δ-octalactone
2-butanol	ethyl butyrate	n-octanoic acid
butanone	ethyl hexanoate	2-octanol
n-butyl acetate	2-heptanone	2,4-pentanediol
2-butyl acetate	n-hexanal	n-pentanoic acid
n-butyl butyrate	n-hexanoic acid	2-pentanol
n-butyric acid	n-hexanol	pentan-2-one
carbon dioxide	2-hexanone	n-propanol
p-cresol	hexanethiol	propanal
γ-decalactone	2-hexenal	propenal
δ-decalactone	isobutanol	n-propyl buryrate
n-decanoic acid	isohexanal	tetrahydrofuran
diacetyl	methane thiol	thiophen-2-aldehyde
diethyl ether	methional	2-tridecanone
		2-undecanone

nant of bitterness rather than any particular amino acid sequence. Although further work suggested that the nature of the terminal amino acids and certain steric parameters influence the perception of bitterness (see Lemieux & Simard, 1991, 1992), the mean hydrophobicity of a peptide appears to be the single most important factor determining its bitterness. Peptides with a molecular weight of 0.1 to 10 kDa and Q less than 1,300 cal per residue are not bitter while those with Q greater than 1,400 cal per residue and a molecular weight of 0.1 to 6 kDa are bitter; above 6 kDa, even peptides with Q greater than 1,400 cal per residue are not bitter. Hydrolysates of proteins with a high mean hydrophobicity are likely to contain bitter peptides, although the distribution of hydrophobic residues along the polypeptide and the specificity of the proteinase used to prepare the hydrolysate also influence the development of bitterness. Since the caseins, especially β-casein, are quite hydrophobic and the hydrophobic residues are clustered (see Chapter 3), casein hydrolysates have a high propensity to bitterness.

As discussed in Chapter 11, cheese contains a great diversity of proteinases and peptidases with different and complementary specificities. Although detailed kinetic studies are lacking, at least some of the peptides in cheese are transient, and hence bitterness may wax and wane as bitter peptides are formed and hydrolyzed or masked by other sapid compounds. It is very likely that all cheeses contain bitter peptides, which probably contribute positively to the overall desirable flavor. Bitterness becomes a problem only when bitter peptides accumulate to an excessive, unbalanced level. Although bitter peptides can originate from either α_{s1}- or β-casein, the action of chymosin and/or lactococcal cell envelope–associated proteinase (CEP) on the very hydrophobic C-terminal region of β-casein may result in the production of bitter peptides. This action

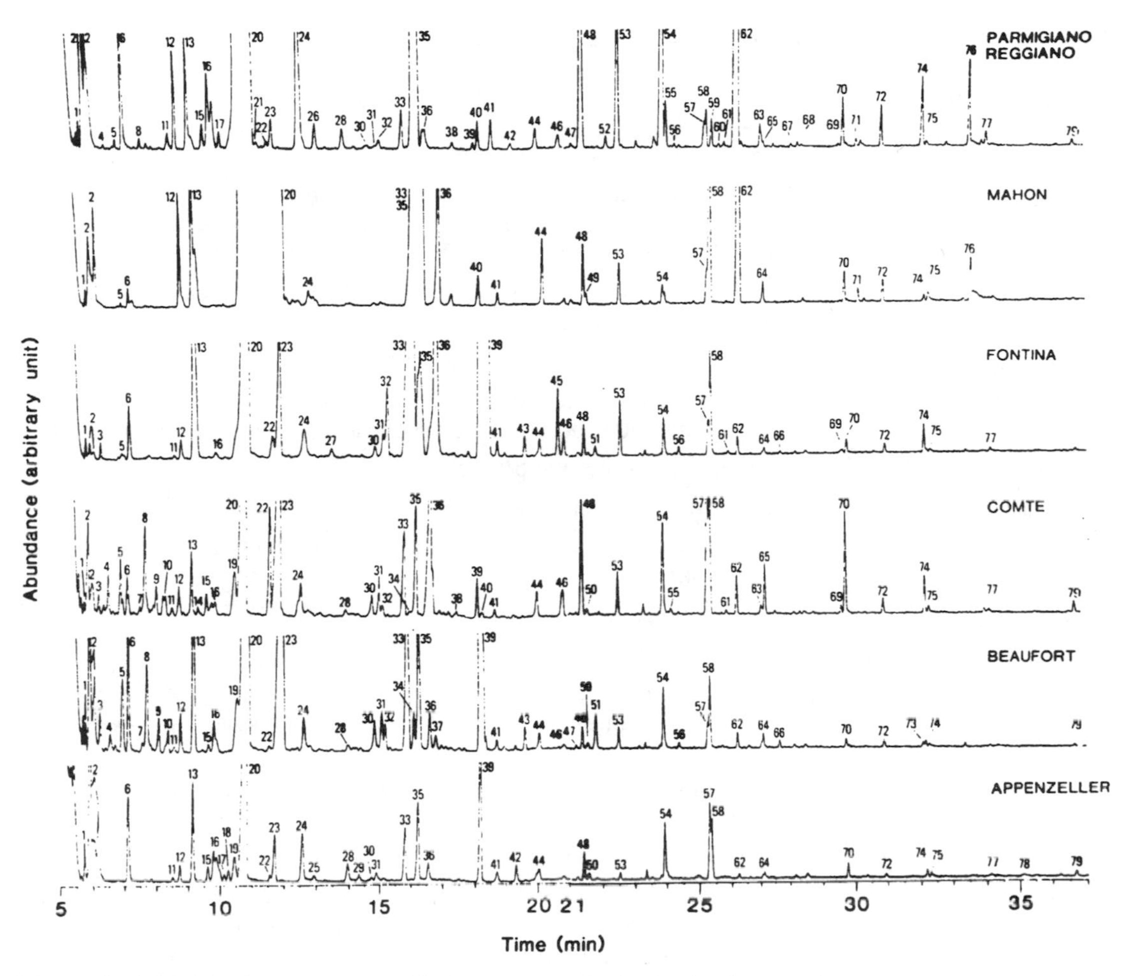

Figure 12–8 GC-MS chromatograms of the headspace volatiles for six cheese varieties.

is very dependent on salt concentration (see Chapter 11). The peptides produced initially from α_{s1}-casein are from the hydrophilic N-terminal region and therefore are less likely to be bitter. The production of bitter peptides also depends on the specificity of the lactococcal CEP. For example, P_{III}-type CEP (*Lc. lactis* ssp. *cremoris* AM_1) produces less bitter casein hydrolysates than P_I-type CEP (*Lc. lactis* ssp. *cremoris* HP), perhaps because more peptides are initially released from the hydrophobic C-terminal region of β-casein by the latter. The concentration of bitter peptides depends on the rate at which they are degraded by lactococcal peptidases and perhaps, in the case of larger bitter peptides, by the CEP. The debittering effect of aminopeptidase N on a tryptic digest of β-casein has been demonstrated.

Factors Affecting the Development of Bitterness in Cheese

Certain starters have a propensity to cause bitterness. Nonbitter cheese can be made using these strains provided they are combined with "nonbitter" strains. Since the coagulant may cause the release of bitter peptides, factors that

Table 12–2 Important Flavor Compounds in Various Cheeses

Cheese	*Compounds*
Cheddar	Methanethiol
Camembert	Nonan-2-one, oct-1-en-3-ol, N-isobutylacetamide, 2-phenylethanol, 2-phenylethyl acetate, 2-heptanol, 2-nonanol, NH_3, isovaleric acid, isobutyric acid, hydroxybenzoic acid, hydroxyphenylacetic acid
Emmental	Methional, 4-hydroxy-2,5-dimethyl-3(2H)-furanone (Furaneol), 5-ethyl-4-hydroxy-2-methyl-3 (2H)-furanone
Romano	Butanoic, hexanoic, and octanoic acids
Parmesan	Butanoic, hexanoic, and octanoic acids, ethyl butyrate, ethyl hexanoate, ethyl acetate, ethyl octanoate, ethyl decanoate, methyl hexanoate
Provolone	Butanoic, hexanoic, and octanoic acids
Goat milk cheese	4-Methyloctanoic acid, 4-ethyloctanoic acid
Pecorino Romano	4-Methyloctanoic and 4-ethyloctanoic acids, *p*-cresol, *m*-cresol, 3,4-dimethylphenol
Limburg	Methanethiol, methyl thioacetate
Surface-ripened cheeses	Methyl thioesters
Pont-l'Evêque	*N*-Isobutylacetamide, phenol, isobutyric acid, 3-methylvaleric acid, isovaleric acid, heptan-2-one, nonan-2-one, acetophenone, 2-phenylethanol, indole
Vacherin	Acetophenone, phenol, dimethyl disulfide, indole, terpineol, isoborneol, linalool
Roquefort	Oct-1-en-3-ol, methyl ketones
Livarot	Phenol, *m*- and *p*-cresols, dimethyl disulfide, isobutyric acid, 3-methylvaleric acid, isovaleric acid, benzoic acid, phenylacetic acid, nonan-2-one, acetophenone, 2-phenylethanol, 2-phenylethyl acetate, dimethyl disulfide, indole
Munster	Dimethyl disulfide, isobutyric acid, 3-methylvaleric acid, isovaleric acid, benzoic acid, phenylacetic acid
Trappist	H_2S, methanethiol
Blue cheeses	Heptan-2-one, nonan-2-one, methyl esters of $C_{4,6,8,10,12}$ acids, ethyl esters of $C_{1,2,4,6,8,10}$ acids
Brie	Isobutyric acid, isovaleric acid, methyl ketones, sulfur compounds, oct-1-en-3-ol
Carré de l'Est	Isobutyric acid, isovaleric acid, 3-methylvaleric acid
Epoisses	2-Phenylethanol
Maroilles	2-Phenylethanol, nonan-2-one, acetophenone, phenol, indole
Langres	2-Phenylethanol, dimethyl disulfide, styrene, indole
Buffalo Mozzarella	Oct-1-en-3-ol, nonanal, indole, component RI 975 with odor of truffles
Bovine Mozzarella	Ethyl isobutanoate, ethyl 3-methylbutanoate

affect its retention in cheese curd (type and quantity used in cheesemaking, pH at whey drainage, and cook temperature) influence the development of bitterness. Certain rennet substitutes produce bitter cheese, owing to excessively high activity and/or incorrect specificity. The pH of cheese also influences the activity of residual coagulant and other enzymes. Cheese with a low salt concentration is very prone to bitterness, perhaps because the susceptibility of β-casein to hydrolysis by chymosin, with the production of the bitter peptide β-CN f193-209, is strongly affected by the NaCl concentration in cheese. Salt also inhibits lactococcal CEP and may promote the aggregation of large, nonbitter hydrophobic peptides that would otherwise be degraded to bitter peptides. Bitterness can be particularly problematic in low-fat cheeses, perhaps owing to reduced partitioning of hydrophobic peptides into the fat phase.

Bitter Peptides Isolated from Cheese

Bitter peptides that have been isolated from cheese are summarized in Figures 12–9 and 12–10. As expected, bitter peptides originate principally from hydrophobic regions of the caseins, including sequences 14 to 34, 91 to 101, and 143 to 151 of α_{s1}-casein and 46 to 90 or 190 to 209 of β-casein. As discussed by McSweeney et al. (1997), the majority of these peptides show evidence of some degradation by lactococcal proteinases and/or peptidases.

12.5.2 Astringency

Astringency is a taste-related phenomenon perceived as a dry feeling in the mouth and a puckering of the oral tissue. In many foods, astringency involves interaction between tannins (polyphenols) and proteins in saliva, but this is probably not the cause of astringency in cheese. An astringent fraction was extracted from Cheddar cheese using chloroform-methanol, but it was not characterized. Since the abstract absorbed light at 280 nm, it probably contained peptides. The aqueous fraction of Comté cheese contains N-propionyl methionine, which is slightly bitter, astringent, and pungent.

12.5.3 Fruitiness

The principal compounds responsible for fruitiness in cheese are esters, especially ethyl butyrate and ethyl hexanoate, formed by esterification of fatty acids with ethanol. Production of ethanol appears to be the limiting factor, as fatty acids are present in cheese at relatively high concentrations. Ethyl esters are present at low concentrations in nonfruity cheeses, and the fruity defect occurs as a result of excessive production of ethanol or its precursors.

12.5.4 "Unclean" Off-Flavors

Phenylacetaldehyde, which is produced via the Strecker reaction, causes "unclean" and related flavors in Cheddar cheese. At higher concentrations (> 500 μg/kg), phenylacetaldehyde imparts astringent, bitter, and stinging flavors to cheese. *p*-Cresol imparts a "utensil-type" flavor when present at high concentrations. Short-chain fatty acids potentiate the flavor impact of *p*-cresol. Phenol contributes to the unclean flavor of cheese but enhances the sharpness of Cheddar flavor.

12.6 FORMATION OF FLAVOR COMPOUNDS

Some of the flavor compounds in cheese, including lactic acid, diacetyl, and methyl ketones, are produced by the metabolic activity of living microorganisms. A few are produced by chemical reactions, for example, via the Maillard and Strecker reactions between amino acids and various carbonyls, especially dicarbonyls, e.g., diacetyl, glyoxal, or methyl glyoxal. Some examples of the Strecker reaction are shown in Figure 12–11. However, most flavor compounds are produced by enzymes that are either indigenous to milk (especially plasmin and lipase), added to the milk (especially chymosin or rennet

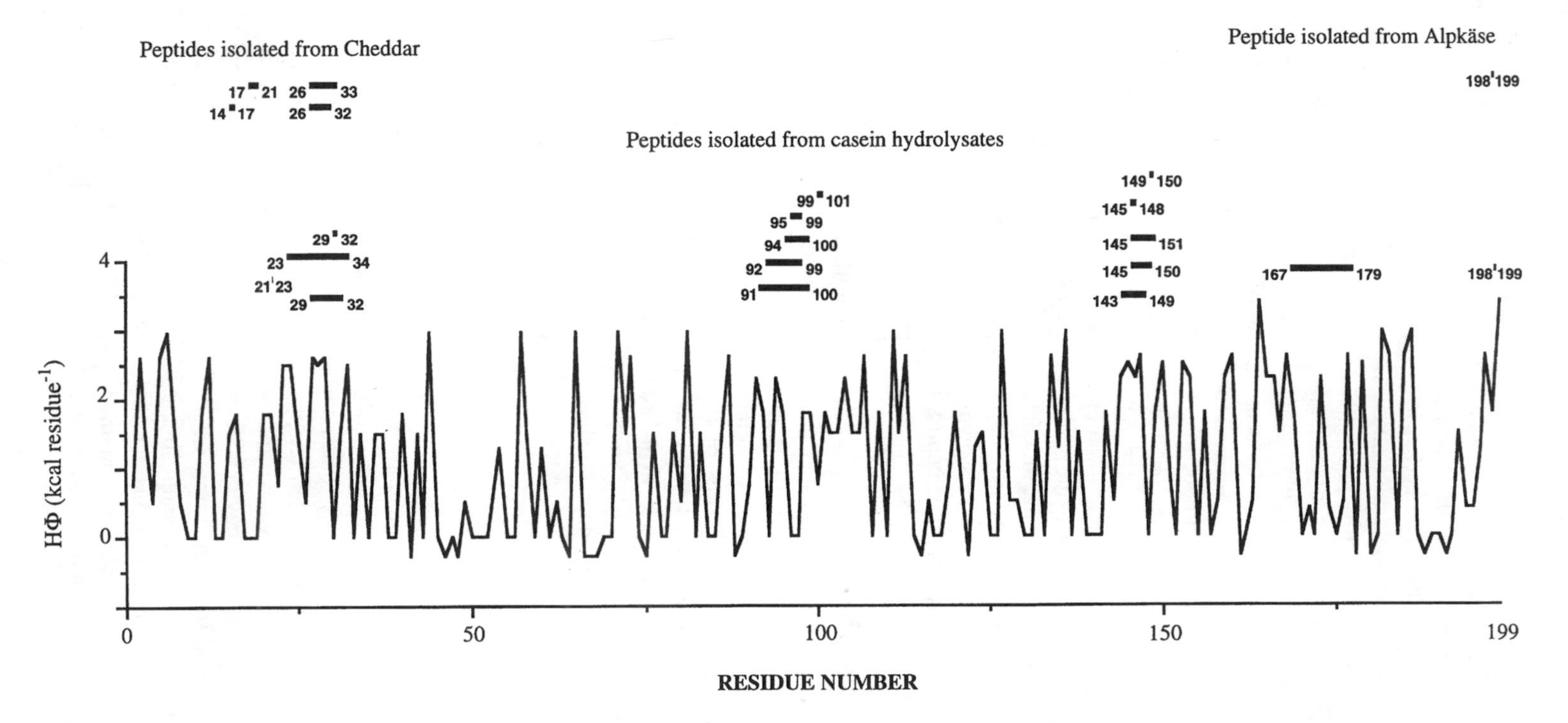

Figure 12–9 Hydrophobicity of bovine α_{s1}-casein plotted as a function of amino acid residue. The locations of bitter peptides originating from α_{s1}-casein and isolated from cheese or casein hydrolysates (as listed by Lemieux and Simard, 1992) are shown approximately to scale.

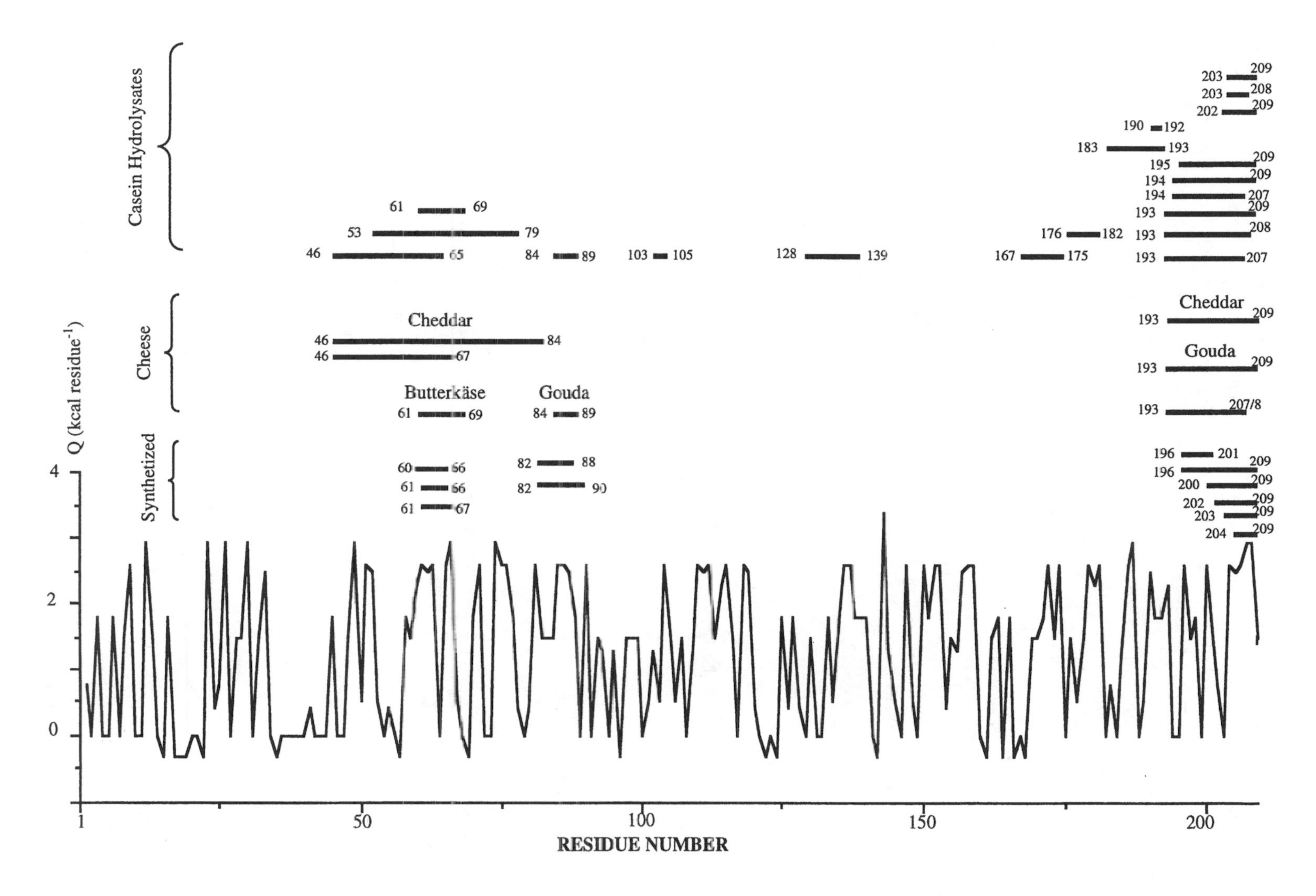

Figure 12–10 Hydrophobicity of bovine β-casein plotted as a function of amino acid residue. The locations of bitter peptides originating from β-casein and isolated from cheese or casein hydrolysates or synthesized (as listed by Lemieux and Simard, 1992) are shown approximately to scale.

substitute and, in some varieties, pregastric esterase), secreted by microorganisms (especially molds and other microorganisms in the surface smear), or released from microbial cells following cell death and lysis.

The pathways leading to the formation of most of the flavor compounds in cheese are known and are described in Chapter 11. The subject has been reviewed by Fox and Wallace (1997). The principal gap in our knowledge on flavor generation is the ability to precisely balance and control the various reactions. Many of the compounds that are considered to be most important in cheese flavor are present at very low concentrations and, in many varieties, are produced mainly via the catabolism of amino acids. The enzymes involved are intracellular bacterial (starter and nonstarter) enzymes present at low levels and are often relatively difficult to assay. Therefore, they have to date been largely neglected, although they are now attracting some attention.

12.7 INTERVARIETAL AND INTRAVARIETAL COMPARISON OF CHEESE RIPENING

As discussed in Section 12.4, most ripened cheeses contain essentially the same sapid and aromatic compounds but at different concentra-

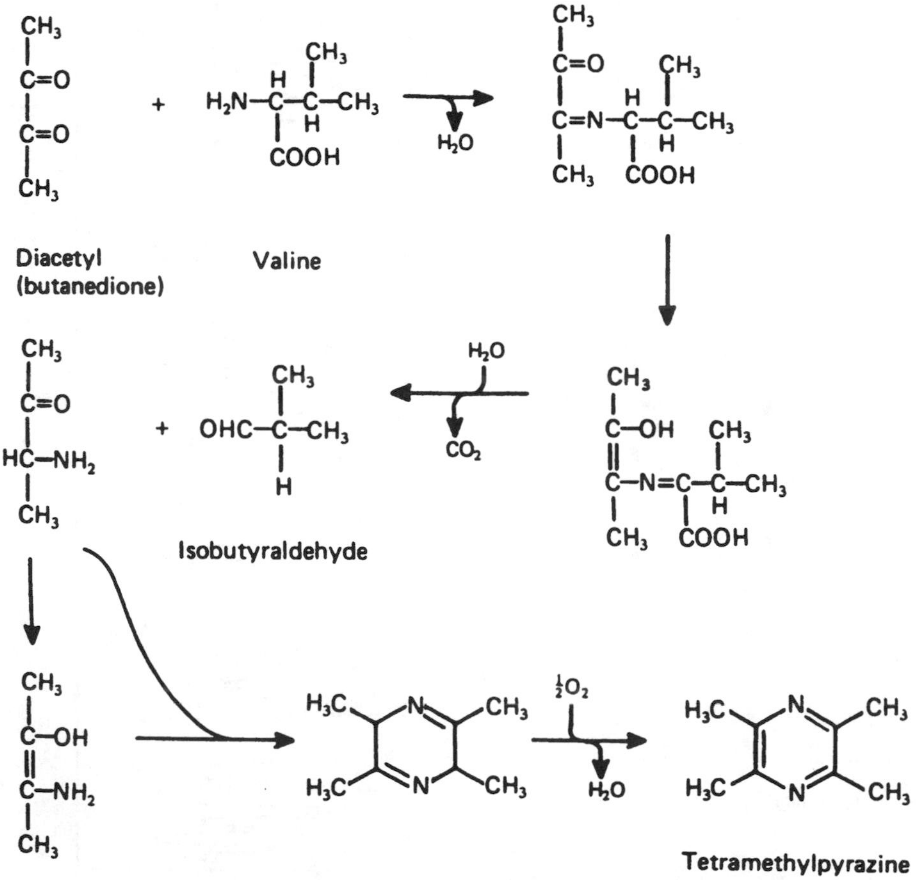

Diacetyl (butanedione)

Valine

Isobutyraldehyde

Tetramethylpyrazine

Figure 12–11 Example of the Strecker degradation reaction.

tions and proportions. Therefore, it appears reasonable to presume that inter- and intravarietal comparison, especially of closely related varieties, might help to identify compounds most likely to contribute to characteristic cheese flavors. However, although both the water-soluble nonvolatile and volatile fractions of several cheese varieties have been analyzed, there are relatively few intervarietal comparisons, especially of the nonvolatile fractions. In this section, the results of some such studies are discussed.

12.7.1 Gel Electrophoresis

Gel electrophoresis, especially alkaline (pH 9.0) urea-PAGE, is the best method for characterizing the large water-insoluble peptides and is also useful for characterizing the larger water-soluble peptides.

Urea-PAGE of water-insoluble and -soluble fractions of cheese indicate variety-specific peptide patterns (Figures 11–16 and 11–17), reflecting mainly the relative activities of chymosin and plasmin (see Chapter 11). The peptides detectable by urea-PAGE are too large to affect cheese flavor, but large peptides signal the presence of certain small peptides that may affect flavor or perhaps, more correctly, off-flavor, especially bitterness. For instance, the presence of a band corresponding to β-CN f1-192 indicates the presence of β-CN f193-209 or fractions thereof, which are bitter.

Urea-PAGE of the water-insoluble and -soluble fractions of cheeses made using single strain starters show very little difference between cheeses differing substantially in quality. This is not surprising, since starter proteinases contribute little to primary proteolysis in cheese, as detected by PAGE. Urea-PAGE also fails to show substantial differences between cheeses made from pasteurized or raw milk, although the flavor of these cheeses differ markedly. Urea-PAGE of the water-insoluble fraction generally reflects the age of the cheese (if ripened at a particular temperature) and is a useful index of its textural quality (if the age is known) but not of its flavor.

12.7.2 High Performance Liquid Chromatography

RP-HPLC is more effective than PAGE (which does not detect peptides less than 3,000 Da) for analysis of the water-soluble peptides. RP-HPLC profiles of the UF permeate and retentate of cheese become more complex as the cheese matures (see O'Shea, Uniacke-Lowe, & Fox, 1996). Unfortunately, it is not possible at present to relate cheese flavor or texture to HPLC chromatograms. HPLC of the 70% ethanol-soluble and -insoluble fractions of the water-soluble peptides shows clear varietal characteristics (see Figures 11–18 and 11–19). The concentration of free amino acids (measured by reaction with Cd-Ninhydrin) in Cheddar is highly correlated with age (Figure 12–12) and hence with the intensity of cheese flavor. The ratio of free amino acids to water-soluble N appears to be characteristic of the variety (Figure 12–13). Cheddar appears to contain a very low level of free amino acids relative to small peptides. Parmesan contains a particularly high concentration of amino acids, which have a major effect on its flavor, but it contains a low concentration of small peptides (Figures 11–18 and 11–19).

The water-soluble fraction also contains short-chain fatty acids (< $C_{9:0}$), which impart a "cheesy" aroma. The few available studies indicate that there are substantial intervarietal differences with respect to short-chain fatty acids. Further work in this area appears to be warranted.

12.7.3 Cheese Volatiles

A number of intra- and intervarietal comparisons of cheese volatiles have been published. An early example is the study of Manning and Moore (1979), who analyzed headspace volatiles of nine fairly closely related varieties. Considerable intervarietal differences were evident, but the four samples of "Cheddar" that were analyzed also differed markedly. The intensity of cheese flavor was reported to be related to the

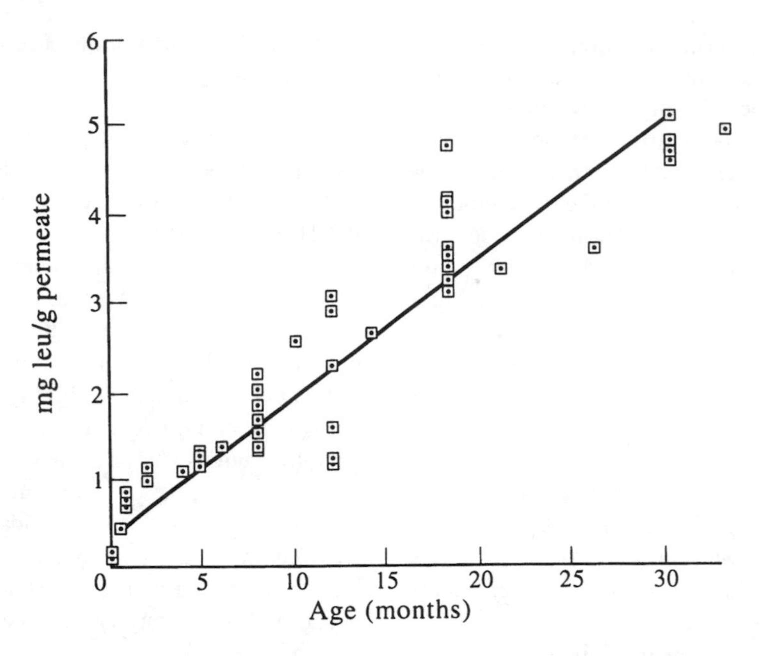

Figure 12–12 Total free amino acid concentrations (Cd-Ninhydrin assay) in Cheddar cheese as a function of age.

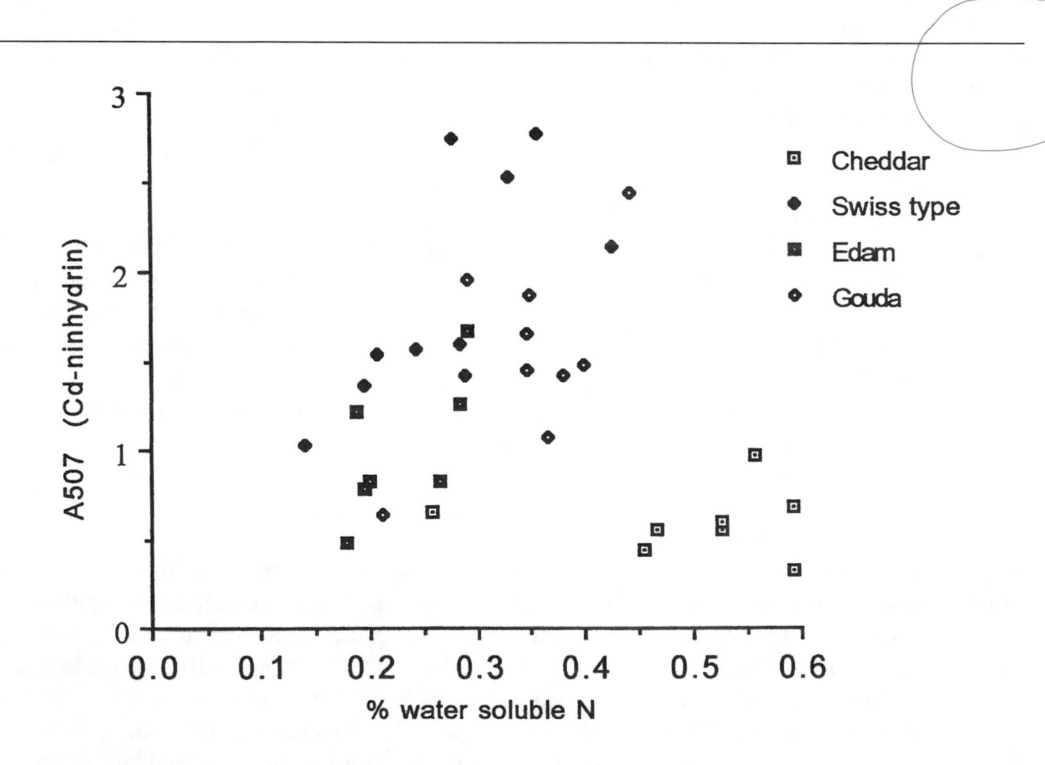

Figure 12–13 Relation between the concentration of amino acids (Cd-Ninhydrin assay) and percentage of water-soluble N in a selection of cheese varieties.

concentration of sulfur compounds; 2-pentanone was also considered to be important for Cheddar cheese flavor.

A comprehensive study on Cheddar, Gouda, Edam, Swiss, and Parmesan (total of 82 samples) was reported by Aishima and Nakai (1987). The volatiles were extracted by CH_2Cl_2 and analyzed by GC. More than 200 peaks were resolved in all chromatograms, 118 of which were selected as variables for discriminative analysis. Expression of the area of each of the 118 peaks as a percentage of total chromatogram area clearly permitted classification of the five varieties. The compounds likely to be responsible for the characteristic flavor of each variety were not discussed.

Bosset and Gauch (1993) concentrated the headspace volatiles from six cheese varieties by a "purge and trap" method for analysis by GC-MS. A total of 81 compounds were isolated and identified (Figure 12–8), 20 of which were found in all six varieties and a further nine in five of the six varieties. The authors concluded that "practically all types of cheese analysed contain more or less the same volatiles, but at different concentrations. Thus, the flavour of these cheeses seems to depend not on any particular key compound, but rather on a critical balance, or a weighted concentration ratio of all components present" (p. 366).

12.8 CONCLUSION

The component balance theory of cheese flavor still applies. While analytical techniques have improved greatly during the last 40 years and data are available on the concentrations of many flavor compounds in cheese, the critical compounds, if any, are unknown in most cases. Further intervarietal comparisons may be useful if quantitative data are obtained, although frequently they are not. Perhaps it would be fruitful to reinvestigate cheeses with a controlled microflora. The last studies on such systems were done in the 1970s and those were concerned mainly or totally with proteolysis. It would seem to be particularly useful to combine studies on controlled microflora cheese with intervarietal comparisons, but perhaps such an undertaking is beyond the capabilities of a single laboratory.

REFERENCES

Adda, J., Gripon, J.-C., & Vassal, L. (1982). The chemistry of flavour and texture development in cheese. *Food Chemistry, 9*, 115–129.

Aishima, T., & Nakai, S. (1987). Pattern recognition of GC profiles for classification of cheese variety. *Journal of Food Science, 52*, 939–942.

Aston, J.W., & Dulley, J.R. (1982). Cheddar cheese flavour. *Australian Journal of Dairy Technology, 37*, 59–64.

Bartlett, P.N., Elliott, J.M., & Gardner, J.W. (1997). Electronic noses and their application in the food industry. *Food Technology, 51(*12), 44–48.

Bosset, J.O., & Gauch, R. (1993). Comparison of the volatile flavour in six European AOC cheeses by using a new dynamic headspace GC-MS method. *International Dairy Journal, 3*, 359–377.

Fox, P.F. (1989). Proteolysis in cheese during manufacturing and ripening. *Journal of Dairy Science, 72*, 1379–1400.

Fox, P.F., Singh, T.K., & McSweeney, P.L.H. (1995). Biogenesis of flavour compounds in cheese. In E.L. Malin & M.H. Tunick (Eds.), *Chemistry of structure-function relationships in cheese.* New York: Plenum Press.

Fox, P.F., & Wallace, J.M. (1997). Formation of flavour compounds in cheese. *Advances in Applied Microbiology, 45*, 17–85.

Gardner, J. (1996). *An introduction to electronic nose technology*. Essex, UK: Neotronics Scientific Ltd.

Hirst, D., Muir, D.D., & Naes, T. (1994). Definition of the sensory properties of hard cheese: A collaborative study between Scottish and Norwegian panels. *International Dairy Journal, 4*, 743–761.

Imhof, R., & Bosset, J.O. (1994). Relationships between microorganisms and formation of aroma compounds in fermented dairy products. *Zeitschrift fur Untersuchung und Forschung, 198,* 267–276.

International Dairy Federation. (1991). *Chemical methods for evaluation of proteolysis in cheese maturation* [Bulletin No. 261]. Brussels; Author.

Kristoffersen, T. (1973). Biogenesis of cheese flavour. *Journal of Agricultural and Food Chemistry*, *21*, 573–575.

Kristoffersen, T. (1985). Development of flavour in cheese. *Milchwissenschaft*, *40*, 147–199.

Lemieux, L., & Simard, R.E. (1991). Bitter flavour in dairy products. I. A review of the factors likely to influence its development, mainly in cheese manufacture. *Lait*, *71*, 599–636.

Lemieux, L., & Simard, R.E. (1992). Bitter flavour in dairy products. II. A review of bitter peptides from the caseins: their formation, isolation and identification, structure masking and inhibition. *Lait*, *72*, 335–382.

Maarse, H., & Visscher, C.A. (1989). *Volatile compounds in food* (6th ed.). Zeist, The Netherlands: TNO Division for Nutrition and Food Research.

Manning, D.J., & Moore, C. (1979). Headspace analysis of hard cheese. *Journal of Dairy Research*, *46*, 539–545.

McEwan, J.A., Moore, J.D., & Colwill, J.S. (1989). The sensory characteristics of Cheddar cheese and their relationship with acceptability. *Journal of the Society of Dairy Technology*, *4*, 112–117.

McGugan, W.A. (1975). Cheddar cheese flavour: A review of current progress. *Journal of Agricultural and Food Chemistry*, *23*, 1047–1050.

McSweeney, P.L.H., & Fox, P.F. (1993). Cheese: Methods of chemical analysis. In P.F. Fox (Ed.), *Cheese: Chemistry, physics and microbiology* (2d ed., Vol. 1). London: Chapman & Hall.

McSweeney, P.L.H., & Fox, P.F. (1997). Chemical methods for characterization of proteolysis in cheese during ripening. *Lait*, *77*, 41–76.

McSweeney, P.L.H., Nursten, H.E., & Urbach, G. (1997). Flavours and off-flavours in milk and dairy products. In P.F. Fox (Ed.), *Advanced Dairy Chemistry* (2d ed., Vol. 3). London: Chapman & Hall.

Muir, D.D., & Hunter, E.A. (1992). Sensory evaluation of Cheddar cheese: The relation of sensory properties to perception of maturity. *Journal of the Society of Dairy Technology*, *45*, 23–30.

Ney, K.H. (1979). Bitterness of peptides: Amino acid composition and chain length. In J.C. Boudreau (Ed.), *Food taste chemistry* (ACS Symposium Series 115). Washington, DC: American Chemical Co.

Nielsen, R.G., Zannoni, M., Beriodier, F., Lavanchy, P., Muir, D.D., & Siverten, H.K. (1998). Progress in developing an international protocol for sensory profiling of hard cheese. *International Journal of Dairy Technology*, *51*, 57–64.

Olson, N.F. (1990). The impact of lactic acid bacteria on cheese flavour. *Federation of European Microbiological Societies Microbiological Reviews*, *43*, 497–499.

O'Shea, B.A., Uniacke-Lowe, T., & Fox, P.F. (1996). Objective assessment of Cheddar cheese quality. *International Dairy Journal*, *6*, 1135–1147.

Reiter, B., Fryer, T.F., Sharpe, M.E., & Lawrence, R.C. (1966). Studies on cheese flavour. *Journal of Applied Bacteriology*, *29*, 231–242.

Schieberle, P. (1995). New developments in methods for analysis of volatile flavor compounds and their precursors. In A.G. Gaonkar (Ed.), *Characterization of food: Emerging methods*. Amsterdam: Elsevier Science.

Urbach, G. (1993). Relations between cheese flavour and chemical composition. *International Dairy Journal*, *3*, 389–422.

Urbach, G. (1995). Contribution of lactic acid bacteria to flavour compound formation in dairy products. *International Dairy Journal*, *5*, 877–903.

Urbach, G. (1997). The flavour of milk and dairy products. II. Cheese: Contribution of volatile compounds. *International Journal of Dairy Technology*, *50*, 79–89.

CHAPTER 13

Cheese Rheology and Texture

13.1 INTRODUCTION

Rheology involves the study of the deformation and flow of materials when subjected to a stress or strain. The rheological properties of cheese are those that determine its response to a stress or strain (e.g., compression, shearing, or cutting) that is applied during processing (e.g., portioning, slicing, shredding, or grating) and consumption (slicing, spreading, masticating, and chewing). These properties include intrinsic characteristics—such as elasticity, viscosity, and viscoelasticity—that are related primarily to the composition, structure, and strength of the attractions between the structural elements of the cheese. The rheological characteristics are determined by the application of a fixed stress or strain to a sample of cheese under defined experimental conditions. The relationship between stress and strain may be described using various rheological terms, including *bulk modulus*, *elastic modulus*, *shear modulus*, *fracture stress or strain*, and *firmness*. In lay language, the behavior of cheese when subjected to a stress or strain can be described by terms such as *hardness*, *firmness*, *springiness*, *crumbliness*, and *adhesiveness*. The rheological properties of cheese are of considerable importance, since they affect its

- handling, portioning, and packing characteristics
- texture and eating quality, which are determined by the effort required to masticate the cheese or, alternatively, the level of mastication achieved for a given level of chewing (the latter may, in turn, influence its flavor and aroma and the suitability of the cheese for different consumer groups, such as children or aged people)
- use as an ingredient, since they determine its behavior when subjected to different size reduction methods (such as shredding, grating, and shearing) and how the cheese interacts with other ingredients in foods
- ability to retain a given shape at a given temperature or when stacked
- ability to retain gas and hence to form eyes or cracks or to swell

In short, the rheological properties of cheese are quality attributes that are important to the manufacturer, packager, distributor, retailer, industrial user, and consumer.

The rheology of cheese is a function of its composition, microstructure (i.e., the structural arrangement of its components), the physicochemical state of its components, and its macrostructure, which reflects the presence of heterogeneities such as curd granule junctions, cracks, and fissures. The physicochemical properties include parameters such as the level of fat coalescence, solid fat:liquid fat ratio, degree of hydrolysis and hydration of the paracasein matrix, and level of intermolecular attractions between paracasein molecules. Hence, the rheological characteristics differ markedly with the cheese variety and its age. The effect of variety on the rheological properties is readily apparent when one compares an almost flowable mature Camembert with a firm, brittle Parmesan or compares a crumbly Cheshire cheese with a springy Emmental-type cheese or String cheese. Similarly, the influence of age is shown by the difference between a young (< 1–2 months) rubbery Cheddar with a fully mature pliable Cheddar.

Cheese texture may be defined as a composite sensory attribute resulting from a combination of physical properties and perceived by the senses of sight, touch, and hearing (Brennan, 1988). The properties of cheese that contribute to its texture may be divided into three principal categories (Szczesniak, 1963): (1) mechanical, (2) geometrical, and (3) others.

Mechanical properties are manifested by the reaction of food to a stress applied during consumption (e.g., squeezing between the fingers, manual cutting, and mastication). They comprise the following characteristics: hardness, cohesiveness, viscosity, springiness, chewiness, brittleness, and gumminess. The mechanical properties are measured organoleptically by the pressure exerted on the cheese by the teeth, tongue, and roof of the mouth during eating. Geometrical properties include the size distribution, shape, and orientation of particles within a food (particles of the whole food material or of components such as occluded air or fat). They are reflected mainly in the visual characteristics of the cheese (e.g., granularity in Cottage cheese or the fibrous nature of String cheese). However, the geometrical characteristics may be sufficiently pronounced to affect the mechanical properties of the cheese. The other properties that contribute to cheese texture include characteristics such as greasiness, oiliness, succulence, and mouth-coating associated with the presence of fat and moisture within the cheese.

An alternative scheme to that of Szczesniak (1963) for classifying textural properties was proposed by Sherman (1969). In this scheme, textural properties are considered to be primary (or fundamental), secondary, or tertiary. The primary characteristics—from which the secondary and tertiary characteristics are derived—include the composition of the food, its micro- and macrostructure, and its molecular properties. The textural characteristics are further subdivided into two broad groups based on sensory perceptions of the material before and during consumption (Figure 13–1). Characteristics that contribute to the initial perception of cheese texture, before eating, include

- visual appearance (e.g., presence of holes, eyes, or granules and surface roughness or smoothness)
- crumbliness, springiness, stickiness, and slicing characteristics
- spreading characteristics (important for pasteurized processed cheese spreads)

The secondary and tertiary categories of textural properties include many characteristics (e.g., hardness, brittleness, softness, and springiness) that are directly related to the intrinsic rheological properties (e.g., elasticity and viscosity) that determine the cheese's response to the stresses applied during biting, chewing, and salivation. Hence, cheese texture and cheese rheology are closely related, in that many of the textural properties of cheese are determined by its rheological properties.

Cheese rheology, which has been studied and reviewed extensively (Culioli & Sherman, 1976; Prentice, Langley, & Marshall, 1993; Rao, 1992; Sherman, 1969, 1988; van Vliet, 1991; Visser, 1991) is the focus of this chapter. For detailed information on cheese texture, the reader is referred to a number of studies and reviews (Brandt, Skinner, & Coleman, 1963; Brennan, 1988; Jack, Paterson, & Piggott, 1993, 1995; Sherman, 1969; Szczesniak, 1963).

13.2 CHEESE MICROSTRUCTURE

The micro- and macrostructure of cheese are major determinants of its rheological and textural properties and are discussed briefly below. Natural rennet-curd cheese is essentially a calcium phosphate–paracasein matrix composed of

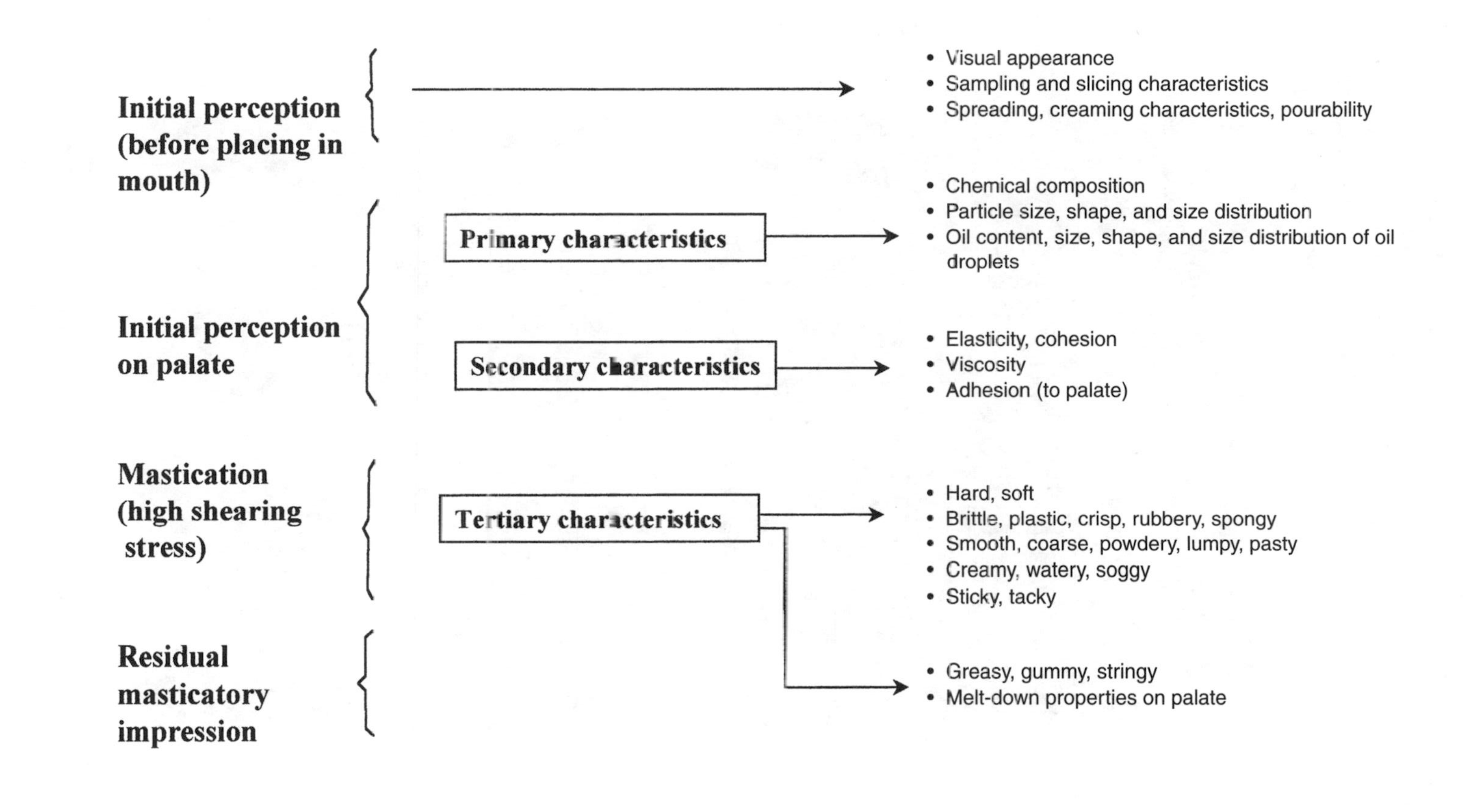

Figure 13–1 Textural characteristics of foods.

overlapping and cross-linked strands of partially fused paracasein aggregates (in turn formed from fused paracasein micelles) (Figure 13–2). The integrity of the matrix is maintained by various intra- and interaggergate hydrophobic and electrostatic attractions. The matrix occludes, within its pores, fat globules (in varying degrees of coalescence), moisture, and dissolved sub-

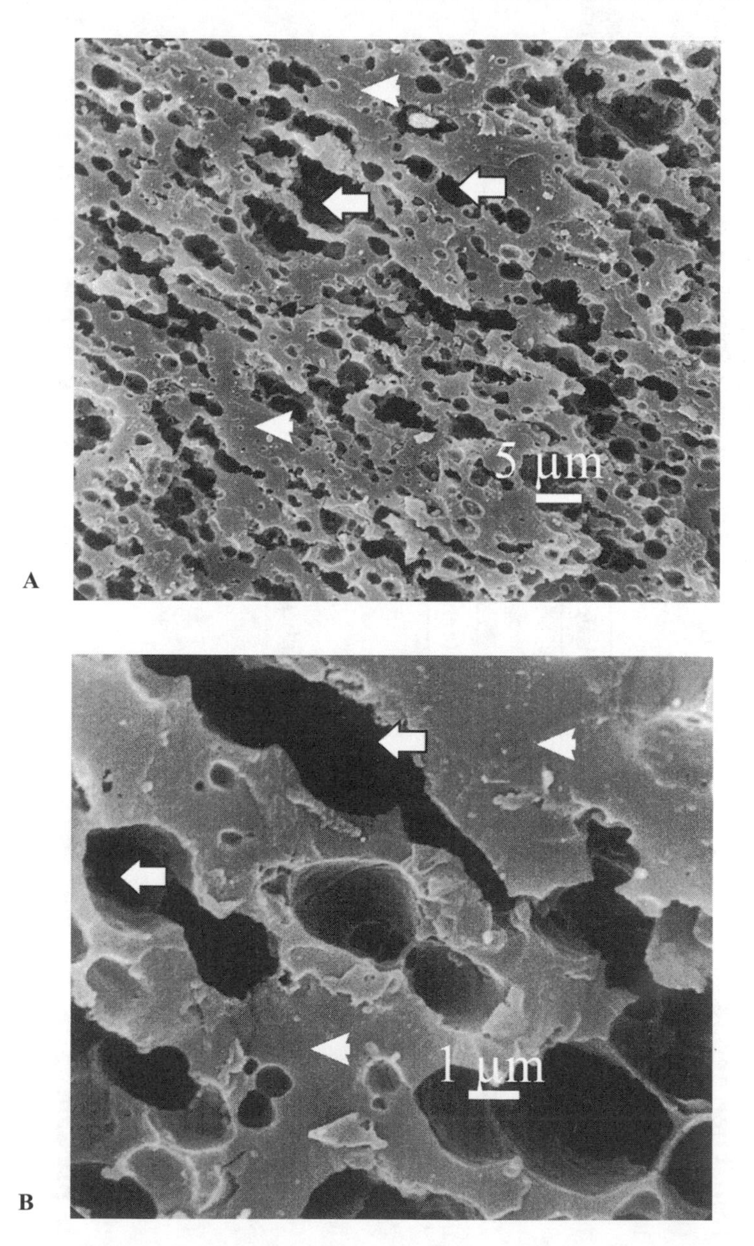

Figure 13–2 Scanning electron micrographs of Cheddar cheese, showing the continuous paracasein matrix (arrow heads) permeated by holes and fissures corresponding to discreet fat globules and/or pools of clumped or coalesced fat globules (solid arrows). Bar in (A) equals 5 µm and in (B) 1 µm.

stances (minerals, lactic acid, peptides, and amino acids), and enzymes (e.g., residual rennet and proteinases from starter and nonstarter microorganisms). The cheese matrix also contains various microorganisms (in most cases starter and nonstarter bacteria and in some cases molds, yeasts, and other bacteria on the surface of the cheese) and their enzymes (e.g., proteinases, peptidases, and lipases), which are released into the cheese matrix at various rates during maturation. A dynamic equilibrium exists between the concentrations of Ca and inorganic phosphate in the paracasein matrix and the cheese serum. The equilibrium is influenced by pH and other factors such as the concentration of Na^+ in the serum phase.

The paracasein network is essentially continuous, extending in all directions, although some discontinuities in the matrix may exist due to the presence of curd granule junctions and/or curd chip junctions (e.g., as occur in Cheddar and related dry-salted varieties). Microscopical examination of low-moisture Mozzarella cheese reveals the presence of well-defined curd granule junctions ($\approx$ 4 µm wide) (Figure 13–3), which appear as veins running along the perimeters of neighboring curd particles. Unlike the interior of the curd particles, the junctions are almost devoid of fat, owing to the expression of fat globules from the surfaces of the curd particles into the whey during cutting of the coagulum and cooking. Similarly, chip junctions in Cheddar and related dry-salted varieties have a higher casein:fat ratio than the interior and are clearly discernible upon examination of the cheese by light microscopy (Figure 13–4) or by visual inspection of cheese, especially in the case of cheeses exhibiting the defect of seaminess. Differences in cheese composition between junctions and the interior of the curd particles probably lead to differences in the molecular attractions between the casein molecules in the interior and exterior of curd particles.

Various physicochemical changes occur in the structural components of the matrix during maturation. These changes are mediated by the residual rennet, microorganisms and their enzymes (see Chapter 11), and changes in mineral equilibria between the serum and paracasein matrix. The type and extent of physicochemical changes depend on the cheese variety, cheese composition, and ripening conditions. The changes include the following:

- The conversion of residual lactose to lactic acid and/or acetic and propionic acids.
- Hydrolysis of the caseins to peptides of varying molecular weights and amino acids and catabolism of amino acids to amines, aldehydes, alcohols, and NH_3 (Fox, O'Connor, McSweeney, Guinee, & O'Brien, 1996; Chapter 11).
- Hydrolysis of triglycerides to free fatty acids, which may be degraded further to ketones and alcohols.
- Increased hydration of the paracasein mediated by factors such as its hydrolysis, the increase in cheese pH, and solubilization of casein bound calcium. Solubilization of Ca^{2+} attached to the casein occurs when they are partially replaced by Na^+, especially when the concentrations of Na^+ and Ca^{2+} in the cheese moisture (serum) are low (i.e., 30 g/L and 4 g/L, respectively). The increase in casein hydration is paralleled by a physical expansion or swelling of the paracasein matrix that is clearly observed upon examination of the cheese by confocal laser scanning microscopy at various intervals during storage (see Figure 19–7).
- Coalescence of fat globules resulting in the formation of fat pools. This appears to occur in all cheeses, as reflected by increases in the level of fat that can be expressed from the cheese when subjected to hydraulic pressure or centrifugation (see Figures 19–7 and 19–14) or exudes from the cheese upon baking (Guinee, Mulholland, Mullins, & Corcoran, 1997; Kindstedt, 1995). The increase in free fat during maturation may be due to the physical swelling of the protein phase into spaces previously occupied by fat, which forces the partially denuded fat globules closer together.

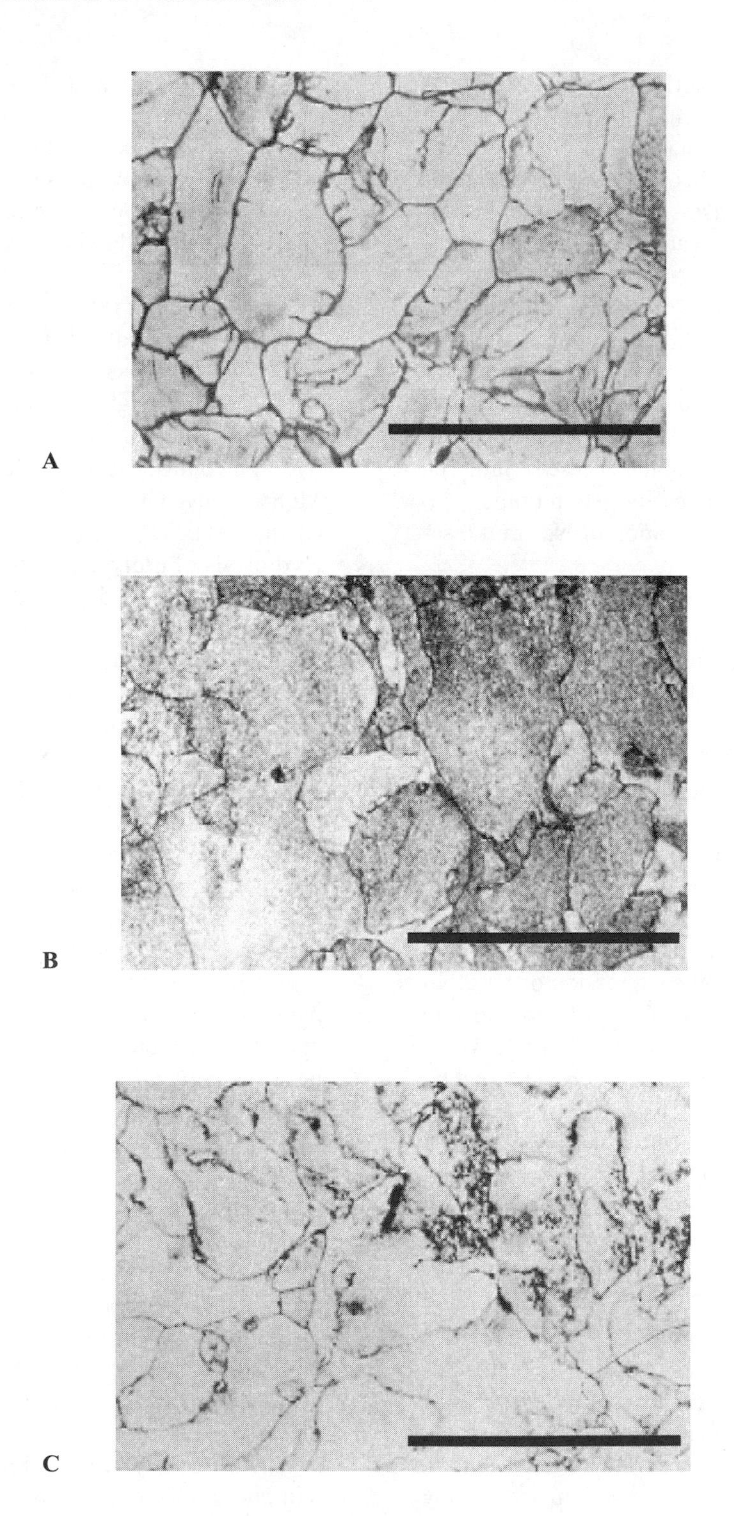

Figure 13–3 Light micrographs of Mozzarella (A), Gouda (B), and Edam (C) cheese showing curd granule junctions, which appear as dark lines. Bar equals 5 mm.

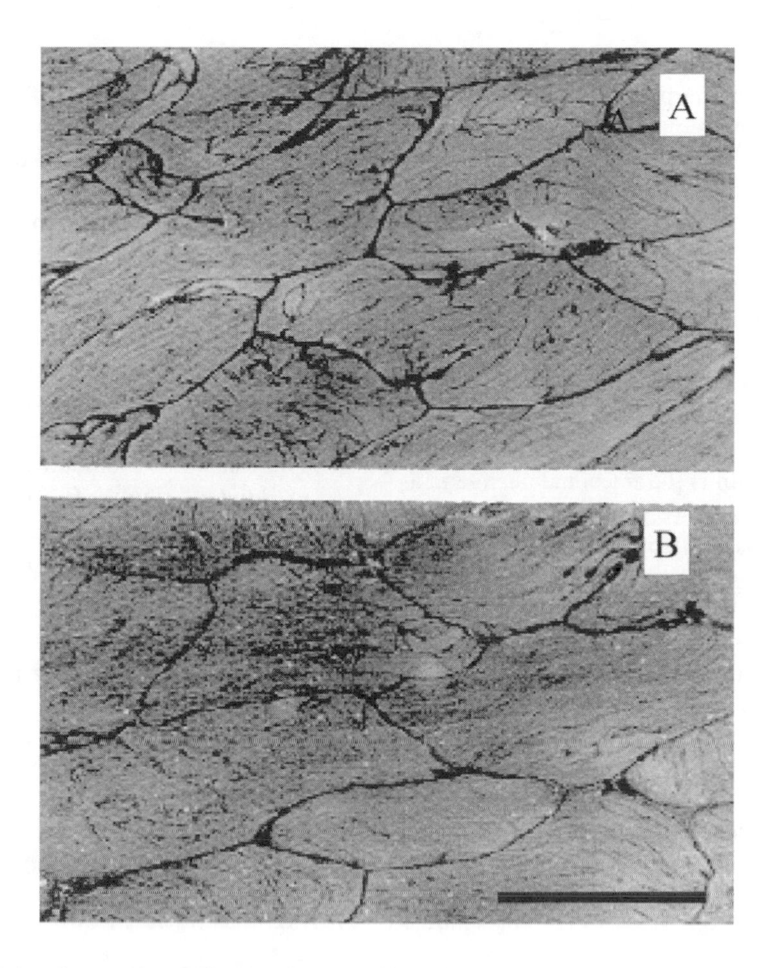

Figure 13–4 Light micrographs of Cheddar cheeses subjected to pressing in a Wincanton block former (A) or in a mold (B), as in traditional Cheddar manufacture. Curd chip junctions appear as heavy dark lines, and the junctions of curd particles within the curd chips are also discernible as fine black lines. Bar equals 10 mm.

Hence, cheese is, chemically, biologically, and biochemically, a dynamic system in which the structural components undergo storage-related physicochemical and microstructural changes such as hydrolysis and hydration of the casein, matrix swelling, fat coalescence, and/or fat hydrolysis. These changes aid in the conversion of fresh curd to a mature cheese and markedly influence its rheological, textural, functional, and flavor characteristics (see Chapters 12 and 19). Thus, a ripening period is generally required for all natural cheeses, apart from fresh acid-curd varieties, before they attain the desired attributes for the particular variety (e.g., flavor, aroma, and degree of meltability).

13.3 RHEOLOGICAL CHARACTERISTICS OF CHEESE

13.3.1 Definitions of Different Types of Rheological Behavior Based on Creep and Recovery Experiments

The application of stress to a solid material results in deformation and strain. The applied stress (denoted σ or τ) is the force per unit surface area of the material. It may be applied to the surface in a normal, perpendicular direction (σ), resulting in compression (e.g., when a weight is placed on top of a piece of material), or in extension (e.g., when a weight is hung from the material) (Figure 13–5, diagram A). Alternatively, the stress may be applied tangentially, or parallel, to the direction of the surface (τ), causing contiguous parts of the material to slide relative to each other in a direction parallel to their plane of contact (Figure 13–5, diagram B). The deformation is the change in distance between two points within the cheese mass as stress is applied, and the strain is the fractional change in a particular dimension (e.g., height or length). When the stress is applied normally, the ensuing strain (denoted by ε) may be defined as $\Delta h/h$ or $\Delta l/l$, where h and l correspond to the original height or length of the sample and Δh and Δl to the change in height or length. When the stress is applied tangentially, the ensuing shear or shear strain (denoted γ) may be defined as the distance (Δl) through which the point of application moves divided by the distance (l) between the moving and stationary planes of the sample (Figure 13–5, diagram B).

A solid material is described as being perfectly elastic (or Hookean) if the strain is directly proportional to the applied stress, with the σ/ε curve passing through the origin (Figure 13–6). Depending on how the stress is applied, two types of moduli (the proportionality constant between σ and ε or the slope of the σ/ε curve) may be obtained for a Hookean solid:

- The modulus of elasticity or Young's modulus (E), where the force is normal to the area bearing the stress, is expressed as $\sigma = E \times \varepsilon$, with $\varepsilon = \Delta l/l$.
- The elastic shear modulus, also termed the modulus of rigidity (G or G′), where the force is parallel to the area bearing the stress, is expressed as $\tau = G \times \gamma$, with $\gamma = \Delta l/l$.

The moduli E and G for a perfectly elastic material are independent of the rate at which the stress is applied. Hence, the stress-strain curve is always linear. A plot, known as a creep curve, shows the variation of ε with time upon the application and removal of a fixed stress (σ) to an ideal elastic material (where σ is low enough not to fracture the material). An ideal elastic mate-

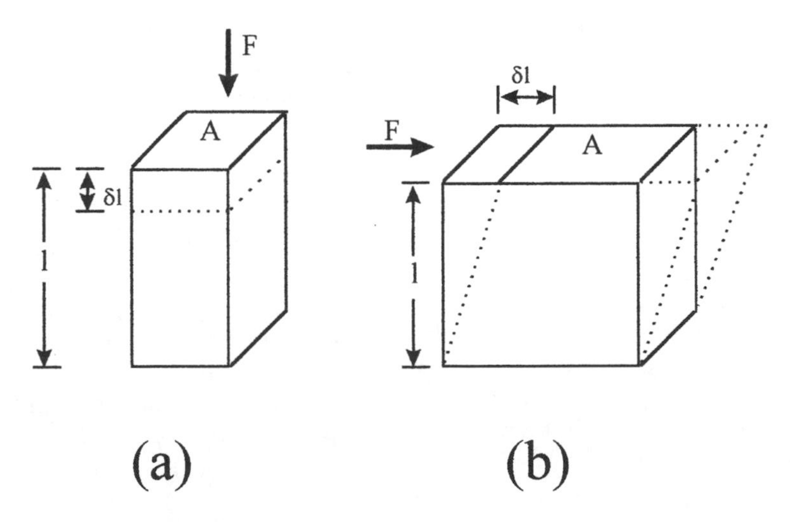

Figure 13–5 Application of stress (i.e., force F per unit area A) to a solid material. Stress may be applied (a) in a direction normal to the surface, resulting in uniaxial compression, or (b) parallel to the surface, resulting in shear compression. In both cases, the deformation is Δl and the strain is $\Delta l/l$.

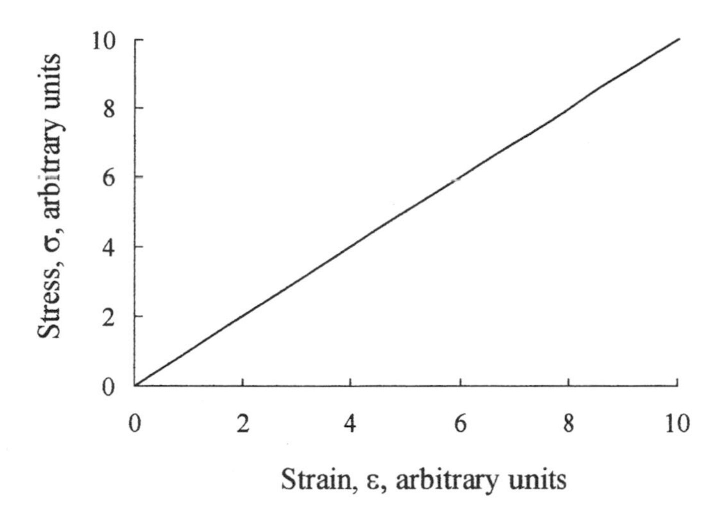

Figure 13–6 Relationship between stress (σ) and strain (ε) for an elastic solid.

rial deforms instantly upon the application of σ and recovers instantly to its original shape and dimensions when σ is removed (Figure 13–7, diagram A). During the application of σ, stress energy is absorbed or stored by the structural elements of the material. Upon removal of σ, the stored energy is released and the material regains its original dimensions instantly. This rheological behavior, which is a hypothetical one, may be represented mechanically by a simple spring (Figure 13–7, diagram B), where the degree of extension (i.e., the strain) is directly proportional to the mass hanging from the spring (and hence the force per unit area).

In contrast to an elastic solid, a fluid does not support a stress. Hence, the strain changes constantly as long as the stress is maintained. A material is defined as an ideal viscous (Newtonian) fluid if the rate of change of the strain (i.e., $d\gamma/dt$, usually denoted $\dot{\gamma}$) is directly proportional to the applied (shear) stress, σ, with the $\sigma/\dot{\gamma}$ curve passing through the origin (Figure 13–8). The proportionality constant—the slope between stress and strain rate (more usually denoted as shear rate)—is known as the coefficient of viscosity or just viscosity (η, where $\eta = \sigma/\dot{\gamma}$). A typical creep curve for a Newtonian fluid shows that it starts to flow instantly at a fixed rate upon the application of a constant shear stress and ceases to flow immediately (but is permanently deformed) upon its removal (Figure 13–9, diagram A). Unlike a solid material, the stress energy is not stored but is dissipated in the form of flow, with the strain being proportional to the time over which the stress is applied. This rheological behavior may be represented mechanically by a dashpot (i.e., a piston enclosed in a cylinder filled with a viscous liquid; Figure 13–9, diagram B), where the rate of movement of the piston (strain rate) is directly proportional to the applied stress.

Like most solid and semisolid foods, cheese exhibits characteristics of both an elastic solid and a Newtonian fluid and therefore is termed *viscoelastic*. The relationship between stress and strain for these materials is not linear except at very low strains, as discussed below. The stress increases less than proportionally with strain, resulting in a curve that is concave downward (Figure 13–10). The rheological properties (i.e., E, G, η) of viscoelastic materials differ from those of perfectly elastic or viscous materials in that they are dependent on time (they are a function of the time over which a fixed stress or strain is applied) and the magnitude of the stress. However, upon the application of a strain that is

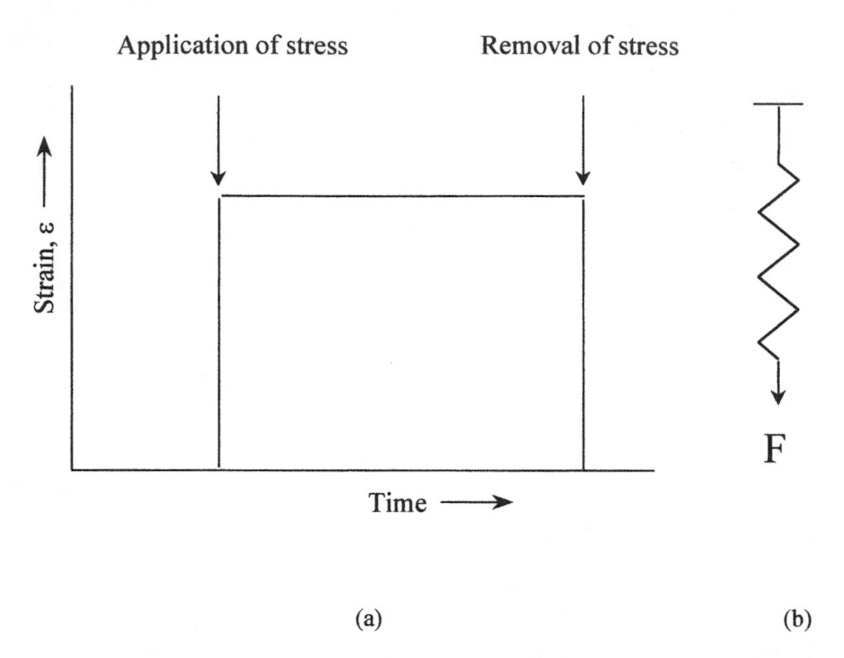

Figure 13–7 Time-related change in the strain of an elastic solid upon the application or removal of a fixed stress (force [*F*] per unit area) (a). The behavior is mechanically represented by a single spring (b).

sufficiently small so as not to induce permanent damage or fracturing of the microstructure, viscoelastic solid and semisolid food materials exhibit elastic behavior. The strain at which linearity between stress and strain is lost is referred to as the *critical strain*, which for most foods, including cheese, is relatively small (e.g., 0.02–0.05). At a strain less than the critical strain, the rheological properties of viscoelastic materials are time dependent. The typical change in strain with time upon the application of a constant stress to a 3-month-old Cheddar cheese is shown in Figure 13–11. The curve is termed a *creep compliance–time behavior curve*, where creep compliance (*J*) is the ratio of the strain to the applied constant stress. Three characteristic regions are evident in the creep curve:

1. elastic deformation (region AB), where the strain is instantaneous and fully reversible (the strain is referred to as *elastic compliance* [J_0])
2. viscoelastic deformation (region BC), where the strain is partly elastic and partly viscous (the strain is referred to as *retarded elastic compliance* [J_R], and the elastic component of the strain recovers slowly upon removal of the stress)
3. viscous deformation (region beyond C), where the deformation increases linearly with time and is permanent (the strain, which is referred to as *Newtownian compliance* [J_N], is not recoverable)

Upon removal of the strain at point D, the recovery follows a similar sequence to strain creep, with three regions evident: an instantaneous elastic recovery (DE), a delayed elastic recovery (EF), and an eventual flattening. The vertical distance from the flat portion of the curve to the time axis is the nonrecoverable strain per unit stress, which is related to the amount of structural damage to the sample during the test.

In the elastic region, the strands of the cheese matrix absorb and store the applied stress en-

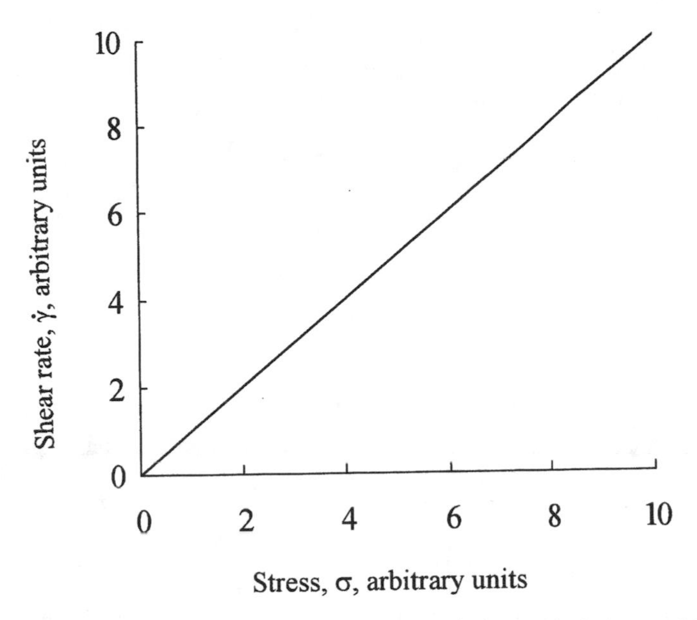

Figure 13–8 Relationship between shear stress (σ) and strain rate($\dot{\gamma}$) for an ideal viscous (Newtonian) liquid.

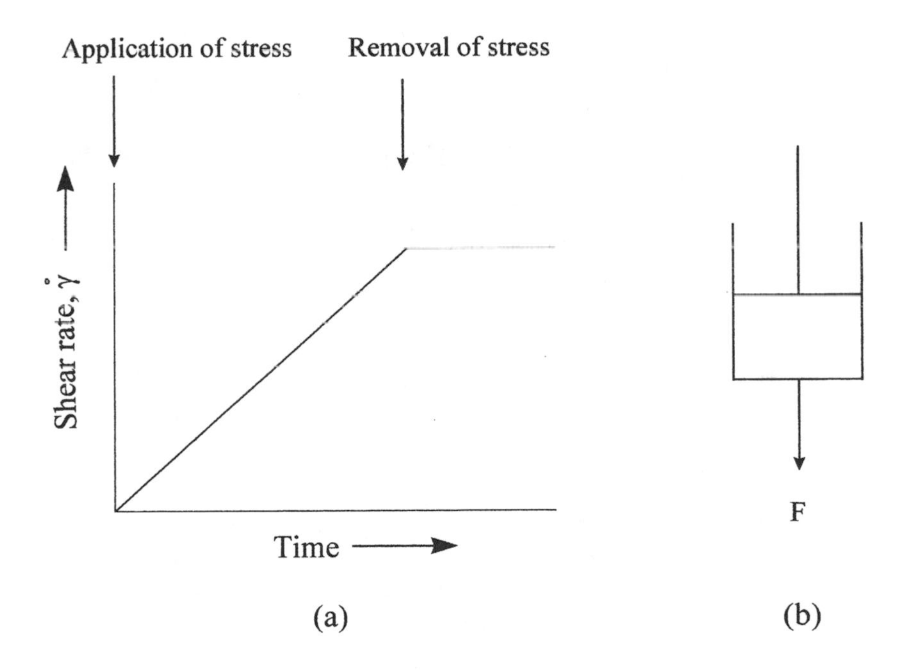

Figure 13–9 Time-related change in the shear of an ideal viscous liquid upon the application or removal of a fixed shear stress (force [*F*] per unit area) (a). The behavior is mechanically represented by a dashpot (b).

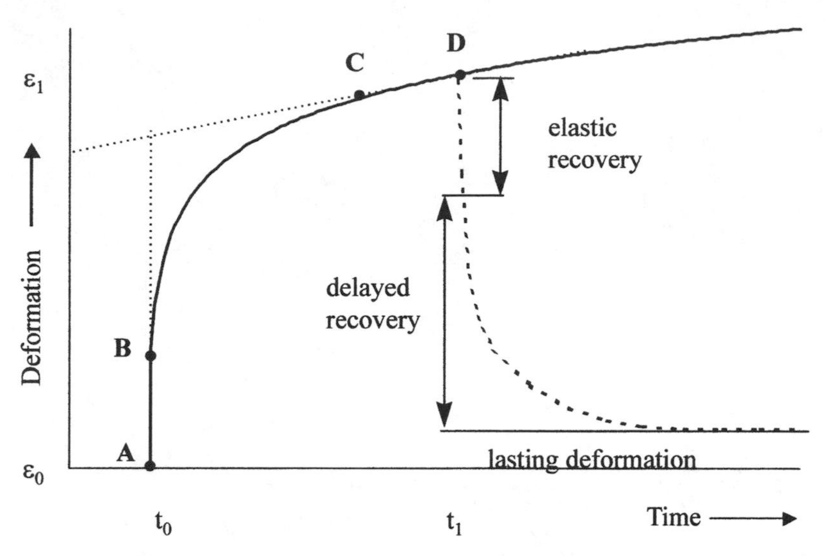

Figure 13–10 Schematic of time-related change in the deformation of a piece of cheese after subjecting it to a constant stress at time t_0 (—) and deformation recovery upon removal of the stress after a fixed time t_1 (- - -). Three types of deformation are apparent during the application of stress: elastic (AB), viscoelastic (BC), and viscous (beyond C). Similarly, upon removal of the stress (at D) three regions of deformation recovery are evident: elastic, viscoelastic (delayed), and lasting deformation as a result of viscous deformation.

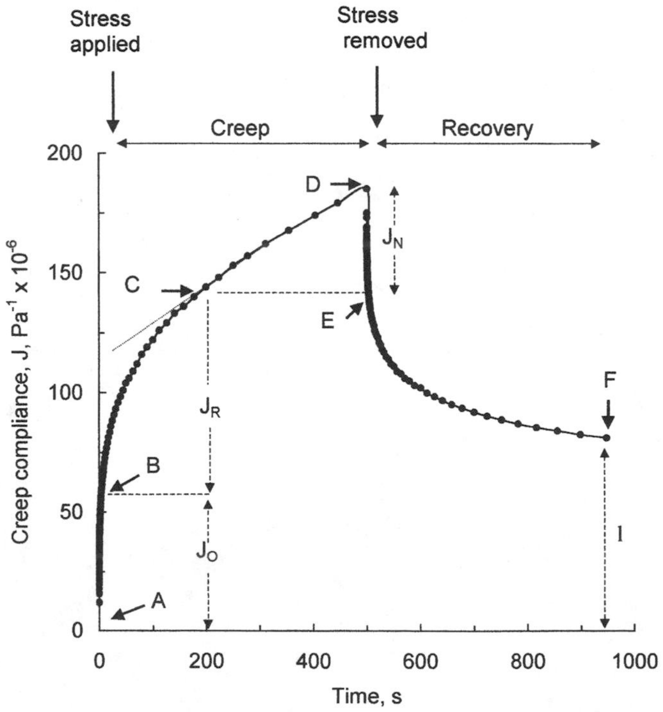

Figure 13–11 Creep compliance and recovery of a 3-month-old Cheddar cheese. The various terms are discussed in Section 13.3.

ergy, which is instantly released upon removal of the stress, enabling the cheese to regain its original dimensions. The extent and duration of the elastic region depend on the magnitude of the applied stress and the structural and compositional characteristics of the cheese. The strain at which elastic behavior is lost for cheese depends on the variety but is generally in the order of 0.02–0.05. At strains greater than the critical strain, the structure of the cheese is altered due to breaking of bonds between structural elements, which are stressed beyond their elastic limit. Eventually, when the stress-bearing structural casein matrix has fractured completely, the cheese flows.

Viscoelastic behavior may be represented by mechanical models that contain different arrangements of dashpots and springs in series and/or in parallel, such as Maxwell (spring and dashpot in series), Kelvin (spring and dashpot in parallel), or Burger bodies. The shape of the creep curve for the latter model, which consists of a Maxwell and a Kelvin body in series (Figure 13–12, diagram C), indicates that it gives a much closer approximation of the rheological behavior of cheese upon the application of a small load or stress than that obtained using the model of either an elastic solid or a Newtonian fluid on its own (see Figures 13–7, 13–9, and 13–12).

The viscoelastic nature of cheese implies that the ratio of elastic to viscous properties depends on the time-scale over which the deformation is applied. At short time scales, cheese is essentially elastic, whereas after a long deformation time, cheese flows, although very slowly in the case of hard cheeses. However, even hard cheeses flow eventually when stressed and will not recover completely upon removal of the stress. Failure to appreciate this characteristic can often lead to alteration of shape, especially bulging, during distribution and retailing, when cheeses of different consistencies are often haphazardly laid upon each other.

The flow of rigid materials is not always readily apparent because of the relatively long time required to produce a notable deformation, and the notion may even appear somewhat abstract. However, there are many natural examples that reveal the slow flow of rigid materials over very long time periods, equivalent to creep experiments where the time approaches infinity—for example, the flow of glass in window panes due to the force of gravity, as indicated by the increase in the thickness of individual window panes of old buildings with increasing distance from the top of the pane.

13.3.2 Measurement of the Rheological Behavior of Cheese

Rheological measurements on cheese may be classified into three main types:

1. Empirical tests. In these tests, the test conditions are arbitrary and the aim is to obtain a number that gives a vague indication of the textural characteristics of the cheese (e.g., its hardness). Often used by cheese graders, empirical tests may be based on the subjection of cheese to a stress or strain that results in visual fracture (e.g., rubbing cheese between the fingers until it becomes pliable or cutting the cheese with a knife). Alternatively, the cheese may be subjected to a stress or stain that causes no visible fracture (e.g., pressing one's thumb or a ball into cheese and recording the resistance either mentally or by an instrument [see Chapter 23]).
2. Creep and stress relaxation experiments. These allow the quantitative determination of precise rheological characteristics, as described in Section 13.3.1.
3. Large-scale deformation tests. These allow determination of the force required to fracture the cheese at given deformation rates. The test conditions may be designed to simulate the conditions to which the cheese is subjected in practice, such as during mastication or during size reduction processes (e.g., portioning). In this case, the test is described as being quantitative in relation to the process it is

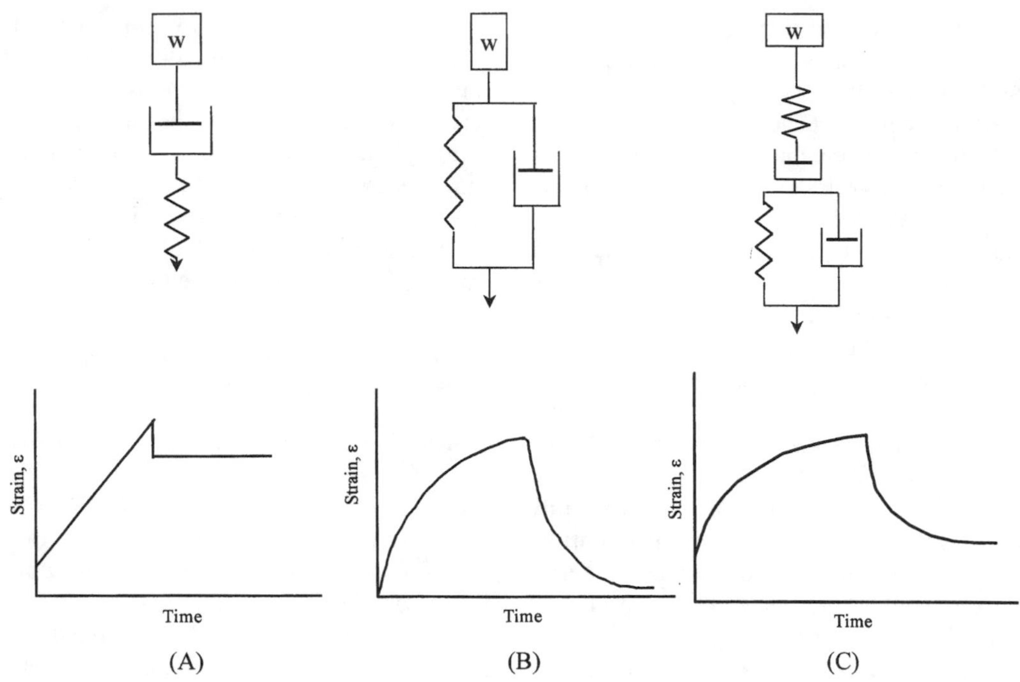

Figure 13–12 Mechanical models (bodies) representing viscoelastic behavior and their corresponding creep and recovery curves: Maxwell body (A), Kelvin body (B), and Burger body (C).

attempting to simulate. Alternatively, the test conditions may not simulate the stress or strain to which the cheese is subjected in a particular application. Although physical quantities (e.g., fracture stress or fracture strain) may be measured during the latter test, the test is described as being nonquantitative for that application, and the quantities obtained are of little use in trying to predict how the cheese will behave rheologically in that application.

Empirical Tests

These tests involve subjecting a cheese sample to a stress or strain by various techniques (e.g., inserting a penetrometer). The different types include

- penetration tests, where the force required to insert a probe a given distance into the cheese or, alternatively, the depth of penetration by a probe under a fixed load for a given time is measured (e.g., penetrometer test)
- compression tests, where the extent of compression under a constant load for a specified time is measured (e.g., ball compressor test)

The penetrometer test measures the depth to which a penetrometer (e.g., needle or cone) can be forced into a cheese under a constant stress. As the needle or cone penetrates the cheese, the cheese in its path is fractured and forced apart. The progress of the penetrometer is retarded to an extent that depends on the hardness of the cheese in its path, the adhesion of the cheese to its surface (which increases with the depth of penetration into the cheese), and its surface area of contact with the cheese (regulated by the thickness of the needle or angle of the cone used). Eventually, the retardation stresses become equal to the applied stress and penetration ceases. The test, which is used to provide an index of hardness (i.e., resistance of a surface to penetration), is suitable for closed-textured cheeses such as Gouda and Mozzarella, which are macroscopically homogeneous. Conversely, it is unsuitable for open-textured cheeses with small mechanical openings or eyes (e.g., Tilsit and Gruyère) or cheeses that are macroscopically nonuniform owing to the presence of chip boundaries (e.g., Cheddar).

The ball compressor test measures the depth of indentation after a given time made by a small ball or hemisphere when placed under a given load (stress) on the cheese surface (Figure 13–13). Hence, the test simulates the action of a grader who, in the course of examination, presses the ball of his or her thumb into the cheese. The depth of penetration has been used directly as an index of firmness. Alternatively, by making a number of simple assumptions, testers may use it to calculate a modulus, analogous to an elastic modulus (G, given by the equation $G = 3M/[16(RD^3)]^{1/2}$, where M is the applied force and R and D are the radius and depth of the indentation, respectively). The above tests, which are described in detail by Prentice et al. (1993), are generally nondestructive and can be performed on the whole cheese at several locations. However, they provide only an overall measure of the many different facets of rheological behavior.

Quantitative Tests (Creep and Stress Relaxation)

In quantitative tests, the cheese is subjected to a very small stress or strain so as to minimize structural alteration. The ensuing stress or strain is measured dynamically over time, and the resultant curves may be used for the measurement of precise rheological properties such as deformation (at a given stress), modulus of elasticity (E), and relaxation time required for stress alleviation to a certain value of the maximum stress applied. Thus, they provide information on the viscoelasticity and structure of the cheese. A high value for E suggests that the cheese matrix is elastic and continuous, with strong intermolecular attractions, whereas a low value of E indicates that the matrix is less elastic and weaker as a consequence of proteolysis and casein hydration, for example.

The viscoelastic behavior of cheese is measured using two quasi-static tests, creep and stress relaxation. In a creep test, as described in Section 13.3.1, the sample is instantaneously subjected to a low, constant stress (e.g., 10–50 Pa), and the

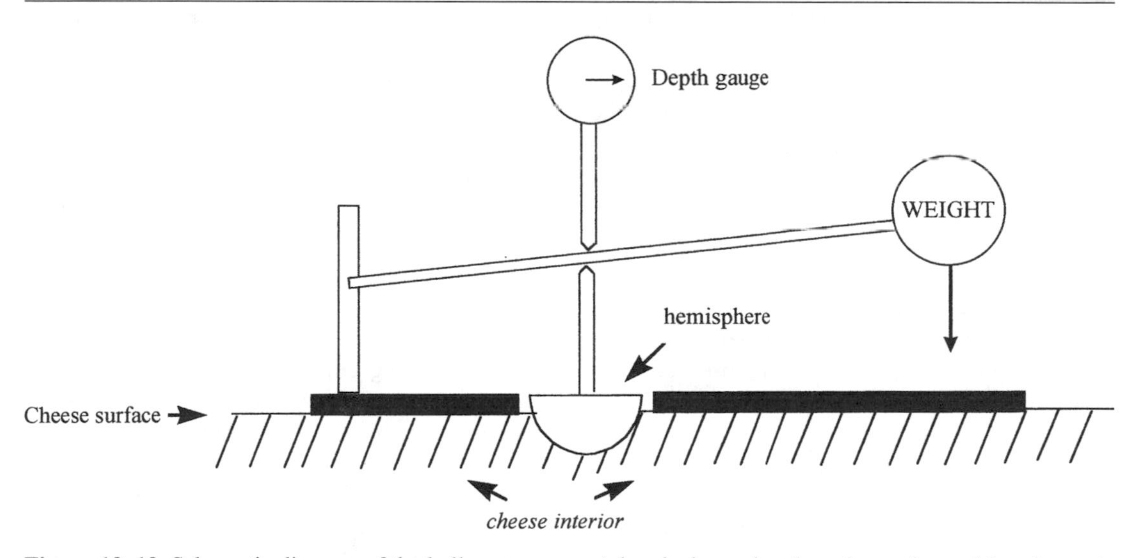

Figure 13–13 Schematic diagram of the ball compressor. A hemisphere placed on the surface of the cheese is forced into the cheese surface under the action of a weight.

ensuing increase in deformation or strain is measured over time (see Figure 13–11). The constant stress may be applied using a control stress rheometer (see Chapter 6), where the test involves placing a cheese sample between two parallel plates and applying a load to the top plate. A stress relaxation test involves instantaneous application of a low constant deformation or strain to a cheese sample (e.g., < 0.05) and measurement of the decrease in stress over time. The constant strain is usually applied by placing the sample between two parallel plates of an instrument such as the Instron Universal Testing Machine and allowing the top plate to compress the sample to a particular deformation (Figure 13–14).

Force Compression Tests between Parallel Plates

In practice, cheese is subjected to large stresses and strains (i.e., > 0.05), which usually result in fracture. The use of cheese as an ingredient may involve precutting of large blocks and/or comminution (e.g., by a conveying auger crushing and forcing precut cheese through die plates with narrow apertures). Other industrial operations that result in the fracture of cheese include shredding (e.g., cutting into thin, narrow cylindrical pieces or shreds, e.g., 2.5 cm in length and 0.4 cm in diameter), dicing (cutting into very small cubes, e.g., 0.4 cm^3), or cubing (2.5 cm^3). In the home, slicing of table cheese ideally results in a clean cut along the path of the knife blade. Also, cheese, when eaten, is submitted to a number of fracture strains that reduce it to a pulp capable of being swallowed. First the cheese is cut by the incisors, then it is compressed by the molars during chewing and sheared between the palate and the tongue. Most of these actions involve a combination of compressive and shear forces. Thus, prediction of how cheese behaves under large stresses and strains is desirable.

Force-compression tests measure the dynamic rheological behavior of cheese as the strain is increased over time to values that generally result in fracture and flow of the cheese being tested. The tests, which are today commonplace for the measurement of fracture properties, are more or less empirical, depending on the magnitude of the stresses and strains involved. At very low strains (< 0.02), the structure of the cheese is effectively unaltered during the measurement, and the values obtained may be then used to quantify the elastic modulus (usually denoted *compression modulus*). Compression tests usually involve large strains (up to 0.8) that cause complete breakdown of the cheese structure. Incidentally, these strains simulate the mastication of cheese, where compression of the cheese between the molars is typically about 70%. The large-strain compression test is useful in that it provides a measure of the stresses and strains at which cheese is effectively elastic, at which it fractures, and at which it flows. These data may be useful for designing size-reduction equipment, optimizing the settings of a piece of equipment for a particular type or batch of cheese, or gauging how difficult or how easily a piece of cheese is degraded when compressed under conditions similar to those in the mouth.

In a large-strain force-compression test, a cylindrical or cubical cheese sample of fixed dimensions and at a fixed temperature is placed between the two parallel plates of an instrument such as the Universal Testing Machine (Figure 13–14). One plate, denoted the *base plate*, is fixed; the other, denoted the *cross-head*, is programmed to move at a fixed rate (typically 50 cm/min) and compresses the cheese sample to a predetermined level (typically to 25% of its original height). The force required to compress the sample to a given level (or percent compression) is a measure of two parameters: the force required to overcome the surface friction of the sample and the force required to actually compress the cheese sample. The force (F) developed during compression is recorded as a function of distance or displacement (ΔL); alternatively, the force may be converted into stress (σ) and the displacement into strain (ε) as follows: σ (Pa) $= F/A$ (N/m^2), where A is the surface area of the cheese sample in square meters, and $\varepsilon = \Delta L/L$, where ΔL (m) is the distance through which the cross-head moves.

Figure 13–14 Compression instrument (Instron Universal Testing Machine, Model 112). A cylindrical sample of cheese is compressed at a fixed rate to a preset deformation between two parallel plates, the bottom (base) plate and the top plate (cross-head). During compression, the force exerted by the cheese on the moving top plate is measured by a load cell interfaced with a personal computer for continuous data capture.

A typical force (stress)/displacement (strain) curve for mature Cheddar reveals a number of distinct regions (Figure 13–15):

- Region AB. The stress increases proportionally with ε. If compression is stopped in this range, the cheese returns to its original dimensions. The slope of this linear region corresponds to the compression modulus, E (i.e., $E = \sigma/\varepsilon$), which is an indication of the texture profile descriptors springiness and elasticity. Alternative (rarely used) expressions for E are *yield stress* and *yield strain* (the stress and strain at which elastic recovery is lost). The value of E decreases with assay temperature and moisture content and increases upon elevation of fat and salt-in-moisture levels and with maturity (Luyten, 1988).
- Region BC. The stress increases less than proportionally with ε (i.e., the curve is concave downward). The slightly lower slope of the curve in this region compared with that in region AB is probably due to the formation of microcracks that allow some stress to be dissipated but do not spread throughout the sample. However, the curve is still relatively linear, indicating that the cheese is still largely elastic and recovers almost completely when the compression ceases.

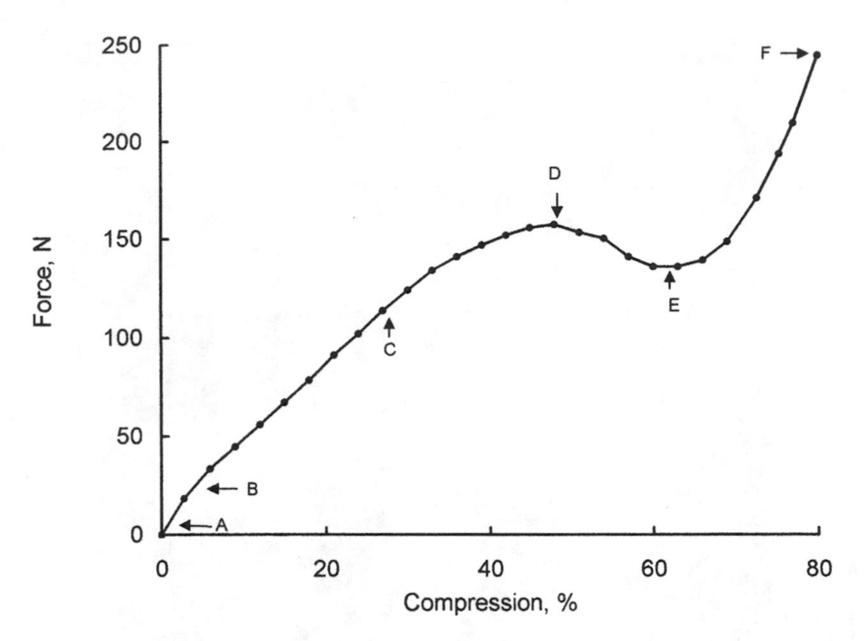

Figure 13–15 Force compression curve for a 6-month-old mature Cheddar cheese compressed to 75% of original height at a rate of 5 cm/min. Several regions are identifiable: AB, BC, CD, DE, EF. (See Section 15.3 for details.)

- Region CD. The slope of the σ/ε curve decreases markedly. The cheese begins to fracture at C, as cracks grow and spread throughout the sample at an increasing rate. Eventually, at D, the rate of collapse of the stress-supporting paracasein matrix overtakes the build-up of stress within the matrix through further compression, and a peak stress, denoted the *fracture stress*, is reached. The fracture stress, σ_f, and strain, ε_f, are measures of the stress and strain, respectively, required to cause complete fracture of the sample into several individual pieces. If the compression is stopped at C, the sample may recover partially and may still be in the form of one mass.
- Region DE. The stress decreases with further compression, reflected by the negative slope of the curve, owing to the collapse of the stress-bearing structure. The sample fractures into pieces that spread, to a greater or lesser degree, over the base plate of the instrument. Hence, the force per unit surface area decreases.
- Region EF. The stress increases as the cross-head begins to compress the fragmented pieces of cheese. The stress at the end of the compression (point F) is a measure of firmness, as judged in the first bite during mastication.

The various quantities obtained from the force-displacement curve and their interpretation are given in Table 13–1.

13.3.3 Relationships Between Rheological Quantities and Cheese Characteristics

The relationships between the outputs from various rheological measurements (e.g., curves from creep, recovery, and large-scale compression tests) and rheological characteristics, as denoted by textural descriptors used in practice, are shown in Table 13–2. An increase in E implies that the cheese is more elastic, springy, or rubbery at low σ. Likewise, a cheese that displays a high σ_f may be referred to as tough, rubbery, firm, or strong. Conversely, a cheese with

Table 13–1 Rheological Parameters Derived from a Stress-Strain Curve Obtained from a Force Compression Test

Rheological Quantity	*Abbreviation*	*Interpretation*	*Rheological Characteristic to Which Quantity Is Related*
Compression modulus	E	Measure of true elasticity at a low strain	Degree of elasticity or springiness, rubberiness
Fracture stress (yield stress)	σ_f	Stress required to cause fracture and collapse the cheese matrix beyond point of recovery	Strength, toughness, brittleness
Fracture strain (compression strain, fracture strain, yield displacement)	ε_f	Strain at which cheese collapses completely	Shortness, longness, crumbliness
Maximum strain (maximum stress, maximum load)	σ_{max}	Stress required to reach a given deformation	Firmness, hardness

Note: Refer to Figure 13–15 for details.

a low σ_f could be described as soft or weak-bodied. Cheeses that exhibit a high ε_f are frequently referred to as being "long," while those with low ε_f are described as being "short" or brittle. A cheese that requires a high σ to achieve a given ε is described as firm.

13.3.4 Factors That Influence the Rheological Characteristics of Cheese as Measured Using Force-Compression Tests

The exact shape of the curve and the magnitude of the various rheological quantities obtained from force-deformation curves depend on test conditions and cheese variety, composition, and degree of maturity (Culioli & Sherman, 1976; Dickinson & Goulding, 1980; Prentice et al., 1993; Shama & Sherman, 1973; Sherman, 1988; Vernon Carter & Sherman, 1978; Visser, 1991).

Test Conditions

Increasing the compression or strain rate (i.e., speed of the cross-head) in the range 5–100 cm/min results in progressive increases in E (σ_f), the stress required to achieve a given level of compression and firmness (Figure 13–16). Moreover, the magnitude of these increases is variety dependent. For example, σ_f increases more rapidly with the compression rate for Cheddar than for Cheshire (Figure 13–17). Hence, comparative studies on two or more cheeses can produce incorrect conclusions if the compressions are performed at different compression rates. However, if the effect of the strain rate for different varieties under standard conditions (e.g., temperature and sample dimensions) is known, then it is possible to compare results obtained at different strain rates by adjusting the results to a common strain rate. The strain rate and the level

Table 13–2 Rheological Properties of Raw Cheese and Their Definitions, Showing the Relationship to Rheological Quantities as Measured Instrumentally

Rheological Property	*Definition*	*Cheese Type Displaying Property*
Elasticity (rubberiness)	Tendency of cheese to recover its original shape and dimensions upon removal of an applied stress	Swiss-type cheese, low-moisture Mozzarella
Springiness	Tendency to recover from large deformation (strain) after removal of deforming stress	Swiss-type cheese, low-moisture Mozzarella
Elastic fracturability	Tendency of hard cheese to crack, with very limited flow (only near the crack) of the cheese (after fracture, the broken surfaces can be fitted closely to each other)	Parmesan, Romano, Gruyère
Brittleness	Tendency of (hard) cheese to fracture at a relatively low permanent deformation	Romano, Parmesan, Gruyère
Firmness (hardness)	High resistance to deformation by applied stress	Cheddar, Swiss-type cheese, Romano, Parmesan, Gouda
Longness	The failure of cheese to fracture until a relatively large deformation is attained	Mozzarella, Swiss
Toughness (chewiness)	A high resistance to breakdown upon mastication	Mozzarella, String cheese
Softness	Low resistance to deformation by applied force	Blue cheese, Brie, Cream cheese
Plastic fracturability	The tendency of cheese to fracture accompanied by flow	Mature Cheddar, Blue cheese, Appenzeller, Chaumes, Raclette
Shortness	The tendency to plastic fracture at a small deformation; low resistance to breakdown upon mastication	Camembert, Brie
Adhesiveness (stickiness)	The tendency to resist separation from another material it contacts (e.g., another ingredient or a surface such as a knife blade or palate)	Mature Camembert
Crumbliness	The tendency to break down easily into small, irregular shaped particles (e.g., by rubbing)	Cheshire, Wensyledale, Blue cheese, Stilton, Feta
Shear thickening	The tendency to increase in apparent viscosity when subjected to an increasing shear rate (especially upon heating)	Cream cheese (when heated)
Shear thinning	The tendency to exhibit a decrease in apparent viscosity when subjected to an increasing shear rate	Quarg (especially at low temperatures, i.e., $< 4°$ C)

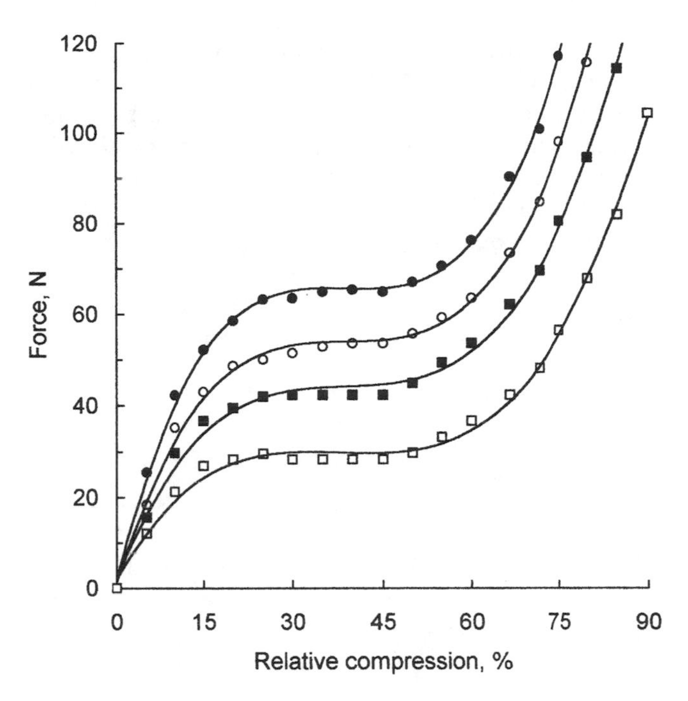

Figure 13–16 Force as a function of compression for samples of Gloucester cheese compressed at 5 (❑), 20 (■), 50 (○), and 100 (●) cm/min.

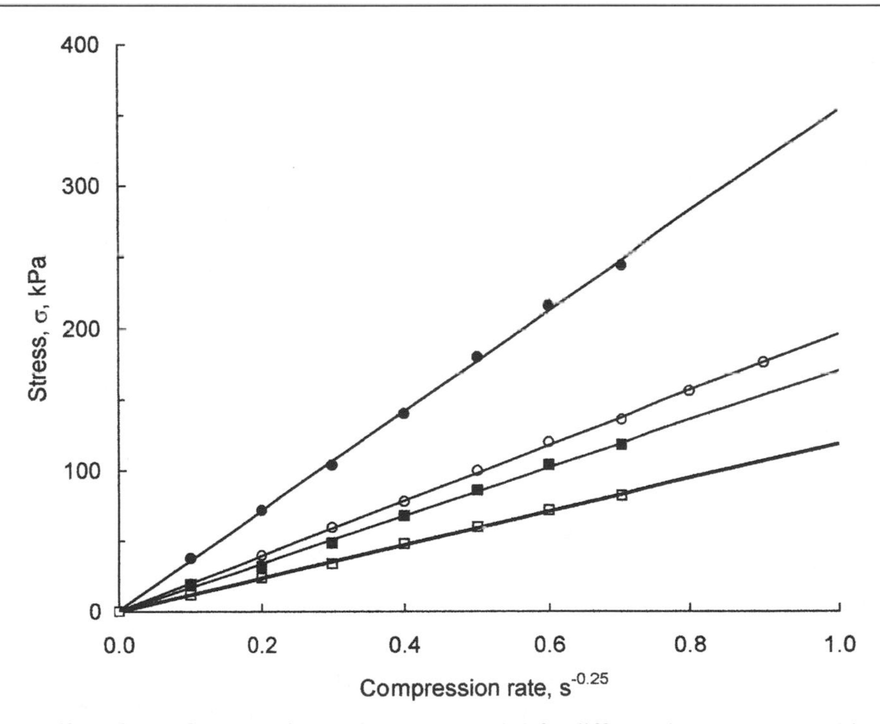

Figure 13–17 Effect of rate of compression on fracture stress (σ_f) for different cheese types: Double Gloucester (●), Cheddar (○), Leicester (■), and Cheshire (❑).

of compression that are chosen depend on what the test is designed to simulate. Comparison of instrumental force-compression data obtained at different compression rates with consumers' evaluations, both orally and using the fingers, of a wide range of solid foods suggests that low compression levels (≤ 50%) correspond to measurement of firmness or hardness by squeezing the cheese between the fingers. Higher compression levels (≥ ≈70%) correspond to firmness as judged by chewing the cheese. Evaluation of food by compressing between the fingers depends more on the elasticity of the cheese, whereas in the mouth it depends on the cumulative contributions of elasticity, fracture, and flow.

Increasing the height of cylindrical samples into the range 0.75–3.5 cm results in reductions in σ_f and the force required to achieve a given compression and an increase in ε_f (Figure

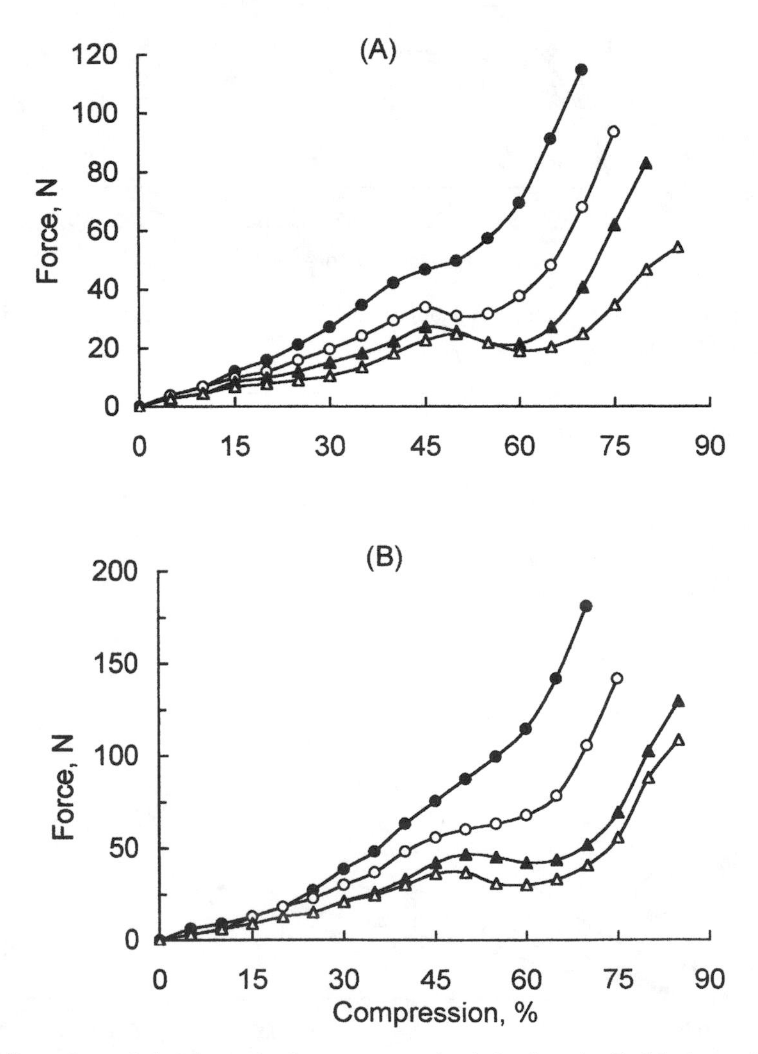

Figure 13–18 Effect of sample height on the force-compression behavior of cylindrical samples (2.5 cm diameter) of Gouda cheese compressed at 5 cm/min (A) and 50 cm/min (B). Sample height: 0.75 (●), 1.5 (○), 2.5 (▲), and 3.5 (Δ) cm.

13–18). The effects are more pronounced as the compression rate is increased into the range 5–50 cm/min. Similarly, the force required to cause a given deformation increases as the diameter:height ratio of the cylindrical samples is increased, a trend that is more accentuated with the level of compression (Figure 13–19). The influence of sample dimensions is related to surface friction (between sample and instrument plates), which is associated with barrel deformation of the cheese as it is compressed. Barrel deformation refers to the shape frequently assumed by a cylindrical sample during compression—progressive increase in diameter from the flat surfaces (in contact with the plates) to a maximum at the central region of the cylinder (Figure 13–20). Experiments with cylindrical samples of Gouda cheese (height, 2.5 cm; diameter, 2.5 cm) showed that barelling, which was first observed at 20% compression, increased with level of compression in the range 20–80%, especially at high compression rates (> 50 cm/min). Barrel deformation restricts the expansion of the flat faces of the sample upon compression and results in the development of a transverse compressive stress in addition to the axial compressive stress and thereby contributes to surface friction. In the absence of surface friction (which can be reduced to nothing by lubricating the flat surfaces of the cheese sample with mineral oil prior to compression), the surface area of the flat sections increases more than in the presence of surface friction. Hence, as the height of the sample decreases during compression, the actual surface area over which the force is applied increases in the absence of surface friction; consequently, the actual stress does not increase as rapidly (Sherman, 1988). The transverse stress

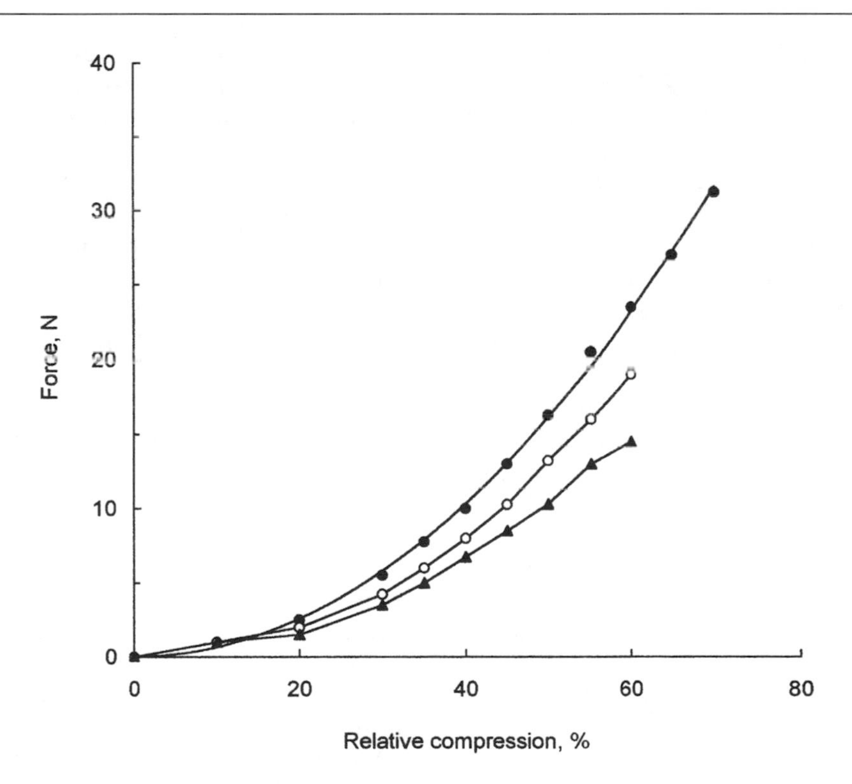

Figure 13–19 Effect of diameter:height ratio on the force-compression behavior of cylindrical samples (1.64 cm diameter) of German loaf cheese. Sample height (cm) and diameter:height ratio are 1.65 and 1.0 (▲), 1.3 and 1.25 (○), and 1.1 and 1.5 (●).

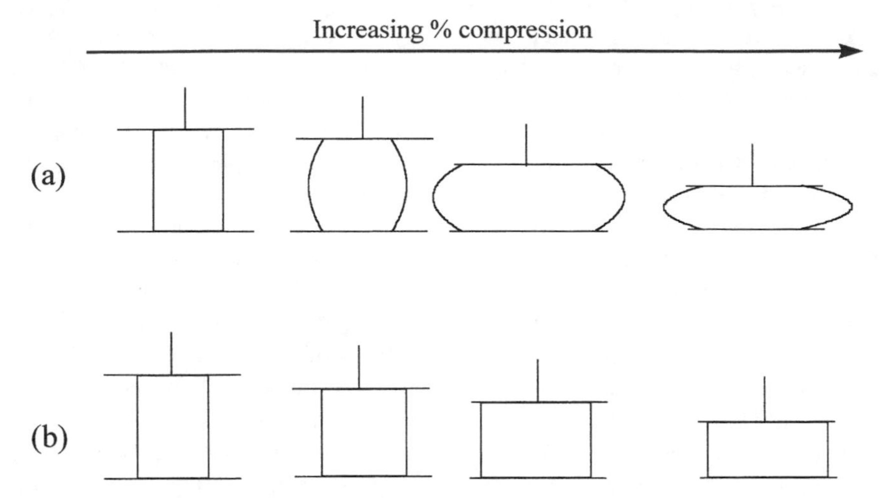

Figure 13–20 (a) Schematic representation of barrel compression of a cylindrical cheese sample during compression, showing a progressive increase in diameter from the surface to the center of the sample. Friction between the cheese and the compression plate (cross-head) prevents the cheese surface from spreading. (b) Schematic representation of ideal compression.

associated with barreling decreases with distance from the flat surface and reaches zero at a distance that is approximately equal to the sample diameter. Hence, the compressive stress decreases as the diameter:height ratio (and hence height) of cylindrical samples is reduced.

Sample shape (i.e., cubical or cylindrical) also has an effect on the shape of the force-compression behavior. Generally, cubes require a higher stress to achieve a given percent compression than cylinders of comparable volume, especially at compression levels greater than 50%. This trend may be related to frictional effects.

Increasing the sample temperature results in a marked decrease in the elastic modulus, fracture stress, and firmness (Figure 13–21). This effect is attributed to liquefaction of the fat fraction, which contributes to lubrication of fracture surfaces, and a probable reduction in the surface friction between the sample and the instrument plates as liquid fat is exuded at the flat surfaces.

Cheese Structure, Composition, and Maturity

The viscoelasticity of cheese results from the interactive rheological contributions of its individual constituents—protein, fat, and moisture. The structure, which represents the way in which the individual constituents coexist, and the physical nature of the constituents (i.e., whether solid or liquid) are major determinants.

Upon the application of a stress to a cheese product, the matrix will at first limit the deformation. As the concentration of casein in the cheese matrix increases, the intra- and inter-strand linkages become more numerous and the matrix displays greater elasticity and is more difficult to deform (Figure 13–22). Hence, reduced-fat Cheddar, which contains a higher concentration of structural matrix per unit volume than full-fat Cheddar, is firmer and has a higher σ_f than the latter (Figure 13–23). Factors that promote weakening of the casein matrix reduce the stress required to cause a given deformation. Hence, the σ_f and firmness of cheese generally decrease with ripening time owing to hydrolysis and hydration of the casein, both of which processes contribute to disintegration of the casein matrix (Figure 13–24). The structure of cheese is markedly weakened by the early hydrolysis of α_{s1}-casein by residual chymosin at the Phe_{23}-Phe_{24} bond. The sequence 1-23 of α_{s1}-casein is strongly hydrophobic and interacts with the hydrophobic regions of other α_{s1}- and β-casein

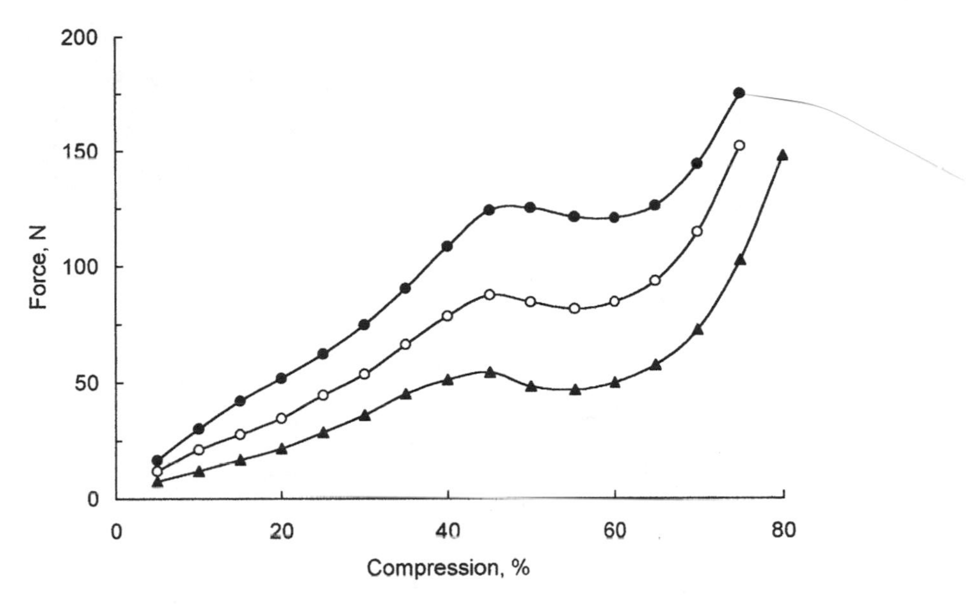

Figure 13–21 Effect of temperature on the force compression behavior of Gouda cheese compressed at 50 cm/min at 10°C (●), 15°C (○), and 20°C (▲).

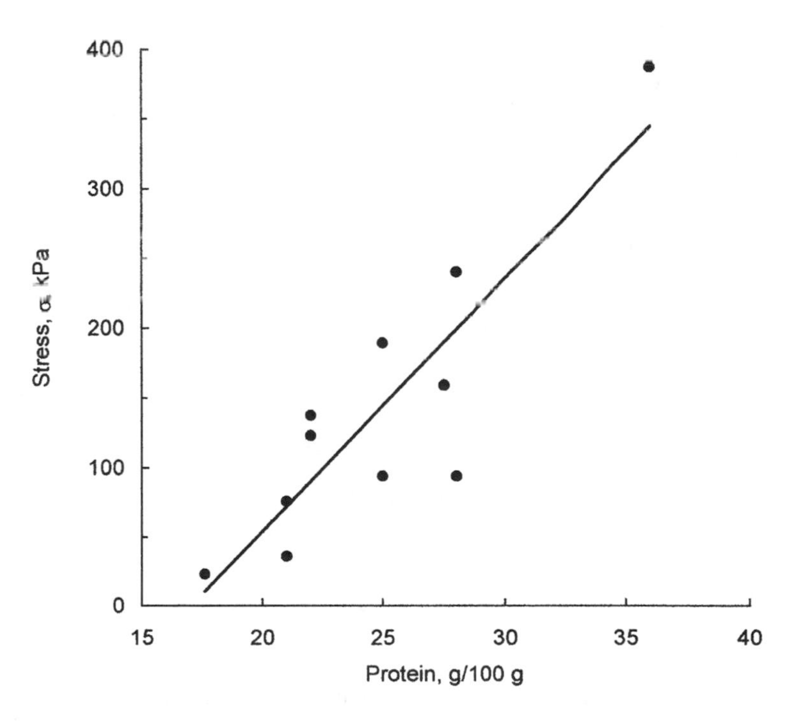

Figure 13–22 Relationship between protein content and cheese firmness. Data derived from the testing of 10 different types of hard cheese and a pasteurized processed Cheddar cheese.

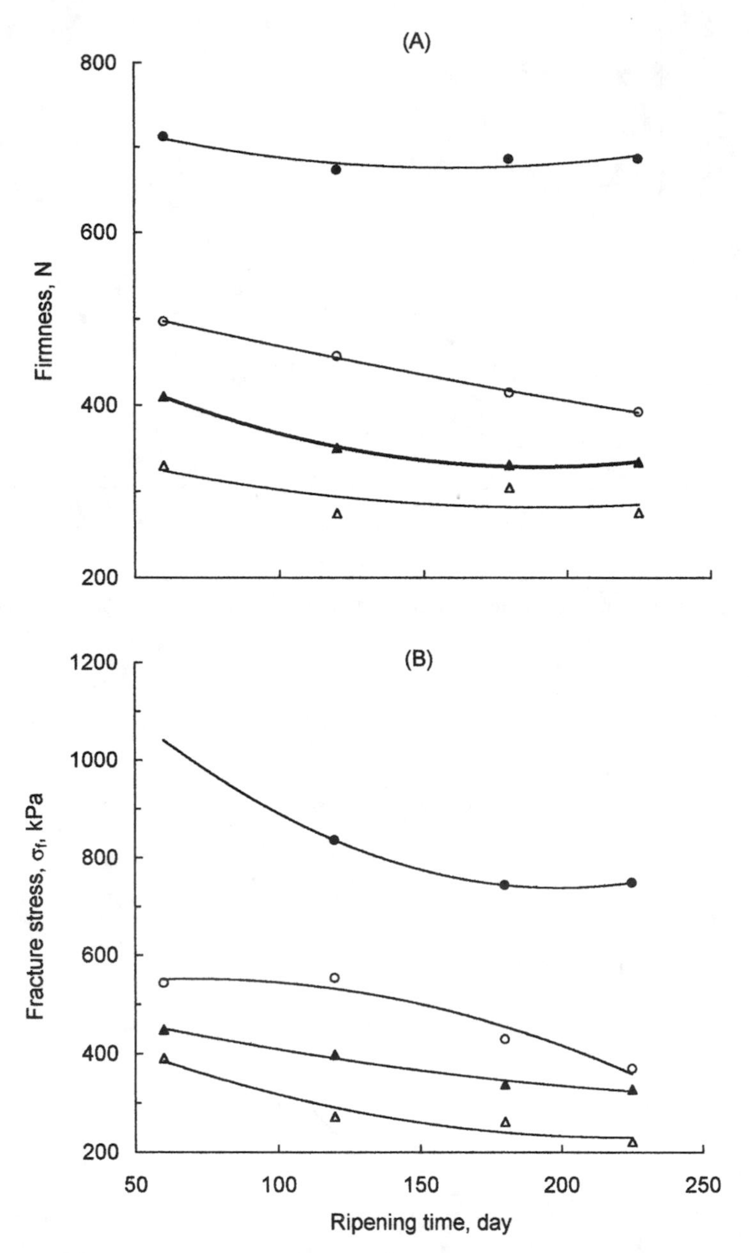

Figure 13–23 Effect of fat level [6 (●), 17 (○), 22 (▲), and 33 (Δ) g/100 g] on the firmness (A) and fracture stress (B) of Cheddar cheese compressed at 5 cm/min.

---

molecules and thus contributes to the overall continuity and integrity of the matrix (Creamer & Olson, 1982).

The contribution of fat to the rheological properties of cheese depends on its physical state and therefore on the temperature, which controls the ratio of solid fat:liquid fat. At low temperatures (< 5°C), where the milk fat is predominantly solid, the fat adds to the elasticity of the casein matrix. The solid fat globules limit defor-

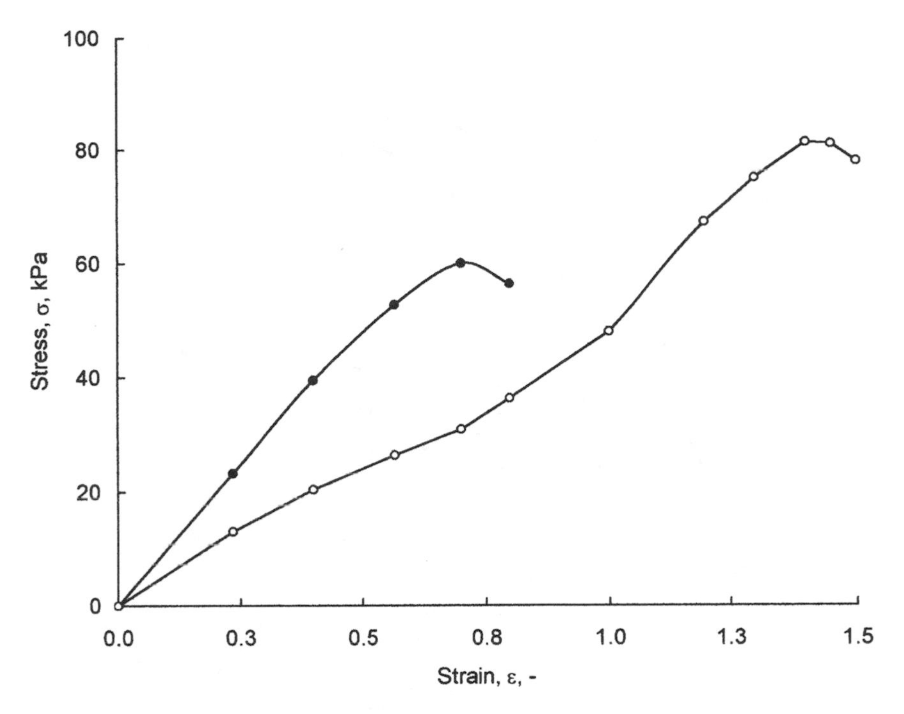

Figure 13–24 Effect of age (1 week, ○; 7.5 month, ●) on the force-compression behavior of Gouda cheese (moisture, 41.6 g/100 g).

mation of the casein matrix, as the deformation of the latter also requires deformation of the fat globules enmeshed within its pores. As the proportion of liquid fat increases, the fat behaves more like a fluid and confers viscosity rather than elasticity or rigidity on the cheese. Moreover, liquid fat acts as a lubricant on fracture surfaces of the casein matrix and thereby reduces the stress required to fracture the matrix. Hence, for a given fat content, increasing the assay temperature during compression results in a marked decrease in the elastic modulus, fracture stress, and firmness (Figure 13–21). An increase in the fat-in-dry-matter level of cheese (while retaining the other compositional parameters constant) is paralleled by a decrease in σ_f, with the effect becoming more pronounced as the temperature is increased. Generally, an increase in the fat content of cheese is accompanied by decreases in the levels of protein and moisture as well as decreases in the σ_f and firmness (Figure 13–23).

The third major component of cheese is moisture, which acts as a plasticizer in the protein matrix, thereby making it less elastic and more susceptible to fracture upon compression. Thus, increasing the moisture content of cheese results in decreases in E, σ_f, and firmness (Figure 13–25). The ε_f increases slightly with moisture content to an extent dependent on cheese pH and maturity (Creamer & Olson, 1982; Luyten, 1988).

Other compositional factors also influence the rheology of cheese. Increasing the pH into the range 5.0–5.2 reduces both E and σ_f (Figure 13–26). In contrast, increasing the pH into the range 5.2–5.6 results in a marked increase in σ_f (to values much higher than those at pH < 5.2) and a slight increase in E. The ε_f for young (1-week-old) Gouda cheese was at its maximum at pH 5.2 and decreased upon lowering or raising the pH to 4.8 and 5.6. However, the pH at which ε_f is maximal increases with ripening time (e.g.,

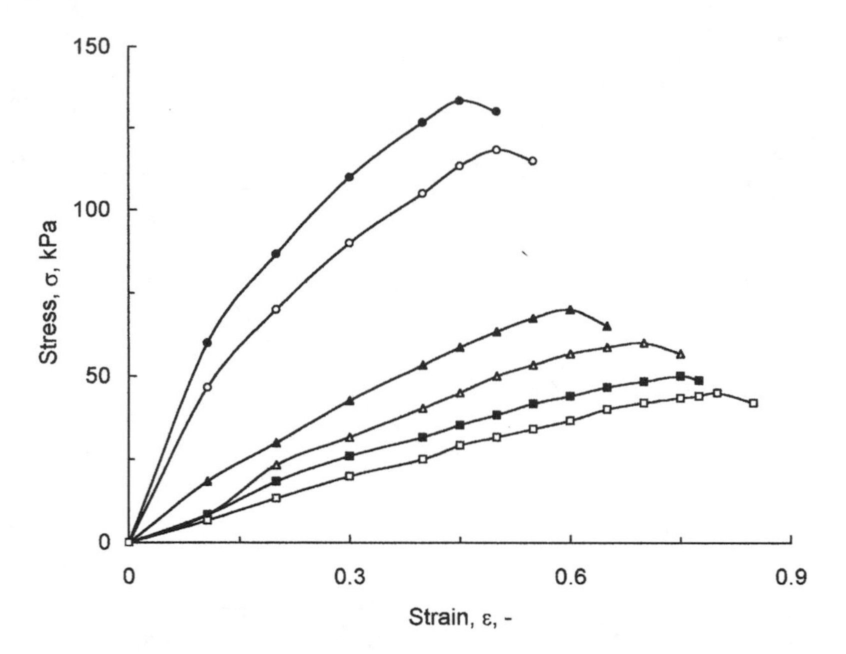

Figure 13–25 Effect of moisture content [32.3 (●), 35.6 (○), 38.6 (▲), 41.4 (Δ), 43.9 (■), and 46.2 (□) g/100 g] on the force-compression behavior of 7.5-month-old Gouda-type cheese.

from ≈ 5.2 for 1-week-old Gouda to ≈ 5.4 for 3-month-old Gouda cheese). Differences in pH may help to explain the different rheological characteristics exhibited by some common cheeses upon compression. Low pH cheeses (e.g., Cheshire and Feta) generally tend to have low σ_f and ε_f values and to crumble into many pieces upon fracturing, whereas relatively high pH cheeses (e.g., Emmental and Gouda) exhibit higher σ_f and ε_f values and tend to fracture into larger pieces (Prentice et al., 1993). The effect of pH probably ensues from its influence on

- the ratio of soluble calcium to colloidal calcium
- the degree of paracasein hydration, which is maximal at about pH 5.2
- the types of intra- and interaggregate bonds

Moreover, the effect of pH appears to be related to other factors, such as the levels of moisture and salt in the cheese and the degree of proteolysis (Walstra and van Vliet, 1982).

The salt content of rennet-curd cheeses varies from about 0.79 g/100 g in Emmental to about 6 g/100 g in Feta; however, because salt is dissolved in the moisture phase, the effective concentration is much higher (≈ 2 and 12 g/100 g, respectively). Increasing the salt-in-moisture in Gouda-type cheese into the range 0–12 g/100 g while maintaining the other compositional parameters relatively constant is associated with increases in E and σ_f (which is relatively constant in the range 3–7 g/100 g S/M) (Figure 13–27, graph A). The fracture strain, ε_f, increases slightly to a maximum at an S/M of 4 g/100 g, then decreases sharply to about half its maximum value, and thereafter plateaus at this value for S/M in the range 5–12 g/100 g (Figure 13–27, graph B).

The rheological characteristics of cheese may also be influenced indirectly by many other factors that influence its composition and/or structure, some of which are listed below:

- Seasonal changes in milk composition and the state of its components. These changes are associated with the stage of lactation, quality of the cows' diet, health status of the cows, and on-farm husbandry practices.

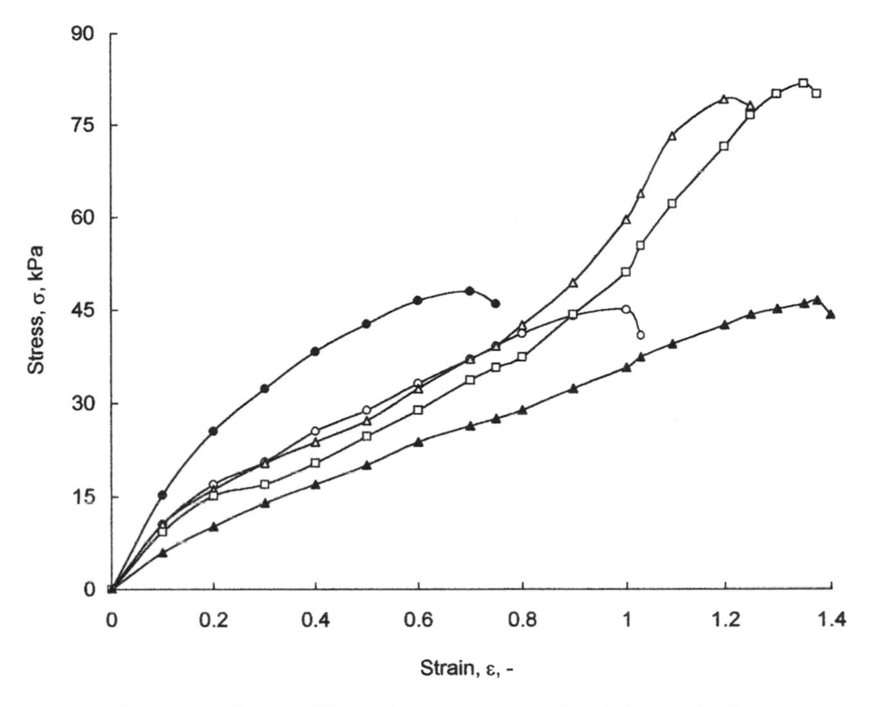

Figure 13–26 Effect of pH [5.02 (●), 5.1 (○), 5.2 (▲), 5.43 (□), and 5.58 (Δ)] on the force-compression behavior of 1-week-old Gouda cheese when composition was otherwise similar.

Hence, poor-quality, late lactation milk from cows on a poor-quality diet often results in cheese with a high-moisture content and low firmness.

- Cheesemaking conditions. For a given cheese variety, composition may vary with cheesemaking conditions, especially if the milk supply is predominantly from spring-calving cows fed on pasture. In the latter situation, the level of casein in milk may vary markedly throughout the cheesemaking season (e.g., in Ireland from ≈ 22 g/kg in March to 31 g/kg in October). Hence, when rennet and starter are added on a volume basis rather than on a casein basis, and certain cheesemaking steps (e.g., setting, cutting, cut programs) are undertaken on the basis of time rather than on the basis of defined criteria (e.g., pH and gel firmness), variations in cheese composition (especially moisture content) may be expected. Increasing the casein content of milk (e.g., by ultrafiltration) enhances its curd-forming characteristics and reduces cheese moisture. Hence, the fracture stress of Cheddar cheese increases linearly with increasing protein content of cheese milk (Figure 13–28).
- Genetic variants of milk protein, such as κ-casein AA or BB variants, owing to their effect on cheese composition (Walsh et al., 1998).
- Cheese-ripening conditions (e.g., temperature and relative humidity, which influence parameters such as the rate of casein degradation and/or moisture loss).

13.4 CHEESE TEXTURE

Cheese texture is a sensory characteristic and therefore is directly measurable only by sensory analysis. Texture perceptions arise from a com-

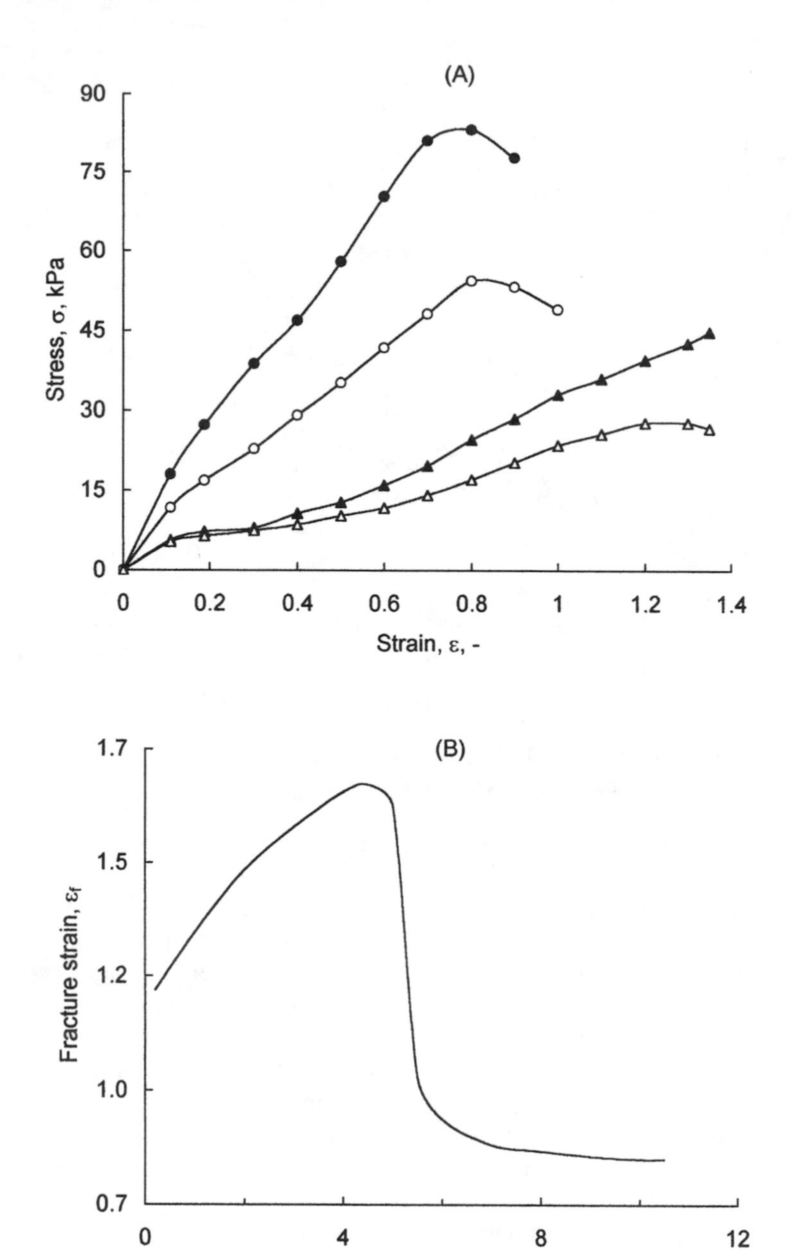

Figure 13–27 Effect of salt-in-moisture level [0.4 (Δ), 3.3 (▲), 7.3 (○), 11.3 (●) g/100 g] on the force compression-behavior (A) and fracture strain (B) of Gouda cheese when composition was otherwise similar.

plex array of sensory inputs that occur both prior to and during food consumption. Consumption of a piece of cheese involves the following series of events:

- Visual assessment (e.g., for eye distribution in a Swiss-type cheese, granularity of Cottage cheese, whiteness of Feta cheese). The visual perception may create an important

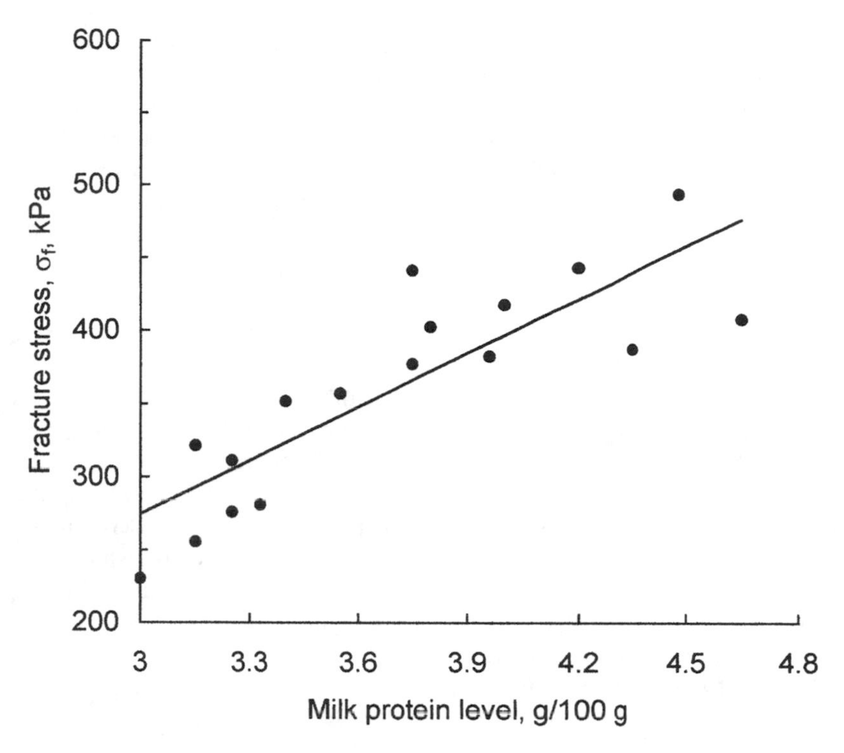

Figure 13–28 Effect of milk protein level on the fracture stress of 6-month-old Cheddar cheese.

first impression about the anticipated taste and texture of the cheese. For instance, the granular, dry (nonglossy) appearance of Parmesan may create the perception that it will likely have a dry, grainy mouthfeel; the surface sheen of a freshly sliced Camembert, or its absence, may be taken as evidence of its state of maturity, flavor, and palatability.

- Assessment by touch (e.g., assessment of the resistance of a piece of String cheese to touch or that of a piece of Cheddar or Camembert when it is cut with a knife or punctured by a fork).
- Eating, which occurs in four phases:
  1. Placement in mouth, including contact with nerve endings on the tongue and cheeeks that contribute to sensations collectively referred to as *somaesthesis* (e.g., sensations of touch, pain, warmth, and cold). The ingested cheese is compressed by the various parts of the oral cavity (palate and inside of lips and cheeks), and concomitantly the counter-stress exerted by the cheese is detected by nerve endings.
  2. An initial bite by the teeth (resistance to cutting by the incisors may be involved).
  3. Chewing and mastication. The cheese is compressed repeatedly, mainly between the molars, moved toward the teeth by the tongue, and mixed with saliva. This results in diminution of the cheese to a pasty bolus ready for swallowing. Evidence suggests that the tooth socket is elastic, permitting vertical and horizontal movement of the teeth (Brennan, 1988). The degree of deformation of the periodontal membrane (the structure that surrounds the tooth in the socket and at-

taches it to the jaw bone) is the stimulus for detection of the mechanical properties of the cheese being eaten.
4. Swallowing of the cheese, during which the bolus is discharged out of the oral cavity.

In the process of eating, the cheese is subjected to cutting, shearing, and compression forces that fracture and reduce it to a state ready for swallowing. The objective of sensory texture evaluation is to translate the cumulative sensations perceived into descriptors or scores.

13.4.1 Sensory Measurement of Cheese Texture

The sensory methods employed to measure food texture are of three general types (Jack, Paterson, & Piggott, 1995; Powers, 1984):

1. attribute (or profiling) methods
2. difference methods
3. preference methods

Attribute methods, which are used most widely include

- the texture profile method (Brandt et al., 1963)
- descriptive analysis
- free choice profiling

In the texture profile method, three categories of texture characteristics were proposed by Szczesniak (1963): (1) mechanical characteristics, which are related to the reaction of food to stress; (2) geometrical characteristics, which are related to size, shape, and orientation of particles within the food; and (3) other characteristics, which are related to the perception of the levels of moisture and fat in the food. The mechanical characteristics in turn were divided into primary characteristics, such as hardness, cohesiveness, viscosity, elasticity, and adhesiveness, and secondary characteristics, such as brittleness, chewiness, and gumminess. The definitions of the mechanical characteristics are summarized in Exhibit 13–1. Each of the above attributes is scored on a nine-point equidistant scale relative to a standard reference cheese.

Descriptive textural analysis involves the scoring of samples in terms of intensity using a fixed consensus vocabulary of textural descriptors that are agreed upon jointly by the sensory panel and the trainers during training sessions. Standard products (of the same generic type as the product being evaluated) considered to exhibit one or more particular descriptors may be used to define characteristics, with the panel members being repeatedly exposed to the standard products during training.

The free choice profiling method of texture evaluation is similar to descriptive analysis, except that the sensory descriptors are proposed by the individual assessor, who quantifies the intensity of a particular attribute by assigning a score on a line scale.

13.4.2 Instrumental Methods for Evaluation of Cheese Texture

Cheese texture, being a sensory property, is ultimately expressed in sensory terms or descriptors. However, trained texture panels may be difficult and costly to establish and maintain. Moreover, instrumental methods are easier to perform, standardize, and reproduce and require the involvement of fewer trained people. Hence, much research effort has focused on the measurement of textural properties using instrumental methods. Textural evaluation by instrumental methods is generally based on force-compression tests that are designed to simulate the compression of cheese between the molars during chewing. The first apparatus of this type, developed for foods in general, was the forced-compression test using the General Foods Texturometer (Bourne, 1978; Friedman, Whitney, & Szczesniak, 1963), which was designed to simulate the compression of food between the teeth (Figure 13–29). Essentially, a food sample is loaded onto a plate attached to a beam and then subjected to a deforming force provided by a tooth-shaped plunger moved by a wheel device in a motion designed to simulate the vertical action of the jaw. When the plunger deforms the

Exhibit 13–1 Definitions of Mechanical Properties of Cheese Using the General Foods Texture Profile.

Mechanical Characteristics

Primary properties

Hardness:	The force required to compress a cheese between the molar teeth (e.g., hard and semi-hard cheese) or between the tongue and palate (e.g., for soft spreadable cheese and process cheese spreads) to a given deformation or to the point of penetration.
Cohesiveness:	The extent to which a cheese can be deformed before it ruptures.
Springiness (elasticity):	The degree of recovery of a deformed (strained) piece of cheese after the deforming force is removed.
Adhesiveness:	The force required to remove cheese that adheres to the mouth (generally the palate) during the normal eating process.

Secondary properties

Fracturability (brittleness):	The force at which a cheese crumbles, cracks, or shatters when deformed. Fracturability is the result of a high degree of hardness and a low degree of adhesiveness.
Chewiness:	The length of time or the number of chews required to masticate a cheese to a state ready for swallowing. Chewiness is the product of hardness, cohesiveness, and springiness.
Gumminess:	A denseness that persists throughout mastication; energy required to disintegrate a piece of cheese to a state ready for swallowing. Gumminess is a product of a low degree of hardness and a high degree of cohesiveness.

sample, strain gauges attached to the sample-holding beam detect the movement of the beam, and a force-time trace is recorded. The sample is subjected to two successive deformations (referred to as *bites*) so as to simulate more closely repetitive chewing action (Figure 13–30). The relationships between the trace and textural descriptors, as described by Szczesniak (1963), are outlined in Exhibit 13–2.

More recently developed instruments (e.g., the Stevens Compressor Response Analyzer, Instron Universal Testing Machine, and TA.XT2 and TA.HD Texture Analyzers) also use the principle of compression for textural evaluation. In addition, these instruments have a range of attachments (e.g., cones, needles, knives, and plungers) that facilitate the determination of the response of the cheese to other forces, such as cutting, at various deformation rates.

Effective simulation of sensory texture evaluation by instrumental compression tests necessitates test conditions (e.g., compression rates) that resemble those during consumption and that are carefully standardized between samples (e.g., temperature, sample size and shape, and size and shape of the compression plates). Research has revealed that most people compress food to around 70% and chew at a rate of 40–80 masticatory strokes per minute, indicating that the duration of the downstroke in a bite is around 5 s, though it can be more or less depending on the food (Sherman, 1988). Yet in many studies that attempted to relate instrumental and sensory evaluations of textural attributes such as firmness or brittleness, lower levels of compression and compression rates were used. As discussed in Section 13.3.4, the quantities derived from compression tests are markedly influenced by compression rate.

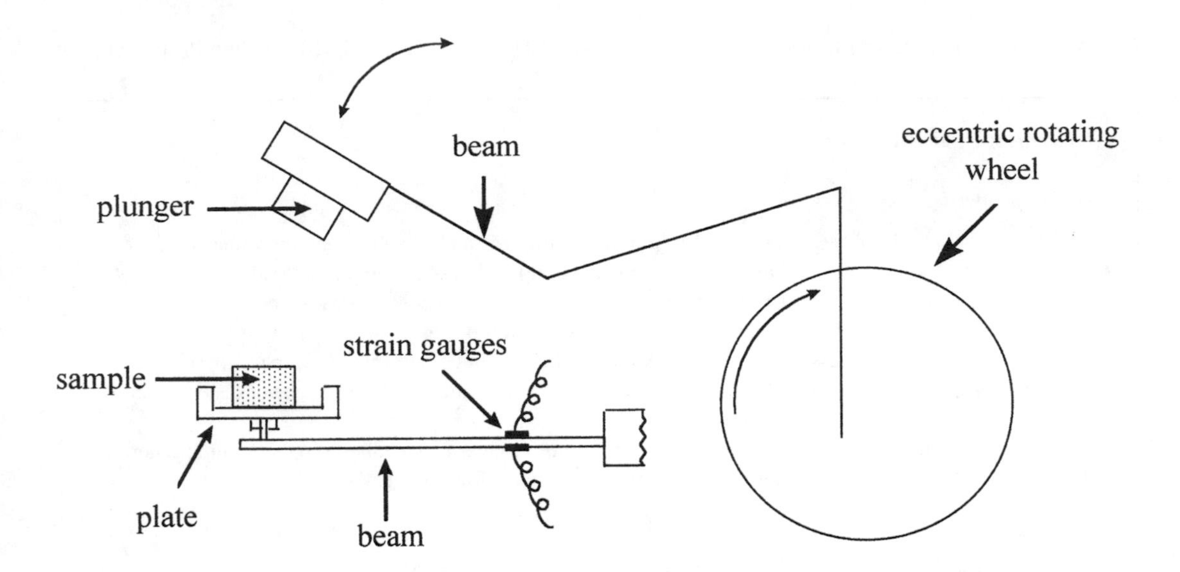

Figure 13–29 Schematic representation of the structure of the General Foods Texturometer. The various terms are discussed in Section 13.4.

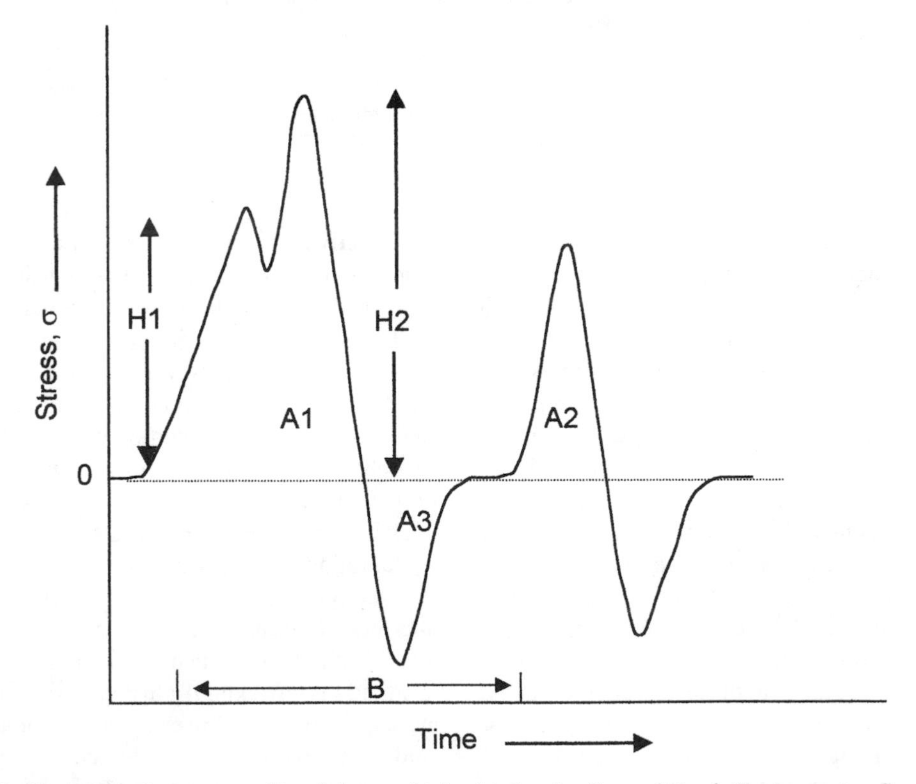

Figure 13–30 A typical texture profile of cheese obtained using the General Foods Texturometer. See Exhibit 13–2 for interpretation of curve.

Exhibit 13–2 Interpretation of the Force-Compression Curve from the General Foods Texturometer

Fracturability:	Height of the first peak (H1) in the first bite (A1).
Hardness:	Height of the second peak (H2) in the first bite (A1).
Cohesiveness:	Ratio of area on second bite to that on first bite (A2/A1).
Adhesiveness:	Area (A3) of the negative peak formed when the plunger is pulled from the sample after first bite, due to cheese adhering to the plunger.
Springiness:	Difference between distance B (measured from the initial point of contact of the plunger with the sample in bite 1 to contact with the sample in bite 2) and distance C (the same measurement made on a completely inelastic material such as clay) (B – C).
Chewiness:	Hardness × cohesiveness × springiness (A1 × [A2/A1] × [B– C]).
Gumminess:	Hardness × cohesiveness × 100 (A1 × [A2/A1] × 100).

13.4.3 Comparison of Instrumental and Sensory Evaluation of Texture

Numerous comparative studies on the evaluation of texture by compression and sensory methods have been undertaken (Green, Marshall, & Brooker, 1985; Jack et al., 1993; Lee, Imoto, & Rha, 1978). Owing to differences in test conditions (temperature, strain rate, level of compression) and sample dimensions, the results from the different studies have varied, including the degree of correlation between the sensory parameters (e.g., firmness and springiness) and the instrumental quantities. However, good correlations have been reported, in general, for hardness, chewiness, adhesiveness, and springiness (Table 13–3).

Table 13–3 Relationships between Textured Parameters Obtained from Sensory Evaluation and Rheological Parameters from a Universal Instron Testing Machine for a Range of Different Cheese Types

	Sensory Parameter			
Rheological Parameters	*Hardness*	*Chewiness*	*Springiness*	*Adhesiveness*
Compression stress (firmness)	0.95*	0.86*	0.27*	0.59
Fracture stress	0.70*	0.82*	0.80*	–0.80*
Elastic recovery	0.17	0.36	0.97*	–0.27
Cohesiveness	0.87*	0.80*	0.21	0.44
Adhesiveness	0.70*	0.82*	0.80*	–0.80*

Note: Cheese types with a wide variety of textures were used: Cream cheese, Camembert, Mozzarella, Muenster, processed Cheddar, Swiss cheese, and Cheddar cheeses of different degrees of maturity. Cheese samples (height, 1.0 cm; diameter, 1.5 cm) were compressed to 20% of original height (80% deformation) at a cross-head speed of 50 mm/min at room temperature, with 2 consecutive cycles applied to the cheese sample (e.g., as in Figure 13–30). The rheological parameters are defined as follows: compression stress is the stress required for a given deformation; elastic recovery is the recovery in the height of the cheese cylinder after the first bite and before the second bite (equivalent to distance B in Figure 13–30); cohesiveness is the ratio of the areas peak 1 to peak 2 (A1/A2); and adhesive force is the force exerted on the ascending motion of the cross-head after the first bite.

* $p < 0.01$.

REFERENCES

Bourne, M.C. (1978). Texture profile analysis. *Food Technology, 32*(2), 63–66, 72.

Brandt, M.A., Skinner, E.Z., & and Coleman, J.A. (1963). Texture profile method. *Journal of Food Science. 4,* 404–409.

Brennan, J.G. (1988). Texture perception and measurement. In J.R. Piggott (Ed.), *Sensory analysis of foods* (2d ed.). London: Elsevier Applied Science.

Creamer, L.K., & Olson, N.F. (1982). Rheological evaluation of maturing Cheddar cheese. *Journal of Food Science, 47,* 631–636, 646.

Culioli, J., & Sherman, P. (1976). Evaluation of Gouda cheese firmness by compression tests. *Journal of Texture Studies, 7,* 353–372.

Dickinson, E., & Goulding, I.C. (1980). Yield behavior of crumbly English cheeses in compression. *Journal of Texture Studies, 11,* 51–63.

Fox, P.F., O'Connor, T.P., McSweeney, P., Guinee, T.P., & O'Brien, N. (1996). Cheese, physical, biochemical and nutritional aspects. *Advances in Food and Nutrition Research, 39,* 163–329.

Friedman, H.H., Whitney, J.E., & Szczesniak, A.S. (1963). The texturometer: A new instrument for objective texture measurement. *Journal of Food Science, 28,* 390–396.

Green, M.L., Marshall, R.J., & Brooker, B.E. (1985). Instrumental and sensory texture assessment of Cheddar and Cheshire cheeses. *Journal of Texture Studies, 16,* 351–364.

Guinee, T.P., Mulholland, E.O., Mullins, C., & Corcoran, M.O. (1997). Functionality of low-moisture Mozzarella cheese during ripening. In T.M. Cogan, P.F. Fox, & P. Ross (Eds.), *Proceedings of the 5th Cheese Symposium.* Dublin: Teagasc.

Jack, F.R., Paterson, A., & Piggott, J.R. (1993). Relationships between rheology and composition of Cheddar cheeses and texture as perceived by consumers. *International Journal of Food Science and Technology, 28,* 293–302.

Jack, F.R., Paterson, A., & Piggott, J.R. (1995). Perceived texture: Direct and indirect methods for use in product development. *International Journal of Food Science and Technology, 30,* 1–12.

Kindstedt, P.S. (1995). Factors affecting the functional characteristics of unmelted and melted cheese. In E.L. Malin & M.H. Tunick (Eds.), *Chemistry of structure-function relationships in cheese.* New York: Plenum Press.

Lee, C.-H., Imoto, E.M., & Rha, C. (1978). Evaluation of cheese texture. *Journal of Food Science, 43,* 1600–1605.

Luyten, H. (1988). *The rheological and fracture properties of gouda cheese.* Unpublished doctoral dissertation, Wageningen Agricultural University, Wageningen, The Netherlands.

Powers, J.M. (1984). Current practices and applications of descriptive methods. In R.J. Piggott (Ed.), *Sensory analysis of foods.* London: Elsevier Applied Science.

Prentice, J.H., Langley, K.R., & Marshall, R.J. (1993). Cheese rheology. In P.F. Fox (Ed.), *Cheese: Chemistry, physics and microbiology* (2d ed., Vol. 1). London: Chapman & Hall.

Rao, M.O. (1992). Classification, description and measurement of viscoelastic properties of solid foods. In M.O. Rao & J.F. Steffe (Eds.), *Viscoelastic properties of foods.* London: Elsevier Applied Science.

Shama, F., & Sherman, P. (1973). Evaluation of some textural properties of foods with the Instron Universal Testing Machine. *Journal of Texture Studies, 4,* 344–353.

Sherman, P. (1969). A texture profile of foodstuffs based upon well-defined rheological properties. *Journal of Food Science, 34,* 458–462.

Sherman, P. (1988). Rheological evaluation of the textural properties of foods. *Progress and Trends in Rheology, 11,* 44–53.

Szczesniak, A.S. (1963). Classification of the textural characteristics. *Journal of Food Science, 28,* 385–389.

van Vliet, T. (1991). Terminology to be used in cheese rheology. In *Rheological and fracture properties of cheese* [Bulletin No. 268]. Brussels: International Dairy Federation.

Vernon Carter, E.J., & Sherman, P. (1978). Evaluation of the firmness of Leicester cheese by compression tests with the Universal Testing Machine. *Journal of Texture Studies, 9,* 311–324.

Visser, J. (1991). Factors affecting the rheological and fracture properties of hard and semi-hard cheese. In *Rheological and fracture properties of cheese* [Bulletin No. 268]. Brussels: International Dairy Federation.

Walsh, C.D., Guinee, T.P., Reville, W.D., Harrington, D., Murphy, J.J., O'Kennedy, B.T., & Fitzgerald, R.F. (1998). Influence of κ-casein genetic variant on rennet gel microstructure, Cheddar cheesemaking properties and casein micelle size. *International Dairy Journal, 8,* 707–714.

Walstra, P., & van Vliet, T. (1982). Rheology of cheese. In Bulletin No. 153. Brussels: International Dairy Federation.

Chapter 14

Factors That Affect Cheese Quality

14.1 INTRODUCTION

As discussed in Chapter 11, the ripening of cheese, and hence its quality, is due to the activity of microorganisms and enzymes from four or five sources. Therefore, it seems like it should be possible to produce premium-quality cheese consistently by controlling these agents. However, in spite of considerable research and quality control efforts, it is not yet possible to do so.

A very wide and diverse range of factors interact to affect the composition of cheese curd and hence the quality of the final cheese; we have attempted to summarize these in Figure 14–1. Some of these factors or agents can be manipulated easily and precisely, whereas others are more difficult, or perhaps impossible, to control. It should be possible to apply the principles of hazard analysis critical control point (HACCP) to cheese production. However, the precise influence of many of the factors included in Figure 14–1 on cheese ripening and quality are not known precisely, and many of the factors are interactive. The interactions between the principal factors that affect cheese quality were reviewed by Lawrence and Gilles (1980). Figure 14–1, which is more comprehensive than the scheme presented by Lawrence and Gilles (1980), is not presented as definitive, but our hope is that it may stimulate others to apply HACCP principles to cheesemaking.

14.2 MILK SUPPLY

It is well recognized that the quality of the milk supply has a major impact on the quality of the resultant cheese. Three aspects of quality must be considered: microbiological, enzymatic, and chemical.

14.2.1 Microbiology

In countries with a developed dairy industry, the quality of the milk supply has improved markedly during the past 30 years. Total bacterial counts are now usually below 20,000 cfu/ml ex-farm. The total bacterial count probably increases during transport and storage at the factory, but growth can be minimized by thermization (65°C × 15 s) of the milk supply, which is standard practice in some countries (see Chapter 4).

Although many cheeses are made from raw milk, most cheese is made from milk pasteurized at or close to 72°C × 15 s. If produced from good-quality raw milk and subsequently handled under hygienic conditions, pasteurized milk should have a very low total bacterial count and therefore represents a very uniform raw material from a microbiological viewpoint.

14.2.2 Indigenous Enzymes

Milk contains as many as 60 indigenous enzymes (see Andrews et al., 1992), but the signifi-

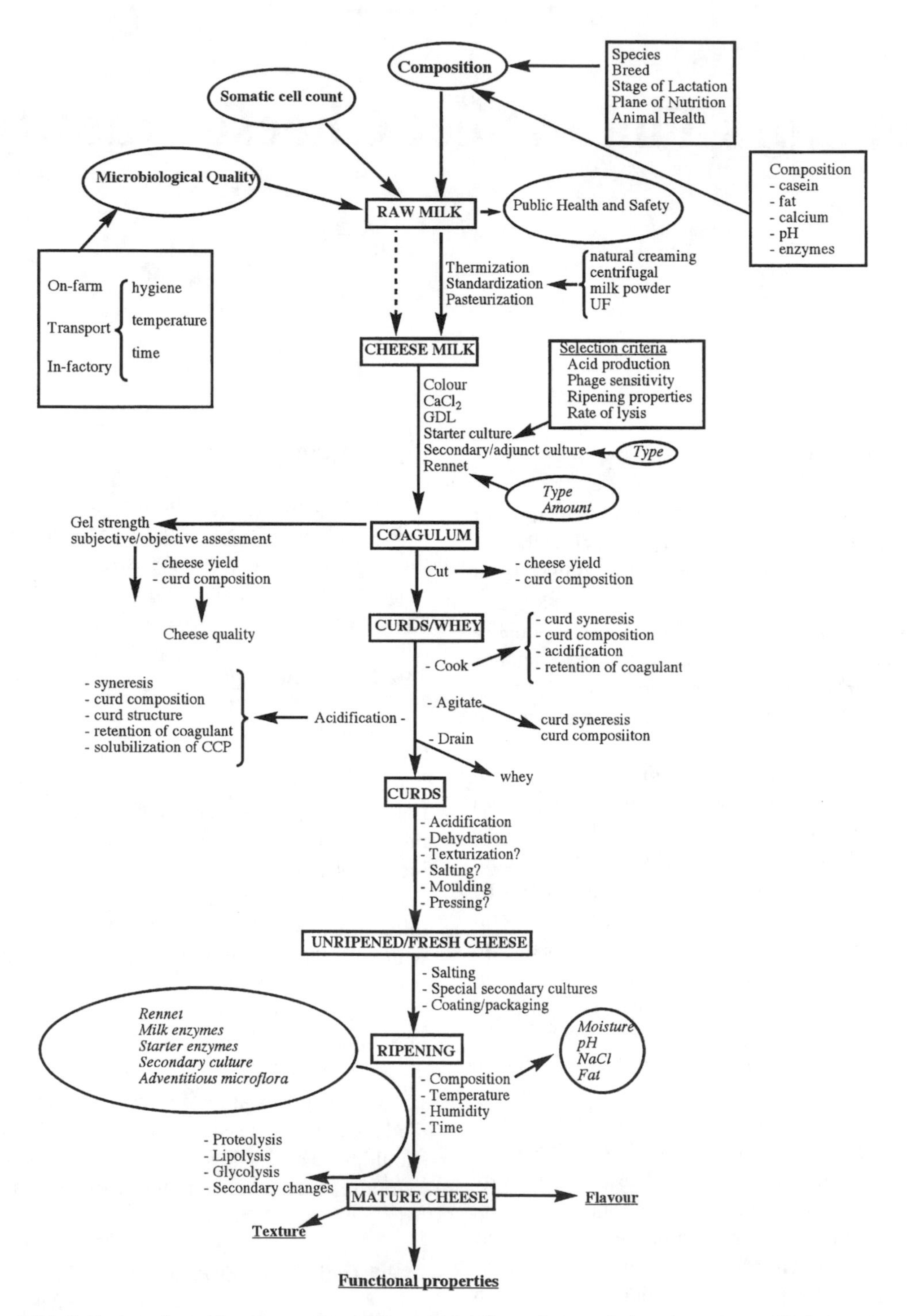

Figure 14–1 Factors that affect the quality of cheese. The figure is intended to show the multiplicity of factors that impact, directly or indirectly, on the quality of rennet-coagulated cheeses. It is possible to standardize and control many of the factors involved. Knowledge of the factors that affect cheese quality should enable a HACCP approach to be applied to cheese production.

cance of these to cheese quality has not yet been researched adequately. Several of these enzymes have the potential to affect cheese quality, especially lipase, proteinase(s), acid phosphatase, and perhaps xanthine oxidase, sulfydryl oxidase, lactoperoxidase, and γ-glutamyl transpeptidase. Many of these survive high-temperature, short-time (HTST; 72°C × 15 s) pasteurization to a greater or lesser extent, and at least some (e.g., plasmin, acid phosphatase, and xanthine oxidase) are active during cheese ripening (see Chapter 11).

Somatic cells are an important source of enzymes, particularly proteinases, in milk. Somatic cell count (SCC) is negatively correlated with cheese yield (see Chapter 9) and quality. A SCC of less than 300,000/ml is recommended.

Although precise information is lacking, it is unlikely that indigenous milk enzymes are a major cause of variability in cheese quality. Some of these enzymes contribute to cheese ripening and may contribute to the superior quality of raw milk cheese, a possibility that warrants investigation.

14.2.3 Chemical Composition

The chemical composition of milk, especially the concentrations of casein, fat, calcium, and pH, has a major influence on several aspects of cheese manufacture, especially rennet coagulability, gel strength, curd syneresis, and hence cheese composition and cheese yield. When seasonal milk production is practiced, as in New Zealand, Ireland, and Australia, milk composition varies widely, and there is some variability even with random calving patterns due to nutritional factors. It is possible to reduce, but not eliminate, the variability in the principal milk constituents by standardizing the concentrations of fat and casein, not just the ratio (protein content can be standardized by adding UF retentate), the pH (using gluconic acid-δ-lactone), and calcium content (by adding $CaCl_2$), as discussed in Chapter 2.

14.3 COAGULANT (RENNET)

It is generally accepted that calf chymosin produces the best-quality cheese. An adequate supply of chymosin from genetically engineered microorganisms is now available (although its use is not permitted in all countries), and therefore rennet quality should not be a cause of variability in cheese quality.

As discussed in Chapter 11, the proportion of added rennet retained in cheese curd varies with rennet type, cook temperature, and pH at draining. These variables should be standardized if cheese of consistent quality is to be produced. Increased retention of the coagulant in the curd results in greater initial hydrolysis of α_{s1}-casein, although this does not appear to be reflected in sensory assessment of cheese texture and flavor. It has been suggested that the activity of chymosin in cheese curd is the limiting factor in cheese ripening. However, excessive rennet activity leads to bitterness. There have been relatively few studies on how chymosin activity affects cheese quality, an issue that appears to warrant further research.

14.4 STARTER

Since the starter plays a key role in cheese manufacture and ripening, it seems that differences between the enzyme profiles of starter strains should affect cheese quality. Modern single-strain starters produce acid very reproducibly and, if properly managed, show good phage resistance. *Lactococcus* strains have been selected mainly on the basis of acid-producing ability, phage resistance, and compatibility. Based on pilot-scale studies and commercial experience, strains that produce unsatisfactory, especially bitter, cheese have been identified and excluded from commercial usage. However, systematic studies on strains with positive cheesemaking attributes are lacking. This probably reflects the lack of information on precisely what attributes of a starter are desirable from a flavor-generating viewpoint. Studies on genetically engineered strains that superproduce proteinase and/or the general aminopeptidase PepN showed that cheese quality was not improved, although proteolysis was accelerated. Since all lactococcal enzymes, except the cell wall–associated proteinase, are intracellular, the cells must lyse before these enzymes can participate in ripening. Therefore, the rate of lysis of *Lactococcus* strains is being studied with the objective of selecting strains with improved cheesemaking properties.

Sulfur compounds have long been considered as contributors to the flavor of many cheese varieties. Some strains of *Lc. lactis* spp. *cremoris* (but not *Lc. lactis* spp. *lactis*) can absorb glutathione (γ-Glu.Cys.Gly; GSH) from the growth medium. Release of GSH into the cheese upon cell lysis may affect the redox potential (E_h) of cheese and hence the concentration of thiol compounds. Comparative cheesemaking studies using starter strains that accumulate glutathione and those that do not are warranted.

Although considerable information is available on the individual enzymes of *Lactococcus* and, to a lesser extent, of *Lactobacillus*, especially on the glycolytic and proteolytic systems, there have been few studies on the proportions of different enzyme activities in starter strains. There have been even fewer studies on the relationship between different starter enzyme profiles and cheese quality. It would appear to be highly desirable that studies should be undertaken to relate cheese quality to the natural enzyme profile of starter strains or genetically engineered starters. The availability of starter strains deficient in or overproducing one or more enzymes will facilitate such studies.

It is very likely that the desirable cheesemaking properties of starters are due to a balance between certain, perhaps secondary, enzymatic activities that have not yet been identified.

14.5 NONSTARTER LACTIC ACID BACTERIA (NSLAB)

The significance of lactobacilli for Cheddar and Dutch cheese quality is controversial (see Chapters 11 and 15). Many researchers consider their contribution to be negative (in the Netherlands, a maximum of 2×10^6 NSLAB/g is specified for Gouda). Although there are several studies on controlled microflora cheeses, we are not aware of studies in which cheese free of NSLAB was compared with "control" cheeses containing "wild" NSLAB. Several comparative studies on cheese made under aseptic or nonaseptic conditions using *Lactococcus* starter alone or with selected *Lactobacillus* adjuncts indicate that inoculation of cheese milk with selected strains of *Lactobacillus* improves cheese flavor and possibly accelerates ripening. Thermophilic *Lactobacillus* spp. are more effective as adjuncts than mesophilic lactobacilli, probably because they die rapidly in cheese, lyse, and release intracellular enzymes. Both mesophilic and thermophilic lactobacilli and *Sc. thermophilus* are being used commercially as adjunct cultures for Cheddar cheese and possibly for other varieties.

Since the numbers and strains of NSLAB in cheese are uncontrolled, it is likely that they contribute to variability in cheese quality. It is impossible to eliminate NSLAB completely, even under experimental conditions. Therefore, it appears worthwhile to determine what factors affect their growth. The number of NSLAB in Cheddar is strongly influenced by the rate at which the curd is cooled and subsequently ripened. Rapid cooling of the curd after molding is the most effective way of retarding the growth of NSLAB, although they will grow eventually to about 10^7 cfu/g. The growth of NSLAB can be prevented by ripening at about 1°C, but all ripening reactions are retarded. The growth of NSLAB does not appear to be influenced by the composition of cheese (moisture, salt, or pH) within the ranges normally found in commercial cheese.

NSLAB grow mainly after the lactose has been metabolized by residual starter activity. Although the growth substrates in cheese for *Lactobacillus* are not known, it is likely that they are limited (NSLAB normally plateau at about 10^7 cfu/g), and hence it might be possible to outcompete wild NSLAB by adding selected strains of *Lactobacillus* to cheese milk, thereby offering better control of the ripening process. NSLAB may also be controlled by including a broad spectrum bacteriocin-producing strain in the starter culture.

14.6 CHEESE COMPOSITION

The quality of cheese is influenced by its composition, especially moisture content, NaCl concentration (preferably expressed as salt-in-moisture [S/M]), pH, moisture in nonfat substances (MNFS; essentially the ratio of protein to moisture), and percentage fat in dry matter (FDM). At least five studies (Fox, 1975; Gilles & Lawrence, 1973; Lelievre & Gilles, 1982; O'Connor, 1971; Pearce & Gilles, 1979) have attempted to relate the quality of Cheddar cheese to its composition. While these authors agree that moisture content, S/M, and pH are the key determinants of cheese quality, they disagree about the relative importance of these parameters (see Figure 14–2).

O'Connor (1971) found that flavor, texture, and total score were not correlated with moisture content but were significantly correlated with the percentage of NaCl and particularly with pH. Salt content and pH were themselves strongly correlated with each other, as were salt and moisture.

Based on the results of a study on experimental and commercial cheeses in New Zealand, Gilles and Lawrence (1973) proposed a grading (selection) scheme that has since been applied commercially in New Zealand for young (14-day-old) Cheddar cheese. The standards prescribed for premium grade were pH, 4.95–5.10; S/M, 4.0–6.02%; MNFS, 52–56%; and FDM, 52–55%. The corresponding values for first-grade cheeses were 4.85–5.20%, 2.5–6%, 50–57%, and 50–56%. Young cheese with a composition outside these ranges was considered unlikely to develop into good-quality mature cheese. Quite wide ranges of FDM are acceptable. Lawrence and Gilles (1980) suggested that since relatively little lipolysis occurs in Cheddar cheese, fat content plays a minor role in determining cheese quality but if FDM falls below about 48%, the cheese is noticeably firmer and less attractive in flavor. Pearce and Gilles (1979) found that the grade of young (14-day-old) cheeses produced at the New Zealand Dairy Research Institute was most highly correlated with moisture content. The optimum compositional ranges were MNFS, 52–54%; S/M, 4.2–5.2%; and pH, 4.95–5.15.

Fox (1975) reported weak correlations between grade and moisture and between salt and pH for Irish Cheddar cheeses, but a high percentage of cheeses with compositional extremes were downgraded, especially those with low salt (< 1.4%), high moisture (> 38%), or high pH (> pH 5.4). Salt concentration seemed to exercise the strongest influence on cheese quality, and the lowest percentage of downgraded cheeses can be expected in the salt range 1.6–1.8% (4.0–4.9% S/M). Apart from the upper extremes, pH and moisture appear to exercise little influence on quality. High salt levels tend to cause curdy textures, probably due to insufficient proteolysis. A pasty body, often accompanied by off-flavors, is associated with low salt and high moisture levels. In the same study, the composition of extra mature cheeses was found to vary less, and the mean moisture content was 1% lower than that of regular cheeses.

A very extensive study of the relationship between the composition and quality of nearly 10,000 cheeses produced at five commercial New Zealand factories was reported by Lelievre and Gilles (1982). As in previous studies, considerable compositional variation was evident, but the variation was less for some factories than others. While the precise relationship between quality and composition varied between plants, certain generalizations emerged:

- Within the compositional range suggested by Gilles and Lawrence (1973) for premium quality cheese, composition does not have a decisive influence on grade, which decreases outside this range.
- Composition alone does not provide a basis for grading as currently acceptable in New Zealand.
- MNFS was again found to be the principal factor affecting quality.
- Within the recommended compositional bands, grades declined marginally as MNFS increased from 51% to 55% and

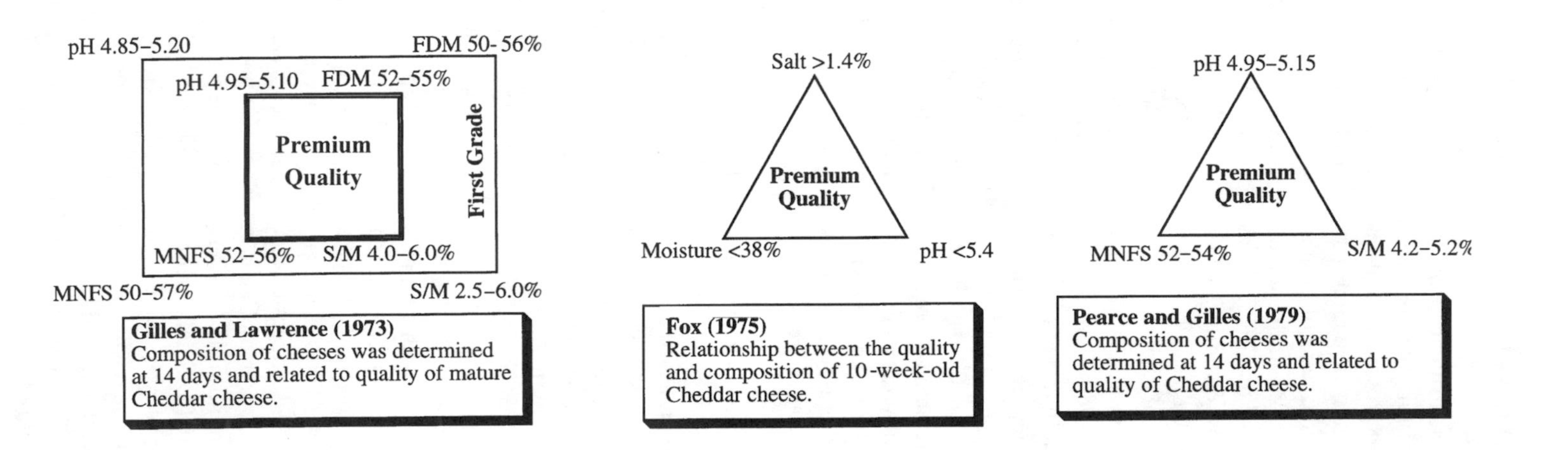

Figure 14–2 Relationships between composition (determined at various stages during ripening) and the quality of mature Cheddar cheese (moisture-in-nonfat substances [MNFS]; fat-in-dry-matter [FDM], and salt-in-moisture [S/M]).

> they increased slightly as S/M decreased from 6 to 4. pH had no consistent effect within the range 4.9–5.2, and FDM had no influence in the range 50–57%.

The authors stress that, because specific interplant relationships exist between grade and composition, each plant should determine the compositional parameters that are optimum for itself.

The results of the foregoing investigations indicate that high values for moisture and pH and a low salt level lead to flavor and textural defects. The desired ranges suggested by Gilles and Lawrence (1973) appear to be reasonable, at least for New Zealand conditions, but within the prescribed zones composition is not a good predictor of Cheddar cheese quality. Presumably, several other factors—such as microflora, activity of indigenous milk enzymes, relatively small variations in cheese composition, and probably other unknown factors—influence cheese quality but become dominant only under conditions where the principal determinants (moisture, salt, and pH) are within appropriate limits.

Although the role of calcium concentration in cheese quality has received occasional mention, its significance was largely overlooked until the work of Lawrence and Gilles (1980), who pointed out that the concentration of calcium in cheese curd determines the cheese matrix and, together with pH, indicates whether proper procedures were used to manufacture a specific cheese variety. As the pH decreases during cheese manufacture, colloidal calcium phosphate dissolves and is removed in the whey. The whey removed after cooking comprises 90–95% of the total whey lost during cheesemaking, and this whey contains, under normal conditions, about 85% of the calcium and about 90% of the phosphorus lost from the cheese curd. Thus, the calcium content of cheese reflects the pH of the curd at whey drainage. There are strong correlations between the calcium content of cheese and the pH at 1 day and between pH at 14 days and the amount of starter used (see Lawrence, Heap, & Gilles, 1984). Since the pH of cheese increases during ripening, the pH of mature cheese may be a poor index of the pH of the young cheese. Therefore, calcium concentration is probably a better record of the history of a cheese with respect to the rate of acidification than the final pH. Reduction in calcium phosphate concentration by excessively rapid acid development also reduces the buffering capacity of cheese, and hence the pH of the cheese will fall to a lower value for any particular level of acid development. Unfortunately, no recent work on the level and significance of calcium in Cheddar cheese appears to be available.

14.7 RIPENING TEMPERATURE

The final factor known to influence the rate of ripening and cheese quality is ripening temperature. Ripening at an elevated temperature is normally done with the objective of accelerating ripening, but it also affects cheese quality. The literature on the accelerated ripening of cheese is discussed in Chapter 15.

14.8 CONCLUSION

Through increased knowledge of the chemistry, biochemistry, and microbiology of cheese, it is now possible to consistently produce cheese of an acceptable quality, although this acceptability is not always achieved, owing to failure to control one or more of the key parameters that affect cheese composition and ripening. Milk is a variable raw material, and although it is possible to eliminate major variations in the principal milk constituents, some variation persists. Variability in milk composition can also be compensated for by manipulating some process parameters in the cheesemaking process. Most large factories operate on a strict time schedule, and hence subtle process manipulation on an individual vat basis may not be possible. Therefore, strict control of milk composition and starter activity are critical.

From a microbiological viewpoint, the milk supplied to modern cheese factories is of very high quality and, after pasteurization, is essentially free of bacteria. In modern factories, where enclosed vats and other equipment is used, the level of contamination from the environment is very low. Cheese curd containing fewer than 10^3 NSLAB/g at day 1 is normal. However, these adventitious NSLAB grow to about 10^7 cfu/g and dominate the microflora of long-ripened cheese. Since the adventitious NSLAB grow slowly in cheese, they are most significant in long-ripened cheese. Although the significance of the adventitious NSLAB in long-ripened cheese is unclear, it would appear to be desirable to control them, either by eliminating them or standardizing them. It is not possible to eliminate NSLAB, even in cheese made on a pilot scale under aseptic conditions. Their growth can be prevented by ripening at about 1°C, but the overall ripening process then becomes unacceptably lengthened. Outcompeting indigenous NSLAB by an adjunct *Lactobacillus* culture, which does not have to contribute to ripening, is a possibility, but this approach has not been investigated.

Although it is now possible to avoid major defects in cheese produced using modern technology, further research on the biochemistry of cheese ripening is required to enable the process of cheese manufacture and ripening to be refined to an extent that will allow the consistent production of premium quality cheese.

The key to successful cheesemaking is a good reliable starter, both from the viewpoint of reproducible acid production and subsequent ripening. If properly managed, modern starters are generally satisfactory, and their performance is being improved progressively.

The use of starter adjuncts, usually mesophilic lactobacilli, for some varieties, especially Cheddar, is increasing, with the objective of intensifying and modifying cheese flavor, accelerating ripening, and perhaps controlling adventitious NSLAB and thus standardizing quality.

REFERENCES

Andrews, A.T., Olivecrona, T., Vilaro, S., Bengtsson-Olivecrona, G., Fox, P.F., Bjorck, L., & Farkye, N.Y. (1992). Indigenous enzymes in milk. In P.F. Fox (Ed.), *Advanced dairy chemistry* (Vol. 1). London: Elsevier Applied Science.

Fox, P.F. (1975). Influence of cheese composition on quality. *Irish Journal of Agricultural Research, 14*, 33–42.

Gilles, J., & Lawrence, R.C. (1973). The assessment of cheese quality by compositional analysis. *New Zealand Journal of Dairy Science and Technology, 8*, 148–151.

Lawrence, R.C., & Gilles, J. (1980). The assessment of the potential quality of young Cheddar cheese. *New Zealand Journal of Dairy Science and Technology, 15*, 1–12.

Lawrence, R.C., Heap, H.A., & Gilles, J. (1984). A controlled approach to cheese technology. *Journal of Dairy Science, 67*, 1632–1645.

Lelievre, J., & Gilles, J. (1982). The relationship between the grade (product value) and composition of young commercial Cheddar cheese. *New Zealand Journal of Dairy Science and Technology, 49*, 1098–1101.

O'Connor, C.B. (1971). Composition and quality of some commercial Cheddar cheese. *Irish Agricultural and Creamery Review, 26*(10), 5–6.

Pearce, K.N., & Gilles, J. (1979). Composition and grade of Cheddar cheese manufactured over three seasons. *New Zealand Journal of Dairy Science and Technology, 14*, 63–71.

Chapter 15

Acceleration of Cheese Ripening

15.1 INTRODUCTION

The original objective of cheese manufacture was to conserve the principal nutrients in milk (i.e., lipids and proteins). This was achieved by a combination of acidification, dehydration, low redox potential, and salting. Although a few minor cheese varieties are dehydrated sufficiently or contain a sufficiently high level of NaCl to prevent microbiological and/or enzymatic changes during storage, the composition of most varieties permits biological and enzymatic activity to occur, which causes numerous changes during storage. These changes are referred to as ripening (maturation) and were described in Chapter 11.

Cheese ripening can be a slow process, ranging from about 3 weeks for Mozzarella to 2 or more years for Parmesan and extra mature Cheddar. In general, the rate of ripening is directly related to the moisture content of the cheese, with low-moisture cheeses ripening more slowly. A time-consuming process, cheese ripening is also an expensive one, owing to inventory costs, the need for controlled-atmosphere ripening rooms, and the risk of defects. It is estimated that the cost of ripening cheese is about $100 per tonne per month. Cheese ripening is also a rather uncontrolled process, although increased knowledge of the microbiology and biochemistry of ripening has increased the probability of producing a good-quality cheese. Consequently, there are economic and technological incentives for accelerating the cheese-ripening process, provided that the flavor and textural properties of the cheese are not altered.

As discussed in Chapter 11, glycolysis of the residual lactose in cheese curd is complete within, at most, a few weeks. The racemization of L-lactic acid to DL-lactic acid, as occurs in Cheddar and Dutch-type cheeses, has no effect on the flavor or texture of cheese, while its catabolism in mold-ripened and Swiss-type cheeses occurs relatively quickly—within a few weeks. Hence, it is not necessary to accelerate the metabolism of lactose or lactic acid.

Only limited lipolysis occurs in most cheese varieties. Major exceptions are Blue and some Italian varieties, such as Romano and Provalone. Blue cheeses ripen relatively rapidly (less than 4 months), and the principal lipase is that secreted by *P. roqueforti*. The characteristic flavor of Blue cheeses is due to methyl ketones, produced by β-oxidation of free fatty acids by the mold. Lipolysis in the above Italian cheeses is due mainly to an exogenous lipase, pregastric esterase, added to the cheese milk. Consequently, the rate and extent of lipolysis can be readily altered if desired.

Proteolysis occurs in all cheese varieties, ranging from limited, such as Mozzarella, to very extensive, such as Blue, Parmesan, and extra mature Cheddar. Proteolysis is largely responsible for the textural changes in most varieties. It also makes a direct contribution to flavor (e.g., peptides and amino acids), produces substrates (amino acids) for the generation of sapid compounds (e.g., amines, acids, thiols, and thioesters), and facilitates the release of sapid compounds from the cheese mass during mastication. Proteolysis is perhaps the most important reaction during cheese ripening, but in Blue and Italian varieties lipolysis and fatty acid oxidation are also very important. Proteolysis appears to be rate limiting in the maturation of most cheese varieties and hence has been the focus of most research on the acceleration of ripening, which is most pertinent for low-moisture, slow-ripening varieties (most published work has been on Cheddar). Techniques for the acceleration of ripening are also applicable to low-fat cheeses, which ripen more slowly than their full-fat counterparts.

Thus, the objective of accelerating cheese ripening should be to accelerate the proteolytic process and related events that occur in naturally ripened cheese as closely as possible. As described in Chapter 11, proteolysis in Cheddar and related cheeses is now fairly well understood at the molecular level, and hence it should be feasible to accelerate the process. Several approaches have been adopted. An extensive literature on the acceleration of cheese ripening has accumulated and has been the subject of several reviews (see Fox et al., 1996; Wilkinson, 1993).

The methods used to accelerate ripening fall into six categories:

1. elevated ripening temperature
2. exogenous enzymes
3. chemically or physically modified bacterial cells
4. genetically modified starters
5. adjunct cultures
6. cheese slurries and enzyme-modified cheeses

These methods, each of which has advantages and limitations (Table 15–1), aim to accelerate ripening either by increasing the level of putative key enzymes or by making the conditions under which the "indigenous" enzymes in cheese operate more favorable for their activity.

15.2 ELEVATED TEMPERATURE

Traditionally, cheese was ripened in caves or cellars, probably at 15–20°C for much of the year. Since the introduction of mechanical refrigeration for cheese-ripening rooms in the 1940s, the use of a controlled ripening temperature has become normal practice in modern factories. Typical ripening temperatures are as follows: Emmental, 22–24°C (for part of ripening, i.e., the critical "hot room" period); mold and smear-ripened cheeses, 12–15°C; Dutch varieties, 12–14°C; and Cheddar, 6–8°C (the ripening temperature for Cheddar is exceptionally low). The ripening temperature for most varieties is profiled. The above temperatures are the "maximum" in the profiles and are usually maintained for 4–6 weeks, usually to induce the growth of a desired secondary microflora, after which the cheese is transferred to a much lower temperature (e.g., 4°C). Again, Cheddar is an exception, since it is normally kept at 6–8°C throughout the ripening process.

About 20°C is probably the upper limit for cheese ripening. Above this temperature, the cheese is very soft and deforms readily. Exudation of fat and excessive evaporation of moisture may also occur in cheeses that are not film wrapped (some brine-salted varieties). Thus, the scope for accelerating the ripening of most cheese varieties by increasing the ripening temperature is quite limited. However, this approach has potential for Cheddar and offers the simplest and cheapest method for accelerating ripening: no additional costs are involved (indeed, savings may accrue from reduced refrigeration costs), and there are no legal barriers. However, considering the numerous complex biochemical reactions that occur during ripening, it is unlikely that all reactions will be accelerated equally at elevated temperatures, and unbalanced flavor or off-flavors may result. We are not aware of data on the effect of ripening temperature on the relative rates of individual reactions during ripening. The growth of nonstarter bacteria is accelerated at elevated temperatures, but at least in the case of Cheddar, the final number of nonstarter lactic

Table 15–1 Principal Methods Used To Accelerate Cheese Ripening

Method	*Advantages*	*Disadvantages*
Elevated temperature	No legal barriers; technically simple; no cost (perhaps saving)	Nonspecific action; increased risk of spoilage
Exogenous enzymes	Low cost; specific action, choice of flavor options	Limited choice of useful enzymes; possible legal barriers; difficult to ensure uniform incorporation of enzymes; risk of overripening
Modified starter cells	Easy to incorporate; natural enzyme balance retained	Technically complex; rather expensive
Genetically engineered starters	Easy to incorporate; choice of options	Possible legal barriers; may experience consumer resistance
Cheese slurries and enzyme-modified cheese	Very rapid flavor development	High risk of microbial spoilage; final product requires processing

acid bacteria (NSLAB) is independent of ripening temperature, at least up to 20°C, provided that composition and microbiological status are satisfactory. Furthermore, the species of NSLAB present may depend on the ripening temperature, but we are not aware of data on this, apart from the fact that the growth of heterofermentative lactobacilli is promoted by high temperatures during the early phase of ripening.

At least three studies on the ripening of Cheddar cheese at elevated temperatures (up to 20°C), alone or in combination with other agents (e.g., exogenous proteinases and/or lactose-negative starter supplements), have been published (see Fox et al., 1996). These studies agree that ripening at an elevated temperature is the single most effective method for accelerating ripening. For example, the duration of ripening can be reduced by 50% by increasing the ripening temperature from 6°C to 13°C, without adverse effects. The highest temperature that can be used continuously is about 16°C, although 20°C could be used for a short period; 12–14°C is probably optimal. Ripening can be accelerated or delayed by raising or reducing the temperature at any stage during the process, thus enabling the supply of mature cheese to be regulated. Cheese intended for ripening at an elevated temperature should be of good chemical composition (cheese with a high pH, low NaCl, or high moisture content is unsuitable) and have a low initial NSLAB count (< 10^3 cfu/g at the end of manufacture has been suggested), which can be controlled by cooling the cheese curd rapidly after pressing.

The quality of Cheddar cheese produced in modern factories is, in most cases, sufficiently high to withstand ripening at 12–14°C without problems. Except in cases where it is desired to retard ripening, for whatever reason, there is no justification for ripening at a temperature as low as 2°C, which is sometimes used. The ripening of all hard and semi-hard cheeses could probably be conducted successfully at 12–14°C. A temperature of 16°C was used successfully to accelerate the ripening of Manchego cheese.

15.3 EXOGENOUS ENZYMES

Since enzymes are directly responsible for most of the changes that occur during ripening, it seems reasonable to assume that ripening could be accelerated by adding exogenous enzymes to the cheese curd. A number of options have been investigated.

15.3.1 Coagulant

Since the coagulant is principally responsible for primary proteolysis in most cheese varieties (see Chapter 11), it might be expected that ripening could be accelerated by increasing the level or activity of rennet in the cheese curd. This can be done by increasing the level of rennet added to the cheese milk (which would necessitate process modification), renneting at a lower pH (the proportion of gastric and fermentation rennets retained in the curd increases as the pH of the milk is reduced), or adding powdered rennet to the cheese curd (e.g., at salting, in the case of Cheddar). Although chymosin appears to be the limiting proteolytic agent in the production of soluble N in cheese, increasing the level of rennet in cheese curd does not accelerate ripening and may cause bitterness. Large chymosin-produced polypeptides accumulate in cheese during ripening, but many of the smaller peptides are further hydrolyzed by microbial proteinases and/or peptidases, and these may in fact be rate-limiting with respect to flavor development. Therefore, it might be necessary to increase the starter population or its proteolytic activity to exploit the benefits that might accrue from higher rennet activity. As far as we are aware, this approach has not been investigated.

The natural function of chymosin is to coagulate milk in the neonatal stomach, thereby increasing the efficiency of digestion. It is fortuitous that chymosin is not only the most efficient milk coagulant but also gives best results in cheese ripening. However, it seems probable that the efficiency of chymosin in cheese ripening could be improved by protein engineering. As discussed in Chapter 6, some modified chymosins have been produced through genetic engineering, but the cheesemaking properties of these have not been studied to date.

As discussed in Chapter 11, chymosin has very little activity on β-casein in cheese, probably because the principal chymosin-susceptible bond in β-casein, Leu_{192}-Tyr_{193}, is in the hydrophobic C-terminal region of the molecule, which is inaccessible in β-casein because of its hydrophobicity. However, *C. parasitica* proteinase preferentially hydrolyzes β-casein in cheese without causing flavor defects (perhaps its preferred cleavage sites are in the hydrophilic N-terminal region, but its specificity is unknown). A rennet containing chymosin and *C. parasitica* proteinase might be useful for accelerating ripening, but, as far as we are aware, this has not been investigated.

15.3.2 Plasmin

Plasmin contributes to proteolysis in cheese, especially in high-cooked varieties in which chymosin is extensively or totally inactivated (see Chapter 11). Plasmin is associated with the casein micelles in milk, which can bind at least 10 times the amount of plasmin normally present in milk, and is totally and uniformly incorporated into cheese curd, thus overcoming one of the major problems encountered with the use of exogenous enzymes to accelerate cheese ripening.

Addition of exogenous plasmin to cheese milk accelerates the ripening of Cheddar cheese made from that milk, without off-flavors. At present, plasmin is too expensive for use in cheese on a commercial scale. Perhaps the gene for plasmin can be cloned in and expressed by a suitable bacterial host, which would reduce its cost. It may also be possible to clone the plasmin gene in *Lactococcus*, which could be used as a starter that would lyse and release plasmin during ripening. A nonstarter mesophilic *Lactobacillus* might also be a suitable host, but these bacteria lyse to only a limited extent during cheese ripening, and therefore the plasmin would either have to be excreted or extracellularly located.

Since milk normally contains four times as much plasminogen as plasmin, an alternative strategy might be to activate indigenous plasminogen by adding a plasminogen activator (e.g., urokinase) that also associates with the casein micelles. Recent work in the authors' laboratory has shown that this approach does work, but it may be too expensive for commercial use.

Preliminary reports have indicated that trypsin, which is relatively cheap and readily available and has a specificity similar to that of plasmin, may also have potential for accelerating cheese ripening. This hypothesis requires confirmation.

15.3.3 Exogenous Proteinases

The possibility of accelerating ripening through the use of exogenous proteinases other than those in rennet has attracted considerable attention for more than 20 years. The principal problems with this approach are that uniform distribution of the enzyme(s) in the curd is difficult to achieve and that exogenous enzymes are prohibited in many countries.

Most proteinases can accelerate proteolysis in cheese, but most have given unsatisfactory results, probably due to inappropriate specificity. The neutral proteinase "neutrase," from *Bacillus subtilis,* has been found to be the most effective, and when used alone or in combination with another means, such as elevated temperature or a lactose-negative culture, it is claimed to accelerate ripening substantially. Initially, neutrase was used alone, but more recently a cocktail of proteinase and peptidases and in some cases a lipase has been used. Commercial preparations include NaturAge, Accelase, FlavourAge, and DCA50. Some investigators have claimed positive results with these preparations, but others have found no substantial acceleration of flavor development and in some cases have observed flavor and textural defects. It is not known how widely these preparations are used commercially.

With the exception of rennet and plasmin (which adsorbs on casein micelles), the incorporation and uniform distribution of exogenous proteinases throughout the cheese matrix poses several problems:

- Since most proteinases are water soluble, most of the enzyme added to cheese milk is lost in the whey, which increases cost.
- Enzyme-contaminated whey must be heat-treated to inactivate the proteinase if the whey proteins are recovered for use as functional proteins. Therefore, the enzyme must be inactivated at a temperature below that which causes denaturation of whey proteins.
- To ensure a sufficient level of enzyme in the curd, a very high level of enzyme must be added to the milk, which causes extensive early proteolysis, leading to a loss of casein-derived peptides in the whey and a reduction in cheese yield.

Consequently, most investigators have added enzyme preparations, usually diluted with salt to facilitate mixing, to the curd at salting. Since the diffusion of large molecules, like proteinases and lipases, in the cheese matrix is very slow or occurs not at all, this method is applicable only to Cheddar-type cheeses, which are salted as chips at the end of manufacture. This method is not suitable for surface-salted cheeses, which include most varieties. Even with Cheddar-type cheeses, the enzyme will be concentrated at the surface of chips, and uneven mixing of the salt-enzyme mixture with the curds may cause "hot spots" in which excessive proteolysis, with concomitant off-flavors, may occur.

Microencapsulation is a fairly widely used technology in the pharmaceutical industry. The compound of interest is encapsulated in some type of membrane, usually with the objective of protecting it in a particular environment. When the microcapsules reach the target site, the membrane dissolves or disintegrates, releasing the entrapped compound. Enzymes have been encapsulated for a number of applications and have attracted the attention of cheese technologists as a means of adding exogenous enzymes to cheese. Usually, the enzyme, encapsulated in a lipid or phospholipid membrane, is added to the cheese milk. The microcapsules are efficiently entrapped in the coagulum formed upon renneting and are retained in the curd. During cooking, the lipid membrane melts, releasing the encapsulated enzymes (Figure 15–1).

Several studies on the microencapsulation of enzymes for use in cheese have been reported.

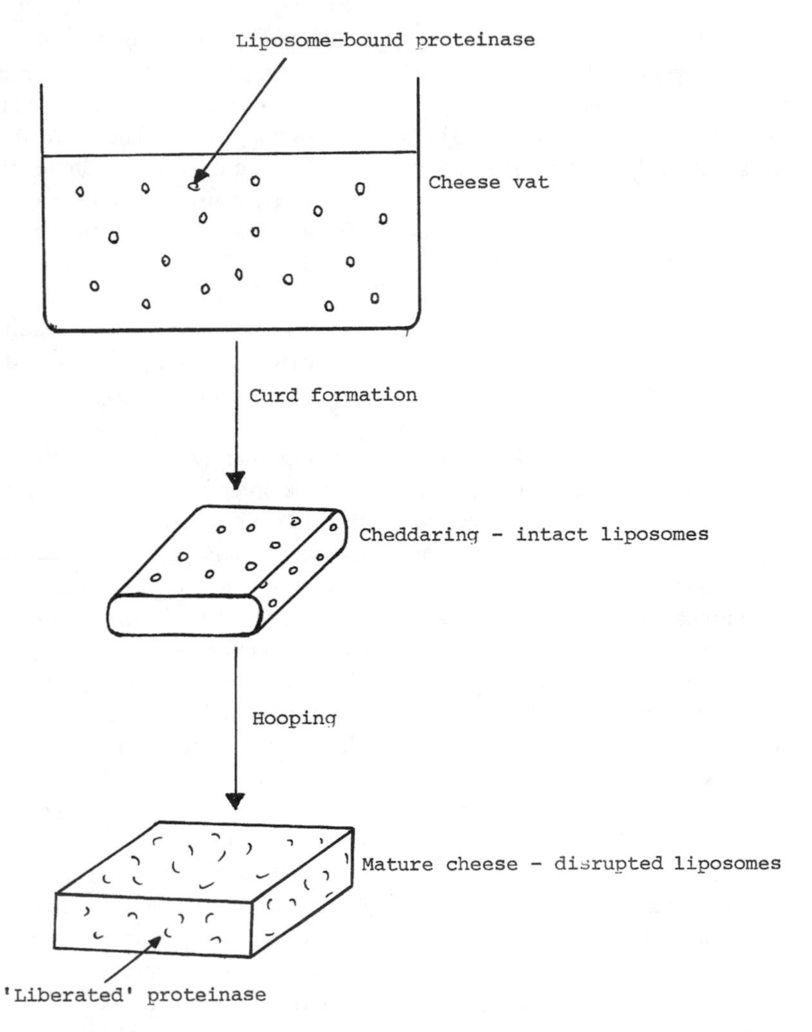

Figure 15–1 Schematic representation of the incorporation of microencapsulated exogenous enzymes into cheese curd.

Although microcapsules added to milk are incorporated efficiently into cheese curd, the efficiency of enzyme encapsulation is low, thus increasing cost. At present, encapsulated enzymes are not being used commercially in cheese production.

15.3.4 Exogenous Lipases

Lipolysis is a major contributor, directly or indirectly, to flavor development in strong-flavored cheeses, such as Romano, Provolone, and Blue varieties. Rennet paste or crude preparations of pregastric esterase (PGE) are normally used in the production of Romano, Provolone, and some other Italian cheeses. *R. miehei* lipase may also be used for Italian cheeses, although it is less effective than PGE. It has been reported that lipases from *P. roqueforti* or *P. candidum* may be satisfactory also. The ripening of Blue cheese may be accelerated and quality improved by the addition of lipases.

Although Cheddar-type and Dutch-type cheeses undergo little lipolysis during ripening, it has been claimed that the addition of rennet paste improves the flavor of Cheddar cheese, especially that made from pasteurized milk. Several authors have claimed that lipases improve the flavor of "American" or "processed American" cheese. In contrast, a number of authors have reported that the addition of PGE or *R. miehei* lipase, with or without neutase, to Cheddar cheese curd has a negative effect on flavor quality.

A strain of *A. oryzae* secretes a lipase that has an exceptionally high specificity for C_6-C_8 acids and forms micelles in aqueous media, as a result of which about 94% of the enzyme added to milk is retained in the cheese curd. This lipase, commercialized as FlavourAge (Chr. Hansen's, Milwaukee, Wisconsin), is claimed to accelerate the ripening of Cheddar cheese. The formation of short-chain fatty acids parallels flavor intensity in Cheddar cheese. In contrast to the free fatty acid profile caused by PGE, which liberates high concentrations of butanoic acid, the profile in cheese treated with FlavourAge is similar to that in the control cheese, except that the levels of free fatty acids are much higher.

Feta-type cheese produced from cow milk with a starter containing *Lc. lactis* and *Lb. casei* and a blend of kid and lamb PGEs developed a body, flavor, and texture similar to those of authentic Feta cheese. It has been reported that the flavor of Ras and Domiati cheeses can be improved and flavor development accelerated by the addition of low levels of PGE or lipases from *R. miehei* or *R. pusillus*.

15.4 SELECTED, ACTIVATED, OR MODIFIED STARTERS

Since the starter bacteria are mainly responsible for the formation of small peptides and amino acids and for flavor development in cheese (Chapter 11), it seems obvious to exploit certain characteristics of the starter to accelerate ripening. At least four approaches have been adopted.

15.4.1 Selected Starters

The primary function of starters is to produce acid at a reliable and adequate rate. As discussed in Chapter 5, highly refined starters, containing only one or a few selected strains, are now widely used, especially by large cheese manufacturers. The principal criteria for the selection of single-strain starters are phage-unrelatedness, the ability to grow well and produce acid over the temperature profile used in cheesemaking, and interstrain compatibility. The selection protocol does not include specific criteria for the ability to produce high-quality cheese, but strains with undesirable cheesemaking properties (e.g., bitterness) have been excluded, and strains that more or less consistently produce high-quality cheese have been selected on the basis of commercial experience. The scientific selection of starter strains with desirable cheesemaking properties is hampered by the lack of precise knowledge as to which enzymes are most important. However, there is strong evidence that *Lactococcus* spp. differ markedly with respect to the activity and specificity of cell wall–associated proteinase and the activity of various intracellular exopeptidases, acid phosphomonoesterase, and esterase. There is also strong evidence that the cheesemaking properties of starter strains differ markedly, but the cheesemaking properties and enzyme activities have not been correlated. Research in this area appears to be warranted. Information is also required on the activity of enzymes involved in the catabolism of amino acids, which is believed to be important in flavor development.

Since the cell envelope–associated proteinase is the only extracellularly located enzyme in *Lactococcus* and *Lactobacillus* and the cells cease to grow in cheese within about 1 day and are therefore unable to transport compounds into the cell, the cells must lyse so that their intracellular enzymes (exopeptidases, esterases, phosphatase, etc.) can encounter their substrates. Thus, the faster cells lyse, the sooner their intracellular enzymes can become involved in cheese ripening. There is natural variation in the susceptibility of *Lactococcus* strains to lysis, and hence the selection of fast-lysing strains may be a useful approach to accelerating ripening. Lysis of some *Lactococcus* strains is thermoinducible (e.g., occurs during cooking). It has been suggested that lysis might be induced via a controlled phage infection, but this would probably be unacceptable to cheese manufacturers owing to the difficulty of controlling the level of infection.

Some strains of *Lactococcus* produce bacteriocins that cause the lysis of other *Lactococcus* strains. By including an appropriate level of a bacteriocin-producing strain in the starter, it appears to be possible to induce early lysis of the starter cells without interfering with acid production. The results of preliminary experiments suggest that this approach accelerates ripening and improves flavor, but larger-scale studies are required.

15.4.2 Attenuated Starters

Since the starter plays a key role in cheese ripening, it might be expected that increasing cell numbers would accelerate ripening. At least in the case of Cheddar, high numbers of starter cells have been associated with bitterness. However, not all authors agree that bitterness is related simply to starter cell numbers. Some suggest that too much proteolytic activity or the wrong specificity is responsible, for example, too little peptidase activity relative to proteinase activity. In fact, a number of authors reported that stimulating starter growth accelerates ripening. Perhaps the significance of starter cell numbers for cheese ripening should be reinvestigated.

An alternative to the use of high starter cell numbers is the addition of attenuated starter cells to the cheese milk. The rationale behind this method is that it would likely destroy the acid-producing ability of the starter (since excessively rapid acid development is undesirable) but cause as little denaturation of the cells' enzymes as possible. The discussion in the preceding paragraph suggests that adding attenuated cells might cause bitterness, but this has not been reported to be a problem with the use of attenuated starters; in fact, the opposite has usually been reported. However, most studies on the use of attenuated starters have been on varieties other than Cheddar.

Five alternative approaches to the production of attenuated starters have been investigated; these are discussed below.

Lysozyme Treatment

Lysozyme is a widely distributed enzyme that hydrolyzes the cell wall of certain bacteria, including *Lactococcus* and *Lactobacillus*, causing the cells to lyse unless the osmotic pressure of the surrounding medium is high. Lysozyme-treated *Lactococcus* cells do not lyse in milk or unsalted cheese curd but do lyse when the curd is salted, releasing their intracellular enzymes into the cheese matrix (if the lysozyme-treated cells lysed in the milk, most of the intracellular enzymes would be lost in the whey). Limited studies on the addition of lysozyme-treated cells to milk for Cheddar cheese indicate that, although the addition of the equivalent of 10^{10} cells/g cheese accelerated proteolysis, flavor development was not accelerated significantly. Lysozyme is rather expensive, and a cheaper source is required before this approach would be commercially successful.

Heat- or Freeze-Shocked Cells

The lactic acid-producing ability of lactic acid bacteria can be markedly reduced by a sublethal heat treatment (e.g., 60–70°C for 15 s) without reducing proteinase and peptidase activity to any significant extent. Heat-shocked cells added as a concentrate to cheese milk are entrapped in the curd (≈ 90% retention). A number of independent studies on several types of cheese have shown that this approach accelerates ripening and intensifies flavor. However, addition of a large number of cells (e.g., 10^9 cfu/g) is necessary to achieve a significant impact, and the cost may be prohibitive.

Freezing and thawing also kill lactic acid bacteria (LAB) without inactivating their enzymes.

Although this technique would appear to be easier to execute than heat-shocking and appears to be equally effective, it has been used to a very limited extent. As far as we know, neither heat-shocked nor freeze-thaw-shocked cells are used commercially. Lactose-negative (Lac$^-$) strains of *Lactococcus*, which do not grow in milk, would appear to be a more attractive alternative to heat-shocked or freeze-shocked cells (see "Mutant Starters" below).

Solvent-Treated Cells

Treatment of starter cells with n-butanol activates some membrane-bound proteinases and peptidases, presumably by increasing accessibility to the substrate. The addition of butanol-treated cells to cheese milk accelerated ripening slightly and reduced the intensity of bitter flavor. The use of solvent-treated cells in cheesemaking may not be permitted by regulatory authorities.

Mutant Starters

Lactose-negative (Lac$^-$) cells cannot grow in milk, probably because they lack the lactose transport system (see Chapter 5), and hence they can be added to cheese milk without interfering with the rate of acid production. The Lac$^-$ cells contain the full complement of enzymes and thus serve as a package of enzymes that is entrapped in the cheese curd and lyses slowly, releasing enzymes. Since the Lac gene is encoded on a plasmid and plasmids are easily lost, Lac$^-$ mutants occur naturally and are easily isolated. The principal characteristics required in a starter (phage resistance, reliable acid production, and strain compatibility) are not important in Lac$^-$ adjuncts, and hence a wider range of strains may be suitable, such as those with high proteinase and/or peptidase activity. Strains may also be engineered to have very high levels of certain enzyme activities that are considered to be important in cheese ripening. The ability of Lac$^-$ or Lac$^-$Prt$^-$ mutants to accelerate ripening and/or intensify flavor has been assessed in several studies, dating back to 1983. In general, they have given satisfactory results. Lac$^-$ *Lactococcus* strains with high exopeptidase activity are commercially available and have given satisfactory results in pilot-scale and commercial-scale studies. The cultures are available as frozen concentrates or freeze-dried preparations. Although such cultures are relatively expensive (the recommended level of usage costs ≈ \$96 per tonne of cheese), the extra cost is affordable if ripening time is reduced by more than 20% and/or cheese flavor is intensified. The extent of their commercial use is not known.

Other Bacteria as Additives

LAB are weakly proteolytic in comparison with other microorganisms. It might be possible to accelerate or modify ripening by adding cells of other genera with a very high level of general or specific peptidolytic activity. The cells, which serve as packages of enzymes, when added to the milk will be efficiently entrapped and uniformly distributed in the curd. As discussed in Chapter 10, cheese is a very selective environment in which few bacterial genera can grow or survive.

Preliminary studies have been reported on the use of *Pseudomonas, Brevibacterium*, and *Propionibacterium* to accelerate or modify ripening. The first two of these bacteria are strictly aerobic and therefore will not grow in the interior of cheese or on its surface when vacuum packed. Significant modification of ripening (i.e., accelerated release of amino acids and slightly accelerated flavor development) was found only when high numbers (10^8–10^9/g) of washed *Pseudomonas* or *Brevibacterium* cells were present in the cheese. Such high numbers may not be economical, especially if the effect obtained is rather small.

Propionibacterium freudenreichii subsp. *shermanii*, the characteristic organism in Swiss-type cheese, is anaerobic but is strongly inhibited by NaCl and does not grow below about 18°C. When it was added to milk for Cheddar cheese, it had a very significant effect on the flavor of the cheese at numbers above 10^8 cfu/g. Proteolysis was accelerated slightly, and the resultant cheese had a very pronounced Swiss-type flavor. Although different from that of

Cheddar, the flavor generated was very attractive. In fact, a new type of cheese was produced (i.e., the ripening of the cheese was modified rather than accelerated). Further work with such genera, especially species of *Propionibacterium*, appears warranted.

15.4.3 Genetically Engineered Starters

Considerable knowledge is now available on the genetics of the cell envelope–associated proteinase (CEP) and of many of the intracellular peptidases of *Lactococcus* spp. and to a lesser extent of *Lactobacillus*. Thus, it may be possible to specifically modify their proteolytic system.

Mutants deficient in CEP or in one or more intracellular peptidases are available and have been used to study the significance of these enzymes for cell growth. The significance of CEP in cheese ripening has also been assessed using Prt^- or CEP superproducing mutants. Proteolysis in the cheese made using the Prt^- mutant was slower than in the control, and the cheese lacked flavor, but the superproducing mutant did not accelerate proteolysis or flavor development, suggesting that CEP is not the limiting factor in cheese ripening.

The gene for the neutral proteinase (neutrase) of *B. subtilis* has been cloned in *Lc. lactis* UC317. Cheddar cheese manufactured with this engineered culture as the sole starter underwent very extensive proteolysis, and the texture became very soft within 2 weeks at 8°C. Since the genetically modified cells were not food grade, the cheese was not tasted, but its aroma was satisfactory. By using a blend of unmodified and neutrase-producing cells as starter, a more controlled rate of proteolysis was obtained and ripening was accelerated. An 80:20 blend of unmodified:modified cells gave best results. The results appear sufficiently interesting to warrant further investigation when a food-grade mutant becomes available.

Since free amino acids are widely believed to make a major contribution, directly or indirectly, to flavor development in cheese, a starter with increased aminopeptidase activity would appear to be attractive. Two studies on a starter genetically engineered to superproduce the general aminopeptidase PepN have been reported. Although the release of total amino acids was accelerated, the rate of flavor development and flavor intensity were not, suggesting that the release of total amino acids is not rate limiting.

15.5 ADJUNCT STARTERS

The fourth group of contributors to the ripening of cheese are NSLAB, which may originate in the milk, especially if raw milk is used, or the cheesemaking environment (equipment, air, and personnel). Their most likely source is the milk. As discussed in Chapter 10, the interior of cheese is a hostile environment for bacteria, and consequently few genera of bacteria can grow within cheese. The principal NSLAB are mesophilic lactobacilli. Cheddar cheese made from good-quality pasteurized milk in modern enclosed equipment with a good active starter contains very few NSLAB initially (< 50 cfu/g), but these multiply to roughly 10^7 cfu/g within about 3 months. The NSLAB population of pasteurized milk cheese is dominated by a few species, usually *Lb. casei* and/or *Lb. paracasei*. The NSLAB population in raw milk Cheddar cheese usually exceeds 10^8 cfu/g and is more heterogeneous.

There is a widely held view, substantiated by comparative studies on cheese made from raw, pasteurized, or microfiltered milk, that cheese made from raw milk ripens faster and develops a more intense (although not always typical or desirable) flavor than cheese made from pasteurized milk and that the indigenous microflora is responsible. The results of these studies have stimulated interest in the selection of *Lactobacillus* cultures for addition to pasteurized milk to simulate the quality of raw milk cheese. Such cultures are now available from commercial starter suppliers. Several studies in which commercial or noncommercial *Lactobacillus* adjuncts were used have been published (see Fox et al., 1996; Fox, McSweeney, & Lynch, 1998). In all of these studies, low numbers of selected mesophilic lactobacilli were added to the cheese milk. There is general agreement that the lactobacilli modify proteolysis; in particular, they result in a higher concentration of free amino acids and improve the sensoric quality.

In contrast to mesophilic lactobacilli, thermophilic lactobacilli die rapidly in cheese, lyse, and release their intracellular enzymes. Consequently, cheeses made with thermophilic *Lactobacillus* spp. as starters contain high concentrations of amino acids (the concentrations are particularly high in Parmesan cheese). Although thermophilic lactobacilli will not grow in Cheddar cheese, their inclusion as a starter adjunct markedly intensifies the flavor of Cheddar. Adjuncts of thermophilic lactobacilli and *Sc. thermophilus* are available commercially.

There appears to be strong evidence that selected lactobacilli have the potential to improve cheese flavor and to accelerate flavor development. It is likely that research on this subject will continue and that improved strains of *Lactobacillus* will be isolated. It is also likely that a cocktail of strains will be more effective than individual strains. It appears that the principal contribution of adjunct NSLAB is to the formation of amino acids. Perhaps superproducing adjuncts can be developed through genetic engineering once the key enzymes have been identified.

15.6 SECONDARY CULTURES

Secondary cultures are involved in the ripening of many cheese varieties, including *Propionibacterium, Brevibacterium, Penicillium*, and some yeasts. As discussed in Chapters 10 and 11, these cultures play key and characterizing roles in the ripening of cheeses in which they are used. With the exception of some Swiss varieties, cheeses in which secondary cultures are used have relatively short ripening times, due to their relatively high moisture content and the very high level of activity of the secondary starter.

Apart from the production of enzyme-modified cheeses (discussed in Section 15.7), there has been little work on accelerating the ripening of cheeses using a secondary starter.

15.7 ENZYME-MODIFIED CHEESE

An extreme form of accelerated ripening is practiced in the production of enzyme-modified cheese (EMC), which has been reviewed by Kilcawley, Wilkinson, and Fox (1998). The basic steps involved in the production of EMCs are shown in Figure 15–2. Fresh curd or young cheese is homogenized (dispersed) and pasteurized, and a cocktail of enzymes (proteinases, peptidases, lipases, and perhaps bacterial cultures) is added. The mixture is incubated for the requisite period, depending on the activity of the enzymes added, and then repasteurized to terminate the microbiological and enzymatic reactions. The product may be spray dried or commercialized as a paste (see Chapter 19).

Although their flavor does not resemble or even approximate that of natural cheeses, EMCs have the ability to potentiate cheese flavor in various food products, including processed cheese, cheese analogues, cheese sauces, cheese dips, and products incorporating cheese, such as crackers and crisps. For such applications, EMCs may be able to replace 20–50 times their weight of natural cheese and therefore provide a cheaper alternative to natural cheese for imparting cheese flavor to formulated foods. Typical costs of natural Cheddar cheese and Cheddar EMC are $3.75 and $12 per kg, respectively. Cheddar EMCs are the most important commercially, but EMCs that simulate several varieties have been developed, such as Blue cheese, Swiss, and Romano (see Chapter 19).

EMCs are based on "cheese slurries," which were developed in the 1970s. Cheese slurries have served as model cheese systems in which to study the pathways involved in cheese ripening, for the selection of enzymes to accelerate the ripening of cheese, and more recently for the selection of starter cultures with superior cheese-

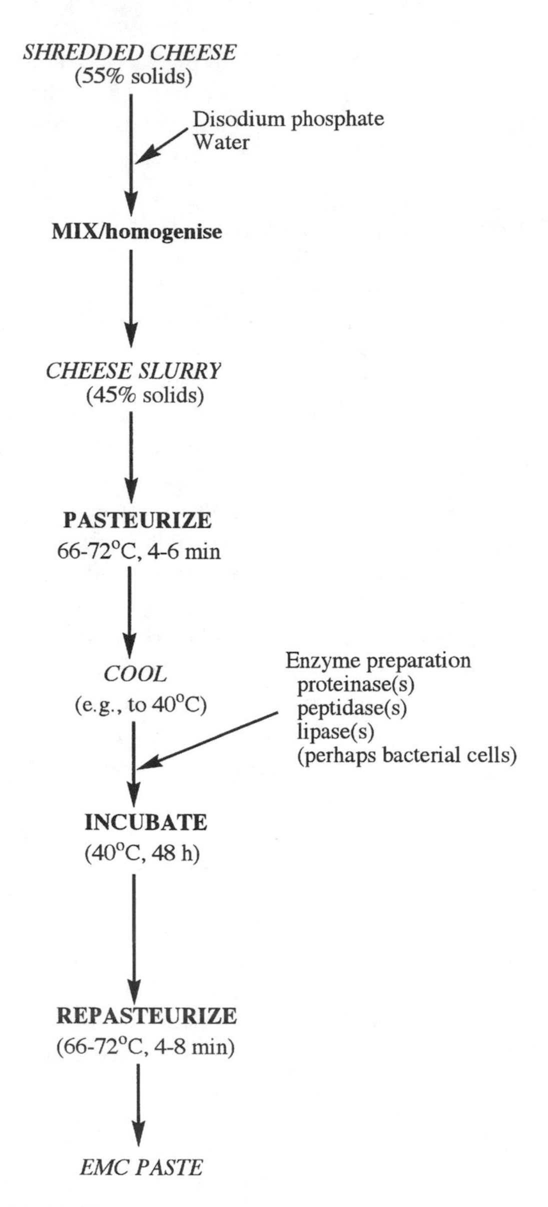

Figure 15–2 Typical protocol for the manufacture of enzyme-modified cheese.

making properties. It is claimed that cheese slurries may be added to natural cheese to accelerate ripening. Presumably, they act as cultures of desirable secondary bacteria. It is also claimed that cheese slurries develop a characteristic flavor in as little as 4–5 days. Again, most of the work on cheese slurries has been related to Cheddar cheese, but slurries mimicking other varieties have also been produced.

15.8 ADDITION OF AMINO ACIDS TO CHEESE CURD

Since amino acids are considered to be important contributors, either directly or indirectly, to cheese flavor and their production in cheese is relatively slow, it seems reasonable to assume that the addition of amino acids to cheese curd might accelerate flavor development. Glutamic acid, leucine, and methionine are considered to be the most important amino acids with respect to cheese flavor. Glutamic acid has a brothy flavor, while methionine is the precursor of several sulfur compounds considered to be very important contributors to the flavor of many cheeses (see Chapter 12).

Preliminary studies by Wallace and Fox (1997) have shown that the addition of intermediate levels of free amino acids to Cheddar cheese curd at salting (5–6 g/kg curd) had a beneficial effect on the development of cheese flavor. Amino acids appear to stimulate proteolysis, particularly secondary proteolysis involving the breakdown of small peptides to amino acids, either due to the activation of peptidases, increased cell lysis, or perhaps increased growth of NSLAB, which was not studied. The products of amino acid catabolism were also not studied, but they may merit study, as they are thought to be major contributors to cheese flavor. The economics of incorporating amino acids into cheese curd also requires evaluation.

15.9 PROSPECTS FOR ACCELERATED RIPENING

There is undoubtedly an economic incentive for accelerating the ripening of low-moisture, highly flavored, long-ripened cheeses. Although consumer preferences are tending toward more mild flavored cheeses, there are considerable niche markets for highly flavored products. While the ideal might be to have cheese ready for consumption within a few days, this is unlikely to be attained, and in any case it would be necessary to stabilize the product after it reaches optimum quality, such as by heat treatment (as is used in the production of EMCs).

Although the possibility of using exogenous (nonrennet) proteinases and in some cases peptidases attracted considerable attention for a period, this approach has not been commercially successful, for which a number of factors may be responsible:

- Although a necessary prerequisite, primary proteolysis is probably not the rate-limiting reaction in flavor development.
- The use of exogenous enzymes in cheese is prohibited in several countries.
- Uniform incorporation of enzymes is still problematic, and the use of encapsulated enzymes is not viable at present.

Plasmin may have potential as a cheese-ripening aid because it can be easily incorporated into cheese curd, is an indigenous enzyme active in natural cheese, and has narrow specificity, producing nonbitter peptides. At present, it is too expensive, but its cost may be reduced via genetic engineering.

Attenuated cells appear to have given useful results in pilot-scale experiments, but, considering the mass of cells required, the cost of such cells would appear to be prohibitive for commercial use, except perhaps in special circumstances. Selected peptidase-rich Lac$^-$/Prt$^-$ *Lactococcus* cells added as adjuncts have given promising results, but further work is required, and they may not be cost effective.

The selection of starter strains according to scientific principles holds considerable potential. Such selection is hampered by the lack of information on the key enzymes involved in ripening. Preliminary studies on the significance of early cell lysis have given promising results, and further studies are warranted. Bacteriocin-induced lysis appears to be particularly attractive.

The ability to genetically modify starters holds enormous potential, but results to date using genetically engineered starters have been disappointing. Identifying the key enzymes in ripening is essential for the success of this approach. It is hoped that current research on cheese ripening will identify the key sapid compounds in cheese and hence the critical rate-limiting enzymes. Genetic manipulation of Lac$^-$/Prt$^-$ adjunct *Lactococcus* will also be possible when key limiting enzymes have been identified. We believe that adjunct starters, especially lactobacilli, hold considerable potential. It appears to be possible to produce cheese of acceptable quality without lactobacilli, but these bacteria do intensify (Cheddar) cheese flavor and offer flavor options. The volume of literature published on starter adjuncts has been rather limited to date. Further work will almost certainly lead to the development of superior adjuncts. There is the obvious possibility of transferring desirable enzymes from lactobacilli to starter lactococci.

At present, an elevated ripening temperature ($\approx$ 15°C) offers the most effective and certainly the simplest and cheapest method for accelerating the ripening of Cheddar, which is usually ripened at an unnecessarily low temperature. However, this approach is less applicable to most other varieties, for which relatively high ripening temperatures are used at present.

The key to accelerating ripening ultimately rests on identifying the key sapid compounds in cheese. This, so far, has been an intractable problem. Work on the subject commenced nearly 100 years ago and has been quite intense since 1960, when gas chromatography was developed. Although as many as 400 compounds that might be expected to influence cheese taste and aroma have been identified, it is not possible to describe cheese flavor precisely (see Chapter 12). Until such information is available, attempts to accelerate ripening will be speculative and empirical.

REFERENCES

Fox, P.F., McSweeney, P.L.H., & Lynch, C.M. (1998). Significance of non-starter lactic acid bacteria in Cheddar cheese. *Australian Journal of Dairy Technology*, *53*, 83–89.

Fox, P.F., Wallace, J.M., Morgan, S., Lynch, C.M., Niland, E.J., & Tobin, J. (1996). Acceleration of cheese ripening. *Antonie von Leeuwenhoek*, *70*, 271–297.

Kilcawley, K.N., Wilkinson, M.G., & Fox, P.F. (1998). Enzyme-modified cheese [A review]. *International Dairy Journal*. *8*, 1–10.

Wallace, J., & Fox, P.F. (1997). Effect of adding free amino acids to Cheddar cheese curd on proteolysis and flavour development. *International Dairy Journal*, *7*, 157–167.

Wilkinson, M.G. (1993). Acceleration of cheese ripening. In P.F. Fox (Ed.), *Cheese: Chemistry, physics and microbiology* (2d ed., Vol. 1). London: Chapman & Hall.

Chapter 16

Fresh Acid-Curd Cheese Varieties

16.1 INTRODUCTION

Fresh acid-curd cheeses comprise those varieties that are produced by the coagulation of milk, cream, or whey via acidification or a combination of acid and heat and that are ready for consumption once the manufacturing operations are complete (Figure 16–1). They differ from rennet-curd cheeses, for which coagulation is induced by the action of rennet at pH 6.4–6.6, in that coagulation occurs close to the isoelectric pH of casein (i.e., pH 4.6) or at a higher value when a higher temperature is used (e.g., pH 6.0 at 80°C for Ricotta). While a very small amount of rennet may be used in the production of Quarg, Cottage cheese, and Fromage frais to give firmer coagula and to minimize casein losses during subsequent whey separation, its addition is not essential.

Annual world production of fresh acid-curd cheeses amounts to about 3.5 million tonnes, which is equivalent to roughly 23% of total cheese production (Sørensen, 1997). Quarg, Cottage cheese, Cream cheese, Fromage frais, and Ricotta are commercially the most important types. Consumption grew by about 2.5% per annum during the 1987–1996 period. Factors contributing to this increase include these:

- They offer a large variety of consistencies and flavors, made possible by changes in cheesemaking protocols; blending of one or more cheese types to create new products; and the addition of sugar, fruit purees, spices, and condiments.
- Their soft, ingestible consistency makes them safe for and attractive to very young children.
- They are perceived as healthy by diet-conscious consumers. In general, the fat content of these cheeses is lower than that of rennet-curd cheeses. Double Cream cheese, an exception in the group, has a fat content (≈ 330 g/kg) similar to that of Cheddar. However, the cheeses are relatively low in calcium (typically < 0.8 g/kg) compared with rennet-curd cheeses such as Cheddar (≈ 7.5 g/kg) or Swiss (≈ 9.5 g/kg) (Table 16–1).

16.2 OVERVIEW OF THE MANUFACTURING PROCESS FOR FRESH ACID-CURD CHEESE PRODUCTS

Production generally involves pretreatment of milk (standardization, pasteurization, and perhaps homogenization), slow quiescent acidification, gel formation, dehydration of the gel (whey separation), and in some cases further treatments of the curd (pasteurization; shearing; addition of salt, condiments, and stabilizers; and homogenization) (Figure 16–2). Acidification is generally slow, 12–16 hr at 21–23°C (long set) or 4–6 hr at 30°C (short set), and is usually brought about by the in situ conversion of lactose to lactic acid, by an added starter culture and/or by the addition of

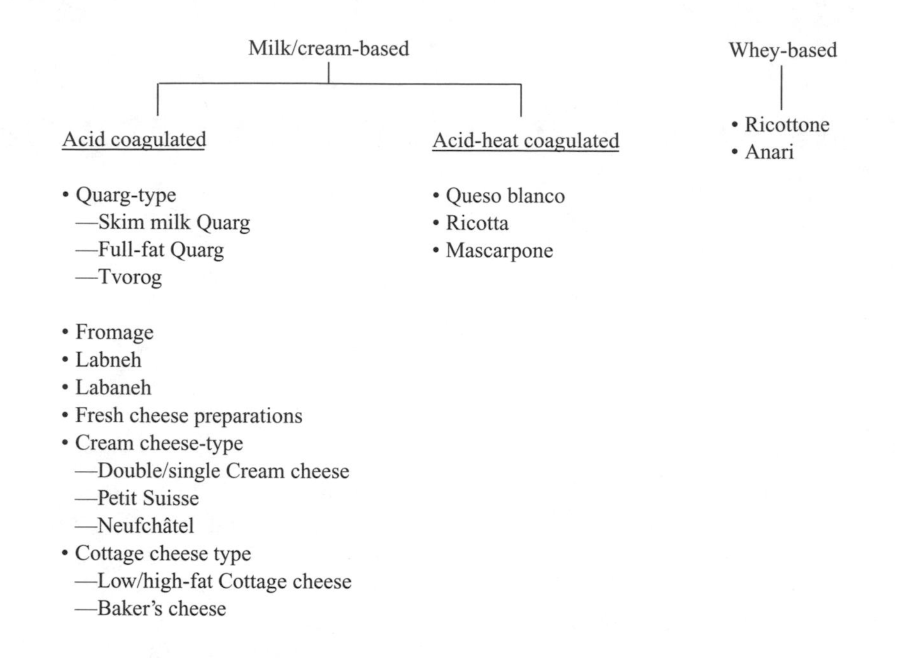

Figure 16–1 Fresh acid-curd cheese varieties.

food-grade acid (e.g., lactic or citric) or acidogen, such as gluconic acid-δ-lactone (which hydrolyzes to gluconic acid). The structure of the gel has a major effect on the texture (e.g., spreadability and firmness) and sensory attributes (smoothness) of the final product and its physicochemical stability (i.e., stability to wheying-off and/or to the development of a chalky/grainy mouthfeel) during storage. The structure of the gel is influenced by many processing factors, such as the protein level, the milk pasteurization treatment, homogenization, the temperature during acidification, and the pH at which the gel is broken and subjected to dehydration. The effect of gel structure on product quality is most pronounced in products whose curd, following whey separation and concentration, is not treated further (e.g., Quarg and Fromage frais). In hot-pack products, such as Cream cheeses and some fresh cheese preparations, curd treatments (pasteurization, homogenization, and hydrocolloid addition) have a major impact on the quality of the final product (Guinee, Pudja, & Farkye, 1993).

16.3 PRINCIPLES OF ACID MILK GEL FORMATION

Slow acidification of milk under quiescent conditions is accompanied by two opposing sets of physicochemical changes:

1. a tendency toward disaggregation of the casein micelles into a more disordered system as a result of
   - solubilization of the internal micellar cementing agent, colloidal calcium phosphate (CCP), which, at 20–30°C, is fully soluble at about pH 5.2–5.3 (Figure 16–3)

Table 16–1 Approximate Composition of Various Fresh Cheeses

Variety	*Dry Matter (%, w/w)*	*Fat (%, w/w)*	*Protein (%, w/w)*	*Lactose (Lactate) (%, w/w)*	*Salt (%, w/w)*	*Ca (mg/100 g)*	*pH*
Cream cheese							
Double	40	30	8–10	2–3	0.75	80	4.6
Single	30	14	20	3.5	0.75	100	4.6
Neufchâtel	35	20	10–12	2–3	0.75	75	4.6
Labneh	25	11.6	8.4	4.4	–	–	4.2
Quarg							
Skim milk	18	0.5	13	3–4	–	120	4.5
Full fat	27	12	10	2–3	–	100	4.6
Cottage cheese							
Low-fat	21	2	14	–	–	90	4.8
Creamed	21	5	13	–	–	60	4.8
Fromage frais							
Skim milk	14	1	8	3.5	–	0.15	4.4
Queso blanco	49	15	23	1.8	3.9	–	5.4
Ricotta							
Whole milk	28	13	11.5	3.0	–	200	5.8
Part skim	25	8	12	3.6	–	280	5.8
Ricottone	18	0.5	11	5.2	–	400	5.3
Brunost							
Gubrandsdalsost	82	30	11	38	–	400	–
Fløtemyost	80	19	11	46	–	–	–

    - a pH- and temperature-dependent dissociation of individual caseins, especially β-casein, from the micelles, with a concomitant increase in the level of serum casein (casein dissociation decreases with decreasing pH to about pH 6.2, then increases to a maximum at pH 5.3–5.6, depending on the temperature, and thereafter decreases to a minimum at the isoelectric pH [Figure 16–4])
    - an increase in casein solvation with pH reduction in the range 6.7–5.3 (Figure 16–5)
2. a tendency for the casein micelles to aggregate into a more ordered system due to
    - the reduction of the negative surface charge on the casein micelles and hence of intermicellar repulsive forces due to the production of lactic acid
    - a decrease in casein hydration in the pH range 5.3–4.6 (Figure 16–5)
    - the increase in the ionic strength of the milk serum (due to the increased concentrations of calcium and phosphate ions), which has a shrinking effect on the casein micelles

At pH values greater than that at the onset of gelation (~ 5.1–5.3 at 20–30°C), disaggregating forces predominate and hence a gel is not formed. At lower pH values, forces that promote aggregation of the casein micelles prevail and gel formation begins. Electron microscopic examination has revealed the presence of a heterogeneous size distribution of casein aggregates (composed of fused casein micelles) at the onset of gelation. Further reduction of pH is paralleled by a touching of aggregates that initiates the formation of loose porous strands. Eventually, as the pH of the milk approaches the isoelectric point, dangling strands touch and cross-link to form a three-dimensional particu-

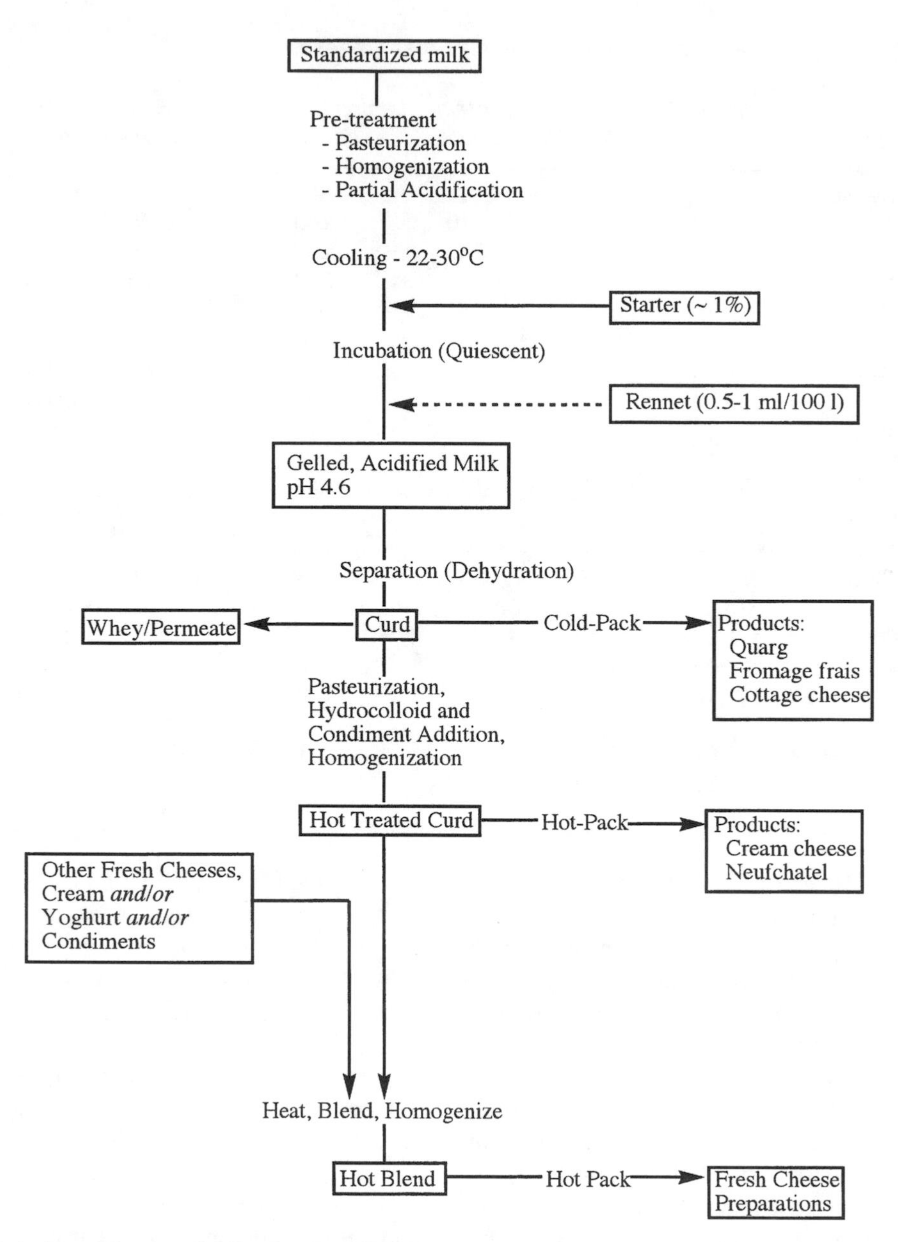

Figure 16–2 Generalized production protocol for fresh cheese products.

late casein network (or gel), which extends continuously throughout the serum phase. The gel is described as particulate because, when viewed by scanning electron microscopy, the individual gel strands are found to be composed of particles (casein aggregates) that undergo limited touching (over part of their surfaces) and are linked together, rather like the beads in a necklace.

Gel formation is accompanied by a marked increase in the elastic shear modulus (index of curd firmness), which increases progressively

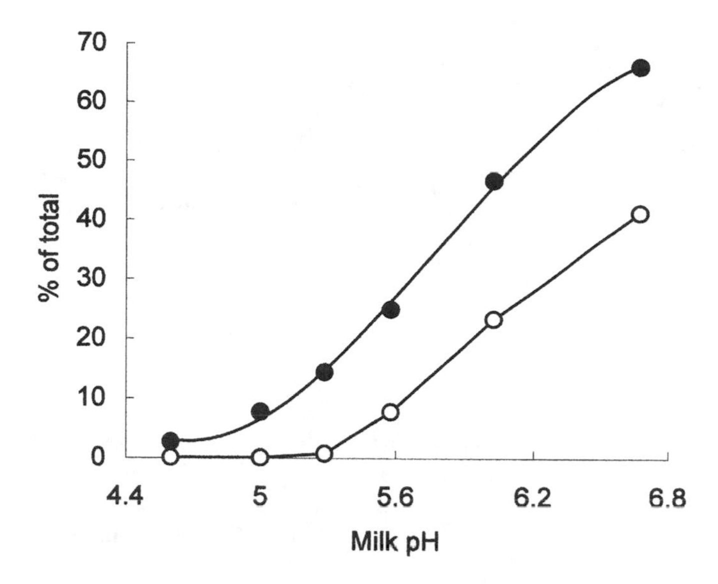

Figure 16–3 Micellar calcium (●) and inorganic phosphate (○) in skim milk as a function of pH at 30°C (determined by ultracentrifugation at 88,000 g for 1.5 hr and expressed as percentage of total concentration in milk.

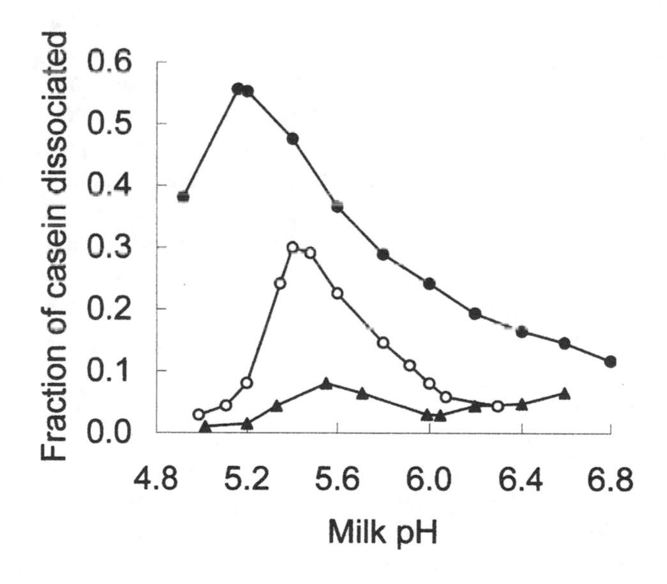

Figure 16–4 Serum (nonsedimentable) casein in skim milk as a function of pH at 4°C (●), 20°C (○), and 30°C (▲), determined by centrifugation at 70,000 g for 4, 2, and 1.75 hr, respectively, and expressed as a percentage of total casein.

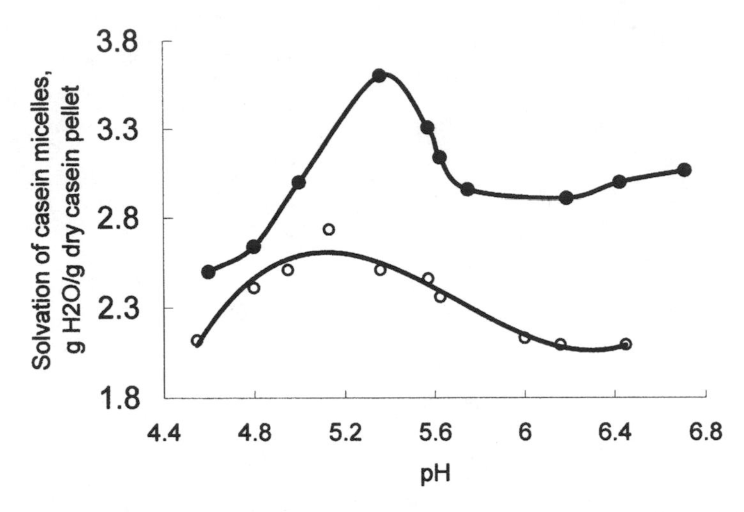

Figure 16–5 Solvation of casein micelles as a function of pH at 20°C for skim milk (●) and rennet-treated skim milk (○).

with further pH reduction (to 4.6) and casein aggregation (Figure 16–6).

The physicochemical and microstructural changes accompanying the conversion of milk to an acid gel are summarized in Figure 16–7.

16.4 PREREQUISITES FOR GEL FORMATION

Acidification of milk may result in the formation of a gel or a precipitate, depending on the rate and extent of casein aggregation. Gelation occurs when forces that promote aggregation of the casein micelles slowly overcome those that promote repulsion of the micelles. These conditions result in the formation of relatively loose, porous, hydrated aggregates of casein micelles, which are only slightly more dense than the serum phase in which they are dispersed. Owing to the relatively small density gradient between the aggregates and the serum phase, the aggregates have sufficient time to link together, via strand formation, to form a continuous casein network, which physically entraps the serum (whey) phase. In contrast, the casein micelles aggregate more rapidly and undergo a high degree of fusion (i.e., touch neighboring micelles over a much larger part of their surface) to form smaller, less porous, and less hydrated aggregates when conditions that promote aggregation of the casein micelles are more extreme, e.g., rapid acidification under nonquiescent conditions at a high temperature, as in the manufacture of acid casein. Owing to their relatively high density, these aggregates sediment as a precipitate. Compared to a gel, the casein in a precipitate is highly aggregated, has a very low water-holding capacity, and occupies a much lower specific volume (i.e., has a low voluminosity). Although the production of both acid casein and fresh acid-curd cheeses, such as Quarg, involves the acidification of skim milk, the conditions of acidification differ markedly (Figure 16–8) and result in the production of two very different types of product—a precipitate from which the moisture is expelled rapidly, enabling the recovery of casein as a food ingredient (as in acid casein), and a cheese (gel), which has superior water-holding capacity.

To obtain a gel rather than a precipitate, the number of attractive forces and hence the surface area of contact between the dispersed particles (casein micelles) must be limited. Conditions conducive to limited aggregation include a

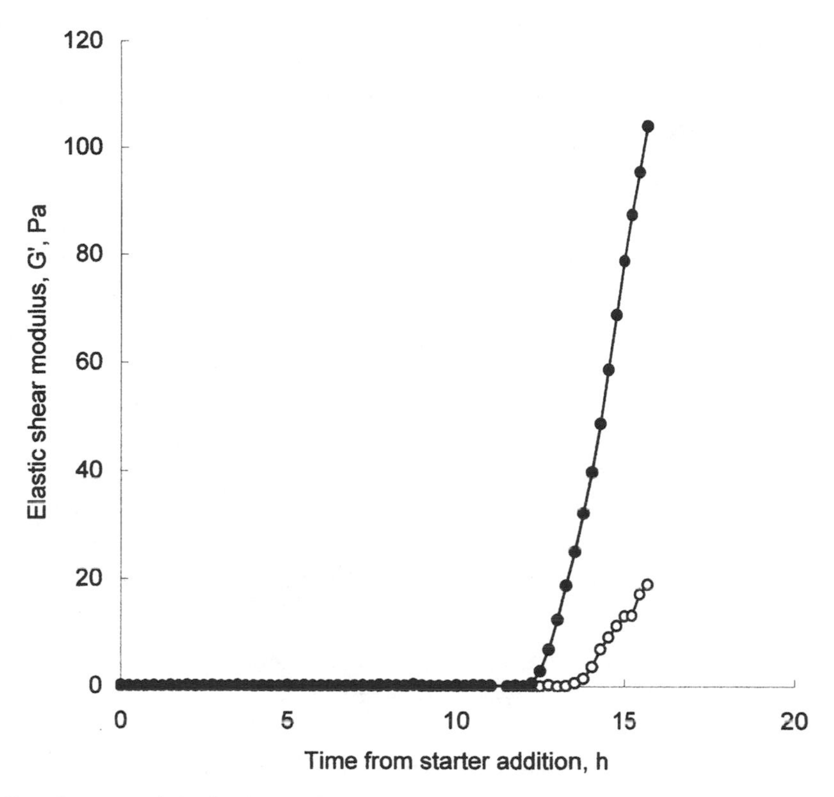

Figure 16–6 Development of elastic shear modulus (G′) during fermentation of skim milk pasteurized at 72°C × 15 s (○) and at 90°C × 300 s (●); see Table 16–2.

slow rate of acidification under quiescent conditions. Slow acidification is promoted by the use of a starter culture and a relatively low temperature (e.g., 22–30°C) during acidification. During the production of fresh acid-curd cheeses, conditions that promote a greater degree of casein attraction and fusion (e.g., when the rate of acidification is increased) lead to the formation of a gel that is less voluminous and closer to a precipitate. The latter type of gel, which has a lower water-holding capacity, is said to be coarser. Alternatively, conditions that promote a lower degree of casein aggregation (e.g., a slow rate of acidification) result in a finer gel network, which has a relatively high water-holding capacity. The structures of fine and coarse gels and of a precipitate in which the concentration of gel-forming protein is equal are illustrated schematically in Figure 16–9.

In extreme situations (e.g., acidification occurs very slowly and it takes more than 16 hr for the pH to fall from ≈ 6.6 to 4.6), if the number of interparticle attraction sites is lower than optimum, slowly forming aggregates may have sufficient time to precipitate before fusing and linking with neighboring aggregates to form into a network. An example of the latter is the defect in Cottage cheese production known as "major sludge formation," whereby phage infection of the starter, after acid development has progressed to an advanced stage (≈ pH 5.2–5.3), leads to casein precipitation rather than gelation.

16.5 EFFECT OF GEL STRUCTURE ON QUALITY

Gel structure is a major determinant of quality attributes, such as the mouthfeel (smoothness or

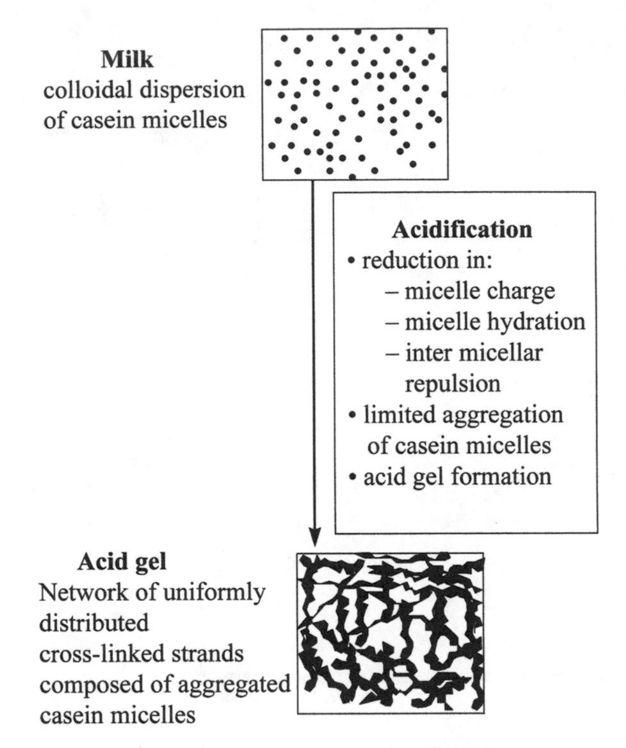

Figure 16–7 Schematic representation of the conversion of milk to a gel by slow quiescent acidification using a starter culture.

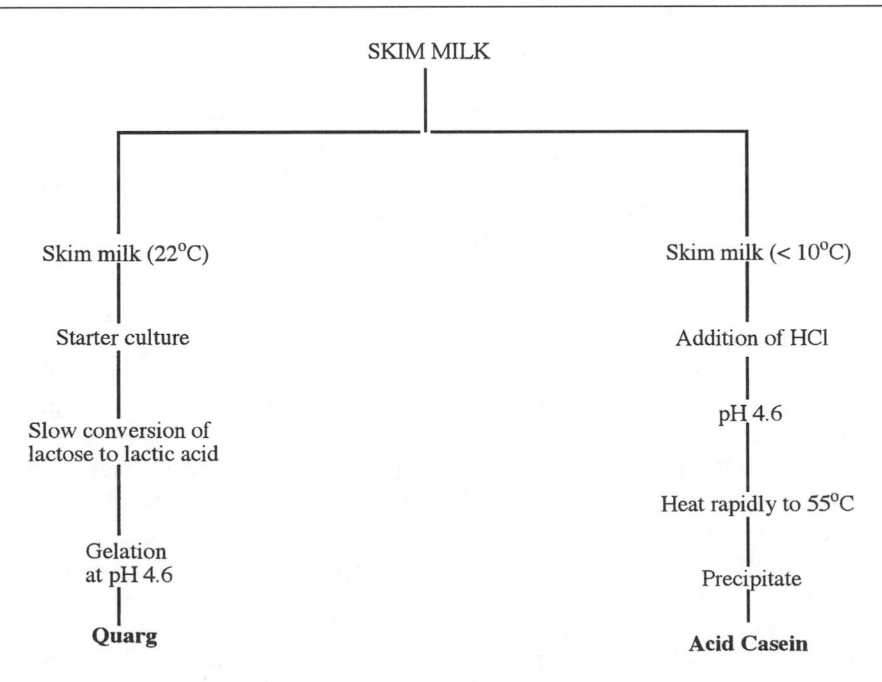

Figure 16–8 Acidification conditions for the production of Quarg and acid casein.

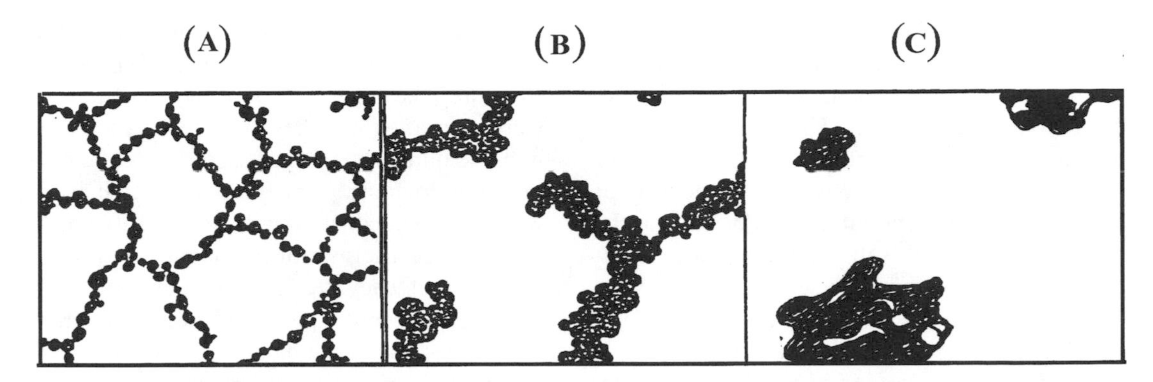

Figure 16–9 Schematic representation of a fine-structured (A) and coarse-structured (B) acid milk gel and a precipitate (C). The progression from a fine gel through a coarse gel to a precipitate is paralleled by an increasing degree of fusion of the milk protein (dark areas). Simultaneously, the protein network in the acidified milk becomes more open and porous.

chalkiness), appearance (coarseness or smoothness), and physicochemical stability (absence of wheying-off and graininess during storage) of fresh acid-curd cheese products, especially cold-pack varieties (e.g., Quarg and Fromage frais), where the curd, following whey separation, is not further treated. Cold-pack varieties are discussed below in more detail.

16.5.1 Syneresis

Syneresis, or whey expulsion, of acid milk gels is necessary for dehydration of the gel during cheese manufacture and may be achieved by subjecting the gel, following incubation, to concentration (i.e., whey removal) by cutting, stirring, cooking, whey drainage, and/or mechanical centrifugation. On the other hand, wheying-off in the final fresh cheese product during storage is undesirable, as it leads to the formation of a whey layer, which consumers view as undesirable. However, syneresis frequently occurs in acid gel–based products (e.g., yogurt and fresh cheese) because of the relatively high moisture:protein ratio compared with rennet-curd cheeses (e.g., ≈ 17.6 g H_2O/g protein in yogurt compared with ≈ 1.44g H_2O/g protein in Cheddar cheese). Moreover, owing to the fact the casein is at its isoelectric pH, the water in acid-curd cheeses is mainly physically imbibed rather than chemically bound by the protein. In contrast, about 15% of the moisture in young (i.e., < 1 week) rennet-curd cheeses is chemically bound by the paracasein, and the level appears to increase during ripening.

Syneresis of acid (and rennet) milk gels requires rearrangement and shrinkage of the casein matrix (gel) into a more compact structure. However, the gel is generally unable to contract to any appreciable degree when left under quiescent conditions, owing to its rigidity. Thus, the initiation of extreme syneresis (e.g., as required to separate the whey and recover the curd during manufacture) necessitates the application of a stress to the gel to break the gel strands. Breaking of the gel strands and hence the gel as a whole enhances syneresis by

- permitting a large portion of the entrapped whey to escape, via the surfaces of the newly created curd particles, from the matrix
- allowing the broken strands to come into closer proximity, enabling them to reknit into a more compact arrangement
- facilitating the physical expulsion of whey, due to the rearrangement and shrinkage of the protein phase, which has the effect of "squeezing out" the entrapped whey

Moreover, the prevailing conditions of low pH and relatively high temperature (usually > 22°C during separation) are conducive to casein dehydration and aggregation. The stress required to initiate syneresis during the manufacture of fresh acid-curd cheeses is applied in the form of external pressure via cutting and/or stirring the gel. The broken gel is then subjected to centrifugal force (e.g., in a nozzle separator) or gravitational force (e.g., by pouring the broken gel onto muslin bags suspended on a frame).

Acid milk gels formed in situ in the package, such as set natural yogurt, show little tendency to synerese if left undisturbed. However, even in this situation, spontaneous syneresis may eventually occur to a greater or lesser degree, depending on the level of fortification and processing conditions, such as preheating of milk, which causes differences in the porosity and structure of the gel. The syneresis may be partly due to slow proteolysis of the casein caused by enzymes of starter bacteria, a decrease in pH, and temperature fluctuations. Hydrolysis of casein influences its hydration and hence its tendency to aggregate. The change in the degree of casein aggregation results in internal stresses within the gel and rearrangement of the casein matrix, which in turn leads to syneresis. It has been suggested that casein hydrolysis may be responsible for the widely different practical experience (day-to-day in-factory and interfactory inconsistencies) regarding syneresis in set fermented milk products (Walstra, van Dijk, & Geurts, 1985). Changes in pH and temperature during storage, which would alter the state of casein aggregation, probably also contribute to spontaneous syneresis (shrinkage) or wheying-off.

Shrinkage of the casein particles of the network caused, for example, by a reduction in pH and/or an increase in temperature following gel formation enhances both induced and spontaneous syneresis.

Once syneresis has started, the outward flow of whey through the gel matrix becomes increasingly impeded over time by the sieve effect exerted on it by the relatively narrow pores of the gel, the porosity of which depends on the gel structure. The impedance of the pores to the outward migration of whey (syneresis) increases with the duration of syneresis, owing to the progressive contraction of the matrix as it loses whey and the reduction in the size of its pores. Hence, once a gel is broken, the rate of outward migration of whey decreases with time.

The structure of a gel has a marked effect on its ability to synerese or undergo wheying-off. For unidimensional flow through a porous medium, such as an acid milk gel, the rate of syneresis, v, may be expressed by Darcy's law:

$$v = B\Delta P/hl$$

where v is the whey flux (i.e., the volume flow rate in the direction of flow divided by the cross-sectional area perpendicular to this direction through which the whey flows, measured in meters per second); B is the permeability coefficient of the gel matrix, which corresponds to the average cross-sectional area of the gel pores; h is the viscosity of the whey flowing through the matrix; ΔP is the pressure gradient arising from syneretic pressure exerted on the entrapped serum by the matrix; and l is the distance over which the serum flows.

The permeability coefficient B depends on the volume fraction of the protein matrix and the spatial distribution of the matrix strands (i.e., gel fineness or coarseness). For a given syneretic pressure, the resistance to the passage of whey through the gel decreases as the permeability coefficient increases. Hence, a fine gel structure has a relatively low porosity and a lower permeability to outflowing whey than its coarse-structured counterpart. Therefore, while it is more difficult to remove whey from fine-structured gels during manufacture, they are much less prone to wheying-off during storage.

The influence of gel structure on its susceptibility to wheying-off may be easily explained by reference to Figure 16–9, which depicts the structural differences between fine and coarse gels in which the concentration of gel-forming protein is equal. In the fine gel, A, the micelles have formed into thin strands (chains), resulting in a highly branched, continuous gel network. Conversely, in the coarse gel, B, the micelles have fused to a much greater degree to produce thicker gel strands and a less continuous, more porous structure, which is more susceptible to syneresis during storage (whey drains easily through the large open channels between the network strands). The gel-forming protein in A is more uniformly distributed, and the gel has smaller interstitial spaces or pores. The relatively low porosity of gel A retards the outflow of whey and so endows the gel with better water-holding capacity than B and a low tendency toward syneresis.

16.5.2 Rheology

The structure of a gel also has a major influence on its rheological properties. For gels of similar composition, gel strength is primarily dependent on the homogeneity of the gel, which determines the number of stress-bearing strands per unit area of the gel. In the case of a gel to which a relatively small stress (i.e., much less than yield stress) is applied in direction x, the elastic shear modulus (G', i.e., ratio of shear stress to shear strain, σ/γ), which is an index of elasticity or strength of the gel, can be related to the number of strands per unit area according to this equation (Walstra & van Vliet, 1986):

$$G' = CN \times d_2F/dx^2$$

where N is the number of stress-bearing strands per unit area of the gel in a cross section perpendicular to x; C is a coefficient related to the characteristic length determining the geometry of the network; and dF is the change in Gibb's free energy when the aggregates in the strands are moved apart by a distance dx upon the application of the stress. The number of strands per unit area of a gel is determined by its fineness or coarseness, with a fine gel network having a greater number of stress-bearing strands than a coarse gel. The thickness and hence the strength of the stress-bearing strands is, on average, greater in the coarser gel because of the greater number of attractions between the aggregates. However, within the normal parameters of fresh acid-curd cheese manufacture, a fine gel generally has a greater gel firmness than a coarser gel with a similar composition and concentration of gel-forming protein. Compared to a gel, a precipitate (and its accompanying expressed whey) with the same level of gel-forming protein has a much lower G' value, as the rheological contribution ensues mainly from the continuous whey phase.

16.5.3 Sensory Attributes

The structure of the gel may also influence the sensory characteristics of fresh fermented products, especially in cold-pack products where, following its formation, the gel is subjected to little further processing (e.g., Quarg and Fromage frais) or none (e.g., set yogurt, where the gel is formed in its package). A smooth mouthfeel is generally an indicator of good quality in fresh cheese products. Cottage cheese is an exception, in that granularity, as imparted by the "chewy" curd particles, is an indicator of quality. Common sensory defects in fresh cheese products include "chalkiness" (perceived as a dry or powdery mouthfeel), grittiness, and graininess. Electron microscopic analyses of cheeses with these defects have revealed the presence of large protein conglomerates (masses of highly fused casein aggregates), suggesting that the defects ensue from excessive protein aggregation during gel formation and/or during whey separation. These defects are more likely to occur in products made from coarse-structured gels than fine-structured-gels, owing to the higher level of casein aggregation in the former. The defects are more prevalent in products where the gel, following fermentation, is concentrated and/or heated (e.g., Cream cheese); these conditions are conducive to protein dehydration and hence aggregation.

16.6 FACTORS THAT INFLUENCE THE STRUCTURE OF ACID GELS AND THE QUALITY OF FRESH CHEESE PRODUCTS

Many factors influence the structure of acid milk gels and hence impact the rheology, susceptibility to wheying-off, and the mouthfeel characteristics of fresh acid-curd cheeses. The principal compositional and processing factors that influence these are discussed below.

16.6.1 Level of Gel-Forming Protein

Higher concentrations of gel-forming protein generally result in denser gel matrices, which have a higher number of strands per unit volume and are more highly branched and less porous. The resultant gels are generally less prone to syneresis and are firmer and more elastic (i.e., higher G′). The lower susceptibility to syneresis is particularly desirable in products where the gel is essentially the final product, such as set and stirred curd yogurts. Wheying-off is a common defect in these products, especially those made without the addition of stabilizers, such as hydrocolloids. A high gel firmness is also desirable in yogurt, as it conveys the impression of being creamier, more viscous, and richer to the consumer. Owing to the fact that fresh cheeses are usually manufactured with a standard level of protein in the final product, it is envisaged that increasing the level of milk protein has a less dramatic impact on their rheological properties (e.g., fracture stress and firmness) than it has on those of yogurt, although the authors are not aware of any studies on this aspect of fresh acid-curd cheese. However, a higher gel firmness (due to a higher level of milk protein) is desirable in the manufacture of cold-pack fresh cheeses, as it reduces the susceptibility of the gel to shattering during subsequent whey separation, minimizes the loss of curd fines in the whey, and contributes positively to cheese yield. Thus, it is widespread practice in the commercial manufacture of yogurt and cold-pack fresh cheese products to increase the level of milk protein prior to acidification by, for example, ultrafiltration of the milk or the addition of dairy ingredients, such as skim milk powder, whey protein concentrate, or blends of dairy ingredients.

The structure of the gel is also markedly influenced by the ratio of casein to whey protein in the milk. Reducing the ratio from 4.6:1 to 3.2:1 results in set yogurt that has a finer, more highly branched, and less porous matrix; a smoother consistency; and a lower susceptibility to wheying-off during storage. Moreover, for a given concentration of protein and a given whey protein:casein ratio in the milk, the syneretic and rheological characteristics of the resultant gel may differ markedly, depending on the type of ingredient used to increase the level of milk protein. Stirred curd yogurt made from milk standardized to 5% protein with whey protein concentrate (75% protein, WPC 75) is markedly more viscous and less prone to syneresis than that made from milk standardized to 5% protein with WPC 35 (Guinee, Mullins, Reville, & Cotter, 1995). Differences in the performance of protein ingredients with the same type of protein, whether casein or whey protein, may be related to differences in the degree of whey protein denaturation, the type and level of minerals, and the level of other materials, such as lactose.

The effective concentration of milk protein may also be increased, while maintaining the actual protein level constant, by

- homogenization of the milk (as practiced in the production of yogurt and Cream cheese), which converts fat globules to pseudoprotein particles
- high heat treatment of the milk (e.g., 95°C × 5 min), which causes denaturation and binding of whey proteins to casein micelles (the denatured whey proteins become part of the ensuing gel, but undenatured whey proteins are soluble at their isoelectric pH, around 4.6, and do not participate in gel formation)

16.6.2 Heat Treatment of Milk

High heat treatment of milk (e.g., 90°C × 5 min compared to 72°C × 15 s) prior to culturing of fermented products, such as yogurt and Quarg, gives a smoother and firmer consistency (Figure 16–6). This effect is due to extensive denaturation of the whey proteins (e.g., > 70% of total) and their binding (especially of β-lactoglobulin), via disulfide interaction, to κ-casein; the denatured whey proteins subsequently become part of the gel. This interaction results in a higher effective concentration of gel-forming protein and a finer-structured gel with a lower permeability coefficient (and hence lower porosity) and a reduced propensity to spontaneous wheying-off (Table 16–2 and Figures 16–10 and 16–11). Electron microscopic analysis of high heat–treated milk shows that these complexes result in the formation of filamentous appendages that protrude from the surface of the micelles and prevent the close approach, and hence extensive fusion, of micelles upon subsequent acidification.

For similar levels of whey protein denaturation, the type of heat treatment has a significant influence on the textural parameters of fermented milks. Thus, the viscosity of natural yogurt produced from ultrahigh temperature–treated milk (130–150°C × 2–15 s) is lower than that of yogurt produced from high temperature-short time–treated milk (≈ 80–90°C × 0.5–5 min), which in turn is lower than that of yogurt produced from batch-heated milk (63–80°C × 10–40 min). The differences in viscosity and firmness at similar levels of whey protein denaturation may be associated with different types of denaturation (e.g., level of unfolding) and/or binding of denatured whey proteins to the casein, which alters the structure (i.e., coarseness or fineness) of the ensuing gel.

16.6.3 Incubation Temperature and Rate of Acidification

Increasing the acidification temperature of milk, in the range 20–43°C, results in

- the onset of gelation at a higher pH value (e.g., pH 5.5 at 43°C compared with pH 5.1 at 30°C)
- a coarser gel structure that is firmer (more elastic) and more prone to wheying-off during storage

These effects are thought to be associated with the faster rate of acidification (when using thermophilic cultures) and the reduced degree of casein dissociation from the micelles at the higher incubation temperature (see Section 16.3).

In an extreme situation, rapid acidification to pH 4.6 promotes rapid aggregation of casein and the formation of large dense aggregates that precipitate rather than form a gel. Gel formation by rapid acidification is, however, possible when the tendency of micelles to coagulate is reduced by acidifying to about pH 4.6 at a low temperature (0–4°C) and then heating the milk up slowly (≈ 0.5°C/min) under quiescent conditions to about 30°C.

16.6.4 pH of the Gel

The firmness (elastic shear modulus) of acid milk gels increases with decreasing pH toward the isoelectric point and is maximal at around pH 4.5. This effect is due to a greater degree of casein aggregation and a concomitant reduction in negative charge. Lowering the pH of acid milk gels at cutting (e.g., from 4.92 to 4.59 in Cottage cheese gels) reduces the level of syneresis. The latter effect may be attributed to the fact that the gel strands are more rigid at the lower pH and hence are less susceptible to breakage upon cutting. A lower degree of strand breakage affords less potential for new bonding sites, and matrix contraction is thus less severe than otherwise. Consequently, the pressure exerted on the entrapped serum by the matrix is relatively low and less wheying-off ensues. As a corollary, a decrease in pH during syneresis of an acid milk gel results in a higher level of syneresis than if the gel is brought to the same pH before cutting (Walstra et al., 1985).

Table 16–2 Effect of Heat Treatment on the Level of Whey Protein Denaturation in Skim Milk and the Permeability Coefficient of the Resultant Skim Milk Gels

	Heat Treatment	
	72°C × 15 s	*90°C × 5 min*
Milk composition		
Dry matter (g/kg)	98.6	98.4
Total protein (g/kg)	36.6	36.5
Casein number	75.2	87.2
NPN (% total N)	7.3	7.4
Whey protein denaturation (% total)	2.5	70.0
Gel Characteristics		
Elastic shear modulus, G′ at 16 hr (Pa)	100.0	20
Permeability coefficient, B (m^2)	2.56×10^{-13}	1.61×10^{-13}

Note: The data presented in Figures 16–6 and 16–11 are for gels obtained from the above milks.

16.6.5 Rennet Addition

It is common practice during the manufacture of some fresh cheese products, such as Quarg and Cottage cheese, to add a small quantity of rennet to the milk shortly after culture addition (e.g., ≈ 1–2 hr), when the pH is roughly 6.1–6.3. Typical levels of addition are 30–60 RU (or 0.5–1.0 ml single strength rennet) per 100 L. The rennet hydrolyzes some κ-casein, and there are concomitant decreases in

- the negative charge on the micelles (i.e., ζ-potential)
- casein dissociation from the micelles
- casein hydration over the pH region 6.6–4.6 (Figure 16–5)

These changes contribute to enhanced aggregation of the micelles, and gelation begins at a higher pH value than otherwise. Hence, a gel sufficiently firm for cutting and whey separation is obtained at a higher pH value (e.g., 4.8 compared with 4.6). In the absence of added rennet, cutting is performed at about pH 4.6 so as to prevent excessive loss of fines upon whey separation.

16.6.6 Added Stabilizers

A wide variety of plant- and animal-derived hydrocolloids (including pectins, pregelatinized starch, cellulose derivatives, alginates, carrageenans, and gelatin) are added to the milk prior to fermentation (e.g., in yogurt) or to the separated curd (e.g., Cream cheese and fresh cheese products) to immobilize water and reduce syneresis. Although these additives are very effective, their inclusion in the product detracts from the natural image, and some may have adverse effects on flavor and consistency. Recent studies have shown that fortification of yogurt milk with various dairy-based proteins (or blends) can be as effective as adding hydrocolloids in retarding wheying-off. The use of slime-producing cultures in yogurt has also been found to reduce syneresis.

16.6.7 Packaging and Retailing

Disturbance of set fermented milk products, such as by movement during cartoning and transport, creates stress for bond breakage and

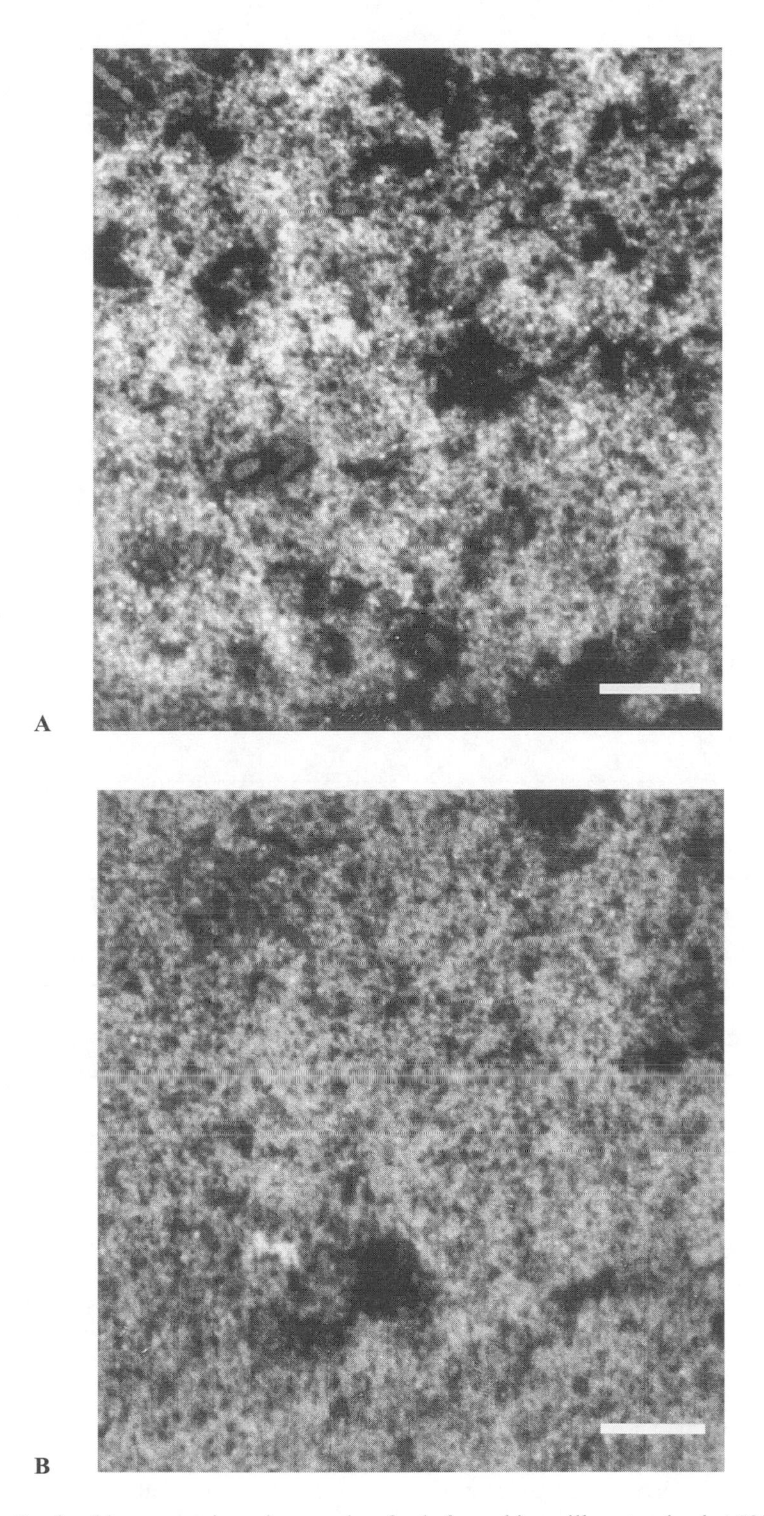

Figure 16–10 Confocal laser scanning micrographs of gels from skim milk pasteurized at 72°C × 15 s (A) and at 90°C × 5 min (B) (described in Table 16–2). Bar equals 10 μm.

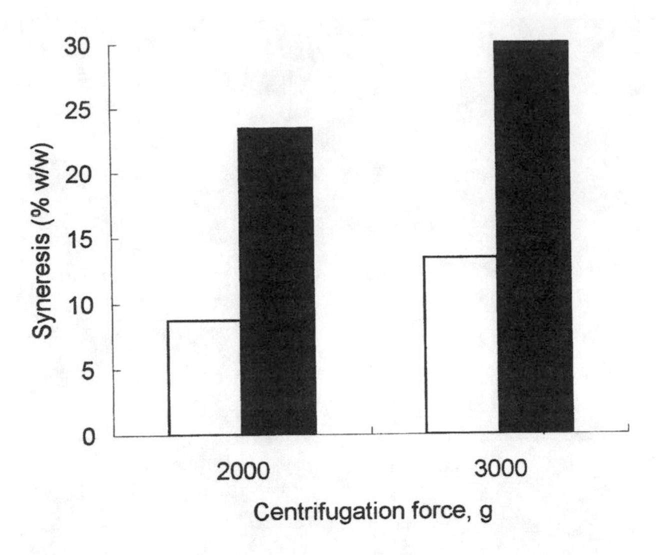

Figure 16–11 Level of syneresis from acid-coagulated gels formed from skim milk pasteurized at 72°C × 15 s (■) and at 90°C × 5 min (❑) (described in Table 16–2). After fermentation of the skim milk by a starter culture at 22°C, the gels (pH 4.6) were stirred gently and samples were weighed in centrifuge tubes and held at 8°C for 36–48 hr. The samples were then centrifuged at 2,000 or 3,000 g. The weight of whey expelled was expressed as a percentage of the original sample weight.

matrix rearrangement and may initiate or accentuate syneresis. For a given level of syneretic pressure, syneresis increases with increasing surface area to volume ratio of the gel. The shape of the package containing the gel may also influence syneresis. For example, in a package with sloping walls, the gel may have a tendency to detach from the walls, which leads to a stress in the gel and breakage. The breakage of the gel strands provides new bonding sites and rearrangement of the gel matrix into a more compact structure.

16.7 TREATMENTS OF THE SEPARATED CURD

In the production of many fresh cheese products, the gel produced upon acidification is subjected to a number of further processing steps, such as stirring, whey separation or concentration, heating, homogenization, agitation, and cooling (Figure 16–2). Various materials, such as cream, sugar, salt, fruit purees, and hydrocolloids, may be added to the curd. Such treatments influence the structural, rheological, and syneretic properties of the final product. The effects of various processing steps are discussed below.

Cutting of the gel into cubes, as in Cottage cheese, initiates syneresis, which is enhanced by cooking and stirring, as in the manufacture of rennet-curd cheese. Stirring of the gel (as in Quarg, Cream cheese, and Fromage frais) breaks the matrix strands to an extent that depends on the severity of agitation. For a given degree of agitation, cooling of the gel to below 20°C (e.g., to retard a further decrease in pH before whey separation) may result in more destruction of the gel matrix. The contribution of hydrophobic bonds to the integrity of the casein matrix decreases upon reducing the temperature (Hayakawa & Nakai, 1985; Kinsella, 1984). Increasing the temperature of the gel (e.g., in the range 25–85°C) enhances whey separation. Any factor that increases the firmness of the gel at separation (e.g., proximity to the isoelectric pH, rennet addition, milk homogenization, higher levels of gel-forming protein, and increased temperature at gelation) renders the gel less susceptible to fracture and disintegration at a given shear. Whey separation, which may be achieved by pouring the hot fluid onto cheese cloth, ultrafiltration, or centrifugation, results in concentration and aggregation of the broken pieces of gel to a greater or lesser degree. Collision of the pieces of gel during concentration forces them into close proximity and thus contributes to further casein aggregation. The moisture content of the curd is inversely related to the degree of aggregation. Factors that enhance casein aggregation and hence casein dehydration (e.g., higher temperature and higher centrifugation force) reduce the moisture content. Indeed, in the manufacture of many fresh acid-curd cheese varieties (such as Cream cheese and Cottage cheese), heating of the curds-whey mixture, after stirring of the gel, is done to induce whey separation and permit efficient recovery of curds with the desired dry matter level.

Homogenization or shearing of the dehydrated gel results in destruction of casein conglomerates to an extent dependent on the magnitude of the shear and thereby contributes to a more homogeneous size and spatial distribution of the matrix-forming material. The holding of hot Cream cheese at 75–85°C (e.g., in a surge tank prior to packaging) may result in a marked increase in elasticity. The thickening of the consistency is rather similar to the thickening process (often referred to as "creaming") observed in pasteurized processed cheese products when held for a long time at a high temperature (e.g., > 75°C) during processing. However, while imparting a more elastic character, holding a product at a high temperature may also lead to the development of chalkiness, powdery mouthfeel, grittiness, or graininess. These quality defects may be attributed to acute protein dehydration and the consequent formation of compact protein conglomerates. It is noteworthy that a high temperature and low pH (generally in the range 4.5–4.8 for fresh acid-curd products) are very conducive to casein aggregation. Hence, as frequently observed in the commercial production of Cream cheese, prolonged holding of the cheese at a high temperature (> 75°C) causes the cheese to be notably more brittle and firmer and to have a tendency toward elastic fracture. Slow cooling probably accentuates this defect, as protein aggregation has more time to proceed unhindered before being arrested by the lower temperature. The matrix of the cooled Cream cheese is more or less continuous, with the degree of continuity being governed by the size and spatial distribution of the matrix-forming material before cooling and the rate of cooling. A finer matrix manifests itself in a product that has a smoother appearance and mouthfeel and that is less susceptible to spontaneous wheying-off during storage. The addition of hydrocolloids to the curd also minimizes syneresis. Stabilizers that interact with casein, particularly κ-carrageenan, may interrupt matrix formation and yield a smoother, softer product.

16.8 MAJOR FRESH ACID-CURD CHEESE VARIETIES

16.8.1 Quarg and Related Varieties

Also referred to as Tvorog in some European countries, Quarg is a cheese of major commercial significance in Germany, where annual per capita consumption is about 7.1 kg. Quarg is a soft, homogeneous, mildly supple white cheese with a smooth mouthfeel and a clean, refreshing, mildly acidic flavor. The product is shelf-stable for 2–4 weeks at below 8°C. Stability refers to the absence of bacteriological deterioration, wheying-off (syneresis), and the development of graininess and overacid or bitter flavors during storage.

Quarg is sometimes loosely referred to as the German equivalent of Cottage cheese. However, while these cheeses are related, in the sense that both are fresh acid-curd products of similar composition, they are quite different from a production viewpoint and in sensory aspects. Cottage cheese is a (dressed) granular cheese, and its granules ideally have a chewy, meat-like texture.

Quarg is normally made from pasteurized (72–85°C × 15 s) skim milk cooled to 20–23°C and inoculated with an O-type culture. The milk, at 20–23°C, is held for 14–18 hr until the desired pH of 4.6–4.8 is reached. Shortly after culture addition (e.g., 1–2 hr), a small quantity of rennet (30–60 RU/100 L) is added when the pH is around 6.3–6.1. Rennet gives a firmer coagulum at a higher pH, and its addition minimizes casein loss upon subsequent whey separation and reduces the risk of overacidity (in the absence of rennet, a lower pH is required to obtain the same degree of curd firmness). The fermented gelled milk is stirred gently (100–200 rpm) into a smooth flowable consistency and pumped to a nozzle centrifuge, where it is separated into curds (Quarg) and whey, containing 0.65% whey protein and 0.19% nonprotein N. The Quarg is cooled immediately (< 10°C) en route to the buffer tank feeding the packaging machine.

Various methods have been employed to reduce the loss of whey proteins and to increase yield:

- In the *Westfalia thermoprocess*, the milk is pasteurized at 95–98°C × 2–3 min and the gelled milk (pH 4.6) is heated to 60°C × ≈ 3 min and then cooled to the separating temperature (25°C). In this process, 50–60% of the whey proteins are recovered in the cheese.
- In the *centriwhey process*, the whey from the separator is heated to 95°C to precipitate the whey proteins. The denatured whey proteins are recovered by centrifugation in the form of a concentrate (≈ 12–14% dry matter), which is added to the milk for the next batch of Quarg.
- In the *lactal process*, the whey from the separator is heated to 95°C to precipitate the whey proteins, which are allowed to settle. A concentrated whey (≈ 7–8% dry matter) is obtained upon partial decantation of the whey. A whey Quarg (17–18% solids), which is blended at a level of about 10% with regular Quarg, is produced upon further concentration using a nozzle centrifuge.
- *Ultrafiltration* of the gelled milk is now being used on a large scale for the commercial production of Quarg and other fresh cheese varieties. This method gives full recovery of whey proteins in the cheese. However, the nonprotein nitrogen fraction, which amounts to 2–3 g/kg of milk, is not concentrated and passes into the permeate.

Quarg cheeses that are made from milk pasteurized at a similar temperature (72°C × 15 s) and have the same level of protein (140 g/kg) and dry matter (180 g/kg) may have different levels of casein and whey protein, depending on the curd separation technique. Hence, the levels of casein and whey protein in skim milk Quarg produced by the separator (centrifuge) or ultrafiltration techniques are approximately 134 and 6 g/kg and 110 and 30 g/kg, respectively. However, whey proteins in the native state do not gel under the cheesemaking conditions used for Quarg made by either the ultrafiltration or centrifugation techniques. Hence, while the levels of total protein in the products produced by these methods are similar, the concentrations of gel-forming protein differ. Therefore, suppliers of ultrafiltration units to the Quarg industry recommend a high milk pasteurization treatment (95°C × 3–5 min). The high heat treatment results in binding of denatured whey proteins with the casein and thereby increases the level of gel-forming protein in ultrafiltration-produced Quarg to the same level as in separator-produced Quarg. Otherwise, ultrafiltration-produced Quarg, although containing the correct level of total protein, has a relatively thin consistency due to the lower level of matrix-building protein (110 g/kg compared to 134 g/kg). Quarg produced by the recommended ultrafiltration procedure (i.e., high heat milk treatment prior to culturing) has sensory characteristics similar to those of Quarg produced using the standard separator process.

Owing to its relatively high moisture (820 g/kg) and low protein (140 g/kg) levels, the shelf-life of Quarg is 2–4 weeks at below 8°C owing to microbial growth, syneresis, and off-flavor defects (especially bitterness). Microbiological quality can be improved by various methods, including the addition of sorbates, modified atmosphere packaging, thermization (58–60°C) of the broken gel prior to separation, and high heat treatment of the product (containing hydrocolloids). Addition of excessive rennet (> 78 RU/100 L), while increasing yield, leads to bitterness in Quarg after storage at 5–10°C for 4 weeks. Addition of rennet at a level of about 39 RU/100 L has been found to give the best compromise between yield and lack of bitter flavor. It is easier to prevent bitter flavor development in ultrafiltration-produced Quarg as the addition of rennet is not necessary, since yield is not affected by its addition. Quarg produced from lactose-hydrolyzed milk is sweeter and has a yellower color than that produced from normal milk.

Further processing (e.g., heating, homogenization, and/or aeration) and the addition of various ingredients (e.g., spices, herbs, fruit purees, cream, sugar, other fresh fermented products of different fat levels, and hydrocolloids) give rise to a range of Quarg-based products such as half-fat (20% FDM) and full-fat (40% FDM) Quarg, fruit and savory Quargs, Shrikhand, dairy desserts, and fresh cheese preparations.

Labneh, Labeneh, Ymer, and Fromage frais are similar to Quarg. These products are versions of concentrated natural stirred-curd yogurt and represent the interface between the classical fresh cheeses (i.e., standard separator Quarg and Cream cheese) and yogurt. As for yogurt, the milk is subjected to a high heat treatment (≈ 95°C × 5 min) in order to cause a high degree of β-lactoglobulin–κ-casein interaction, which in turn leads to a finer gel network, manifesting itself in the form of a product with a smoother mouthfeel and the ability to occlude more water. Unlike in the case of yogurt, the milk is not normally fortified, and the coagulated milk is concentrated by various means (pouring into cloth bags, as is done in traditional Labeneh manufacture, or use of a Quarg-type separator or ultrafiltration).

Production of these products generally involves

- standardization and heat treatment of the milk
- acidification by a yogurt-type starter culture to pH 4.6
- concentration of the coagulated milk
- homogenization of the curd

They may be flavored by the addition of sugar, fruit purees, or other condiments, which are blended in prior to homogenization. While acceptable products in their own right, they are, like Quarg and Cream cheese, often blended with yogurt and other fresh cheeses for the production of "new" fresh cheese preparations with different compositional, textural, and flavor attributes.

16.8.2 Cream Cheese and Related Varieties

Cream cheese (hot pack) is a cream-colored, clean, and slightly acid tasting product with a mild diacetyl flavor. Its consistency ranges from brittle (especially double Cream cheese) to spreadable (e.g., single Cream cheese). The product, which is most popular in North America, has a shelf-life of around 3 months at below 8°C.

Cream cheese is produced from standardized, homogenized, pasteurized (72–75°C × 30–90 s) milk (typically with a fat:protein ratio of 2.85:1 for double and 1.2:1 for single Cream cheese). Homogenization is important for the following reasons:

- It reduces creaming of fat during the fermentation or acidification stage and therefore prevents compositional heterogeneity of the resultant gel.
- It reduces fat losses upon subsequent whey separation.
- It converts, via the coating of fat with casein and whey protein, naturally emulsified fat

globules to pseudoprotein particles, which participate in gel formation upon subsequent acidification. The incorporation of fat into the gel structure by this means gives a smoother, firmer curd and therefore is especially important to the quality of cold-pack Cream cheese, in which the curd is not treated further.

Following pasteurization, the milk is cooled (20–30°C), inoculated with a D-type starter culture, and held at this temperature until the desired pH (≈ 4.5–4.8) is reached. The resulting gel is agitated gently, heated, and concentrated by various methods:

- draining through muslin bags at 60–90°C over 12–16 hr, as in the traditional batch method
- continuous concentration using a centrifugal curd separator at 70–85°C
- ultrafiltration at 50–55°C

In the batch method, the curd is cooled to around 10°C, and salt (5–10 g/kg) and hydrocolloid (< 5 g/kg; e.g., sodium alginate and carrageenan) are added. Then the treated curd may be packaged directly as cold-pack Cream cheese, which has a somewhat spongy, aerated consistency and a coarse appearance, or heated (70–85°C) and sheared by batch cooking (in a process cheese–type cooker at a relatively high shear rate for 4–15 min) or continuous cooking (in scraped-surface heat exchangers). The degree of heat and shear and the duration of cooking have a major influence on the consistency of the final product. Increasing the latter two parameters while keeping the temperature constant generally results in an increasingly more elastic and brittle texture. The hot, molten product, known as hot-pack Cream cheese, has a shelf-life of about 3 months at 4–8°C.

In the continuous production method, curd from the separator is treated continuously with stabilizer via an online metering and mixing device, pumped through a scraped-surface heat exchanger, homogenized online, and fed to the buffer tank feeding the packaging machine.

Owing to the thick, viscous consistency of the curd, concentration by ultrafiltration necessitates a two-stage process in order to maintain satisfactory flux rates and obtain the correct dry matter level. Stage 1 involves standard modules with centrifugal or positive displacement pumps, and stage 2 involves high-flow modules with positive displacement pumps.

The flavor diversity of Cream cheese may be increased by adding various flavors, spices, herbs, and sterilized, slurried, deboned fish. Cream cheese–type products may also be prepared by blending two or more acid-curd products (e.g., fermented cream, Ricotta, Quarg, and cultured buttermilk), then pasteurizing and homogenizing the blend and hot packing. These cheeses compare well with commercial double Cream cheese in all quality aspects.

The manufacture and sensory attributes of other cream cheese–type products, such as Neufchâtel and Petit Suisse, are similar to double Cream cheese. They differ mainly with respect to composition. Mascarpone, however, differs from other Cream cheese–type products in that acidification and coagulation are brought about by a combination of chemical acidification (using food-grade organic acids, such as lactic or citric) to about pH 5.0–5.6 and heat (90–95°C) rather than by starter fermentation at 20–45°C. The hot, acidified cream (400–500 g/kg fat), which is Mascarpone cheese, is packed in cartons or tubs and stored at around 5°C. The product, which has a shelf-life of 1–3 weeks, has a soft homogeneous texture and a slightly buttery, slightly tangy flavor.

16.8.3 Cottage Cheese

Cottage cheese is a soft granular unripened cheese in which the curd granules are lightly coated with a salted cream dressing. The flavor ranges from a cream-like blandness to mildly acidic with overtones of diacetyl.

Cottage cheese is made from skimmed, pasteurized (72°C × 15 s) milk inoculated with a lactic acid–producing starter at a level depending on the set time. Long-set Cottage cheese is inoculated with 5–10 g/kg starter and incubated at 21–23°C for 14–15 hr, whereas short-set Cottage cheese is inoculated with 5–10 g/kg starter and incubated at 30–32°C for 4–5 hr. The starter normally consists of lactic acid–producing bacteria (*Lc. lactis* subsp. *cremoris* or *Lc. lactis* subsp. *lactis*) and flavor-producing bacteria (citrate-positive lactococci or *Leuconostoc mesenteroides* subsp. *cremoris*). The metabolism of citrate by the latter results in the production of the flavor compounds diacetyl and acetate as well as CO_2, which vaporizes upon subsequent cooking (≈ 55°C) and forms gas bubbles that tend to cause floating of curd particles to the top of the whey. Excessive CO_2 production gives rise to the defect known as "floating curd," which reduces the yield. The curd is fragile and shatters upon cutting and stirring to give fines that are lost in the whey and wash water. However, selection of a suitable starter with the correct balance of acid and flavor producers gives cheese of a satisfactory flavor while avoiding the above defect. Much of the diacetyl produced (> 3.2 mg/kg) is lost in the whey. The risk of floating curd is minimized by removing the flavor-producing strains from the culture. Instead, diacetyl may be added directly to the cream dressing or the creaming mixture may be cultured with a diacetyl-producing starter.

Another starter-related problem in Cottage cheese manufacture is agglutination, and its associated defect is known as "minor sludge formation" (i.e., the formation of a layer of fragile, discolored [yellowish] material at the bottom of the vat during acidification). Agglutination of starter lactococci is caused by immunoglobulins that occur naturally in milk as part of the whey protein fraction and are at a particularly high level in colostrum and mastitic milk. Upon agglutination, the starter bacteria clump together and settle to the bottom of the cheese vat. Lactic acid production becomes localized, and a pH difference of around 0.5 unit between the milk at the top and the bottom of the vat occurs after about 4 hours of incubation. Consequently, precipitation of casein (≈ 4–8% of the total casein) results in the formation of a sludge, which shatters upon subsequent cutting and stirring to produce fines that are lost during whey drainage and washing. The risk of starter agglutination is reduced by homogenizing the skim milk (e.g., at a pressure of ≈ 155 bar) or the bulk culture (≈ 176 bar) or by the addition of lecithin to the bulk culture. Homogenization of skim milk destroys agglutins, while homogenization or addition of lecithin to the culture causes fragmentation of starter chains without affecting cell numbers or acid production. The defect known as "major sludge formation," in which all the casein forms a precipitate (which cannot be made into satisfactory cheese) rather than a gel during acidification, is thought to be due to phage infection of starter after acid development is well advanced (i.e., at pH 5.2–4.9).

As in the manufacture of Quarg, a small amount of rennet is added to the milk when the pH reaches 6.3 with the aim of increasing gel firmness at cutting. The optimum pH at cutting is about 4.8. At a constant cooking temperature in the range 50–60°C, the firmness and dry matter of the final Cottage cheese increase with pH at cutting in the range 4.6–4.9. At a cutting pH above 4.8, the curds tend to mat upon cooking, giving rise to clumping of the curd granules. However, if the heat treatment is more severe than normal pasteurization (i.e., > 72°C), the curds synerese poorly during subsequent stirring and cooking, resulting in a soft, mushy product. Increasing the cutting pH to above 4.9 (e.g., to 5.1) and increasing the level of rennet added enhances the syneretic properties of curd from high heat–treated milk without the risk of matting. In addition to heat treatment of the milk, other factors that influence the firmness and syneretic properties of the curd during cooking determine the desired cutting pH:

- Milk composition (stage of lactation). At a given cutting pH, higher casein levels give firmer gels, which synerese better than those from low-protein milk.
- Level of added rennet. Higher levels of rennet addition promote firmer gels at a given

pH. However, as in Quarg, excess rennet causes bitterness.
- Grain size. Large curd granules, because they require a longer time to dehydrate upon cooking and stirring, tend to be more susceptible than smaller grains to shattering and therefore generally necessitate a higher cutting pH and hence a coagulum that firms more rapidly upon cooking.

All factors being equal, reducing the pH from 4.8 to 4.6 tends to give a softer, more fragile coagulum, an effect that may be due to increased loss of casein-bound calcium, which impairs the ability of the curd to synerese and become firm during stirring and cooking. The cut size depends on whether a large curd (≈ 2 cm cut) or small curd (≈ 1 cm cut) end product is desired.

After cutting, the curds are allowed to settle for 5–15 min (depending on their firmness) in order to undergo "healing" of the cut surfaces, then gently agitated and cooked slowly (at the rate of increase of 1°C per 5 min to 40°C and 2°C per 5 min to 55°C). Higher cooking temperatures enhance syneresis and consequently give a product in which the curd granules are more defined and stronger and have a more chewy, meatlike texture. When the curd particles have acquired the correct degree of resilience and firmness, the whey is drained off and the curds are washed to

- prevent them from matting together
- remove lactose and minimize the growth of spoilage bacteria in the final product
- remove lactic acid and therefore prevent the likelihood of overly acid tasting cheese

Washing involves the addition of water (at a volume equivalent to the volume of whey drained) at about 25°C to the curds, stirring for 2–3 min, and then draining. This process is repeated two or three times using ice-water chlorinated to a level of 5–25 mg/kg. The wash water is drained, and the cooled curd grains are trenched and allowed to stand for at least 1 hr until all wash water has drained away.

A homogenized, pasteurized, salted (10–40 g/kg salt) cream dressing (90–180 g/kg fat) is mixed with the curds in the proper amount to create a finished product with the desired level of salt (8–10 g/kg) and fat. In dry-curd Cottage cheese, the level of fat is below 5 g/kg; in low-fat Cottage cheese, it is between 5 g/kg and 20 g/kg; and in Cottage cheese, it is not less than 40 g/kg. The dressing may be cultured or contain added starter distillate. The use of such a dressing is advocated when a plant is experiencing production difficulties as a result of "floating curd," as discussed above.

Recent developments in Cottage cheese manufacture include direct-acid-set coagulation using food-grade acid or acidogen and a transition from all-cheese-vat batch operations to continuous production systems, where washing, cooling, and draining (pressing) of the curds occur on rotating belts.

16.8.4 Ricotta and Ricottone

Ricotta is a soft, cream-colored unripened cheese with a sweet-cream and somewhat nutty or caramel flavor and a delicate aerated texture. The cheese, which was produced traditionally in Italy from cheese whey from ewe milk, now enjoys more widespread popularity, particularly in North America and Western Europe, where it is produced mainly from whole or partly skimmed bovine milk or whey–skim milk mixtures.

In the traditional batch production method, the milk or milk-whey blend is directly acidified to around pH 5.9–6.0 by the addition of food-grade acid (e.g., acetic, citric, or lactic acid), starter culture (≈ 200 g/kg inoculum), or acid whey powder (≈ 25% addition). Heating of the milk to about 80°C by direct steam injection induces coagulation of the casein and whey proteins and thus results in the formation of curd flocs in the whey after about 30 min, at which point direct steam heating is discontinued. The curd particles, now under quiescent conditions, begin to coalesce and float to the surface, where they form into a layer. Indirect steam (applied to the vat jacket), together with manual movement of curd from the vat walls toward the center, initiates the process of "rolling," whereby the curds roll from the walls toward the center of the vat and there form into a layer that is easily recovered by scooping (using perforated scoops). The curds are filled into perforated molds and allowed to drain for 4–6 hr at below 8°C.

The above procedure gives only partial recovery of the whey proteins. A secondary precipitation, in which the whey from Ricotta cheese manufacture is acidified to pH 5.4 with citric acid, heated to 80°C, and treated as for Ricotta, is therefore sometimes practiced in order to recover remaining whey proteins in the form of Ricottone cheese. Ricottone has a relatively hard and tough consistency and therefore is normally blended with Ricotta in an attempt to moderate its undesirable features.

Owing to its relatively high pH, its high moisture content (Table 16–1), and the manual method of filling, Ricotta produced by the traditional method is very susceptible to spoilage by yeasts, molds, and bacteria and hence has a relatively short shelf-life—1 to 3 weeks at 4°C. However, significant advances have been made in the automation of Ricotta cheese production with the objective of improving curd separation, cheese yield, and shelf-life. Excellent quality Ricotta has been produced using an ultrafiltration-based production method. Whole milk is acidified with acid whey powder to pH 5.9 and ultrafiltered at 55°C to 11.6% protein (≈ 29% dry matter). The retentate is heated batchwise at 80°C for 2 min to induce coagulation (without whey separation). The coagulum is hot packed and has a shelf-life of at least 9 weeks at 9°C. In another process based on ultrafiltration, milk and/or whey is standardized, pasteurized at pH 6.3, cooled to 50°C, and ultrafiltered to 30% dry matter. The retentate is heated to 90°C and continuously acidified to pH 5.75–6.0 at a pressure of 1.0–1.5 bar. The pressure is reduced to induce coagulation without whey separation, and the curds are cooled to 70°C and hot packed. In a process developed by Modler (1988), a 20:80 blend of whole milk and concentrated whey (neutralized to pH 6.9–7.1) is heated from 4°C to 92°C, pumped to a 10 min holding tube (to induce whey protein denaturation), and acidified, by on-line dosing with citric acid (250 g/kg), to induce coagulation. The curds are separated from the "deproteinated" whey on a nylon conveyor belt. This process gave excellent recoveries of fat and protein (99.6 and 99.5 g/kg, respectively). Other methods employed to increase yield and automate the production of Ricotta include filtration of whey after curd removal and the use of perforated tubes or baskets in the bottom of the curd-forming vat to collect the curds after whey drainage.

Ricotta cheese, in addition to being an acceptable product itself, has many applications, including use as a base for whipped dairy desserts, Cream cheese, and pasteurized processed cheese products and use in confectionery fillings and cheesecake.

16.8.5 Queso Blanco

Queso blanco (white cheese) is the generic name for white, semi-hard cheeses produced in Central and South America. These cheeses can be consumed fresh but some cheeses may be held for 2–8 weeks before consumption (Torres & Chandan, 1981a, 1981b). Elsewhere in the world, similar cheeses include Chhana and Paneer in India, Armavir in the Western Caucasus, Zsirpi in the Himalayas, and low-salt (< 10 g/kg), high-moisture (> 600 g/kg), unripened cheeses in the Balkans (e.g., Beli sir types). Beli sir–type cheeses may also be salted and ripened in brine for up to 2 months to give white pickled cheeses usually known by local names, such as Travnicki sir and Sjenicki sir.

In Latin America, Queso blanco covers many white cheese varieties, which differ from each other by the method of production (i.e., acid/heat or rennet coagulated), composition, size, shape, and region of production. Examples include Queso de Cincho, Queso del Pais, and Queso Llanero, which are acid/heat coagulated, and Queso de Matera and Queso Pasteurizado, which are rennet coagulated. The use of a high temperature (80–90°C) during the production of acid/heat-coagulated white cheeses was, traditionally, very effective for improving the keeping quality in warm climates.

In general, Queso blanco–type cheeses are creamy, highly salted, and acid in flavor. Their texture and body resemble those of very young high-moisture Cheddar, and they have good slicing properties. The average composition of a fresh cheese is 40–50% moisture, 22–25% protein, and 15–20% fat (Kosikowski & Mistry, 1997; Torres & Chandan, 1981a).

The production method for acid/heat-coagulated Queso blanco varies but generally involves the following steps:

- Standardization of milk to the required protein:fat ratio to achieve the desired end-product composition.
- Heat treatment of the milk to about 82–85°C, followed by holding for about 5 min. This heat treatment achieves partial denaturation (≈ 600–700 g/kg) of whey proteins, which complex with the caseins and are therefore recovered with the casein upon subsequent coagulation.
- Acidification of the hot milk to pH 5.3 by adding food-grade acid (acetic, citric, or tartaric acid, lime juice, or lactic culture) to the milk while stirring gently. Citric and acetic acid are used most frequently, sometimes jointly. The acids are diluted prior to addition, typically to a concentration of 50–100 g/L, to facilitate dispersion and prevent localized coagulation.
- Curd formation. Protein aggregation occurs rapidly under nonquiescent conditions owing to the low pH and high temperature of the milk, resulting in the formation of curd particles and whey.
- Curd recovery, salting, molding, and pressing. The curd particles are separated from the whey and dry stirred, dry-salted (at a level sufficient to result in ≈ 20–40 g/kg in the final cheese), and pressed. The pressed cheese is cut into consumer-size portions, which are vacuum packed and stored at 4–8°C. The product is shelf-stable at this temperature for 2–3 months.

Queso blanco is traditionally consumed fresh because, as a result of high heat treatment during curd formation, very few biochemical changes occur during storage. However, starter bacteria (*Lactobacillus* spp) and/or exogenous lipases may be added to the curd before salting and pressing to improve the flavor of the cheese during storage. Major volatile compounds that contribute to the flavor and aroma of Queso blanco include acetaldehyde, acetone, isopropanol, butanol and formic, acetic, propionic, and butyric acids. The pH of Queso blanco decreases from about 5.2 to 4.9 during ripening, an effect that may be due to fermentation of residual lactose to lactic acid by heat-stable indigenous bacteria in milk that survive cheesemaking or by post-manufacture contaminating bacteria (Torres & Chandan, 1981b).

One of the interesting properties of the cheese is its flow resistance upon heating, owing to the inclusion of whey proteins that gel upon heating. This enables the cheese to be deep-fat fried in the preparation of many savory snack foods, such as cheese sticks in batter.

16.8.6 Whey Cheeses

Brunost, meaning "brown cheese," refers to a distinctive group of Norwegian unripened "cheese" varieties made from sweet whey (rennet casein or cheese whey) or skim milk, to which cream may be added (Kosikowski & Mistry, 1997; Otterholm, 1984). The best known members of the group include Mysost and Gudbrandsdalsost (≥ 350 g/kg fat-in-dry-matter [FDM]; from bovine and caprine milk), Ektegeitost (≥ 33% FDM; from goat milk components), Flotemysost (≥ 33% FDM; from bovine milk), and Primost.

In the classical sense, Mysost is not a cheese but rather a fat-protein–enriched concentrated heated whey. However, being unripened, it may be defined as a "fresh cheese." The cheese, characterized by a light golden to a dark brown color, is produced using these steps:

- In this step, known as "standardization," the whey, milk, and/or cream are blended to give the correct end-product FDM. The whey is first filtered or decanted to remove casein particles, which otherwise occur in the product as black-brown specks.

- The standardized whey is preconcentrated in a multistage film evaporator to 50–60% total solids. The viscous concentrate is then transferred to special steam-jacketed, conical kettles in which it is further concentrated to 800–820 g/kg dry matter by being heated, while agitated vigorously, under a vacuum.
- The molten viscous mass is transferred to a vessel with a strong rotary, swept metal agitator. Kneading over a 20 min period while slow atmospheric cooling occurs helps to give the product its butterlike, plastic consistency and prevents the formation of large lactose crystals (and therefore grittiness).

REFERENCES

Guinee, T.P., Mullins, C.G., Reville, W.J., & Cotter, M.P. (1995). Physical properties of stirred-curd unsweetened yoghurts stabilized with different dairy ingredients. *Milchwissenschaft, 50*, 196–200.

Guinee, T.P., Pudja, P.D., & Farkye, N.Y. (1993). Fresh acid-curd cheese varieties. In P.F. Fox (Ed.), *Cheese: Chemistry, physics and microbiology* (2d ed., Vol. 1). London: Chapman & Hall.

Hayakawa, S., & Nakai, S. (1985). Relationships of hydrophobicity and net charge to the solubility of milk and soy proteins. *Journal of Food Science, 50*, 486–491.

Kinsella, J.E. (1984). Milk proteins: Physicochemical and functional properties. *Critical Reviews in Food Science and Nutrition, 21*, 197–262.

Kosikowski, F.V., & Mistry, V.V. (1977). *Cheese and fermented milk food: Vol. 1. Origins and principles.* Westport, CT: F.V. Kosikowski, LLC.

Modler, H.W. (1988). Development of a continuous process for the production of Ricotta cheese. *Journal of Dairy Science, 71*, 2003–2009.

Oterholm, A. (1984). *Cheesemaking in Norway* [Bulletin No. 171]. Brussels: International Dairy Federation.

Sørensen, H.H. (1997). *The world market for cheese* [Bulletin No. 326]. Brussels: International Dairy Federation.

Torres, N., & Chandan, R.C. (1981a). Latin American white cheese [A review]. *Journal of Dairy Science, 64*, 552–557.

Torres, N., & Chandan, R.C. (1981b). Flavor and texture development in Latin American white cheese. *Journal of Dairy Science, 64*, 2161–2169.

Walstra, P., van Dijk, H.J.M., & Geurts, T.J. (1985). The syneresis of curd: 1. General considerations and literature review. *Netherlands Milk and Dairy Journal, 39*, 209–246.

Walstra, P., & van Vliet, T. (1986). The physical chemistry of curd-making. *Netherlands Milk and Dairy Journal, 40*, 241–259.

CHAPTER 17

Principal Families of Cheese

17.1 INTRODUCTION

The diversity of cheese types is truly breathtaking. Despite the limited range of raw materials (bovine, ovine, caprine, or buffalo milk), approximately 500 varieties of cheese recognized by the International Dairy Federation (Burkhalter, 1981) are produced. Numerous other minor local cheeses are also manufactured.

In order to facilitate their study, a number of attempts have been made to classify cheese varieties into meaningful groups or families. As discussed by Fox (1993a), traditional classification schemes have been based principally on moisture content, such as hard, semi-hard, or soft. Although this is a widely used basis for classification, it suffers from a serious drawback: it groups together cheeses with widely different characteristics and manufacturing protocols. For example, Cheddar, Parmesan, and Emmental are often grouped together as hard cheeses, although they have quite different flavors and the methods for their manufacture are quite different. Attempts have been made to make this scheme more discriminating by including factors such as origin of the cheese milk, moisture content, texture, principal ripening microorganisms, and cooking temperature (Exhibit 17–1). A classification scheme proposed by Walstra (see Fox, 1993a) differentiates between varieties based on the type of primary and secondary starter used and moisture content (expressed as the moisture:protein ratio). Walter and Hargrove (1972), who classified cheeses on the basis of manufacturing technique, suggested that there are only 18 distinct types of natural cheese, which they grouped into 8 families under the headings very hard, hard, semi-soft, and soft:

1. Very hard (grating)
   - 1.1 Ripened by bacteria (e.g., Parmesan)
2. Hard
   - 2.1 Ripened by bacteria, without eyes (e.g., Cheddar)
   - 2.2 Ripened by bacteria, with eyes (e.g., Emmental)
3. Semi-soft
   - 3.1 Ripened principally by bacteria (e.g., Gouda)
   - 3.2 Ripened by bacteria and surface microorganisms (e.g., Limburger)
   - 3.3 Ripened principally by blue mold in the interior (e.g., Roquefort)
4. Soft
   - 4.1 Ripened (e.g., Brie)
   - 4.2 Unripened (e.g., Cottage)

Unfortunately, none of these schemes is completely satisfactory and thus none is universally accepted. Fox (1993a) proposed a number of "superfamilies" into which all cheeses would be grouped based on the method of milk coagulation:

- rennet-coagulated cheeses (most major international cheese varieties)
- acid-coagulated cheeses (e.g., Cottage and Quarg)

Exhibit 17–1 Classification of Cheese According to Source of Milk, Moisture Content, Texture, and Ripening Agents

1.	COW MILK										
1.1	*Hard (< 42% H_2O)*	*1.2*	*Semi-hard/semi-soft soft (43–55% H_2O)*	*1.3*	*Soft (> 55% H_2O)*	*1.4*	*Fresh, rennet*	*1.5*	*Fresh, acid*	*1.6*	*Fresh*
1.1.1	Grating cheese (extra hard)	1.2.1	Small round openings	1.3.1	Blue veined						
1.1.2	Large round openings	1.2.2	Irregular openings	1.3.2	White surface mold						
1.1.3	Medium round openings	1.2.3	No openings	1.3.3	Bacterial surface smear						
1.1.4	Small round openings	1.2.4	Blue veined	1.3.4	No rind						
1.1.5	Irregular openings										
1.1.6	No openings										
2.	SHEEP MILK Hard; semi-hard; soft; blue-veined; fresh										
3.	GOAT MILK										
4.	BUFFALO MILK										

Note: Unless otherwise stated, the cheeses are internally bacterial ripened.

- heat/acid coagulated (e.g., Ricotta)
- concentration/crystallization (e.g., Mysost)

All ripened cheeses are coagulated by rennet (≈ 75% of total world production). Acid-curd cheeses (see Chapter 16) are the next most important group. Coagulation by a combination of heat and acid is used for a few minor varieties, including Ricotta. Concentration/crystallization is used in Norway to produce "whey cheeses" (e.g., Mysost).

There is a great diversity of rennet-coagulated cheeses, and therefore they must be classified further. One such classification scheme is proposed in Figure 17–1. Rennet-coagulated varieties are subdivided into relatively homogeneous groups based on the characteristic ripening agents and/or manufacturing technology. The most diverse family of rennet-coagulated cheeses is that containing the internal bacterially ripened varieties, which include most hard and semi-hard cheeses. The term *internally bacterially ripened* may be somewhat misleading, since indigenous milk enzymes and residual coagulant also play important roles in the ripening of these cheese varieties. This group may be subdivided based on moisture content (extra hard, hard, and semi-hard) and on whether the cheese has eyes. Many varieties produced on a large industrial scale are included in this group. Parmesan (extra hard) is used as a grating cheese, and its manufacture is characterized by a high cook temperature. Cheddar and British territorial varieties (for which the curds are often textured and dry-salted) are classified as hard or semi-hard internal bacterially ripened cheeses. Internal bacterially ripened cheeses with eyes are further subdivided on the basis of moisture content into hard varieties (e.g., Emmental), in which the numerous large eyes are formed by CO_2 produced on fermentation of lactate by *Propionibacterium freudenreichii* subsp. *shermanii*, and semi-hard varieties (e.g., Edam and Gouda), in which a few small eyes develop due to the formation of CO_2 by fermentation of citrate by a component of the starter (see Chapters 10 and 11).

Most of the varieties classified in groups other than internal bacterially ripened cheeses are soft or semi-hard. Pasta filata cheeses (e.g., Mozzarella) are characterized by stretching in hot water, which texturizes the curd prior to salting. Mold-ripened cheeses are subdivided into surface mold–ripened varieties (e.g., Camembert and Brie), in which ripening is characterized by the growth of *Penicillium camemberti* on the surface, and internal mold–ripened cheeses (Blue cheeses), in which *P. roqueforti* grows in fissures throughout the mass of the cheese. Surface smear–ripened cheeses are characterized by the development of a complex microflora consisting initially of yeasts and ultimately of bacteria (particularly coryneforms) on the cheese surface during ripening. White, brined cheeses, including Feta and Domiati, are ripened under brine and have a high salt content, and consequently they are grouped together in a separate category.

The classification scheme proposed in Figure 17–1 is not without inconsistencies. A cursory glance will show that cheeses made from the milk of different species are grouped together (e.g., Roquefort and Gorgonzola are Blue cheeses but the former is made from sheep milk and the later from cow milk) and that the subdivision between hard and semi-hard cheeses is rather arbitrary. There is also some crossover between categories. Gruyère is classified as an internal bacterially ripened variety with eyes, but it is also characterized by the growth of a surface microflora, while some cheeses classified as surface ripened (e.g., Havarti and Port du Salut) are often produced without a surface flora and therefore are, in effect, soft internal bacterially ripened varieties. Likewise, Pasta filata and high-salt varieties are considered as separate families because of their unique technologies (stretching and ripening under brine, respectively), but they are actually ripened by the same agents as internal bacterially ripened cheeses. However, we believe that the scheme proposed in Figure 17–1 is a useful basis for classification, and therefore the diversity of cheeses will be discussed under these headings. Ultrafiltration technology is used for the manufacture of some

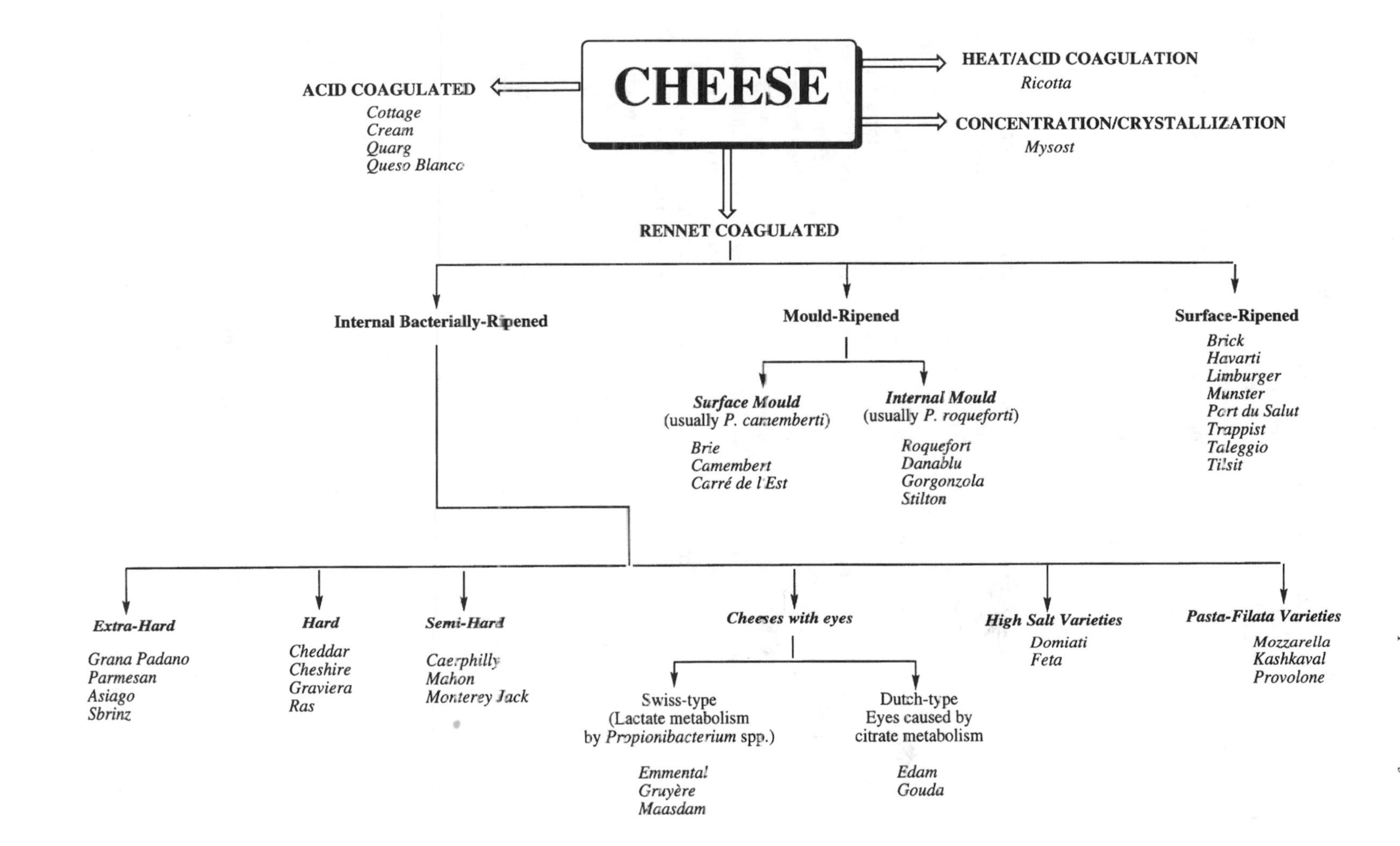

Figure 17–1 The diversity of cheese. Cheese varieties are classified into superfamilies based on the method of coagulation and further subdivided based on the principal ripening agents and/or characteristic technology.

cheese varieties, which are discussed separately (Section 17.6). The following discussion is based largely on descriptions of cheeses given in Fox (1993b), Kosikowski and Mistry (1997), Robinson (1995), Robinson and Wilbey (1998), and Scott (1986), to whom the reader is referred for detailed manufacturing protocols. In some cases, the manufacturing protocols described in these sources for certain varieties are inconsistent, and therefore should be treated with due caution. This divergence is particularly true for minor varieties, which are probably ill defined and variable in any case. The typical composition of a selection of cheeses is shown in Appendix 17–A.

A number of cheese varieties have Protected Designation of Origin (PDO) (or "Appellation d'Origine Contrôlée") status, which recognizes a specific heritage and provides consumers with a guarantee of authenticity. Unlike commercial trademarks, PDO denomination reflects a collective heritage and may be used by all producers of a particular cheese in a particular geographical area. PDO cheeses are protected by the European Union under various international agreements (Bertozzi & Pandri, 1993). PDO denomination also certifies that the cheese has been made using specified (usually traditional) technology. A list of cheeses with PDO status is given in Table 17–1.

17.2 RENNET-COAGULATED CHEESES

17.2.1 Internal Bacterially Ripened Varieties

Internal bacterially ripened cheese varieties form a very diverse group of cheeses characterized by the absence of a surface microflora or internal mold growth. The agents that contribute to the ripening of internal bacterially ripened varieties originate from the milk (plasmin and other enzymes), the rennet (chymosin and/or other proteinases and, in certain cases, lipases), and the internal bacterial microflora (starter and nonstarter bacteria in all cases and adjunct starter bacteria in some cheeses, particularly those in which eyes develop). Some internal bacterially ripened varieties are easily classified into homogeneous groups based on some distinctive technology (e.g., Pasta filata cheeses or cheeses ripened under brine). However, the classification used here for most varieties is based on texture (extra hard, hard, and semi-hard) and is therefore somewhat arbitrary.

Extra Hard Varieties

The majority of extra hard internal bacterially ripened cheese varieties originated in Italy. Guinee and Fox (1987) grouped extra hard Italian-type cheeses into 3 subcategories: Parmesan and related varieties, Asiago, and Romano. These cheeses are usually matured for a long period (2 years or more) and often have a hard, grainy texture. They may be consumed as table cheeses when young or in grated form when mature. The hard texture of these cheeses results from the use of semi-skimmed milk in their manufacture, a high cooking temperature, and evaporation of moisture during ripening.

Parmigiano-Reggiano and Grana Padano are important members of this group. They are produced from raw milk in the Po valley in Northern Italy and are protected by designation of origin. A grainy texture in the mature cheese is desirable (hence the name "Grana"; Figure 17–2). Parmesan-type cheeses are made worldwide, particularly in the United States, from pasteurized milk. These cheeses are often smaller than Italian Grana-type cheeses, are cooked to a lower temperature (≈ 50°C) than the traditional product (54°C), are salted more heavily, are ripened for a shorter period, and are usually used in grated form.

The manufacturing protocol for Parmigiano-Reggiano is summarized in Figure 17–3. The major features of its manufacture are the use of semi-skimmed raw milk produced by gravity creaming. In traditional manufacture, the evening milk is creamed in shallow vats and the lower skimmed milk layer is drawn off and mixed with whole morning milk. Traditionally, the milk was coagulated and the curds cooked in copper vats that were shaped like an inverted bell and heated by a fire underneath. Modern vats are

Table 17–1 Cheeses with Protected Designations of Origin

Milk	*Variety*
France	
Bovine	Abondance
	Beaufort
	Bleu d'Auvergne
	Bleu des Causses
	Bleu de Gex-Haut Jura-Septmoncel
	Brie de Meaux
	Brie de Melun
	Camembert de Normandie
	Cantal
	Chaource
	Comté
	Epoisse de Bourgogne
	Fourme de'Ambert ou Montbrison
	Laguiole
	Langres
	Livarot
	Maroilles or Marolles
	Mont d'Or/Vacherin du Haut Doubs
	Münster or Münster Géromé
	Neufchâtel
	Pont l'Evêque
	Reblochon and Petit Reblochon
	Saint Nectaire
	Salers
Caprine	Chabichou du Poitou
	Crottin de Chavignol
	Picodon de l'Ardèche/Drome
	Pouligny Saint Pierre
	Sainte Maure de Touraine
	Selles sur Cher
Ovine	Ossau-Iraty-Brebis-Pyrénées
	Roquefort
Caprine-ovine whey	Brocciu Corse or Brocciu
Spain	
Bovine	Cantabria
	Mahón
Ovine	Idiazábal
	Manchego
	Roncal

continues

Table 17–1 continued

Milk	*Variety*
Mixed (cow, goat, ewe)	Cabrales
	Liebana
Portugal	
Bovine	San Jorge
Ovine	Azeitao
	Serpa
	Serra da Estrela
	Castelo Blanco
	Picante da Beira Baixa Amarelo
Italy	
Bovine	Asiago
	Bra
	Castelmagno
	Fontina
	Formai de Mut
	Gorgonzola
	Grana Padano
	Montasio
	Murazzano
	Parmigiano-Reggiano
	Raschera
	Robiola di Roccaverano
	Taleggio
Ovine	Canestrato Pugliese
	Casciotta di Urbino
	Fiore Sardo
	Pecorino Romano
	Pecorino Siciliano
	Pecorino Toscano
	Pecorino Sardo

often made from copper-plated steel and have steam jackets for heating, but the traditional shape is retained. Acidification is by a whey culture prepared by incubating whey from the previous day's manufacture. Calf rennet is used to coagulate the milk, and the coagulum is broken by means of a wire basketlike implement (spino) (see Chapter 7). The curds are cooked to 53–55°C in 10–12 min and then transferred to a mold large enough to produce a cheese of 25–40 kg. The cheeses may be subjected to light pressure and turned frequently to encourage whey expulsion. The cheeses are brine-salted for 20–23 days and ripened at 16–18°C and 85% equilibrium relative humidity (ERH) for 18–24 months. The rind of these cheeses is cleaned frequently.

The manufacturing protocol for Grana Padano cheese is generally similar to that for

Figure 17–2 Grana cheese.

Parmigiano-Reggiano, except that it is made from raw milk from a single batch of milk that is partially skimmed after creaming for about 8 hours. Grana Padano is brine-salted for about 25 days and ripened for 14–16 months.

Asiago is produced in the province of Vicenza, Italy, from partly skimmed raw cow milk. Rennet paste is used to coagulate the milk, and a natural whey starter is added for acidification. The cheeses, weighing 8–12 kg, are matured for various lengths of time, depending on the intensity of flavor desired. Mature Asiago (≈12 months old) has a hard, granular texture. Montasio, which originated in northeastern Italy, is made from raw cow milk and coagulated using calf rennet extract. Montasio may be consumed as a table cheese after 2–3 months or it may be matured for a longer period (14–18 months), during which time the cheese hardens and becomes suitable for grating. Sbrinz is a hard grating cheese that originated in Switzerland. It is made from full-fat cow milk using rennet extract and a natural thermophilic whey starter. The curds are cooked to about 57°C, placed in molds, pressed for 2–3 days, and either dry- or brine-salted. The cheese may be consumed as a table cheese after a short ripening period or matured for up to 3 years and used for grating.

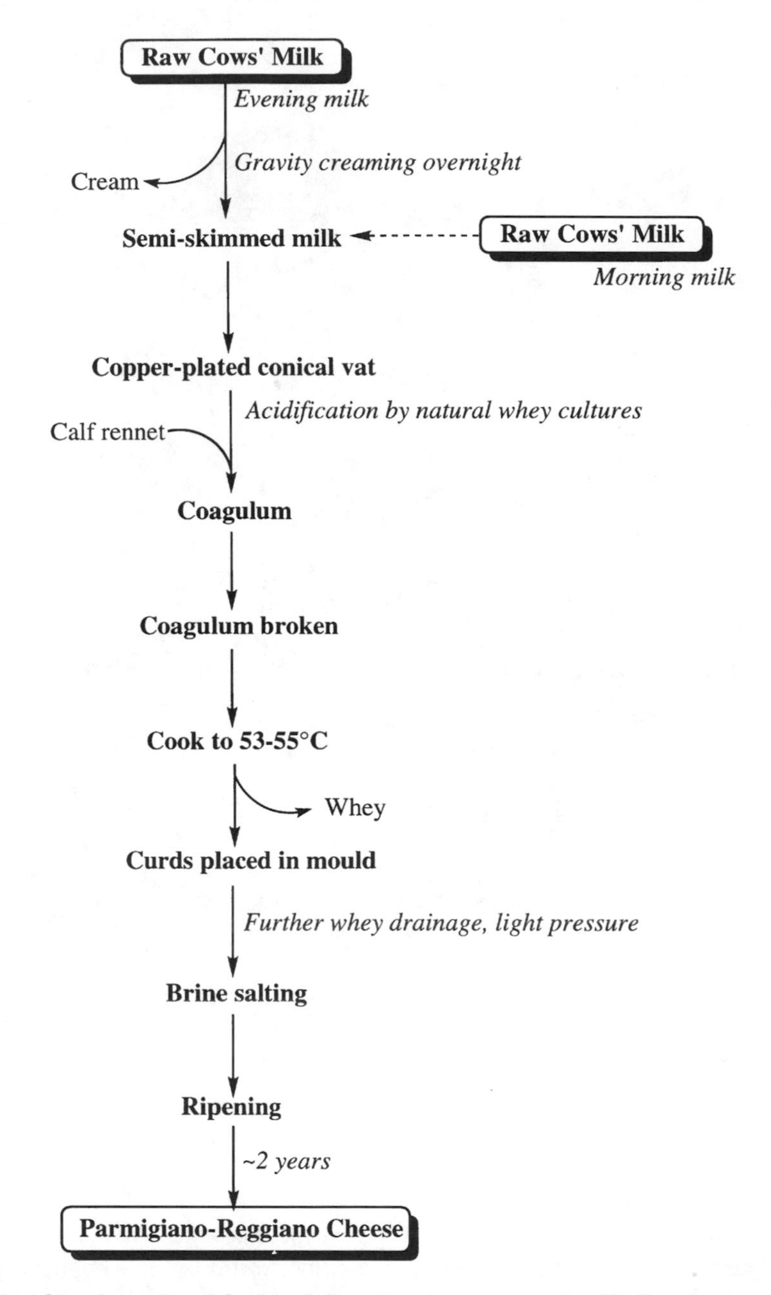

Figure 17–3 Manufacturing protocol for Parmigiano-Reggiano, an extra hard Italian cheese variety.

Romano-type cheeses are important members of the extra hard group. Cheese manufactured in Sardinia is called *Sardo* and that in Sicily is called *Siciliano*. The adjectives *Pecorino*, *Vacchino*, and *Caprino* indicate whether the cheese was made from ewe, cow, or goat milk, respectively. Italian Pecorino Romano cheese is made from sheep milk using a thermophilic starter (commercial or whey based). Rennet paste is used as coagulant. The high lipolytic activity of the rennet paste results in the development of a strong, slightly rancid flavor in the mature cheese (the fat in ewe milk has a high content of short-chain, middle-chain, and branched fatty acids, which give it a characteristic flavor). The curds-whey mixture is cooked to 45–48°C, and the whey is then drained off. Blocks of curd are placed in molds and pressed lightly before brine- or dry-salting. The cheeses are ripened for about 8 months. Pecorino Romano cheese is usually grated and used as a condiment.

Hard Varieties

Hard pressed varieties include some of the most commercially important cheeses produced worldwide (e.g., Cheddar; Figure 17–4). There is some heterogeneity in manufacturing technology for cheeses within this group, and whether a variety should be considered a hard cheese is not always clear. However, hard varieties usually have a moisture content in the range 30–45% and are subjected to high pressure during manufacture to give a hard, uniform, close texture. According to Robinson (1995), the manufacture of these cheeses has a number of features in common, including renneting at about 30°C, cutting the coagulum into small pieces and cooking to 39–40°C, and whey drainage. In the case of some hard cheeses, such as Cheddar and other British varieties, the curds are textured in the vat (cheddared) and are milled and dry-salted when sufficient acidity has developed. The salted curds are then molded and pressed at a high pressure for 12–16 hr or longer and matured for 3–12 months. Hard cheese varieties include Cheddar; Cheshire, Derby, Gloucester, and Leicester (British Territorial varieties); Cantal (French); Friesian Clove cheese and Leiden (The Netherlands); Graviera and Kefalotiri (Greece); Manchego, Idiazábal, Roncal, and La Serena (Spain); and Ras (Egypt).

Cheddar cheese originated around the village of Cheddar, England, and is now one of the most important cheese varieties worldwide. Cheddar is produced on a large scale in most English-speaking countries, particularly in the United States, United Kingdom, Australia, New Zealand, Canada, and Ireland. Cheddar cheese is usually made from pasteurized whole cow milk standardized to a casein:fat ratio of 0.67–0.72:1 and coagulated using calf rennet or a rennet substitute. The starter used is *Lc. lactis* subsp. *cremoris* or *Lc. lactis* subsp. *lactis*. Defined strain starter systems are now used in large Cheddar factories in New Zealand, Australia, Ireland, and the United States, but mixed, undefined cultures are also used (see Chapter 5). The milk is renneted at about 30°C, and the coagulum is cut and cooked to 37–39°C over 30 min and held at this temperature for about 1 hr. The whey is then drained and the curds are cheddared. The traditional cheddaring process involves piling blocks of curd on top of each other, with regular turning and stacking of the curd blocks. The cheddaring process allows time for acidity to develop in the curd (pH decreases from around 6.1 to 5.4) and subjects the curds to gentle pressure, which assists in whey drainage. During the cheddaring process, the curd granules fuse and the texture changes from soft and friable to quite tough and pliable. The curd should have a texture similar to cooked chicken breast meat at the end of cheddaring. When the pH has reached around 5.4, the curd blocks are milled into small pieces and dry-salted. A "mellowing" period follows, during which the salt dissolves in moisture on the surface of the curd chips. The curds are then molded and pressed overnight at up to 200 kN/m^2. Cheddar is matured at 6–10°C for a period ranging from 3–4 months to 2 years, depending on the maturity desired.

Although this traditional manufacturing process (Figure 17–5) is still practiced on a farmhouse scale, most Cheddar cheese is now manufactured in highly automated factories (e.g., Figure 17–6). The principal features of automated Cheddar production include the use of a number of cheese vats in which cheesemaking commences at 30 min intervals to provide a semicontinuous supply of curd. Whey drainage is mechanized and automated, as is the cheddaring process, which occurs as the curd passes

Figure 17–4 Cheddar cheese.

continuously through a large tower in which the curds are pressed gently by the weight of the column of curds above or, alternatively, on a belt system. The curds are then milled and salted mechanically on a belt system. Pressing and molding are done automatically by pneumatically conveying the salted curds to the top of a block former, which is a large (Wincanton) tower in which the curds are compressed by their own weight. A close texture is ensured by applying a vacuum. As the curds exit the block former, 20 kg blocks are cut off by a guillotine and vacuum packaged in plastic bags, placed in cardboard boxes, stacked on a pallet, and transferred to the cheese store. In many large factories, the boxed cheeses are cooled rapidly by being passed through a forced air cooling tunnel before palleting. The objective is to retard the growth of nonstarter lactic acid bacteria (NSLAB), which may cause defects in flavor and texture. Most Cheddar is now produced in block form, although traditional Cheddar has a cylindrical shape. Annatto (see Chapter 2) or another colorant may be added to milk for Cheddar cheese; the resulting product is known as Red Cheddar.

Cheshire cheese is a British Territorial variety (such varieties originated in various parts of Britain) with a hard texture and perhaps some mechanical openings. Its manufacture is characterized by rapid acidification and a lower cook temperature (32–35°C) than Cheddar, which results in a lower pH, a higher moisture content, and a shorter manufacturing time. The curd mat that develops while the curd mass stands on the bottom of the vat after whey drainage is broken frequently to prevent the development of an extensive structure in the curd mass. Cheshire curds are dry-salted, placed in molds, allowed to drain overnight, and then pressed. The cheeses are packaged and matured at 6–8°C. Leicester is similar to Cheshire but is normally colored with

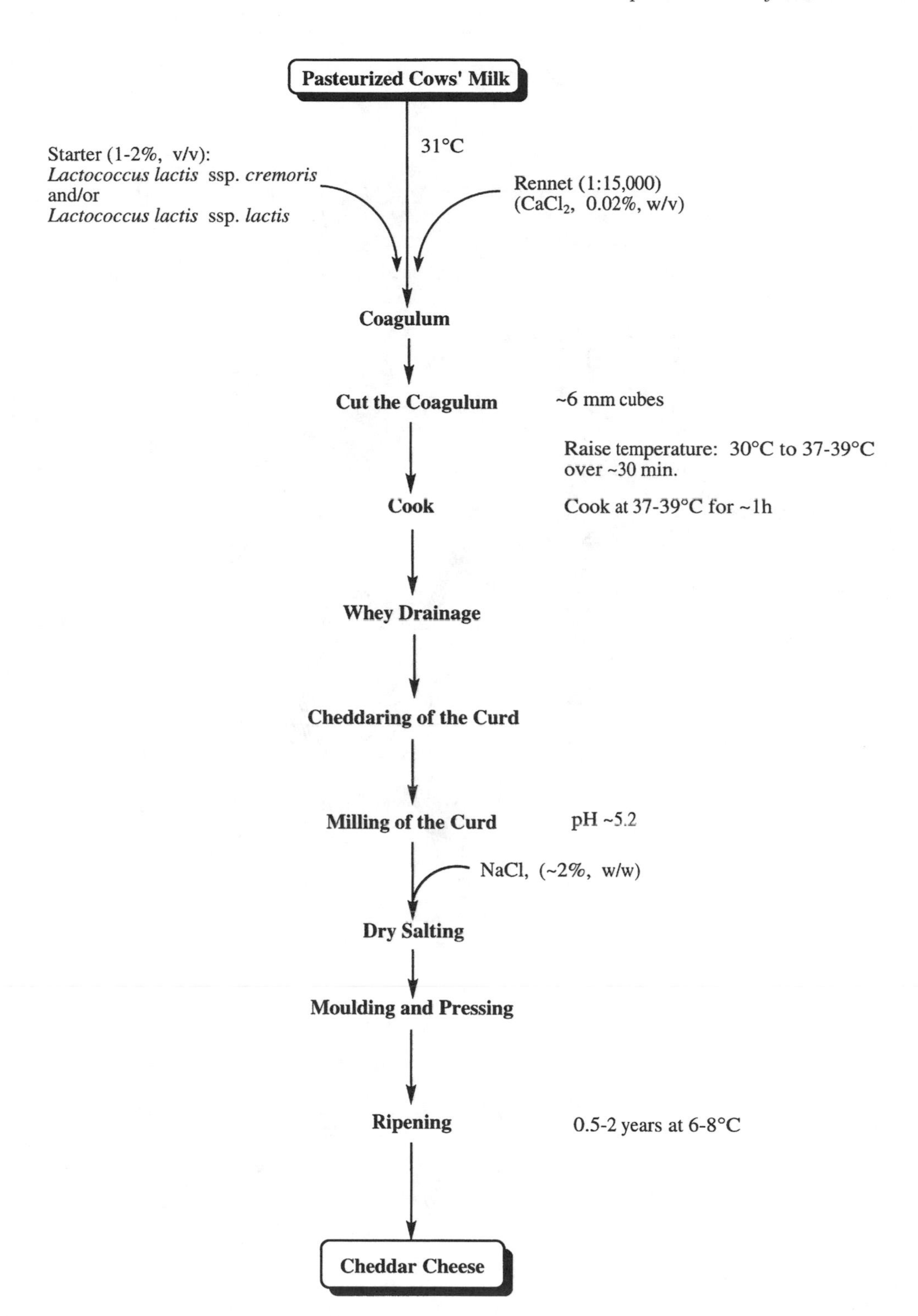

Figure 17–5 Traditional protocol for the manufacture of Cheddar cheese.

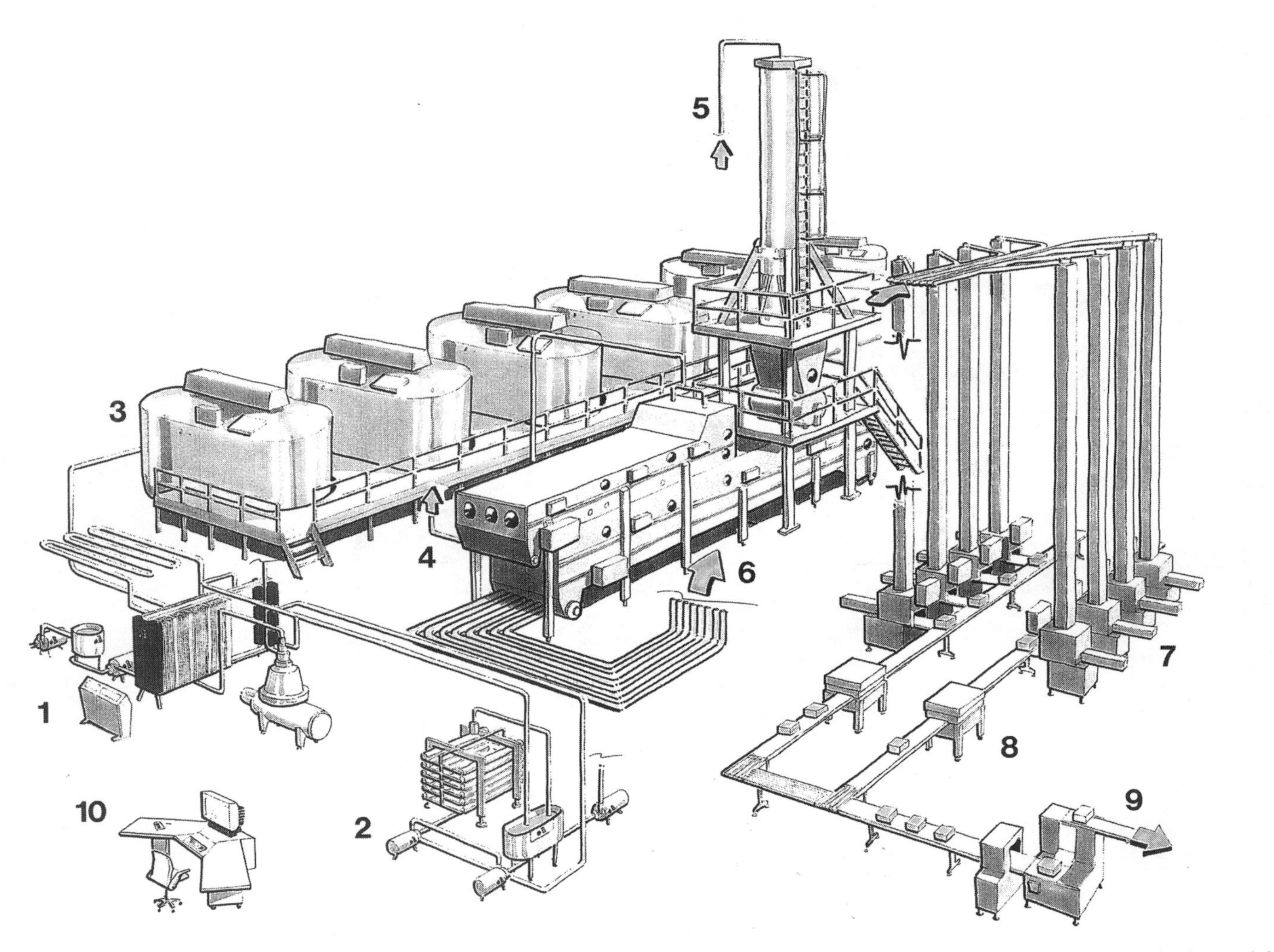

Figure 17–6 Large-scale Cheddar cheesemaking factory incorporating the CheddarMaster 3 system with a Cheddaring tower. (1) Pasteurization and fat standardization, (2) protein standardization using UF, (3) cheesemaking, (4) draining conveyor, (5) cheddaring tower, (6) salting/mellowing conveyor, (7) block former, (8) vacuum packaging, (9) cheese block packing, and (10) main process control panel.

annatto. It is made from cow milk using a mesophilic starter and is cooked at about 37°C. The curds are pressed after whey drainage, and blocks of curd are placed on a draining rack and turned and cut to promote further whey drainage. The cheeses are dry-salted and matured for 4–8 months at 10–15°C. Derby is similar to Leicester and is somewhat softer and flakier than Cheddar. Gloucester, another British Territorial variety, has a cylindrical shape and is about 40 cm in diameter. Single Gloucester is 6–8 cm high while a cylinder of Double Gloucester is 15–20 cm high. The manufacturing procedure for Gloucester is similar to that for Cheddar, and annatto is added to the milk to color the curds, which are cooked to 35–38°C. The curds are textured, milled twice, dry-salted, pressed, and matured for 4–6 months.

Cantal is a hard French cheese from the Auvergne region and is manufactured by a process somewhat similar to that used for British Territorial varieties. The milk is coagulated using standard calf rennet and acidified by a mesophilic lactic starter. The curds-whey mixture is not cooked, but the whey is drained and the curds are cheddared. Weights may be placed on the bed of curd to assist in whey drainage. The blocks of curd are then milled, dry-salted, molded, pressed, and matured at 8–10°C for 3–6 months.

Kefalotiri is a Greek cheese made from pasteurized sheep or goat milk standardized to about 6.0% fat. The milk is inoculated with a thermophilic culture (usually *Sc. thermophilus* and *Lb. delbrueckii* subsp. *bulgaricus*) and coagulated by calf rennet. The curds are cooked to 43–45°C, transferred to molds lined with cheesecloth and subjected to a low pressure, which is increased slowly. Upon removal from the molds, the cheeses are dried overnight and brine-salted. After brining, dry salt is rubbed onto the surface of the cheese over the next few days to give a final salt content close to 4%. During this time, the cheeses are washed with a brine-soaked cloth to control microbial growth on the surface and ripened for about 3 months. According to Robinson (1995), Kefalotiri has a hard texture and a strong, salty flavor.

Graviera is a relatively recently developed Greek variety made principally from ewe milk. It is acidified by a mixed mesophilic culture (1%) containing *Lc. lactis* subsp. *lactis* or *Lc. lactis* subsp. *cremoris* and a smaller amount (0.1%) of a thermophilic culture containing *Sc. thermophilus* and *Lb. helveticus*. The curds are cooked first to about 50°C, and after whey drainage the curds are molded and pressed at an increasing pressure. The cheeses are then salted by frequent application of dry salt to the surface for 2–3 weeks and ripened for 3–4 months. In some factories, the early stages of dry-salting may be replaced by brining.

The principal hard cheese produced in Egypt is Ras, the production process for which is similar in many respects to that for Kefalotiri. It is made from cow milk standardized to 3% fat. The curds are cooked to 45°C, salted at a level of 1% after whey drainage, molded, and pressed in a manner similar to Kefalotiri. The cheeses are brined for 24 hr and rubbed with a small quantity of dry salt daily for several weeks. During this period, the cheeses are also washed with brine.

Manchego cheese, probably the most important Spanish variety, is made from ewe milk, although generally similar cheeses (without PDO status) are manufactured from milks of other species. Two types of Manchego are produced: artisanal (made from raw milk without culture addition) and commercial (made from pasteurized milk inoculated with a mesophilic starter; see Chapter 5). The milk is coagulated with standard calf rennet. The curds are cooked to about 38°C, transferred to molds, and pressed for 12–16 hr. Manchego has characteristic side markings made by binding the cheeses in basketwork wrappings for about 30 min after pressing. The cheeses are then brine-salted and matured for at least 2 months at 10–15°C and 85% ERH.

Idiazábal cheese is produced in the Basque region of northern Spain from raw ewe milk coagulated at 38°C. The coagulum is allowed to cool to 25°C and then broken, and the curds are ladled into molds. The cheeses are salted by brining or by the application of dry salt and matured in caves for 2 months. They are then smoked in beechwood kilns and further matured for up to 1 year. Roncal is made in northern Spain from raw ewe milk and acidified by the indigenous flora of the milk. The curd-whey mixture is cooked to about 37–40°C, and the curds are then allowed to settle to the bottom of the vat. The whey is removed slowly, and the curds are pressed against the sides of the vat, molded, and pressed before being dry-salted. Roncal is smoked and ripened at 6–8°C and 100% ERH for 45–50 days. La Serena is a hard cheese made in western Spain from ewe milk. Traditionally, raw milk is used, although pasteurized milk is used in large-scale production. The milk is coagulated with rennet extracted from the cardoon thistle (*Cynara cardunculus*).

Other hard cheese varieties include Leiden and Friesian Clove cheeses from the Netherlands, which are characterized by the addition of cumin seeds (Leiden) or cloves and cumin seeds (Freisian Clove).

Semi-Hard Varieties

The description of a cheese as semi-hard is arbitary. The semi-hard group of cheeses is thus heterogeneous, and the distinction between this and other groups of cheeses (e.g., hard cheeses, smear-ripened varieties, and Pasta filata cheeses) may not be clear. Semi-hard cheeses include Colby and Monterey (stirred-curd Cheddar-type cheeses), a number of British Territorial varieties (Caerphilly, Lancashire, and Wensleydale), and cheeses such as Bryndza (Slovakia) and Mahón and Majorero (Spain).

Caerphilly, which originated in Wales, is a crumbly acid cheese. It is made from pasteurized cow milk using calf rennet and a mesophilic starter. The curds are cooked to 32–34°C and held at this temperature for about 1 hr. The whey is drawn off and the curds are collected at the bottom of the vat, where rapid acid production occurs. Some dry salt (1%) is added to the curds before molding and pressing overnight. The pressed curds are then brine-salted for 24 hr and packaged. Caerphilly matures rapidly and is ready for sale after 10–14 days. Lancashire, another British Territorial variety, is made from cow milk using rennet and a mesophilic starter. The curds and whey are not cooked, but after draining, the curds are cheddared and held overnight, during which time extensive acid production occurs. The next day, fresh curds are mixed with the acidified curds, and the mixture is milled to ensure homogeneity. The curds are dry-salted, placed in molds overnight at room temperature, and pressed for 3 days. Lancashire is ripened at 13–18°C for 3–12 weeks.

Wensleydale cheese, which originated in Yorkshire, England, is made from cow milk inoculated with a mesophilic starter. The curds are cooked at 32–34°C. The whey is drained off, and the curd mat is broken into pieces to assist whey drainage. The pieces of curd are dry-salted and molded, held overnight at about 21°C without pressing, and then pressed lightly for 5 hr. Wensleydale, which is matured for about 1 month, has mechanical openings and a mild, acidic taste.

Stirring Cheddar-type cheese curd inhibits the development of curd structure and results in a cheese with a higher moisture content and thus a softer texture. Two stirred-curd variants of Cheddar cheese are recognized: Colby and Monterey. The manufacture of Colby, which originated in the United States, follows a protocol similar to Cheddar until after cooking, when some whey is removed and replaced by cold water. The curds and whey are stirred, and most of the whey is removed. The curds and remaining whey are stirred vigorously. The remaining whey is then drained off, and the curds are stirred further. Stirring prevents the development of an extensive structure while the curds are in the vat. Salt is added to the curds, which are molded and pressed. Colby is ripened for 2–3 months at 3–4°C. Monterey (Monterey Jack) cheese was first made in California and is similar to Colby. The whey is removed and the curds are left on the bottom of the vat, with occasional stirring until the pH reaches 5.3. The curds are dry-salted and pressed lightly overnight. The cheeses are allowed to form a rind before waxing or packaging in films. Monterey, which is ripened for 5–7 weeks, has many mechanical openings.

Bryndza is made from sheep milk coagulated with rennet (sometimes with significant lipase activity) and acidified by the indigenous microflora of the milk or by a mesophilic starter. The curds are allowed to settle to the bottom of the vat, most of the whey is removed, and the curds are consolidated into lumps by hand. The lumps of curd are placed into cloth bags and stored for 3 days while sufficient acidity develops. At this stage, the cheese is known as *Hrudka* and may be sold locally. However, most Hrudka is transported to a central factory, where it is broken into pieces, salted, and passed between granite rollers to produce a smooth paste, which is placed in polyethylene-lined wooden tubs and matured.

Mahón is produced in the Balearic island of Minorca from raw cow milk acidified by its indigenous microflora. Mahón is brine-salted and ripened at 18°C for about 2 months. Although the texture of Mahón is semi-hard, its moisture content is reported to be roughly 32%. Majorejo is made from goat milk on Fuerteventura Island, one of the Canary Islands. In commercial practice, milk is acidified by a mesophilic starter and coagulated using rennet, although artisanal cheesemakers rely on the indigenous microflora of the milk and use rennet paste as a coagulant. The curds are molded in braided palm leaves (which impart to its surface a characteristic pattern), pressed lightly, and dry-salted. Majorejo develops a strong flavor during ripening.

Cheeses with Eyes

Mechanical openings, resulting from the incomplete fusion of curd pieces, are common in many cheese varieties and may be considered desirable (e.g., Monterey) or a defect (e.g., Cheddar). However, some internal bacterially ripened varieties are characterized by the development of eyes caused by the entrapment within the curd of gas produced by bacterial metabolism. The development of eyes in cheese is governed by the rate of gas production by bacteria and the ability of the curd to retain the gas. There are two main families of cheese with eyes: Dutch types (Edam, Gouda [Figure 17–7], and related varieties), which have small eyes, and Swiss types, which are characterized by large eyes. In the case of Edam and Gouda, CO_2 is produced from citrate by the DL culture (see Chapter 5), while in Swiss varieties, CO_2 is produced by *Propionibacterium freudenreichii* spp. *shermanii* from lactate during ripening (see Chapters 10 and 11).

Gouda (Figure 17–8) originated in the Netherlands but is now produced worldwide from pasteurized cow milk coagulated using calf rennet and acidified by a mesophilic DL starter. Nitrate may be added to the milk to suppress the growth of *Clostridium* spp., which produce gas (H_2 and CO_2) from lactate during ripening, causing a defect known as "late gas blowing." Butanoic acid is also produced from lactate, and it causes off-flavors. The coagulum is cut and the curd-whey mixture is stirred for 20–30 min before a portion (about 30%) of the whey is drained off and replaced by hot water, which raises the temperature to 36–38°C. This washing step removes some of the lactose and consequently reduces the development of acidity after the curds are molded. The curds are cooked at this temperature and then allowed to settle to the bottom of the vat where they are pressed under the whey. The bed of curd is cut into blocks, which are transferred to molds (wheel-shaped or rectangular, producing a cheese of 4–20 kg) and pressed for 5–6 hr. Gouda is brine-salted, coated with yellow wax (traditionally), and ripened at 15°C for 2–3 months or longer (up to 2 years for extra mature cheese).

Edam, a Dutch variety similar to Gouda but with a distinctive spherical shape, is coated with red wax and is made from semi-skimmed milk (≈ 2.5% fat). A few small eyes develop in Edam, which may be sold after ripening for 6–30 weeks. Maribo is a similar variety produced in Denmark from cow milk, and, in addition to eyes caused by a mesophilic DL starter, it has numerous mechanical openings. Other Dutch-type cheeses include Danbo (Denmark), Colonia and Hollanda (Argentina), and Svecia(ost) (Sweden).

Figure 17–7 Gouda cheese.

Swiss-type cheeses are characterized by large eyes produced by *P. freudenreichii* subsp. *shermanii*, which metabolizes lactate to propionate, acetate, and CO_2 (see Chapters 10 and 11). Propionibacteria do not grow in the milk during cheesemaking but grow in the cheese during maturation, when it is transferred to a hot room (≈ 20–22°C). The curd of these cheeses is quite rubbery and is able to trap the CO_2 (which migrates through the curd until it reaches a fissure or weakness, at which place an eye develops). The texture of these cheeses is influenced by a high cook temperature (≈ 55°C), which inactivates most of the coagulant, and a high pH at draining (which leads to a high concentration of calcium in the curd). Swiss-type cheeses are traditionally made as large wheels and are brine-salted. The relatively slow diffusion of salt in a large cheese allows the salt-sensitive propionibacteria time to grow.

Emmental (Figure 17–9) is a typical Swiss cheese variety that is now made worldwide. Traditional Emmental is made from raw cow milk that is acidified with a mixed thermophilic starter consisting of *Sc. thermophilus* and a *Lactobacillus* species (Figure 17–10). *Lb. helveticus* was used traditionally but *Lb. delbrueckii* subsp. *lactis* is now common also. Propionibacteria may be added to the milk or may contaminate milk from the environment. The milk, at 30°C, is coagulated using calf rennet, and the coagulum is cut into small pieces and cooked to about 55°C until the curd grains are of the desired firmness. The curds and whey are transferred to molds, where the whey is separated. The molds are sufficiently large to give a wheel of cheese weigh-

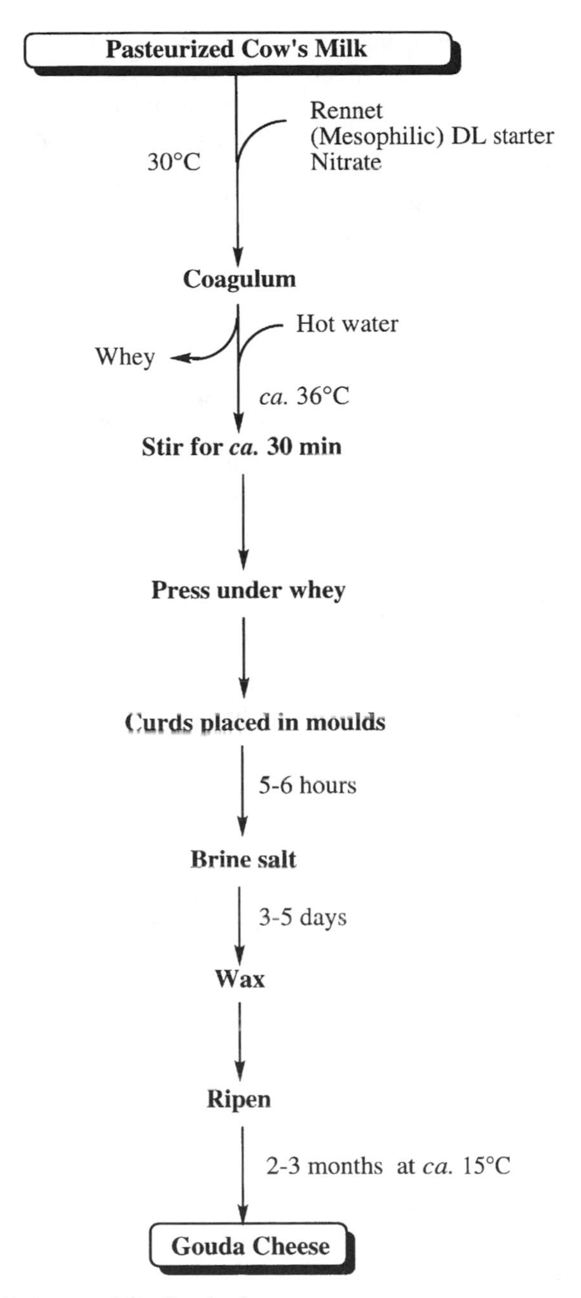

Figure 17–8 Manufacturing protocol for Gouda cheese.

ing up to 100 kg and as much as 1 m in diameter. The size of the Emmental wheel is significant because it determines the rate of cooling of the curd (and thus the activity of the starter; see Chapter 10), determines the diffusion of salt throughout the cheese mass, and helps to trap gas within the cheese. Over the next 1–2 days, the wheels are pressed and turned frequently. During this time, the curds cool and acid production by the starter organisms (which were dormant during cooking) recommences. Complete fermentation of lactose and its constituent monosaccharides in Emmental takes about 24 hr. After pressing, the cheese wheels are brine-salted, stored in a cool room (10–15°C, 90% ERH) for 10–14 days, and brushed, dry-salted, and turned daily until a smooth rind develops. The cheeses are then transferred to a hot room (20–22°C, 80–83% ERH) and held there for 3–6 weeks, until adequate eye formation has occurred (which is indicated by a drum-like sound when a cheese is tapped with a cheese trier). The cheeses are then matured at about 7°C for a further 1 or 2 months. Rindless Emmental in block form is produced by a protocol generally similar to that for Emmental, but milk with a lower fat content and a lower cooking temperature are used. The cheese is wrapped in a plastic film, and therefore no rind develops during maturation.

Gruyère is another popular Swiss-type cheese. It differs from Emmental in being smaller and having a somewhat stronger flavor and fewer eyes, and it is characterized by the development of a surface flora (similar to that which develops on smear-ripened varieties, Section 17.2.3). The surface flora is encouraged by ripening for 2–3 weeks at 10°C and then for 2–3 months at 15–18°C and 90–95% ERH, during which time the cheeses are rubbed with a brine-soaked cloth. Further ripening at 12–15°C is required before sale at 8–12 months of age.

Similar varieties include Raclette (which is manufactured from raw milk and is acidified by the indigenous milk microflora) and Gruyère de Comté, which is produced in eastern France from raw bovine milk. Beaufort is a French variety similar to but larger than (≈ 45 kg) Gruyère, but only *Lb. helveticus* is used as starter. Appenzeller, which originated in Switzerland, is a small cheese (≈ 30 cm in diameter) with a soft texture. It undergoes propionic acid fermentation, resulting in the development of a few eyes. The curds are cooked at 43–45°C. Appenzeller

Figure 17–9 Emmental cheese.

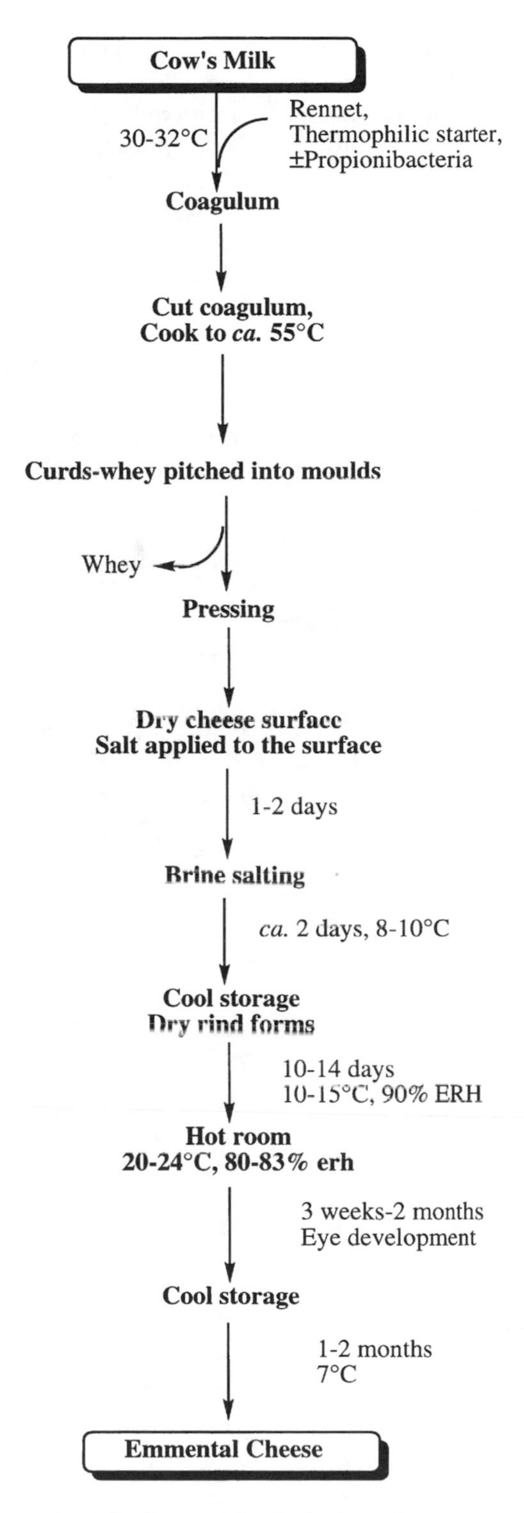

Figure 17–10 Manufacturing protocol for Emmental, a Swiss-type cheese.

is immersed in cider or spiced wine or rubbed with a mixture of salt and spices during ripening, which impart a distinctive flavor to the cheese. Maasdammer, a variety developed recently in the Netherlands, is characterized by the use of a mesophilic starter and extensive propionic acid fermentation, which causes large eyes and gives a domed appearance to the cheese wheels. Jarlsberg is a Swiss-type cheese produced in Norway using a mesophilic starter.

High-Salt Varieties

White brined cheeses originated around the eastern Mediterranean. According to Robinson (1995), the original characteristic features of the manufacture of these cheeses were the use of sheep milk (which yields a very white curd), a high ambient temperature, and storage in brine (leading to a high salt content) for purposes of preservation. The principal white brined varieties today are Feta (Figure 17–11), Telemes (Greece), and Domiati (Egypt). White brined cheeses are now made worldwide and are major industrial products.

Feta cheese is made from sheep milk or mixed sheep and goat milk. Strenuous efforts on the part of the Greek government have resulted in PDO status for Feta, although similar cheeses

Figure 17–11 Feta cheese.

are manufactured worldwide. Milk for most Feta is pasteurized and standardized to a casein:fat ratio of about 0.7–0.8:1 (Figure 17–12). A thermophilic or mesophilic starter culture is added to ensure rapid acidification. The rennet used is often a mixture of standard calf rennet and a local rennet extract with some lipase activity. $CaCl_2$ may be added to the milk before renneting. The rennet-induced coagulum is cut (into 2–3 cm cubes), and the soft curds are ladled directly into

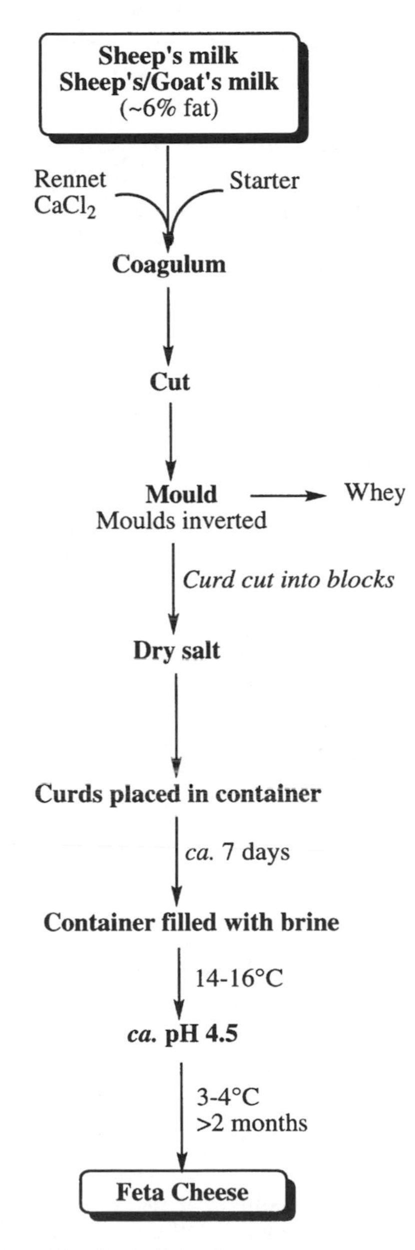

Figure 17–12 Manufacturing protocol for Greek Feta cheese.

molds. The curd-whey mixture is not cooked. Whey drainage occurs in the molds, which are inverted after 2–3 hr. The curd mass is then firm enough to be removed from the molds and is cut into blocks, which are dry-salted (or brined) before being transferred to a barrel or other container. The container is filled with brine (≈14% NaCl) after around 7 days and held at 14–16°C until the pH of the cheese has decreased to about pH 4.5, at which point the cheeses are stored at 3–4°C for at least 2 months.

Feta-type cheese is now also manufactured industrially from cow milk concentrated by ultrafiltration (concentrated approximately fivefold using membranes with a cutoff of 10–20 kDa). Less rennet is used in the manufacture of Feta-type cheese from ultrafiltration retentate, and the yield is higher than for Feta made from unconcentrated milk owing to the incorporation of whey proteins into the curd (see also Section 17.6).

Telemes cheese is made by a protocol similar to that for Feta, except that some pressure is applied to the curds in the molds to aid in the expulsion of whey and the curd blocks are brined after removal from the molds.

Domiati is made from cow or buffalo milk or a mixture of both containing 2%, 4%, or 8% fat (giving reduced- or full-fat cheese). Domiati may be made from pasteurized milk to which NaCl is added to a level of 5–15%. Alternatively, about one-third of the milk may be heated to around 80°C and salt added to the remainder. At the level used, NaCl has a strong antibacterial effect, and halotolerant lactobacilli are used as starter. Domiati may be consumed fresh or ripened in brine for a number of months.

Halloumi is a brined cheese made in Cyprus from sheep milk. The curds are cooked to 38–42°C during manufacture and the cheeses pressed. Blocks of curd are scalded by immersion in hot whey (90–92°C) for 30 min but are not stretched. The cheeses are dry-salted and consumed fresh or after storage in brine.

Other white pickled cheeses include Lightvan (Iran), Beda (Egypt), and Bulgarian white brined cheeses.

Pasta Filata Varieties

Pasta filata cheeses are semi-hard varieties, the curds for which are heated to 55°C and above and mechanically stretched during manufacture. Stretching causes the curds to become fibrous and malleable. Most Pasta filata cheeses originated in the Mediterranean region.

By far the most important member of this group is Mozzarella, which originated in southern Italy and was originally manufactured from buffalo milk. Mozzarella di bufala (Figure 17–13) is hand-molded into round pieces (100–300 g) during manufacture. This cheese is still manufactured in Italy, but the type of Mozzarella now widely manufactured around the world is made from pasteurized, partly skimmed cow milk (Figure 17–14) and is often referred to as pizza cheese or, in the United States, low-moisture, part-skimmed Mozzarella. This type of Mozzarella has a higher salt concentration (1.5–1.7%) than Mozzarella di bufala (<1.0%). The production of low-moisture, part-skimmed Mozzarella cheese has increased greatly in recent years as a result of the increased popularity of pizza pie. Some is consumed as a table cheese or as a component of salads.

The manufacturing process for Mozzarella for use as pizza topping (Figure 17–14) involves standardizing pasteurized cow milk to around 1.8% fat. A higher fat content (≈ 3.6%) is used for Mozzarella intended to be consumed as a table cheese. A thermophilic starter (1–2%) containing a combination of *Lactobacillus* spp. and *Sc. thermophilus* is used in the manufacture of pizza cheese. The *Lactobacillus* is often omitted when Mozzarella is intended as a table cheese, since the rate of acidification need not be as fast as in pizza cheese. Proteolytic enzymes of the *Lactobacillus* may make a minor contribution to the functionality of the final product by causing slight hydrolysis of the caseins. The milk is renneted after some acidity has developed, and the coagulum is cut and cooked to around 41°C. The whey is then usually drained off, and texture is developed in the curds (usually by cheddaring) until the pH drops

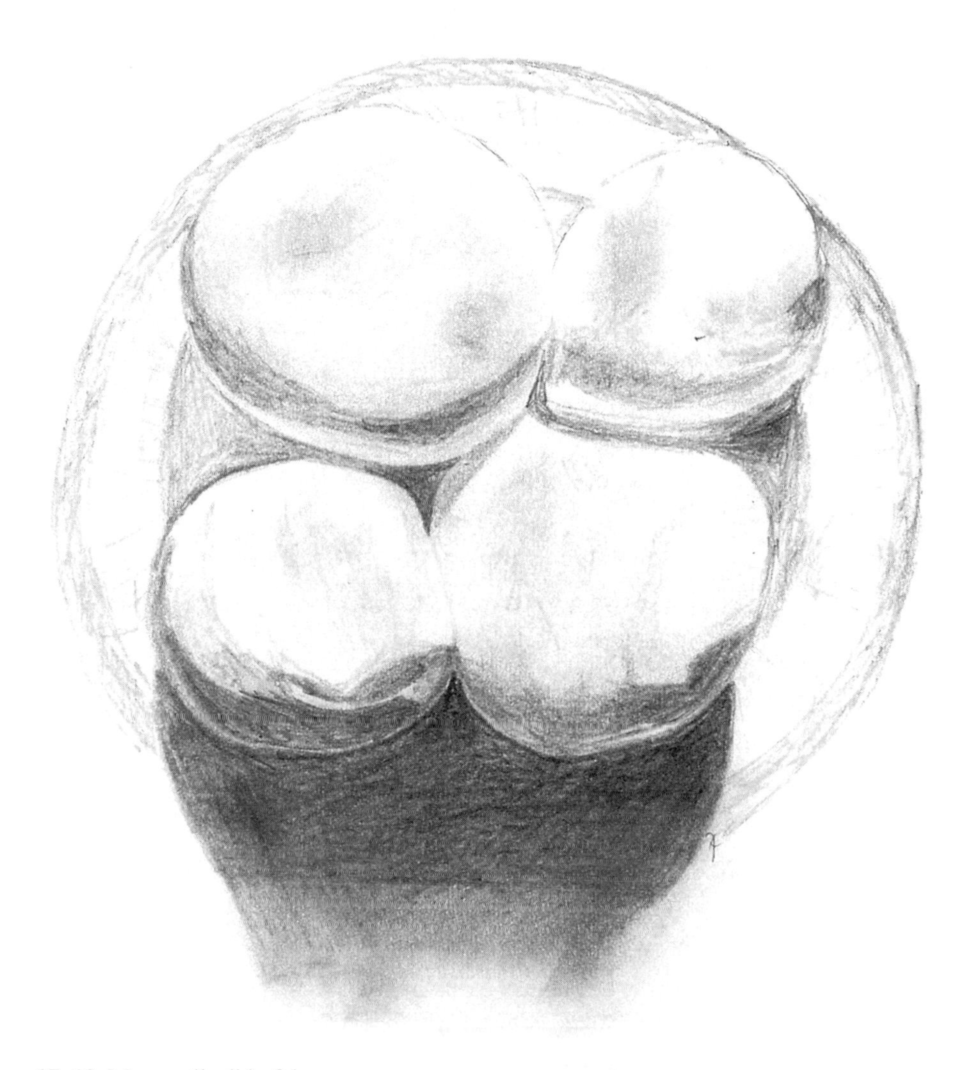

Figure 17–13 Mozzarella di bufala.

to around pH 5.1–5.3. Because the production and the treatment of the curds are quite similar to those used for Cheddar up to this stage, Cheddar plants may be easily modified to produce Mozzarella by altering the starter and temperature profile used and by the use of an appropriate stretcher. The next stages in Mozzarella manufacture are stretching and kneading, which are characteristic of pasta filata varieties. The cheddared curds are placed in hot water (≈ 70°C) and kneaded, stretched, and folded until the desired texture has been developed. The curds for pizza cheese are stretched more extensively than those for table Mozzarella. The former may also be salted during the stretching and forming stages. The hot, plastic curds are molded (usually into rectangular blocks) and cooled quickly in cold water or brine, and if salt was not added during the cooking and stretching process, the cheeses are then brine-salted. Mozzarella is usually consumed within a few weeks of manufacture. Extensive ripening is undesirable, since the functional properties of the cheese deteriorate.

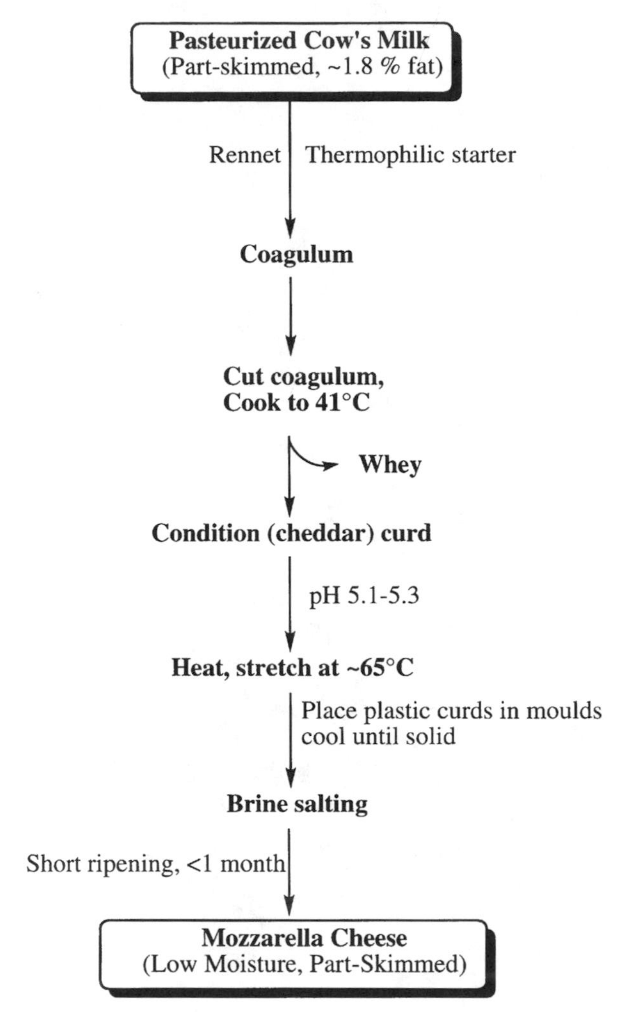

Figure 17–14 Manufacturing protocol for low-moisture, part-skimmed Mozzarella (pizza) cheese.

So-called string cheese is produced from Mozzarella, Cheddar, or similar cheese curd by cooking and extruding the plastic curd as long rods (1–2 cm in diameter) and then brining the rods (brining occurs rapidly because of the small cross-sectional area of the rods). The rods are then cut into convenient lengths and packaged. Strings of cheese may be torn from these rods, and this novelty feature is the major selling point for string cheese, whose target market consists of young children.

Kashkaval is a stretched-curd variety from the Balkans and was traditionally made from sheep milk, although cow milk Kashkaval is now common. The cheese is usually brine-salted, but some is dry-salted or stored in brine. Typically, the cheese is matured for 2–3 months before consumption. Kasseri, a Greek cheese similar to Kashkaval, is made from sheep milk or a mixture of sheep and goat milk. Provolone (Figure 17–15), which is characteristically pear shaped, originated in southern Italy, where it is made from cow milk. Rennet paste may be used in its manufacture, giving the resulting cheese (Provolone piccanti) a stronger flavor than normal Provolone (Provolone dulce), which is manufac-

Figure 17–15 Provolone cheese.

tured using rennet extract. Provolone is ripened for 2–6 months. Caciovallo is a hard Italian cheese manufactured from cow milk by a process somewhat similar to that used for Provolone. The curds are stretched in hot water and brine-salted. Caciovallo is ripened for 3–4 months—or longer (>12 months) if the cheese is to be grated. Ostiepok, a stretched-curd cheese from central Europe (Czech Republic and Slovakia), is made from sheep milk (although cow milk is used sometimes), brine-salted, and smoked heavily.

17.2.2 Mold-Ripened Cheeses

Cheese varieties on which molds grow during ripening fall into two broad categories: surface mold–ripened cheeses (e.g., Brie and Camembert; Figure 17–16), in which the mold grows as a mat on the surface, and blue-veined varieties (Figure 17–17), which are characterized by the growth of *P. roqueforti* in fissures throughout the cheese. Although these two groups have mold growth in common, the methods used for their manufacture and the flavor and texture of the mature cheese are quite different. Hybrid mold-ripened cheeses are also produced (e.g., Cambazola); these have a white mold growth on the surface and blue mold in the interior.

Surface Mold–Ripened Varieties

Surface mold–ripened cheeses are generally soft varieties characterized by the growth of the white mold *Penicillium camemberti* on the surface of the cheese. The surface flora is often more complex, particularly in cheeses made

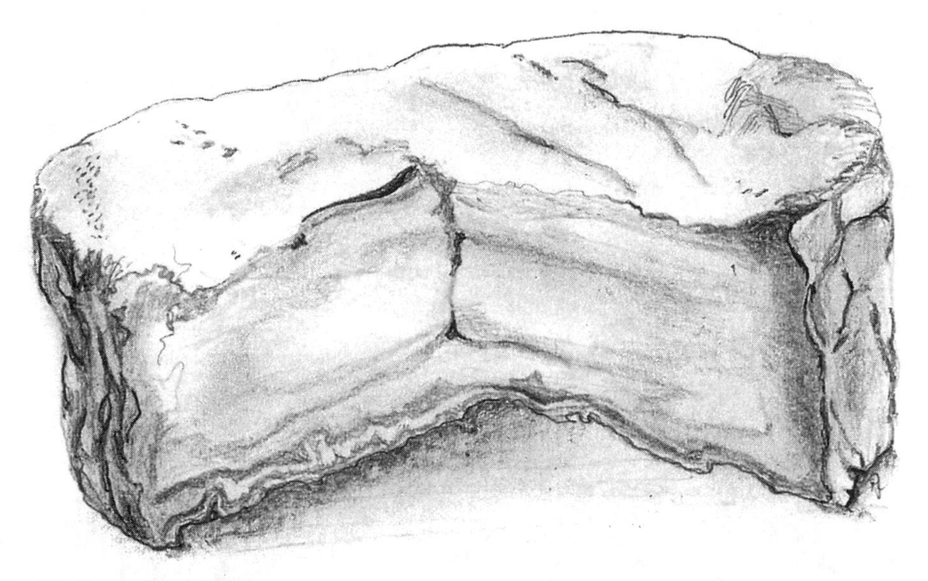

Figure 17–16 Camembert cheese.

Figure 17–17 Blue-veined cheese.

from raw milk by traditional technology. These curds are acidified to around pH 4.6 during manufacture using a mesophilic starter. Lactic acid produced by the starter is metabolized by the mold, which also produces ammonia, and therefore the pH of the surface layer of the cheese increases to around 7.0. If present, yeast also catabolize lactic acid. An important consequence of the increase in the pH of these cheeses is that there is considerable migration of calcium phosphate to the surface layer (which contains ≈ 80% of the calcium and ≈ 55% of the phosphorus of the mature cheese). As in all cheeses containing active rennet, the coagulant plays a role in the development of texture. However, in surface mold–ripened cheeses, the role of rennet is relatively minor. These cheeses soften from the surface toward the center during ripening, owing, not to proteolysis by mold enzymes, which diffuse only a few millimeters into the cheese, but to the establishment of pH and calcium phosphate gradients. Softening in these cheeses is often quite extensive, leading to a spreadable, almost liquid consistency.

Many surface mold–ripened varieties originated in France. They are usually manufactured from cow milk acidified by a mesophilic starter. The first microorganisms to become established on the surface are yeasts, including *Debaromyces* spp. and *Kluyveromyces* spp. *Geotrichum candidum* also becomes established at this time, although its growth may be limited if the level of salt is high. *P. camemberti* is observed after 6–7 days of ripening and forms the characteristic mat on the cheese surface. Once the surface of the cheese has been neutralized (after 15–20 days), aerobic bacteria (particularly micrococci and coryneforms), which are inhibited by the low initial pH, begin to grow.

Camembert, the most important surface mold–ripened variety, originated in Normandy, France, in the 18th century. It is a small cheese (≈ 10 cm diameter and 200–250 g) manufactured from cow milk (Figure 17–18). Raw milk is used traditionally but industrial Camembert is now produced from pasteurized milk. A mesophilic starter (≈ 0.1%) is used, and when the pH of the milk has fallen to about 6.1, rennet is added. Traditionally, the coagulum is not cut but is ladled into molds, where drainage occurs. To facilitate manufacture, the coagulum for industrial Camembert is first cut into large cubes and then transferred to molds without cooking. Traditionally, the cheeses are dry-salted and *P. camemberti* spores are sprayed on the surface, although it is now industrial practice to inoculate the milk with mold spores and to brine-salt the cheeses. The surface of the cheese is allowed to dry at ambient temperature in a well-ventilated room, after which the cheeses are transferred to a storeroom at about 12°C for 10–12 days for mold development. The cheeses are then packaged in waxed paper and placed in wooden or cardboard boxes prior to final ripening at 7°C for 7–10 days.

Brie is a flat, cylindrical surface mold–ripened cheese with a larger diameter than Camembert, which it resembles closely in flavor, texture, and manufacturing protocol. Carré de l'Est is a square surface mold–ripened cheese that originated in eastern France. Neufchâtel originated near Rouen, France, where it is still produced, mainly on farms. Raw cow milk is inoculated with a small quantity of an artisanal starter and renneted. The coagulum that forms overnight is transferred to muslin bags through which the whey is drained. After most of the whey has drained, the curds are removed from the bags, mixed, salted, and placed in molds. When the curds are firm enough to remove from the molds, their surfaces are salted, dusted with *P. camemberti* spores, and ripened for 2–3 weeks. In addition to the above-mentioned varieties, St. Marcellin (which was manufactured originally from goat milk but is now made from cow or sheep milk) and a number of minor goat milk cheeses also develop a surface mold growth during ripening. The microflora of these minor varieties is often uncontrolled.

Blue-Veined Cheeses

Blue-veined cheeses are characterized by the growth of *P. roqueforti* in fissures throughout the cheese (Figure 17–17). These cheeses usually have a soft texture and a flavor dominated by alkan-2-ones (methyl ketones), which are produced from free fatty acids by the mold via the β-oxidation pathway (see Chapter 11). The pH of blue mold cheeses increases during ripening, from 4.6–5.0 after molding to 6.0–6.5 when mature.

Since *P. roqueforti* requires O_2 for growth, the manufacture of Blue cheese is dominated by the need to provide a suitable environment for its growth. This is achieved by encouraging large mechanical openings in the cheese (by not pressing the cheese when molded) and by piercing the cheese to allow air into its center and to allow CO_2 produced by the mold to escape. Lipolysis is often encouraged in Blue cheeses by separation of the raw cheese milk and homogenization of the cream, which encourages the action of the indigenous milk lipase. The level of starter used (usually a DL starter) is normally

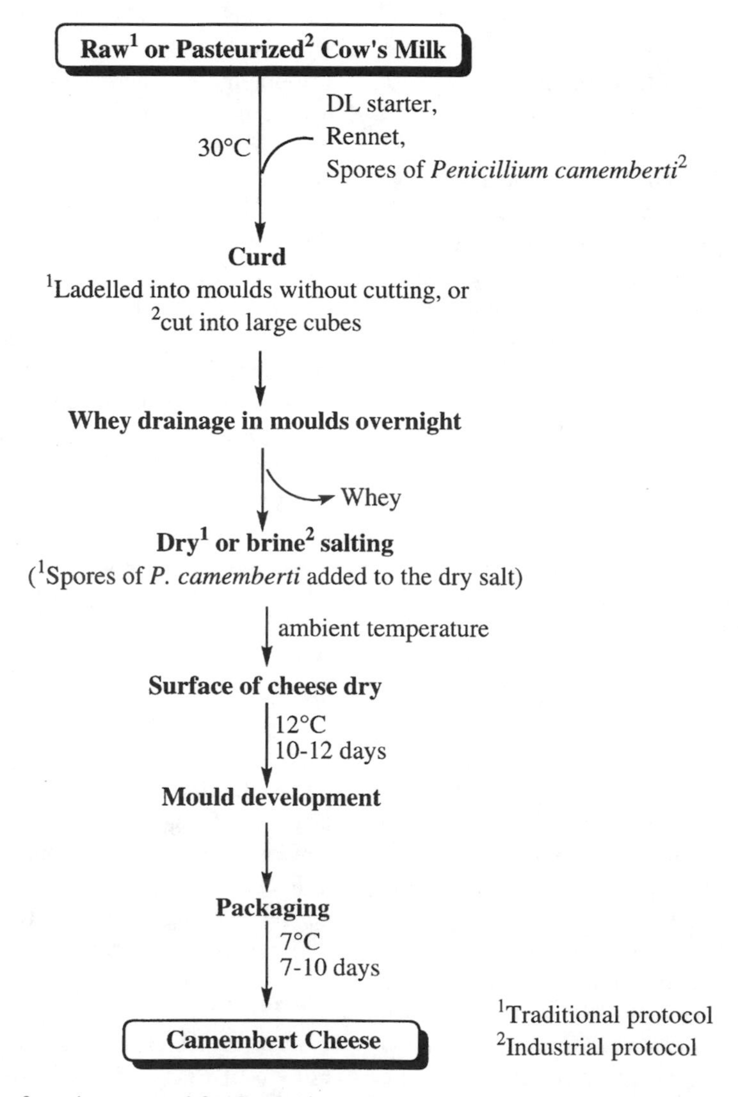

Figure 17–18 Manufacturing protocol for Camembert, a surface mold–ripened cheese.

very low, and in some cases the curds are acidified by the indigenous microflora, since the curds are usually held for an extended period in molds or special drainers, during which time acidity develops. Spores of *P. roqueforti* are added either to the milk or to the curds during manufacture. The curds for Blue cheese are cooked at a low temperature and transferred to drainers or molds, where the whey is separated from the curds. Blue cheeses are dry-salted, either by the application of salt to the surface of the cheese or by milling the curds after whey drainage and mixing with salt prior to molding. The cheeses are ripened under conditions of temperature (usually 10–12°C) and relative humidity that favor mold growth. They are pierced during ripening to facilitate uniform mold growth, they are turned, and their surfaces are cleaned regularly.

Roquefort is a Blue cheese with PDO status manufactured from raw ewe milk. Cheeses must be ripened in caves in a defined area of southeastern France. The absence of carotenoids from ewe milk results in a very white cheese that highlights the contrast in color between the curd and the mold. The manufacturing protocol for Roquefort is shown in Figure 17–19. Usually, no starter is added to the milk, which is acidified by its indigenous microflora. In addition to homofermentative lactococci, this microflora contains leuconostocs and other heterofermentative lactic acid bacteria (LAB) that produce CO_2 as a byproduct. The CO_2 causes small openings in the curd that favor the growth of the mold. Lamb rennet is added to the cheese milk, and coagulation is complete in about 2 hr, when the coagulum is cut. The curds are not cooked but are mixed with spores of *P. roqueforti* and placed in perforated metal molds for whey drainage. The whey is allowed to drain for 4–5 days, during which the cheeses are inverted periodically and acidity develops. The cheeses are then removed from the molds and dry-salted over a period of 1 week, after which they are pierced and placed in limestone caves that have the correct temperature and relative humidity to encourage mold growth. The cheeses are matured for 3–5 months, during which their surface is cleaned to remove adventitious molds or smear-forming bacteria.

Bleu d'Auvergne is a Blue cheese manufactured in France from cow milk. The Spanish Blue cheese, Cabrales, is usually made from cow milk and is ripened in local caves. Cabrales cheeses are often covered with sycamore (*Acer pseudoplatanus*) leaves to retain humidity around the cheese. The cheeses are not inoculated with *P. roqueforti* spores but become contaminated with mold spores from the environment during ripening. Thus, the degree of mold development in this variety can be variable.

Danablu (Danish blue) is perhaps the most commercially important Blue cheese. It is manufactured from cow milk, the cream from which is homogenized (to encourage lipolysis) prior to pasteurization. Pasteurized skim milk and cream are mixed and inoculated with starter and mold spores. The manufacturing protocol for Danablu is broadly similar to that for Roquefort, although chlorophyll is sometimes added to mask the yellow color of the carotenoids in cow milk, thus giving a whiter cheese. Danablu is ripened under controlled temperature (14–18°C) and relative humidity (90–95%) for up to 3 months. Edelpilkäse is a Blue cheese produced in Austria and Germany. Mycella, a Blue cheese larger than Danablu produced in Denmark, is characterized by a yellow-white color and intense mold growth. Gorgonzola, the traditional Blue cheese of Italy, is also manufactured from cow milk. The traditional protocol for the manufacture of Gorgonzola involves the separate production of curds from evening and morning milk. *P. roqueforti* spores are added to both batches of curd. The evening curds are stored overnight in cloth bags and used for cheesemaking the following morning. The morning curds are placed on the bottom and around the sides of the mold while still warm, and the cool evening curds are placed in the center. The top of the mold is then filled with warm morning curds. This layering encourages the development of mechanical openings in the cheese and yields a cheese with a smooth, firm surface.

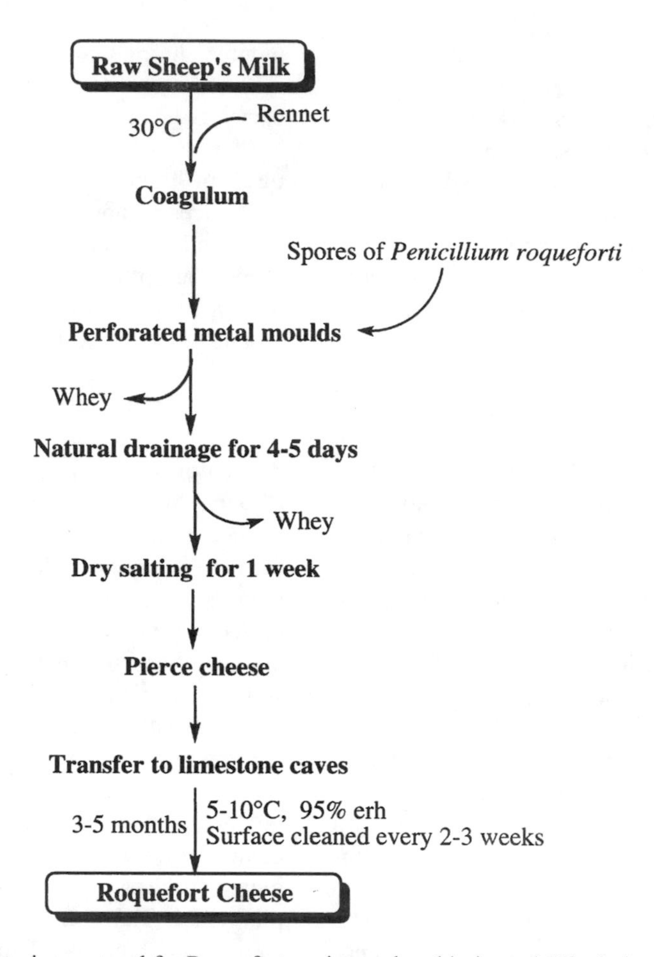

Figure 17–19 Manufacturing protocol for Roquefort, an internal mold–ripened (Blue) cheese.

Stilton is a Blue cheese that legally can only be produced in the English counties of Leicestershire, Derbyshire, and Nottinghamshire from pasteurized cow milk. A mesophilic DL starter (< 0.04%) and *P. roqueforti* spores are added to the milk. After renneting, the curds are allowed to settle on the bottom of the vat, and the whey is withdrawn slowly over the next 12–18 hr. The curd mass is cut to facilitate drainage, and the curd pieces are milled, dry-salted, and placed in molds. Whey drainage continues for about 7 days and is facilitated by frequent turning. During this time, the cheeses are kept warm (26–30°C, 90% relative humidity) so that the starter bacteria can produce sufficient acid in the cheese curd. The cheeses are then placed in a cooler room (13–15°C, 85–90% relative humidity) for 6–7 weeks, during which time the cheeses cool and a rind develops on the surface. The cheeses are pierced, and after sufficient mold growth has occurred (2–3 weeks), they are moved to a cold storeroom at 5°C.

17.2.3 Surface Smear–Ripened Cheeses

Cheeses ripened with a mixed surface microflora perhaps constitute the most hetereogeneous group of rennet-coagulated cheeses. Although most varieties in this group are soft or semi-hard, a surface flora may also develop on hard cheeses such as Gruyère. However, in the latter case, the contribution of the surface flora to cheese ripening is relatively minor.

The distinguishing feature of surface-ripened cheeses is the development of a mixed microflora on the cheese surface, forming a red-orange smear. These cheeses are manufactured using a mesophilic starter (most varieties) or a thermophilic starter (Gruyère and similar cheeses) and are usually brine-salted. Manufacturing protocols usually result in curds with a high moisture content. After manufacture, a range of salt-tolerant yeasts (*Kluyveromyces*, *Debaromyces*, *Saccharomyces*, *Candida*, *Pichia*, *Hansenula*, and *Rhodotorula*), together with *Geotrichum candidum,* become established on the cheese surface, where they metabolize lactate to CO_2 and H_2O. This change in environment favors the growth of other microorganisms. The microflora is complex and consists of Gram-positive bacteria, including *Micrococcus*, *Staphylococcus*, and various coryneform bacteria (which are responsible for the color of the smear). One component of the surface smear is the coryneform *Brevibacterium linens*, which is widely used in smear inocula, although recent research has suggested that it may be only a relatively minor component of the complex surface flora.

The surface microflora may reach 10^{11} cfu/cm^2 and thus the enzymatic activities of the smear microorganisms contribute significantly to the flavor of the cheese. Since enzymes do not diffuse through cheese curd, patterns of proteolysis in smear-ripened cheeses are similar to those in internal bacterially ripened varieties except at the cheese surface. However, products of the metabolic activities of the smear diffuse into the cheese and influence its flavor. Soft surface-ripened cheeses usually have a strong flavor whereas semi-hard varieties have a milder flavor.

Soft surface-ripened cheeses mature quite rapidly. The rate at which they ripen is governed by the size of the cheese, its moisture content, the ripening conditions, and the composition of the surface microflora. The high moisture content of these cheeses accrues from cutting the coagulum into large pieces and cooking the curds at a low temperature. Much whey is retained, and therefore the curds have a relatively high lactose content, favoring the growth of the starter, which acidifies the cheese to a low pH ($\approx$ 5). The ratio of surface area to volume (and thus the size and shape of the cheese) is very important in surface-ripened varieties. The smaller the cheese (and thus the greater this ratio), the greater the influence of the surface flora on the flavor of the cheese. The relative humidity (> 95%) and temperature (12–20°C) of ripening rooms are controlled so as to favor the growth of the surface smear. In some cases, cheeses are held in an environment with a lower relative humidity to encourage the development of a rind.

The smear develops initially as a series of colonies on the surface of the cheese. During ripening, the surface of these cheeses may be "massaged" (washed) with a brine solution, which distributes the microorganisms evenly over the surface of the cheese and results in the development of a uniform smear. Although it is common practice to inoculate cheeses with *Br. linens*, the principal source of the surface microflora is the cheesemaking environment. To encourage development of the microflora, young cheeses are often smeared with the same brine used previously to smear older, high-quality cheeses.

The ripening period for smear-ripened cheeses depends on the desired flavor intensity but is usually relatively short. For some cheeses (e.g., Brick), the smear is washed from the surface and the cheeses are coated with protective material before being transferred to a ripening room at a lower temperature ($\approx$ 10°C) for further maturation. The degree to which the smear is permitted to develop varies greatly among varieties. In some cheeses, smear development is desired principally to color the cheese surface and is therefore very limited. These cheeses mature in a similar manner to internal bacterially ripened varieties. In some cases, the surface is painted with dye to mimic smear color.

The majority of bacterial surface-ripened cheeses originated in northern Europe, but many are now produced worldwide. Limburger (Figure 17–20), one of the most important smear-ripened varieties, is named after the region of Limburg, now divided between the Netherlands and Belgium, but the cheese is currently manufactured widely in Germany and North America. The manufacturing protocol (Figure 17–21) is similar to those for other smear-ripened varieties. Limburger cheese is produced from pasteurized cow milk acidified using a mesophilic DL starter and coagulated using calf rennet. After cutting the coagulum, the curds and whey are cooked slowly to around 37°C. Most of the whey is then drained off and in some cases replaced by dilute brine, which firms the curds and removes lactose, thereby reducing the level of acid produced. The curds are then transferred to block-shaped molds, and whey drainage is assisted by frequent turning or the application of low pressure. During this time, the pH decreases and the curds mat sufficiently to retain the shape of the cheese when it is removed from the mold. The cheese is then either salted by the application of dry salt to the surface or by immersion in brine (10–15°C) for 1–2 days. The salted cheeses are transferred to ripening rooms kept at 10–15°C and above 95% relative humidity, where the characteristic surface microflora de-

Figure 17–20 Limburger cheese.

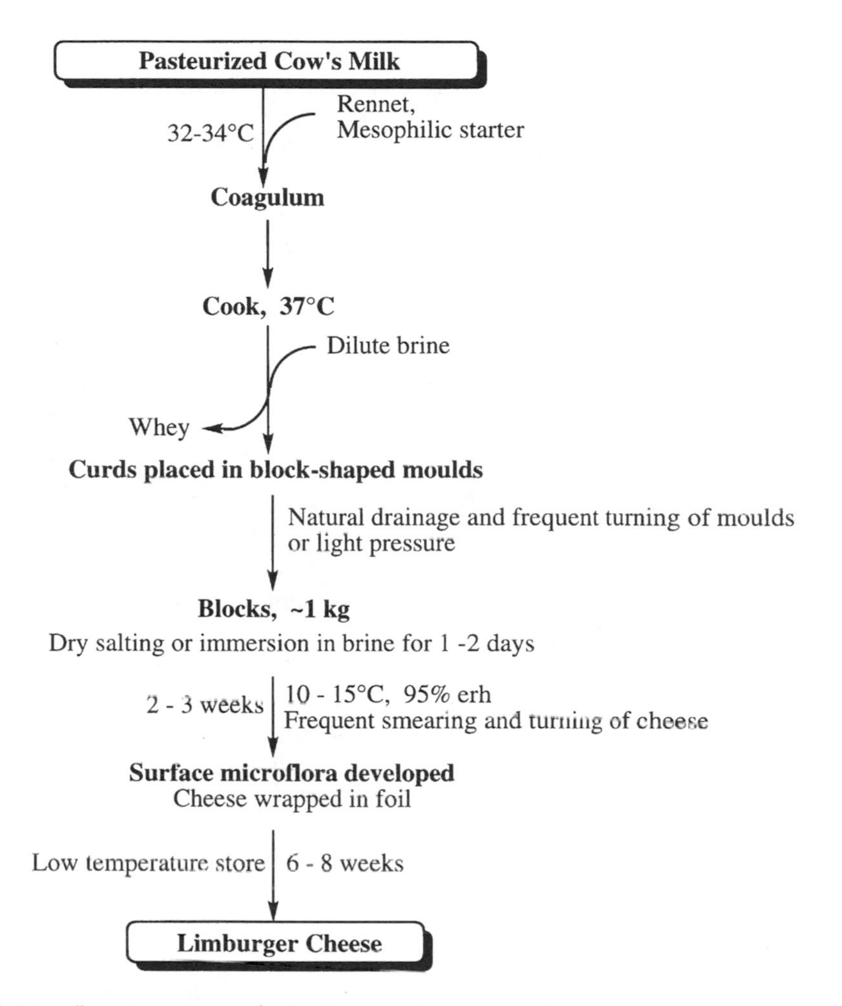

Figure 17–21 Manufacturing protocol for Limburger, a bacterial surface-ripened cheese.

velops during the next 2–3 weeks. During this time, the cheeses are turned frequently and smeared with a brine solution containing desirable microorganisms. After the development of the surface microflora, the cheeses are wrapped in foil and ripened for a further 3–8 weeks at around 4°C to complete the development of flavor. Limburger is a strong-flavored, soft, rindless cheese with mechanical openings. A number of cheeses similar to Limburger are produced, including low-fat Limburger, Romadour, and Weisslacker.

Pont l'Evêque is a square (10–11 cm sides) smear-ripened cheese that originated in Normandy and is manufactured from pasteurized cow milk. Curds are placed in cloth bags for initial draining and then in metal molds standing on rush mats. The cheeses are dry-salted. The surface flora is dominated initially by *G. candidum*, which gives the cheese a white appearance. Excessive mold growth is prevented by daily turning and washing with dilute brine. Later during ripening, Pont l'Evêque develops a bacterial microflora characteristic of smear-ripened cheeses. Port du Salut and related varieties (e.g., Saint Paulin) are semi-hard cheeses with an elastic body. They are made from cow milk, ripened at a relative humidity above 90%, and washed periodically with brine to restrict the development of the smear. Brick is a semi-hard smear-ripened cheese that originated in the United States. Smear development on this cheese is terminated after about 2 weeks by waxing or wrapping in film, and ripening is completed at a low temperature during a further 2–3 months. Butterkäse is another surface-ripened variety with limited growth of smear. The cheese is brine-salted (which encourages the development of a rind) and smear growth is encouraged for 2–3 weeks. As suggested by its name, Butterkäse is a soft cheese with a butter-like consistency.

Trappist cheese is reputed to have originated in a Trappist monastery in Bosnia. Smear growth is limited by the short ripening time (2–3 weeks). Tilsit is an important smear-ripened variety that originated in Prussia. The cheese is somewhat similar to Limburger, but its texture is firmer and there are more mechanical openings. The cheeses are brine-salted, and a strong smear develops during maturation. Taleggio is a soft smear-ripened variety that originated in Lombardy in the 1920s and has a characteristic square shape. Serra da Estrela is manufactured in Portugal from sheep milk coagulated using an extract from cardoon flowers of the thistle, *Cynara cardunculus*. After manufacture, cheeses are first ripened at a high humidity, which promotes the development of a yeasty smear ("reima"). The smear is removed about 14 days after manufacture and the cheeses are then ripened at a lower humidity to promote rind development. Münster cheese, which originated in Germany, is brine-salted, and smear growth is encouraged to give a color to the surface of the cheese. Livarot is produced in Normandy. Some is sold as fresh cheese but most is matured. During ripening an intense smear develops. Characteristic reed bands are placed around the cheese. Havarti, which originated in Denmark, has numerous irregular openings. The cheese is brine-salted, and a dry surface develops when cheeses are held at room temperature for 1–2 days. A surface flora is then encouraged to contribute to flavor development.

Variants of some of the above smear-ripened cheeses (e.g., Havarti and Saint Paulin) are also produced in which a smear is not allowed to develop. These cheeses are semi-soft and internally bacterially ripened. Such cheeses are sometimes covered with a red or orange coating to give the impression of smear growth.

17.3 ACID-COAGULATED CHEESES

Acid-coagulated cheeses (e.g., Cream cheese, Cottage cheese, Quarg, and some types of Queso blanco) are produced from milk or cream by acidification to around pH 4.6, which causes the caseins to coagulate at their isoelectric point (≈ pH 4.6). Acid-curd cheeses were probably the first cheeses produced, since such products may result from the natural souring of milk. Acidification is usually achieved by the action of a mesophilic starter, but direct acidification is also practiced. A small amount of rennet may be used in certain varieties (e.g., Cottage cheese and Quarg) but is not essential and serves to increase the firmness of the coagulum and to minimize casein losses in the whey. The coagulum may or may not be cut or cooked during manufacture, but the curds are not pressed. Acid-coagulated cheeses are characterized by a high moisture content and are usually consumed soon after manufacture. They are discussed in detail in Chapter 16.

17.4 HEAT/ACID-COAGULATED CHEESES

A small group of cheeses are coagulated by a combination of heat and acid. The most important member of this subgroup is Ricotta, an Italian cheese originally produced from whey. Ricotta (the name derives from the Italian *ricottura*, meaning "reheated") was produced originally from cheese (Mozzarella or Provolone) whey, perhaps with some milk added, by heat-induced coagulation (85–90°C) and some acidifying agent (e.g., lemon juice or vinegar). Ricotta curds were then transferred to molds surrounded by ice, where drainage occurred.

However, much Ricotta cheese is now produced from full-cream or skim milk. The milk is acidified to about pH 6.0 by the addition of a large amount (typically > 20%) of bulk starter. Unlike in the case of other varieties, the starter does not produce acid in the cheese vat and is used simply as a source of preformed lactic acid. Alternatively, the milk may be acidified with acetic acid (white wine vinegar) or citric acid. The acidified milk is heated to about 80°C by direct steam injection, during which time NaCl (0.2%, w/v) and stabilizers may be added to the milk. Precipitation occurs in about 30 min at about 80°C, after which the curds and whey are held for 15–20 min. During holding, the curd particles become firm, coalesce, and float to the surface due to entrapped air. They are scooped into perforated molds, which are cooled with crushed ice. Ricotta has a high moisture content (≈ 73%) and thus has a short shelf-life. It is normally consumed soon after manufacture as a table cheese or as an ingredient in lasagne, ravioli, or desserts.

Ricottone cheese is manufactured roughly like Ricotta but from sweet whey to which milk, skim milk, or buttermilk is added. Since the pH of sweet cheese whey is about 6.2, little additional acidification is usually necessary. Dry Ricotta, which is a grating cheese, is produced by pressing Ricottone curd and drying the cheese at 10–16°C for several months (or 4 weeks at 21°C). Mascarpone is similar to Ricotta except that cream is added to the milk, and a slightly higher cooking temperature is used. The resulting cheese is more creamy than Ricotta and is usually salted at a low level, whipped, and formed into a cylindrical shape. Impastata is made like Ricotta except that the curds are agitated gently as they form, which causes them to sink to the bottom of the vat, where they are cooked more efficiently than Ricotta curd, which remains on the surface. The result is a drier cheese, with a coarse texture, which is often ground to give a smooth, dough-like texture. Impastata is used mainly in the confectionery industry as an ingredient in pastry.

Other recooked cheese varieties and their country of origin include Bruscio (Corsica), Cacio-ricotta (Malta), Mizthra and Manouri (Greece), and Ziger (Yugoslavia).

17.5 CONCENTRATION AND CRYSTALLIZATION

A few Norwegian cheeses are produced from whey by concentration and crystallization of lactose and concentration of other solids in the whey. One could argue that such varieties are not cheeses at all but rather byproducts of cheese manufacture made from whey. These cheeses (Brunost, meaning "brown cheese") are characterized by a smooth creamy body and a sweet, caramel-like flavor. Sweet whey is the usual starting material, although acid whey may be used for some varieties. Sometimes skim milk or cream is added to the whey to give a lighter product (which would otherwise be dark brown).

The manufacturing protocol for Primost is shown in Figure 17–22. Primost ("premium quality cheese") differs from the otherwise similar Gjetost by the addition of cream to whey derived from a mixture of goat and cow milk. The whey-cream mixture is first concentrated to around 60% total solids (often in a multistage vacuum evaporator). A second concentration step (to > 80% total solids) follows, and it requires a higher vacuum. The resulting plastic mass is heated to around 95°C. The Maillard reaction is encouraged during the manufacture of these cheeses and is important for the final color and flavor. The concentrate is then cooled, kneaded, and packaged. Crystallization of lactose is controlled so as to avoid sandiness in the product. Several varieties (including Mysost, Gjetost, Niesost, Fløtemyost, and Gudbrandsdalsost) are produced using this basic process. Differences arise from the origin of the whey (cow or goat milk); the addition of skim milk, milk, or cream to the mix; and the use of sweet or acid whey. These cheeses have a high total solids content (< 18% moisture), are high in calories, and are characterized by a long shelf-life.

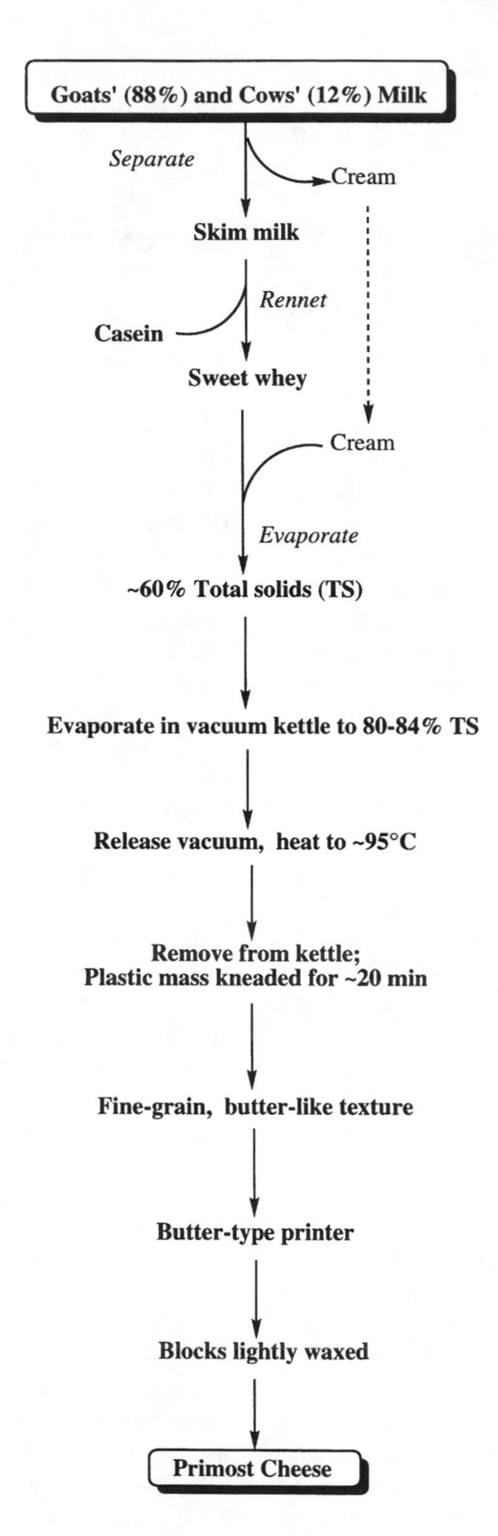

Figure 17–22 Protocol for the manufacture of Primost from whey by concentration and crystallization.

17.6 ULTRAFILTRATION TECHNOLOGY IN CHEESEMAKING

Ultrafiltration as a technology for cheese manufacture was introduced in the early 1970s and has been investigated extensively and reviewed (Ernstrom & Anis, 1985; Lawrence, 1989; Lelievre & Lawrence, 1988; Ottosen, 1988; Spangler, Jensen, Amundson, Olson, & Hill, 1991; Zall, 1985). It has attracted the attention of cheese and equipment manufacturers, primarily because of its potential to increase yield through the recovery of whey proteins in the cheese curd (see Chapter 9). Other advantages include its potential to reduce production costs and to produce new cheese varieties with different textural and functional characteristics. In this section, some of the more important aspects of ultrafiltration in cheesemaking are highlighted.

The most successful commercial applications of ultrafiltration in cheese manufacture to date have been in the production of cast Feta in Denmark, the production of fresh acid-curd varieties (Quarg, Ricotta, and Cream cheeses) in Germany and other European countries, and the standardization of milk protein, to 4–5%, for the production of Camembert and other varieties.

Based on the degree of concentration and whether whey expulsion following concentration is necessary, ultrafiltration in cheesemaking may be classified into three general areas:

1. *Low concentration factor ultrafiltration*, followed by cheesemaking and whey removal using conventional equipment. The main application of this type of ultrafiltration is the standardization of milk to a fixed protein level to obtain a more consistent end product. Variations in gel strength at cutting, buffering capacity, and rennet:casein ratio are minimized. However, when using conventional cheesemaking vats, concentration appears limited to a maximum concentration factor of around 1.5, or 4–5% protein, because of difficulties in handling the curd and yield losses.
2. *Medium concentration factor ultrafiltration* (2–6 ×) to the final solids content of the cheese without whey expulsion. The main attraction of this type of technology is the increased cheese yield associated with retention of whey proteins and the increased moisture when whey proteins are denatured prior to ultrafiltration. The main commercial application is in the production of high-moisture, unripened cheeses (e.g., Quarg and Cream cheese) and of cheeses that are not very dependent on proteolysis during ripening for flavor development (e.g., Feta). Feta produced by this method (by addition of rennet to a concentrate, i.e., pre-cheese, without cutting the coagulum) has a smoother, more homogeneous texture than the more "curdy textured" traditional product, hence the name "cast" Feta.

   There are numerous reports on using ultrafiltration concentration to attain the final cheese dry matter level for the production of soft or semi-hard rennet-curd cheeses, including Camembert, Blue cheese, Havarti, and Mozzarella. Manufacture essentially involves preacidification, ultrafiltration or diafiltration, starter addition, rennet addition, coagulation and automated cutting of the coagulum using specialized equipment (e.g., Al-Curd or Ost Retentate coagulators), molding, pressing, and brining. To date, the use of ultrafiltration technology by the dairy industry for the production of the latter cheeses has been limited. Apart from uncertainties concerning the regulatory status of such cheeses and the relatively small reported increases in yield, the main drawbacks include changes in cheese texture, flavor, and functionality (i.e., meltability and stretchability).
3. *High concentration factor ultrafiltration* followed by whey expulsion in novel equipment. Because the upper limit of concentration by ultrafiltration is about 7:1 for whole milk, it is not possible to

achieve the dry matter levels required for hard cheeses such as Cheddar and Gouda. Hence, further whey must be expelled following coagulation of the retentate and cutting of the coagulum. Owing to the high curd:whey ratio, efficient curd handling (i.e., stirring and heat transfer) is not feasible in conventional systems. The only continuous system capable of handling such concentrates is the Siro-Curd, which was used for the production of Cheddar cheese in Australia during the 1980s, although its use has been discontinued. The cheese produced by this process, which gives a yield increase of around 4–6%, was claimed to be indistinguishable from Cheddar manufactured using standard equipment.

When renneting is done at a fixed dosage level, increasing the protein level in milk results in a reduced rennet coagulation time, an increase in the level of soluble (nonaggregated) casein at the point of gelation, an increased rate of curd firming, a reduced set-to-cut time when cutting at a given curd strength, a decrease in the degree of aggregation at cutting, and a coarser gel network. Micelles that are not modified, or aggregated, at the onset of gelation are presumably modified later and incorporated into the gel to a greater or lesser degree.

Owing to the rapid rate of curd firming, it becomes increasingly difficult, as the milk protein level is increased, to cut the coagulum cleanly (without tearing) before the end of the cutting cycle. Reflecting the tearing of the coagulum and consequent shattering of curd particles, fat losses in the whey are greater than those predicted on the basis of volume reduction (due to ultrafiltration) for milks with protein concentrations less than 5%. Similar findings have been attributed partly to the poorer fat-retaining ability of higher protein curds, which have coarser, more porous protein networks. Reducing the setting temperature, in the range 31–27°C, and reducing the level of rennet added results in set-to-cut times and curd-firming rates for concentrated milk closer to those for the control milk.

Increasing the concentration of protein also results in slower proteolysis during ripening when equal quantities of rennet on a milk volume basis are used. The slower rate of proteolysis in cheeses made from ultrafiltered milks may be attributed to a number of factors, including

- the lower effective rennet concentration (i.e., rennet:casein ratio) and hence activity in the cheese
- the inhibition of the indigenous milk proteinase plasmin by retained β-lactoglobulin in cheeses containing a significant quantity of whey proteins
- the concentration during ultrafiltration of indigenous proteinase or peptidase inhibitors
- the resistance of undenatured whey proteins to proteolysis in cheese in which they represent a substantial portion (≈ 18%) of the proteins

However, at equal rennet:casein ratios, the level of α_{s1}-casein hydrolysis is higher in control Cheddar cheese than in that made from milk concentrated fivefold by ultrafiltration. The reduced surface area:volume ratio (SA:V) of the protein network in cheeses made from concentrated milks, resulting from their coarser network, may also contribute to the observed reduction in proteolysis. It is conceivable that for a given level of enzyme activity in the cheese curd, casein degradation decreases as the SA:V of the matrix decreases.

Cheese becomes progressively firmer (i.e., requires a higher compression force to induce fracture), more cohesive, mealier, and drier and the structure of the protein matrix becomes coarser and more compact (fused) with an increasing concentration factor. The reduced rate of proteolysis results in slower softening and flavor development during maturation.

REFERENCES

Bertozzi, L., & Panari, G. (1993). Cheeses with Appellation d'Origine Contrôlée (AOC): Factors that affect quality. *International Dairy Journal*, *3*, 297–312.

Burkhalter, G. (1981). *Catalogue of cheeses* [Document No. 141]. Brussels: International Dairy Federation.

Ernstrom, C.A., & Anis, S.K. (1985). Properties of products from ultrafiltered whole milk. In *New dairy products via new technology* [Proceedings of International Dairy Federation Seminar, Atlanta]. Brussels: International Dairy Federation.

Fox, P.F. (1993a). Cheese: An overview. In P.F. Fox (Ed.), *Cheese: chemistry, physics and microbiology* (2d ed., Vol. 1). London: Chapman & Hall.

Fox, P.F. (Ed.) (1993b). *Cheese: Chemistry, physics and microbiology: Vol. 2. Major cheese groups* (2d ed.). London: Chapman & Hall.

Guinee, T.P., & Fox, P.F. (1987). Salt in cheese: Physical, chemical and biological aspects. In P.F. Fox (Ed.), *Cheese: Chemistry, physics and microbiology* (Vol. 1). London: Elsevier Applied Science.

Kosikowski, F.V., & Mistry, V.V. (1997). *Cheese and fermented milk foods* (3d ed., Vols. 1, 2). Westport, CT: F.V. Kosikowski, LLC.

Lawrence, R.C. (1989). *The use of ultrafiltration technology in cheesemaking* [Bulletin No. 240]. Brussels: International Dairy Federation.

Lelievre, J., & Lawrence, R.C. (1988). Manufacture of cheese from milk concentrated by ultrafiltration. *Journal of Dairy Research*, *55*, 465–478.

Ottosen, N. (1988). *Protein standardization: Technical information*. Silkeborg, Denmark: APV Pasilac.

Robinson, R.K. (Ed.). (1995). *A colour guide to cheese and fermented milks*. London: Chapman & Hall.

Robinson, R.K., & Wilbey, R.A. (1998). *Cheesemaking practice* (3d ed.). Gaithersburg, MD: Aspen Publishers.

Scott, R. (1986). *Cheesemaking practice*. London: Elsevier Applied Science.

Spangler, P.L., Jensen, L.A., Amundson, C.H., Olson, N.F., & Hill, G.G., Jr. (1991). Ultrafiltered Gouda cheese: Effects of preacidification, diafiltration, rennet and starter concentration and time to cut. *Journal of Dairy Science*, *74*, 2809–2819.

Walter, H.E., & Hargrove, R.C. (1972). *Cheeses of the world*. New York: Dover.

Zall, R.R. (1985). On-farm ultrafiltration. In *New dairy products via new technology* [Proceedings of International Dairy Federation Seminar, Atlanta]. Brussels: International Dairy Federation.

Appendix 17–A

Compositions of Selected Cheese Varieties

Cheese	*Fat (%)*	*Total Solids (%)*	*Total Protein (%)*	*Salt (%)*	*Ash (%)*	*pH*
Asiago	30.8	72.5	30.9	3.6	6.6	5.3
Blue	29.0	58.0	21.0	4.5	6.0	6.5
Blue Stilton	33.0	61.7	24.8	3.5	3.2	5.2
Brick	30.0	60.0	22.5	1.9	4.4	6.4
Bulgarian White	32.3	68.0	22.0	3.5	5.3	5.0
Caciocavallo Siciliano	27.5	70.9	33.1	4.0	7.0	6.0
Caerphilly	34.0	67.7	27.2	1.5	3.4	5.4
Camembert	23.0	47.5	18.5	2.5	3.8	6.9
Cheddar (American)	32.0	63.0	25.0	1.5	4.1	5.5
Cheshire	33.0	66.7	26.7	1.8	3.9	5.3
Comté	30.0	66.5	30.0	1.1	4.1	5.7
Cottage	4.2	21.0	14.0	1.0	1.0	5.0
Cream	33.5	50.0	10.0	0.75	1.3	4.6
Edam	24.0	57.0	26.1	2.0	3.0	5.7
Emmental (Swiss)	30.5	64.5	27.5	1.2	3.5	5.6
Feta	20.3	40.3	13.4	2.2	2.3	4.2
Gouda	28.5	59.0	26.5	2.0	3.0	5.8
Grana (Parmesan)	25.0	69.0	36.0	2.6	5.4	5.4
Gruyère	30.0	66.5	30.0	1.1	4.1	5.7
Havarti	26.5	56.5	24.7	2.2	2.8	5.9
Leicester	33.0	64.7	25.5	1.6	3.5	6.5
Limburger	28.0	55.0	22.0	2.0	4.8	6.8
Manchego	25.9	62.1	28.1	1.5	3.6	5.8
Mitzithra	25.0	56.3	18.4	1.6	2.5	5.0
Mozzarella	18.0	46.0	22.1	0.7	2.3	5.2
Mozzarella–low moisture	23.7	53.0	21.0	1.0	3.0	5.2
Muenster	29.0	57.0	23.0	1.8	4.4	6.2
Pont L'Eveque	25.8	57.2	26.5	2.8	2.4	7.0
Provolone	27.0	57.5	25.0	3.0	4.0	5.4
Quarg	0.2	21.0	15.0	0.70	1.0	4.5
Queso blanco	15.0	49.0	22.9	2.0–3.9	5.4	5.3
Ricottone (whey Ricotta)	0.5	27.5	11.0	<0.5	4.0	4.9
Ricotta	12.7	28.0	11.2	<0.5	–	5.9
Romano	24.0	77.0	35.0	5.5	10.5	5.4
Roquefort	31.0	60.0	21.5	3.5	6.0	6.4
Samsoe	27.0	59.9	26.5	1.8	3.7	5.5
Serra da Estrela	27.5	51.3	21.3	1.9	2.8	6.5
Svecia	28.3	56.0	21.8	2.5	4.1	5.5

Chapter 18

Processed Cheese and Substitute or Imitation Cheese Products

18.1 INTRODUCTION

Products in this group of cheese products differ from natural cheeses in that they are not made directly from milk (or dehydrated milk) but rather from various ingredients such as natural cheese, skim milk, water, butter oil, casein, caseinates, other dairy ingredients, vegetable oils, vegetable proteins, and minor ingredients. The two main categories, namely pasteurized processed cheese products and substitute or imitation products, may be further subdivided depending on composition and the types and levels of ingredients used (Figure 18–1). The individual categories are discussed below.

18.2 PASTEURIZED PROCESSED CHEESE PRODUCTS

Pasteurized processed cheese products (PCPs) are produced by comminuting, melting, and emulsifying into a smooth homogeneous molten blend one or more natural cheeses and optional ingredients using heat, mechanical shear, and (usually) emulsifying salts. Optional ingredients permitted are determined by the product type (processed cheese, processed cheese food, and processed cheese spread) and include dairy ingredients, vegetables, meats, stabilizers, emulsifying salts, flavors, colors, preservatives, and water (Table 18–1 and Exhibit 18–1).

Although a product of recent origin (first developed in 1911), processed cheese products have had a growth rate similar to that of natural cheese (≈ 2.3% per annum) in Europe and North America during the period 1987–1996 (Sørensen, 1997). Current global production is 1.9 million tonnes per annum, equal to 17% of natural cheese production (Sørensen, 1997). Factors contributing to the continued growth and success of these products are as follows:

- They offer almost unlimited variety in flavor, consistency, functionality (e.g., sliceability, meltability, flowability), and consumer appeal as a result of differences in formulation, processing conditions, and packaging into various shapes and sizes.
- They cost less than natural cheese because they incorporate low-grade natural cheese and cheaper noncheese dairy ingredients.
- They are adaptable to the fast-food trade. The most notable example is the inclusion of cheese slices in burgers and the use of dried processed cheeses (cheese powder) as snack and popcorn coatings.
- They have a relatively long shelf-life, and waste is minimal.
- Some companies specialize in the manufacture of equipment, emulsifying salts, and other ingredients tailor-made to industry's need for new products and consistent quality.

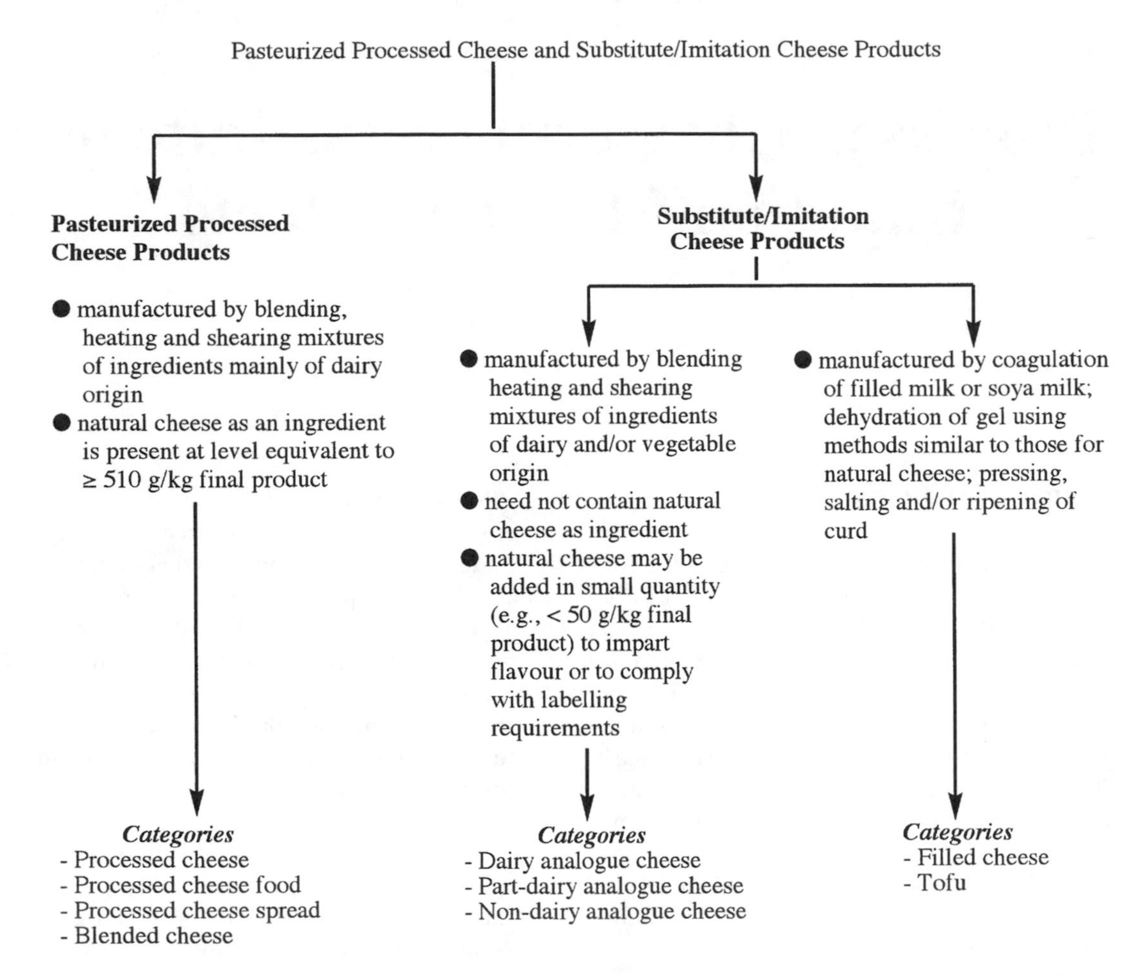

Figure 18–1 Generalized classification scheme for pasteurized processed cheese and substitute or imitation cheese products based on manufacturing procedure and ingredients used.

18.2.1 Classification of Processed Cheese Products

The named categories of PCPs and their standards of identity (composition and levels and types of permitted ingredients) vary somewhat between countries. Hence, in the United Kingdom there are two categories of PCPs, namely, processed cheese and cheese spread (Cheese and Cream Regulations 1995 SI 1995/3240), whereas in Germany there are four categories, namely, Schmelzkäse (processed cheese), Schmelzkäsezubereitung (processed cheese preparation), Käsezubereitung (cheese preparation), and Käsekomposition (cheese composition), as described in the Deutsche Käseverordnung of 12 November 1990. Currently, the International Dairy Federation, under the auspices of the Codex Alimentarius Commission, is endeavoring to draft a single standard for pasteurized processed cheese products that will be accepted globally. It is expected that a Codex standard will assume increased importance because such a standard will be used by the World Trade Organization in the resolution of trade disputes. In this chapter, the scheme used in the

Table 18–1 Permitted Ingredients in Pasteurized Processed Cheese Products

Product	*Ingredients*
Pasteurized blended cheese	Cheese; cream, anhydrous milk fat, dehydrated cream (in quantities such that the fat derived from them is less than 5% [w/w] in finished product); water; salt; food-grade colors, spices, and flavors; mold inhibitors (sorbic acid, potassium/sodium sorbate, and/or sodium/calcium propionates), at levels ≤ 0.2% (w/w) of finished product
Pasteurized process cheese	As for pasteurized blended cheese, but with the following extra optional ingredients: emulsifying salts (sodium phosphates, sodium citrates; ≤ 3% [w/w] of finished product); food-grade organic acids (e.g., lactic, acetic, or citric) at levels such that pH of the finished product is ≥ 5.3
Pasteurized process cheese foods	As for pasteurized process cheese, but with the following extra optional dairy ingredients: milk, skim milk, buttermilk, cheese whey, whey proteins—in wet or dehydrated forms
Pasteurized process cheese spread	As for pasteurized process cheese food but with the following extra optional ingredients: food-grade hydrocolloids (e.g., carob bean gum, guar gum, xanthan gums, gelatin, carboxymethylcellulose, and/or carrageenan) at levels < 0.8% (w/w) of finished products; food-grade sweetening agents (e.g., sugar, dextrose, corn syrup, glucose syrup, hydrolyzed lactose)

Note: For more detail, see Code of Federal Regulations (1988).

United States to classify PCPs will be used. Under this system, which is described in the Code of Federal Regulations, Food and Drugs, Part 133 (Edition 4-1-93), four main categories of PCPs are identified:

1. pasteurized process cheese
2. pasteurized process cheese food
3. pasteurized process cheese spread
4. pasteurized blended cheese

A fifth category comprises nonstandarized pasteurized cheese-type products such as dips and sauces. The criteria for classification include permitted ingredients and compositional parameters. The main aspects of the different categories are summarized in Tables 18–1 and 18–2.

Pasteurized process cheese is usually sold in the form of sliceable blocks (e.g., processed Cheddar) or slices. Spreads and foods may be in the form of blocks, slices, spreads, dips, sauces, or pastes (e.g., in tubes). Pasteurized blended cheese, which is the least common type, is usually sold in forms that give a natural cheese image.

18.2.2 Manufacturing Protocol for Processed Cheese Products

Manufacture involves the following steps (Figure 18–2):

- formulation of blend, which involves selection of the correct type and quantity of natural cheeses, emulsifying salts, water, and optional ingredients
- shredding or comminuting of cheese and blending with optional ingredients

Exhibit 18–1 Optional Ingredients Permitted in Pasteurized Processed Cheese Products

Dairy ingredients
- Anhydrous milk fat, cream, milk, skim milk solids, whey solids, milk proteins, co-precipitates, milk ultrafiltrates

Stabilizers
- Emulsifying salts, including sodium phosphates and sodium citrates
- Hydrocolloids: guar gum, xanthan gum, carrageenans
- Organic emulsifiers: lecithin, mono- and diglycerides

Acidifying agents
- Various food-grade organic acids, including lactic, acetic, phosphoric, and citric acids

Sweetening agents
- Sucrose, dextrose, corn syrup, hydrolyzed lactose

Flavors
- Enzyme modified cheese (EMC), artificial flavors, smoke extracts, starter distillate

Flavor enhancers
- NaCl, autolyzed yeast extract

Colors
- Annatto, oleoresin, paprika, artificial colors

Preservatives
- Potassium sorbate, calcium/sodium propionates, nisin

Condiments
- Cooked meats/fish
- Cooked or dried fruit or vegetables

- processing of the blend
- homogenization of the hot molten blend (this step is optional and implementation depends on the fat content of the blend, type of cooker used, and body characteristics of the end product)
- packaging and cooling

Processing refers to the heat treatment of the blend by direct or indirect steam, with constant agitation. Application of a partial vacuum during cooking is optional. It may be used to regulate moisture content when direct steam injection is used, and it is also beneficial in removing air and thus preventing air openings in the finished set product. In batch processing, the temperature-time combination varies (70–95°C for 4–15 min), depending on the formulation; extent of agitation; and desired product texture, body, and shelf-life characteristics. At a given temperature, the processing time generally decreases with agitation rate, which may vary, depending on the kettle (cooker) type, from 50 to 3000 rpm. In continuous cookers, which are used mainly for dips and sauces, the blend is mixed and heated to 80–90°C in a vacuum mixer, from which it is pumped through a battery of tubular heat exchangers and heated to 130–145°C for a few seconds and then flash cooled to 90°C. The cooked product is then pumped to a surge tank that feeds the packaging machine.

In the manufacture of slices, the hot molten cheese is pumped through a manifold with 8–12 nozzles that extrude ribbons of cheese onto the first of 2 or 3 counter-rotating chill rolls, which cool the ribbons from 70–80°C to 30°C. The ribbons are cut automatically into slices, which are stacked and packed.

18.2.3 Principles of Manufacture for Processed Cheese Products

Microstructurally, natural cheese may be viewed as a three-dimensional paracasein network composed of overlapping and cross-linked chains of partially fused aggregates (in turn formed from fused paracasein micelles). Moisture and fat, in the form of discrete or coalesced globules, are entrapped within the pores of the network, which visually resembles a loose semirigid sponge (Figure 18–3). The integrity of the paracasein network is maintained by various inter- and intra-aggregate bonds, including hydrophobic and electrostatic attractions (e.g., calcium cross-linking via casein phosphoserine and ionized carboxyl residues).

Application of heat (70–90°C) and mechanical shear to natural cheese, as in processing, in the absence of stabilizers, usually results in a

Table 18–2 Compositional Specifications for Pasteurized Processed Cheese Products

Product Category	*Moisture (%, w/w)*	*Fat (%, w/w)*	*Fat in Dry Matter (%, w/w)*
Pasteurized blended cheese	≤ 43	–	≥ 47
Pasteurized process cheese	≤ 43	–	≥ 47
Pasteurized process cheese foods	≤ 44	≥ 23	–
Pasteurized process cheese spread	40–60	≥ 20	–

Note: Minimum temperature and time specified for processing is 65.5°C for 30 s. The compositional specifications for pasteurized process cheese may differ from those given, depending on the type of product. For more detail, see the Code of Federal Regulations (1988).

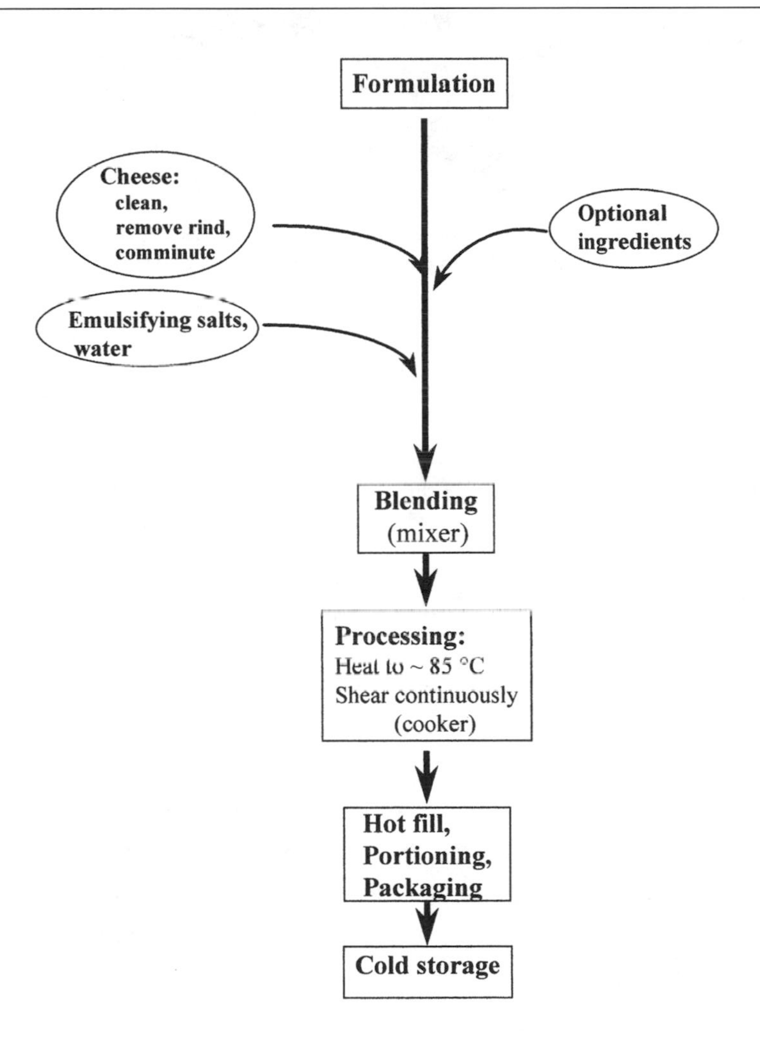

Figure 18–2 Batch process for the manufacture of pasteurized processed cheese products.

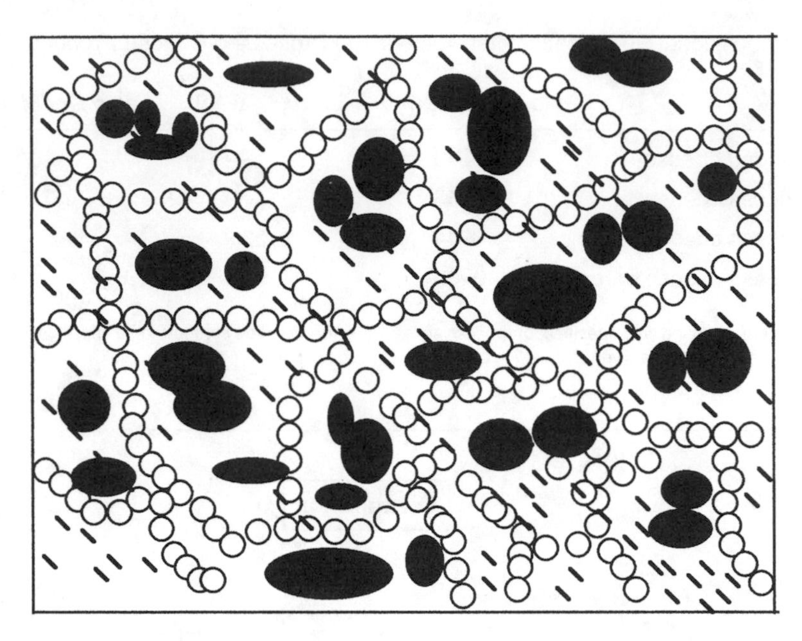

Figure 18–3 Schematic representation of the structure of natural cheese, showing network of fused paracasein micelles (❍) that occludes within its pores the fat phase, consisting of individual and partly coalesced globules (●) and moisture (- -).

heterogeneous, gummy, puddinglike mass that undergoes extensive oiling-off and exudation of moisture during manufacture, especially upon cooling. These defects arise from

- the coalescence of liquified fat due to shearing of the fat globule membranes
- partial dehydration or aggregation and shrinkage of the paracasein matrix induced by the relatively low pH of cheese (for most cheeses < 5.7) and the high temperature applied during processing

The modified structure, consisting of a shrunken paracasein matrix with large pools of free oil and free moisture, has an impaired ability to occlude fat and free moisture. Consequently, free moisture and de-emulsified liquified fat seep through the more porous, modified structure.

The addition of emulsifying salts (10–30 g/kg) during processing promotes emulsification of free fat and rehydration of protein and thus contributes greatly to the formation of a smooth, homogeneous, stable product. The emulsifying salts most commonly used for pasteurized processed cheese manufacture include sodium citrates, sodium orthophosphates, sodium pyrophosphates, sodium tripolyphosphates, sodium polyphosphates (e.g., Calgon), basic sodium aluminum phosphates (e.g., Kasal), and phosphate blends (e.g., Joha, Solva blends). These salts generally have a monovalent cation (i.e., sodium) and a polyvalent anion (e.g., phosphate). While these salts are not emulsifiers, they promote, with the aid of heat and shear, a series of concerted physicochemical changes in the cheese blend that result in rehydration of the aggregated paracasein matrix and its conversion into an active emulsifying agent. These changes include calcium sequestration, upward adjustment and stabilization (buffering) of pH, paracasein hydration (solvation) and dispersal, emulsification of free fat, and structure formation. They are discussed briefly below.

Calcium Sequestration

This involves the exchange of the divalent Ca^{2+} of the paracasein matrix (attached to casein via the carboxyl groups of acidic amino acids and/or by phosphoseryl residues) for the monovalent Na^{+} of the emulsifying salt. The removal of Ca^{2+}, which is referred to as calcium sequestration or chelation, results in

- partial breakdown of the paracasein matrix due to disintegration of the intra- and inter-aggregate bonds and consequently of the links between the strands of the paracasein matrix
- an ensuing conversion of the calcium paracasein gel network into a sodium phosphate paracaseinate dispersion (sol), to a greater or lesser degree, depending on the processing conditions and type of salt (calcium chelating strength, pH, and buffering capacity)

Displacement and Stabilization (Buffering) of pH

The use of the correct blend of emulsifying salts usually shifts the pH of the cheese upward, typically from around 5.0–5.5 in the natural cheese to 5.6–5.9 in the processed cheese product, and stabilizes it by virtue of the high buffering capacity of the salts. This change contributes to the formation of a stable product by increasing

- the calcium-sequestering ability of the emulsifying salts
- the negative charge on the paracaseinate, which in turn promotes further disintegration of the calcium paracasein network and a more open, reactive paracaseinate conformation, with superior water-binding and emulsifying properties

Hence, the extent of pH buffering is a critical factor controlling the textural attributes of processed cheese products.

Dispersion and Water Binding of Paracasein

Dispersion of paracasein, also referred to as *peptization* or *swelling*, involves the disintegration of the cheese matrix and conversion of the calcium paracasein into a charged, hydrated sodium (phosphate) paracaseinate; it is caused by the above-mentioned emulsifying salt-induced changes in combination with the mechanical and thermal energy inputs of processing.

The conversion of calcium paracasein to sodium (phosphate) paracaseinate during processing is the major factor affecting the water-binding capacity of the protein. The increase in casein hydration during processing is consistent with the inverse relationship found between casein-bound calcium and casein hydration.

Emulsification

Under the conditions of cheese processing, the dispersed hydrated paracaseinate contributes to

- emulsification by coating the surfaces of dispersed free fat droplets, resulting in the formation of recombined fat globule membranes
- emulsion stability by immobilization of a large amount of free water

18.2.4 Structure Formation upon Cooling

During the cooling of processed cheese products, the homogeneous, molten, viscous mass sets to form a characteristic body, which, depending on the blend formulation, processing conditions, and cooling rate, may vary from firm and sliceable to semi-soft and spreadable. Factors that contribute to structure formation (setting) during cooling include solidification (crystallization) of fat and protein-protein interactions, which result in the formation of a new matrix. It is envisaged that the newly formed emulsified fat globules become an integral part of the matrix owing to interaction of their paracaseinate membrane with the paracaseinate matrix.

Electron microscopy studies of processed cheese products indicate the following:

- The protein phase exists in the form of relatively short strands connected to varying degrees, resulting in a matrix with different degrees of continuity depending on product type. The matrix strands are much finer than those in natural cheese.
- The fat globules are uniformly distributed (unlike natural cheese) within the protein matrix and generally range from 0.3 to 5.0 μm in diameter. Fat globule size varies with the degree of emulsification, which in turn is regulated by the formulation (type and quantity of emulsifying salt and other ingredients and age of the cheese) and processing conditions (shear rate, temperature, and time).
- The paracaseinate membranes of the emulsified fat globules attach to the matrix strands. The ensuing anchoring of the relatively short strands by the recombined fat globules probably contributes to the continuity and elasticity of the matrix in the cooled product.

The size of the fat globules is important, as it influences the firmness of the final product and the ability of the fat to become free and contribute to oiling-off when the processed cheese product is subsequently cooked (e.g., processed cheese slices on cheeseburgers and processed cheese insets in burgers). When cheese is baked or grilled, some oiling-off is desirable, as it limits drying out of the cheese and thus contributes to the desired flowability, succulence, and surface sheen of the melted product (see Chapter 19). Generally, for a given formulation, a reduction in the mean diameter of the emulsified fat globules results in processed cheese products that are firmer and exhibit a low tendency to oil-off and poor flowability upon cooking. Comparative studies on the effect of different emulsifying salts indicate that, for a given processing time, the mean fat globule diameter is generally smallest when tetrasodium pyrophosphate (TSPP) is used, largest with basic sodium aluminum phosphate (SALP), and intermediate with trisodium citrate (TSC) or disodium phosphate (DSP). Hence, SALP is generally claimed to produce processed cheeses with good melting properties. Increasing the concentration of emulsifying salt (10–40 g/kg) and processing temperature (80–140°C) results in a progressive decrease in mean fat globule diameter and a concomitant increase in firmness. Increasing the processing time for a given formulation results in final products that are firmer, more elastic, and less flowable (Rayan, Kalab, & Ernstrom, 1980). These trends undoubtedly reflect decreases in the mean fat globule diameter and the level of paracasein hydration (or alternatively an increase in protein aggregation) upon prolonged holding or shearing of the hot molten blend at a high temperature. The degree of paracasein hydration in cheese is a major factor influencing its rheology and functionality upon cooking (see Chapter 19).

18.2.5 Properties of Emulsifying Salts

The emulsifying salts most commonly used are sodium citrates, sodium hydrogen orthophosphates, sodium polyphosphates, and sodium aluminum phosphates (Table 18–3). Other potential emulsifying agents include gluconates, lactates, malates, ammonium salts, gluconic acid, lactones, and tartarates. Nowadays, emulsifying salts are generally supplied as blends of phosphates (e.g., Joha C special and Solva 35S) or phosphates and citrates (e.g., Solva NZ 10), tailor-made to impart certain functionalities (e.g., different degrees of meltability, sliceability, spreadibility) to different pasteurized products (e.g., blocks, slices, spreadable products) manufactured under different conditions (e.g., with cheeses of varying degrees of maturity, in cookers with varying degrees of shear input). The properties of different emulsifying salts have been studied and reviewed extensively (Caric & Kalab, 1993; Cavalier-Salou & Cheftel, 1991; Fox, O'Connor, McSweeney, Guinee, & O'Brien, 1996; Rayan, Kalab, & Ernstrom, 1980; Scharf, 1971; Tanaka et al., 1986; van Wazar, 1971).

Sodium citrates are used widely. While the use of potassium citrates for the manufacture of low-sodium PCPs has been reported, they are not normally used commercially, as they tend to impart a bitter flavor (relative to sodium citrate). Trisodium citrate is used most commonly. The mono- and disodium forms ($NaH_2C_6H_5O_7$ and $Na_2HC_6H_5O_7$), when used alone, tend to result in overacidic processed cheese products that are mealy, acidic, and crumbly and show a tendency toward oiling-off due to poor emulsification. The dissociation constants (pK_a's) of citric acid, at the ionic strength of milk, are 3.0, 4.5, and 4.9. Owing to their acidic properties, mono- and disodium citrates may be used to correct the pH of a processed cheese blend, for example, when a high proportion of very mature, high-pH cheese or skim milk solids is used.

The phosphates used in cheese processing include sodium monophosphates (sodium orthophosphates), which contain 1 P atom ($n = 1$), and linear condensed phosphates such as pyrophosphates ($n = 2$), and polyphosphates ($n = 3$–25; e.g., tripolyphosphate, $n = 3$). Of the orthophosphates, disodium hydrogen orthophosphate (Na_2HPO_4) is the form normally used. When used alone, the mono- and trisodium salts tend to produce overacidic and underacidic products, respectively. Comparative studies have shown that the potassium salts of orthophosphates, pyrophosphates, and citrates produce processed cheeses with textural properties similar to those made with the equivalent sodium salts at similar concentrations. Hence, potassium emulsifying salts may have potential in the preparation of reduced-sodium formulations. Owing to their aluminum content and the possible association of aluminum with Alzheimer's disease, sodium aluminum phosphates (e.g., kasal) are used to only a limited extent.

The effectiveness of the different salts in promoting the various physicochemical changes that occur during processing has been studied extensively in both pasteurized processed cheese products and analogue cheeses. Discrepancies exist between these studies as regards the influence of emulsifying salts on different physicochemical changes, probably due to differences in product formulation (e.g., levels of total protein and intact protein, pH), the level of emulsifying salts added, and processing conditions (e.g., cooker type, degree of shear, and time-temperature treatment). However, these studies indicate definite trends, which are summarized in Table 18–4, and discussed below.

Calcium Sequestration

Ion exchange is best accomplished by salts that contain a monovalent cation and a polyvalent anion, and effectiveness generally increases with the valency of the anion. The general ranking of the calcium sequestration ability of the common emulsifying salts used in cheese processing is in the following order: polyphosphates $>$ pyrophosphates $>$ orthophosphates $>$ sodium aluminum phosphates $\approx$ citrates. However, the sequestering ability, especially of the shorter chain phosphates, is strongly influenced by pH. The increased ion exchange function at higher pH values is attributed to more complete dissociation of the sodium phosphate molecules, resulting in the formation of a higher valency anion. Thus, for the shorter chain phosphates, calcium binding increases in the following order: $NaH_2PO_4 < Na_2HPO_4 < Na_2H_2P_2O_7 < Na_3HP_2O_7 < Na_4P_2O_7 < Na_4P_2O_7$.

Displacement and Buffering of pH

The buffering capacity of sodium phosphates in the pH range normally encountered in processed cheese products (5.5–6.0) decreases with increasing chain length and is effectively zero for the longer chain phosphates ($n > 10$). This decrease in buffering capacity with chain length is due to the corresponding reduction in the number of acid groups per molecule, which oc-

Table 18–3 Properties of Emulsifying Salts for Processed Cheese Products

Group	*Emulsifying Salt*	*Formula*	*Solubility at 20°C (%)*	*pH Value (1% Solution)*
Citrates	Trisodium citrate	$2Na_3C_6H_5O_7.1H_2O$	High	6.23–6.26
Orthophosphates	Monosodium phosphate	$NaH_2PO_4.2H_2O$	40	4.0–4.2
	Disodium phosphate	$Na_3HPO_4.12H_2O$	18	8.9–9.1
Pyrophosphates	Disodium pyrophosphate	$Na_2H_2P_2O_7$	10.7	4.0–4.5
	Trisodium pyrophosphate	$Na_3HP_2O_7.9H_2O$	32.0	6.7–7.5
	Tetrasodium pyrophosphate	$Na_4P_2O_7.10H_2O$	10–12	10.2–10.4
Polyphosphates	Pentasodium tripolyphos-phate	$Na_5P_3O_{10}$	14–15	9.3–9.5
	Sodium tetrapolyphosphate	$Na_6P_4O_{13}$	14–15	9.0–9.5
	Sodium hexametaphosphate (Graham's salt)	$Na_{n+2}PnO_{3n+1}$ (n = 10–25)	Very high	6.0–7.5
Aluminum phosphates	Sodium aluminum phosphate	$NaH_{14}Al_3(PO_4)_8.4H_8O$	–	8.0

Table 18–4 General Properties of Emulsifying Salts in Relation to Cheese Processing

Property	*Citrates*	*Orthophosphates*	*Pyrophosphates*	*Polyphosphates*	*Aluminum Phosphate*
Ion exchange (calcium sequestration)	Low	Low	Moderate	High–very high	Low
Buffering action in the pH range 5.3–6.0	High	High	Moderate	Low–very low	–
Paracaseinate dispersion (peptization)	Low	Low	High	Very high	–
Emulsification	Low	Low	Very high	Very high (n = 3–10)	Very low
Bacteriostatic effects	Nil	Low	High	High–very high	–

cur singly at each end of the polyphosphate chain. The ortho- and pyrophosphates possess high buffering capacities in the pH ranges 2–3, 5.5–7.5, and 10–12. Thus, in cheese processing they are suitable not only as buffering agents but also as pH correction agents. Within the citrate group, only the trisodium salt has buffering capacity in the pH range 5.3–6.0. The more acidic mono- and disodium citrates produce overacidic, crumbly cheese with a propensity to oiling-off.

The pH of processed cheese products is related to the pH of the solution of emulsifying salt and to its buffering capacity. The pH of analogue cheese made with trisodium citrate or different sodium phosphate emulsifying salts, at equal concentrations, decreases in the following order: tetrasodium pyrophosphate ≈ trisodium citrate ≈ pentasodium tripolyphosphate > disodium hydrogen phosphate > sodium polyphosphate. The pH of processed cheese generally increases linearly with emulsifying salt concentration in the range 0–30 g/kg for trisodium citrate, tetrasodium pyrophosphate, sodium tripolyphosphate, and disodium hydrogen phosphate.

Hydration and Dispersion of Casein

The ability of the different groups of emulsifying salts to promote protein hydration and dispersion during cheese processing is in the following general order: polyphosphates ($n = 3$–10) > pyrophosphates > monophosphates ≈ citrates. The greater hydrating effect of polyphosphates over citrates and monophosphates can be explained in terms of the greater calcium-sequestering ability of the former.

Ability To Promote Emulsification

The effectiveness of different emulsifying salts in promoting emulsification, as indicated by electron microscopy and oiling-off studies, in processed cheese is in the following general order: sodium tripolyphosphates > pyrophosphates > polyphosphates (P > 10) > citrates ≈ orthophosphates ≈ basic sodium aluminum phosphates. Their ability to promote emulsification generally parallels their effectiveness in promoting hydration of the paracaseinate complex.

Hydrolysis (Stability)

During processing and storage of processed cheese products, linear condensed phosphates undergo varying degrees of hydrolysis to orthophosphates. The extent of degradation increases with processing time and temperature, product storage time and temperature, and phosphate chain length. Other influencing factors include the type of cheese, quantity of emulsifying salt, and type of product being produced. In experiments with pasteurized processed Emmental, the level of polyphosphate ($n > 4$) breakdown during melting at 85°C varied from 7% for block cheese (processed for 4 min) to 45% for spreadable cheese (processed for 10 min). While the breakdown of condensed phosphates to monophosphates was complete in the spreadable cheese after 7 weeks, low levels were detectable in block processed cheese even after 12 weeks. The greater degradation of polyphosphates in spreadable processed cheeses is also believed to be due to their higher pH and moisture content.

Bacteriostatic Effects

Cheese processing normally involves temperatures (70–95°C) that are lower than those used for sterilization. Thus, processed cheese products may contain viable spores, especially of the genus *Clostridium*, which originate in the raw materials. Germination of spores during storage often leads to problems such as blowing of cans, protein putrefaction, and off-flavors. While bacterial spoilage is minimized through the addition of preservatives, some of the emulsifying salts also possess bacteriostatic properties. Polyphosphates inhibit many microorganisms, including *Staphylococcus aureus*, *Bacillus subtilis*, *Clostridium sporogenes*, and various *Salmonella* species. Orthophosphates have been found to inhibit the growth of *Cl. botulinum* in processed cheese. Citrates possess no bacteriostatic effects and may even be degraded by bacteria, thus reducing product-keeping quality (Caric & Kalab, 1993). The inhibitory effect of sodium orthophosphates on *Cl. botulinum*, which has been found to be superior to that of sodium citrates in pasteurized processed cheese spreads with moisture levels in the range of 520–580 g/k, depends on the levels of moisture and NaCl and the pH of the processed cheese product (Tanaka et al., 1986). The general bacteriostatic effect of phosphates, which increases with chain length, may be attributed to their interactions with bacterial proteins and sequestration of calcium, which generally serves as an important cellular cation and cofactor for some microbial enzymes (Stanier, Ingraham, Wheelis, & Painter, 1987).

Flavor Effects

It is generally recognized that sodium citrates impart a "clean" flavor, whereas phosphates may impart off-flavors such as soapiness (in the case of orthophosphates) and bitterness. Potassium citrates also tend to cause bitterness.

18.2.6 Influence of Various Parameters on the Consistency of Processed Cheese Products

Numerous investigations have been undertaken to assess the effects of different variables (e.g., levels and types of ingredients, changes in processing conditions, and composition) on the textural and functional characteristics of PCPs. Some discrepancies occur with regard to the conclusions of different studies in which similar variables were investigated, probably due, in part, to differences in formulation and processing conditions. However, certain trends emerge, which are discussed below.

Blend Ingredients

Cheese is the major constituent of processed cheese products. Its proportion ranges from a minimum of around 51% in cheese spreads and cheese foods to around 98% in processed cheeses. Hence, both the type and degree of maturity of the cheese used have a major influence on the consistency of the product. Block processed cheese with good sliceability and elasticity requires predominantly young cheese (75–90% intact casein), whereas predominantly medium-ripe cheese (60–75% intact casein) is required for spreads.

There is an inverse relationship between the age (and hence degree of proteolysis) of the cheese and its emulsifying capacity. Therefore, it is not surprising that the meltability and firmness of PCPs generally increase and decrease, respectively, with the maturity of the cheese blend, since there is a positive relationship between the degree of emulsification and firmness or hardness. Owing to intervarietal differences in microstructure, composition, and level of proteolysis, different types of cheese give processed products different consistency characteristics. It is generally recognized that hard and semi-hard cheese varieties, such as Cheddar, Gouda, and Emmental, which have a relatively high level of intact casein, give firmer, longer bodied (high fracture strain), more elastic processed products than mold-ripened varieties, such as Camembert and Blue cheese. The latter cheeses undergo more extensive proteolyis during ripening and have a low Ca:casein ratio.

Rework refers to processed cheese that is not packaged for sale. It is obtained from "leftovers" in the filling and cooking machines—damaged packs and batches that have overthickened and are too viscous to pump. When added at a maximum level of about 200 g/kg, rework cheese increases the viscosity of the molten blend during processing, especially in blends with a high moisture content (e.g., cheese spreads) or a high proportion of overripe cheese. Overripe cheese tends to give poor emulsification due to the very low level of intact casein. Addition of rework cheese generally produces PCPs that are firmer and less spreadable and have poor flowability when remelted.

Cheese base is being used increasingly as a cheese substitute in processed cheese manufacture. Its main advantages are its lower cost and more consistent quality (i.e., intact casein content). Production generally involves ultrafiltration and diafiltration of milk, inoculation of the retentate with a lactic culture, incubation to a set pH (5.2–5.8), pasteurization, and scraped-surface evaporation to 600 g/kg dry matter. Increasing the level of cheese substitution with cheese base generally results in PCPs that are longer bodied, firmer, and less flowable upon remelting. However, the effects vary depending on the method of cheese base preparation and subsequent heat treatment during processing:

- Decreasing the pH of milk, in the range 6.6–5.2, prior to ultrafiltration gives a lower Ca concentration in the cheese base and yields PCPs with improved meltability.
- Rennet treatment of the ultrafiltration retentate results in poorer flowability of the PCPs, an effect that may be attributed to greater interaction of β-lactoglobulin with para-κ-casein than with native casein during subsequent processing. This interaction contributes to the gelation of whey proteins, which impairs the flowability of the PCP upon remelting.
- Treatment of the retentate with other proteinases (e.g., Savourase-A, proteinases from *Aspergillus oryzae* and *Candida cylindracea*) leads to higher levels of proteolysis in the cheese base, which in turn yields PCPs that are softer and more flowable than those containing untreated base.
- For a given level of cheese base inclusion in the processed cheese blend, increasing the processing temperature in the range 66–82°C results in PCPs with reduced flowability, an effect attributed to the heat gelation of whey proteins at the higher temperatures, especially when rennet-treated base is used. In this context, it is noteworthy that flow-resistant PCPs may be prepared by adding a heat-coagulable protein (30–70 g/kg whey protein or egg albumin), at a temperature below 70°C, to the cheese blend upon completion of processing.

Noncheese dairy ingredients may account for a maximum of about 150 g/kg of the PCP blend. Addition of skim milk powder at a level of 30–50 g/kg of the blend results in a softer, more spreadable PCP but increases the propensity to undergo nonenzymatic browning during storage. Higher levels (70–100 g/kg) are conducive to the development of textural defects, such as crumbliness.

The addition of milk protein coprecipitates (produced by high heat treatment of milk followed by acidification and calcium addition), at levels up to 50 g/kg of the blend, yields pasteurized processed Cheddar that is firmer and less flowable upon remelting. However, the level at which flowability becomes noticeably impaired varies from 2.5 to 30 g/kg and varies with the source of the coprecipitate.

Hydrocolloids, including carob bean gum, guar gum, carrageenan, sodium alginate, karaya gum, pectins, and carboxymethylcellulose, are permitted in pasteurized processed cheese spreads at a maximum level of 8 g/kg. Owing to their water-binding and/or gelation properties, they increase viscosity and thus find application in PCPs that have a high water content or a high proportion of overripe cheese. These materials, along with polysaccharides and polysaccharide derivatives (e.g., inulin), are finding increasing application as fillers and texturizers in the manufacture of low-fat products, including PCPs.

Processing Conditions

Increases in the shear and temperature (in the range 70–90°C) during processing generally result in a higher degree of emulsification and PCPs that are firmer, less spreadable, and less flowable on remelting. Hence, high moisture formulations, such as processed cheese spreads, are generally subjected to conditions (higher temperature and more vigorous agitation) that promote a higher degree of emulsification than block processed cheese.

Composition

Although the rheological attributes of processed cheese products with the same moisture level can differ significantly owing to variations in blend composition and processing conditions, increasing the moisture content generally yields products that are softer, less elastic, and more adhesive and spreadable. Product pH has a major effect on texture. Low pH (4.8–5.2), caused, for example, by use of monosodium citrate, monosodium phosphate, or sodium hexametaphosphate alone, produces short, dry, crumbly cheeses with a tendency toward oiling-off. High pH values (> 6.0) yield products that tend to be soft and exhibit excessive flow upon remelting.

18.3 IMITATION AND SUBSTITUTE CHEESE PRODUCTS

Cheese substitutes or imitation cheeses may be generally defined as products that are intended to partly or wholly substitute for or imitate cheese and in which milk fat, milk protein, or both are partially or wholly replaced by non-milk-based alternatives, principally of vegetable origin. However, their designations and labeling should by law clearly distinguish them from cheese or PCPs. The labeling requirements for imitation and substitute cheeses have been reviewed by McCarthy (1991). In the United States, "an imitation cheese is defined as a product which is a substitute for, and resembles, another cheese but is nutritionally inferior, where nutritional inferiority implies a reduction in the content of an essential nutrient(s) present in a measurable amount but does not include a reduction in the caloric or fat content" (Food and Drug Administration Regulation 101.3, Identity Labeling of Food in Packaged Form (e)). A substitute cheese is defined as a product that is a substitute for and resembles another cheese and is not nutritionally inferior. Outside the United States, there is little specific legislation covering imitation or substitute cheeses. For pertinent information regarding designation and labeling, the reader should consult International Dairy Federation (1989), McCarthy (1991), and current national regulations and the Codex Alimentarius.

There are few, if any, standards relating to permitted ingredients or manufacturing procedures for imitation cheese products. The products may be arbitrarily classified into three categories—analogue cheeses, filled cheeses, and tofu—based on the ingredients used and the manufacturing procedure (Figure 18–1). The effects of various ingredients, various processing conditions, and low temperature storage on the quality of imitation cheese products have been reported extensively (see Cavalier-Salou & Cheftel, 1991; Marshall, 1990; Mulvihill & McCarthy, 1994; Yang & Taranto, 1982). However, in many of these studies, model product formulations that bear little resemblance to those used in commercial practice have been used. Despite the fact that substitute Mozzarella cheese is the principal substitute cheese product used commercially, research has concentrated principally on composition and texture, with little focus on functionality and viscoelastic properties during melting and comparison with those of natural cheese. Therefore, most of the relevant information on these products is of a proprietary nature. Pertinent reviews include Shaw (1984) and International Dairy Federation (1989). The individual products are discussed below.

18.3.1 Cheese Analogues

Analogue cheeses, which were introduced in the United States in the early 1970s, constitute by far the largest group of imitation or substitute cheese products. The manufacture of analogues of a wide variety of natural cheeses (e.g., Cheddar, Monterey Jack, Mozzarella, Parmesan, Romano, Blue cheese, and Cream cheese) and pasteurized processed cheese products has been reported in the trade literature. Based on feedback from the marketplace, current annual production of analogue cheese in the United States, the major producing region, amounts to around 300,000 tonnes. The major products are substitutes for or imitations of low-moisture Mozzarella, Cheddar, and pasteurized processed Cheddar. These products find application mainly as cheese topping for frozen pizza pie and as slices in beef burgers. Other applications include use in salads, sandwiches, spaghetti sprinkling, cheese sauces, cheese dips, and ready-made meals. European production is estimated to be relatively small (≈ 20,000 tonnes/annum), a fact that may be attributed to the lack of a common effective legislation policy, the efforts of groups with the objective of protecting the designation of milk and dairy products, the lower level of pizza consumption compared with the United States, and the fact that the flavor systems used for analogues are still not developed to a point where the analogues have the same flavors and textures as the corresponding table cheese products. The success of analogue cheese products in the United States may be attributed to a number of factors:

- Natural cheeses cost more than substitutes. The low cost of analogues is due to the low cost of vegetable oils compared with butterfat, the low cost of casein imported from New Zealand and Europe (price-subsidized), the absence of a maturation period (for natural cheeses maturation costs about US$1.4/week), and the relatively low cost of manufacturing equipment compared with that required for natural cheese.
- They offer a diverse functionality range (e.g., flowability, melt resistance, and shredability) made possible by tailor-made formulations, and they exhibit high functional stability during storage.
- Fast food and ready-made meals have become extremely popular.
- They can be designed to meet special dietary needs through formulation changes (e.g., products can be lactose free, low calorie, low in saturated fat, and vitamin enriched).

Classification

Cheese analogues may be arbitrarily categorized as dairy, partial dairy, or nondairy, depending on whether the fat and/or protein components are from dairy or vegetable sources

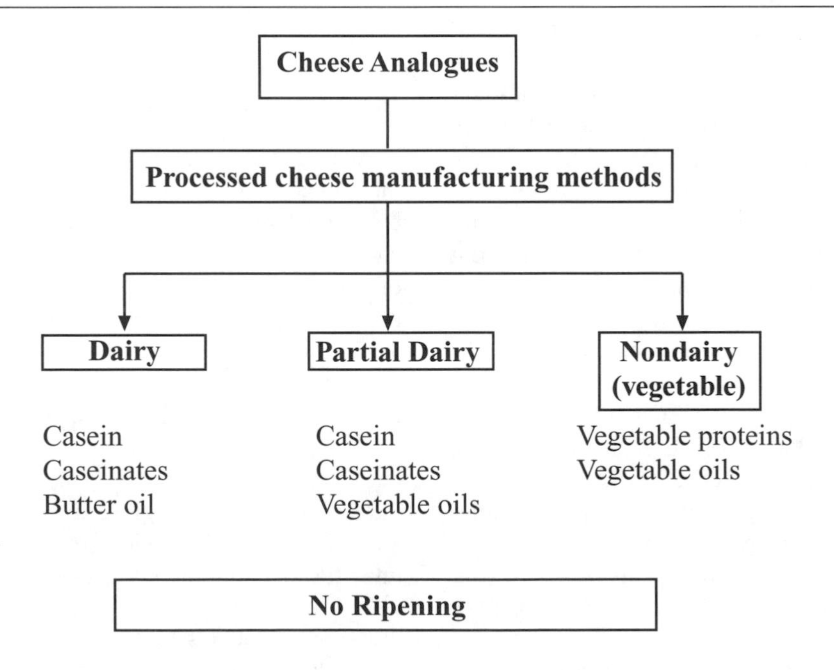

Figure 18–4 Classification of analogue cheeses based on the sources of proteins and oils used in product formulation.

(Figure 18–5). Partial dairy analogues, in which the fat is mainly vegetable oil (e.g., soya oil, palm oil, rapeseed, and their hydrogenated equivalents) and the protein is dairy based (usually rennet casein and/or caseinate) are the most common. Nondairy analogues, in which both fat and protein are vegetable derived, have little or no commercial significance and to the authors' knowledge are not commercially available. The preparation of experimental substitute or imitation cheese products (e.g., Mozzarella cheese and PCPs) from various vegetable proteins (e.g.,

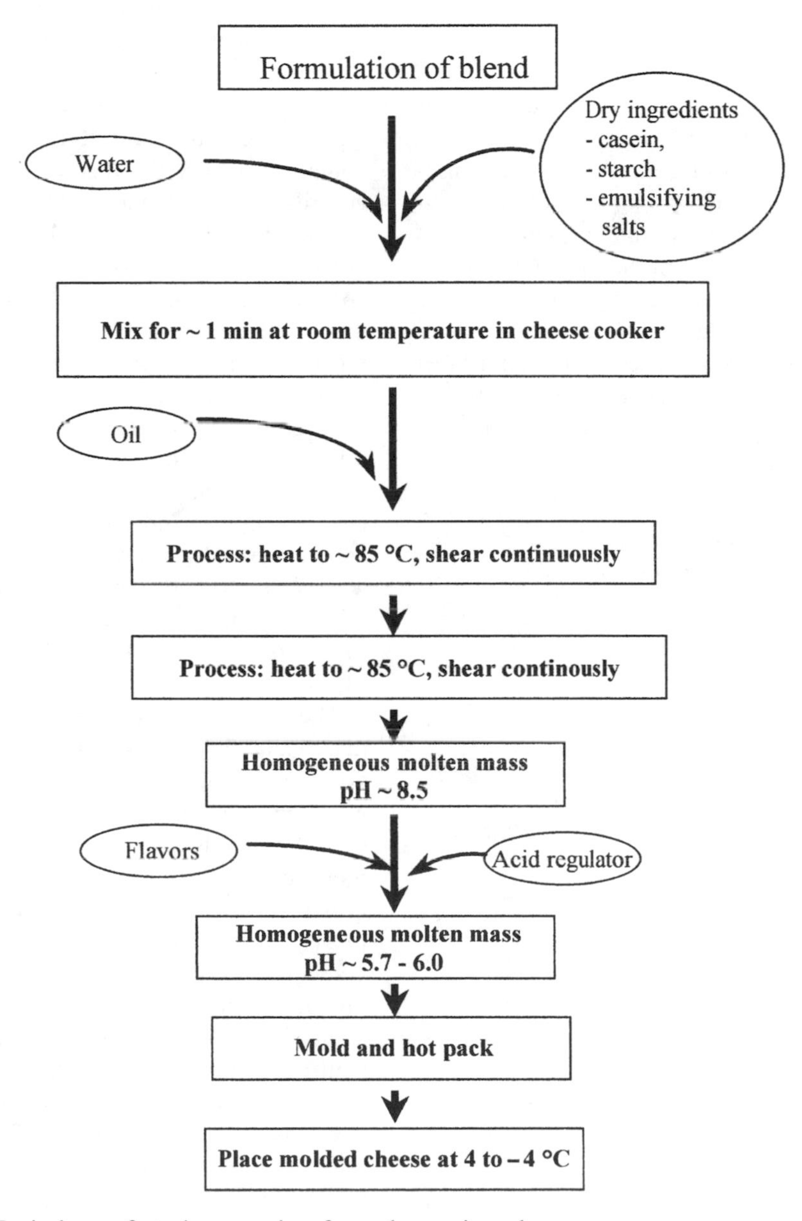

Figure 18–5 Typical manufacturing procedure for analogue pizza cheese.

peanut and soya proteins or blends of these proteins with casein) has generally shown that the substitution of casein by vegetable proteins results in impaired texture. Dairy analogues are not produced in large quantities because their cost is prohibitive. The following discussion pertains to partial dairy analogue cheese (e.g., low-moisture Mozzarella type) for use in pizza pie (analogue pizza cheese, APC).

Analogue Pizza Cheese

Manufacturing Protocol. The manufacture of analogue pizza cheese (APC), which is similar to the manufacture of PCPs, involves the formulation, processing, and packing of a hot molten product. A typical formulation (Table 18–5) shows that it differs from that for PCPs in that cheese is not normally included, although some cheese may be introduced as a flavoring agent or as required by customer specifications for label declaration. While production methods vary somewhat, a typical manufacturing procedure (Figure 18–6) involves

- simultaneous addition of the required amount of water and of dry ingredients (e.g., casein and emulsifying salts)
- addition of oil and cooking to about 85°C using direct steam injection while continuously shearing until a uniform homogeneous molten mass is obtained (typically 5–8 min)
- addition of flavoring materials (e.g., enzyme-modified cheese or starter distillate) and acid regulator (e.g., citric acid) and blending the mixture for a further 1–2 min
- packing the hot molten blend

Table 18–5 Typical Formulation for Analogue Pizza Cheese

Ingredient	*Level Added (g/100 g Blend)*
Casein and caseinates	23.00
Vegetable oil	25.00
Starch	2.00
Emulsifying salts	2.00
Flavor	2.00
Flavor enhancer	2.00
Acid regulator	0.40
Color	0.04
Preservative	0.10
Water	38.50
Condensate[a]	7.00

[a] Upon cooking the blend to about 85°C using direct steam injection, condensate equivalent to about 7.0 g is absorbed by the blend.

Horizontal twin screw cookers operating at typical screw speeds of 40 rpm are used in the manufacture of APC. This design of the cooker ensures adequate blending and a relatively low degree of mechanical shear, compared with the homogenizing effects of some processed cheese cookers. These process conditions, together with the correct formulation, promote a low degree of fat dispersion and hence a relatively large fat globule size (e.g., 5–25 μm). Upon subsequent baking of the analogue cheese on pizza pie, the relatively large fat globules ensure a sufficient degree of oiling-off, limit dehydration of the cheese topping, and thereby are conducive to achieving the desired flow and succulence characteristics. It is noteworthy that there is generally an inverse relationship between the degree of fat emulsification and the flowability of PCPs.

The addition of flavors toward the end of the manufacturing process minimizes the loss of flavor volatiles in the dissipating steam. In the manufacture of PCPs, the pH of the final product (e.g., 5.5–5.9) is regulated by adding the correct blend of emulsifying salts, which adjust and buffer the pH of the blend during processing to this pH value. In contrast, the addition of food-grade acids toward the end of the process (following casein hydration and oil dispersion or emulsification) is the normal procedure used to adjust the pH of the cooked APC to that required in the finished product. This protocol is essential to ensure a high pH (typically > 7.0) during processing of the product when rennet casein is the major protein ingredient. A high pH at this stage gives a higher negative charge to the casein and

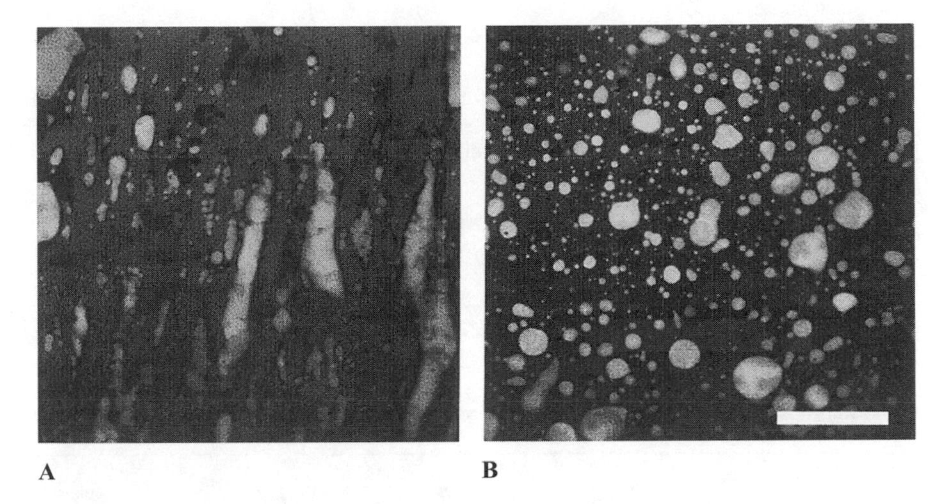

Figure 18–6 Confocal laser scanning micrographs showing the microstructures of commercial samples of low-moisture Mozzarella (a) and analogue pizza cheese (b). The bar = 25 μm; protein is shown in black, and fat is shown in white/gray. In the low-moisture Mozzarella, the protein is in the form of elongated fibers and the fat is in the form of pools trapped between the protein fibers. In the analogue pizza cheese, the protein is not organized into fibers and the fat is mainly in the form of discrete globules.

---

is conducive to greater calcium sequestration by sodium phosphate emulsifying salts. Both factors contribute to the efficient hydration of rennet casein and hence the emulsification of added vegetable oil. Two factors necessitate this procedure:

- Rennet casein has a higher (e.g., 35 mg/g casein) calcium content than natural cheese (e.g., ≈ 28 mg/g casein for Cheddar).
- Rennet casein is in a dehydrated state whereas the casein in natural cheese is hydrated to a degree dependent on the extent of proteolysis, pH, and concentrations of Na and Ca in the moisture phase.

Principles of Manufacture. The principles of manufacture of APC from rennet casein are similar to those for PCPs. The combined effects of emulsifying salts, heat, and shear promote sequestration of Ca from the rennet casein (dehydrated paracasein, which in effect is equivalent to dehydrated cheese protein), pH adjustment and buffering of the blend, casein hydration, fat dispersion and fat emulsification by the hydrated paracaseinate, and setting of the molten mass upon cooling.

Composition and Functionality. Comparison of the mean composition of commercial samples of low-moisture Mozzarella cheese (LMMC) and APC (Table 18–6) indicates that while many of the gross compositional parameters of APC are similar to those of LMMC, the latter generally has a lower level of protein and higher levels of fat-in-dry-matter, Ca, and P. Although intravarietal differences in composition occur for both cheese types, they are more pronounced in APC. Moreover, the sum of the mean values for moisture, fat, protein, and ash account for only about 930 g/kg dry matter (compared with ≈ 990 g/kg in LMMC), suggesting the addition of carbohydrate-based ingredients (e.g., lactose, maltodextrins, starch) during formulation. These materials may be added to impart certain functional characteristics to the end product and/or as partial substitutes for rennet casein, thereby reducing formulation costs. The relatively large compositional variations exhibited by APCs probably reflect deliberate differences in formu-

Table 18–6 Typical Compositions of Low-Moisture Mozzarella and Analogue Pizza Cheese

	Low-Moisture Mozzarella	*Analogue Pizza Cheese*
Moisture (g/100 g)	46.4	48.8
Protein (g/100 g)	26.0	18.5
Fat (g/100 g)	23.2	25.0
Fat-in-dry-matter (g/100 g)	44.6	49.0
Salt-in-moisture (g/100 g)	3.1	3.5
Ash (g/100 g)	3.9	4.2
Ca (mg/100 g)	27.5	34.4
pH	5.5	6.1

Note: Values presented are means of 8 samples of each cheese type sourced in Ireland, the United Kingdom, and/or Denmark.

lation so as to achieve customized functionalities in the finished products.

Some important functional attributes of melted cheese on a cooked pizza pie are as follows:

- Melt time. An index of how rapidly the shredded cheese on a pizza pie melts and flows into a homogeneous molten mass showing no traces of shred identity.
- Flowability. A measure of the degree of flow.
- Stretchability. A measure of the tendency to form cohesive strings or sheets when extended.
- Apparent viscosity. A measure of chewiness (Table 18–7 and Chapter 13).

Upon baking, a good-quality pizza cheese melts relatively quickly, flows adequately to give the desired degree of surface coverage, and possesses the desired degrees of chewiness and stretchability, which, perhaps more than other functional properties, endow pizza pie with its unique culinary qualities (see Chapter 19). Comparison of the functional characteristics of commercial LMMC and APC indicates that both cheeses have similar mean values for melt time, flowability, and apparent viscosity. However, the stretchability of APC is generally inferior to that of LMMC. The differences in stretchability between LMMC and APC may be related primarily to differences in the degree of aggregation and microstructure of the paracasein caused by differences in the procedures used to manufacture the two products. During the manufacture of LMMC and other Pasta filata cheeses, such as Provolone or Kashkaval, the cheese curd, at around pH 5.15, is subjected to a plasticization process, which involves heating to around 55–60°C and kneading the curd in hot (e.g., 70°C) water or dilute brine. These conditions promote a limited degree of aggregation and contraction of the paracasein gel matrix and thereby lead to the formation of paracasein fibers with a high tensile strength (see Figure 18–7 and Figure 19–7). The cheese fat is physically entrapped between the paracasein fibers. In contrast, the conditions used in the manufacture of APC are designed to disaggregate and hydrate the paracasein aggregates of rennet casein and caseinate. The hydrated paracaseinate immobilizes large quantities of added water and emulsifies the added vegetable oil, thereby contributing to formation and physicochemical stability of the product. Hence, unlike low-moisture Mozzarella, the casein in the APC is in the form of a partially hydrated dispersion rather than paracasein fibers (Figure 18–7). Both LMMC and APC exhibit marked intravarietal differences attributable to designed differences in formulation (for APC) or processing conditions and degree of maturity (for LMMC). The intravarietal dif-

Table 18–7 Functionality of Low-Moisture Mozzarella and Analogue Pizza Cheese

Functional Attributes	*Low-Moisture Mozzarella*	*Analogue Pizza Cheese*
Aggregation index	3.95	3.74
Melt time (seconds)	108	105
Flowability (%)	53	42
Stretchability (cm)	87	28
Apparent viscosity (Pa × s)	630	650

ferences in functionality allow manufacturers to customize their cheeses to the requirements of different pizzerias.

Functional Stability during Storage. The functional properties of LMMC change markedly during storage at 4°C (see Chapter 19). Initially, during the first 1–5 days, depending on manufacturing procedure and composition, the cheese is nonfunctional and burns or crusts during baking. After 5–10 days of ripening, the cheese acquires functionality, as reflected by the decreases in melt time and apparent viscosity (chewiness) and the increases in flowability and stretchability (see Chapter 19). Thereafter, these changes occur more slowly, and the cheese retains desirable functionality for around 40–50 days. However, prolonged aging of LMMC (e.g., to 75 days) is associated with excessive flowability, loss of chewiness, and a "soupy" consistency in the grilled or baked cheese. Moreover, the uncooked shredded cheese develops an increased susceptibility to clumping or sticking, an undesirable change, as it leads to the blocking of cheese-dispensing units on pizza pie production lines and to nonuniform distribution of the cheese topping on the pizza pie. The functional changes that occur during ripening are mediated by proteolysis of paracasein (by plasmin and possibly residual coagulant), hydration and swelling of the paracasein matrix, and coalescence of the fat phase (see Chapters 11 and 19). The increase in the hydration of the paracasein is thought to result from a number of factors: proteolysis and the concomitant increase in the number of amino and carboxyl groups, the increase in pH during early ripening, and the low concentration of NaCl in the moisture phase of the cheese (which is conducive to hydration of the paracasein). Few studies have considered the changes in casein-based APCs during ripening. Mulvihill and McCarthy (1994) reported a progressive increase in proteolysis (e.g., pH 4.6–soluble N increased from ≈ 35 at 0 days to 195 g/kg total N after 51 weeks) and decreases in elasticity and chewiness during storage at 4°C for 51 weeks. However, the changes during the first 6 weeks were relatively small. Normally, analogues are used within 1 month after manufacture. Kiely, McConnell, and Kindstedt (1991) reported that casein-based APCs were more functionally stable than LMMCs during storage at 4°C for 28 days. In the authors' experience, casein-based analogues containing a high level of starch (> 40 g/kg) may lose their functionality relatively rapidly (e.g., after 4 weeks) during storage at 4°C, an effect possibly associated with the retrogradation of amylose. The loss of functionality is reflected by the increase in loose moisture upon shredding, the loss of meltability and flowability, and burning or crusting upon baking. Added starch may undergo postmanufacture retrogradation during cold storage to form gels to an extent depending on processing conditions (e.g., the level of heat and shear) and the level and type of starch used (e.g., amylose:amylopectin ratio and whether the starch is native or modified). These gels contract during storage, resulting in the expulsion of moisture. It is envisaged that these changes in the starch component of APCs during storage result in products with higher levels of unbound water and impede flowability upon remelting.

18.3.2 Filled Cheeses

Filled cheeses generally differ from natural cheeses in that the milk fat is partly or fully replaced by vegetable oils, which may be partially hydrogenated to impart a melting profile similar to that of milk fat. However, filled cheeses may be categorized into two types depending on whether the base material is native skim milk or reformed skim milk; the latter is prepared by dispersing dairy ingredients, such as whey and total milk protein, in water. In all cases, preparation of the filled milk involves dispersion of the vegetable oil in the native or reformed skim milk using a high speed mixer and subsequent homogenization of the blend. Dispersion and homogenization ensure emulsification of the added vegetable oils and thus prevent phase separation and/or excessive creaming during cheesemaking.

The filled milk is then subjected to the conventional in-vat cheesemaking procedure used for the variety being substituted for or imitated.

Homogenization of milk generally results in curd that synereses poorly and therefore tends to yield cheeses with a higher moisture content, lower yield stress and firmness, and lower flowability upon remelting than those made from nonhomogenized milk (Fenelon and Guinee, unpublished results).

18.3.3 Tofu or Soybean Cheeses

Tofu, a stable food in the Orient for centuries, is a tough, rubbery curd made from soya (soybean) milk. Manufacture essentially involves soaking the soybeans in water for a long period (during which they swell), adding extra water, grinding and milling the bean-water mixture into a smooth slurry, and filtering the slurry to obtain soya milk. The soya milk is boiled to induce protein denaturation, cooled to about 37°C, and coagulated by adding a divalent salt (e.g., calcium lactate) and adjusting the pH to 4.5–5.0 (McCarthy, 1991; Tharp, 1986). Following coagulation, the whey is drained off and the curd is molded and lightly pressed to give tofu, in which the levels of dry matter, protein, fat, and carbohydrate are typically 152, 77, 42, and 24, respectively. The molded curd may be subjected to a high pressure and brine-salted to yield soybean cheeses with a higher dry-matter level than Tofu (e.g., 530 g/kg dry matter; Abou El-Ella, 1980). Ras cheese made from soya milk was found to have a higher moisture level and received lower sensory scores for color, flavor, and body and texture characteristics than that made from cow milk by conventional cheesemaking procedures (Abou El-Ella, 1980).

REFERENCES

Abou El-Ella, W.M. (1980). Hard cheese substitute from soy milk. *Journal of Food Science*, *45*, 1777–1778.

Caric, M., & Kalab, M. (1993). Processed cheese products. In P.F. Fox (Ed.), *Cheese: Chemistry, physics and microbiology* (2d ed., Vol. 2). London: Chapman & Hall.

Cavalier-Salou, C., & Cheftel, J.C. (1991). Emulsifying salts' influence on characteristics of cheese analogs from calcium caseinate. *Journal of Food Science*, *56*, 1542–1547, 1551.

Code of Federal Regulations. (1988). Part 133: Cheese and related products. In *Food and Drugs 21. Code of Federal Regulations, Parts 100 to 169.* Washington, D.C.: US Government Printing Office.

Fox, P.F., O'Connor, T.P., McSweeney, P.L.H., Guinee, T.P., & O'Brien, N.M. (1996). Cheese, physical, biochemical and nutritional aspects. *Advances in Food and Nutrition Research*, *39*, 163–329.

International Dairy Federation. (1989). *The present and future importance of imitation dairy products.* [Bulletin No. 239]. Brussels: Author.

Kiely, L.J., McConnell, S.L., & Kindstedt, P.S. (1991). Observations on the melting behaviour of imitation mozzarella cheese. *Journal of Dairy Science*, *74*, 3568–3592.

Marshall, R.J. (1990). Composition, structure, rheological properties and sensory texture of processed cheese analogues. *Journal of the Science of Food and Agriculture*, *50*, 237–252.

McCarthy, J. (1991). *Imitation cheese products* [Bulletin No. 249]. Brussels: International Dairy Federation.

Mulvihill, D.M., & McCarthy, A. (1994). Proteolytic and rheological changes during ageing of cheese analogues made from rennet casein. *International Dairy Journal*, *4*, 15–23.

Rayan, A.A., Kalab, M., & Ernstrom, C.A. (1980). Microstructure and rheology of process cheese. *Scanning Electron Microscopy*, *3*, 635–643.

Scharf, L.G., Jr. (1971). The use of phosphates in cheese processing. In J.M. Deman & P. Melnchyn (Eds.), *Phosphates in food processing*. Westport, CT: AVI Publishing Co.

Shaw, M. (1984). Cheese substitutes: Threat or opportunity? *Journal of the Society of Dairy Technology*, *37*, 27–31.

Sørensen, H.H. (1997). *The world market for cheese* [Bulletin No. 326]. Brussels: International Dairy Federation.

Stanier, R.Y., Ingraham, J.L., Wheelis, M.L., & Painter, P.R. (1987). *General microbiology* (5th ed.). London: Macmillan Press.

Tanaka, N., Traisman, E., Plantinga, P., Finn, L., Flom, W., Meske, L., & Guggisberg, J. (1986). Evaluation of factors involved in antibotulinal properties of pasteurized press cheese spreads. *Journal of Food Protection*, *49*, 526–531.

Tharp, B. (1986, September). Frozen desserts containing tofu. *Dairy Field*, pp. 38–42, 59.

van Wazer, J.R. (1971). Chemistry of the phosphates and condensed phosphates. In J.M. Deman & P. Melnchyn (Eds.), *Phosphates in food processing*. Westport, CT: AVI Publishing Co.

Yang, C.S.T., & Taranto, M.V. (1982). Textural properties of Mozzarella cheese analogs manufactured from soya beans. *Journal of Food Science*, *47*, 906–910.

CHAPTER 19

Cheese as a Food Ingredient

19.1 INTRODUCTION

Owing to its numerous varieties, cheese offers the consumer a very wide diversity of flavors, aromas, and textures. Hence, as a product, cheese has been enjoyed since antiquity, as testified by the numerous references to it in early writings (see Chapter 1).

While it is generally assumed that cheese was originally eaten on its own or with bread, humans probably soon realized that it enhanced the organoleptic qualities of other foods to which it was added. The Romans were the first to record the use of cheese as an ingredient (Ridgway, 1986). Typical applications included the blending of hard cheese with oil, flour, and eggs in the preparation of cakes and the mixing of soft cheeses with meat or fish, boiled eggs, and herbs in the making of pies. Cheese has long been used as a culinary ingredient, along with other foods and condiments, to create an extensive array of dishes (Figure 19–1). Today, natural cheese continues to be used as a major ingredient in the hotel and catering industry. Typical cheese dishes include omelets, quiches, sauces, chicken cordon-bleu, and pasta. Natural cheese is also used extensively by the industrial catering sector in the mass production of ready-to-use grated cheeses, shredded cheeses, cheese blends, combination products, and cheese-based ingredients, such as pasteurized processed cheese products (PCPs), cheese powders, and enzyme-modified cheeses (EMCs). Combination products contain two or more types of food, such as cheese, meat, fish, and vegetables, each of which retains its own distinct identity (e.g., as a layer in the product). They are generally produced by coextrusion of the different foods or by the dipping of one food (e.g., cooled salami) into a hot molten form of another food (e.g., pasteurized processed cheese). Commercially, cheese-based ingredients are used by the catering industry (e.g., burger outlets, pizzerias, and restaurants) and by the manufacturers of formulated foods such as soups, sauces, and ready-made meals.

In this chapter, the functional properties of natural cheese, cheese powders, and enzyme-modified cheese as an ingredient are discussed. Cheese base, a concentrated cultured ultrafiltered milk retentate ($\approx$ 550 g/kg dry matter; pH $\approx$ 5.3), is an important cheeselike ingredient that finds applications as a substitute for young cheese in PCPs. The use of cheese base in PCPs is discussed in Chapter 18.

19.2 OVERVIEW OF THE REQUIREMENTS OF CHEESE AS AN INGREDIENT

When used as an ingredient in food applications, cheese is required to perform one or more functions, some of which are listed in Table 19–1. In its natural state, cheese is required to exhibit a number of rheological properties so as to facilitate its use in the primary stages of preparation of various dishes, such as the ability

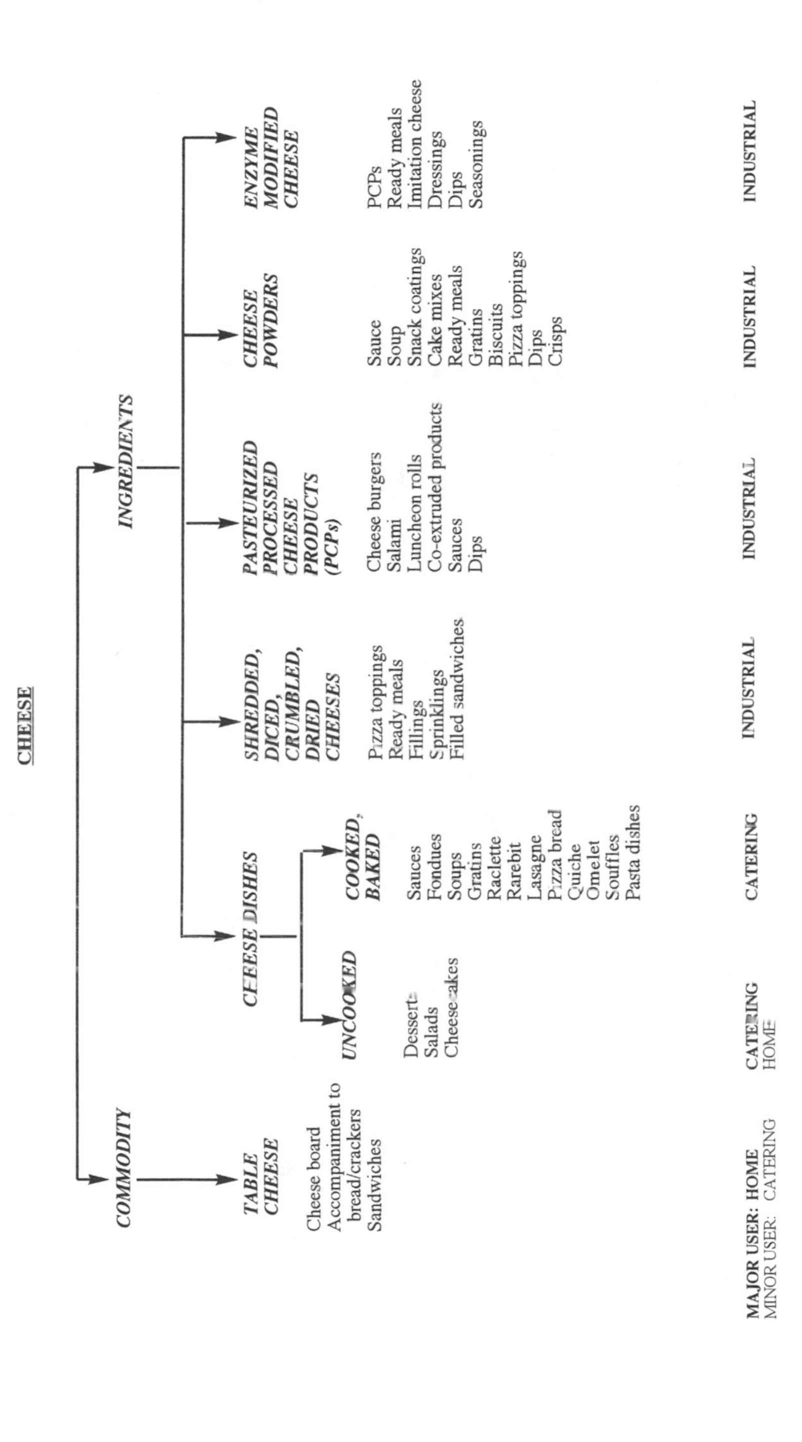

Figure 19–1 Uses of cheese as an ingredient

Table 19–1 Typical Requirements of Cheese as a Food Ingredient

Requirement	*Examples of Food Applications*	*Examples of Cheese or Cheese-Based Ingredient*
Ability to crumble when rubbed	Mixed salads	Feta, Cheshire, Stilton
	Soup	Stilton
Sliceability	Filled cheese rolls (finger foods)	Swiss-type, Gouda, Edam
	Sandwiches (filled, open, toasted)	Swiss-type, Cheddar, Mozzarella
	Cheese slices in burgers	Cheddar
	Cheese slices on crackers	Cheddar
Shreddability	Consumer packs of shredded cheese	Swiss-type, Cheddar, Mozzarella
	Pizza pie (frozen/fresh baked)	Mozzarella, Provolone, Cheddar, analogue pizza cheese, Monterey
	Pasta dishes (lasagne, macaroni and cheese)	Cheddar, Romano, Provolone
Free-flow when shaken	Cheese sprinklings (e.g., on lasagne)	Grated Parmesan and Romano
	Snack coating (e.g., popcorn)	Cheese powders
	Dry soup/sauce mixes	Cheese powders, enzyme-modified cheeses
Ability to flow when blended with other materials in the raw state	Fresh cheese desserts	Quarg, Fromage frais, Cream cheese
Ability to "cream" or to form a paste when sheared	Cheesecake	Cream cheese, Ricotta
	Tiramisu	Mascarpone
	Home-made desserts	Cream cheese
Nutritional value	Baby foods	Dried cheeses, especially rennet-curd varieties (high in calcium)
Meltability upon grilling/baking	All cooked dishes (including sauces, fondues, pizza pie)	Mozzarella, Cheddar, Raclette, Swiss, Romano, analogue pizza cheese, PCPs

Flowability upon grilling/baking	Most cooked dishes (e.g., pizza pie, cheese slices on burgers)	Mozzarella, Cheddar, Swiss, analogue pizza cheeses
	Chicken cordon-bleu	PCPs, Cream cheese
Flow resistance upon deep-frying	Deep-fried breaded cheese sticks	PCPs, analogue pizza cheese, custom-made Mozzarella or string cheese
	Deep-fried burgers containing cheese insets	PCPs, analogue pizza cheese
	Fried cheese dishes	Paneer, acid-coagulated Queso blanco
Stretchability when baked/grilled	Pizza pie	Mozzarella, Kashkaval, young Cheddar, analogue pizza cheese
Chewiness when baked/grilled	Pizza pie	Halloumi, Mozzarella, Provolone, Kashkaval, young Cheddar
Limited oiling-off when baked/grilled	Pizza pie	Mozzarella, Kashkaval
Limited browning when baked/grilled	Macaroni and cheese	Cheddar, Romano
	Lasagne	Cheddar, Romano, Parmesan
	Pizza pie	Mozzarella, analogue pizza cheese
Viscosity	Soups	Cheese powders, PCPs
	Sauces	Cheese powders, Cheddar, Blue cheese, PCPs
	Cheesecake	Cream cheese
Flavor	Most cheese dishes, soups	Cheddar, Romano, Swiss-type, Parmesan
	Baked products	Cheese powders, enzyme-modified cheeses
	Snack coatings	Cheese powders
	Dressings	Cheese powders
	Baby food	Dried cheeses
	Ready-made meals	Cheese powders

Key: PCPs = pasteurized processed cheese products.

to crumble easily, to slice or to shred cleanly, or to bend when in sliced form. The rheological properties also determine the textural properties of the cheese during mastication. Cheese is generally required to contribute to the organoleptic characteristics (flavor, aroma, and texture) of the food in which it is an ingredient. Upon grilling or baking, the cheese may be required to melt, flow, stretch, brown, blister, oil-off, and/or stretch to varying degrees. The baked cheese may also be expected to be chewy (as in pizza pie) and contribute to certain mouth-coating characteristics (as in sauces and soups). In many dishes, including sauces, the cheese is required to be able to interact with other food components such as water, carbohydrates, proteins, and fats during food preparation.

19.3 FUNCTIONAL PROPERTIES OF CHEESE AS AN INGREDIENT

The ability of cheese to fulfill its expected requirements as an ingredient is related to its functional properties. These may be defined as those rheological, physicochemical, microstructural, and organoleptic properties that affect the behavior of the cheese in food systems during preparation, processing, storage, cooking, and/or consumption and that therefore contribute to the quality and organoleptic attributes of the food in which cheese is included.

The functional properties of cheese can be classified into four main types:

1. Rheology-related properties of the raw cheese. These properties are exhibited when the cheese is subjected to a stress (such as cutting, shearing, and mastication) or strain (compression and extension). They may include sensoric properties such as consistency, fracturability, crumbliness, stickiness, firmness, hardness, and softness (see Chapter 13).
2. Rheology-related properties of the heated cheese. These properties are exhibited when the cheese is subjected to a stress throughout its mass as a result of heat-induced physicochemical and microstructural changes such as liquefaction of the fat, protein dehydration, fat coalescence, and matrix collapse. Included among them is the ability of the cheese to melt, flow, and stretch.
3. Physicochemical and microstructural properties induced by heating. These properties include oiling-off, browning, blistering, fat coalescence and exudation or separation, interaction of free amino groups with reducing sugars, moisture evaporation, and paracasein aggregation and precipitation.
4. Flavor- and aroma-related properties. These include properties such as cheddariness, saltiness, fruitiness, piquancy, and sweetness.

19.3.1 Functional Properties of Raw Cheese

The functional properties of raw cheese are related to its taste and aroma and its rheological characteristics, discussed in Chapters 12 and 13, respectively. Although the rheological properties do not directly affect the taste and aroma, their influence on the rate and extent of breakdown during mastication may alter the latter characteristics indirectly. For example, a cheese with low values for fracture stress and strain is expected to deform rapidly and release its fat more quickly after a given residence time in the mouth. Free liquid oil quickly coats parts of the mouth, allowing the aroma and taste of its volatile and nonvolatile flavor compounds to be perceived rapidly.

Rheology-Related Characteristics

The primary stage of preparation of any food containing cheese requires that the cheese mass be reduced in size so as to facilitate dispersion, mixing, and/or layering onto the food. Size reduction is usually achieved by cutting into relatively large pieces and then crumbling, slicing, shredding, dicing, grating, and/or shearing. These actions usually involve a combination of compressive and shear stresses. The behavior of the cheese when subjected to different size reduction methods constitutes a group of important functional properties, which are listed in Table 19–2. These properties are related to the rheological characteristics of the cheese, which determine the magnitude of the strain (e.g., change in dimensions and fracture) upon application of the stresses applied by the different size-reduction methods.

The rheology-related functional properties of the raw cheese determine its suitability for a particular application (Table 19–1), as shown by the examples below. Mature Camembert and Chaumes, which are soft, short, and adhesive (see Chapter 13), are not used in shredded or diced cheese applications, such as pizza pie, because of their tendency to ball and clump. However, the ability of these cheeses to undergo plastic fracture and flow under shear (i.e., spread) makes them ideal for blending with other materials such as butter, milk, or flour in the preparation of fondues and sauces (Tables 19–1 and 19–2). The brittleness and tendency of hard cheeses, such as Parmesan and Romano, with low levels of moisture and fat-in-dry-matter (FDM) to undergo elastic fracturability endow them with excellent gratability, such as when crushed between rollers, and they are suitable for sprinkling onto dishes such as spaghetti Bolognase. However, these properties render the latter cheeses unsuitable in food applications that require cheese slices, such as filled sandwiches or cheeseburgers. Conversely, other hard cheeses, such as Cheddar and Gouda, are unsuitable for grating owing to their lack of brittleness and to their elasticity and relatively high fracture stress and strain, which enables a relatively high degree of recovery to their original shape and dimensions following crushing. Moreover, the relatively high moisture and fat-in-dry matter levels of the latter cheeses are conducive to a higher degree of flow following fracture than Romano or Parmesan and hence to the development of tackiness following crushing. Owing to its springiness, elasticity, and "long" body, Swiss-type cheese is ideal for slicing very thinly and therefore is particularly well suited for applications such as filled sandwiches and rolled cheese slices containing fillings. Similarly, the springiness of low-moisture Mozzarella cheese (LMMC) endows it with good shreddability (a low tendency to fracture and form fines or curd dust) and nonstick properties and facilitates uniform distribution on the surface of pizza pies (Tables 19–1 and 19–2). Owing to their crumbliness, cheeses such as Feta, Cheshire, and Caerphilly are particularly well suited for inclusion in mixed salads. While it could be argued that shreddable cheese such as Mozzarella or Gouda could also be included in mixed salads, crumbly cheeses are more desirable, as the irregularly shaped, curdlike particles are more visually appealing to the consumer than cheese shreds (they convey an image of "real" cheese).

Factors That Influence the Functionality of Raw Cheese

Little or no information is available on factors that influence the various rheology-related functional properties of raw cheese, apart from shreddability.

The suitability of cheese for shredding may be quantitatively assessed by determining the tendency of the shredded cheese to aggregate or clump when vibrated under controlled conditions similar to those used on commercial pizza production lines. Cheese, after storing at 4°C for 12 hr or more, is cut into cubes of fixed dimensions (e.g., 2.5 cm), and a fixed weight (W_1) of shredded cheese is placed immediately on the top sieve of a stack of sieves ranging in aperture from 9.5 to 1 mm. The stack is vibrated at a fixed amplitude for a given time, resulting in the cheese shreds passing through the stack to a degree dependent on their susceptibility to stick or clump on the one hand or fracture on the other. The cheese on each sieve is then weighed (W) and an aggregation index (AGI) is calculated:

$$\text{AGI} = \Sigma\ (W \times SA)/W_1$$

where SA is the sieve aperture. A higher AGI value corresponds to a higher susceptibility to

Table 19–2 Functional Properties of Raw Cheese That Influence Its Functionality as an Ingredient

Property	*Definition*	*Cheeses Generally Displaying the Property*	*Positively Associated Rheological Parameters*
Shreddability	The ability of a cheese block to shred into thin strips of uniform dimensions, resist fracture during shredding, and resist clumping/balling during shredding	Low-moisture Mozzarella, Swiss-type cheese, medium-aged Cheddar, Gouda, Provolone	Elasticity, springiness, firmness, longness
Sliceability	The ability to be cut cleanly into thin slices without fracturing or crumbling or sticking to cutting implement	Low-moisture Mozzarella, Swiss-type cheese, Provolone, analogue pizza cheese, PCPs (some)	Elasticity, springiness, firmness, longness
Gratability	The ability to fracture (elastically) into small particles, with a low tendency to stick, upon shearing and crushing	Parmesan, Romano	Elastic fracturability
Spreadability	The ability to spread easily when subjected to a shear stress	Mature Camembert, Cream cheese, mature Blue cheese	Plastic fracturability, softness, adhesiveness, shortness
Crumbliness	The ability to break down into small irregular shaped pieces when rubbed (at low deformation)	Blue cheese, Cheshire	Elastic fracturability at low deformation, low cohesiveness

aggregation and clumping. Similar approaches are also used commercially (e.g., hand vibrating cheese in a colander and determining the quantity retained after a fixed time).

Young Mozzarella (1–5 days old) generally does not shred well owing to the large amount of free moisture on its surface and within the body of the cheese (Kindstedt, 1995). Thereafter, shreddability improves, and it becomes optimal after about 3 weeks' storage at 4°C, but it deteriorates progressively during further storage, and the cheese becomes soft and sticky. Free water and stickiness are undesirable, as they promote clumping of the cheese shreds, which leads to blocking of cheese-dispensing units on pizza pie production lines, poor distribution of cheese on pizza pies, and matting of shredded cheese when placed in retail packs.

The trend in the shreddability of LMMC as a function of maturation time is associated with increases in the levels of primary proteolysis, the water-binding capacity of the cheese proteins, and the free (nonglobular) fat in the cheese during ripening (see Section 19.3). The effect of free fat on shreddability depends on the ratio of solid to liquid fat, which decreases as the temperature is raised. At temperatures where milk fat is largely in the liquid state (60% of total fat at 20°C), free fat exudes to the surface of the cheese shreds, where it acts as an adhesive for other shreds. This defect is compounded by temperature fluctuations during storage. For example, cooling after holding at ambient temperature (≈ 20°C) leads to solidification of exuded fat, which makes the breaking up of clumps of cheese shreds more difficult. Such a problem is sometimes encountered in small retail units where stacked packs of grated cheese are subjected to temperature fluctuations, leading to clumping of the entire contents of the packs, especially those at the bottom of the pile. Hence, in practice, cheeses are cooled to about 2°C prior to shredding and distributing on pizza pies to avoid blockage of cheese shredding and dispensing machines. Other factors that are conducive to clumping of shredded cheese include

- longer shred length and shred diameter, which increase the chance of shred entanglement
- increasing moisture content, although the effect appears to be related to the method of Mozzarella production (e.g., whether acidified by a starter culture or food-grade acid), composition, and level of primary proteolysis
- increasing fat content (in the range 50–330 g/kg for Cheddar cheese)

The AGI value for a range of commercial cheeses indicates that cheese variety has a marked influence on shreddability (Figure 19–2). No intervarietal correlation was found between AGI and individual gross compositional parameters, pH, or level of primary proteolysis (as measured by N solubility at pH 4.6).

Although little or no published information is available on factors that affect the other rheology-related functional properties of raw cheese, it may be assumed that these are influenced by conditions that impact the rheological characteristics. The factors that influence the latter are discussed in Chapter 13 and include

- cheese macrostructure, which determines the discontinuity of the cheese matrix due to curd granule junctions, chip boundaries, cracks and fissures, gas holes, and eyes
- cheese composition (e.g., the levels of moisture, protein, fat, salt, and pH)
- temperature of the cheese, which influences the ratio of solid fat to liquid fat
- the extent of cheese maturation, which causes certain physical changes in the structural components during ripening (e.g., changes in the ratio of intact casein to hydrolyzed casein, the level of casein hydration, and/or the degree of fat coalescence)

Organoleptic Characteristics

Cheese is generally required to contribute to the organoleptic characteristics of most foods in which it is incorporated by possessing certain taste, aroma, texture, and/or mouth-coating characteristics. The importance of the contribu-

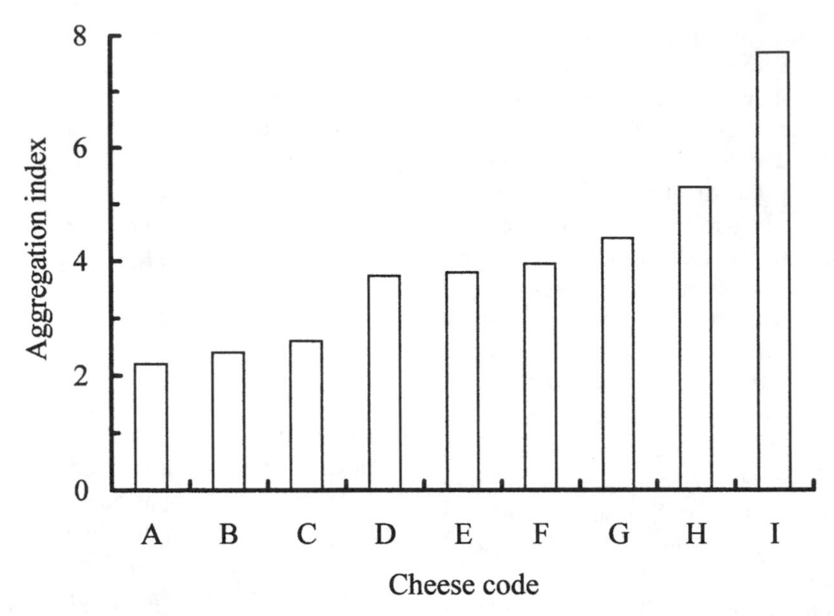

Figure 19–2 Susceptibility of different types of shredded cheese to clumping (as measured by aggregation index): Gruyère (A), Emmental (B), Appenzeller (C), analogue pizza cheese (D), Kashkaval (E), low–moisture Mozzarella (F), Cheddar (G), Tetilla (H), and Fontina (I).

tion of the cheese ingredient to overall flavor is highlighted by the development of and growth in the use of highly flavored enzyme-modified cheeses for a range of products, such as PCPs, imitation cheese products, and cheese powders. These products, in turn, are used to impart cheesy flavor to other products, such as ready-made meals, snacks, soups, and sauces. The increasing importance of cheese flavor is also highlighted by the increasing use of cheeses, such as mature Cheddar and Colby, which have poor stretchability compared to LMMC, in pizza cheese toppings. Cheese flavor is discussed in Chapter 12.

19.3.2 Functional Properties of Heated Cheese

Comparison of Different Cheese Types

The properties displayed by cheese upon cooking are of importance in many applications of cheese, especially grilled cheese sandwiches, pizza pie, cheeseburgers, pasta dishes, and sauces. The various terms used to describe the functional properties of baked or grilled cheese are defined in Table 19–3. Upon grilling or baking, the cheese may be required to melt, flow, stretch, brown, blister, oil-off, and/or stretch to varying degrees; it may also be expected to be chewy and contribute to certain mouth-coating characteristics. The cooking application determines whether one or more of these functions is necessary. Despite the increasing use of different cheese varieties in cooked dishes, surprisingly little is known about the functionality of cheeses other than LMMC, which has been studied extensively and reviewed (Guinee, Mulholland, Mullins, & Corcoran, 1997; Kindstedt, 1995).

Guinee (unpublished study) compared the functional properties of commercial samples of different natural cheese varieties and analogue pizza cheeses upon heating. There were large inter- and intravarietal differences in melt time, flowability, stretchability, and apparent viscosity, which is an index of chewiness (Table 19–4). This trend undoubtedly reflects intervarietal differences in the conditions of manu-

Table 19–3 Functional Properties of Grilled or Baked Cheese That Influence Its Functionality as an Ingredient

Property	*Definition*	*Cheeses Generally Displaying This Property*	*Property Related to Physicochemical State*
Meltability	The ability of cheese to soften to a molten cohesive mass on heating	Most cheeses after a given storage period, PCPs, APCs, Cream cheese	Fat liquifaction, fat coalescence
Flowability	The ability of the melted cheese to flow	Most cheeses after a given storage period, PCPs, OACs, Cream cheese	Fat liquifaction, casein hydration, high degree of fat coalescence, limited oiling-off
Stretchability	The ability of the melted cheese to form cohesive fibers, strings, or sheets when extended uniaxially	Low-moisture Mozzarella, Kashkaval, young Cheddar (15 days)	Moderate degree of casein hydration and casein aggregation, level and type of molecular attractions between paracasein molecules
Flow resistance (often referred to as melt-resistance)	The resistance to flow of melted cheese	Paneer, PCPs, OACs, natural cheeses made from high heat-treated milk	Absence of fat coalescence, heat-induced gelation of particular component(s) (e.g., whey proteins), thermo-irreversibility of gel system in the uncooked product upon heating
Chewiness (rubbery, tough, elastic)	High resistance to breakdown upon mastication	Low-moisture Mozzarella, Kashkaval, young Cheddar (15 days)	As for stretchability
Viscous (soupy)	Low resistance of melted cheese to breakdown upon mastication	Mature Cheddar, aged Mozzarella, Cream cheese, PCPs, OACs	Relatively high level of casein hydration
Limited oiling off	Ability of cheese to express a little free oil upon heating so as to reduce cheese dehydration and thereby to maintain succulence of and impart surface sheen to melted cheese	Most natural cheeses (if not very mature or very young), PCPs, APCs	Limited degree of fat coalescence
Desirable surface appearance	Desired degree of surface sheen with few, if any, dry, scorched black or brown patches	Mature Cheddar, aged Mozzarella, Cream cheese, PCPs, OACs (depending on formulation)	Adequate degrees of casein hydration and of oiling off during baking; low level of residual reducing sugars (e.g., lactose galactose) and Maillard browning

Key: PCP = processed cheese product; APC = analogue pizza cheese; OAC = other analogue cheese.

Table 19–4 Functional Characteristics of Different Types of Natural Cheeses

Cheese Type	Sample Size	Melt Time(s)	Flowability (%)	Stretchability (cm)	Apparent Viscosity (Pa × s)
Pasta filata-type					
Low-moisture Mozzarella	8	108 (6)	53 (8)	83 (21)	623 (303)
Kashkaval	2	96 (11)	67 (4)	87 (13)	522 (330)
Provolone dolce	3	86 (6)	64 (21)	80 (13)	950 (-)
Provolone fumica	1	92	71	76	–
Provolone	1	92	71	76	–
Cheese with eyes					
Gruyère	1	105	78	67	391
Jarlsberg	1	82	52	35	371
Emmental	1	81	74	35	269
Cheddar	8	100 (7)	69 (9)	23 (10)	349 (129)
Analogue pizza cheese	8	105 (13)	42 (19)	27 (8)	668 (307)

Note: Where the sample size ≥ 2, mean values are presented. Values in parentheses are standard deviations.

facture, composition (Table 19–5), degree of maturity, and/or formulation (e.g., levels and types of added ingredients) in the case of the APCs.

The pasta filata cheeses were differentiated from all other varieties by their superior stretchability (Figure 19–3), relatively high apparent viscosity (Figure 19–4), moderate flowability (Figure 19–5), and melt time (Figure 19–6). These functional properties endow these cheeses with attributes typically associated with the melted cheese on pizza pie; sufficiently rapid melt and desirable levels of stringiness, chewiness, and flow. Some of the pasta filata cheeses (Provolone dolce and string cheese) had a very high viscosity, which would undoubtedly be associated with overchewiness on pizza pie. The functional requirements of the global pizza market tend to be region specific. However, while the functional requirements of different pizzerias vary somewhat within and between countries, cheese with the following characteristics is generally acceptable: melt time less than 120 s; flowability, 40–55%; stretchability, greater than 75 cm; apparent viscosity, 800–400 Pa × s; AGI, 3.5–4.5. Hence, the higher flow and/or apparent viscosity of Provolone dolce and string cheese may be better suited to the customized requirements of local pizza markets. While the degree of browning required on pizza pie appears to depend very strongly on the pizzeria, generally a low degree is more desirable.

The superior stringiness of pasta filata cheeses upon baking, compared with other natural cheeses, may be attributed primarily to plasticization of the curd during the kneading and stretching process. In this process, which is unique to the manufacture of pasta filata–type cheeses, the milled curd is heated to about 57–60°C and kneaded in hot water or brine at about 78–82°C when the curd pH is about 5.2. The combined effect of the high temperature and low pH during kneading results in casein aggregation and contraction of the strands of the para-casein gel, resulting in the formation of para-casein fibers of high tensile strength. The curd kneading and stretching operations of the plasticization process, which are carried out in equipment that simulates the traditional hand stretching (Figure 7–6), give the newly formed curd fibers a linear orientation (parallel to the direction of stretching). Confocal laser scanning micrographs of the curd before and after texturization clearly demonstrate the formation and

Table 19–5 Compositional Analyses of Different Commercial Cheese Varieties

Cheese Type	*Source*	*Sample Size*	*Moisture (g/kg)*	*Protein (g/kg)*	*Fat (g/kg)*	*FDM (g/kg)*	*MNFS (g/kg)*	*S/M (g/kg)*	*Ash*	*Ca (mg/g cheese protein)*	*P (mg/g cheese protein)*	*pH 4.6 SN (g/100 g total N)*	*PTAN (g/100 g total N)*	*pH*
Pasta filata type														
Low moisture Mozzarella	Ireland, UK, Denmark	8	464	260	232	446	604	31	38	27.2	20.6	4.7	0.5	5.53
Kashkaval	Yugoslavia	2	441 (7)	253 (3)	256 (12)	458 (17)	596 (4)	49 (2)	43 (0)	31.0 (5)	22.2 (18)	6.1 (3.9)	0.7 (0.2)	5.37 (0.02)
Provolone dolce	Italy	2	386 (32)	276 (5)	294 (37)	477 (36)	546 (17)	57 (7)	–	–	–	10.2 (9.8)	–	5.54 (0.14)
Provolone fumica	Italy	1	434	278	243	428	573	47	–	–	–	7.5	–	5.29
Provolone	Italy	1	402	281	285	476	562	44	–	–	–	10.8	–	5.29
Cheese with eyes														
Gruyère	Switzerland	1	341	277	368	558	540	49	–	–	–	10.8	–	5.83
Jarlsberg	Norway	1	404	277	313	524	587	28	–	–	–	8.6	–	5.72
Emmental	Switzerland	1	343	243	380	578	553	13	–	–	–	9.2	–	5.64
Cheddar	Ireland	8	372 (11)	254 (8)	331 (20)	526 (24)	556 (10)	45 (7)	37 (3)	28.0 (1.4)	20.6 (1.3)	20.3 (3.7)	4.6 (2.5)	5.14 (0.12)
Analogue pizza cheese	Ireland	8	489 (37)	184 (19)	250 (19)	490 (34)	651 (44)	35 (4)	42 (3)	34.4 (21)	28.1 (44)	2.3 (0.8)	0.2 (0.1)	6.3 (0.1)

Note: Where sample size ≥ 2, mean values are presented. Values in parenthesis are standard deviations.

Key: FDM = fat-in-dry matter; MNFS = moisture-in-non-fat substance; S/M = salt-in-moisture; pH4.6 SN = nitrogen soluble at pH 4.6; PTAN = nitrogen soluble in 5% phosphotungstic acid.

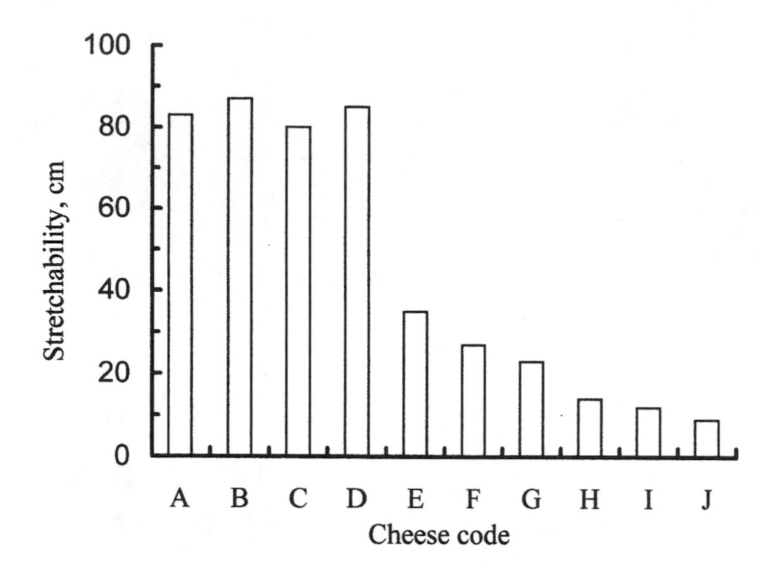

Figure 19–3 Stretchability of different cheese types after baking at 280°C for 4 min: low-moisture Mozzarella (A), Kashkaval (B), Provolone (C), string cheese (D), Emmental (E), analogue pizza cheese (F), Cheddar (G), Parmesan (H), Raclette (I), and Appenzeller (J).

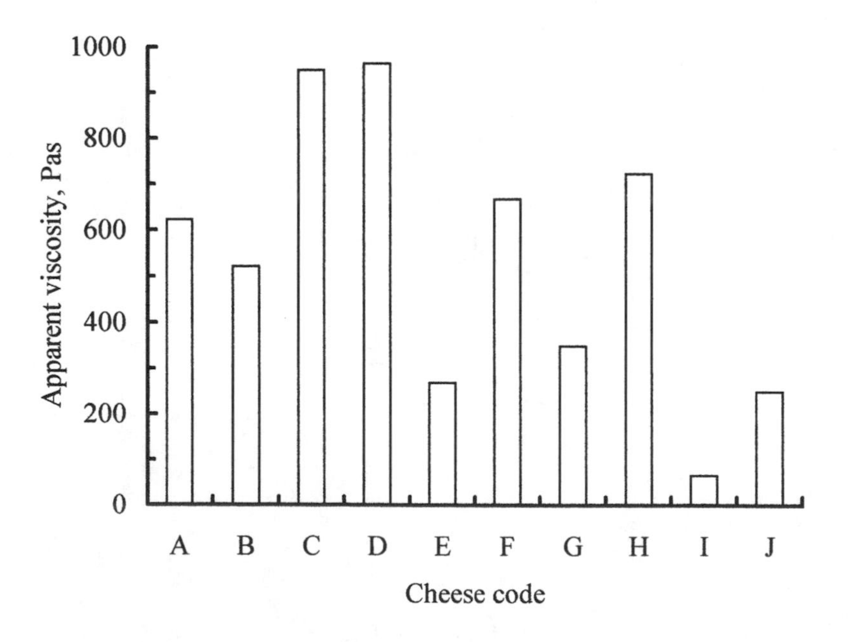

Figure 19–4 Apparent viscosity, at 70°C, of different cheese types: low-moisture Mozzarella (A), Kashkaval (B), Provolone (C), string cheese (D), Emmental (E), analogue pizza cheese (F), Cheddar (G), Parmesan (H), Raclette (I), and Appenzeller (J).

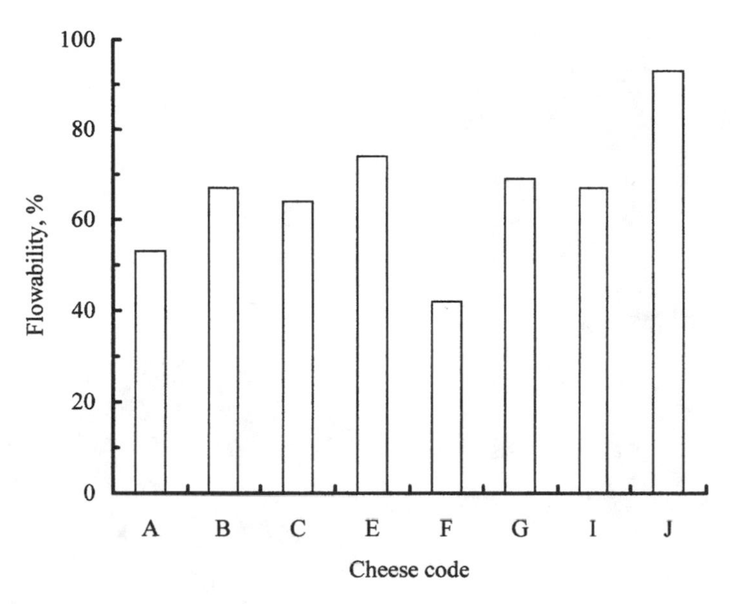

Figure 19–5 Flowability of different cheese types after baking at 280°C for 4 min: low-moisture Mozzarella (A), Kashkaval (B), Provolone (C), Emmental (E), analogue pizza cheese (F), Cheddar (G), Parmesan (H), Raclette (I), and Appenzeller (J).

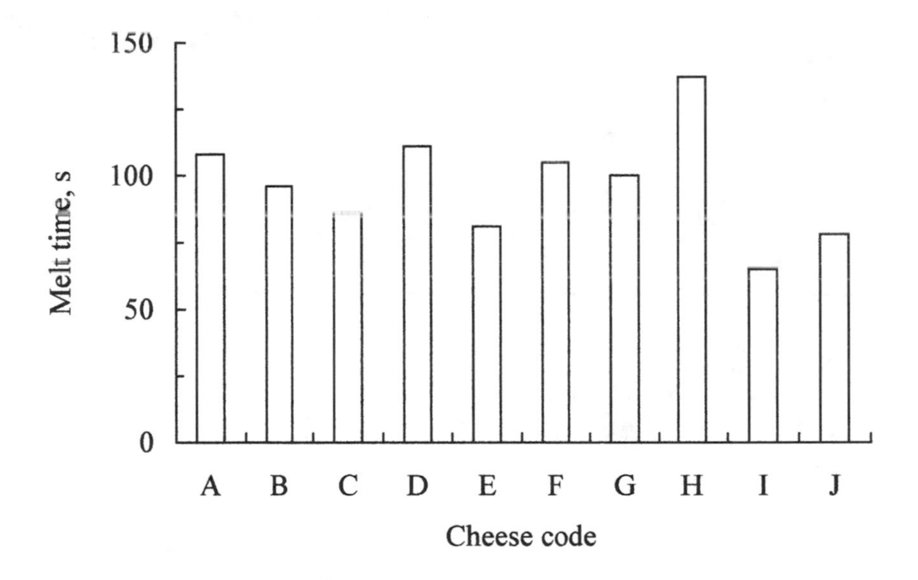

Figure 19–6 Melt time of different cheese varieties at 280°C: low-moisture Mozzarella (A), Kashkaval (B), Provolone (C), string cheese (D), Emmental (E), analogue pizza cheese (F), Cheddar (G), Parmesan (H), Raclette (I), and Appenzeller (J).

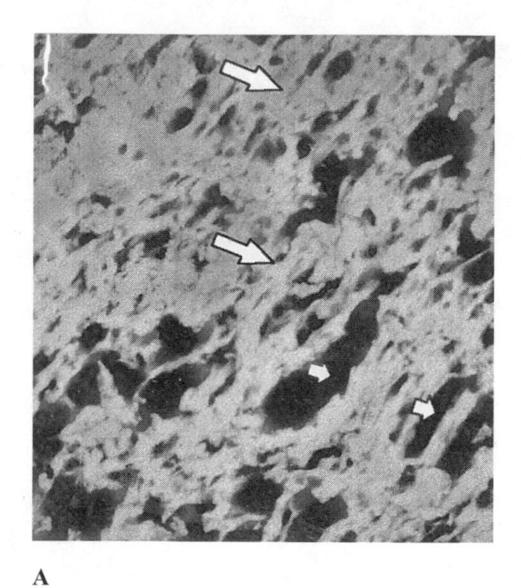

A

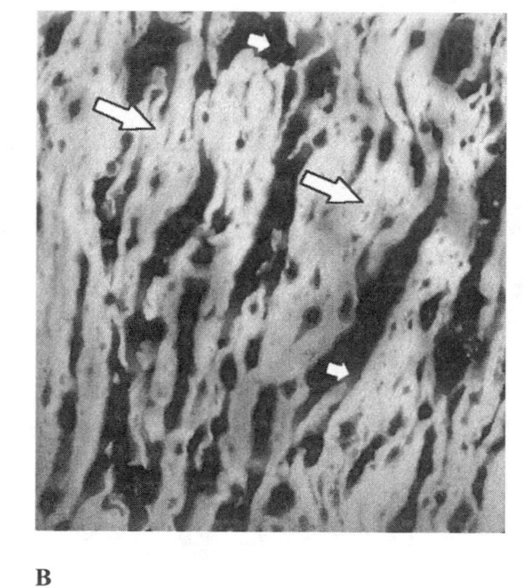

B

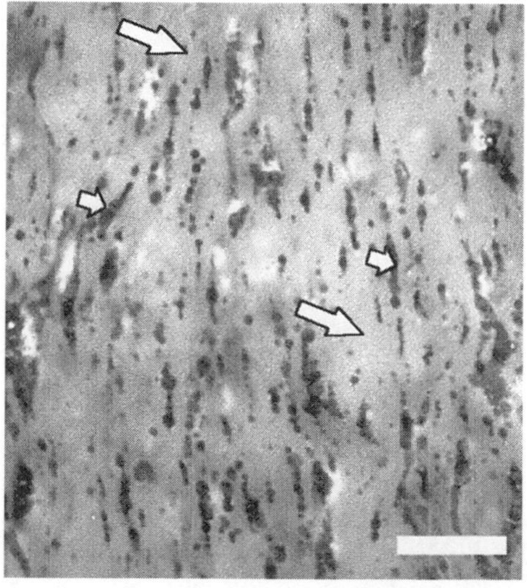

C

Figure 19–7 Confocal laser scanning micrographs of low-moisture Mozzarella cheese at different stages of production and storage at 4°C. (a) Before plasticization: salted and milled curd showing paracasein matrix strands (long arrows) and void spaces containing fat globules (short arrows). (b) After plasticization (24 hr storage at 4°C): stretched curd showing extensive linearization of paracasein into fibers (long arrows) and fat globules or pools (short arrows). (c) After plasticization (43 days storage at 4°C): aged cheese showing hydrated swollen paracasein fibers forming a continuous protein phase occluding fat mainly in the form of pools. Bar = 25 μm.

linearization of protein fibers (Figure 19–7).

In contrast to the pasta filata cheeses, most other cheese types, including analogue pizza cheeses, Cheddar, and Emmental, had relatively low stretchability, low apparent viscosity (< 400 Pa × s) and in some cases excessive (> 55%) or impaired (19%) flowability characteristics. If such cheeses were used on pizza pie, the melted cheese would lack the desired stringiness, would flow excessively or insufficiently, and would lack the desired chewiness. Conversely, stringiness, which is typical of LMMC and other pasta filata cheeses such as Kashkaval and Provolone, is an undesirable attribute for applications such as sauces, fondues, and toasted sandwiches. Cheeses such as mature Cheddar, Emmental, Raclette, and Gouda are more satisfactory because of their excellent flowability and flavor and the absence of stringiness upon baking or grilling.

The findings of the above study indicate that, whereas many cheeses have good flow and viscosity characteristics, few, apart from LMMC and other pasta filata cheeses, possess the good stretchability associated with cheese topping on pizza pie. However, other cheeses, such as Cheddar, are being used increasingly on pizza pie to impart more cheese flavor. Relatively little is known on how the blending of LMMC with other cheeses affects the functionality of the cheese topping on pizza pie. The intervarietal differences in functional properties probably arise from variations in proteolysis, the water-binding capacity of the cheese proteins, and the concentrations of structural components, namely, fat, protein, and moisture (as discussed in the following section). The preliminary findings suggest that more knowledge is required to identify the effects of the different factors that contribute to the functionality and viscoelasticity of baked or grilled cheeses to enable an efficient approach to the blending of different cheese varieties for pizza cheese toppings.

In other applications, melt is essential but very limited flow is required so as to preserve the shape and identity of the cheese. Examples of the latter include fried Paneer, grilled or fried burgers containing cheese insets, and deep-fried breaded cheese sticks. Most mature natural cheeses are unsuitable for these applications owing to excessive flow and oiling-off during cooking. In the case of cheese insets in deep-fried burgers, such attributes would cause the cheese to permeate the interstices of the coarse meat emulsion and thereby cause the cooked cheese to lose its shape and visual effect. For these applications, pasteurized processed cheese products, which are specially formulated to endow them with varying degrees of flow resistance, are more suitable than most natural cheeses (Figure 19–8). Selective manipulation of formulation and processing conditions enables the structural and physicochemical properties of natural cheese to be modified readily by heating and shearing processes and by blending with other ingredients, such as emulsifying salts and whey proteins (see Chapter 18). However, some natural cheeses, such as Queso blanco and Cream cheeses (manufactured using a particular protocol), exhibit excellent flow resistance and are, therefore, often used in applications such as deep-fried battered cheese sticks or cheese insets in various dishes. The manufacture of these cheeses generally involves a high heat treatment of the milk and/or the curd (e.g., > 85°C for 5–15 min) following whey drainage, as in Cream cheese produced by the ultrafiltration method. This results in high levels of whey protein denaturation (e.g., ≈ 600 g/kg total whey protein) and the complexation of whey protein with casein. Owing to the heat gelation of the included whey proteins during subsequent baking or grilling of the cheese, they impede flow of the cheese.

In other meat products, such as cordon-bleu poultry products, extensive flow is required, with little or no oiling-off or stringiness.

Factors That Influence the Functionality of Cheeses upon Cooking

Little information is available on the factors that affect the functional characteristics of natural cheeses during cooking, apart from LMMC. Most natural cheeses are consumed principally as table cheeses, and the functionality of the cooked cheese is of only secondary importance

Figure 19–8 Pasteurized processed Cheddar cheeses exhibiting varying levels of flowability in grilled cheese–meat burgers.

compared to the taste and texture of the natural cheese.

19.3.3 Ripening Time

Changes in the functional characteristics of LMMC during storage have been reported extensively (Guinee et al., 1997; Kindstedt, 1995). Comparatively little information is available on the age-related changes in other cheeses such as other pasta filata cheeses, Cheddar, Parmesan, or Emmental. LMMC made by the conventional procedure (see Chapter 17) generally is nonfunctional upon cooking during the first 5–10 days of storage at 4°C after manufacture. This is reflected by drying out and crusting of the cheese on the pizza pie during baking owing to the low water-binding capacity of paracasein (and the concomitant extensive dehydration of the cheese during baking) and the low propensity of the cheese to express free oil. Both factors are conducive to excessive evaporation of moisture at the high temperature (typically ≈ 90°C) in the mass of melting cheese when heated in a convection oven at 280°C for 4 min). The dried-out, crusted cheese lacks succulence and fails to melt, flow, or stretch. Moreover, it is extremely tough and chewy, as reflected by a high apparent viscosity of more than 1,000 Pa × s.

The functionality of LMMC improves markedly during the first 2 weeks of ripening at 4°C, as indicated by decreases in melt time and apparent viscosity and increases in flowability and stretchability. This status is maintained until about 40–50 days (Figures 19–9 to 19–12). The improved functionality may be attributed to increases in protein hydration and free fat during aging of the cheese (Figures 19–13 and 19–14) (Guinee et al., 1997; Kindstedt & Guo, 1997). The water vapor pressure of water bound by the paracasein is lower than that of free water and thus has a lower propensity to evaporate during baking. The exu-

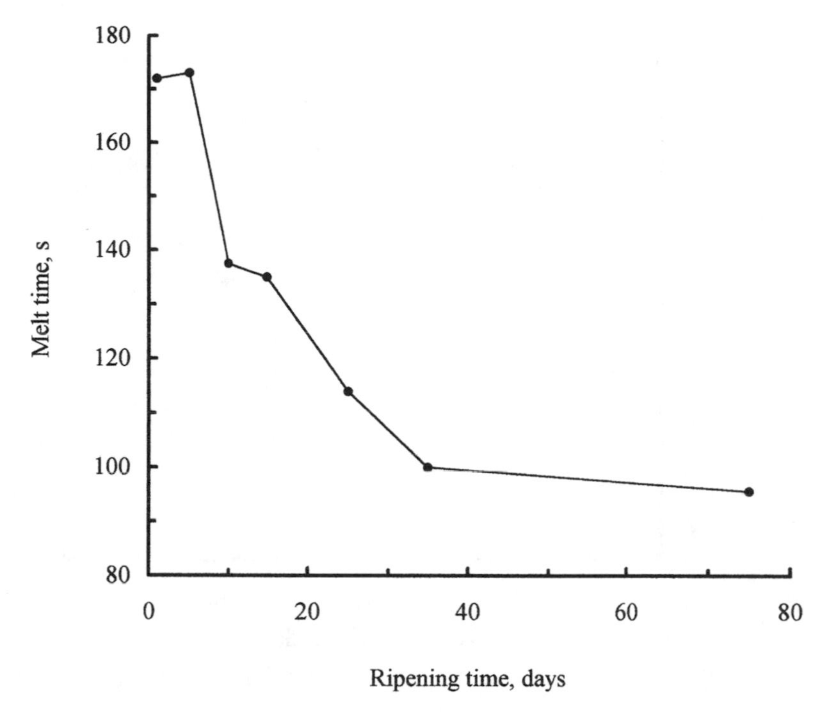

Figure 19–9 Typical changes during storage at 4°C in the melt time of low-moisture Mozzarella cheese baked at 280°C for 4 min.

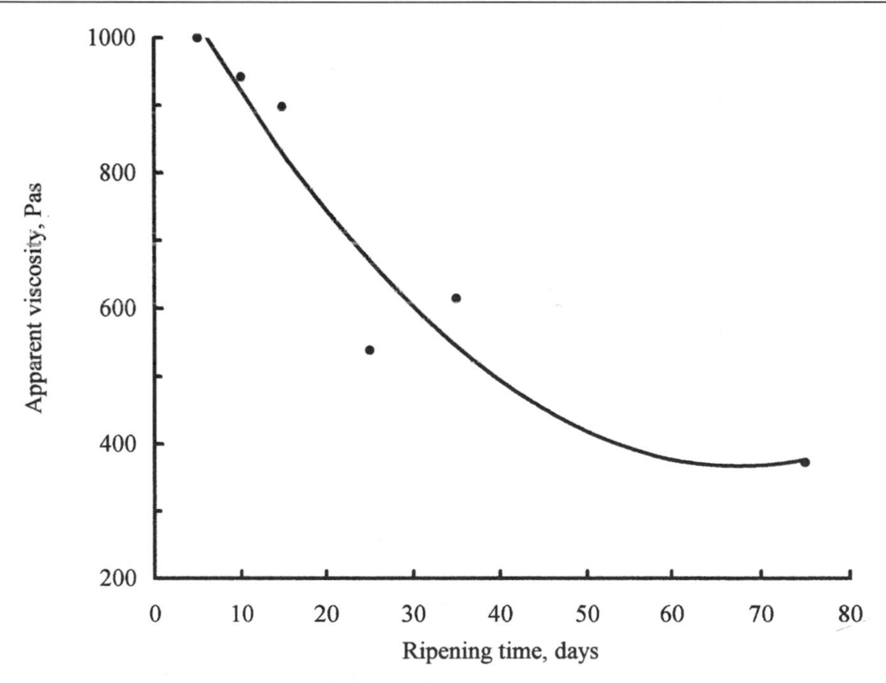

Figure 19–10 Typical changes during storage at 4°C in the apparent viscosity of low-moisture Mozzarella cheese heated at 70°C.

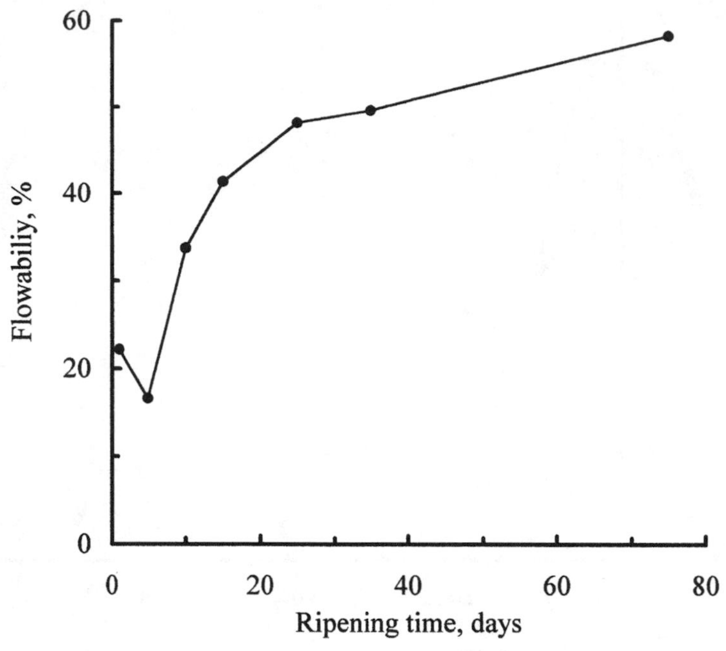

Figure 19–11 Typical changes during storage at 4°C in the flowability of low-moisture Mozzarella cheese baked at 280°C for 4 min.

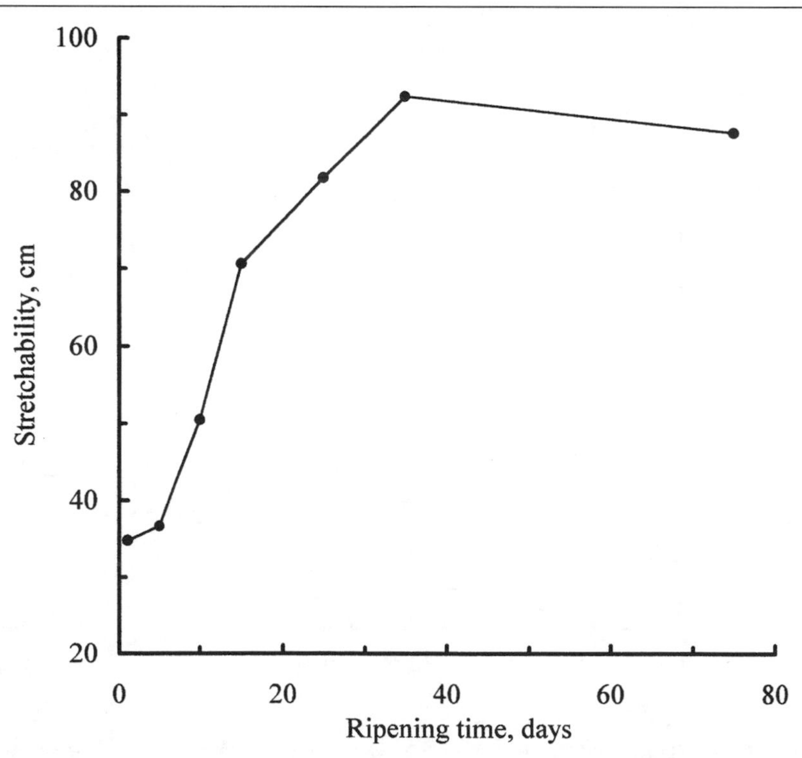

Figure 19–12 Typical changes during storage at 4°C in the stretchability of low-moisture Mozzarella cheese baked at 280 °C for 4 min.

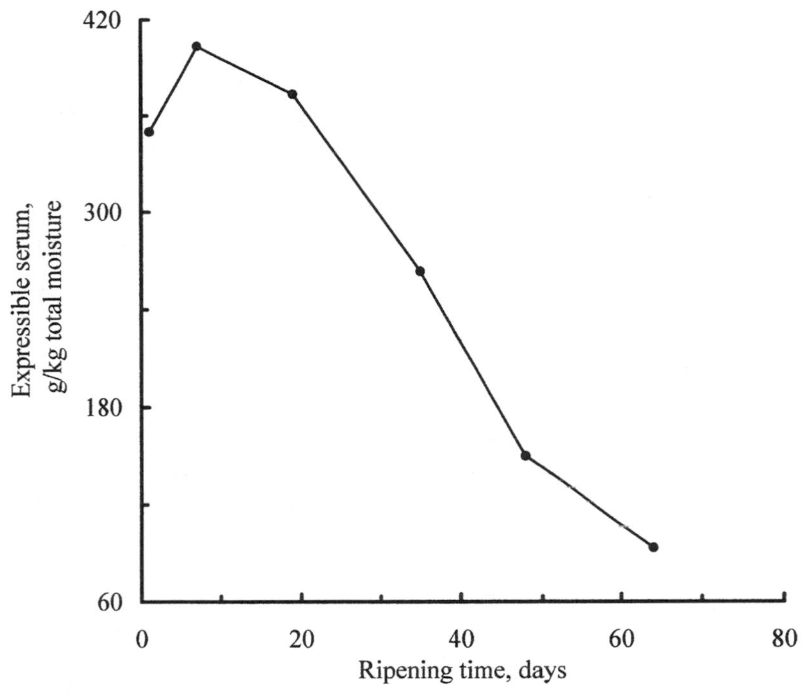

Figure 19–13 Typical changes during storage at 4°C in the level of serum expressed from low-moisture Mozzarella upon hydraulic pressing (3.2 MPa at 20°C for 3 hr). A decrease in the level of expressible serum is an index of increased water-binding capacity, or hydration, of the paracasein in the cheese.

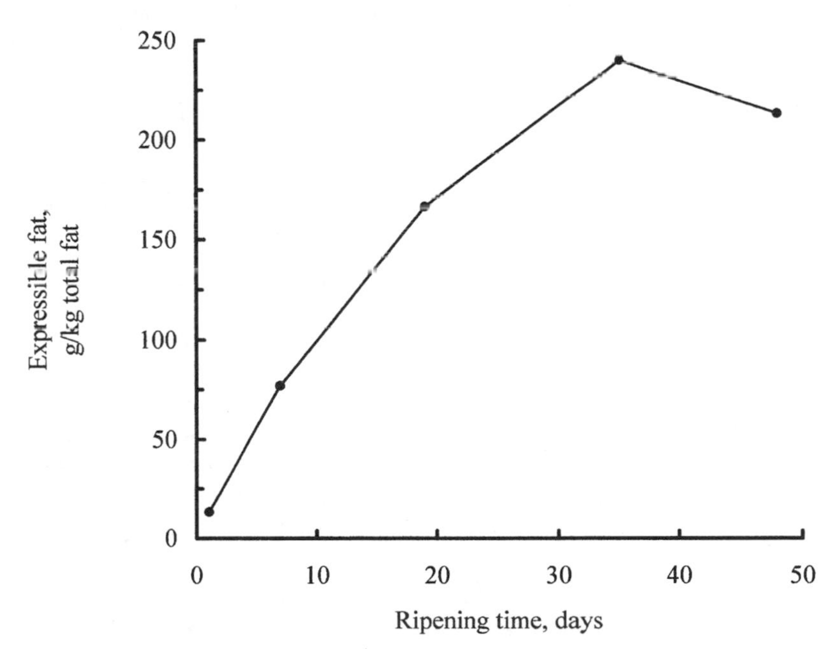

Figure 19–14 Typical changes during storage at 4°C in the level of oil expressed from low-moisture Mozzarella upon hydraulic pressing (3.2 MPa at 20°C for 3 hr). An increase in expressible oil is an index of the increased potential of the cheese to oil-off during baking.

dation of free oil from the shredded cheese during baking also limits dehydration; the free oil forms an apolar surface layer, which impedes the escape of water vapor. The changes in protein hydration and free oil formation during storage are mediated by the following:

- The small increase in pH (from ≈ 5.15 before plasticization to ≈ 5.35 at 5 days). The increase in pH during plasticization is due to the loss of lactic acid and soluble calcium phosphate in the water used to heat the curd during plasticization. The loss of soluble Ca results in the solubilization of colloidal calcium phosphate in the curd during subsequent cooling of the curd, to restore equilibrium between the soluble and colloidal phases. The resulting phosphate anions (PO_4^{3-}) scavenge free H^+ from the aqueous phase and thereby reduce the [H^+] (increase pH). Paracasein has maximum hydration at around pH 5.35 (see Chapter 16).
- The increase in primary proteolysis (Figure 19–15) of paracasein by residual rennet and/or plasmin, as reflected by the increase in pH 4.6 soluble N. Compared with other varieties, such as Cheddar and Gouda, LMMC has a low level of proteolysis, apparently because of the relatively low level of active rennet in the curd due to its thermal inactivation during plasticization (see Chapter 11). A comparative study of 12-week-old Cheddar, Gouda, and LMMC showed that the latter had the highest level of intact α_{s1}-casein and intermediate levels of α_{s1}-CN f24-199 and γ-caseins. The presence of these breakdown products suggests contributions from residual chymosin and plasmin to primary proteolysis in LMMC.
- The solubilization of casein-bound calcium that occurs when the calcium attached to the casein matrix is partially replaced by sodium. Conditions in the cheese conducive to this ion exchange effect are 2.0–4.0% NaCl and 0.4% Ca in the serum phase (Kindstedt & Guo, 1997).
- The increased propensity of the cheese to oil-off upon cooking, which may be associated with the age-related degradation of the fat globule membrane and/or the continued coalescence of the partially denuded fat globules after plasticization. The latter effect may result from the hydration and concomitant physical swelling of the protein matrix, an occurrence that physically forces the partially denuded globules into close proximity.

During prolonged storage (up to 75 days), the unbaked cheese generally becomes too soft and sticky whereas the baked cheese becomes excessively flowable and "soupy" and lacks the desired chewiness, reflected by the relatively low apparent viscosity. These changes in functionality are attributed to excessive proteolyis. However, stretchability remains relatively constant even when the product is stored for up to 4 months at 4°C, suggesting that the level of primary proteolysis in the cheese at this time (i.e., pH 4.6–soluble N ≈ 120 g/kg total N) is insufficient to significantly impair stretchability. Indeed, preliminary experiments show that young Cheddar cheese (i.e., 15–35 days, with a pH 4.6–soluble N level of < 120 g/kg total N) has good stretchability, similar to that of LMMC (Guinee, unpublished results). In contrast, Cheddar (with a pH 4.6–soluble N level > 150 g/kg total N) has inferior stretchability compared with LMMC.

Maillard browning on pizza pie results from heat-induced reactions between the carbonyl group of reducing sugars (lactose and galactose) and the amino groups of peptides and amino groups. The degree of browning is related to the sugar-fermenting and proteolytic characteristics of the starter culture used (Kindstedt, 1993). Most strains of *Streptococcus thermophilus* and *Lacobacillus delbrueckii* spp. *bulgaricus*, which are commonly used in the manufacture of LMMC, are unable to metabolize galactose, and hence cheese made solely with such cultures is susceptible to browning. However, *Lb. helveticus*, which is frequently used as a component of the starter culture, ferments lactose and galactose (resulting from the incomplete metabolization of lactose by the former cultures) completely to lactic acid (see Chapter 5). Attempts to

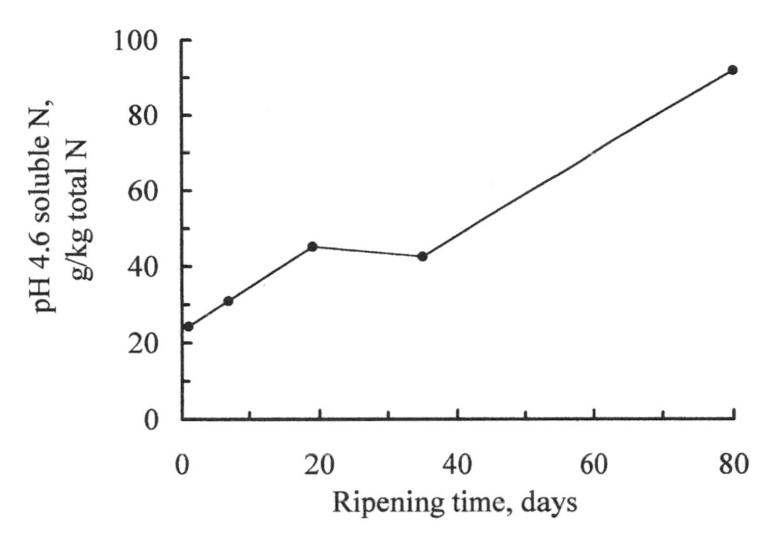

Figure 19–15 Typical changes in pH 4.6–soluble N in low-moisture Mozzarella cheese during storage at 4°C.

control the level of browning in pizza cheese include control of residual sugars and/or proteolysis products via

- adjustment of the ratio of *Lb. helveticus* to *Sc. thermophilus* in the starter culture
- use of galactose-positive strains of *Lb. delbrueckii* spp. *bulgaricus*
- the use of proteinase-negative starter strains to limit the formation of free amino groups
- the use of curd washing to remove lactose from the cheese curd

The propensity of LMMC to brown upon baking changes markedly during ripening. Fresh cheese curd (< 2–5 days) shows a high propensity to brown. Its propensity to brown decreases markedly during the first few weeks of ripening due to the metabolism of lactose and/or galactose by galactose-fermenting starters but increases progressively thereafter owing to the accumulation of small peptides and amino acids.

19.3.4 Cheese Composition and Proteolysis

The composition of cheese has a major effect on its functionality, as shown by inter- and intravarietal correlations between composition and functionality. A recent study involving commercial samples of 21 different natural hard and semi-hard cheese varieties showed several across-varietal correlations between cheese composition and functionality upon heating (Table 19–6). Stretchability was negatively correlated with the levels of fat, FDM, and pH 4.6–soluble N and positively correlated with moisture content. In contrast, flowability was negatively correlated with the levels of moisture and moisture-in-non-fat-substance (MNFS). A similar across-varietal relationship between flowability and MNFS has been found for a range of hard and semi-hard cheeses. Like stretchability, the apparent viscosity was negatively correlated with the level of pH 4.6–soluble N. Hence, Cheddar and other cheeses that had higher levels of pH 4.6–soluble N and relatively lower levels of moisture than pasta filata cheeses had significantly higher flowability and lower apparent viscosity and stretchability than the former.

The effects of intravarietal compositional differences on the functionality of LMMC have been reviewed by Kindstedt (1993). Increasing the moisture content of LMMC results in lower apparent viscosity and higher flowability. Reducing the level of fat in LMMC results in a lower level of free oil and flowability upon baking. Reducing the levels of calcium and phosphate, by reducing the pH at coagulation, results

Table 19–6 Relationships between Compositional and Functionality Variables for Different Natural Commercial Cheese Types

Functional Variable	*Compositional Variable*	*Degrees of Freedom*	*Correlation Coefficient (r)*	*Significance*
Raw cheese				
Aggregation index	pH 4.6SN	27	+0.36	$p < 0.05$
Cooked or baked cheese				
Stretchability (cm)	Fat	36	–0.66	$p < 0.001$
	FDM	36	–0.57	$p < 0.001$
	Moisture	36	+0.68	$p < 0.001$
	pH 4.6 SN	34	–0.67	$p < 0.001$
Flowability (%)	MNFS	34	–0.53	$p < 0.001$
	Moisture	32	–0.40	$p < 0.02$
Melt time (s)	pH	38	–0.45	$p < 0.05$
Apparent viscosity (Pa × s)	pH 4.6 SN	28	–0.43	$p < 0.02$

Note: Correlation coefficients obtained by linear regression of the data. Data refer to cheeses in Tables 19–4 and 19–5.

Key: FDM = fat-in-dry-matter (g/kg); pH 4.6 SN = nitrogen soluble at pH 4.6 (g/100g total N); MNFS = moisture-in-non-fat-substance (g/kg).

in higher flowability in LMMC made by direct acidification, that is, by the addition of food-grade acid rather than acid development by a starter culture. Increases in the levels of primary and secondary proteolysis in LMMC, through the use of a more proteolytic coagulant than chymosin (e.g., *C. parasitica* proteinase), reduces the apparent viscosity and increases free oil and flowability. Increasing the salt content of LMMC from 11 to 17.8 g/kg results in cheese that is less stringy and less flowable and has a higher apparent viscosity.

Increasing the fat content of Cheddar cheese from 50 to 330 g/kg is associated with shorter melt times, greater flowability, greater stretchability, and lower apparent viscosity.

19.3.5 Cheesemaking Conditions

The composition and functionality of LMMC during ripening may be altered by varying cheesemaking conditions and factors, including

- the type of coagulant
- pH at whey drainage, which influences the level of rennet retained in the curd
- temperature of the curd during cooking and stretching, which influences the stability of rennets
- pH at stretching, which influences the level of casein-bound calcium and its water-binding capacity
- salting method

Hence, Mozzarella cheese with customized functionalities may be produced by varying the cheesemaking conditions. Other factors that influence the composition and functionality of LMMC include

- the stage of lactation and diet of the cows
- homogenization of cheese milk
- design and type of stretching equipment
- whether a starter or direct acidification is used

The composition and functionality of Mozzarella acidified by the addition of acids is influenced markedly by the calcium-chelating properties of the acids used and the pH and temperature at coagulation. The latter parameters affect the structure of the gel, its ability to synerese following cutting or cooking, the level of casein-bound calcium, and the water-binding capacity of the casein matrix during subsequent stretching.

19.4 DRIED CHEESE PRODUCTS

Dehydrated cheese products are industrially produced cheese-based ingredients that were developed for the U.S. Army during World War II as a means of preserving cheese solids under conditions to which natural cheese would not normally be subjected, such as temperature above 21°C for long periods. Since then, they have become ingredients of major economic importance owing to their widespread use as flavoring agents and/or nutritional supplements in a wide range of foods. These include bakery products, biscuits, dehydrated salad dressings, sauces, snack coatings, soups, pasta dishes, savory baby meals, cheese dips, au gratin potatoes, and ready-made dinners. They are also included in processed and analogue cheese products as flavoring agents or as functional ingredients in powdered instant cheese preparations, which can be reconstituted by the consumer for the preparation of instant functional cheeses, such as pizza-type cheese, for domestic use. Advantages over natural cheeses as an ingredient in the above applications include these:

- Convenience of use in fabricated foods. Cheese powders can be easily applied to the surface of snack foods, such as popcorn, potato crisps, and nachos, or easily incorporated into food formulations by dry mixing with other dry ingredients such as skim milk powder (e.g., dried soup, sauce, and cake mixes) or blending into wet formulations. In contrast, natural cheeses require size reduction prior to their use in these applications.
- Longer shelf-life because of their lower water activity (a_w) than natural cheese. The water activity (a_w) of natural cheeses ranges from about 0.99 for Quarg to 0.917 for Parmesan (see Chapter 8), it ranges from about 0.93 to 0.97 for processed cheese products, but it only ranges from 0.2 to 0.3 for various dairy powders. Owing to their relatively high stability, cheese powders may be stored for a long period without alteration or deterioration of quality. In contrast, the changes that occur in natural cheese during storage may influence its processability (e.g., the ease with which it can be size-reduced or its interaction with other ingredients) and its flavor profile and intensity. Hence, cheese powders are more amenable to inventory management and set manufacturing methods, and they yield end products with more consistent quality in large-scale manufacturing operations than natural cheese.
- Greater diversity of flavor and functional characteristics. The wide range of characteristics is made possible by the use of different types of cheese, EMCs, and other ingredients in the preparation of cheese powders.

Dehydrated cheese products may be classified into four categories, depending on the ingredients used:

1. dried grated cheeses, such as Parmesan and Romano
2. natural cheese powders, made using natural cheeses, emulsifying salts, and natural cheese flavors (optional)
3. extended cheese powders, incorporating natural cheese and other ingredients such as dairy ingredients (e.g., skim milk solids, whey, lactose), starches, maltodextrins, flavors, flavor enhancers, and/or colors
4. dried EMCs

19.4.1 Dried Grated Cheeses

Dried grated cheeses are normally used as highly flavored sprinklings, on pasta dishes, for example, and in bakery products (e.g., biscuits). Essentially, the production of these products involves finely grating hard cheeses and drying the ground cheese, usually in a fluidized bed drier by exposure to low humidity air (15–20% relative humidity) at an inlet temperature below 30°C. Under these conditions, the cheese is dehydrated rapidly and evaporatively cooled, thereby reducing the risk of fat exudation and the tendency to ball or clump. The dried, grated cheese (typically containing 17% moisture) is usually pulverized and packed under nitrogen to reduce the risk of oxidative rancidity during distribution and retailing.

Certain properties are required for the production of dried grated cheeses, such as relatively low levels of moisture (300–340 g/kg) and fat-in-dry-matter (FDM; 390 g/kg), brittleness, and elastic fracture characteristics. These properties foster efficient size reduction upon grinding, minimal susceptibility to fat exudation and sticking of the cheese particles, and efficient drying to a homogeneous product free of clumps. An intense cheese flavor is also a desirable characteristic. Generally, dried grated cheeses are used in small quantities, as sprinklings, to impart strong cheese notes to pasta dishes, soups, and casseroles. The cheeses that best meet these criteria are Parmesan and Romano, because of their composition; fracture properties; and strong, piquant, lipolyzed flavor. The flavor of Romano-type cheeses results mainly from the addition of pregastric esterase (from kid, goat, or lamb) to the cheese milk, which preferentially hydrolyzes the short-chain fatty acids (especially butanoic acid) from milk fat triglycerides during maturation. A high level of butanoic acid (1,500–2,000 mg/kg cheese) is responsible for the peppery, piquant flavor of Romano cheese.

Owing to their lower firmness and higher levels of moisture and FDM, cheeses such as mature Cheddar (moisture ≈ 370 g/kg and FDM ≈ 520 g/kg) or Gouda (moisture ≈ 410 g/kg and FDM ≈ 480 g/kg) are unsuitable for drying. These characteristics render the cheese susceptible to fat exudation and clumping during grinding and drying. However, they can be dried if they are first shredded and blended with Parmesan- or Romano-type cheeses before grinding. The moisture content of dehydrated grated cheese may be reduced further by using the Sander's drying process. The grated cheese powder is placed on trays that are conveyed through a drying tunnel, where it is exposed to hot air. The hot air heats the cheese particles to 63°C and reduces the moisture content from 170 g/kg to about 35 g/kg. High moisture cheeses (820 g/kg), such as Cottage cheese, may also be dried directly to a 30–40 g/kg moisture level by first pulverizing and then drying them in specialized spray driers (e.g., silo spray drying using the Birs Dehydration Process; Kosikowski & Mistry, 1997). These low-moisture, dried natural cheeses are generally used for nutritional supplementation of foods (e.g., dried baby meals).

19.4.2 Cheese Powders

Manufacture

The manufacture of cheese powders essentially involves the production of a pasteurized processed cheese slurry (400–450 g/kg dry matter), which is then spray-dried (Figure 19–16). The production steps include formulation of the blend, processing of the blend to form a slurry, homogenization, and drying of the slurry.

The blend usually consists of comminuted natural cheese, water, emulsifying salts, flavoring agents, flavor enhancers, colors, antioxidants, and perhaps filling materials, such as whey or skim milk solids, starches, maltodextrins, and milk fat. The type of cheese powder (e.g., natural or extended), flavor required, and application (e.g., whether intended for use in a sauce, soup, snack coating, or cheese dip) determine the type of ingredients included. Antioxidants, such as propyl gallate and butylated hydroxyanisole, may be added at a level of 0.5–1.0 g/kg fat to retard oxidative rancidity. Typical formulations of the slurries required for the production of natural and extended cheese powders with different levels of cheese solids are given in Table 19–7.

The flavor profile and intensity of the final cheese powder is determined by the types of

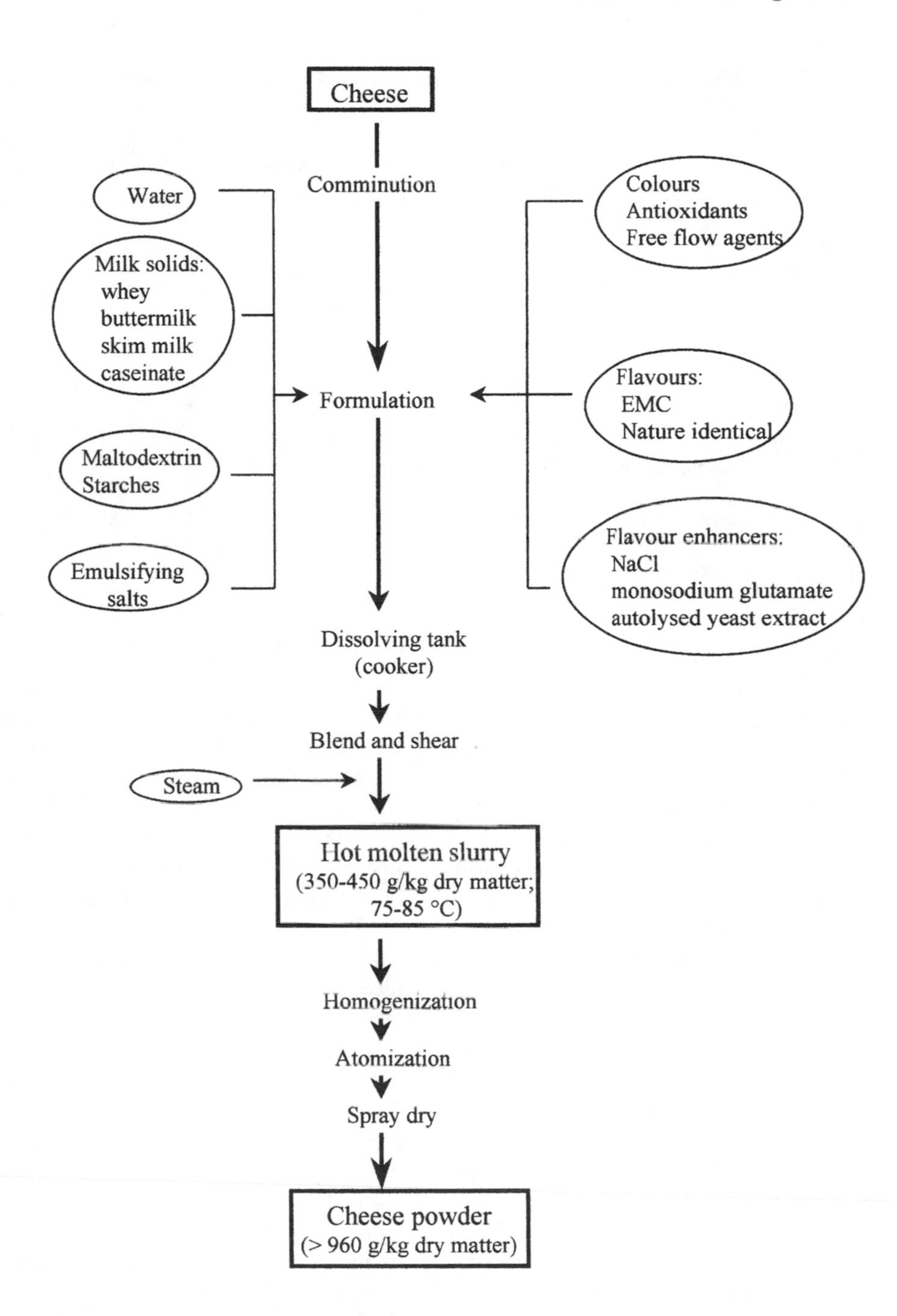

Figure 19–16 Production process for cheese powder.

Table 19–7 Typical Formulations of Cheese Slurries for the Production of Cheese Powder with Different Levels of Cheese Solids

	Low (262 g/kg)	*Medium (530 g/kg)*	*High (950 g/kg)*
Medium aged Cheddar	–	190	–
Mature Cheddar	200	170	635
EMC paste	5	2	10
EMC powder	5	20	–
Whey powder	120	50	–
Skim milk powder	80	38	–
Maltodextrin (DE 17)	165	110	–
Emulsifying salts	15	25	15
Butylated hydroxyanisole (BHA)	0.5	0.5	0.5
Sodium chloride	15	10	5
Water and condensate	390	385	335

Key: EMC = enzyme-modified cheese.

cheese used and the types and levels of other flavoring agents (e.g., EMC, hydrolyzed milk fat, and starter distillate) and flavor enhancers (e.g., NaCl, monosodium glutamate, and autolyzed yeast extract). Generally, mature cheese with an intense flavor is used so as to impart a strong flavor to the final product. Apart from their lack of flavor-imparting properties, young cheeses with a high level of intact casein are unsuitable, as they result in slurries that are very viscous and are difficult to atomize and dry efficiently. Filling materials in extended cheese powders are usually added to replace cheese solids and reduce formulation costs. However, they may influence flavor, wettability, and mouth-coating characteristics of the product in which the cheese powder is used.

Processing principles and technology are similar to those used for the manufacture of processed cheese products. Processing involves heating the blend to around 80°C in a processed cheese–type cooker using direct steam injection or in large "dissolving tanks" (e.g., 5,000 L) containing shearing blades using indirect steam injection and continuous shearing at 1,500–3,000 rpm. The blend is worked until the hot molten slurry is homogeneous in color and consistency and free of lumps or nonhydrated material. The maximum processing temperature should be maintained below 85°C to minimize the loss of volatile flavor compounds in the dissipating steam and to minimize the risk of browning, especially for formulations containing a high level of lactose or high dextrose (glucose) equivalent (DE) maltodextrins.

The viscosity of the cheese slurry has a major influence on its tendency to foam and therefore on the level of air in the resultant powder. Owing to its effects on the air content of the powder, the viscosity of the cheese slurry influences the physical characteristics (bulk density and wettability) of the resultant powder and its susceptibility to oxidative rancidity and flavor deterioration during storage. High-viscosity slurries ($\geq$ 3.0 Pa $\times$ s) have a lower propensity to foam and therefore yield a powder with a lower level of air than low-viscosity slurries ($<$ 0.3 Pa $\times$ s). The viscosity of the cheese slurry is determined by its dry matter content and the characteristics of its ingredients, such as density of the different ingredients, levels of fat and protein, pH, and degree of ingredient hydration. The air content of the cheese powder is also influenced by the levels of formulation ingredients that tend to promote foaming of the slurry during preparation and drying (undenatured whey proteins) or to depress such foaming (fat and food-grade antifoaming agents).

Homogenization of the slurry is optional but is commonly practiced to ensure homogeneity of the slurry. The pressures applied (typically 15 and 5 MPa in the first and second stages, respectively) have a major effect on the viscosity of the slurry, with higher pressures generally imparting higher viscosity for a given level of dry matter.

Several spray-drying processes (e.g., single stage or two stage) and dryer configurations (e.g., tall-form, filtermat, silo-form) may be used. The design and operation of the dryer (e.g., atomizer type) and the pressure, direction of air flow, air inlet and outlet temperatures, and air humidity influence the physical characteristics (e.g., bulk density, wettability, and solubility) and the flavor characteristics of the cheese powder. The physical properties are important in applications that require reconstitution of the cheese powder, such as ready-made soups, sauces, and baby foods.

In all cases, the homogenized cheese slurry is pumped to the dryer, where it is atomized and dried, typically at an inlet air temperature of 180–200°C and an outlet air temperature of 85–90°C, depending on the type of dryer. The powder is then cooled from about 55°C to 20°C, separated from the air, and packaged. The moisture content of the dried powder is typically 30–40 g/kg and generally decreases with increasing outlet air temperature. However, an elevated outlet temperature, above 95°C, for example, may be detrimental to product quality owing to

- increased Maillard browning
- reduced product wettability and solubility (due to denaturation of ingredients)
- loss of volatile flavor compounds
- greater susceptibility to oiling-off and free-fat formation (free fat in the cheese powder leads to lumpiness, flow problems, and flavor deterioration)

Commercially, cheese powders are normally manufactured using two-stage drying systems. Filtermat (box) dryers are used frequently in the United States, whereas tall-form dryers with an integrated fluidized bed are widely used in Europe. While the operating conditions of these dryers influence the quality of the cheese powder, the tall-form dryer is generally considered to give better flavor retention, larger powder particles, and better powder flowability. Conventional single-stage tall-form dryers are rarely used because of the high outlet air temperature (e.g., > 95°C) necessary to achieve the low moisture content required. However, single-stage silo dryers (with a 70 m drying tower, compared with the 10 m tower used in tall-form dryers) may be used, as in the Birs Dehydration Process. In this process, the drying air is dehumidified but not heated. The main advantages over conventional two-stage drying are improved color stability and enhanced flavor retention, especially in mildly flavored products, the flavor of which is dominated by a few compounds (e.g., Cottage cheese).

Composition

The composition of cheese powders varies considerably, depending on the formulation ingredients; typical values are shown in Table 19–8.

Applications

Cheese powders are generally used as flavoring ingredients in a wide variety of foods, especially snack coatings (e.g., popcorn, nachos, tortilla shells), cheese sauces, soups, savory dressings, and savory biscuits. In snack foods, the powder is dusted after the snack has been sprayed lightly with vegetable oil. In cheese sauces, the level of cheese powder is typically 50–100 g/kg, depending on the flavor intensity of the cheese powder and the types and levels of other flavoring ingredients in the formulation. Generally, at these levels, the cheese powder has little influence on the rheological properties of the sauces, which are controlled mainly by the types and levels of starchy materials used (Guinee, O'Brien, & Rawle, 1994).

19.4.3 Enzyme-Modified Cheeses

Enzyme-modified cheeses (EMCs) are used principally as flavoring agents in industrially produced cheese products and ingredients, such as pasteurized processed cheese products,

Table 19–8 Composition of Cheese Powders with Different Levels of Cheese Solids

	Low (262 g/kg)	*Medium (530 g/kg)*	*High (950 g/kg)*
Dry matter (g/kg)	970	970	960
Protein (g/kg)	201	230	361
Fat (g/kg)	145	219	388
Lactose (g/kg)	264	123	3
Ash (g/kg)	89	104	104
pH	6.35	6.50	6.3

cheese substitutes or imitations, cheese powders, and ready-made meals. Natural cheeses have certain limitations as flavor ingredients:

- low flavor stability due to ongoing biochemical and microbiological changes during storage
- flavor inconsistency, due to changes, for example, in cheese composition
- insufficient flavor intensity (large quantities are required to impart a strong cheese flavor)
- high cost, due to the relatively long ripening time required for most cheese varieties
- the need to comminute cheese prior to its incorporation into foods and the fact that comminuted cheese is not suited to the bakery and snack food industries, which are large users of cheese

These limitations led to the development, in the 1960s, of enzyme-modified cheeses with flavors 5–20 times more intense than those of the corresponding natural cheeses.

The production of EMCs generally involves the following production steps (Figure 19–17):

- *Production of a cheese curd,* as for conventional cheese.
- *Formation of a paste* (typically 400–500 g/kg dry matter) by blending the curd with water and emulsifying salts. The addition of emulsifying salts assists in adjusting the pH of the paste to a value that is optimal for subsequent enzymatic reactions.
- *Pasteurization* of the cheese paste to inactivate the cheese microflora and enzymatic activities. Pasteurization reduces the risk of flavor inconsistencies due to variations in strain composition and populations of starter and nonstarter lactic acid bacteria and variations in enzyme (e.g., residual chymosin) activity in curd obtained from different suppliers.
- *Treatment of the pasteurized curd* with the desired cocktail of enzymes and perhaps starters to produce the required flavor profile and intensity. Added enzymes may include proteinases, peptidases, and lipases chosen based on knowledge of the enzymology and flavor profile of the cheese being simulated. Some commercial EMCs are prepared using a combination of added enzymes and bacterial culture systems. The advantage of using starter culture systems is that each starter cell is essentially a sack of enzymes that are known to contribute to balanced flavor production in any given cheese variety. Hence, it is generally easier to simulate a particular cheese flavor by using cultures than enzyme cocktails.
- *Incubation of the slurry* at 30–40°C for 24–72 hr. During this period, the added enzymes, or those released from starter cells during growth and/or autolysis, act on the casein and fat in the paste to produce the correct balance of peptides, amino acids, amines, aldehydes, ammonia, fatty acids, ketones, and alcohols.

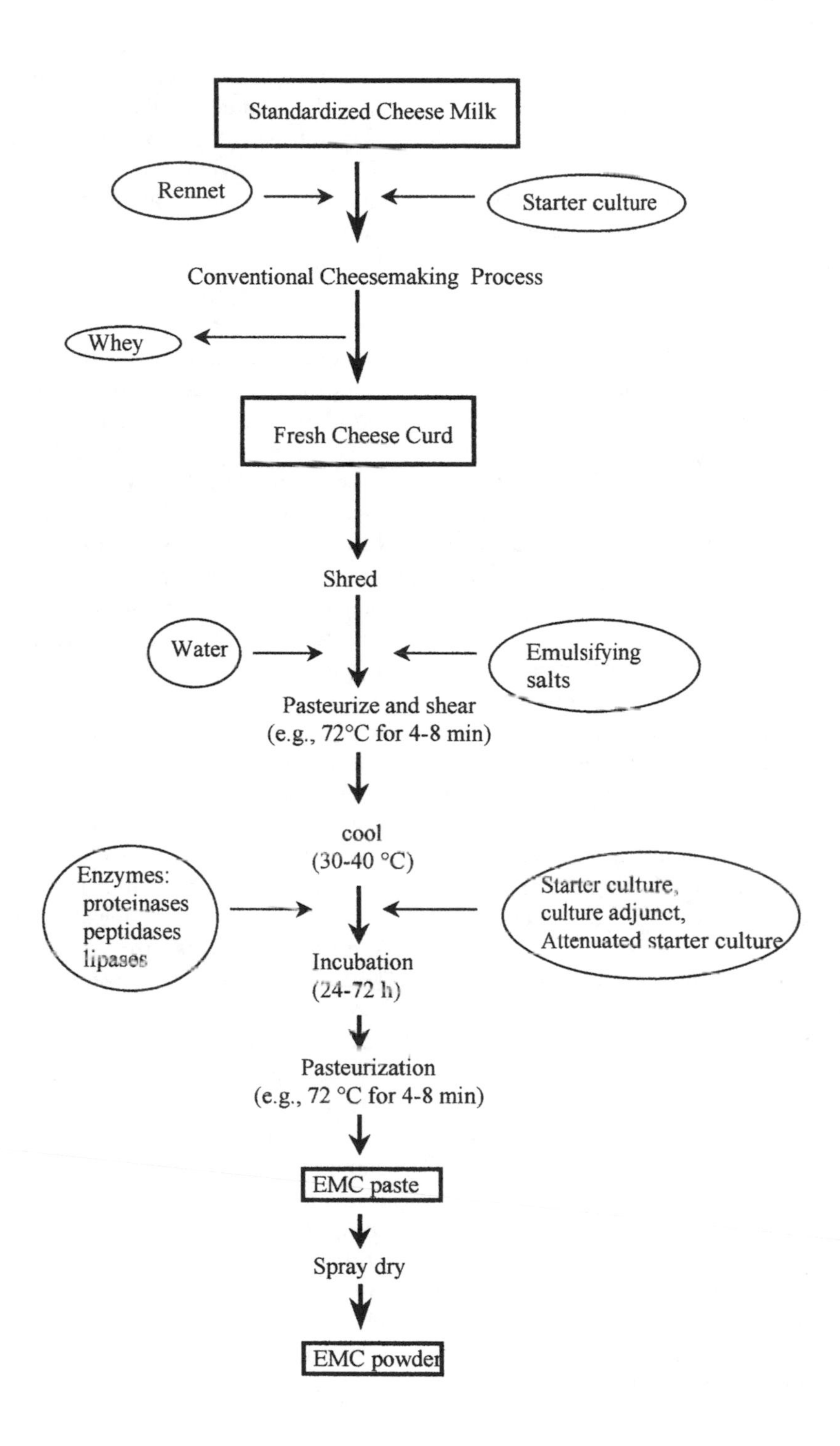

Figure 19–17 Production process for enzyme-modified cheese.

- *Pasteurization of the enzyme-treated paste* to inactivate enzymes and thereby preserve the flavor characteristics generated with minimum change during storage.
- *Homogenization* of the hot pasteurized paste to reduce the risk of phase separation during storage and ensure product homogeneity. The homogenized cheese paste, known as EMC paste, may be packaged and stored at refrigeration temperatures, usually in opaque materials to minimize the risk of oxidative rancidity and off-flavor development.
- *Drying*. The paste may be dried to produce an EMC powder, which has a longer shelf-life than the paste and is better suited for applications involving dry-blending with other ingredients.

EMC variants of many natural cheeses, such as Cheddar, Blue cheese, Romano, and Emmental, are commercially available (see Kilcawley, Wilkinson, & Fox, 1998). Little information has been published on the production of EMCs, as most of it is proprietary and guarded by the manufacturers. The development of EMCs requires elucidation of the flavor compounds and/or enzyme activities in the cheese being simulated, followed by testing of the various enzyme cocktails under various conditions (e.g., various levels of paste dry matter, pH, and incubation temperature) until the desired flavor characteristics are obtained. The composition, flavor-forming reactions, and flavor components of EMCs have been reviewed extensively (see Kilcawley et al., 1998).

19.5 CONCLUSION

Cheese is a highly versatile dairy ingredient that can be used directly in an array of culinary dishes, formulated food products, and ready-made meals. In these applications, added cheese performs a number of functions; it contributes to structure, texture, flavor, mouth-feel, melt properties, and nutrition. However, natural cheese is an expensive ingredient, may be functionally unstable, and may have somewhat variable functionality. Moreover, the consistency of natural cheese is unsuitable for some applications, such as those involving dry-blending with other ingredients in the manufacture of cake mixes, dried soups, and baby meals. Hence, cheese may be dried in the original or an extended form to yield cheese powder or be used as a substrate for the development of EMCs.

Little information is available on the quantities of natural cheese and cheese products consumed as ingredients in other foods. Sutherland (1991) reported that about 30% of Australian cheese is used by the food industry and food service sectors and predicted that the consumption of cheese as an ingredient of other foods will grow rapidly because of the greater demand for prepared meals. The Australian trend probably reflects trends in the United States and Europe. In addition, cheese is used as an ingredient in the home. Indeed, it is estimated that about 25% of total cheese consumed is incorporated into various homemade dishes and that, in developed countries, about 50% of cheese is used as an ingredient by home cooks and by the food industry. It is envisaged that the industrial use of cheese as an ingredient will become a major driving force in increasing the per capita consumption of cheese in developed countries, where the demand for convenience and prepared foods is growing.

In cheese-containing prepared foods, the cheese is expected to exhibit functional characteristics such as flowability, mouth-feel, flavor, and stretchability. As the production and consumption of "fast foods" and ready-made meals grow, the demand for greater functionality and for customized cheeses is increasing. Functionality is a major factor contributing to the increase in cheese consumption, which is clearly reflected by the recent dramatic growth in the consumption of pizza cheese, especially in the United States, where current annual production is about 1 million tonnes (Lavoie & Mertz, 1995; Sørensen, 1997). Production of pizza cheese is also increasing in Europe (100,000 tonnes per annum) and New Zealand and Australia (100,000 tonnes per annum), where consumption is following the number of outlets of the major pizzeria chains. Moreover, it is becoming apparent that the taste and functionality expected of pizza cheese show regional variations. Hence, it is conceivable that as pizzeria chains expand into new markets, there will be a demand for cheeses with different flavors and functionality than those currently required in order to meet local taste preferences.

REFERENCES

Guinee, T.P., Mulholland, E.O., Mullins, C., & Corcoran, M.O. (1997). Functionality of low moisture Mozzarella cheese during ripening. In T.M. Cogan, P.F. Fox, & R.P. Ross (Eds.), *Proceedings of the Fifth Cheese Symposium*. Dublin: Teagasc.

Guinee, T.P., O'Brien, N., & Rawle, D.F. (1994). The viscosity of cheese sauces with different starch systems and cheese powders. *Journal of the Society of Dairy Technology, 47*, 132–138.

Kilcawley, K.N., Wilkinson, M.G., & Fox. P.F. (1998). Enzyme-modified cheese [A review]. *International Dairy Journal, 8*, 1–10.

Kindstedt, P.S. (1993). Effect of manufacturing factors, composition, and proteolysis on the functional characteristics of Mozzarella cheese. *Critical Reviews in Food Science and Nutrition, 32*, 167–187.

Kindstedt, P.S. (1995). Factors affecting the functional characteristics of unmelted and melted Mozzarella cheese. In E.L. Malin & M.H. Tunick (Eds.), *Chemistry of structure-function relationships in cheese.* New York: Plenum Press.

Kindstedt, P.S., & Guo, M.R. (1997). Recent developments in the science and technology of pizza cheese. *Australian Journal of Dairy Technology, 52*, 41–43.

Kosikowski, F.V., & Mistry, V.V. (1997). Drying and freezing of cheese. In *Cheese and fermented milk foods: Vol. 1. Origins and principles*. Westport, CT: F.V. Kosikowski, LLC.

Lavoie, C., & Mertz, T. (1995). *The U.S. market for pizza* [Report]. Commack, NY: Business Trend Analysts, Inc.

Ridgway, J. (1986). *The complete cheese book.* London: Judy Piatkus Ltd.

Sørensen, H.H. (1997). *The world market for cheese* [Bulletin No. 326]. Brussels: International Dairy Federation.

Sutherland, B.J. (1991). New cheese products as food ingredients. *Food Research Quarterly, 51*(1, 2), 114–119.

Chapter 20

Pathogens and Food-Poisoning Bacteria in Cheese

20.1 INTRODUCTION

Milk is a highly nutritious medium of almost neutral pH and, therefore, many bacteria, including spoilage and pathogenic bacteria, can grow in it. Numerous outbreaks of food poisoning have been traced to milk. Although cheese is equally nutritious, it has been shown to be responsible for relatively few food-poisoning outbreaks. These are summarized in Table 20–1. There were 21 confirmed outbreaks of food poisoning in Western Europe during the years 1970–1997, 7 outbreaks in the United States from 1948 to 1997, and 4 outbreaks in Canada from 1970 to 1997 due to consumption of cheese ("Food Safety and Cheese," 1998; Johnson, Nelson, & Johnson, 1990a, 1990b, 1990c). During the period 1970–1997, an estimated 235,000,000 tonnes of cheese were produced in Western Europe, the United States, and Canada. Such data imply that cheese is a relatively safe food product, but 28% of the outbreaks involved cheeses made from raw milk.

Several organisms were involved in cheese-related outbreaks of food poisoning, but *Salmonella* spp., *Staphylococcus aureus*, and *Listeria monocytogenes* were the most common (Table 20–1). The species of organism involved in the European outbreaks were more diverse than those in the United States, which may reflect the greater diversity of cheese varieties made in Europe compared with the United States and Canada. The primary reasons for these cheese-related food-poisoning outbreaks were poor starter activity (due to phage, antibiotic residues in the milk, etc.), poor hygiene in the plant, gross environmental contamination, and faulty pasteurization.

In recent years, the most important pathogens found in cheese have been *L. monocytogenes* and enteropathogenic *Escherichia coli*. The former is the more important since the outbreaks involved several deaths. The outbreaks due to *E. coli* O157 involved cheese made from raw milk, and only a few cases were involved. Nevertheless, a major outbreak due to *E. coli* O157 has the potential to be very serious. The major cheeses involved were soft surface-ripened cheese and cheese with a low acidity (e.g., Mexican-style cheese). There is no indication that *Mycobacterium bovis* and *Brucella abortus* and the so-called emerging pathogens, *Campylobacter jejunii, Yersinia entercolitica*, and *Aeromonas hydrophilia* grow during manufacture of cheese. *M. bovis*, which causes tuberculosis in cows and sometimes in humans, and *B. abortus*, which causes abortion in cows and undulant fever in humans, are sometimes excreted in the milk of infected cows. These microorganisms were probably important causes of human disease in the past but are not important now, since TB and brucellosis are controlled in almost all dairy herds in developed countries. An outbreak of food poisoning in Canada involving

Table 20–1 Food-Poisoning Outbreaks Associated with Cheese in the United States (1948–1997), Canada (1970–1997), and Europe (1970–1997)

Country of Origin	*Variety of Cheese*	*Year*	*No. of Cases*	*No. of Deaths*	*Causative Organism(s)*	*Reference*
US	Colby	1956	200		*Staph. aureus*	Allen and Stovall, 1960
US	Cheddar	1958	200		*Staph. aureus*	Hendricks et al., 1959
US	Cheddar, Kuninost, and Monterey	1965			*Staph. aureus*	Zehren and Zehren, 1968
US	Cheddar	1976	>339		*Salmonella*	Fontaine et al., 1980
US	Homemade	1983	16		*Sc. zooepidemicus*	Sharpe, 1987
US	Mexican style	1985	142	48	*L. monocytogenes*	Linnan et al., 1988
US	Mozzarella	1989	164	0	*Salmonella javiana; Salmonella oranienberg*	Hedberg et al., 1992
Mexico	Mexican style[a]	1975	3		*Brucella*	Eckman, 1975
Canada	Emmental	1977	12		*Staph. aureus*	Todd et al., 1979, 1981b
Canada	Cheese curd	1980	62		*Staph. aureus*	Todd et al., 1981a
Canada	Cheddar, other types	1982			*Salmonella*	Sharpe, 1987
Canada	Cheddar	1984	>2700	1	*Salmonella typhimurium*	D'Aoust et al., 1985; D'Aoust, 1994
					Salmonella	Sharpe, 1987
Finland	Farmhouse	1983	35		*E. coli* O124:B17	Marier et al., 1973
France	Camembert, Brie, Coulommiers[a]	1971	387	0		
France	Brie[b]	1983	>3000		*E. coli* O27:H20	MacDonald et al., 1985
France	Goat milk cheese[c]	1993	273	1	*Salmonella paratyphi B*	Desenclos et al., 1996
France	Brie de Meaux[c]	1995	20	4	*Listeria monocytogenes*	Goulet et al., 1995
France	Fromage frais[c]	1992	NR	1	Veratoxic *E. coli*	Public Health Laboratory Service, London, 1994
France, Switzerland	Soft cheese	1974	77		*Cl. botulinum*	Kauf et al., 1974; Sebald et al., 1974
Greece	Homemade, unripened	1983	23	0	*Brucella*	Sharpe, 1987
Ireland	Soft cheese[c]	1989	42	0	*Salmonella dublin*	Maguire et al., 1992

continues

Table 20–1 continued

Country of Origin	*Variety of Cheese*	*Year*	*No. of Cases*	*No. of Deaths*	*Causative Organism(s)*	*Reference*
Italy	Mascarpone	1996	8	1	*C. botulinum*	Aureli et al., 1996
Malta	Soft cheese	1995	135	1	*Br. melitensis*	Public Health Laboratory Service, London, 1995
Scandinavia	Brie	1982	>50		*Bacillus; Shigella sonnei*	Sharpe, 1987
Switzerland	Vacherin Mont d'Or[c]	1983/87	122	34	*L. monocytogenes*	Bille, 1990; Malinverni et al., 1985
Switzerland, France	Vacherin Mont d'Or[c]	1985	>40	0	*Salmonella typhimurium*	Sadik et al., 1986
Switzerland, France	Doubs[c]	1995	25	5	*Salmonella dublin*	Valliant et al., 1996
UK	Cheddar	1983	2		*Staph. aureus*	Sharpe, 1987
UK	Cheddar	1994	>84	0	*Salmonella gold-coast*	Public Health Laboratory Service, London, 1997a
UK	Sheep milk cheese[c]	1984	>13	0	*Staph. aureus*	Bone et al., 1989
UK	Farmhouse cheese	1992	>20	0	*E. coli* O157	Public Health Laboratory Service, Edinburgh, 1994
UK	Lancashire[c]	1997	2	0	*E. coli* O157	Public Health Laboratory Service, London, 1997b

[a] Consumed in the United States.
[b] Consumed in the United States, Sweden, and The Netherlands.
[c] These cheeses were made from raw milk.

Cheddar cheese made from raw milk was traced to a farm where one cow was shedding about 200 cfu of salmonella per milliliter of milk.

The infective dose of pathogens can be low. For instance, *S. heidelberg* has an infective dose of 100–500 cells, and the infective dose for *E. coli* O157:H7 is thought to be 10 cells. In contrast, the infective dose for *S. aureus* is high because the actual cause of the food poisoning is not the organism itself but a number of closely related, heat-stable protein toxins (able to withstand 100°C for > 30 min) produced by it. These toxins are called staphylococcal enterotoxin A (SEA), SEB, SEC, etc., and some of them (e.g., SEC) are further subdivided into SEC1, SEC2, SEC3, etc. The amino acid sequences of all the enterotoxins are very similar. Some of them (e.g., SEA and SEE) are produced during exponential growth, while others (e.g., SEB and SEC) are produced mainly during the stationary phase of growth. Those produced in the exponential phase are the more common causes of staphylococcal food poisoning. The minimum number of staphylococci and the level of toxin required to cause food poisoning are thought to be 10^5 cells/g and 1 ng/g of food ingested. The strains of *S. aureus* present in raw milk are primarily those that cause mastitis, and about 20% of these strains produce enterotoxin.

20.2 PATHOGENS IN RAW MILK

Recent surveys of raw milk quality show that the incidence of pathogens in raw milk is low. For example, about 5% of individual Irish, English, and French farm milks are contaminated with *Listeria* and less than 1% with *Salmonella* (Desmasures, Bazin, & Guéguen, 1977; Ministry of Agriculture, Fisheries and Food, 1997; O'Donnell, 1995; Rea, Cogan, & Tobin, 1992). The microbiological quality of farm milks in Normandy (France), around the area where raw milk Camembert is made, is generally very good; 83% of 69 samples had total bacterial counts below 20,000 cfu/ml, and the samples had an average somatic cell count of 176,000/ml (Table 20–2). The average number of coliforms, enterococci, and *S. aureus* was 77, 79, and 350 per milliliter, respectively, in summer milk. Winter-produced milk was also good, with counts of 57, 74, and 450 per milliliter for the above bacterial groups, respectively. The incidence of *Yersinia entercolitica* was relatively high and that of *Campylobacter* low. All these data suggest that hygiene and cooling of milk were effective, although 14 of the 43 milks examined did not meet the European Union criterion for *S. aureus* in milk destined for cheesemaking (< 500/ml) (Table 20–3).

Table 20–2 Microbiological Quality (cfu/ml) of Milk Produced in Normandy, France

	n	*Winter*	*n*	*Spring/Summer*
Total count	39	71,000 ± 27,000	30	86,000 ± 21,000
Enterococci	37	74 ± 150	25	79 ± 400
Coliforms	29	57 ± 2,400	19	77 ± 5,000
S. aureus	25	450 ± 1,700	18	350 ± 280
L. monocytogenes	39	4 positive samples	30	No positive samples
Salmonella	39	1 positive sample	30	1 positive sample
Y. entercolitica[a]	39	19 positive samples	30	6 positive samples
Campylobacter	39	1 positive sample	30	No positive samples

[a] Only 1 of 61 isolates was a potential pathogen.

Table 20–3 European Union Guidelines (Standards) for Milk for Cheesemaking and Cheese

	Soft Cheese from Raw Milk	*Soft Cheese from Pasteurized Milk*	*Raw Milk*
Listeria	Absent in 25 g*	Absent in 25 g	
m	5	5	
c	0	0	
Salmonella	Absent in 25g	Absent in 25 g	
m	5	5	
c	0	0	
S. aureus			
m†	1,000/g	100/g	500/ml‡
M	10,000/g	1000/g	2,000/ml
n	5	5	5
c	2	2	2
Coliforms			
m		10,000/g	
M		100,000/g	
n		5	
c		2	
E. coli			
m	10,000/g	100/g	
M	100,000/g	1,000/g	
n	5	5	
c	2	2	
Total plate count			≤100,000/ml§
Somatic cells			≤400,000/ml

Key: n = number of samples. m = threshold value; the result is satisfactory if the number of bacteria in all sample units does not exceed m. M = maximum value; the result is unsatisfactory if the number of bacteria in ≥ 1 sample units exceeds M. c = number of samples for which the counts may lie between m and M; the sample is considered acceptable if the numbers are ≤ m in the other sample units.

* The 25 g sample should consist of five 5 g portions taken from different parts of the same product.

† For products labeled "made from raw milk."

‡ Geometric mean of 2 results, with ≥ 2 samples per unit.

§ Geometric mean of 3 results, with ≥ 1 sample per unit.

20.3 PATHOGENS IN CHEESE

Soft cheeses contain high levels of moisture, and the pH, particularly at the surface, increases during ripening. In addition, some of them are made from raw milk, and smear cheeses are frequently handled during ripening. Consequently, soft cheeses are more prone to microbial growth than hard or semi-hard varieties. A recent survey of the microbiological quality of soft cheeses in the United Kingdom market is summarized in Table 20–4. Cheeses from several countries, in-

Table 20–4 Microbiological Quality of Soft Cheese Made from Raw and Pasteurized Milk

	*ND**	*<10†*	*10 to <10²*	*10² to <10³*	*10³ to <10⁴*	*10⁴ to <10⁵*	*>10⁵*
Raw milk cheeses (72)							
Coliforms	32	8	5	3	9	2	13
E. coli	48	7	10	4	1	0	2
S. aureus	70	0	1	1	0	0	0
L. monocytogenes	71	1	0	0	0	0	0
Other *Listeria* spp.	68	3	0	0	1	0	0
Pasteurized Milk Cheeses (405)							
Coliforms	284	13	38	19	23	9	19
E. coli	383	7	9	4	1	0	2
S. aureus	401	0	2	0	1	1	0
L. monocytogenes	403	2	0	0	0	0	0
Other *Listeria* spp.	400	5	0	0	0	0	0
Unlabeled cheeses (960)							
Coliforms	611	37	71	54	96	30	61
E. coli	891	22	25	9	6	2	5
S. aureus	958	0	0	1	1	0	0
L. monocytogenes	947	13	0	0	0	0	0
Other *Listeria* spp.	926	30	0	1	1	0	2

* Not detected.

†Total per gram.

cluding Scandinavia, France, Germany, Greece, Cyprus, Italy, and the United Kingdom, were analyzed. *Salmonella* were not found in any of the 1,437 cheeses examined. Overall, the micro biological quality of both raw and pasteurized milk cheese was quite good. One (1.4%) of the raw milk cheeses failed to meet the European Union criteria (Table 20–3) for *Listeria* in raw milk cheese, and 2 (3%) failed to meet the criteria for *E. coli*. There are no standards for coliform numbers in raw milk cheese, but 28 of the 405 (7%) samples of pasteurized milk cheese examined failed the coliform standard, and 7 (2%) failed the *Listeria* standard (Table 20–3). Coliforms are killed by pasteurization, and the high numbers in some of the pasteurized milk cheeses probably reflect the spread of contamination during the smearing of surface-ripened cheese (see Chapter 10).

20.4 LISTERIOSIS

Listeriosis is caused by *L. monocytogenes*, especially serotypes 1/2a and 4b, and it affects mainly pregnant women, immunocompromised people (e.g., those who are HIV positive or are recovering from chemotherapy after treatment for cancer), and the elderly. Prominent symptoms include vomiting and diarrhea, which may lead to meningitis and bacteremia. Infection of the blood stream, the central nervous system, the fetus in utero, and infants by mothers who show no obvious signs of infection during birth are the most common forms of listeriosis.

Two relatively recent major outbreaks of listerosis have been traced to cheese, one in the United States, involving Mexican-style cheese, and one in Switzerland, involving Vacherin Mont d'Or cheese, which is a soft cheese made from raw milk. Both outbreaks involved fatalities, 48 in the United States (20 fetuses, 10 infants, and 18 nonpregnant adults) and 34 in Switzerland (Table 20–1). Poor hygiene was the major factor in both outbreaks, and improper pasteurization was also implicated in the case of the Mexican-style cheese. However, the fact that both cheeses also had low salt levels and that the Mexican-style product is a low-acid cheese, made without the deliberate addition of starter cultures, while Vacherin is a surface-ripened variety, in which the pH increases during ripening, contributed to the outbreaks.

L. monocytogenes is a Gram-positive rod that can grow at temperatures from –0.4°C to 45°C, at pH values of 4.4 to 9.4, and in the presence of 10% NaCl. Generation times at 0°C and 1°C are 131 and 62 hr, respectively, and the lag times are 33 and 3 days, respectively. Therefore, holding cheese as close to 0°C as possible will help to prevent the growth of listeria. The optimum pH and temperature are 7.0 and 37°C, respectively. These properties make this microorganism particularly problematical. It is an ubiquitous organism and is found in soil, water, silage, and so on. Sometimes the organism is found inside phagocytes (neutrophils and macrophages) in milk, and this location was thought to protect the cells from inactivation during pasteurization. The general consensus now, however, is that the organism is inactivated by pasteurization whether the cells are inside phagocytes or not.

20.5 PATHOGENIC *ESCHERICHIA COLI*

The normal habitat of *E. coli* is animal (and human) feces whence it can contaminate raw milk, particularly if the animals have been lying in their own dung and the udders have not been properly washed before milking. Most strains of *E. coli* are harmless, commensal organisms, but some are pathogenic. They are differentiated from each other on the basis of the serological detection of somatic (O), flagellae (H), and capsular (K) antigens. To date, 174 O antigens, 56 H antigens, and 80 K antigens have been detected. Pathogenic strains are generally subdivided into enteropathogenic (EPEC), enterotoxigenic (ETEC), enteroinvasive (EIEC), and enterohemorrhagic (EHEC), depending on how they cause infection.

Five outbreaks of foodborne disease due to pathogenic *E. coli* have been traced to the consumption of soft cheeses. These involved ETEC O27:H20, EIEC O124:B17, and EHEC O157 (Table 20–1). The outbreak due to *E. coli* O124:B17 occurred in the United States but involved French Camembert, Brie, and Coulommiers cheese made in the same plant over a 2-day period. Despite this, the cheese was widely distributed, because outbreaks occurred in 14 states from Connecticut in the east to California in the west; 389 people developed food poisoning, but no deaths were reported. In the outbreak involving *E. coli* O27:H20, the cheese involved was from two different lots made 46 days apart, suggesting that contamination was intermittent. It is not clear if the cheeses in these outbreaks were made from raw or pasteurized milk. *E. coli* O157:H7 is an emerging food pathogen that has been associated with several severe food-poisoning outbreaks involving meat products and also a small outbreak involving a raw milk cheese (Table 20–1).

The symptoms of the *E. coli* O124 infection included diarrhea, fever, and nausea; cramps, chills, vomiting, aches, and headaches were less common. The median time before onset was 18 hr (range, 2–48 hr) and the median duration of the illness was 2 days (range, less than 1 day to 15 days). The symptoms of the *E. coli* O27:H20 infection were fairly similar; the average time before onset was 44 hr (range, 6–144 hr), and the symptoms lasted for 4.4 days (range, 1–14 days). The typical symptoms of O157:H7 food poisoning include production of stools containing blood and mucus, as a result of hemorrhagic colitis, and acute renal failure, particularly in children. Generally, a long incubation period (up to 9 days, average 4 days) and duration of illness (up to 9 days, average 4 days) elapse before the onset of symptoms and the occurrence of deaths. In contrast, in food poisoning due to other pathogenic strains of *E. coli,* the symptoms occur within 2 days of ingestion of the contaminated food, and no deaths have been reported. Oral challenges using human volunteers suggest that 10^5–10^{10} cells of EPEC, 10^8–10^{10} cells of ETEC, and 10^8 cells of EIEC are required to cause diarrhea. In contrast, only 10–100 cells of EHEC are required to cause illness.

The precise mechanism by which *E. coli* O157:H7 causes disease has not been fully elucidated, but isolates produce 1 or 2 toxins that are cytotoxic to Vero cells, an African green monkey kidney cell line. For this reason, *E. coli* O157:H7 is also called verotoxigenic *E. coli* (VTEC). One of these toxins is structurally and immunologically indistinguishable from the shiga toxin of *Shigella dysenteriae*. How *E. coli* O157:H7 acquired the shiga toxin is not clear.

There is relatively little information on the incidence in *E. coli* in raw milk or cheese, but the evidence suggests that it is low. A relatively recent survey of milks throughout the production season (Rea, Cogan, & Tobin, 1992) showed that more than 60–100% of samples contained fewer than 10 *E. coli* per ml, depending on the date of sampling (Figure 20–1). Very few milks contained more than 100 *E. coli* per mililiter. None of 568 raw milks examined in the United Kingdom contained *E. coli* O157:H7 (Neaves, Deacon, & Bell, 1994). In a limited survey, *E. coli* O157:H7 was not detected in 19 soft and semi-soft American-made cheese, while other strains of *E. coli* were found in 11 samples (Ansay & Casper, 1997). In a Spanish study of 221 raw milk cheeses and 75 pasteurized milk cheeses, 3 cheeses (1.4%), each of which had been produced from raw milk, showed the presence of toxigenic *E. coli* (Quinto & Cepeda, 1997).

E. coli O157:H7 has a minimum growth temperature of 8°C, an optimum of 37°C, a maximum of 44°C, and it does not withstand pasteurization. *E. coli* strains normally do not tolerate low pH values, but *E. coli* O157:H7 is an exception and can grow at pH 4.5 in media adjusted with HCl—but not if the pH is adjusted with lactic acid. The organism does not grow in cheese at pH values at or below 5.4.

20.6 GROWTH OF PATHOGENS DURING CHEESE MANUFACTURE

The major factors responsible for the control of microbial growth in cheese have been described in Chapter 10. These factors are also involved in controlling the growth of pathogens. The main reason for the low incidence of food-poisoning outbreaks caused by cheese is that most milk for cheesemaking is pasteurized, which kills all pathogens in the raw milk. This is probably the most important factor in controlling the growth of potential pathogens in cheese. There is some evidence that phagocytosis of *L. monocytogenes* by somatic cells present in raw milk increases the heat resistance of *L. monocytogenes*, but the consensus is that this organism is still killed by normal pasteurization. Significant amounts of cheese are made from raw milk in countries bordering the Mediterranean (≈ 15% of all French cheese is made from raw milk), and in some cases no starter is used (e.g., artisanal production of the Spanish cheeses Manchego and Cabrales), which implies that cheese made from raw milk is also safe, since few outbreaks have been attributed to such products. Pasteurization is normally carried out at 72°C for 15 s, but lower heat treatments (e.g., 65°C for 16–18 s) will destroy all the likely pathogenic microorganisms commonly found in milk except, perhaps, *L. monocytogenes*. Subpasteurization heat treatments are also used in some countries (e.g., Canada) for milk for cheesemaking. The reason for this practice is that stronger flavored cheeses are produced from milk that has not been pasteurized at all or has been subpasteurized than from fully pasteurized milk (see Chapter 15).

M. paratuberculosis causes paratuberculosis (Johne's disease) in cattle, and there is some evidence that this organism also may be involved in the etiology of Crohn's disease in humans. Because of this evidence, milk has been suggested as a possible vehicle for the transmission of the organism from cattle to man. Whether the organism withstands normal pasteurization is important. There is conflicting evidence on this point

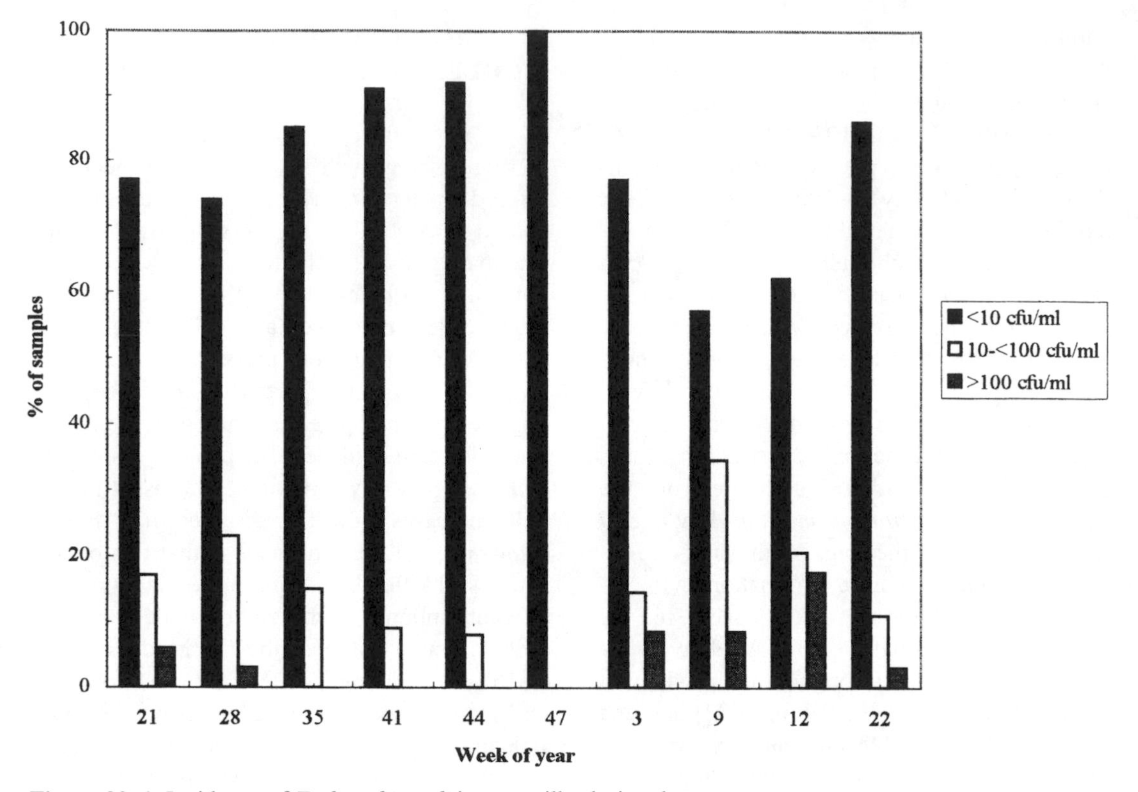

Figure 20–1 Incidence of *Escherichia coli* in raw milks during the year.

in the literature, and some cheese manufacturers have recently increased the temperature of pasteurization by a few degrees and/or the time of pasteurization by a few seconds to ensure that the milk is free of this organism. These more severe conditions may damage the rennetability of the milk.

The presence and survival of pathogens in cheese is influenced by several factors, including

- species
- initial numbers
- physiological condition of the microorganism
- rate of acid production by the starter and consequent decrease in pH
- tolerance of the pathogen to acid and salt
- tolerance to the cooking temperature to which the cheese curd is subjected
- postmanufacturing contamination
- biochemical changes that occur in the cheese during ripening
- ripening and storage temperature of the cheese
- composition of the cheese

The time-temperature profile during cheesemaking, the initial low rate of decrease in pH, and the cooking temperature, which can vary from 33°C for Camembert (essentially no cooking) to 36°C for Dutch-type cheese, 38°C for Cheddar, and 54°C for many Swiss and Italian cheeses, play major roles in promoting the growth of pathogens in cheese during manufacture. A temperature of 25–40°C is conducive to the growth of pathogens, if they are present. The cooking temperature (54°C) and the length of time for which Swiss-type cheese curds are held at this temperature (≈ 60 min) will kill most, if not all, pathogens. In soft cheese, a cooking temperature of 35°C or lower is used, which is ideal for the growth of pathogens. Obviously, if the starter is active, the pH will decrease quickly and the growth of the pathogens will be retarded. The reverse also occurs. That is, if growth of the starter is slow due to phage contamination and/or antibiotic residues in the milk, considerable growth of pathogens can occur. Therefore, a fast acid-producing starter is one of the best means of controlling the growth of pathogens in cheese.

In studies on the growth of pathogens in cheese, the milk is normally inoculated with the pathogens, pregrown under favorable conditions. The numbers found in naturally contaminated milk would be much lower and probably in a more stressed physiological state than those grown in laboratory media. In such a stressed state, bacteria are probably less resistant to the effects of pH and temperature, but experimental evidence for this hypothesis is not available, primarily because of the difficulty of obtaining naturally contaminated milk. Many raw milks are, in fact, free of pathogenic microorganisms.

Data for the growth of *E. coli* O157:H7, *L. monocytogenes*, and *S. aureus* in Cheddar cheese curd during manufacture are shown in Figure 20–2. *E. coli* and *S. aureus* multiplied during manufacture but *L. monocytogenes* did not. When interpreting these data, one must remember that the moisture content in the curd decreases at each stage of manufacture, resulting in an apparent increase in bacterial numbers due to the concentration effect. Based on this phenomenon, it appears that a small decrease in the number of listeria occurred during manufacture. Considerable growth of *E. coli* occurred between the beginning of manufacture and cutting the coagulum, when little acid production would have occurred. No data were reported for *S. aureus* between these two stages, but growth of this organism is also likely to occur. These data refer to Cheddar cheese curds, in which acid production is relatively rapid (acid production will also have a major rate-limiting effect on growth). Growth of pathogens in the curd of most other varieties would probably be greater than in Cheddar because of slower acid production.

The growth of most pathogens will slow down and eventually cease owing to the decrease in pH (as significant amounts of lactic acid are produced) and also to the increase in temperature during cooking.

20.7 GROWTH OF PATHOGENS IN CHEESE DURING RIPENING

What happens during ripening of a cheese depends on the variety. Each cheese is a unique microbial ecosystem and should probably be considered individually. Nevertheless, broad generalizations can be made. Hard and semi-hard cheeses, if made properly, are safe, since almost all pathogens die off during ripening; in contrast, significant growth of pathogens can occur in soft cheese.

20.7.1 Hard and Semi-Hard Cheeses

Coliform bacteria die off at a rate of 0.3 log per week in Cheddar and 0.7 log per week in Gouda. The fate of several pathogens in Emmental and Cheddar cheese is shown in Figures 20–3 and 20–4. Both Emmental and Cheddar cheeses are hard cheeses with similar pH values (≈ 5.2) immediately after manufacture. None of the pathogens, except *S. aureus* at very low levels, was detected in the Emmental cheese within 1 day of manufacture. This is most likely a result of the high cook temperature (≈ 52°C) used in the manufacture of this cheese. In Cheddar cheese, *S. aureus, E. faecalis, E. coli*, and a *Salmonella* species all decreased during ripening, and the Gram-negative bacteria decreased at a faster rate than the Gram-positive organisms (Figure 20–4). One of the problems with *S. aureus* is that, even though the numbers of the organism decrease significantly during ripening, sufficiently high numbers may have been present during the early stages of ripening to produce the small amounts of enterotoxin necessary to cause food poisoning. Therefore, it is possible that a cheese with a low level of *S. aureus* may contain a high level of enterotoxin. Enterotoxins are proteins, and whether they are hydrolyzed by

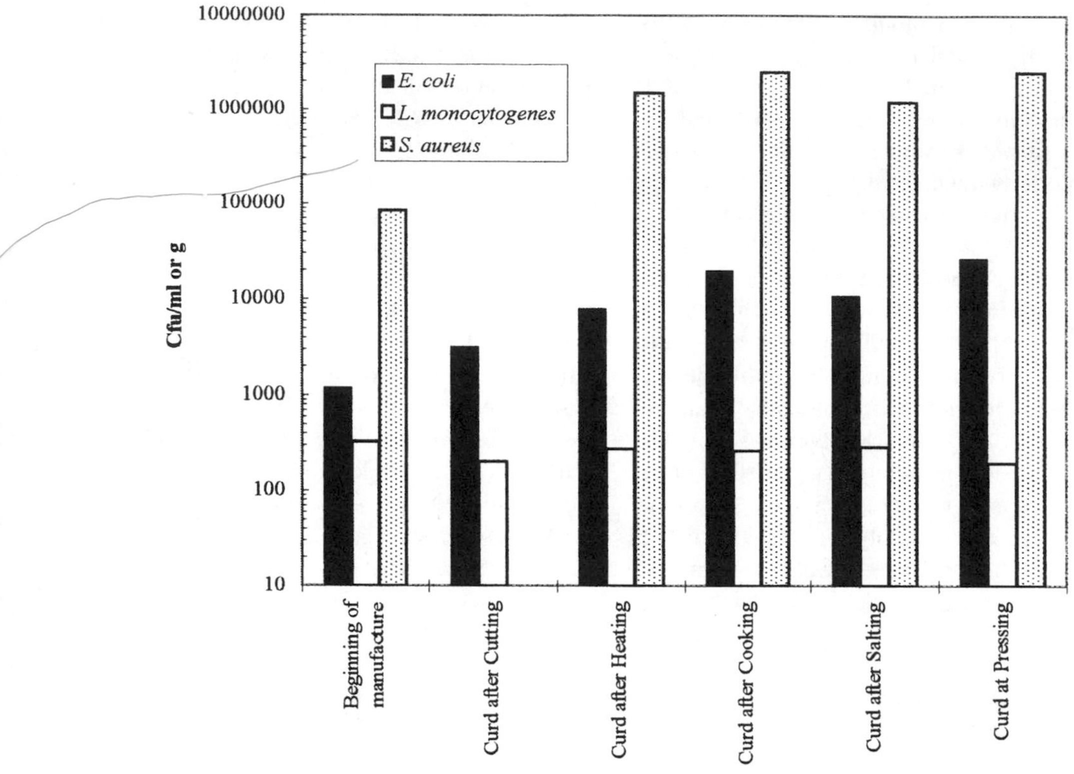

Figure 20–2 Growth of *E. coli* O157, *L. monocytogenes*, and *S. aureus* in Cheddar cheese curd during manufacture.

chymosin or bacterial proteinases during cheese ripening does not appear to have been studied. In the United States, storage of cheese at 2°C for 60 days may be used instead of pasteurization.

The number of *L. monocytogenes* in Cheddar cheese also decreases during ripening (Figure 20–5), but some variations in individual trials occur. There was also some variation in the rate of die-off of *L. monocytogenes* in cheese ripened at 13°C and in cheese ripened at 6°C, but generally the differences were small. *E. coli* O157:H7 died off relatively rapidly (2 log cycles in 25 days) in Cheddar cheese during ripening at 6.5°C (Figure 20–6).

The effect of the pH of the cheese is also critical. Data for *Salmonella* in Cheddar cheese are shown in Figure 20–7. At pH 5.03 and 5.23, they died off quite quickly, but at pH 5.7 they did not die at all. A pH of 5.23 is typical of a well-made Cheddar, and a pH of 5.7 indicates poor starter activity, either as a result of phage contamination or antibiotic residues in the milk.

In Tilsit, a semi-hard cheese, the numbers of all the pathogens tested decreased during ripening, except *L. monocytogenes*, which remained fairly constant during ripening for 30 days, after which a gradual decrease of about 1 log cycle occurred over the following 2 months (Figure 20–3). The stability of *L. monocytogenes* in this cheese was attributed to the relatively low cooking temperature and short cooking time (42°C for 15 min), which were bacteriostatic rather than bactericidal. pH may also be important. The pH of the Tilsit cheese increased during ripening from 5.2 at day 1 to 5.8 at day 90. Commercial samples normally have a pH of about 6.2 at 90 days. *L. monocytogenes* can grow over a wide range of temperature, from –1°C to 45°C, and

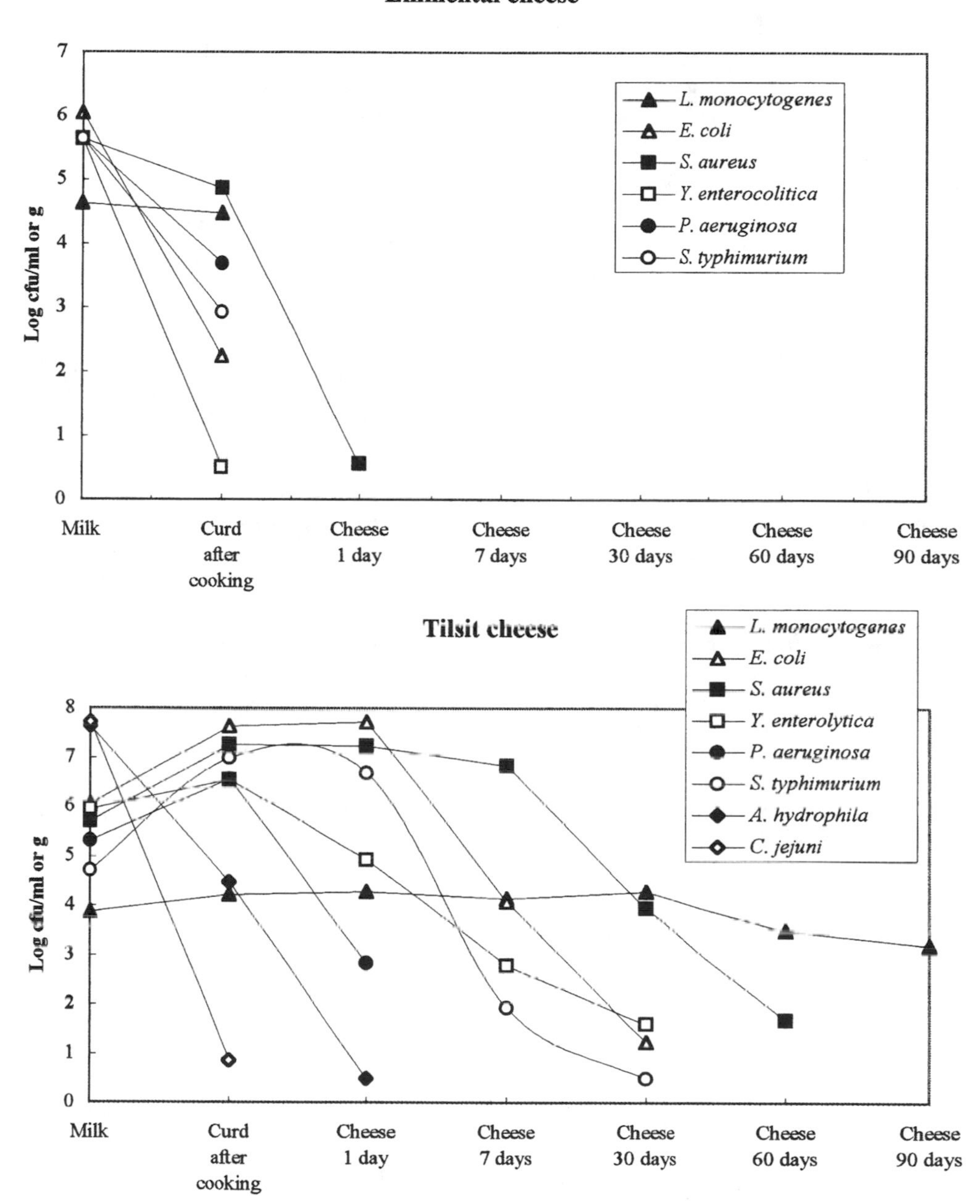

Figure 20–3 Growth of *L. monocytogenes, E. coli, S. aureus, Y. enterocolitica, P. aeruginosa, S. typhimurium, Aeromonas hydrophila*, and *Campylobacter jejuni* in Emmental and Tilsit cheese during ripening.

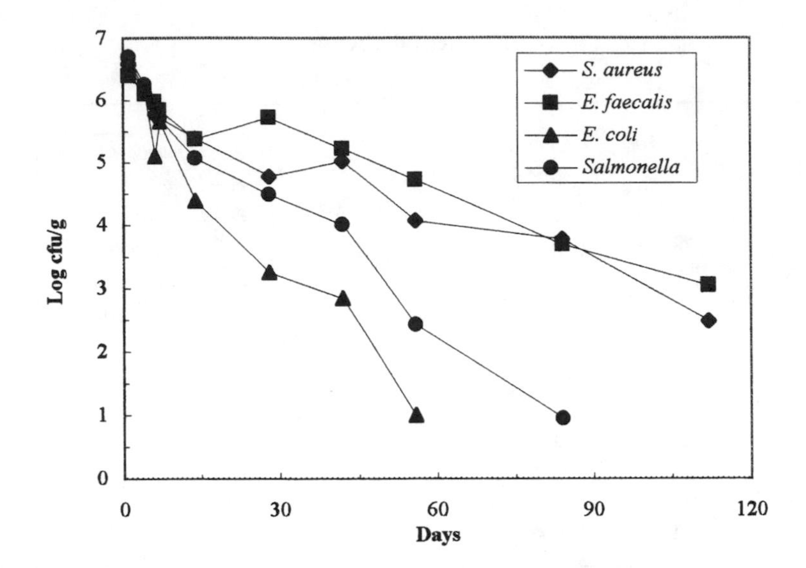

Figure 20–4 Decrease in numbers of *S. aureus, E. faecalis, E. coli*, and *Salmonella* in Cheddar cheese during ripening at 12°C.

this ability may be important for its survival in Tilsit cheese.

20.7.2 Soft Cheeses

The situation in soft mold- and smear-ripened varieties like Camembert, Brie, and Limburger is quite different, and many pathogens can grow readily in such cheeses. The reasons for this are as follows:

- These cheeses have a relatively high moisture content.
- They are ripened at a temperature (10–15°C) at which bacterial growth can occur.

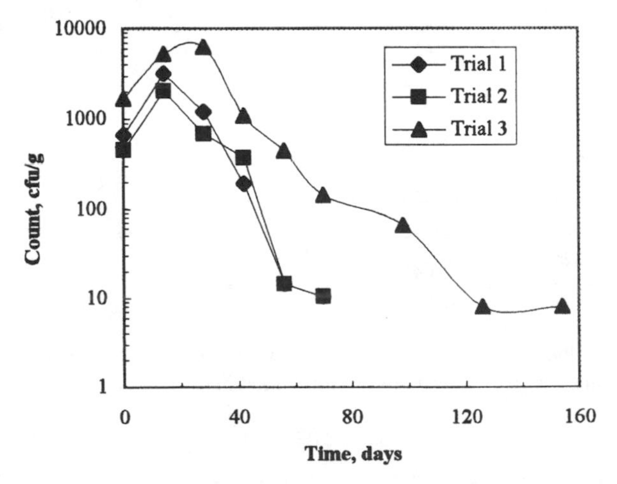

Figure 20–5 Growth of *L. monocytogenes* Scott A in three trials of Cheddar cheese during ripening at 6°C. The variation that occurred in different trials is clearly seen.

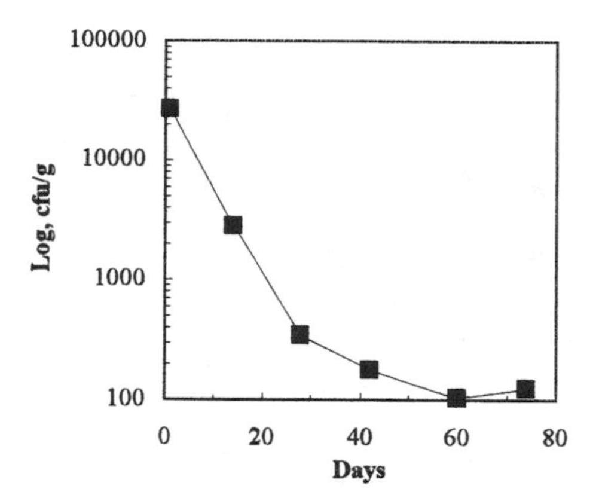

Figure 20–6 Survival of *E. coli* O157:H7 in Cheddar cheese during ripening.

- The pH increases during ripening, especially at the surface, owing to metabolism of lactate by *Geotrichum candidium* and the *Penicillium* spp., to a point where growth of bacterial contaminants can occur (see Chapter 10).

The increase in the pH of soft cheese during ripening is significantly greater at the surface than in the interior of the cheese (see Chapter 10) and is conducive to the growth of some pathogens. Growth of *L. monocytogenes* Scott A, *E. coli* B_2C, *Enterobacter aerogenes* MF1, and *Hafnia* strain 14-1 occurred in Camembert cheese during manufacture. The strain of *E. coli* used was an enterotoxigenic strain. *Hafnia* are closely related to coliform bacteria, and only one species, *H. alvei*, appears to occur in water and the feces of humans and animals. What happens to these organisms during ripening is shown in Figure 20–8. The number of *L. monocytogenes* decreased initially during ripening but increased again once the pH rose above 6. The number also increased in the core but not to the same extent, probably because the pH increased more slowly. In contrast, the number of *E. coli* and *Ent. aerogenes* increased during manufacture but began to decrease once the pH of the curd reached 5.0, and the number continued to decrease during ripening. This pattern of response probably characterizes all coliforms, with *Hafnia* strain 14-1 as an exception to the rule. The number of *Hafnia* strain 14-1 increased until the pH fell to around 5; then it remained constant and decreased to 10 cfu/g during the first week

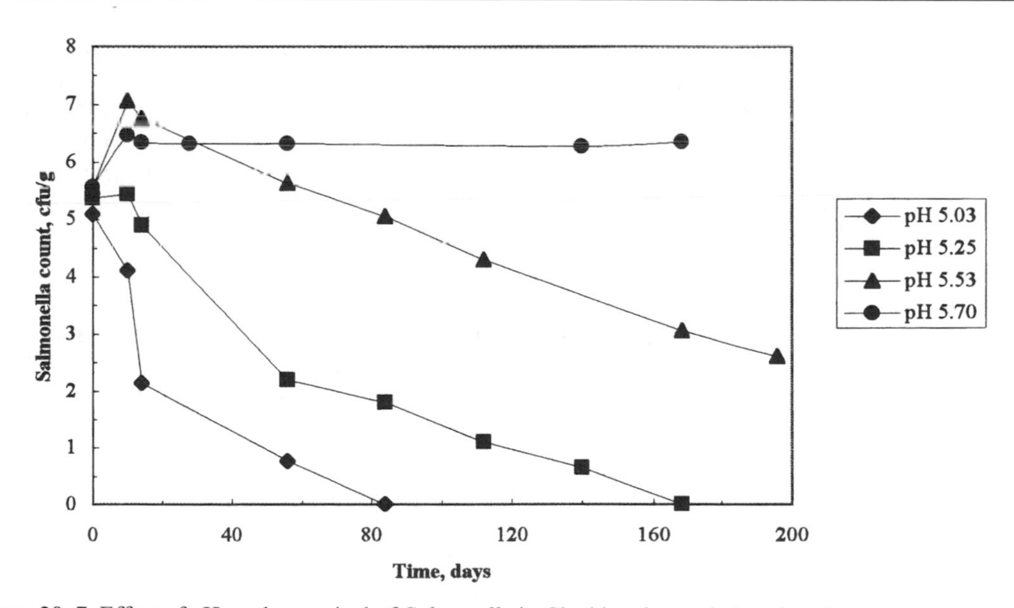

Figure 20–7 Effect of pH on the survival of *Salmonella* in Cheddar cheese during ripening.

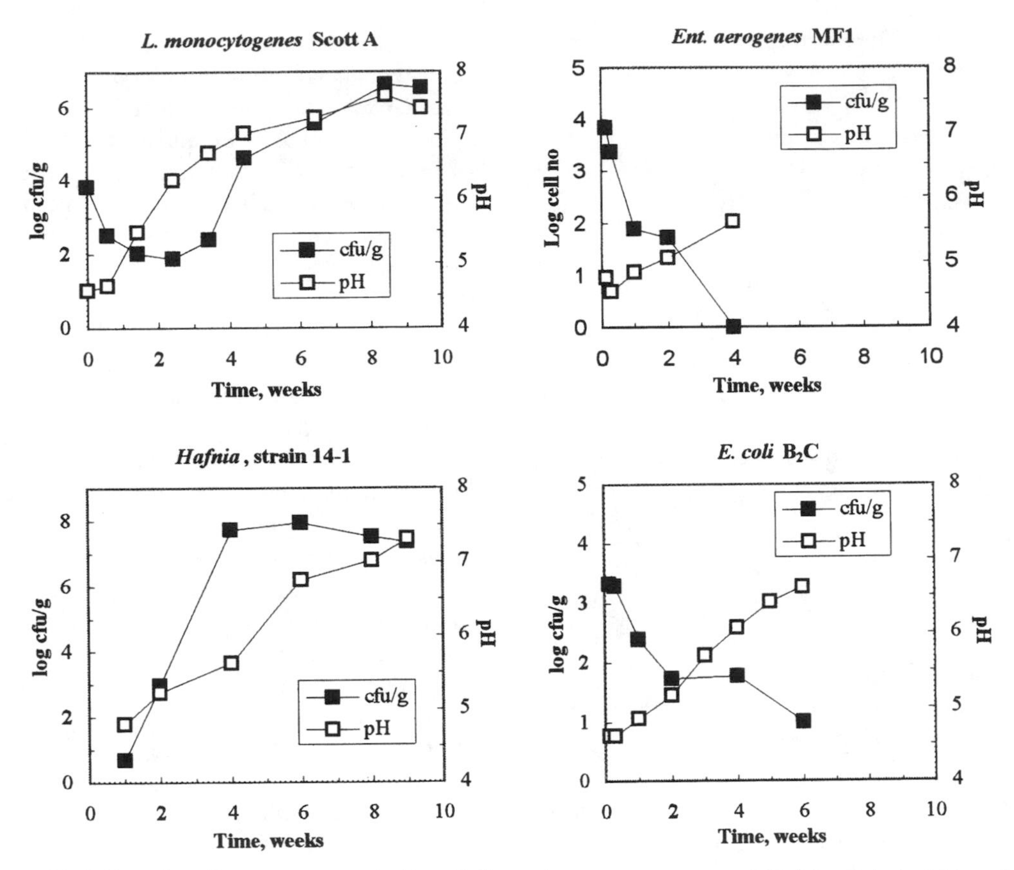

Figure 20–8 Growth of *L. monocytogenes*, *E. coli*, *Enterobacter aerogenes*, and *Hafnia* of strain 14-1 and the increase in pH in Camembert cheese during ripening.

of ripening. The number began to increase again as soon as the pH began to increase, reaching a final cell number of 10^8 cfu/g. The rate of increase in the pH of the three Camembert cheeses varied. This was probably due to differences in the manufacturing procedures and in the rate of growth of the different strains of yeast and *P. camemberti* used. Of course, it is the combined effect of the temperature during ripening, the salt concentration, and the decrease in pH that really determine the extent of growth of pathogens.

Blue cheese is also a soft variety, but *L. monocytogenes* dies out during ripening even though the pH increases from about 4.6 at day 1 to 6.2 after 10 days. The death of *Listeria* in Blue cheese has been attributed to inhibition of their growth by the high level of salt ($\approx$ 10% salt-in-moisture) in these cheeses. The growth of other pathogens on the surface of soft cheeses does not appear to have been investigated.

Significant differences in pH between the core and the surface of soft cheese develop during ripening. This difference creates problems in obtaining representative samples of these cheeses for analysis. Wedge-shaped samples are the most representative of soft cheese (see Chapter 23).

20.8 RAW MILK CHEESES

Cheese made from raw milk has a much stronger flavor than the same cheese made from pasteurized milk, and this is an important marketing advantage for raw milk cheeses. Nevertheless, from the foregoing it is clear that soft cheeses can be problematic, and those made from raw milk are particularly so. *S. aureus* is a common cause of mastitis in dairy cows and therefore is probably present in most raw milks. Approximately 20% of the *S. aureus* strains present in raw milk produce enterotoxin. Growth of such strains to high numbers could therefore cause problems in cheeses made from raw milk. It is also likely that *E. coli* O157:H7 is present in raw milk, as its major source is bovine feces. Despite this, only two food-poisoning outbreaks—and these were small ones—have been traced to cheese containing *E. coli* O157:H7 (Table 20–1). Small numbers of *L. monocytogenes* may also be present in raw milk and may subsequently grow in the cheese. *S. aureus* and *E. coli* grow during cheese manufacture and are potential problems in cheeses made from raw milk. In addition, soft mold- and smear-ripened cheeses with a high moisture content and in which the pH increases, especially at the surface, during ripening are potentially hazardous, especially when they are made from raw milk.

To produce a good-quality raw milk cheese, free from pathogens, the raw milk should

- be of good quality, with bacterial counts below 20,000 cfu/ml and somatic cell counts below 400,000/ml
- be produced under extremely hygienic conditions
- be free of pathogens
- be held at a low temperature (< 4°C)
- be made with an active (fast acid-producing) starter
- have good quality control procedures in place, including hazard analysis critical control points (HACCP)

HACCP was developed by the U.S. space program to ensure the microbiological safety of foods for astronauts. It involves identifying the microbiological hazards and preventive measures that can be taken at each step in the manufacture of the product. The points at which control is critical to managing the safety of the product are then identified and limits are set. These critical control points are then monitored and recorded during the manufacture of each subsequent batch of product. In cheese manufacture, pasteurization and the pH of the curd at a predetermined time after addition of the starter, which estimates the rapidity of acid development, are obvious critical control points.

The European Union standards for different pathogens in raw milk and in soft cheese made from raw or pasteurized milk are shown in Table 20–3. *Listeria* and *Salmonella* must be absent in 25 g, and a distinction is made between soft cheeses produced from raw milk and those made from pasteurized milk, with more stringent standards being set for the former. This reflects the propensity of pathogens to grow in raw milk cheeses during manufacture.

20.9 CONTROL OF THE GROWTH OF PATHOGENS

To prevent the growth of pathogens, it is imperative to prevent contamination of the milk and cheese and to be meticulously hygienic. Today, much cheese is made in automated systems, but small-scale artisanal production involves manual manipulation of the curd during manufacture, molding, and ripening. Good hygiene is critical at each of these steps. Implementation of HACCP systems is also very effective in preventing the growth of pathogens in cheese. An active, phage-free starter and pasteurization are major critical control points. The activity of the starter should be assessed by determining the pH of every batch of cheese at a preset time after starter inoculation, such as 10 hr and 24 hr in the case of soft cheeses and Cheddar, respectively. Comparisons of the data on a daily basis will indicate if starter activity is normal. Soft cheeses are small and will cool quickly. Therefore, keeping the ambient temperature high is important when the curds are in the molds.

Good hygiene is particularly important in the manufacture of smear cheeses, especially where old smear is used to inoculate the fresh cheeses. Old smear may be contaminated with *L. mono-*

cytogenes and may infect all cheeses. The use of old smear is traditional in the production of these cheeses, and efforts are being made to develop defined-strain smear starters to overcome the problem. For example, much attention is being focused on identifying smear bacteria that produce bacteriocins active against *L. monocytogenes*. Application of such cultures to the surface of a cheese should be very useful in helping to prevent the growth of *Listeria* spp. The direct application of bacteriocins produced by lactic acid bacteria (LAB) that inhibit *Listeria* on the surface of cheese is also being advocated as an effective method for controlling listerial growth on cheese.

As already indicated, several factors are involved in controlling the growth of pathogens (and other organisms) in cheese: pH, temperature, and the level of salt are probably the most important. The combined effect of these factors on growth is much greater than each factor's individual effect, and in recent years there has been a major effort to develop models for predicting the growth of pathogens in foods based on their growth responses to combinations of salt, temperature, and pH. These predictive models have been developed mainly from experiments carried out in complex media, and the results are felt to reflect the worst-case scenario in foods, since growth in foods at the same temperature, salt concentration, and pH value is generally less than in model systems. For example, soft cheeses often have a salt concentration of 1.5%, have a pH of 6.5, and are stored at 5°C. Using these parameters, the model predicts that an initial level of 10 *L. monocytogenes* cells/g would multiply to 10,000/g in 10 days (a generation time of about 1 day).

20.10 ENTEROCOCCI

Enterococci are found at high numbers in many cheeses, particularly those made around the Mediterranean. Many of these are artisanal raw milk cheeses made at the farmhouse level. Enterococci are considered to be important in the development of flavor in these cheeses, where they comprise a significant proportion of the microflora of the raw milk and starter. Their ability to metabolize lactose and their tolerance to salt and heat make them ideal candidates as starters. Enterococci at levels above 10^7/g have been found in such cheeses, and these high numbers would almost certainly play a role in flavor development.

There is considerable debate as to whether enterococci should be considered pathogens. During the past few decades, they have been implicated in several diseases, including bacteremia, urinary tract infections, and endocarditis. Many strains are very promiscuous and easily pick up plasmids that encode vancomycin resistance. Many of these plasmids also are conjugative and can be transferred naturally from cell to cell by sexual combination. Vancomycin is a glycopeptide antibiotic that acts by inhibiting cell wall biosynthesis. The incidence of vancomycin-resistant enterococci (VRE) in hospitals has increased dramatically. The use of avoparcin, which is also a glycopeptide antibiotic, as a growth promoter in animal feed has been implicated in the increased occurrence of VREs in farm animals, including pigs and poultry. Because of this, the use of avoparcin has been banned recently in several European countries. Many VREs are difficult to deal with because they are also resistant to other therapeutic antibiotics, implying that alternative antibiotic therapy may not be available. However, many bacteria, including starter LAB like *Lactobacillus, Pediococcus*, and *Leuconostoc,* are intrinsically resistant to vancomycin.

There is little information on how rapidly *Enterococcus* spp. grow in milk, but in Cheddar cheese they remain fairly constant during ripening. Data for other commercial cheeses are sparse in the literature, but data for some artisanal Spanish and Italian cheeses are shown in Figure 20–9. Casar de Cáceres and La Serena are made from raw ewe milk and Afuega'l Pitu from raw cow milk. No starters are used for any of the three cheeses. Pecorino Umbro is made from pasteurized ewe milk and a mesophilic starter is also used. A surface microflora develops on some of these cheeses, but the counts in Figure 20–9 are for the internal part of each cheese. The first point on each line in the figure is the number of enterococci present in the milk at the beginning of manufacture. The data show that considerable growth occurred during manufacture and during the first days of ripening, after which the numbers remained constant, except for Afuega'l Pitu cheese, in which the number decreased. The numbers of enterococci in Casar de Cáceres and La Serena cheese were well in excess of 10^6/g and probably contribute to ripening. Sometimes, it is difficult to distinguish between lactococci and enterococci. However, the counts of enterococci shown in Figure 20–9 are reliable, as selective media were used to enumerate them.

20.11 BIOGENIC AMINES

Biogenic amines can be formed through decarboxylation of amino acids by some strains of nonstarter LAB, particularly *Lb. buchneri,* during cheese ripening. Tyramine, produced from tyrosine, is probably the most important. These amines, which can cause food intoxication within a few hours of ingestion, are discussed in Chapter 21.

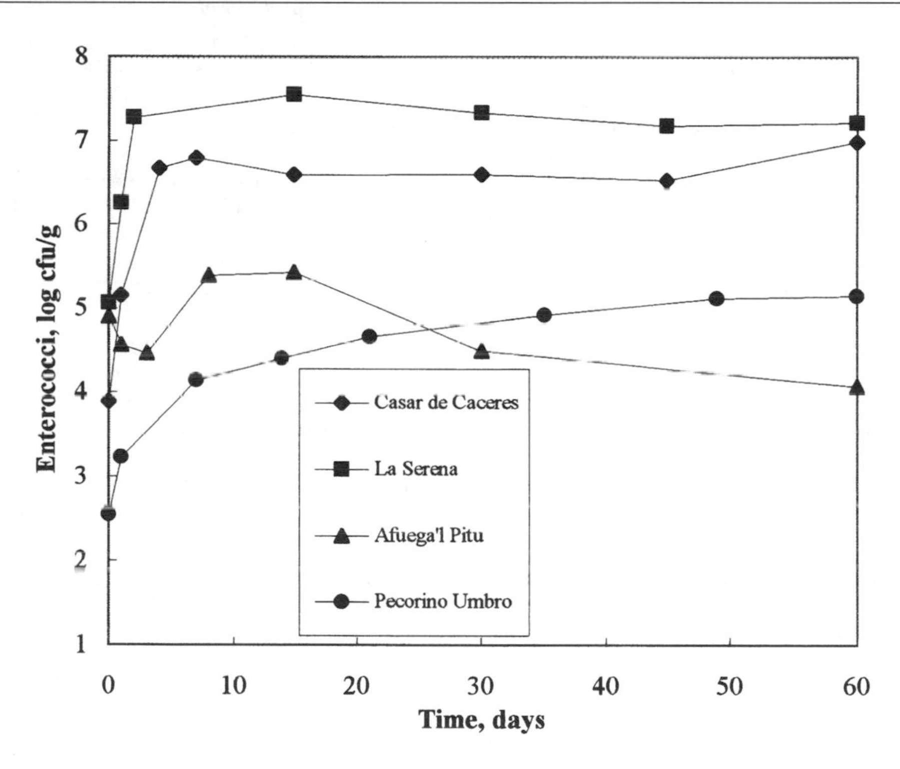

Figure 20–9 Growth of enterococci in Casar de Cáceres (◆), La Serena (■), Afuega'l Pitu (▲), and Pecorino Umbro (●) cheeses during ripening. The first point on each line is the count in the milk.

REFERENCES

Allen, V.D., & Stovall, W.D. (1960). Laboratory aspects of staphylococcal food poisoning from Colby cheese. *Journal of Milk and Food Technology*, *23*, 271–274.

Ansay, S.E., & Casper, C.W. (1997). Survey of retail cheeses, dairy processing environments and raw milk for *Escherichia coli* O157:H7. *Letters in Applied Microbiology*, *25*, 131–134.

Aureli, P., Franciosa, G., & Pourshaban, M. (1996). Foodborne botulism in Italy. *Lancet*, *348*, 1594.

Bille, J. (1990). Epidemiology of human listeriosis in Europe with special reference to the Swiss outbreak. In A.J. Miller, J.L. Smith, & G.A. Somkuti (Eds.), *Foodborne listerosis*. Amsterdam: Elsevier.

Bone, F.J., Bogie, D., & Morgan-Jones, S.C. (1989). Staphylococcal food poisoning from sheep milk cheese. *Epidemiology and Infection*, *103*, 449–458.

D'Aoust, J.Y. (1994). *Salmonella* and international trade. *International Journal of Food Microbiology*, *24*, 11–31.

D'Aoust, J.Y., Warburton D.W., & Sewell, A.M. (1985). *Salmonella typhimurium* phage type 10 from Cheddar cheese implicated in a major Canadian foodborne outbreak. *Journal of Food Protection*, *48*, 1062–1066.

Desenclos, J.C., Bouvet, P., Benz-Lemoine, E., Grimont, F., Desqueyroux, H., Rebiere, I., & Grimont, P.A. (1996). Large outbreak of *Salmonella enterica* serotype *paratyphi B* infection caused by a goats' milk cheese, France, 1993: A case finding and epidemiological study. *British Medical Journal*, *312*, 91–94.

Desmasures, N., Bazin, F., & Guéguen, M. (1997). Microbiological composition of raw milk from selected farmers in the Camembert region of Normandy. *Journal of Applied Microbiology*, *83*, 53–58.

Eckman, M.R. (1975). Brucellosis linked to Mexican cheese. *Journal of the American Medical Association*, *232*, 636–637.

Fontaine, R.E., Cohen, M.L., Martin, W.T., & Vernon, T.M. (1980). Epidemic salmonellosis from Cheddar cheese: Surveillance and prevention. *American Journal of Epidemiology*, *111*, 247–253.

Food safety and cheese. (1998). *International Food Safety News*, *7*, 6–10.

Goulet, V., Jacquet, C., Vaillent, V., Rebiere, I., Mouret, E., Lorente, C., Maillot, E., Stanier, F., & Rocourt, F. (1995). Listeriosis from consumption of raw-milk cheese. *Lancet*, *345*, 1581–1582.

Hedberg, C.W., Korlath, J.A., D'Aoust, J.Y., White, K.E., Schell, W.L., Miller, M.R., Cameron, D.N., MacDonald, K.L., & Osterholm, M.T. (1992). A multistate outbreak of *Salmonella javiana* and *Salmonella oranienberg* infections due to contaminated cheese. *Journal of the American Medical Association*, *268*, 3203–3207.

Hendricks, S.L., Belknap, R.A., & Hausler, W.J. (1959). Staphylococcal food intoxication due to Cheddar cheese: 1. Epidemiology. *Journal of Milk and Food Technology*, *22*, 313–317.

Johnson, E.A., Nelson, J.H., & Johnson, M. (1990a). Microbiological safety of cheese made from heat-treated milk. 1. Executive summary, introduction and history. *Journal of Milk and Food Technology*, *53*, 441–452.

Johnson, E.A., Nelson, J.H., & Johnson, M. (1990b). Microbiological safety of cheese made from heat-treated milk: 2. Microbiology. *Journal of Milk and Food Technology*, *53*, 519–540.

Johnson, E.A., Nelson, J.H., & Johnson, M. (1990c). Microbiological safety of cheese made from heat-treated milk: 3. Technology, discussions, recommendations, bibliography. *Journal of Milk and Food Technology*, *53*, 610–623.

Kauf, C., Lorent, J.P., Mosimann, J., Schlatter, I., Somanini, B., & Velvart. J. (1974). Botulismusepidemic vom type B. *Schweizerische Medizinische Wochenschrift*, *104*, 677–685.

Linnan, M.J., Mascola, M., Lou, X.O., Goulet, V., May, S., Salminen, C., Hird, D.W., Yonekura, L., Hayes, P., Weaver, R., Andurier, A., Plikaytis, B.D., Fannin, S.L., Kleks, A., & Broome, C.V. (1988). Epidemic listeriosis associated with Mexican-style cheese. *New England Journal of Medicine*, *319*, 823–828.

MacDonald, K.L., Eidson, M., Strohmeyer, C., Levy, E., Wells, J.G., Puhr, N.D., Wachsmuth, K., Nargett, N.T., & Cohen, M.L. (1985). A multistate outbreak of gastrointestinal illness caused by enterotoxigenic *Escherichia coli* in imported semisoft cheese. *Journal of Infectious Diseases*, *151*, 716–720.

Maguire, H.C.F., Boyle, M., Lewis, M.J., Pankhurst, J., Wieneke, A.A., Jacob, M., Bruce, J., & O'Mahony, M. (1992). An outbreak of *Salmonella dublin* infection in England and Wales associated with a soft, unpasteurised cow's milk cheese. *Epidemiology and Infection*, *109*, 389–396.

Malinverni, R., Bille, J., Perret, C., Regli, F., Tanner, F., & Glauser, M.P. (1985). Listériose épidémique. *Schweizerische Medizinische Wochenschrift*, *115*, 2–10.

Marier, R., Wells, J.G., Swanson, R.C., Callahan, W., & Mehlman, I.J. (1973, December 15). An outbreak of enteropathogenic *Escherichia coli* foodborne disease traced to imported French cheese. *Lancet*, *2*, 1376–1378.

Ministry of Agriculture, Fisheries and Food. (1997). *Microbiological survey: Raw cow's milk on retail sale.* Food Safety Information Bulletin, 12–13.

Neaves, P., Deacon, J., & Bell, C. (1994). A survey of the incidence of *E. coli* O157 in the UK dairy industry. *International Dairy Journal*, *4*, 679–696.

O'Donnell, E.T. (1995). The incidence of *Salmonella* and *Listeria* in raw milk from farm bulk tanks in England and Wales. *Journal of the Society of Dairy Technology, 48*, 25–29.

Public Health Laboratory Service, Edinburgh. (1994). *E. coli* O157 phage type 28 infections in Grampian. *Communicable Diseases and Environmental Health, 28* (No. 94/46), 1.

Public Health Laboratory Service, London. (1994). Two clusters of haemolytic uremic syndrome in France. *Communicable Disease Report, 4*(7), 29.

Public Health Laboratory Service, London. (1995). Brucellosis associated with unpasteurised milk products abroad. *Communicable Disease Report, 5*(32), 151.

Public Health Laboratory Service, London. (1997a). *Salmonella gold-coast* and Cheddar cheese [Update]. *Communicable Disease Report, 7*(11), 93, 96.

Public Health Laboratory Service, London. (1997b). Vero cytotoxin producing *Escherichia coli* O157. *Communicable Disease Report, 7*(46), 409, 412.

Quinto, E.J., & Cepada, A. (1997). Incidence of toxigenic *Escherichia coli* in soft cheese made from raw milk. *Letters in Applied Microbiology, 24*, 291–295.

Rea, M.C., Cogan, T.M., & Tobin, S. (1992). Incidence of pathogenic bacteria in raw milk in Ireland. *Journal of Applied Bacteriology, 73*, 331–336.

Sadik, C., Krending, M.J., Mean, F., Aubort, J.D., Schneider, P.A., & Roussianos, D. (1986). An epidemiological investigation following an infection by *Salmonella typhimurium* due to the ingestion of cheese made from raw milk. In *Proceedings of the Second World Congress on Foodborne Infections and Intoxications* (Vol. 1). Berlin.

Sebald, M., Jouglard, J., & Gilles, G. (1974). Type B botulism in man due to cheese. *Annales de l'Institut Pasteur, 125A*(3), 349–357.

Sharpe, J.C.M. (1987). Infections associated with milk and dairy products in Europe and North America, 1980–85. *Bulletin of the World Health Organization, 65*, 397–406.

Todd, E.C.D., Shelley, D., Szabo, R., Robern, H., Gleeson, T., Durante, A., Marcoux, A., Entis, P., Morrison, D., Purvis, U., Foster, R., Burgener, D.M., Wright, W.W., Maharaja, R.S., Brodsky, M., Magus, M., Ruf, F.W., & Shab, H.W. (1979). Staphyloccal intoxication from Swiss-type cheese: Quebec and Ontario. *Canadian Diseases Weekly Report, 5*(26), 110–112.

Todd, E.C.D., Szabo, R., Gardiner, M.A., Akhtar, M., Delorme, L., Tourillon, P., Moisan, S., Rochefort, J., Roy, D., Loit, A., Lamontagne, Y., Gosselin, L., Martineau, G., & Breton, J.P. (1981a). Staphylococcal intoxication from cheese curds: Quebec. *Canadian Diseases Weekly Report, 7*(34), 171–172.

Todd, E.C.D., Szabo, R., Robern, H., Gleeson, T., Park, C., & Clark, D.S. (1981b). Variation in counts, enterotoxin levels and TNase in Swiss-type cheese contaminated with *Staphylococcus aureus*. *Journal of Food Protection, 44*, 839–846, 851–852, 856.

Vaillant, V., Haeghebaert, S., & Desenclos, J.C. (1996). Outbreak of *Salmonella dublin* infection in France, November-December 1995. *Eurosurveillance, 1*, 2, 9–10.

Zehren, V.L., & Zehren, V.F. (1968). Examination of large quantities of cheese for staphylococcal enterotoxin A. *Journal of Dairy Science, 51*, 635–644.

SUGGESTED READINGS

Bell, C., & Kyriakides, A. (1998a). *E. coli: A practical approach to the organism and its control in foods*. London: Blackie Academic and Professional.

Bell, C., & Kyriakides, A. (1998b). *Listeria: A practical approach to the organism and its control in foods*. London: Blackie Academic and Professional.

Doyle, M.P., Beuchat, L.R., & Montville, T.J. (1997). *Food microbiology*. Washington, DC: American Society for Microbiology.

Microorganisms in foods 5. (1996). In *Microbiological specifications of food pathogens*. London: Blackie Academic and Professional.

Mortimore, S., & Wallace, C. (1994). *HACCP: A practical approach.* London: Chapman & Hall.

CHAPTER 21

Nutritional Aspects of Cheese

Thomas P. O'Connor and Nora M. O'Brien

21.1 INTRODUCTION

While the nutritional merits of any food should be considered in the context of overall dietary intake, nevertheless, it is accurate to describe cheese as a nutritious and versatile food that can play an important role in a healthy diet, as outlined in current guidelines (Exhibit 21–1). Although per capita consumption of most dairy products has declined worldwide, cheese is a notable exception. Current consumption in a number of countries is shown in Table 1–3. The popularity of cheese is enhanced by its healthy and positive image, the variety of cheeses available, and the compatibility of cheese and cheese-containing products with modern trends toward greater consumption of convenience and prepared foods.

Cheese is a nutrient-dense food, the precise nutritional composition of which is determined by multifactorial parameters, including the type of milk used (species, breed, stage of lactation, and fat content) and the manufacturing and ripening procedures. In general, cheese is rich in the fat and casein constituents of milk, which are retained in the curd during manufacture, and it contains relatively small amounts of the water-soluble constituents (whey proteins, lactose, and water-soluble vitamins), which partition mainly into the whey. The composition of selected cheeses is shown in Table 21–1.

Exhibit 21–1 Dietary Guidelines for Americans Issued by the U.S. Departments of Agriculture and Health and Human Services

- Eat a variety of foods.
- Balance the food you eat with physical activity; maintain or reduce your weight.
- Choose a diet containing plenty of grain products, vegetables, and fruits.
- Choose a diet low in fat, saturated fat, and cholesterol.
- Choose a diet moderate in sugars.
- Choose a diet moderate in salt and sodium.
- If you drink alcoholic beverages, do so in moderation.

21.2 FAT AND CHOLESTEROL

Fat plays several important functions in cheese: it affects, for example, cheese firmness, adhesiveness, mouth-feel, and flavor (see Chapters 12 and 13). It also contributes significantly to the nutritional properties of cheese, as most cheeses contain significant amounts of fat. For example, a 50 g serving of Cheddar cheese provides 17 g fat, in which approximately 66% of the fatty acids are saturated, 30% are monounsaturated, and 4% are polyunsaturated. A typical Western diet providing 2,000 kcal (8,400 kJ) per day, with 40% of energy derived from fat, contains approximately 88 g fat. Thus, cheese contributes a significant amount of both saturated fat and total fat to the diet.

Table 21–1 Composition of Selected Cheeses

Cheese Type	*Water (g/100 g)*	*Protein (g/100 g)*	*Fat (g/100 g)*	*Carbohydrate (g/100 g)*	*Cholesterol (mg/100 g)*	*Energy*	
						kcal	*kJ*
Brie	48.6	19.2	26.9	Trace	100	319	1,323
Caerphilly	41.8	23.2	31.3	0.1	90	375	1,554
Camembert	50.7	20.9	23.7	Trace	75	297	1,232
Cheddar (normal)	36.0	25.5	34.4	0.1	100	412	1,708
Cheddar (reduced fat)	47.1	31.5	15.0	Trace	43	261	1,091
Cheshire	40.6	24.0	31.4	0.1	90	379	1,571
Cottage cheese	79.1	13.8	3.9	2.1	13	98	413
Cream cheese	45.5	3.1	47.4	Trace	95	439	1,807
Danish blue	45.3	20.1	29.6	Trace	75	347	1,437
Edam	43.8	26.0	25.4	Trace	80	333	1,382
Emmental	35.7	28.7	29.7	Trace	90	382	1,587
Feta	56.5	15.6	20.2	1.5	70	250	1,037
Fromage frais	77.9	6.8	7.1	5.7	25	113	469
Gouda	40.1	24.0	31.0	Trace	100	375	1,555
Gruyère	35.0	27.2	33.3	Trace	100	409	1,695
Mozzarella	49.8	25.1	21.0	Trace	65	289	1,204
Parmesan	18.4	39.4	32.7	Trace	100	452	1,880
Processed cheese[a]	45.7	20.8	27.0	0.9	85	330	1,367
Ricotta	72.1	9.4	11.0	2.0	50	144	599
Roquefort	41.3	19.7	32.9	Trace	90	375	1,552
Stilton	38.6	22.7	35.5	0.1	105	411	1,701

[a] Variety not specified.

Many expert groups worldwide have issued dietary guidelines recommending reductions in the intake of both total and saturated fat by Western populations (Exhibit 21–1). Although some experts dispute the merit and efficacy of these guidelines, the vast body of expert opinion supports the concept that dietary intakes do significantly influence the risk of chronic disease. This message has been, in large part, accepted by consumers, and the food industry has responded by producing foods low in fat and cholesterol to meet market trends. A range of "light" cheese products with a reduced level of fat has been developed.

The cholesterol content of cheese varies from approximately 10 to 100 mg/100 g, depending on the variety (Table 21–1). It is well established that dietary cholesterol intake exerts a much smaller influence than the intake of dietary saturated fat on a person's blood cholesterol level, which is a significant risk indicator for coronary heart disease (Keys, 1984). Most individuals (80%) show little change in their blood cholesterol level in response to a change in dietary cholesterol intake in the range 250–800 mg/day. However, a minority of adults do exhibit an increased level of blood cholesterol in response to increased dietary intake of cholesterol (McNamara, 1987).

In recent years, there has been considerable research interest in the role of ingested cholesterol oxidation products (COPs) in the etiology of chronic diseases. However, under normal conditions of manufacture, ripening, and storage, negligible amounts of COPs are formed in cheese.

21.3 PROTEIN AND CARBOHYDRATE

The concentration of protein in cheese varies from approximately 3% to 40%, depending on the variety (Table 21–1). Cheese protein is predominantly casein, as the vast majority of the whey proteins are lost in the whey. As casein is slightly deficient in sulfur-containing amino acids, the biological value of cheese protein is slightly less than that of total milk protein. If the essential amino acid index of total milk protein is assigned a value of 100, the corresponding value for cheese protein varies from 91 to 97, depending on the variety. If whey proteins are incorporated into cheese, such as by use of ultrafiltration, the biological value of cheese protein is similar to that of total milk protein.

Cheese ripening typically involves the progressive breakdown of casein by indigenous milk enzymes, rennet, and bacterial enzymes into water-soluble and -insoluble peptides and amino acids. This process, which is essential for the development of flavor and texture (see Chapters 11–13), also increases the digestibility of cheese protein to almost 100%.

Cheese contains only trace amounts of residual carbohydrate, primarily lactose. The residual lactose in cheese curd is, normally, fermented to lactic acid by starter bacteria during manufacture and ripening. Thus, cheese can be safely consumed by persons deficient in the intestinal enzyme β-galactosidase, which is involved in the digestion of lactose.

21.4 VITAMINS AND MINERALS

Since most of the milk fat is retained in the cheese curd, it follows that the fat-soluble vitamins in milk also partition into the curd. Most of the vitamin A in milk fat (80–85%) is present in cheese fat. Conversely, most of the water-soluble vitamins in milk partition into the whey during curd manufacture. However, some microbial synthesis of B vitamins may occur in cheese during ripening. Significant quantities of vitamin B_{12} are produced in Swiss cheeses by propionic acid bacteria. The vitamin content of a range of cheeses is indicated in Table 21–2. In general, most cheeses are good sources of vitamin A, riboflavin, vitamin B_{12}, and, to a lesser extent, folate. Cheese contains negligible amounts of vitamin C.

Cheese is also an important source of several nutritionally important elements, including calcium, phosphorus, and magnesium (Table 21–3). It is a particularly good source of bioavailable calcium, with most hard cheeses containing approximately 800 mg calcium/100 g cheese. Acid-coagulated cheeses (e.g., Cottage cheese) contain significantly lower levels of calcium than rennet-coagulated varieties. Recker, Bammi, Barger-Lux, and Heaney (1988) reported that the bioavailability of calcium from cheese is comparable to that from milk. Osteoporosis, which may lead to debilitating bone fractures, is a common condition in Western societies. Although it is a disease of multifactorial etiology, there is widespread agreement that adequate calcium intake during childhood and teenage years, especially by girls, is important in ensuring the development of optimum peak bone mass and reducing the risk of subsequent osteoporotic fractures. Cheese can play a positive role in the context of overall diet in supplying highly bioavailable calcium.

As discussed in Chapter 8, sodium chloride plays several important roles during cheese manufacture. The amount of salt added during the manufacture of different cheese varies significantly, resulting in large differences in the concentration of sodium in cheese (Table 21–3). There is substantial evidence that adults in Western societies consume, on average, above optimum levels of sodium. Elevated sodium intake is recognized as a risk factor for hypertension, particularly in those members of the population who are genetically salt sensitive. Hypertension, in turn, is an important risk factor for coronary heart disease. Most dietary guidelines worldwide recommend moderate salt intake. However, even among populations with a high cheese intake, cheese contributes only about 5–8% to total sodium intake.

Table 21–2 Vitamin Content of Selected Cheeses

Cheese Type	*Retinol (μg/100 g)*	*Carotene (μg/100 g)*	*Vitamin D (μg/100 g)*	*Vitamin E (mg/100 g)*	*Thiamine (mg/100 g)*	*Riboflavin (mg/100 g)*	*Niacin (mg/100 g)*	*Vitamin B_6 (mg/100 g)*	*Vitamin B_{12} (μg/100 g)*	*Folate (μg/100 g)*	*Pantothenate (mg/100 g)*	*Biotin (μg/100 g)*
Brie	285	210	0.20	0.84	0.04	0.43	0.43	0.15	1.0	58	0.35	5.6
Caerphilly	315	210	0.24	0.78	0.03	0.47	0.11	0.11	1.1	50	0.29	3.5
Camembert	230	315	0.18	0.65	0.05	0.52	0.96	0.22	1.1	102	0.36	7.6
Cheddar (normal)	325	225	0.26	0.53	0.03	0.40	0.07	0.10	1.1	33	0.36	3.0
Cheddar (reduced fat)	165	100	0.11	0.39	0.03	0.53	0.09	0.13	1.3	56	0.51	3.8
Cheshire	350	220	0.24	0.70	0.03	0.48	0.11	0.09	0.9	40	0.31	4.0
Cottage cheese	44	10	0.03	0.08	0.03	0.26	0.13	0.08	0.7	27	0.40	3.0
Cream cheese	385	220	0.27	1.0	0.03	0.13	0.06	0.04	0.3	11	0.27	1.6
Danish blue	280	250	0.23	0.76	0.03	0.41	0.48	0.12	1.0	50	0.53	2.7
Edam	175	150	0.19	0.48	0.03	0.35	0.07	0.09	2.1	40	0.38	1.8
Emmental	320	140	N	0.44	0.05	0.35	0.10	0.09	2.0	20	0.40	3.0
Feta	220	33	0.50	0.37	0.04	0.21	0.19	0.07	1.1	23	0.36	2.4
Fromage frais	100	Tr	0.05	0.02	0.04	0.40	0.13	0.10	1.4	15	N	N
Gouda	245	145	0.24	0.53	0.03	0.30	0.05	0.08	1.7	43	0.32	1.4
Gruyère	325	225	0.25	0.58	0.03	0.39	0.04	0.11	1.6	12	0.35	1.5
Mozzarella	240	170	0.16	0.33	0.03	0.31	0.08	0.09	2.1	19	0.25	2.2
Parmesan	345	210	0.25	0.70	0.03	0.44	0.12	0.13	1.9	12	0.43	3.3
Processed cheese[a]	270	95	0.21	0.55	0.03	0.28	0.10	0.08	0.9	18	0.31	2.3
Ricotta	185	92	N	0.03	0.02	0.19	0.09	0.03	0.3	N	N	N
Roquefort	295	10	N	0.55	0.04	0.65	0.57	0.09	0.4	45	0.50	2.3
Stilton	355	185	0.27	0.61	0.03	0.43	0.49	0.16	1.0	77	0.71	3.6

Key: Tr = trace; N = nutrient is present in significant quantities but reliable information on the amount is lacking.

[a] Variety not specified.

Table 21–3 Mineral Content of Selected Cheeses

Cheese Type	*Na (mg/100 g)*	*K (mg/100 g)*	*Ca (mg/100 g)*	*Mg (mg/100 g)*	*P (mg/100 g)*	*Fe (mg/100 g)*	*Zn (mg/100 g)*
Brie	700	100	540	27	390	0.8	2.2
Caerphilly	480	91	550	20	400	0.7	3.3
Camembert	650	100	350	21	310	0.2	2.7
Cheddar (normal)	670	77	720	25	490	0.3	2.3
Cheddar (reduced fat)	670	110	840	39	620	0.2	2.8
Cheshire	550	87	560	19	400	0.3	3.3
Cottage cheese	380	89	73	9	160	0.1	0.6
Cream cheese	300	160	98	10	100	0.1	0.5
Danish blue	1,260	89	500	27	370	0.2	2.0
Edam	1,020	97	770	39	530	0.4	2.2
Emmental	45	89	970	35	590	0.3	4.4
Feta	1,440	95	360	20	280	0.2	0.9
Fromage frais	31	110	89	8	110	0.1	0.3
Gouda	910	91	740	38	490	0.1	1.8
Gruyère	670	99	950	37	610	0.3	2.3
Mozzarella	610	75	590	27	420	0.3	1.4
Parmesan	1,090	110	1,200	45	810	1.1	5.3
Processed cheese[a]	1,320	130	600	22	800	0.5	3.2
Ricotta	100	110	240	13	170	0.4	1.3
Roquefort	1,670	91	530	33	400	0.4	1.6
Stilton	930	130	320	20	310	0.3	2.5

[a] Variety not specified.

Cheese, along with other dairy products, is a poor source of dietary iron. Iron deficiency anemia is a major worldwide nutrition-related problem, both in developed and developing countries. In an attempt to alleviate this problem, there is considerable interest in fortifying commonly consumed foods with iron. Cheddar and processed cheeses have been successfully fortified with iron.

21.5 ADDITIVES IN CHEESE

Preservatives are added occasionally during the manufacture of certain cheeses. Growth of yeasts and molds on hard and semi-hard cheeses may be inhibited by the addition of sorbic acid or its salts.

Nitrate may be added to milk prior to the manufacture of certain types of cheese. It is reduced to nitrite, which inhibits growth of *Clostridium* species, a possible cause of late gas blowing and flavor defects (see Chapters 10 and 11). However, nitrite does not persist well in cheese, and the contribution from cheese to total intake of nitrite is negligible.

In recent years, there has been considerable research interest in the efficacy of natural bacterially produced preservatives in cheese. Most interest has focused on the bacteriocin nisin, a peptide produced by some strains of *Lactococcus lactis*. It has been exploited commercially in Swiss-type and processed cheeses to prevent late blowing by *Clostridium* species, the spores of which survive pasteurization.

21.6 CHEESE AND DENTAL CARIES

Dental caries are a commonly occurring problem that, in simple terms, involves degeneration of tooth enamel due to acid produced by oral microorganisms during the metabolism of sugars. The progress and extent of dental caries are influenced by a variety of dietary parameters and nutrient interactions, including the composition, texture, solubility, and retentiveness of food and by its ability to stimulate saliva flow. In recent years, considerable work has been conducted on the cariostatic effects of cheese.

Early work demonstrated that dairy products reduced the development of dental caries in rats and also in vitro. These effects were attributed to the high concentrations of calcium and phosphate in milk and to the protective effects of casein. This early work was supported by further detailed work on rats that indicated that both casein and whey proteins had protective effects, the former being the more effective, and also confirmed the protective effects of calcium and phosphate.

The first evidence that cheese had an anticariogenic effect in humans was reported by Rugg-Gunn, Edgar, Geddes, and Jenkins (1975). Consumption of Cheddar cheese after sweetened coffee or a sausage roll increased plaque pH, possibly due to increased output of saliva, which can buffer the effect of acids formed in plaque. Some early work suggested that the consumption of cheese resulted in reduced numbers of *Streptococcus mutans,* which is involved in acid production, in the mouth. However, later work suggested that the cariostatic effects of cheese may not be directly related to its effect on *Sc. mutans* but could be explained primarily by mass action effects on soluble ions, particularly calcium and phosphate (Jenkins & Harper, 1983).

Further investigation of the cariostatic effects of cheese in humans was reported by Silva, Jenkins, Burgess, and Sandham (1986), who measured the demineralization and hardness of enamel slabs fastened to a prosthetic appliance made specifically for each human subject to replace a missing lower first permanent molar. Each subject chewed 5 g of cheese immediately after rinsing his or her mouth with a 10% (w/v) solution of sucrose. Chewing cheese resulted in a 71% decrease in demineralization of the enamel slabs and an increase in plaque pH but did not significantly affect the oral microflora.

Further trials on humans have confirmed that the consumption of hard cheese results in significant rehardening of softened enamel surfaces (Gedalia, Ionat-Bendat, Ben-Mosheh, & Shapira, 1991; Jenkins & Hargreaves, 1989). While more research is required to define precisely the mechanisms involved in the cariostatic effects of cheese, it is reasonable to recommend the consumption of cheese at the end of a meal as an anticaries measure.

21.7 MYCOTOXINS

Mycotoxins (Figure 21–1) are fungal metabolites that have been shown to be cytotoxic, mutagenic, teratogenic, and carcinogenic in animals. Certain mycotoxins (e.g., aflatoxin) are among the most potent animal toxins known, hence giving rise to concerns regarding their potential effects in the human food supply.

The consumption of aflatoxin-contaminated food, particularly in conjunction with hepatitis B infection, is a key risk factor for liver cancer, which is the principal form of cancer reported in less developed countries. Mycotoxins may be present in milk and dairy products, such as cheese, owing to indirect contamination (contamination of the cows' feedstuff) or direct contamination (growth of mycotoxin-producing fungi in the milk and dairy products).

21.7.1 Indirect Contamination

It has been known for almost 40 years that the intake of feedstuff contaminated with aflatoxin B_1 by dairy cows may result in the excretion of toxic factors (principally aflatoxin M_1) in their milk within a few hours (Allcroft & Carnaghan, 1962). On average, 1–2% of ingested aflatoxin B_1 is excreted in milk as aflatoxin M_1. It has been shown subsequently that indirect contamination of milk with other mycotoxins, such as ochra-

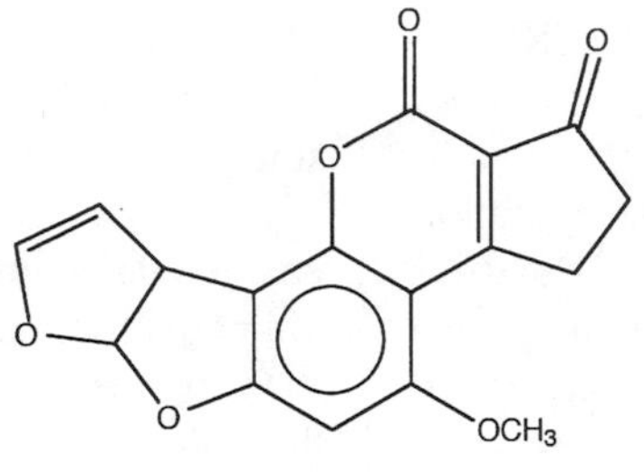

Aflatoxin B1

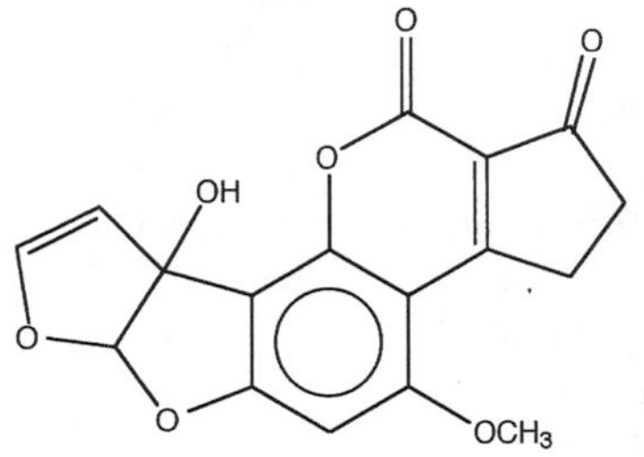

Aflatoxin M1

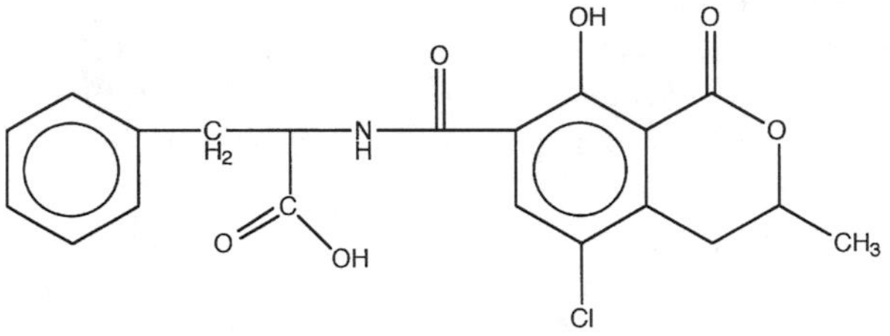

Ochratoxin A

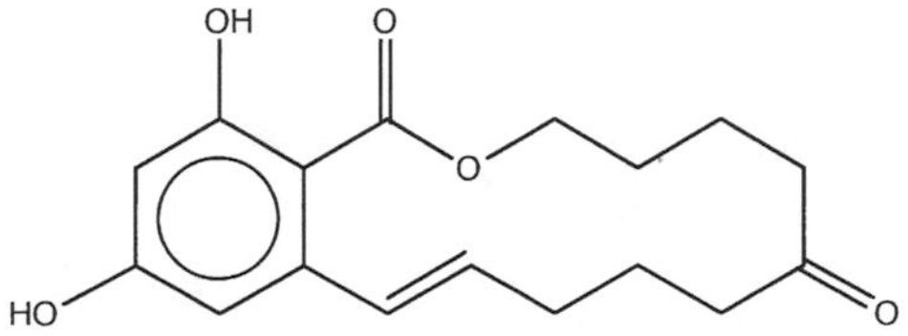

Zearalenone

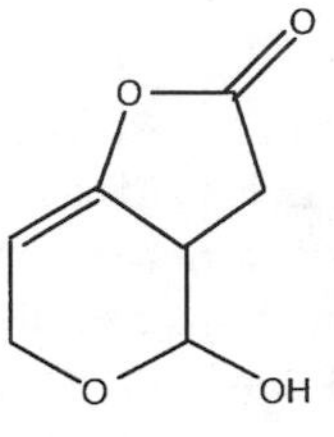

Patulin

Figure 21–1 Structures of selected mycotoxins.

toxin A, zearalenone, T-2 toxin, sterigmatocystin, and deoxynivalenone, does not represent a major public health issue (for a comprehensive review of mycotoxins in dairy products, see van Egmond, 1989).

Van Egmond (1989) summarized the results of surveillance programs in many countries for aflatoxin M_1 in milk and milk products. The incidence and levels of aflatoxin M_1 in milk and milk products have decreased significantly over the years, which can be attributed primarily to implementation of regulatory limits on the contamination of feedstuff with aflatoxin B_1. A notable exception was the significant increase in the level of aflatoxin M_1 in U.S. dairy products in 1988–89. This resulted from feeding aflatoxin B_1–contaminated maize products due to the severe drought in the U.S. Midwest in 1988 which created ideal conditions for the growth of and aflatoxin production by the producing organism, *Aspergillus flavus*. Results reported in surveillance programs for cheese have, in general, indicated that aflatoxin M_1 was not detected or occurred at concentrations below legal limits (0.2–0.25 μg/kg).

Studies have been conducted on the fate and stability of aflatoxin M_1 in milk during cheese manufacture and ripening. These studies indicate that aflatoxin M_1 partitions between the curd and whey in both acid-coagulated and rennet-coagulated cheeses and that aflatoxin M_1 is very stable during cheese manufacture. The partition coefficient for aflatoxin M_1 in water suggests that most of the toxin should partition into the whey. This anomaly can be explained by findings that aflatoxin M_1 tends to associate hydrophobically with casein micelles, resulting in a greater than expected partitioning into the cheese curd.

In general, it appears that aflatoxin M_1 is stable in several cheese varieties during ripening.

21.7.2 Production of Toxic Metabolites in Mold-Ripened Cheese

Penicillium roqueforti and *P. camemberti* are used in the manufacture of various types of blue-veined and white surface-mold cheeses. These molds can produce a range of toxic metabolites. Some strains of *P. roqueforti* can produce PR toxin, patulin, mycophenolic acid, penicillic acid, roquefortine, cyclopiazonic acid, isofumigaclavine A and B, and festuclavine. *P. camemberti* strains produce cyclopiazonic acid, which has been detected in commercial samples of Camembert and Brie. It occurs primarily in the rind at levels below 0.5 mg/kg whole cheese but may occur at levels up to 5 mg/kg if the storage temperature is too high. However, evaluation of available toxicological data for cyclopiazonic acid, together with potential human exposure estimated from consumption data for Camembert and Brie, suggests that this metabolite causes no appreciable public health risk (Engel & Teuber, 1989).

P. roqueforti can produce a range of toxins, as outlined above. Patulin, penicillic acid, and PR toxin have not been detected in commercial samples of cheese. Mycophenolic acid has been detected in commercial cheese samples but at levels well below those that pose a risk to human health. Roquefortine and isofumigaclavine A and B have been detected at low levels in commercial Blue cheese, and their toxicity is low. Compelling evidence that the consumption of mold-ripened cheeses is not hazardous to human health was provided by studies on rats and rainbow trout that consumed levels of mold equivalent to a daily human intake of 100 kg cheese with no apparent signs of toxicity. Mold-ripened cheeses have been consumed for several hundred years without apparent ill effects.

21.7.3 Direct Contamination of Cheese with Mycotoxins

Cheese is a good substrate for the growth of adventitious molds given suitable conditions of temperature, humidity, and oxygen. Mycotoxin-producing molds require oxygen for growth and hence are very unlikely to grow on vacuum-packed or wax-coated cheese, particularly if there is good plant sanitation during manufacture and handling and if the storage temperature is low.

Unintentional growth of molds on cheese during ripening and/or storage results in financial loss, reduces consumer appeal, and may necessitate trimming. However, the production of mycotoxins may also represent a health risk. Cheeses on which unintentional growth of mold had occurred have been reported to contain mycotoxins that are nephrotoxic (ochratoxin A, citrinin), teratogenic (ochratoxin A, aflatoxin B_1), neurotoxic (penitrem A, cyclopiazonic acid), and carcinogenic (aflatoxin B_1 and G_1, ochratoxin A, patulin, penicillic acid, sterigmatocystin) (Ueno, 1985).

Penicillium species are the mycotoxigenic fungi most frequently isolated from cheese; *Aspergillus* and other species are encountered only occasionally. However, the presence of mold growth does not imply that mycotoxins are present in cheese. A large body of work (see van Egmond, 1989) has been conducted on the occurrence of mycotoxins in moldy cheese. The overall incidence was low in most studies. Furthermore, less than 50% of *Penicillium* species were toxicogenic in animal studies. Work has also been conducted on the incidence of mycotoxins in cheeses contaminated with *Aspergillus* species. There is very little evidence that significant levels of aflatoxins are produced in cheese by these species.

Work has also been undertaken on the ability of mycotoxins to migrate from the surface of cheese into the interior. The data are significant for decisions on whether to trim or discard mold-contaminated cheese. The Health Protection Branch of the Department of Health and Welfare, Canada, has recommended that if a hard cheese is contaminated with a patch of mold, the cheese can be salvaged by removing the infected portion to a depth of 2.5 cm.

21.8 BIOGENIC AMINES IN CHEESE

The term *biogenic amines* refers to nonvolatile, low–molecular weight aliphatic, acyclic, and heterocyclic amines, such as histamine, tyramine, tryptamine, putrescine, cadaverine, and phenylethylamine, that may be present in cheese or other foods. In cheese, biogenic amines are produced by decarboxylation of amino acids during ripening by enzymes released by the microorganisms present. Levels produced vary as a function of ripening period and microflora, with the highest levels most likely in cheeses heavily contaminated with spoilage microorganisms. Renner (1987) reported average values of histamine and tyramine in some cheeses (Table 21–4).

Consumption of foods containing significant levels of biogenic amines may cause food poisoning. However, for most individuals, consumption of even large amounts of biogenic amines does not elicit toxicity symptoms, since they are converted rapidly to aldehydes by mono- and diamine oxidases and then to carboxylic acid by oxidative deamination. However, if these enzymes are impaired, owing to a genetic defect or inhibitory drugs, toxic symptoms may result.

Histamine is a normal body constituent formed from histidine by a pyridoxal phosphate–dependent decarboxylase. Its concentration in blood is tightly regulated, and orally administered histamine results in toxicity only following ingestion of a very high dose or impairment of histidine metabolism. Toxic symptoms generally become apparent within 3 hr of ingestion and include, initially, a flushing of the face and neck, followed by an intense, throbbing headache. Other symptoms are observed occasionally, including cardiac palpitations, dizziness, faintness, rapid and weak pulse, gastrointestinal complaints, bronchospasms, and respiratory distress.

Consumption of fish, particularly of the Scombroidae family, has been associated with most cases of histamine poisoning. However, some instances have been reported to be related to cheese consumption. Gouda cheese containing 85 mg histidine/100 g was implicated in an outbreak in the Netherlands. Incidences in the United States resulting from the consumption of contaminated Swiss cheese have also been reported (Taylor, Kiefe, Windham, & Howell, 1982).

Table 21–4 Average Tyramine and Histamine Contents of Some Cheese Types

Cheese Type	*Tyramine (μg/g)*	*Histamine (μg/g)*
Cheddar	910	110
Emmental	190	100
Blue	440	400
Edam, Gouda	210	35
Camembert, Brie	140	30
Cottage	5	5

Tyramine is normally present at low levels in the body. In humans, monoamine oxidase (MAO)–catalyzed oxidative deamination to *p*-hydrophenylacetic acid is the main degradative pathway for tyramine. However, if a genetic deficiency of MAO exists or if MAO-inhibitory drugs are administered, toxicity symptoms may be manifest. These include a hypertensive crisis, often accompanied by severe headache and, in certain cases, intercranial hemorrhage, cardiac failure, and pulmonary edema. The main dietary sources of tyramine, besides cheese, include marinated herring, dry sausages, and marmite. The tyramine content of cheese is generally greater in long-ripened varieties, such as extra mature Cheddar, than in young cheese. Patients prescribed MAO-inhibitory drugs should be advised to avoid intake of tyramine-rich foods. Tyramine poisoning in the absence of MAO-inhibitory drugs has not been reported. The toxicity threshold for tyramine has been estimated at 400 mg, and therefore healthy individuals can tolerate intakes of large amounts of tyramine-rich cheese.

REFERENCES

Allcroft, R., & Carnaghan, R.B.A. (1962). Groundnut toxicity: *Aspergillus flavus* toxin (aflatoxin) in animal products. *Veterinary Record, 74*, 863–864.

Engel, G., & Teuber, M. (1989). Toxic metabolites from fungal cheese starter cultures. In H.P. van Egmond (Ed.), *Mycotoxins in dairy products*. London: Elsevier Applied Science.

Gedalia, I., Ionat-Bendat, D., Ben-Mosheh, S., & Shapira, L. (1991). Tooth enamel softening with a cola type drink and rehardening with hard cheese or stimulated saliva. *Journal of Oral Rehabilitation, 18*, 501–506.

Jenkins, G.N., & Hargreaves, J.A. (1989). Effect of eating cheese on Ca and P concentrations of whole mouth saliva and plaque. *Caries Research, 23*, 159–164.

Jenkins, G.N., & Harper, D.S. (1983). Protective effect of different cheeses in an in vitro demineralization system [Abstract]. *Journal of Dental Research, 62*, 284.

Keys, A. (1984). Serum cholesterol response to dietary cholesterol. *American Journal of Clinical Nutrition, 40*, 351–359.

McNamara, D.J. (1987). Effects of fat-modified diets on cholesterol and lipoprotein metabolism. *Annual Review of Nutrition, 7*, 273–290.

Recker, R.R., Bammi, A., Barger-Lux, M.J., & Heaney, R.P. (1988). Calcium absorbability from milk products, an imitation milk and calcium carbonate. *American Journal of Clinical Nutrition, 47*, 93–95.

Renner, E. (1987). Nutritional aspects of cheese. In P.F. Fox (Ed.), *Cheese: Chemistry, physics and microbiology* (Vol. 1). London: Elsevier Applied Science.

Rugg-Gunn, A.J., Edgar, W.M., Geddes, D.A.M., & Jenkins, G.N. (1975). The effect of different meal patterns upon plaque pH in human subjects. *British Dental Journal, 139*, 351–356.

Silva, M.D. de A., Jenkins, G.N., Burgess, R.C., & Sandham, H.J. (1986). Effect of cheese on experimental caries in human subjects. *Caries Research, 20*, 263–269.

Taylor, S.L., Kiefe, T.J., Windham, E.S., & Howell, J.F. (1982). Outbreak of histamine poisoning associated with consumption of Swiss cheese. *Journal of Food Protection, 45*, 455–457.

Ueno, Y. (1985). The toxicology of mycotoxins. *CRC Critical Review of Toxicology, 14*, 99–132.

van Egmond, H.P. (1989). Aflatoxin M_1: Occurrence, toxicity, regulation. In H.P. van Egmond (Ed.), *Mycotoxins in dairy products*. London: Elsevier Applied Science.

Chapter 22

Whey and Whey Products

22.1 INTRODUCTION

The liquid remaining after removal of the fat and casein from milk by isoelectric or rennet coagulation of the casein is called whey. The term *milk serum* is used for the supernatant liquid obtained by ultracentrifuging skimmed (fat-free) milk; 95% of the casein micelles are sedimented by ultracentrifugation at 100,000 g for 1 hr. Milk serum represents the aqueous phase of milk, unchanged by the process of separation, although it does contain small casein micelles.

The wheys prepared by isoelectric precipitation or rennet coagulation are called acid whey and sweet (rennet) whey, respectively. They differ in composition from each other (Table 22–1) and from milk serum because of changes that occur during their preparation. For example, acid whey contains a much higher concentration of calcium, magnesium, phosphate, and citrate than sweet whey or milk serum owing to the solution of the colloidal milk salts upon acidification. If properly prepared, acid whey should be free of casein. Commercially, acid whey is usually prepared from efficiently skimmed milk in the manufacture of acid-coagulated cheeses (see Chapter 16) or acid casein and is therefore essentially free of fat, although it does contain some phospholipids. Sweet whey is a byproduct of the manufacture of rennet-coagulated cheese or rennet casein and its composition varies depending on its source (e.g., pH 6.2–6.6), depending on the extent of acidification that had occurred prior to whey separation (hence the concentration of some salts varies somewhat). Most cheeses are made from full-fat or partially skimmed milk, and typically about 10% of the fat in such milk is lost in the whey as a result of the formation of free (nonglobular) fat during pasteurization and pumping of the milk and the loss of fat globules from the curd pieces during cutting and cooking. Rennet casein is produced from skimmed milk, and therefore the resulting whey is essentially fat-free. As discussed in Chapter 6, rennet coagulation involves cleavage of κ-casein, and the resulting macropeptides formed are present in rennet whey. Other casein-derived peptides may be present in whey if an excessively proteolytic rennet substitute is used. If the rennet coagulation process is incomplete when the gel is cut, the whey may contain some uncoagulated casein as well as small particles of curd. The compositional data for acid and rennet wheys shown in Table 22–1 are typical values, although the values vary considerably.

It is apparent from Table 22–1 that whey contains about 50% of the total solids of milk, including essentially all of the lactose and whey proteins (provided that the whey proteins were not denatured by heat treatment prior to coagulation), 50–100% of the milk salts (depending on whether it was produced by acid or rennet coagulation), and some fat (depending on whether skimmed or whole milk was used). Thus, whey is a valuable source of food constituents from which numerous food products are now pro-

Table 22–1 Typical Composition and pH of Whole Milk, Sweet (Rennet Casein and Cheddar Cheese) Wheys, and Acid (Lactic and Mineral Acid) Wheys

	Sweet Wheys		*Acid Wheys*		
Component	*Rennet Casein (g/L)*	*Cheddar[a] Cheese*	*Lactic Acid Casein (g/L)*	*Mineral Acid Casein (g/L)*	*Whole Milk (g/L)*
Total solids	66.0	67.0	64.0	63.0	122.5
Total protein (N x 6.38)	6.6	6.5	6.2	6.1	33.0
Nonprotein nitrogen (NPN)	0.37	0.27	0.40	0.30	–
Lactose	52.0	52.0	44.0	47.0	47.0
Milk fat	0.20	3.0	0.30	0.30	35.0
Minerals (ash)	5.0	5.2	7.5	7.9	7.5
Calcium	0.50	0.40	1.6	1.4	1.2
Phosphate	1.0	0.50	2.0	2.0	2.0
Sodium	0.53	0.50	0.51	0.50	0.5
Lactate	–	2.0	6.4	–	–
pH	6.4	5.9	4.6	4.7	6.7

[a] Unseparated whey.

duced. In this chapter, the principal products produced from whey are described briefly. The reader should refer to Sienkiewicz and Riedel (1990) for a detailed discussion on whey and whey utilization. Certain aspects of whey proteins are covered in Fox (1992) and certain aspects of lactose and its derivatives in Fox (1997) and International Dairy Federation (1993).

Traditionally, whey was regarded as a waste product and was disposed of by the cheapest possible method—fed to animals (especially pigs), spray irrigated onto land, dumped in waterways, or treated as effluent. Some whey is still disposed of by such methods, but dumping of whey is unacceptable today for environmental reasons, and improved technology makes it possible to recover whey constituents in a cost-effective manner. World production of whey is about 160 million tonnes per annum, representing about 7 million tonnes of lactose and 1 million tonnes of whey protein.

22.2 CLARIFICATION OF WHEY

The curd fines may be removed using a vibrating screen separator but are removed more often using a centrifugal separator (clarifier). The fines are recovered as a concentrate (e.g., 50% dry matter), which is pressed and may be used in processed cheese or similar products. Removal of fines facilitates further processing of whey (e.g., by ultrafiltration).

Fat, which is typically present at a level of about 0.3% (w/w) in bulk cheese whey, is recovered from the clarified whey using centrifugal separators, typically to levels of 0.07%, w/w. The resultant whey cream (≈ 50% fat) is normally used for the manufacture of whey butter, which is used as a food ingredient in, for example, processed cheese products.

The phospholipids in whey exist as lipoprotein particles, which block ultrafiltration membranes, reducing the flux rate of the plant. A number of methods have been developed to aggregate the lipoprotein particles, such as by adding $CaCl_2$ and raising the pH to around 7.5. The flocculated calcium phosphate–lipoprotein particles may be removed by sedimentation, by centrifugation, or preferably by microfiltration. The lipoproteins have good emulsification properties and may be used in a number of food applications. The clarified whey is subjected to further processing.

22.3 CONCENTRATED AND DRIED WHEY PRODUCTS

Whey powders have been produced for many years and have several applications in the food industry, including as ingredients for bakery and meat products and ice cream. The value of whey powders can be increased and their range of applications extended by one of several process modifications.

22.3.1 Nonhygroscopic Whey Powder

Lactose, which represents about 70% of the total solids in whey, is difficult to crystallize, and if the lactose is not properly crystallized, the whey powder is hygroscopic, making it unstable during storage. Nonhygroscopic whey powder is produced by concentrating the whey to 50–60% total solids, seeding the concentrate with lactose crystals to induce crystallization, and, when crystallization is complete, drying the concentrate.

22.3.2 Demineralized Whey Powder

One of the important applications of whey solids is in the manufacture of infant formulae. Human milk contains more lactose (≈ 7%) and less casein (≈ 1% total protein and a whey protein:casein ratio of 60:40, compared with 20:80 for bovine milk) than cow milk. Many modern baby formulae based on cow milk are humanized; that is, their lactose content and casein:whey protein ratio are adjusted to approximate those in human milk. This adjustment is usually made by blending bovine whey and skimmed milk. However, the concentration of salts in bovine milk is 3–4 times higher than that in human milk and places an undesirably high renal load on the baby. The problem may be resolved by reducing the concentration of ions in whey by electrodialysis and/or ion exchangers.

22.3.3 Delactosed and Delactosed-Demineralized Whey Powder

For many food applications, it is desirable to use a whey product with a higher than normal protein content. This may be achieved by the processes described below for the production of whey protein products or alternatively by crystallizing out some of the lactose. The latter method involves concentrating the whey, seeding it with lactose to induce crystallization, and removing the lactose crystals by centrifugation or filtration. The mother liquor may or may not be demineralized (see Section 22.3.2) and spray-dried to yield a protein-rich whey powder.

22.4 LACTOSE

Lactose is a sugar unique to milk (see Chapter 3). Among commercially available sugars, lactose has many unusual properties:

- low solubility
- difficult to crystalize
- a tendency to form supersaturated solutions
- low sweetness
- low hygroscopicity when properly crystalized
- a tendency to adsorb flavors and pigments

These characteristics create problems for the dairy industry, but methods have been developed for managing and controlling the problems. In fact, some of these characteristics are exploited in the production of improved dairy products, such as instant milk powders and low-hydroscopicity icing sugar mixtures (see Fox, 1997). Consequently, a substantial market has developed for lactose, although very small in comparison with that for sucrose (≈ 250,000 tonnes per annum of pure lactose in comparison with 95 million tonnes per annum of sucrose).

As noted above, lactose is produced by concentrating whey to 50–60% total solids, seeding it with lactose crystals, and recovering of the crystals by centrifugation or filtration. If extra high purity lactose is required, the first crop of crystals may be dissolved and recrystallized.

The market for lactose appears to be relatively limited, but lactose can be converted, enzymatically, chemically, or physically, to a range of useful derivatives:

- *Lactulose* (Figure 22–1). Lactulose is formed when the glucose moiety of lactose is isomerized to fructose by a mild alkaline treatment. Lactulose is not hydrolyzed by β-galactosidase in the human intestine and passes to the lower intestine, where it may act as a laxative or promote the growth of bifidobacteria, which can have beneficial effects on the microbial ecology of the lower intestine (Figure 22–2).

- *Lactitol* (Figure 22–3). The carbonyl group of lactose may be reduced to an alcohol, lactitol, by chemical or electrolytic methods. Lactitol is not hydrolyzed in the human intestine and hence may be used as a nonnutritive sweetener. It is claimed to have anticholesterolemic and anticariogenic properties. Lactitol may be esterified with one or more fatty acids to produce a range of food-grade emulsifiers.
- *Lactobionic acid.* Lactose may be oxidized to lactobionic acid, which may be dehydrated to lactobionic acid lactone (Figure 22–4). Both derivatives have a number of food and industrial applications, but the volumes used are small.
- *Glucose-galactose syrups.* Lactose may be hydrolyzed by β-galactosidase (lactase) or by free acid or cation exchangers to produce glucose-galactose syrups, which are sweeter and more soluble than lactose. Such syrups have several applications in food products but in most cases are not cost competitive with glucose, glucose-fructose, or sucrose.
- *Galacto-oligosaccharides.* β-galactosidase normally functions as a hydrolase, but under certain conditions it may function as a transferase, with the production of galacto-oligosaccharides (Figure 22–5) involving various bond types not digested in the human intestine. These oligosaccharides pass into the large intestine, where they serve to promote the growth of *Bifidobacterium* spp.

22.5 WHEY PROTEINS

Bovine whey contains two main proteins, β-lactoglobulin and α-lactalbumin, with lesser amounts of blood serum albumin and immunoglobulins (mainly IgG_1) and trace amounts of several proteins, especially lactotransferrin, and several enzymes (see Chapter 3). Many of these proteins have desirable nutritional, functional, and, in some cases, pharmaceutical properties. Numerous methods are available for the recovery of whey proteins in toto and, more recently, for the isolation of individual proteins.

The first and simplest of these is heat denaturation and recovery of the aggregated protein, known as *lactalbumin*. The product is insoluble, has very poor functional properties and is used mainly in nutritional fortification of foods.

Whey protein concentrates (WPC, 30–80% protein), prepared by ultrafiltration, are widely used as functional ingredients, such as for the preparation of gels, foams, and emulsions (see Fox, 1992).

Products with a higher protein content (up to 95%; known as *whey protein isolates*) have better functionalities and are produced by ion exchange chromatography. Alternatively, separated whey is ultrafiltrated to 15% dry matter and microfiltered to remove fat. The defatted permeate is further concentrated by ultrafiltration and diafiltration to about 20% dry matter and spray dried. The product typically contains 92% protein, less than 0.5% fat, and 96% dry matter. Since production costs are high, whey protein isolates are produced on a small scale.

The properties of some whey proteins make them more suitable for certain applications than others. For example, the gelation properties of β-lactoglobulin are superior to those of α-lactalbumin, but β-lactoglobulin is less suitable for the fortification of infant formulae, since it does not occur in human milk and many human infants are allergic to it. Several methods have been developed for the fractionation of whey proteins, but most are not amenable to industrial-scale production and none is used on a truly commercial scale. Presumably, the efficacy of these methods will be improved, costs will probably decrease, and demand will increase, and consequently the fractionation of whey proteins on a commercial scale may become a reality.

Some of the minor whey proteins are potentially very valuable as nutraceuticals. Much in-

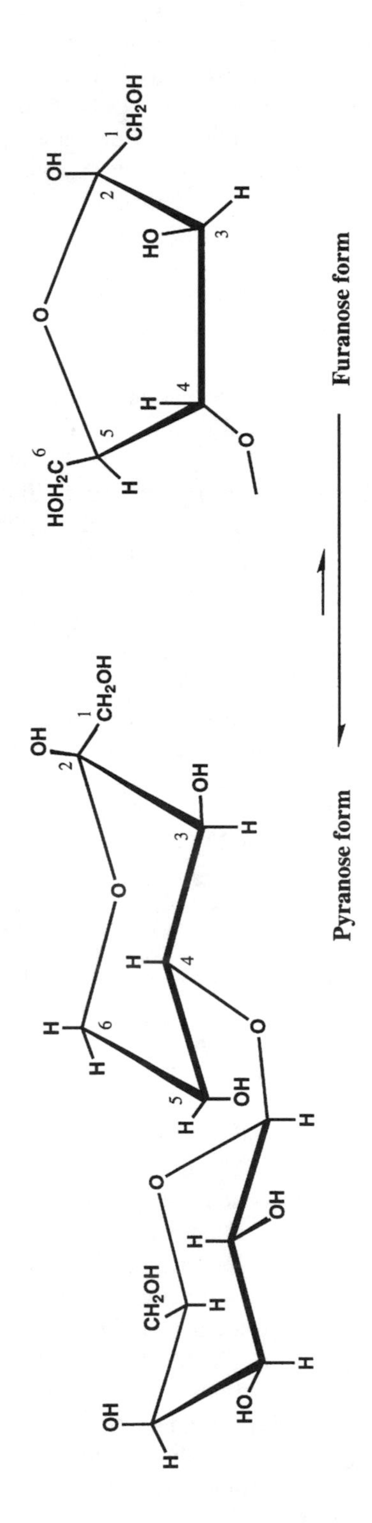

Figure 22–1 Chemical structure of lactulose.

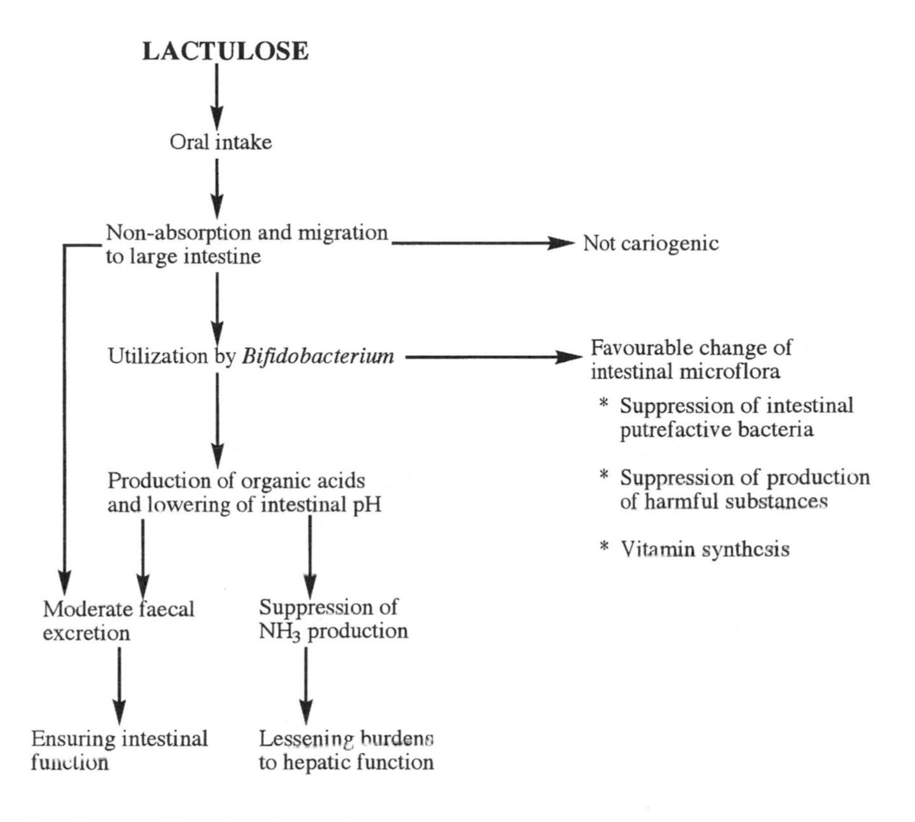

Figure 22–2 Significance of lactulose in health.

terest has focused on lactoferrin (lactotransferrin), which is present at a very much higher concentration in human milk than in bovine milk. Lactoferrin is a nonheme iron-binding protein that has bacteriostatic and other physiological properties and serves as a source of biologically available iron. Because lactoferrin is cationic at the pH of milk (at which most other milk proteins are anionic), it can be easily isolated from milk or whey and has been used to supplement infant formulae and other special dietary products.

Lactoperoxidase is also cationic at the pH of milk and may be readily isolated. In the presence of H_2O_2 and the thiocyanate anion, lactoperoxidase is a very effective bactericidal agent and has been used as an additive in milk replacers for calves and piglets.

The (glyco)macropeptide (GMP) produced from κ-casein upon the renneting of milk contains no aromatic amino acids and therefore is suitable for patients suffering from phenylketonurea. It is also claimed to have several interesting physiological effects. Methods have been developed for the preparation of GMP on a pilot scale.

22.6 WHEY CHEESE

The whey proteins are recovered as soft cheese by heating a mixture of whey and skim or whole milk, adjusted to pH 6.0, at 90°C. Ricotta and variants thereof, Anari and Manouri, are examples of this type of cheese; they are discussed in Chapters 16 and 17.

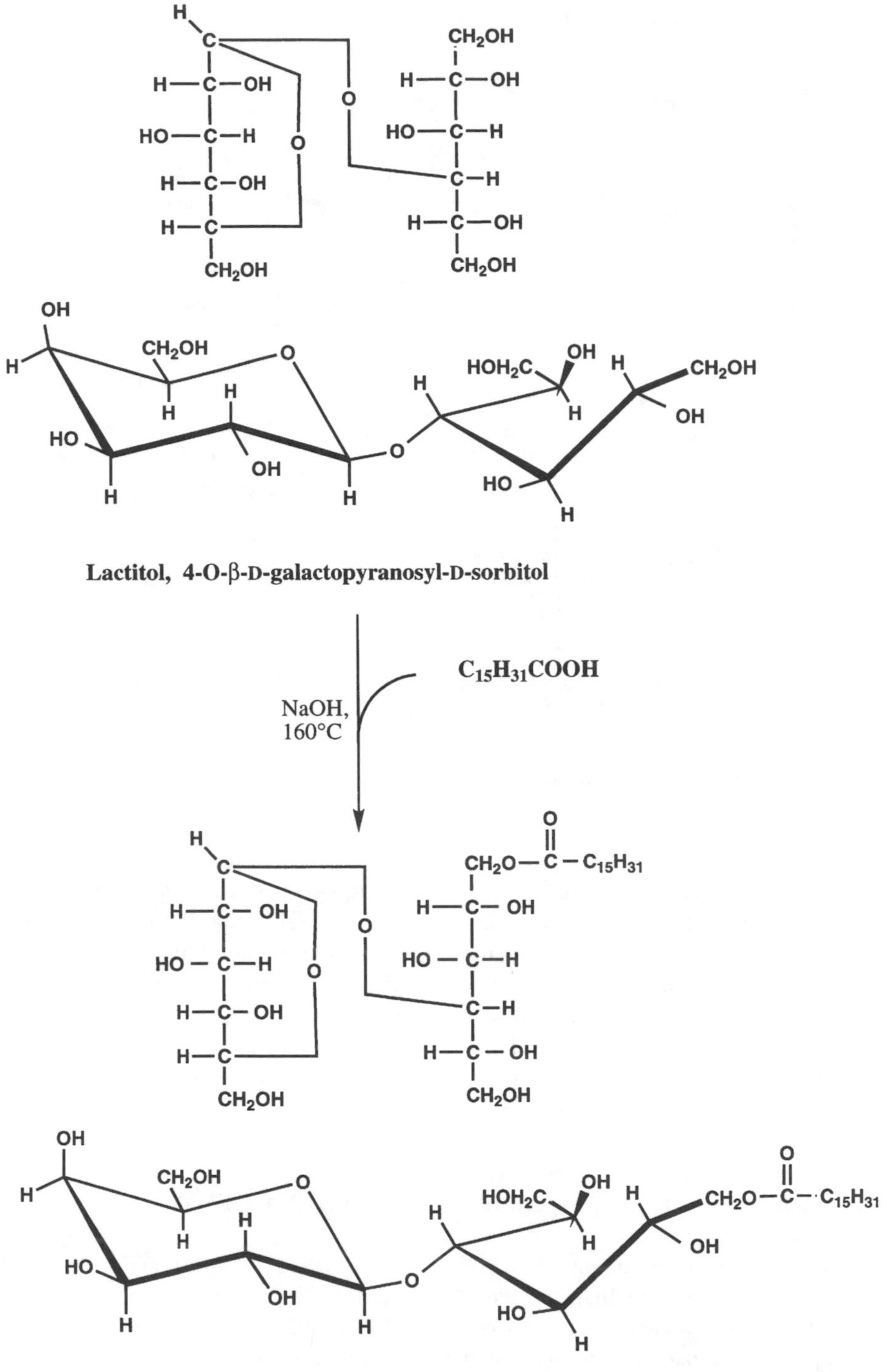

Lactitol, 4-O-β-D-galactopyranosyl-D-sorbitol

NaOH,
160°C

$C_{15}H_{31}COOH$

Lactitol monoester (Lactyl palmitate)

Figure 22–3 Structure of lactitol and its conversion to lactyl palmitate.

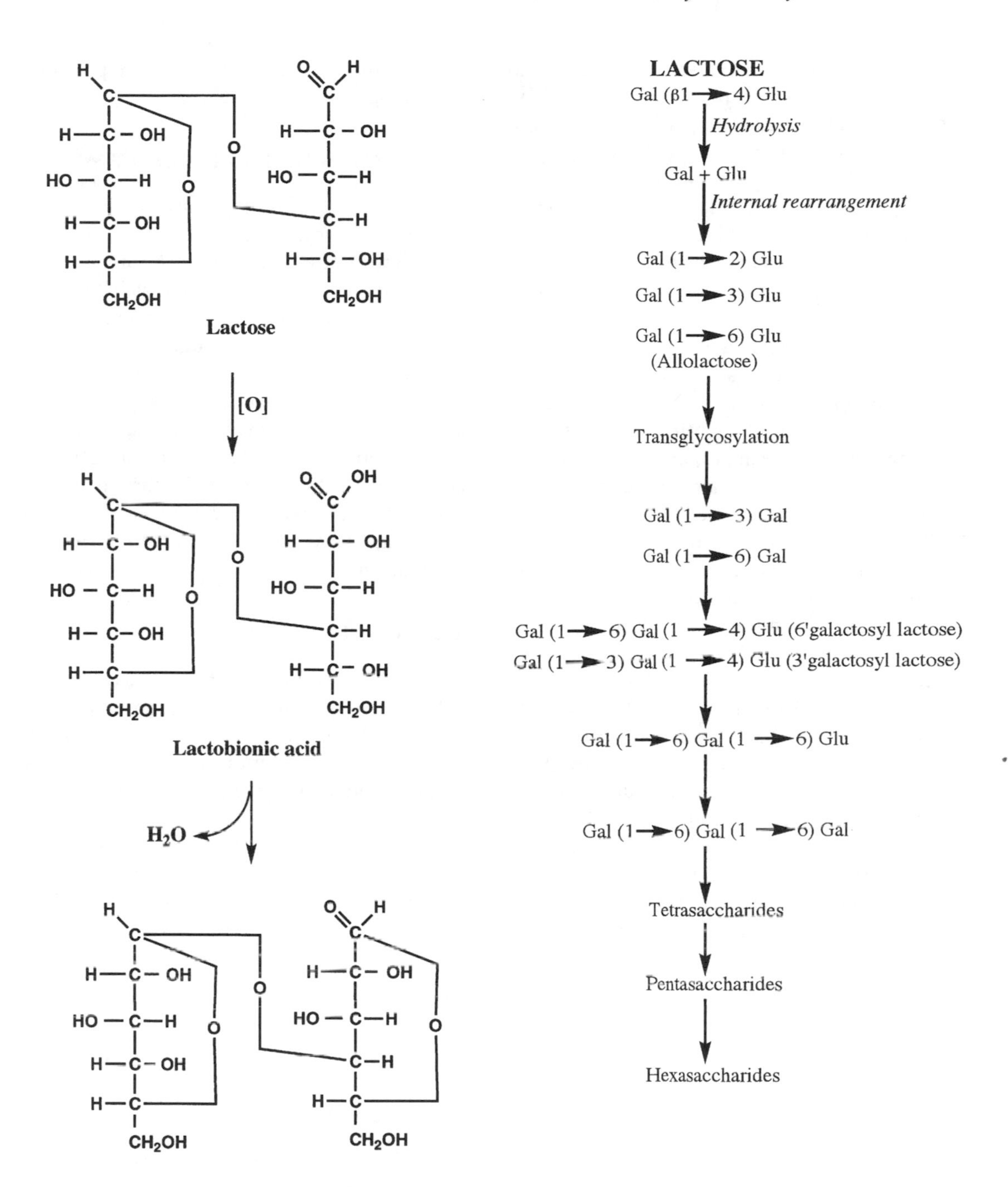

Figure 22–4 Structure of lactobionic acid and its δ-lactone.

Figure 22–5 Possible reaction products from the action of β-galactosidase on lactose.

The whey proteins are also incorporated into some forms of Queso blanco produced from whole milk and acidified to pH 5.4 by heating at 90°C. They may be incorporated into Quarg by using ultrafiltration technology or the Centri-Whey process (whey is heated at 90°C to denature the whey proteins, which are then recovered by centrifugation, added to milk for the next batch of cheese, and become incorporated into the Quarg).

Heating milk damages or destroys its rennet coagulation properties (see Chapter 6), but these properties can be restored by acidifying the cheese milk or supplementing it with $CaCl_2$. This approach has been proposed as a means for increasing the yield of rennet-coagulated cheese but is used to a very limited extent.

The whey proteins can be incorporated into rennet-coagulated cheese by preconcentrating the milk to the total solids content of the particular variety by ultrafiltration and coagulating the pre-cheese by rennet. This technology has been quite successful for soft cheeses but not, to date, for semi-hard or hard cheeses (see Chapter 17).

Finally, whey, usually mixed with whole milk, may be concentrated by thermal evaporation to about 87% solids to produce a unique family of cheeses, examples of which are Mysost and Gjetost (see Chapters 16 and 17). These cheeses are quite different from all other cheeses—they have a sweet taste (owing to the high level of lactose), a brown color (due to the Maillard reaction between lactose and proteins), and a fudgelike consistency.

22.7 FERMENTATION PRODUCTS

Lactose in whey or, more usually, in ultrafiltration permeate may be used in various fermentation processes. The most widespread of the fermentation products is ethanol, which is being produced on a commercial scale in several factories. Other fermentation products include acetic acid (from ethanol), lactic acid, and propionic acid (produced from lactic acid by *Propionibacterium* spp.). Lactose may also be used as a more general fermentation substrate but is not cost-competitive with sucrose in the form of molasses. The production of yeast biomass from whey fermentation has been considered but it is not economical.

22.8 CONCLUSION

Whey, which contains about 50% of the total solids of milk and was regarded as a waste stream until recently, can serve as the raw material for the production of a wide range of food products and food ingredients. Some of these are being produced profitably on a commercial scale. It is likely that as new technologies are developed, new and improved food ingredients derived from or based on whey will be developed.

REFERENCES

Fox, P.F. (1992). *Advanced dairy chemistry: Vol. 1. Proteins* (2d ed.). London: Elsevier Applied Science.

Fox, P.F. (1997). *Advanced dairy chemistry: Vol. 3. Lactose, water, salts and vitamins* (2d ed.). London: Chapman & Hall.

International Dairy Federation. (1993). *Proceedings of the IDF Workshop on Lactose Hydrolysis* [Bulletin No. 289]. Brussels: Author.

Sienkiewicz, T., & Riedel, C.-L. (1990). *Whey and whey utilization* (2d ed.). Gelsenkirchen-Buer, Germany: Verlag Thomas Mann.

CHAPTER 23

Analytical Methods for Cheese

23.1 INTRODUCTION

Cheese is analyzed for a variety of reasons: to ascertain its composition for nutritional purposes, to ensure compliance with standards of identity, to assess the efficiency of production, and to assess the microbial safety of the product or the influence of enzymes and/or microorganisms on cheese quality, among others. Reliable microbiological, physical, chemical, and sensory analyses are of critical importance to the dairy scientist involved in cheese research, to analysts working on quality assurance, and for regulating the production process.

23.2 METHODS OF SAMPLING CHEESE

Unless the entire cheese is sampled, the reliability of the results of any analytical procedure is dependent on how representative the sample analyzed is. Correct sampling is, therefore, of paramount importance, and standard procedures have been published (International Dairy Federation, 1985). In general, samples should be taken by an experienced and responsible person who is familiar with the methods. Since traceability may be an important consideration in industry, the samples should be sealed, appropriately labeled, and accompanied by a sampling report.

Subsamples must be taken from cheese for analysis, unless the cheese is sufficiently small (or has been previously subdivided), in order for the entire cheese to be used as the sample. The basic apparatus for sampling most cheese varieties is the cheese trier (Figure 23–1), although a suitable knife or a cutting wire may also be used. The sampling procedure recommended by the International Dairy Federation (IDF, 1985) involves removing by trier a shallow plug of cheese (15–20 mm deep), which may be retained as a surface sample, followed by the insertion of a smaller trier into the resulting hole, through which an internal sample is taken. Another common practice is to take a plug using a trier and to retain the outer 15–20 mm of the plug as the surface sample. Cheese triers should be made from stainless steel and should be sterilized (by dipping in 70% [v/v] alcohol and flaming) before sampling for microbiological or sensory analysis. Equipment for obtaining samples for chemical or physical analyses should be clean. Cheese samples should be stored in a suitable container (e.g., a plastic container or bag or aluminum foil), and containers for microbiological samples must be sterile. In general, duplicate samples (100–200 g) should be taken and stored at 0–4°C until analyzed, and the analysis should be performed as soon as possible after sampling (preferably within 24 hr). Care should also be taken when sampling fresh cheeses to avoid whey separation. Suitable sampling techniques are shown in Figure 23–2.

Care should be taken when sampling cheeses with gradients from center to surface (e.g., varieties ripened with a surface microflora or brine-

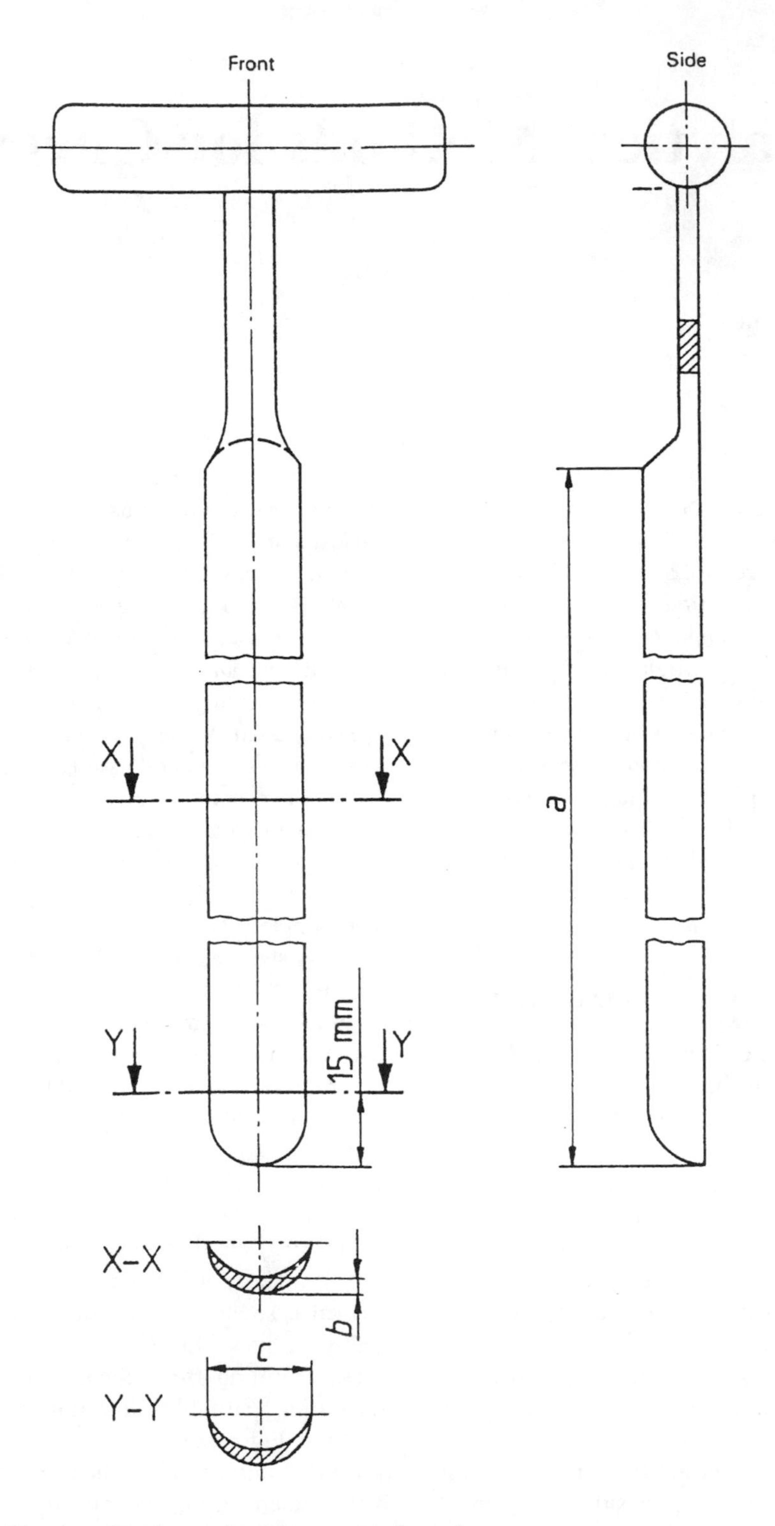

Figure 23–1 Diagram of a trier used to take samples from cheese.

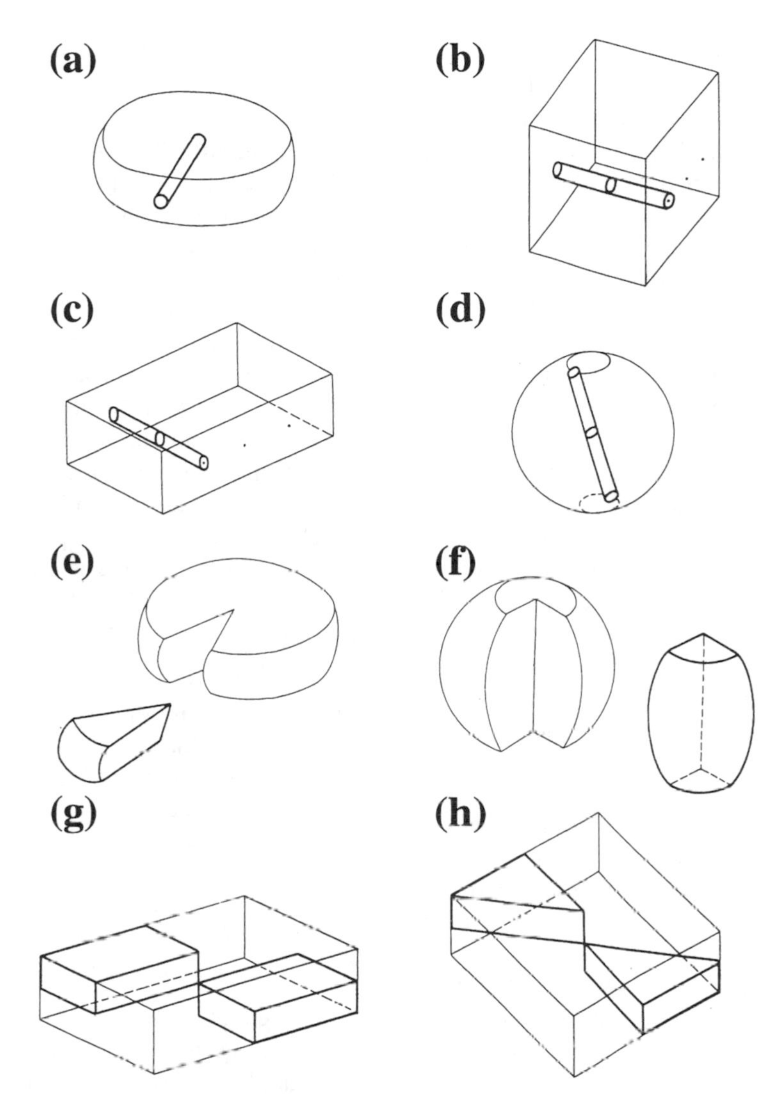

Figure 23–2 Suggested sampling techniques for cylindrical (a), cubic (b), block-shaped (c), and spherical (d) cheeses using a trier; suggested sampling technique for cheeses with a circular cross section and a mass greater than 2,500 g (e) or between 1,100 and 2,500 g (f); suggested sampling techniques for block-shaped cheeses the largest face of which is rectangular (g) or square (h).

salted cheeses). It is recommended that such varieties should not be sampled using a trier but rather by a technique that involves cutting the cheese, because a trier will not give a representative sample from cheeses with radial salt and moisture gradients. For research purposes, it is common to analyze separately surface and internal samples from these cheeses.

23.3 COMPOSITIONAL ANALYSIS

Gross compositional analysis (moisture, ash, protein, fat, acidity, and salt) of cheese is conducted according to standard methods published by the International Dairy Federation (see IDF, 1995) or the Association of Official Analytical Chemists (Cunniff, 1995). Moisture (total solids) is usually determined gravimetrically by drying a sample to constant weight in an oven at 100°C or in a microwave oven with an integral weighing apparatus. Certain cheese varieties (e.g., Blue cheese) contain significant amounts of substances other than water that are volatile at below 100°C. Moisture in such cheeses may be determined by reflux distillation in the presence of an imiscible solvent (e.g., *n*-amyl alcohol and xylene, 1:2) and collection of the water separated from the cheese in a calibrated tube. Ash is determined gravimetrically by heating a sample in a furnace at or below 550°C until completely ashed. Protein is estimated by measuring the N content of cheese by the Kjeldahl method and multiplying by a conversion factor (6.38). Rapid determination of fat in cheese is often made by the Gerber (butyrometric) or Babcock methods, although the standard reference method is based on the Schmid-Bondzynski-Ratzlaff (SBR) technique. The SBR technique, in which fat is extracted by a mixture of diethyl ether and light petroleum after digestion of protein with HCl, is recommended in preference to the Röse-Gottlieb method, since the NH_3 solution used in the latter does not completely dissolve many types of cheese and because free fatty acids released by lipolysis are not fully extracted from the ammoniacal solution.

The chloride (and thus NaCl) content of cheese is usually determined by titration with $AgNO_3$, with potentiometric or colorimetric endpoint determination. There appears to be no standard method for determining the pH of cheese. The method used in our laboratories is as follows: grated cheese (10 g) is thoroughly blended with 10 ml of H_2O using a mortar and pestle, and the pH of the resulting slurry is measured potentiometrically. However, it may be preferable to measure the pH of grated cheese directly (using reinforced electrodes) to minimize changes in pH caused by alternating the balance between colloidal and soluble calcium phosphate upon dispersion in water.

The AOAC standard method for determining the titratable acidity of cheese involves mixing grated cheese with water (40°C) and filtering and titrating an aliquot of the filtrate with 0.1 M NaOH, using phenolphthalein as the indicator. Results are usually expressed as lactic acid percentage (Cunniff, 1995). Although the method is probably suitable for curd or young cheese, it would appear to be unsatisfactory for mature cheese since the buffering capacity of the filtrate and hence its titratable acidity will increase as proteolysis progresses. It is even less suitable for cheeses the pH of which increases during ripening (e.g., mold- and smear-ripened varieties), owing to the catabolism of lactic acid and the production of ammonia (see Chapter 11). Consequently, the concentration of lactic acid decreases but the titratable acidity may increase as the level of water-soluble peptides increases. Measurement of titratable acidity is not common in cheese analysis; it may have potential as an index of ripening.

Calcium can be quantified by

- titration with ethylenediamine tetraacetic acid (EDTA), using ammonium purpurate (murexide) as the indicator
- precipitation as calcium oxalate and weighing
- atomic absorption spectrophotometry
- ion-specific electrodes (it should be noted that this method measures calcium ion activity and not total concentration)

The concentration of Na can be quantified specifically using an ion-selective electrode, atomic absorption spectroscopy, or flame spectrophotometry. Phosphorus may be determined by a colorimetric assay using molybdovanadate or molybdate-ascorbate reagents.

The water activity (a_w) of cheese can be determined by a variety of methods, including psychrometry, cryoscopy, dew-point hygrometry, and isopiestic equilibration. A number of regression equations have been developed to predict a_w from chemical composition for various cheeses (see Chapter 8; Marcos, 1993).

Near infrared reflectance spectroscopy may also be used to determine fat, protein, moisture, and moisture in nonfatty substances in a range of cheese varieties. Infrared transmittance spectrophotometry of solvent extracts of cheese may also be used for gross compositional analysis (see McSweeney & Fox, 1993, for references).

Residual alkaline phosphatase activity in cheese is used to indicate the use of raw or underpasteurized milk for cheese manufacture. Alkaline phosphatase activity may be assayed using several substrates, including disodium phenylphosphate, *p*-nitrophenyl phosphate, phenolphthalein monophosphate, and fluorophos.

Residual coagulant activity, which is important for the development of flavor precursors and texture in cheese (see Chapter 11), may be assessed by monitoring the production of specific peptides (e.g., α_{s1}-CN f24–199) by gel electrophoresis (see Chapter 11), by measuring the coagulation time of milk to which an extract from the cheese is added, by diffusion assays on casein-containing agar, by determining the hydrolysis of various peptide substrates, and by immunochemical methods (see Baer & Collin, 1993).

23.4 BIOCHEMICAL ASSESSMENT OF CHEESE RIPENING

23.4.1 Determination of Products of Lactose, Lactate, and Citrate Metabolism

The metabolism of lactose to lactic acid by the starter bacteria is fundamental to most, if not all, varieties of cheese (see Chapters 5, 10, and 11). L-Lactate is produced by mesophilic and some thermophilic starters, while a mixture of the D- and L-isomers is produced by other thermophiles (see Chapter 5). In certain varieties (e.g., Swiss) the lactate serves as a substrate for further microbial metabolism, and in many varieties the L-isomer produced by the starter is converted to a racemic mixture by the nonstarter flora of the cheese (see Chapters 10 and 11). Many of the products of sugar metabolism may be quantified by enzyme-based assays, and in these cases the enzymatic procedure has replaced earlier methods because of increased sensitivity and specificity. For some other compounds, instrumental or colorimetric methods are also widely used.

Enzyme assay kits (e.g., from Boehringer Mannheim, Mannheim, Germany) for lactose, lactic acid, and citrate are available. Lactose, D-glucose, D-galactose, and lactate may also be determined by high-performance liquid chromatography (HPLC). Lactose in processed cheese may be determined by a modified Fehling's titration (Cunniff, 1995). The enzymatic assay for lactate is particularly useful, since it distinguishes between D- and L-isomers. In this assay, L- and/or D-lactate dehydrogenase (LDH) and glutamic-pyruvic transaminase (GPT) are used. Lactate is oxidized to pyruvate by LDH in the presence of NAD^+. The equilibrium of this reaction normally lies in favor of lactate formation, but the pyruvate formed is trapped by reacting it with L-glutamate in the presence of GPT, shifting the equilibrium in favor of pyruvate and NADH. The concentration of NADH is measured spectrophotometrically (at 334, 340, or 365 nm) and is stoichiometrically related to the concentration of D- or L-lactate present.

The metabolism of citrate leads to the production of flavor (diacetyl) and nonflavor compounds (acetoin and 2,3-butanediol) in some cheeses. Citrate can be measured chemically or enzymatically. The chemical determination of citrate in cheese involves dispersing the cheese in NaOH, precipitating the protein by trichloroacetic acid, filtering the reaction mixture, and reacting an aliquot of the filtrate with pyridine and acetic anhydride. The result is a yellow complex that is quantified spectrophotometrically at 428 nm (Marier & Boulet, 1958). This is a good method, but a standard curve must be included with each assay, as the relationship between citrate and color intensity shows day-to-day variation. Also care should be taken, as pyridine is carcinogenic. In the enzymic assay, citrate is converted to oxaloacetate and acetate by citrate lyase. In the presence of malate dehydrogenase and L-LDH, oxaloacetate and its decarboxylation product, pyruvate, are reduced to L-malate and L-lactate, respectively, with the concomitant oxidation of NADH to NAD^+, which is quantified spectrophotometrically at 340 nm.

Diacetyl and acetoin can be quantified using colorimetric or gas chromatographic (GC) procedures. In the colorimetric procedures, a dispersion of the cheese is steam distilled and the first two 10 ml fractions are collected. Diacetyl is estimated in the first 10 ml fraction and acetoin in the second (Walsh & Cogan, 1974). For the determination of diacetyl, aliquots of the steam distillate are heated with hydroxylamine to form dimethylglyoxime, which is converted to a pink ammonoferrous dimethylglyoxmate complex upon reaction with $FeSO_4$ in acidic solution. The absorbance is then read at 525 nm. Diacetyl can also be quantified by reaction with HCl and 3,3'-diamino-benzidine tetrahydrochloride but care is required since the latter is carcinogenic. Headspace analysis by gas chromatography (GC) can also be used to quantify diacetyl using an electron capture detector.

Acetoin is generally measured in the second 10 ml of distillate by reaction with 2-naphthol and creatine. The red complex formed is quantified at 525 nm. Diacetyl also reacts with this reagent, but no diacetyl is present in the second 10 ml of distillate.

Acetolactate (AL) is produced by some starter cultures and will break down to diacetyl and/or acetoin during steam distillation. When AL is present, the above method for measuring diacetyl must be modified, and steam distillates of samples cannot be used for determination of acetoin. Diacetyl is measured in steam distillates of samples adjusted to pH 0.5 before distillation. Under these conditions, no breakdown of AL to diacetyl occurs (in fact, it is completely converted to acetoin). Acetoin can be measured by solubilizing the cheese with NaOH, followed by direct estimation. Care should be taken not to increase the temperature to an extent that could cause the breakdown of AL to acetoin.

To measure AL, samples are adjusted to pH 3.5, and $CuSO_4$ is added before steam distillation. Under these circumstances, the AL is stoichiometrically converted to diacetyl (Mohr, Rea, & Cogan, 1997). The values obtained must then be corrected for the amount of "free" diacetyl present.

A similar method for measurement of AL was described by Richelieu, Hoalberg, and Nielsen (1997), except that $FeCl_3$ was used to convert AL to diacetyl at pH 3.1, which was then measured by headspace capillary GC. To measure diacetyl, samples were adjusted to pH 7.0 before analysis.

2,3-Butanediol can be quantified by extraction with methylene chloride. The extract is separated from the residue and dried with anhydrous Na_2SO_4, and its volume is reduced by rotary evaporation. Upon equilibrating the extract with water, the 2,3-butanediol passes into the aqueous phase, which is clarified with a mixture of $BaCl_2$, NaOH, and $ZnSO_4$; the butanediol is then estimated by GC.

Propionate and acetate are important in Swiss-type cheeses, which undergo a propionic acid fermentation. These can be extracted from the cheese by low concentrations of H_2SO_4 and analyzed by HPLC. A better method is to acidify and steam distill samples of cheese and then analyze the acids by GC or HPLC. Steam distillation gives good cleanup, but it is important that standards should be included, as each acid is recovered to a different extent by distillation. Pyruvic, lactic, acetic, and propionic acids may also be quantified in cheese extracts by ion-exchange HPLC using an Aminex 87H column.

23.4.2 Assessment of Lipolysis

The degree of lipolysis in cheese depends on the variety and ranges from slight to very extensive (see Chapter 11). Extensive lipolysis in internal bacterially ripened cheeses (e.g., Cheddar, Gouda, and Swiss) is undesirable, whereas in mold-ripened and some hard Italian cheeses lipolysis is essential for flavor development. A number of procedures have been developed to quantify lipolysis. The most effective methods use GC or HPLC but are tedious. Two rapid methods, the copper soaps method and the acid degree value method, have been used to assess lipolysis in cheese. Unfortunately, both have drawbacks, and thus there does not appear to be a simple, rapid method for accurately determining the level of free fatty acids in cheese.

The copper soaps assay is a spectrophotometric method used to estimate free fatty acids in milk and cheese. The copper soaps of free fatty acids in cheese, formed by reaction with $Cu(NO_3)_2$, are extracted using a mixture of chloroform, heptane, and methanol as solvent, followed by centrifugation. An aliquot of the solvent layer is added to a solution of sodium diethyldithiocarbamate in butan-1-ol. The color of the resulting complex is measured spectrophotometrically at 440 nm. The copper soaps of short-chain fatty acids ($< C_{10}$) may partition between the aqueous and apolar solvent phases and therefore may not be extracted completely. The technique has been criticized because of its poor recovery of short-chain fatty acids (which have the greatest effect on cheese flavor) and because it may have low reproducibility in the hands of an inexperienced technician.

The acid degree value (ADV) has been used for many years as an index of lipolysis in dairy products. Fat is released by the combined action of detergent, ion exchange, and heat and is separated from the aqueous phase of the cheese by centrifugation. Aliquots of the fat are weighed and dissolved in solvent, and the free fatty acids are titrated with alcoholic KOH (≈ 0.02 M) using methanolic phenolphthalein as the indicator. Variations of this method are also used, including the Bureau of Dairy Industry (BDI) method. The ADV method is tedious, but it is reliable in the hands of a competent operator for foods containing more than 2% fat.

An extraction and preparation procedure is generally necessary for gas chromatographic techniques, and in many cases esters of the free fatty acids are chromatographed. A wide variety of methods are available. The most critical differences between them are not the GC conditions but rather the extraction and derivitization procedures used for sample preparation. Reliable extraction and derivatization techniques are often tedious, and their tediousness and the capital cost of GC equipment are disadvantages of this technique for routine analysis.

Extraction with aqueous or organic solvents is used but has been criticized owing to its partitioning effects and the extraction of compounds that interfere with analysis. Fatty acids may also be adsorbed on various resins (e.g., Amberlyst A-26) during analysis, but such methods may result in incomplete recovery, glyceride hydrolysis, and low throughput. Free fatty acids are usually chromatographed as their methyl or butyl esters and are identified and quantified by reference to standards. Because complex procedures are often involved in sample preparation, an internal standard (usually a fatty acid with an odd number of C atoms) should be included early in the analytical procedure.

In addition to GC, fatty acids may be quantified by HPLC or, in certain cases, by enzyme-based assays. Specific chromatographic techniques are also available for triglycerides and partial glycerides.

Methyl ketones (alkan-2-ones), the characteristic flavor compounds in Blue cheese, may be analyzed by GC or their 2,4-dinitrophenylhydrazone derivatives may be analyzed by chromatographic or spectrophotometric techniques. Hydroxyacids, fatty acid lactones, and other volatile products of fatty acid catabolism may also be identified and quantified by GC or GC-mass spectrometry (GC-MS), which is discussed in Section 23.5 and in Chapter 12.

23.4.3 Assessment of Proteolysis

As discussed in Chapter 11, proteolysis is the principal and most complex biochemical event that occurs during the ripening of most cheese varieties. Therefore, a wide range of techniques for its assessment have been developed (see Fox, McSweeney, & Singh, 1995; McSweeney & Fox, 1993, 1997). Such techniques can be grouped into two main classes: nonspecific methods (which measure the formation of nitrogenous compounds soluble in various extractants or precipitants or the liberation of reactive groups) and specific methods (which resolve individual peptides). Nonspecific techniques are normally relatively straightforward, and some are suitable for assessment of ripening in quality control laboratories. However, more information about proteolysis is provided by techniques that resolve individual peptides (i.e., various types of electrophoresis and chromatography).

The amount of cheese nitrogen soluble in water or buffers at pH 4.6 is widely used as an index of proteolysis. These extracts contain numerous small and medium-sized peptides, amino acids and their degradation products, organic acids and their salts, and NaCl. Extraction with water efficiently separates the small peptides in cheese from proteins and large peptides. In Cheddar and most other cheeses that are not cooked to a high temperature, the principal proteolytic agents responsible for the production of water-soluble nitrogen (WSN) are the coagulant (principally chymosin) and to a lesser extent plasmin. Starter proteinases and peptidases play a relatively minor role in the hydrolysis of intact caseins and large polypeptides to water-soluble peptides, although they are active on many of the peptides produced by chymosin or plasmin.

WSN, usually expressed as a percentage of total N, varies with variety and increases throughout ripening; a typical value for mature Cheddar cheese is about 25% (see Chapter 11). Water is a suitable extractant only for cheeses the pH of which changes little during ripening. However, many cheese varieties are characterized by an increase in pH during ripening, and since the extractability of N varies with pH, determination of the amount of N soluble in pH 4.6 buffers is preferred (Figure 23–3). Other extractants used to fractionate cheese N include solutions of $CaCl_2$ and a mixture of chloroform and methanol. The latter solvent is particularly suitable for the extraction of hydrophobic peptides, which are often bitter.

Peptides in the primary extract from Cheddar cheese may be fractionated by a number of reagents, including 70% (w/v) ethanol or 2% (w/v) trichloroacetic acid (TCA), which precipitates large and intermediate-sized peptides; 12% (w/v) TCA, which precipitates all but quite short peptides (Figure 23–3); and 5% (w/v) phosphotungstic acid (PTA), in which only free amino acids (except lysine and arginine) and very short peptides (less than ≈ 600 Da) are soluble. PTA (5%)–soluble N is commonly used as an index of total free amino acids (Figure 23–3). WSN may also be fractionated by dialysis or by ultrafiltration (e.g., through 10 kDa membranes).

A number of rapid techniques can also be used to assess proteolysis. Cleavage of a peptide bond results in the liberation of an amino ($-NH_2$) and a carboxylic acid group. Amino groups may be quantified by reaction with trinitrobenzenesulfonic acid (TNBS), ninhydrin, fluorescamine, or *o*-phthaldialdehyde (OPA). The Cd-ninhydrin reagent is particularly sensitive for the amino group of free amino acids and therefore is used to estimate the liberation of amino acids in cheese (Figure 23–4; Folkertsma & Fox, 1992). Cd-ninhydrin–reactive groups correlate well with cheese age (Figure 23–5). Other rapid assays of proteolysis less widely used include measurement of absorbance at 280 nm (and thus peptides containing tryptophan and/or tyrosine residues) of TCA-soluble fractions of cheese or the formation of ammonia or enzyme-based assays for glutamic acid (see McSweeney & Fox, 1997).

Although the above techniques are rapid and suitable for routine analysis, the information obtained from nonspecific assays of proteolysis is limited. Since proteolysis results in the formation of variously sized peptides, techniques that resolve individual peptides provide much more information. They are also more time consuming. Electrophoresis in polyacrylamide gels with alkaline urea–containing buffers (urea-PAGE) is the most commonly used electrophoretic technique for studying cheese (Figure 23–6). Sodium-dodecylsulfate (SDS)-PAGE, which separates proteins based on their molecular mass, is widely used in biochemistry laboratories, but since the caseins have similar molecular masses, the resolution of intact caseins in cheese is poor. However, SDS-PAGE has been used successfully to separate the peptides produced from the caseins by proteolysis. Urea-PAGE with direct staining using Coomassie Brilliant Blue G250 is suitable for assessing the initial hydrolysis of the

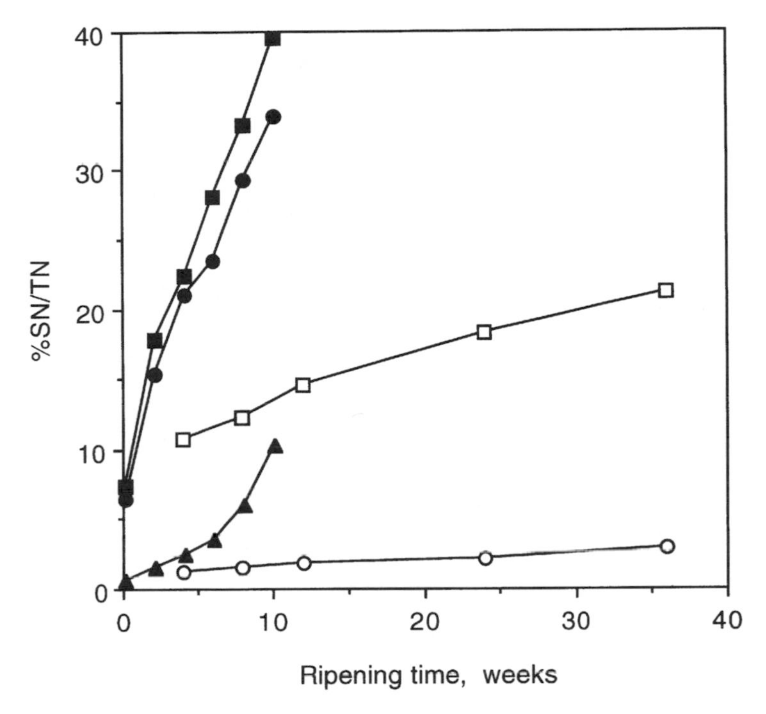

Figure 23–3 Formation of pH 4.6–soluble N (■), 12% trichloroacetic acid–soluble N (●), and 5% phosphotungstic acid (PTA)–soluble N (▲) in an Irish farmhouse Blue cheese and formation of water-soluble N (❑) and PTA-soluble N (❍) in Cheddar cheese.

caseins by chymosin or plasmin and for monitoring the subsequent degradation of the large peptides produced. It is possible to approximately quantify bands on urea-PAGE by densitometry. Capillary electrophoresis (CE) involves the separation of compounds in a buffer-filled capillary under the influence of an electric field (Figure 23–7). CE is likely to be a valuable research tool in the future for studying proteolysis in cheese and is currently being applied by a number of laboratories.

Chromatography is widely used to resolve peptides in cheese or cheese extracts. Anion-exchange chromatography on diethylaminoethyl (DEAE) cellulose in the presence of urea is very useful for separating the larger peptides in Cheddar cheese. High-performance liquid chromatography (e.g., FPLC, Pharmacia, Uppsala, Sweden) using a Mono-Q column is equivalent to DEAE cellulose and has the advantage that peak areas may be quantified easily. Size-exclusion chromatography is also used to separate shorter peptides in cheese extracts. Reverse-phase (RP)-HPLC on C_8 or C_{18} columns using an acetonitrile or water gradient with trifluoroacetic acid as the ion pair reagent and detection at about 214 nm is very successful and is the method of choice for separating the smaller water-soluble peptides from cheese (Figure 23–8).

The isolation of individual peptides from cheese has necessitated the development of various schemes to fractionate cheese N into relatively homogeneous subfractions. One such fractionation scheme, which was developed for Cheddar cheese, is shown in Figure 23–9. The isolation and identification (usually by N-terminal sequencing and/or mass spectrometry) of peptides from cheese have greatly increased the understanding of proteolysis at the molecular level in a number of varieties, particularly

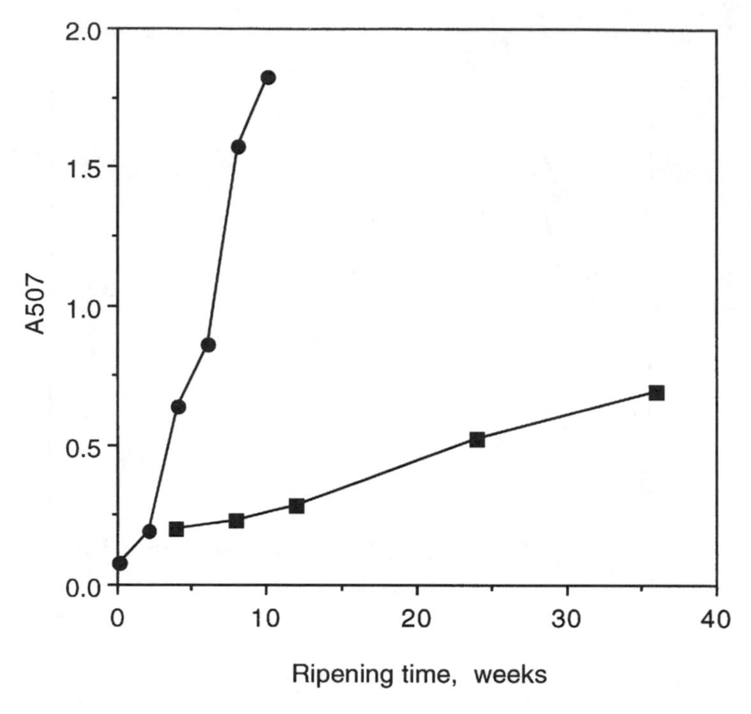

Figure 23–4 Liberation of Cd-ninhydrin–reactive amino groups in Cheddar (■) and an Irish farmhouse Blue cheese (●) during ripening. Water-soluble (Cheddar) and pH 4.6–soluble (Blue cheese) fractions were analyzed.

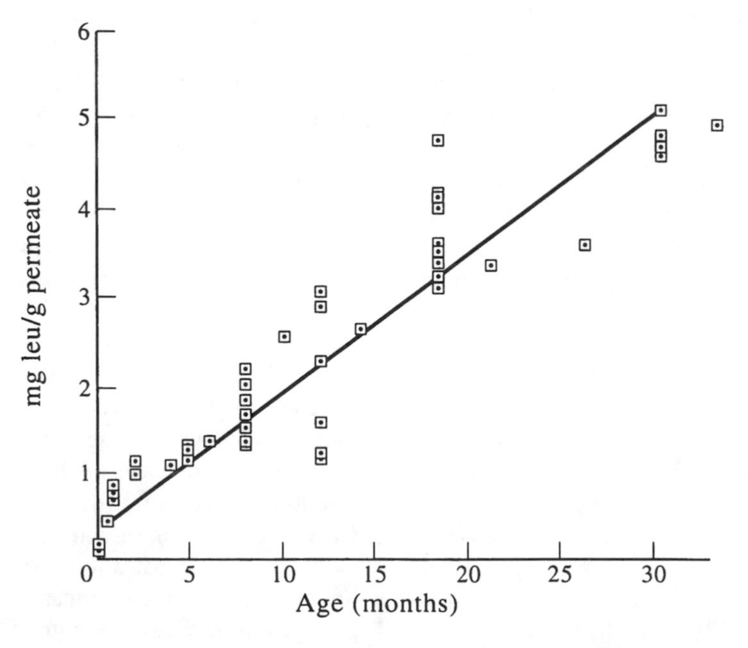

Figure 23–5 Total free amino acid concentrations (Cd-ninhydrin assay of 10 kDa untrafiltration permeates of water-soluble extracts) in Cheddar cheese as a function of age.

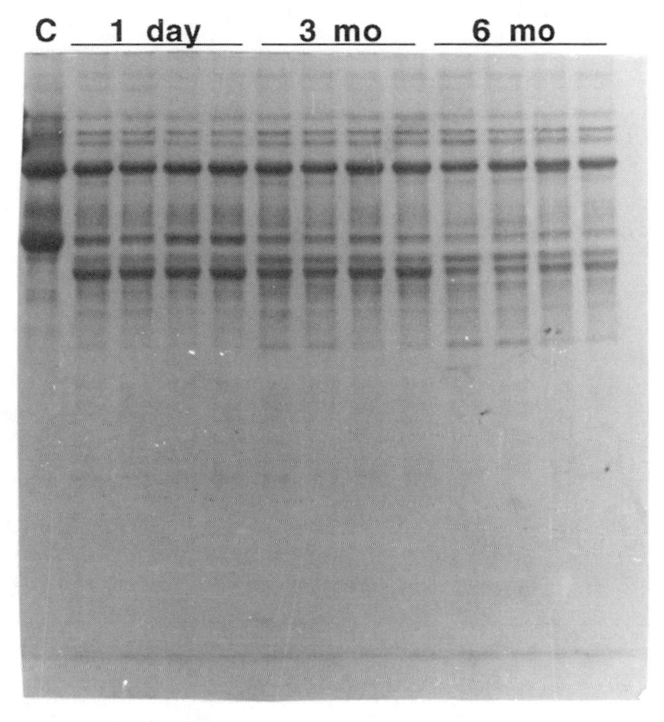

Figure 23–6 Urea-polyacrylamide gel electrophoretograms of Na caseinate (c) and Cheddar cheeses at 1 day, 3 months, and 6 months of ripening.

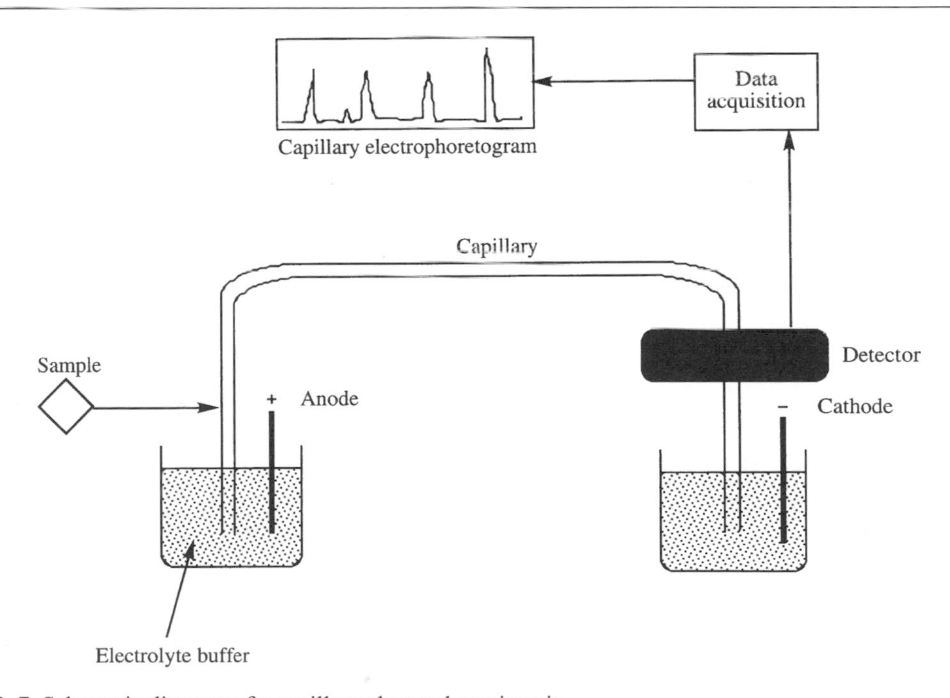

Figure 23–7 Schematic diagram of a capillary electrophoresis unit.

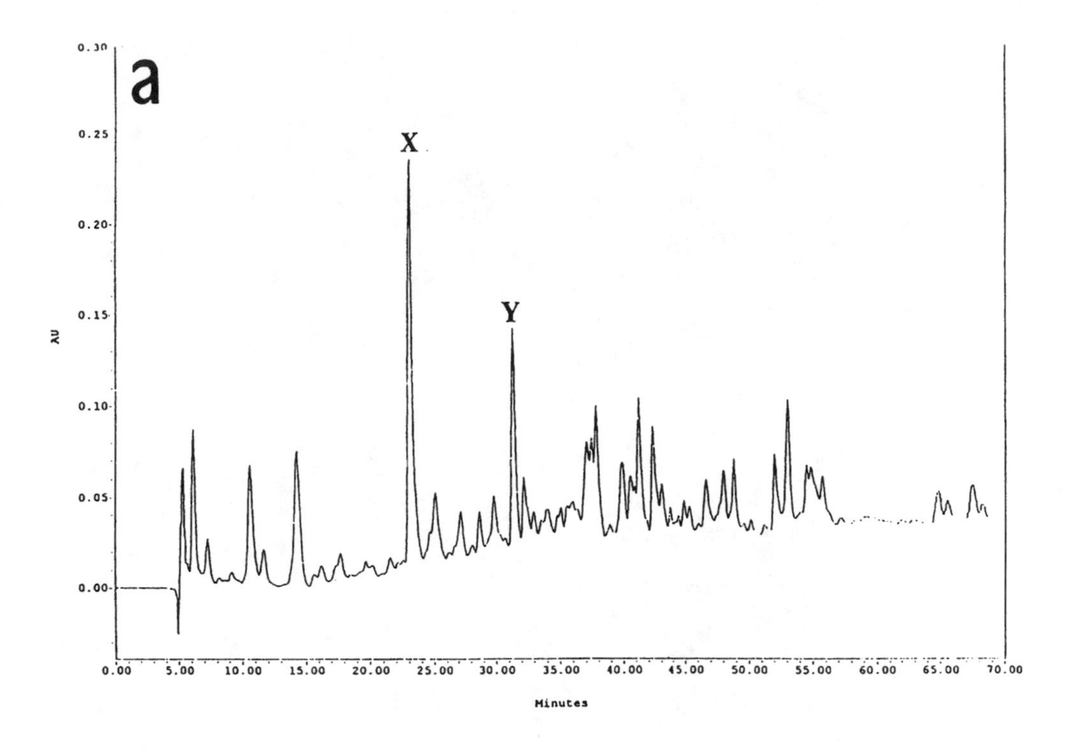

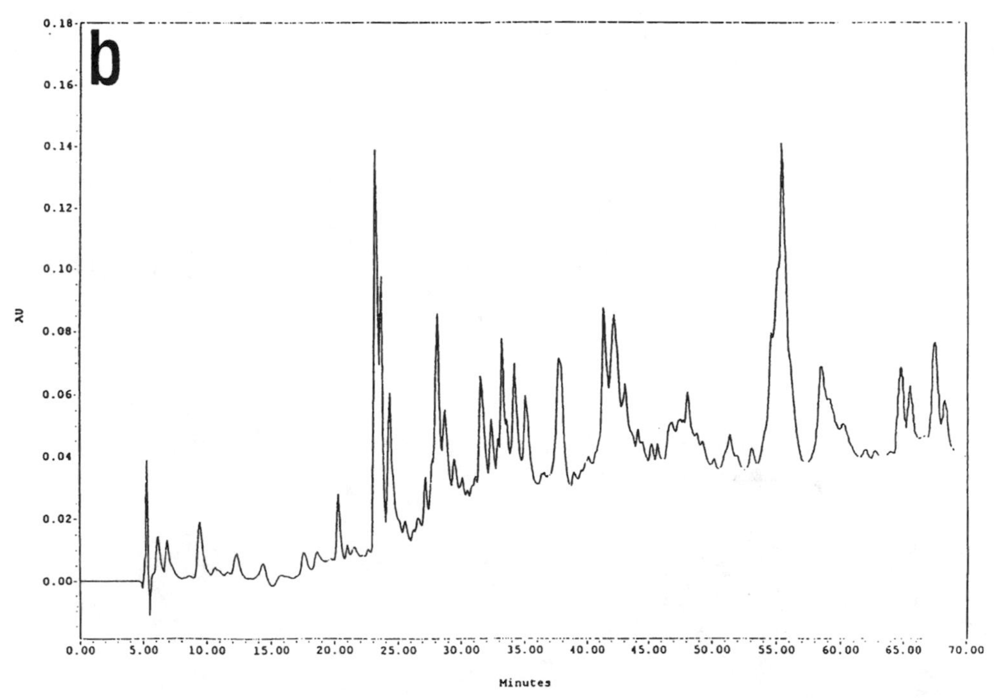

Figure 23–8 Reverse-phase (C_8) high-performance liquid chromatograms of the 70% ethanol-soluble (a) and -insoluble (b) fractions of a water-soluble extract from a 9-month-old Cheddar cheese. The peptides α_{s1}-CN f1-9 (X) and f1-13 (Y) are indicated.

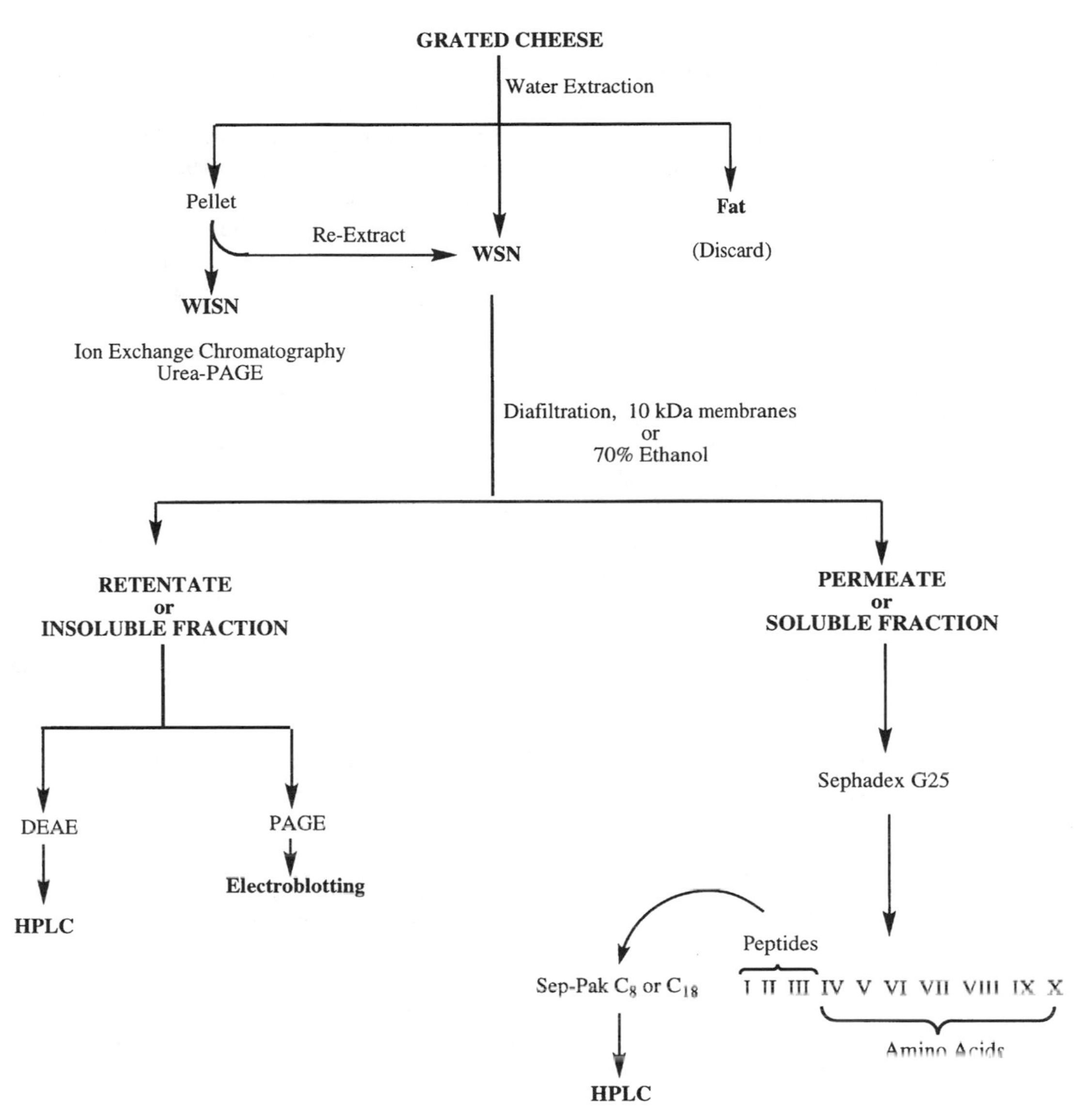

Figure 23–9 Fractionation scheme for cheese N. WSN, water-soluble N; WISN, water-insoluble N; DEAE, ion-exchange chromatography on DEAE-cellulose; PAGE, alkaline urea polyacrylamide gel electrophoresis; HPLC, reverse phase (C_8) high-performance liquid chromatography.

Cheddar, Gouda, and Parmesan (see Fox & McSweeney, 1996, and Chapter 11).

The ultimate products of proteolysis are amino acids. Many amino acids contribute directly to the flavor of cheese but are perhaps more important as precursors of other flavor compounds. As discussed above, total free amino acids can be estimated by rapid assays, while individual amino acids may be quantified by automated amino acid analyzers, which separate amino acids using an ion-exchange resin, followed by postcolumn derivitization (usually with ninhydrin) and spectrophotometric detection. Alternatively, fluorescent amino acid derivatives (e.g., dansyl, OPA, or *N*-(9-fluorenyl-methoxycarbonyl)-) may be prepared and then separated and quantified by RP-HPLC. GC techniques have also been developed to quantify free amino acids but are rarely used (see McSweeney & Fox, 1997).

Decarboxylation of amino acids leads to the formation of amines, many of which are biologically active and have physiological importance. The principal amines in cheese are histamine, tyramine, tryptamine, putrescine, cadaverine, and phenylethylamine (see Chapter 21). Biogenic amines may be determined spectrofluorometrically or by thin-layer chromatography but are now usually quantified by HPLC (often as their fluorescent derivatives) or by GC. Volatile products of amino acid catabolism (see Chapter 12) are usually quantified by GC interfaced with mass spectrometry (GC-MS).

23.5 TECHNIQUES TO STUDY VOLATILE FLAVOR COMPOUNDS

The most important aroma compounds in most cheeses are volatile (see Chapter 12). With the exception of volatile fatty acids and some volatile products of lactate or citrate metabolism (see Sections 23.4.1 and 23.4.2), cheese volatiles are almost universally quantified by various forms of GC-MS. GC-MS is a very sensitive technique with impressive resolving power (see Figure 12–8). Unfortunately, it is not suitable for use in an industrial context owing to the high running and capital cost of the instrument and slow throughput of samples. A number of different techniques are used to prepare samples for GC-MS, including vacuum distillation (at ≈ 70°C) and trapping in liquid N_2 (an approach that has been criticized because of artefact formation) and by trapping volatiles released into the cheese headspace.

23.6 MICROBIOLOGICAL ANALYSIS OF CHEESE

Cheese is a complex microbial ecosystem that may contain starter bacteria (*Lactococcus* spp., *Leuconostoc* spp., *Streptococcus* spp., and *Lactobacillus* spp.), nonstarter lactic acid bacteria (*Lactobacillus* spp., *Pediococcus* spp., and *Enterococcus* spp.), non-lactic-acid bacteria (coliforms, *Staphylococcus aureus*, *Brevibacterium* spp., *Micrococcus* spp., *Corynebacterium* spp., *Microbacterium* spp., and *Arthrobacter* spp.), yeast, and molds. Many cheese varieties contain most of these organisms. Only a limited number of effective selective media are available for enumerating the various microorganisms in cheese. Some of the more common media used in cheese microbiology are listed in Table 23–1.

Some workers have determined total numbers of bacteria in cheese using plate count agar (PCA). Such procedures are not very useful, since cheese contains many different types of bacteria, some of which die out (the starter) whereas others (the nonstarter LAB [NSLAB]) grow during ripening. In addition, many starter and NSLAB either do not grow or grow only poorly on PCA. The reason for this is that the level of nutrients in PCA is too low to sustain good growth of LAB, and its buffering capacity is poor. LAB are quite fastidious and require several amino acids and vitamins for growth. Therefore, media must contain rich sources of amino acids and peptides (peptones and yeast extract) and vitamins (yeast extract) and high levels of buffer to neutralize the large amounts of lactate produced from sugar metabolism during growth.

Many media have been developed to enumerate starter bacteria. Today, the medium of choice for enumerating lactococci is medium 17, containing lactose (LM-17), which was initially developed by Terzaghi and Sandine (1975) to estimate lactococcal phage. It contains sufficient amounts of all the nutrients necessary to support the growth of lactococci and a high concentration of β-glycerophosphate (19 g/L) as a buffer, which, unlike phosphate, does not chelate the Ca^{2+} required for adsorption of phage to its host. Although this medium is nonselective, it is used to count lactococci in cheese because these bacteria outnumber all other microorganisms in the cheese, especially during the early stages of ripening. As the cheese ripens, the medium becomes less selective. This is shown in Table

Table 23–1 Media and Incubation Conditions Used To Enumerate Different Types of Bacteria in Cheese

Group	*Medium*	*Temperature (°C)*	*Incubation Time and Conditions*
Lactococcus	LM-17 Agar[a]	30	3 days; spread or pour plate, aerobic
Leuconostoc	MRS + 20 μg vancomycin/ml[b]	30	3 days; spread or pour plate, aerobic
Sc. thermophilus	LM-17 Agar[a]	45	2 days; spread or pour plate, aerobic
Lb. helveticus	MRS agar, pH 5.4 agar[c]	45	3 days; spread plate, anaerobic
Lb. lactis	MRS agar, pH 5.4 agar[c]	45	3 days; spread plate, anaerobic
Nonstarter LAB	Rogosa agar (RA)[c]	30	5 days; pour plate, aerobic with overlay
Citrate utilizers	Calcium-citrate agar[d] (KCA)	30	2–3 days; spread plate, anaerobic
S. aureus	Baird Parker agar	37	72 hr; spread plate, aerobic
Coliforms	Violet red bile agar	30	18 hr; pour plate, aerobic
Enterococcus spp.	Kanamycin aesculin azide agar[e]	37	24 hr; pour plate, aerobic
Smear bacteria	Plate count agar containing 70 g/L NaCl	25	7–10 days; spread plate, aerobic
Yeast/molds	Yeast extract glucose chloramphenicol agar	25	3–5 days; pour plate, aerobic

[a] Terzaghi and Sandine (1975).
[b] Mathot, Kihal, Prevost, and Divies (1994).
[c] de Man, Rogosa, and Sharpe (1960).
[d] Nickels and Leesment (1964).
[e] Mossel, Baber, and Eeldering (1978).

23–2, based on a study in which the medium was used to enumerate the lactococci in the Spanish cheese, Armada, during ripening (Tornadijo, Fresno, Bernardo, Martin-Sarmiento, & Carballo, 1995). In a study of the Portuguese cheese, Picante de Beira Baixa, 66% of isolates from LM-17 were identified as enterococci (Freitas, Pais, Malcata, & Hogg, 1996).

Incubation of LM-17 plates at 45°C makes it relatively selective for *Sc. thermophilus*, since lactococci do not grow at this temperature and thermophilic lactobacilli grow poorly, if at all, in this medium. However, enterococci will grow at this temperature, and colonies should be examined to determine if they are enterococci or *Sc. thermophilus*.

Thermophilic cultures are often used today as adjuncts in commercial cheeses made with mesophilic cultures. If high counts of *Sc. thermophilus* (on LM-17 at 45°C; see Table 23–2) are found in a cheese made with a mesophilic culture, the count of *Lactococcus* (on LM-17 at 30°C) must be adjusted, since *Sc. thermophilus* will also grow on LM-17 at 30°C. Sometimes, the colonies of *Sc. thermophilus* are much smaller than those of *Lactococcus* spp., and the difference in size could be used to differentiate between them, but a smaller size is by no means an absolute feature of these bacteria.

MRS agar is a general purpose medium developed by de Man, Rogosa, and Sharpe (1960) for the enumeration of lactobacilli. Most other LAB can grow in it, but reducing the pH to 5.4 and increasing the temperature of incubation to 45°C make it more or less selective for the thermophilic lactobacilli found in starters. The mesophilic lactobacilli found in cheese are usually counted on Rogosa agar (RA) (Rogosa,

Table 23–2 Selectivity of Various Agars for the Isolation of Different Lactic Acid Bacteria (Number of Isolates) from Cheese

	Lactose Medium 17 agar (LM-17)[a]					*Mayeaux, Sandine, & Elliker Agar (MSE)*[b]					*Rogosa Agar (RA)*[c]					*Kanamycin Aesculin Agar (KAA)*[d]				
	Ripening Time (Weeks)					*Ripening Time (Weeks)*					*Ripening Time (Weeks)*					*Ripening Time (Weeks)*				
Genus	1	2	4	8	16	1	2	4	8	16	1	2	4	8	16	1	2	4	8	16
Lactococcus	28	25	15	5	7	16	5	4	–	–	–	–	–	–	–	–	–	–	–	–
Lactobacillus	1	11	7	23	13	2	15	21	31	38	36	34	37	38	31	–	–	–	–	6
Leuconostoc	1	1	–	–	1	15	8	6	7	–	–	4	–	–	–	–	–	–	–	–
Enterococcus	–	3	11	5	9	2	4	7	1	–	–	1	–	–	–	38	40	40	37	20
Non–lactic-acid bacteria	–	–	2	3	3	–	1	–	1	1	–	–	–	–	–	–	–	–	–	–

[a] Terzaghi and Sandine (1975).
[b] Mayeaux, Sandine, and Elliker (1962).
[c] de Man, Rogosa, and Sharpe (1960).
[d] Mossel, Baber, and Eeldering (1978).

Mitchell, & Wiseman, 1954). This medium contains a high concentration of acetate (0.225 M) and has a low pH (5.4), which make it quite selective for mesophilic lactobacilli. Some leuconostocs and pediococci may grow on it, but the thermophilic lactobacilli present in starters generally do not grow on this medium. It is currently widely used in cheese microbiology and appears to be quite selective for cheese containing different LAB (Table 23–2).

In the past, the medium of Mayeux, Sandine, and Elliker (1962) (MSE) was used to enumerate leuconostocs. This medium contains sucrose as the energy source, and many authors consider it to be selective for *Leuconostoc*. This is not so. The medium is nutritionally rich, and many, if not all, LAB will grow on it (Table 23–2). It may be selective if only dextran producers (very large colonies) are counted, since dextran formation is generally confined to *Leuconostoc* spp. However, not all leuconostocs produce dextrans. Since most leuconostocs are resistant to the antibiotic vancomycin, the addition of vancomycin (20 μg/ml) to an otherwise nutritionally adequate medium (e.g., MRS) makes it selective for these microorganisms. This method is acceptable when applied to starters, but mesophilic lactobacilli and pediococci are also naturally resistant to vancomycin. Both mesophilic lactobacilli and pediococci are often found in large numbers in cheese, which limits the usefulness of the medium, unless colonies are also examined microscopically and any doubtful ones examined by additional tests.

Calcium citrate agar (Nickels & Leesment, 1964) is very useful for enumerating citrate utilizers present in mesophilic cultures and dairy products. This is a differential, nonselective medium that is opaque owing to the presence of insoluble calcium citrate. As the citrate is metabolized, a clear halo develops around the colony. This medium can also be used to estimate total starter numbers if triphenyltetrazolium chloride (TCC) is added to it (1 ml of a filter-sterilized 1% [w/v] TTC solution per 100 ml of medium). TTC is reduced by the starter bacteria from a colorless soluble form to a red or pink insoluble form that precipitates around the starter colonies, making them more readily apparent. Further details can be found in International Dairy Federation (1997). If this medium is used to enumerate citrate-utilizing bacteria in cheese, colonies surrounded by halos should be examined microscopically, since many mesophilic lactobacilli, which are found in high numbers in ripened cheese, metabolize citrate and will produce halos around the colonies.

Numerous selective media have been proposed for enumerating enterococci, but none is completely reliable. Three commonly used media are m-*Enterococcus*, KF, and kanamycin aesculin azide (KAA) agars. *Ec. faecalis* produces dark red colonies on m-*Enterococcus* agar owing to reduction of TTC and precipitation of formazan around the colony. *Ec. durans*, *Ec. hirae*, *Ec. mundtii*, and *Streptococcus bovis* reduce TTC less strongly than *Ec. faecalis* and produce pale pink colonies, which are counted as enterococci. KF agar also contains TTC as an indicator of growth and NaN_3 as a selective agent. *Ec. faecalis* and *Ec. faecium* produce red colonies, but many enterococci of nonfecal origin (*Ec. mundtii*, *Ec. casseliflavus*, *Ec. pseudoavium*, *Ec. malodoratus*, and *Ec. raffinosus*) and *Sc. bovis* also do so. KAA is a selective, differential medium. Kanamycin is the selective agent, and the differential reaction is the hydrolysis of aesculin to glucose and aesculetin, which gives a black color in the presence of ferric citrate. *Ec. faecalis* and *Ec. faecium* hydrolyze aesculin and produce black colonies in the medium. Its selectivity for the newer *Enterococcus* spp. and for enterococci of nonfecal origin has not been studied. However, some lactobacilli do grow on KAA (Table 23–2).

There are no specific selective media for the bacteria found in the smear on cheese. All these bacteria are salt tolerant, so the addition of NaCl (e.g., 70 g/L) to an otherwise suitable medium (e.g., plate count agar) will inhibit the starter bacteria without affecting the smear bacteria. These bacteria grow slowly, and therefore plates must be incubated for at least 7 days.

Good selective media are available for *Staph. aureus* (Baird Parker agar) and coliforms (violet red bile agar). Yeasts and molds can be enumerated on potato dextrose agar acidified to pH 3.5 with lactic acid or on yeast glucose chlorotetracycline agar. The antibiotic (chlorotetracycline) inhibits the bacteria without affecting yeast and molds. Yeasts and molds are easily distinguished, since the latter produce large fluffy colonies rather than the small, opaque, sometimes glistening colonies of yeast.

23.7 OBJECTIVE ASSESSMENT OF CHEESE TEXTURE

The structure of cheese is formed by a casein network in brine containing fat globules (see Chapter 13 and Prentice, Langley, & Marshall, 1993). The moisture in cheese acts as a plasticizer, and its fat contributes to the rheological properties of cheese to an extent determined by the liquid fat:solid fat ratio. The uniformity of structure varies greatly between cheese varieties due to differences in the manufacturing protocol.

Since texture is a mechanical property of the cheese, it is more amenable to instrumental analysis than is cheese flavor, which results from the combination of a large number of sapid and volatile compounds (see Chapter 12). However, any instrumental technique used to measure cheese texture should involve fracture and mimic the mechanisms involved in sensoric grading or consumption. Obtaining representative samples for physical measurement is a problem when studying cheese texture. Techniques used to study cheese texture are discussed in Chapter 13.

23.8 SENSORY ANALYSIS OF CHEESE FLAVOR AND TEXTURE

The most important indices of cheese ripening are those on which the consumer decides to purchase the product: flavor, texture, and appearance. Ultimately, the quality of cheese is best assessed by sensory analysis (see Delahunty & Murray, 1997, and Muir, Banks, & Hunter, 1995, for reviews). A useful guide for the sensory evaluation of the texture of hard cheeses was prepared by Lavanchy et al. (1994). Texts on general sensory assessment of food products include Stone and Sidel (1993) and Piggott (1988).

Sensory analysis of cheese has, traditionally, been performed by individual cheese graders or by small groups of "experts" with the objective of assessing the potential of a cheese to develop a mature flavor or determining the quality at the point of sale. In such systems, marks are normally deducted for defects according to their intensity. Although characteristics sought by expert judges are usually those desired by consumers, grades awarded do not always correlate with consumer preference. Expert grading systems are widely used in the cheese industry and are, presumably, sufficiently reliable to ensure consistent product quality, but they are less useful in research.

To assess whether there is a difference between cheeses, simple difference tests, such as the paired comparison or triangle tests, may be adequate. However, such tests always leave questions unanswered about the attributes of the cheese and to what extent they vary. The International Dairy Federation (1987) method for cheese grading involves comparing the cheese to a reference cheese selected as a "standard" and rating the cheese in terms of a list of defects. However, the selection of the standard cheese is subjective, and it is difficult to maintain a consistent reference standard.

Descriptive evaluation by a panel of trained assessors (10–20 persons) is the method of choice for determining the sensory profile of a cheese and can determine the influence of processing changes on individual sensory characteristics. Such panels are established by selecting potential assessors based on their ability to perceive sensory stimuli. Assessors taste a range of cheeses and during group discussions agree on and then define a list of descriptors that encompass the sensory attributes of the cheese (Exhibit 23–1). In an alternative approach (free choice profiling), assessors use a personal list of

Exhibit 23–1 A Descriptive Vocabulary Used To Characterize Cheddar Cheese Flavor

Cheddary: Flavor you associate with a typical Cheddar
Creamy: Containing cream, resembling cream
Buttery: Of the nature of or containing butter
Pungent: Physically penetrating sensation in the nasal cavity; sharp smelling or tasting irritant
Moldy: The combination of aromatics generally associated with molds; they are usually earthy, dirty, stale, musty, and slightly sour
Caramel: Burnt sugar or syrup; toffee made with sugar that has been melted further
Burnt caramel: Sweet flavor notes, dominated by a taste not unlike burnt milk
Soapy: Detergentlike, similar to that of food tainted with a cleansing agent
Smoky: The penetrating acrid aromatic of charred wood; tainted by exposure to smoke
Fruity: The aromatic blend of different fruity identities
Mushroom: Organic; the aromatics associated with raw mushrooms
Rancid: Sour milk, fatty, oxidized; having the rank unpleasant taste or smell characteristic of oils and fats when no longer fresh
Nutty: The nonspecific nutlike aromatic that is characteristic of several different nuts (e.g., peanuts, hazelnuts, pecans)
Sweaty: The aromatics reminiscent of perspiration-generated food odor; sour, stale, slightly cheesy, moist, stained or odorous with sweat
Balanced: Mellow, smooth, clean; in equilibrium, well arranged or disposed, with no constituent lacking or in excess
Processed: Tastes of plastic, packaging, shallow; to taste artificial; made by melting, blending, and frequently emulsifying other cheeses
Salty: Fundamental taste sensation of which sodium chloride is typical
Sweet: Fundamental taste sensation of which sucrose is typical
Acidic: Sour, tangy, sharp, citruslike; the fundamental taste sensations of which lactic acid and citric acids are typical
Bitter: Chemical-like, aspirin; taste sensations of which caffeine and quinine are typical
Strength: Intensity of flavor, mildness, and maturity
Astringent: Mouth-drying, harsh; the complex of drying, puckering, and shrinking sensations in the lower cavity causing contraction of the body tissues

descriptors. The assessors are then trained to quantify the intensity of each descriptor and score their perceptions by placing marks on a line scale with defined upper and lower limits. Trained assessors should be able to differentiate between attributes, should give reproducible scores, and should agree with other panelists. Assessors who cannot meet these criteria should be removed from the panel. Descriptive sensory evaluation generates much data concerning the product and computer-assisted data capture is therefore beneficial. Data are analyzed statistically by analysis of variance or, preferably, by multivariate techniques such as principal component analysis, which presents data as a multidimensional matrix that explains as much of the variance as possible (Figure 23–10). Considerable work is involved in establishing a reliable

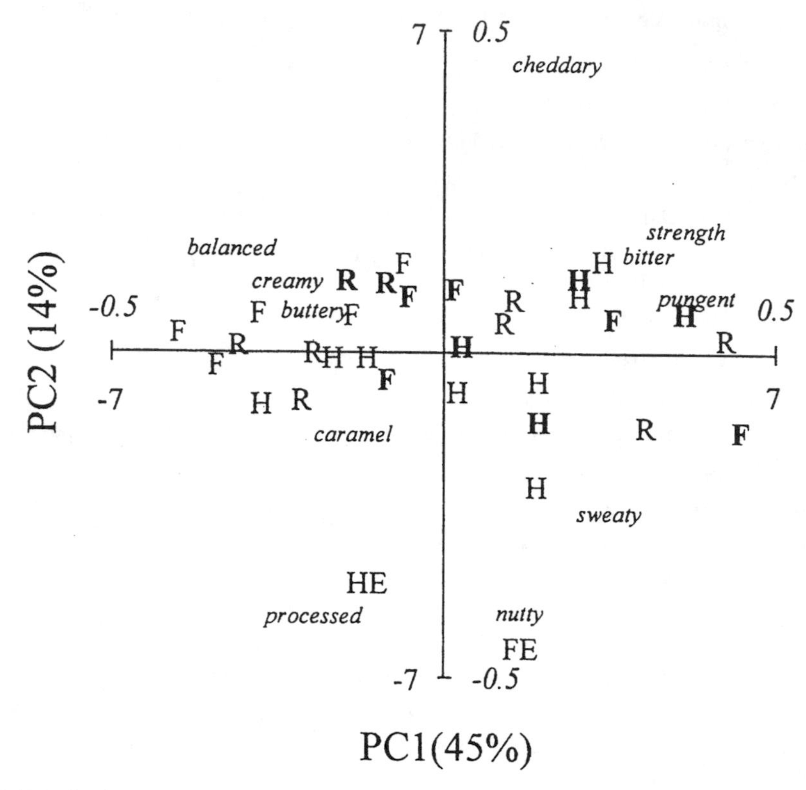

Figure 23–10 Principal component scores for 33 cheeses of varying fat content and loadings of descriptors (in italics and plotted on the scale between –0.5 and 0.5) on the first two components from principal components analysis of descriptive sensory data. F, full-fat Cheddar; R, reduced-fat Cheddar; H, half-fat Cheddar; FE, full-fat Edam; HE, half-fat Edam. Cheeses in bold were mature.

descriptive panel for the sensory analysis of cheese and the effort involved is often underestimated by persons unfamiliar with sensory science.

The question of which cheeses consumers *prefer* is a very important but subjective consideration and can be answered only by an untrained consumer panel, representative of the target market for the cheese. Since a random and untrained group of individuals will vary widely in their preferences and their ability to perceive stimuli, it is essential to have sufficient people on such a panel to obtain reliable results. A minimum number of 50–60 targeted consumers has been recommended.

The power of any sensory analysis technique will be increased if external variables are excluded. Sensory assessment should be performed in a sensory laboratory equipped with individual booths with controlled lighting and ventilation. The order in which cheeses are tasted is very important: the first cheese a panelist tastes is likely to receive a higher score than subsequent cheeses and the perception of a mild-flavored cheese could be influenced by a stronger-tasting cheese tasted immediately before (or vice versa). Order of tasting effects should be eliminated by randomizing the order in which the cheeses are presented and balancing the presentation so that each cheese is tasted an equal number of times by the panel. Assessor fatigue must also be considered; there is a limit (undefined) to the number of cheeses that may be tasted at one session. Assessors should not have prior knowledge of the cheeses, nor of the expected or desired results. However, such a situation may be difficult to achieve with in-house panels.

Obtaining reliable results from sensory analysis of cheese is far from straightforward. It is deceptively simple to establish a taste panel, but practical problems may arise that negate the best efforts of persons inexperienced in sensory analysis. It is therefore desirable that a person thoroughly familiar with sensory science and its pitfalls be responsible for the sensory assessment of cheese.

23.9 DETECTION OF INTERSPECIES ADULTERATION OF MILKS AND CHEESES

Adulteration of ewe or goat milk for cheesemaking with less expensive bovine milk has led to the development of techniques that permit the detection of mixed-species milks (see McSweeney & Fox, 1993). Various authors have used differences in fatty acid profiles to detect adulteration. Such methods have limited sensitivity and will not detect adulteration with skim milk, but they can be applied to mature cheese, since relatively small changes occur in fatty acid composition during maturation. Differences in triglycerides may also be used to identify milks from different species. The absence of β-carotene from sheep and goat milk may permit the detection of bovine milk in caprine or ovine milk.

Differences between HPLC profiles of tryptic hydrolysates of caseins from different species have been demonstrated. However, the effect of proteolysis during cheese ripening may interfere with this approach. Electrophoretic techniques have also been developed to detect mixtures of milks from different species. However, proteolysis during cheese ripening may interfere with these procedures. Interspecies differences between para-κ-caseins (which are degraded only slightly or not at all during ripening) or γ-caseins detected by isoelectric focusing have also been used as indices of adulteration. Interspecies differences in the electrophoretic mobility of whey proteins may be used to identify mixed-species milks but not cheeses.

The higher xanthine oxidase activity of bovine milk has been used to detect the adulteration of goat milk with cow milk. The test is simple and rapid and can be applied to raw or pasteurized ($< 75°C \times 20$ s) milk or fresh cheese; it allows detection of 2% added bovine milk. Differences in the Ca:Mg ratio in cheeses made from cow or sheep milk have also been used as an index of adulteration.

Immunological methods are well suited for analysis of mixed-species milks owing to their sensitivity and specificity and their availability in kit form. Antisera prepared against proteins from a particular milk may react not only with proteins from that milk but with the milks of other species also, but techniques have been developed to overcome this. Caseins have relatively low antigenicity, but immunodotting techniques have been used to overcome this problem. An indirect enzyme-linked immunosorbent assay (ELISA) for bovine caseins has been developed and used to assay for bovine milk in sheep milk and cheese.

REFERENCES

Baer, A., & Collin, J.C. (1993). *Determination of residual activity of milk-clotting enzymes in cheese: Specific identification of chymosin and its substitutes in cheese* [Bulletin No. 284]. Brussels: International Dairy Federation.

Cunniff, P. (Ed.). (1995). *Official methods of analysis of AOAC International* (16th ed., Vols. 1, 2). Arlington, VA: Association of Official Analytical Chemists International.

Delahunty, C.M., & Murray, J.M. (1997). Organoleptic evaluation of cheese. In *Proceedings of the Fifth Cheese Symposium*. Dublin: Teagasc.

de Man, J.C., Rogosa, M., & Sharpe, M.E. (1960). A medium for the culturation of lactobacilli. *Journal of Applied Bacteriology*, *23*, 130–135.

Folkertsma, B., & Fox, P.F. (1992). Use of Cd-ninhydrin reagent to assess proteolysis in cheese during ripening. *Journal of Dairy Research*, *59*, 219–224.

Fox, P.F., & McSweeney, P.L.H. (1996). Proteolysis in cheese during ripening. *Food Reviews International*, *12*, 457–509.

Fox, P.F., McSweeney, P.L.H., & Singh, T.K. (1995). Methods for assessing proteolysis in cheese during ripening. In E.L. Malin & M.H. Tunick (Eds.), *Chemistry of structure-function relationships in cheese*. New York: Plenum.

Freitas, A.C., Pais, C., Malcata, F.X., & Hogg, T.A. (1996). Microbiological characterisation of Picante de Beira Baixa cheese. *Journal of Food Protection*, *59*, 155–160.

International Dairy Federation. (1985). Milk and milk products. Methods of sampling [Standard 50B]. Brussels: Author.

International Dairy Federation. (1987). Sensory evaluation of dairy products [Standard 99A]. Brussels: Author.

International Dairy Federation. (1995). Standards. In *Catalogue of IDF publications*. Brussels: Author.

International Dairy Federation. (1997). Dairy starter cultures of lactic acid bacteria (LAB): Standard of identity [Standard 149Ax]. Brussels: Author.

Lavanchy, A., Berodier, F., Zannoni, M., Noël, Y., Adamo, C., Sequella, J., & Herrero, L. (1994). *A guide to the sensory evaluation of texture of hard and semi-hard cheeses*. Paris: Institut National de la Recherche Agronomique.

Marcos, A. (1993). Water activity in cheese in relation to composition, stability and safety. In P.F. Fox (Ed.), *Cheese: Chemistry, physics and microbiology* (2d ed., Vol. 1). London: Chapman & Hall.

Marier, J.R., & Boulet, M. (1958). Direct determination of citric acid in milk with an improved pyridine-acetic anhydride method. *Journal of Dairy Science*, *41*, 1683–1692.

Mathot, A.G., Kihal, M., Prevost, H., & Divies, C. (1994). Selective enumeration of *Leuconostoc* on vancomycin agar media. *International Dairy Journal*, *4*, 459–469.

Mayeux, J.V., Sandine, W.E., & Elliker, P.R. (1962). A selective medium for detecting *Leuconostoc* organisms in mixed strain starter cultures. *Journal of Dairy Science*, *45*, 655–656.

McSweeney, P.L.H., & Fox, P.F. (1993). Cheese: Methods of chemical analysis. In P.F. Fox (Ed.), *Cheese: Chemistry, physics and microbiology* (2d ed., Vol. 1). London: Chapman & Hall.

McSweeney, P.L.H., & Fox, P.F. (1997). Chemical methods for the characterization of proteolysis in cheese during ripening. *Lait*, *77*, 41–76.

Mohr, B., Rea, M.C., & Cogan, T.M. (1997). A new method for the determination of 2-acetolactic acid in dairy products. *International Dairy Journal*, *7*, 701–706.

Mossel, D.A.A., Baber, P.G.H., & Eeldering, J. (1978). Streptokokken der Lancefield Gruppen D in Lebensmittel und Trinkwasser: Ihre Bedeutung, Erfassung und Bekämpfung. *Archiv für Lebensmittelhygiene*, *29*, 121–127.

Muir, D., Banks, J.M., & Hunter, E.A. (1995). Sensory properties of cheese. In *Proceedings of the Fourth Cheese Symposium*. Dublin: Teagasc.

Nickels, C., & Leesment, H. (1964). Methode zur differenziearung und quantitativen Bestimmung von Säurewecberbakterien. *Milchwissenschaft*, *19*, 374–378.

Piggott, J.R. (1988). *Sensory analyses of foods* (2d ed.). London: Elsevier Applied Sciences.

Prentice, J.H., Langley, K.R., & Marshall, R.J. (1993). Cheese rheology. In P.F. Fox (Ed.), *Cheese: Physics, chemistry and microbiology* (2d ed., Vol. 1). London: Chapman & Hall.

Richelieu, M., Hoalberg, U., & Nielsen, J.C. (1997). Determination of α-acetolactic acid and volatile compounds by head-space gas chromatography. *Journal of Dairy Science*, *80*, 1918–1925.

Rogosa, M., Mitchell, J.A., & Wiseman, R.F. (1954). A selective medium for the isolation and enumeration of oral lactobacilli. *Journal of Dental Research*, *10*, 682–689.

Stone, H., & Sidel, J.L. (1993). *Sensory evaluation practices* (2d ed.). New York: Academic Press.

Terzaghi, B.E., & Sandine, W.E. (1975). Improved medium for lactic streptococci and their bacteriophages. *Applied and Environmental Microbiology*, *29*, 807–813.

Tornadijo, M.E., Fresno, J.M., Bernardo, A., Martin-Sarmiento, R., & Carballo, J. (1995). Microbiological changes throughout the manufacturing and ripening of a Spanish goat's raw milk cheese (Armada variety). *Lait*, *75*, 551–570.

Walsh, B., & Cogan, T.M. (1974). Separation and estimation of diacetyl and acetoin in milk. *Journal of Dairy Research*, *41*, 25–30.

Table of Sources

CHAPTER 1

Table 1–1. *Source:* Data from Scott, 1986.
Table 1–2. Courtesy of the Food and Agricultural Organization of the United Nations, 1997, Rome, Italy.
Table 1–3. Courtesy of the International Dairy Federation, 1995, Brussels, Belgium.

CHAPTER 2

Figure 2–2. *Source:* Reprinted with permission from P.F. Fox and P.H.L. McSweeney, *Dairy Chemistry and Biochemistry*, pp. 226–227, © 1998, Aspen Publishers, Inc.
Figure 2–3. *Source:* Reprinted with permission from P.F. Fox and P.H.L. McSweeney, *Dairy Chemistry and Biochemistry*, p. 460, © 1998, Aspen Publishers, Inc.

CHAPTER 3

Table 3–1. *Source:* Reprinted with permission from P.F. Fox and P.H.L. McSweeney, *Dairy Chemistry and Biochemistry*, p. 2, © 1998, Aspen Publishers, Inc.
Table 3–2. *Source:* Reprinted with permission from P.F. Fox and P.H.L. McSweeney, *Dairy Chemistry and Biochemistry*, p. 21, © 1998, Aspen Publishers, Inc.
Figure 3–1. *Source:* Reprinted with permission from P.F. Fox and P.H.L. McSweeney, *Dairy Chemistry and Biochemistry*, p. 22, © 1998, Aspen Publishers, Inc.
Figure 3–2. *Source:* Reprinted with permission from P.F. Fox and P.H.L. McSweeney, *Dairy Chemistry and Biochemistry*, p. 24, © 1998, Aspen Publishers, Inc.
Figure 3–3. *Source:* Reprinted with permission from P.F. Fox and P.H.L. McSweeney, *Dairy Chemistry and Biochemistry*, p. 25, © 1998, Aspen Publishers, Inc.
Figure 3–4. *Source:* Data from Jenness and Patton, 1959.
Table 3–3. *Source:* Reprinted with permission from W.W. Christie, Composition and Structure of Milk Lipids, in *Advanced Dairy Chemistry, Vol. 2, Lipids,* P.F. Fox, ed., pp. 1–36, © 1995, Aspen Publishers, Inc.
Table 3–4. *Source:* Reprinted with permission from W.W. Christie, Composition and Structure of Milk Lipids, in *Advanced Dairy Chemistry, Vol. 2, Lipids,* P.F. Fox, ed., pp. 1–36, © 1995, Aspen Publishers, Inc.
Table 3–5. *Source:* Reprinted with permission from W.W. Christie, Composition and Structure of Milk Lipids, in *Advanced Dairy Chemistry, Vol. 2, Lipids,* P.F. Fox, ed., pp. 1–36, © 1995, Aspen Publishers, Inc.
Table 3–6. *Source:* Reprinted with permission from P.F. Fox and P.H.L. McSweeney, *Dairy Chemistry and Biochemistry*, p. 95, © 1998, Aspen Publishers, Inc.
Table 3–7. *Source:* Reprinted with permission from P.F. Fox and P.H.L. McSweeney, *Dairy Chemistry and Biochemistry*, p. 148, © 1998, Aspen Publishers, Inc.
Figure 3–5. *Source:* Data from Bernhart, 1961.
Table 3–8. *Source:* Reprinted with permission from P.F. Fox and P.H.L. McSweeney, *Dairy Chemis-*

try and Biochemistry, p. 164, © 1998, Aspen Publishers, Inc.

Figure 3–6. *Source:* Reprinted with permission from H.E. Swaisgood, Chemistry of the Caseins, in *Advanced Dairy Chemistry, Vol. 1, Proteins*, P.F. Fox, ed., pp. 63–110, © 1992.

Figure 3–7. *Source:* Reprinted with permission from H.E. Swaisgood, Chemistry of the Caseins, in *Advanced Dairy Chemistry, Vol. 1, Proteins*, P.F. Fox, ed., pp. 63–110, © 1992.

Figure 3–8. *Source:* Reprinted with permission from H.E. Swaisgood, Chemistry of the Caseins, in *Advanced Dairy Chemistry, Vol. 1, Proteins*, P.F. Fox, ed., pp. 63–110, © 1992.

Figure 3–9. *Source:* Reprinted with permission from H.E. Swaisgood, Chemistry of the Caseins, in *Advanced Dairy Chemistry, Vol. 1, Proteins*, P.F. Fox, ed., pp. 63–110, © 1992.

Figure 3–10. *Source:* Reprinted with permission from H.E. Swaisgood, Chemistry of the Caseins, in *Advanced Dairy Chemistry, Vol. 1, Proteins*, P.F. Fox, ed., pp. 63–110, © 1992.

Figure 3–11. *Source:* Reprinted with permission from P.F. Fox and P.H.L. McSweeney, *Dairy Chemistry and Biochemistry*, p. 161, © 1998, Aspen Publishers, Inc.

Table 3–9. *Source:* Reprinted with permission from J. McMahon and R.J. Brown, Composition and Integrity of Casein Micelles: A Review, *Journal of Dairy Science*, Vol. 67, pp. 499–512, © 1984, American Dairy Science Association.

Figure 3–12. *Source:* P. Walstra and R. Jenness, *Dairy Chemistry and Physics*, © 1984, John Wiley and Sons, Inc. Reprinted by permission of John Wiley and Sons, Inc.

Figure 3–13. *Source:* Data from Holt, 1994.

Table 3–10. *Source:* Reprinted with permission from P.F. Fox and P.H.L. McSweeney, *Dairy Chemistry and Biochemistry*, p. 254, © 1998, Aspen Publishers, Inc.

Table 3–11. *Source:* Reprinted with permission from P.F. Fox and P.H.L. McSweeney, *Dairy Chemistry and Biochemistry*, p. 437, © 1998, Aspen Publishers, Inc.

CHAPTER 4

Figure 4–2. *Source:* Reprinted with permission from A.J. Bramley and C.H. McKinnon, The Microbiology of Raw Milk, *Dairy Microbiology, 2nd Ed.*, Vol.1, pp. 163–208, © 1990.

Figure 4–3. *Source:* Reprinted with permission from R.K. Robinson, *Modern Dairy Technology 2nd Ed.*, Vol.1, © 1994.

Figure 4–6. *Source:* From Principles of Biochemistry by Lehninger, Nelson, Cox, © 1993, 1982 by Worth Publishers, Inc. Used with permission.

Figure 4–7. Courtesy of Westfalia Separator, Inc., Northvale, New Jersey.

CHAPTER 5

Figure 5–1. *Source:* Reprinted with permission from T. Lodics and L. Steenson, Characterisation of Bacteriophages and Bacteria Indigenous to a Mixed-strain Cheese Starter, *Journal of Dairy Science*, Vol. 73, pp. 2685–2696, © 1990, American Dairy Science Association.

Figure 5–2. *Source:* Reprinted with permission from Accolas *et al*, Etude des Iteractions Entre Diverse Bacteries Lactiques Thermophiles et Mesophiles, en Realation avec la Fabrication des Fromages a Pate Cuite, *Le Lait*, Vol. 51, pp. 249–272, © 1971, Editions Scientifiques et Medicales Elsevier.

Figure 5–3. *Source:* Reprinted with permission from K. Cooper and E.B. Collins, Influence of Temperature on Growth of *Leuconostoc cremoris*, *Journal of Dairy Science*, Vol. 61, pp. 1085–1088, 1978; D.A. Lee and E.B. Collins, Influence of Temperature on Growth of *Streptococcus cremoris* and *Streptococcus lactis, Journal of Dairy Science*, Vol. 59, pp. 405–409, © 1976; and F.G. Martley, Temperature Sensitivities of Thermophilic Starter Strains, *New Zealand Journal of Dairy Science Technology*, Vol. 18, pp. 191–196, © 1983.

Figure 5–4. *Source:* Reprinted with permission from V. Bottazzi and F. Bianchi, *I Microrganismi Lattiero-caseari al Micrscopio Elettronico a Scansione*, © 1984, Edi-ermes.

Figure 5–5. Courtesy of M. Vancanneyt, University of Ghent, Belgium.

Table 5–5. *Source:* Reprinted from Kunji *et al*, The Proteolytic Systems of Lactic Acid Bacteria, *Antonie van Leeuwenhoek*, Vol. 70, pp. 187–221, © 1996, with kind permission from Kluwer Academic Publishers.

Figure 5–7. *Source:* Reprinted from *FEMS Microbiology Review*, Vol. 12, B. Poolman, Energy Transduction in Lactic Acid Bacteria, pp. 125–148, © 1993, with permission from Elsevier Science.

Table 5–7. *Source:* Data from Cogan, 1972.
Figure 5–14. Courtesy of H. Neve, Institute of Microbiology, Federal Dairy Research Center, Kiel, Germany.
Figure 5–15. Courtesy of H. Neve, Institute of Microbiology, Federal Dairy Research Center, Kiel, Germany.
Figure 5–16. Courtesy of H. Neve, Institute of Microbiology, Federal Dairy Research Center, Kiel, Germany.
Figure 5–17. *Source:* Reprinted with permission from Pearce *et al*, Bacteriophage Multiplication Characteristics in Cheddar Cheesemaking, *New Zealand Journal of Dairy Science Technology,* Vol. 5, pp. 145–150, © 1970, New Zealand Dairy Research Institute.
Figure 5–18. *Source:* Reprinted with permission from R. Klaenhammer, Genetic Characterization of Multiple Mechanisms of Phage Defense from a Prototype Phage-insensitive Strain, *Lactococcus lactis ME2*, *Journal of Dairy Science*, Vol. 72, pp. 3429–3443, © 1989, American Dairy Science Association.
Figure 5–20. Courtesy of H. Neve, Institute of Microbiology, Federal Dairy Research Center, Kiel, Germany.
Figure 5–21. Courtesy of the Netherlands Institute for Dairy Research.

CHAPTER 6

Figure 6–2. *Source:* Reprinted with permission from P.F. Fox and P.H.L. McSweeney, *Dairy Chemistry and Biochemistry*, p. 383, © 1998, Aspen Publishers, Inc.
Table 6–1. *Source:* Reprinted with permission from P.F .Fox and P.H.L. McSweeney, *Dairy Chemistry and Biochemistry*, p. 384, © 1998, Aspen Publishers, Inc.
Figure 6–4. *Source:* Reprinted with permission from P.F. Fox and P.H.L. McSweeney, *Dairy Chemistry and Biochemistry*, p. 386, © 1998, Aspen Publishers, Inc.
Figure 6–8. *Source:* Reprinted with permission from Guinee *et al,* Ultrafiltration in Cheesemaking, *Proceedings of the 3rd Cheese Symposium*, T.M. Cogan, ed., pp. 49–59, © 1992, Dairy Products Research Center, Moorepark, Fermoy, Ireland.
Figure 6–9. *Source:* Reprinted from *International Dairy Journal*, Vol. 41, vanHooydonk *et al,* The Renneting Properties of Heated Milk, pp. 3–18, © 1987 with permission from Elsevier Science.
Figure 6–10. *Source:* Reprinted with permission from P.F. Fox and P.H.L. McSweeney, *Dairy Chemistry and Biochemistry*, p. 389, © 1998, Aspen Publishers, Inc.
Figure 6–11. *Source:* Reprinted with permission from P.F. Fox and P.H.L. McSweeney, *Dairy Chemistry and Biochemistry*, p. 390, © 1998, Aspen Publishers, Inc.
Figure 6–13. *Source:* Data from Bohlin Rheological VOR Manual, Gloucestershire, England.
Figure 6–14. *Source:* Reprinted with permission from J.J. Mayes and B.J. Sutherland, Further Notes on Coagulum Firmness and Yield in Cheddar Cheese Manufacture, *Australian Journal of Dairy Technology*, Vol. 44, pp. 47–48, © 1989.
Figure 6–15. *Source:* Reprinted with permission from P.F. Fox and P.H.L. McSweeney, *Dairy Chemistry and Biochemistry*, p. 390, © 1998, Aspen Publishers, Inc.
Figure 6–16. *Source:* Reprinted with permission from F.A. Payne, Automatic Control of Coagulation Cutting in Cheese Manufacture, *Applied Engineering in Agriculture*, Vol. 11, pp. 691–697, © 1995, American Society of Agricultural Engineers.
Figure 6–17. *Source:* Reprinted with permission from F.A. Payne, Automatic Control of Coagulation Cutting in Cheese Manufacture, *Applied Engineering in Agriculture*, © 1995, American Society of Agricultural Engineers.
Figure 6–18. *Source:* Reprinted with permission from F.A. Payne, Automatic Control of Coagulation Cutting in Cheese Manufacture, *Applied Engineering in Agriculture*, © 1995, American Society of Agricultural Engineers.
Figure 6–19. *Source:* Guinee *et al,* Effect of Milk Protein Standardization by Ultrafiltration on the Manufacture, Composition and Maturation of Cheddar Cheese, *Journal of Dairy Research*, Vol. 61, pp. 117–131, © 1994. Reprinted with the permission of Cambridge University Press.
Figure 6–20. *Source:* Reprinted with permission from Guinee *et al*, The Effects of Composition and Some Processing Treatments on the Rennet Coagulation Properties of Milk, *International Journal of Dairy Technology*, Vol. 50, pp. 99–106, © 1997, Society of Dairy Technology.
Figure 6–21. *Source:* Reprinted with permission from Guinee *et al*, The Effects of Composition

and Some Processing Treatments on the Rennet Coagulation Properties of Milk, *International Journal of Dairy Technology*, Vol. 50, pp. 99–106, © 1997, Society of Dairy Technology.

Figure 6–23. *Source:* Data from Fox, 1969 and Phelan, 1973.

Figure 6–24. *Source:* Reprinted with permission from Emmons *et al*, Milk Clotting Enzymes.1. Proteolysis During Cheese Making in Relation to Estimated Losses of Yield, *Journal of Dairy Science*, Vol. 73, pp. 2007–2015, © 1990, American Dairy Science Association.

Figure 6–25. *Source:* Reprinted from P.F. Fox, Milk-clotting and Proteolytic Activities of Rennet, and of Bovine Pepsin and Porcine Pepsin, *Journal of Dairy Research*, Vol. 36, pp. 427–433, © 1969. Reprinted with the permission of Cambridge University Press; and J.A. Phelan, Laboratory and Field Tests on New Milk Coagulants, *Dairy Industries International*, Vol. 38, pp. 419–424, © 1973, Wilmington Publishing Ltd.

Figure 6–26. *Source:* Reprinted with permission from Thunell *et al,* Thermal Inactivation of Residual Milk Clotting Enzymes in Whey, *Journal of Dairy Science*, Vol. 62, pp. 373–377, © 1979, American Dairy Science Association.

CHAPTER 7

Figure 7–3. *Source:* Reprinted with permission from P.F. Fox and P.H.L. McSweeney, *Dairy Chemistry and Biochemistry*, p. 393, © 1998, Aspen Publishers, Inc.

CHAPTER 8

Figure 8–1. *Source:* Reprinted from O.R. Fennema, *Food Chemistry, 3rd Ed.*, © 1996, by courtesy of Marcel Dekker, Inc.

Table 8–3. *Source:* Reprinted with permission from Y. Roos, Water Activity in Milk Products, in *Advanced Dairy Chemistry*, Vol. 3, pp. 306–346, © 1997, Aspen Publishers, Inc.

Figure 8–2. *Source:* Reprinted with permission from Y. Roos, Water Activity in Milk Products, in *Advanced Dairy Chemistry*, Vol. 3, pp. 306–346, © 1997, Aspen Publishers, Inc.

Figure 8–4. *Source:* Reprinted with permission from R.C. Lawrence and J. Gilles, Factors that Determine the pH of Young Cheddar Cheese, *New Zealand Journal of Dairy Science Technology*, Vol. 17, pp. 1–14, © 1982, New Zealand Dairy Research Institute.

Figure 8–5. *Source:* Reprinted with permission from T.P. Guinee and P.F. Fox, Influence of Cheese Geometry on the Movement of Sodium Chloride and Water During Brining, *Irish Journal of Food Science Technology*, Vol. 10, pp. 73–96, © 1986a, Teagasc.

Figure 8–7. *Source:* Reprinted from *Food Chemistry*, Vol. 19, T.P. Guinee and P.F. Fox, Transport of Sodium Chloride and Water in Romano Cheese Slices During Brining, pp. 49–64, © 1986b, with permission from Elsevier Science.

Figure 8–8. *Source:* Reprinted with permission from H.A. Morris, T.P. Guinee and P.F. Fox, *Journal of Dairy Science*, Vol. 68, p. 1851, © 1985, American Dairy Science Association.

Figure 8–9. *Source:* Reprinted from *International Dairy Journal*, Vol. 28, T.J. Geurts, P. Walstra, and H. Mulder, Transport of Salt and Water During Salting of Cheese. 1. Analysis of the Processes Involved. p. 106, © 1974, with permission from Elsevier Science.

Figure 8–10. *Source:* Reprinted with permission from T.P. Guinee, Studies on the Movements of Sodium Chloride and Water in Cheese and the Effects Thereof on Cheese Ripening, PhD. Thesis, © 1985.

Figure 8–11. *Source:* Data from Thomas and Pearce, 1981.

Figure 8–12. *Source:* Data from O'Connor, 1974.

CHAPTER 9

Figure 9–2. *Source:* Reprinted with permission from J. Gilles and R.C. Lawrence, The Yield of Cheese, *New Zealand Journal of Dairy Science Technology*, Vol. 20, pp. 205–214, © 1985, New Zealand Dairy Research Institute.

Figure 9–4. *Source:* Guinee *et al*, Effect of Milk Protein Standardization by Ultrafiltration on the Manufacture, Composition and Maturation of Cheddar Cheese, *Journal of Dairy Research*, Vol. 61, pp. 117–131, © 1994. Reprinted with the permission of Cambridge University Press.

Figure 9–5. *Source:* Guinee *et al*, Effect of Milk Protein Standardization by Ultrafiltration on the

Manufacture, Composition and Maturation of Cheddar Cheese, *Journal of Dairy Research*, Vol. 61, pp. 117–131, © 1994. Reprinted with the permission of Cambridge University Press.

Figure 9–6. *Source:* Reprinted from Guinee *et al*, Milk Protein Standardization by Ultrafiltration for Cheddar Cheese Manufacture, *Journal of Dairy Research*, Vol. 63, pp. 281–293, © 1996. Reprinted with the permission of Cambridge University Press.

Figure 9–7. *Source:* Reprinted with permission from I. Politis and K.F. Ng-Kwai-Hang, Association Between Somatic Cell Count of Milk and Cheese-yielding Capacity, *Journal of Dairy Science*, Vol. 71, pp. 1720–1727, © 1988a, American Dairy Science Association.

Figure 9–8. *Source:* Reprinted with permission from I. Politis and K.F. Ng-Kwai-Hang, Effects of Somatic Cell Counts and Milk Composition on the Coagulating Properties of Milk, *Journal of Dairy Science*, Vol. 71, pp. 1740–1746, © 1988b, American Dairy Science Association.

Figure 9–9. *Source:* Reprinted with permission from I. Politis and K.F. Ng-Kwai-Hang, Association Between Somatic Cell Count of Milk and Cheese-yielding Capacity, *Journal of Dairy Science*, Vol. 71, pp. 1720–1727, © 1988a, American Dairy Science Association.

Figure 9–10. *Source:* W.J. Donnelly and J.G. Barry, Casein Compositional Studies.III. Changes in Irish Milk for Manufacturing and Role of Milk Proteinase, *Journal of Dairy Research*, Vol. 50, pp. 433–441, © 1983. Reprinted with the permission of Cambridge University Press.

Figure 9–11. *Source:* Reprinted with permission from Politis *et al*, Plasmin and Plasminogen in Bovine Milk: A Relationship with Involution, *Journal of Dairy Science*, Vol. 72, pp. 900–906, © 1989, American Dairy Science Association.

Figure 9–12. *Source:* Reprinted from *International Dairy Journal*, Vol. 8, Influence of κ-Casein Genetic Variant on Rennet Gel Microstructure, Cheddar Cheesemaking Properties and Casein Micelle Size, pp. 707–714, © 1998, with permission from Elsevier Science.

Figure 9–13. *Source:* Reprinted with permission from Hicks *et al*, Psychrotrophic Bacteria Reduce Cheese Yield, *Journal of Food Protection*, Vol. 45, pp. 331–334, © 1982, International Association of Milk, Food and Environmental Sanitarians.

Table 9–3. *Source:* Data from Weathercup *et al*, 1988.

Figure 9–14. *Source:* Reprinted with permission from Guinee *et al*, The Influence of Milk Pasteurization Temperature and pH at Curd Milling on the Composition, Texture and Maturation of Reduced-fat Cheddar Cheese, *International Journal of Dairy Technology*, Vol. 51, pp. 1–10, © 1998, Society of Dairy Technology.

Table 9–5. *Source:* Data from Lemay *et al*, 1994.

Figure 9–15. *Source:* Reprinted with permission from D.B. Emmons and D.C. Beckett, Milk Clotting Enzymes.1.Proteolysis During Cheesemaking in Relation to Estimated Losses of Yield, *Journal of Dairy Science*, Vol. 73, pp. 8–16, © 1990, American Dairy Science Association.

Figure 9–16. *Source:* Reprinted with permission from D.M. Barbano and R.R. Rasmussen, Cheese Yield Performance of Various Coagulants, *Cheese Yield and Factors Affecting its Control*, pp. 255–259, IDF Seminar, Cork, Ireland, © 1994, International Dairy Federation.

Figure 9–17. *Source:* Reprinted with permission from J. J. Mayes and B. J. Sutherland, Coagulum Firmness and Yield in Cheddar Cheese Manufacture-The Role of the Curd Firmness Instrument in Determining Cutting Time, *Australian Journal of Dairy Technology*, Vol. 39, pp. 69–73, © 1984.

Figure 9–18. *Source:* Reprinted with permission from J.J. Mayes and B.J. Sutherland, Further Notes on Coagulum Firmness and Yield in Cheddar Cheese Manufacture, *Australian Journal of Dairy Technology*, Vol. 44, pp. 47–48, © 1989.

Figure 9–19. *Source:* Johnston *et al*, Effects of Speed and Duration of Cutting in Mechanised Cheddar Cheesemaking on Curd Particle Size and Yield, *Journal of Dairy Research*, Vol. 58, pp. 345–354, © 1991. Reprinted with the permission of Cambridge University Press.

Figure 9–20. *Source:* Reprinted with permission from J.A. Phelan, Standardisation of Milk for Cheesemaking at Factory Level, *International Journal of Dairy Technology*, Vol. 34, pp. 152–156, © 1981, Society of Dairy Technology.

CHAPTER 10

Table 10–1. *Source:* Reprinted with permission from J. Stadhouders and L.P.M. Langeveld, *The*

Microflora of the Surface of Cheese. Factors Affecting its Composition, 17th International Dairy Congress, © 1966, International Dairy Federation.

Figure 10–1. *Source:* Reprinted with permission from M. Ruegg and B. Blanc, Influence of Water Activity on the Manufacture and Aging of Cheese, *Water Activity: Influences on Food Quality*, L.B. Rockland and G.F. Stewart, eds., pp. 791–811, © 1981, Academic Press, Inc.

Figure 10–2. *Source:* Reprinted with permission from M. Ruegg and B. Blanc, Influence of Water Activity on the Manufacture and Aging of Cheese, *Water Activity: Influences on Food Quality*, L.B. Rockland and G.F. Stewart, eds., pp. 791–811, © 1981, Academic Press, Inc.

Table 10–2. *Source:* Reprinted with permission from M. Ruegg and B.Blanc, Influence of Water Activity on the Manufacture and Aging of Cheese, *Water Activity: Influences on Food Quality*, L.B. Rockland and G.F. Stewart, eds., pp. 791–811, © 1981, Academic Press, Inc.

Figure 10–4. *Source:* Reprinted with permission from M. Ruegg and B. Blanc, Influence of Water Activity on the Manufacture and Aging of Cheese, *Water Activity: Influences on Food Quality*, L.B. Rockland and G.F. Stewart, eds., pp. 791–811, © 1981, Academic Press, Inc.

Figure 10–5. *Source:* Reprinted with permission from F.G. Martley and R.C. Lawrence, Cheddar Cheese Flavour. 2. Characteristics of Single Strain Starters Associated with Good or Poor Flavour Development, *New Zealand Journal of Dairy Science Technology*, Vol. 7, pp. 38–44, © 1972, New Zealand Dairy Research Institute.

Figure 10–7. *Source:* Data from Nunez, 1978; del Pozo *et al*, 1985; Poullet *et al*, 1991 and Cuesta *et al*, 1996.

Figure 10–8. *Source:* Data from delPozo *et al*, 1985; Poullet *et al*, 1991; Demarigny *et al*, 1996 and Beresford and Cogan, unpublished.

Table 10–3. *Source:* Data from Valdes-Stauber *et al*, 1997 and Eliskases-Lechner and Gringinger, 1995a.

Figure 10–9. *Source:* Reprinted with permission from Samson *et al*, *Introduction to Food-borne Fungi*, © 1995, Centraalbureau voor Schmmel-cultures.

Figure 10–10. *Source:* Reprinted with permission from K.W. Turner and T.D. Thomas, Lactose Fermentation in Cheddar Cheese and the Effect of Salt, *New Zealand Journal of Dairy Science Technology*, Vol. 15, pp. 265–276, © 1980, New Zealand Dairy Research Institute.

Figure 10–11. *Source:* Reprinted with permission from K.W. Turner and T.D. Thomas, Lactose Fermentation in Cheddar Cheese and the Effect of Salt, *New Zealand Journal of Dairy Science Technology*, Vol. 15, pp. 265–276, © 1980, New Zealand Dairy Research Institute.

Figure 10–12. *Source:* Reprinted with permission from Accolas *et al*, Evolution de la Flore Lactique Thermophile au Courss du Pressage des Fromage a Pate Cuite, *Le Lait*, Vol. 58, pp. 118–132, © 1978, Editions Scientifique et Medicales Elsevier.

Figure 10–14. *Source:* Reprinted with permission from Turner *et al*, Swiss-type Cheese: II. The Role of Thermophilic Lactobacilli in Sugar Fermentation, *New Zealand Journal of Dairy Science Technology*, Vol. 18, pp. 117–124, © 1983.

Figure 10–15. *Source:* Reprinted with permission from J. Lenoir, La Flore Microbienne du Camembert et Son Evolution au Cours de la Maturation, *Le Lait*, Vol. 43, pp. 262–270, © 1963, Editions Scientifiques et Medicales Elsevier.

Figure 10–16. *Source:* M. Nunez, Microflora of Cabrales Cheese: Changes During Maturation, *Journal of Dairy Research*, Vol. 45, pp. 501–508, © 1978, Cambridge University Press. Reprinted with the permission of Cambridge University Press.

Figure 10–17. *Source:* Reprinted with permission from F. Eliskases-Lechner and W. Ginzinger, The Yeast Flora of Surface-ripened Cheese, *Milchwissenschaft,* Vol. 50, pp. 458–462, © 1995 a,b.

CHAPTER 11

Table 11–1. *Source:* Reprinted with permission from T.D. Thomas and V.L. Crow, Mechanism of D (-)-Lactic Acid Formation in Cheddar Cheese, *New Zealand Journal of Dairy Science Technology*, Vol. 18, pp. 131–141, © 1983, New Zealand Dairy Research Institute.

Figure 11–2. *Source:* Reprinted with permission from T.D. Thomas, Acetate Production from Lactate and Citrate by Non-starter Bacteria in Cheddar Cheese, *New Zealand Journal of Dairy Science Technology*, Vol. 22, pp. 25–38, © 1987, Dairy Technology.

Figure 11–3. *Source:* Reprinted with permission from C. Karahadian and R.C. Lindsay, Integrated

Roles of Lactate, Ammonia, and Calcium in Texture Development of Mold Surface-ripening Cheese, *Journal of Dairy Science*, Vol. 70, pp. 909–918, © 1987, American Dairy Science Association.

Figure 11–4. *Source:* Reprinted with permission from C. Karahadian and R.C. Lindsay, Integrated Roles of Lactate, Ammonia, and Calcium in Texture Development of Mold Surface-ripening Cheese, *Journal of Dairy Science*, Vol. 70, pp. 909–918, © 1987, American Dairy Science Association.

Figure 11–5. *Source:* Reprinted with permission from I.V. Zarmpoutis, *Proteolysis in Blue-veined Cheese Varieties*, MSc Thesis, © 1995, National University of Ireland.

Figure 11–6. *Source:* Reprinted with permission from S.A. Lowney, *Characterization of Proteolysis in the Italian Smear-ripened Cheese, Taleggio*, MSc Thesis, © 1997, National University of Ireland.

Figure 11–7. *Source:* Reprinted with permission from P.F.Fox and P.H.L McSweeney, *Dairy Chemistry and Biochemistry*, p.406, © 1998, Aspen Publishers, Inc.

Figure 11–9. *Source:* Reprinted with permission from McSweeney *et al*, Contribution of the Indigenour Microflora to the Maturation of Cheddar Cheese, *International Dairy Journal*, Vol. 3, © 1993, pp. 613–634, with permission from Elsevier Science.

Table 11–2. *Source:* Data from Woo *et al*, 1984, Woo and Lindsay, 1984, and Fox and McSweeney, 1998.

Figure 11–10. *Source:* Reprinted with permission from P.F. Fox and P.H.L. McSweeney, *Dairy Chemistry and Biochemistry*, p. 470, © 1998, Aspen Publishers, Inc.

Figure 11–12. *Source:* Reprinted with permission from D.M. Rea, *Comparison of Cheddar Cheese made with Chymosin, Rhizomucor miehei Proteinase or Chryphonectria parasitica Proteinases*, MSc Thesis, © 1997, National University of Ireland.

Figure 11–13. *Source:* Reprinted with permission from P.F. Fox and P.H.L. McSweeney, *Dairy Chemistry and Biochemistry*, p. 320, © 1998, Aspen Publishers, Inc.

Figure 11–14. *Source:* Reprinted with permission from Kunji *et al*, the Proteolytic System of Lactic Acid Bacteria, *Antonie van Leeuwenhoek*, Vol. 70, p. 97, Table 2, © 1996, with kind permission from Kluwer Academic Publishers.

Table 11–3. *Source:* Reprinted with permission from Fox *et al*, Cheese: Physical, Biochemical and Nutritional Aspects, *Advances in Food Nutrition Research*, Vol. 39, pp. 163–328, © 1996, Academic Press, Inc.

Figure 11–16. *Source:* Reprinted with permission from M. McGoldrick, MSc Thesis, © 1996, University College, Cork.

Figure 11–17. *Source:* Reprinted with permission from M. McGoldrick, MSc Thesis, © 1996, University College, Cork.

Figure 11–18. *Source:* Reprinted with permission from M. McGoldrick, MSc Thesis, © 1996, University College, Cork.

Figure 11–19. *Source:* Reprinted with permission from Mooney *et al*, Identification of the Principal Water-insoluble Peptides in Cheddar Cheese, *International Dairy Journal*, Vol. 8, © 1998, pp. 813–818, with permission from Elsevier Science.

Figure 11–20. *Source:* Reprinted with permission from Mooney *et al*, Identification of the Principal Water-insoluble Peptides in Cheddar Cheese, *International Dairy Journal*, Vol. 8, © 1998, pp. 813–818, with permission from Elsevier Science.

Figure 11–22. *Source:* Reprinted with permission from Mooney *et al*, Identification of the Principal Water-insoluble Peptides in Cheddar Cheese, *International Dairy Journal*, Vol. 8, © 1998, pp. 813–818, with permission from Elsevier Science.

Figure 11–23. *Source:* Reprinted with permission from P.F. Fox and P.H.L. McSweeney, Proteolysis in Cheese During Ripening, *Food Reviews International*, Vol. 12, pp. 457–509, Marcel Dekker, Inc. New York, © 1996.

Figure 11–24. *Source:* Reprinted with permission from P.F. Fox and J.M. Wallace, Formation of Flavour Compounds in Cheese, *Advances in Applied Microbiology*, Vol. 45, pp. 53–58, © 1997, Academic Press, Inc.

CHAPTER 12

Exhibit 12–1. *Source:* Reprinted with permission from Nielson *et al*, Progress in Developing an International Protocol for Sensory Profiling of Hard Cheese, *International Journal of Dairy Technology*, Vol. 51, pp. 57–64, © 1998, Society of Dairy Technology.

Figure 12–1. *Source:* Reprinted with permission from Nielson *et al*, Progress in Developing an International Protocol for Sensory Profiling of Hard Cheese, *International Journal of Dairy Technology*, Vol. 51, pp. 57–64, © 1998, Society of Dairy Technology.

Figure 12–3. *Source:* D.J. Manning and C. Moore, Headspace Analysis of Hard Cheese, *Journal of Dairy Research*, Vol. 46, pp. 539–545, © 1979. Reprinted with the permission of Cambridge University Press.

Figure 12–5. *Source:* Reprinted from P. Schieberle, New Developments in Methods for Analysis of Volatile Flavor Compounds and Their Precursors, *Characterization of Food: Emerging Methods*, A.G. Gaonkar, ed., © 1995, pp. 403–431, with permission from Elsevier Science.

Figure 12–7. *Source:* Reprinted with permission from Aston *et al*, Proteolysis and Flavor Development in Cheddar Cheese, *Australian Journal of Dairy Technology*, Vol. 38, pp. 55–65, © 1983, Dairy Industry Association of Australia.

Exhibit 12–2. *Source:* Reprinted with permission from G. Urbach, Relations Between Cheese Flavour and Chemical Composition, *International Dairy Journal*, Vol. 3, © 1993, pp. 389–422, with permission from Elsevier Science.

Table 12–2. *Source:* Reprinted with permission from G. Urbach, The Flavour of Milk and Dairy Products. II. Cheese: Contribution of Volatile Compounds, *International Journal of Dairy Technology*, Vol. 50, pp. 79–80, © 1997, Society of Dairy Technology.

Figure 12–8. *Source:* Reprinted from *International Dairy Journal,* Vol. 3, J.O. Bosset and R. Gauch, Comparison of the Volatile Flavour in Six European AOC Cheeses by Using a New Dynamic Headspace GC-MS Method, pp. 359–377, © 1993, with permission from Elsevier Science.

Figure 12–11. *Source:* Reprinted with permission from T.P. Coultate, *Food: The Chemistry of Its Components, 2nd Edition*, p. 24, © 1989, The Royal Society of Chemistry.

Figure 12–12. *Source:* Reprinted from *International Dairy Journal*, Vol. 6, O'Shea *et al*, Objective Assessment of Cheddar Cheese Quality, pp. 1135–1147, © 1996, with permission from Elsevier Science.

CHAPTER 13

Figure 13–2. *Source:* Reprinted with permission from Guinee *et al*, The Influence of Milk Pasteurization Temperature and pH at Curd Milling on the Composition, Texture and Maturation of Reduced-fat Cheddar Cheese, *International Journal of Dairy Technology*, Vol. 60, pp. 1–12, © 1998, Society of Dairy Technology.

Figure 13–3. *Source:* Reprinted with permssion from M. Kalab, Milk Gel Structure. VI. Cheese Texture and Microstructure, *Milchwissenschaft*, Vol. 32, pp. 449–457, © 1977, Milk Science International.

Figure 13–4. *Source:* Reprinted with permission from Lowrie *et al*, Curd Granule and Milled Curd Junction Patterns in Cheddar Cheese Made by Traditional and Mechanized Processes, *Journal of Dairy Science*, Vol. 65, pp. 1122–1129, © 1982, American Dairy Science Association.

Figure 13–10. *Source:* Reprinted with permission from P. Walstra and T. van Vliet, Rheology of Cheese, *Bulletin No. 153*, pp. 22–27, © 1982, International Dairy Federation.

Figure 13–11. *Source:* Data from Guinee, unpublished results.

Figure 13–13. *Source:* Reprinted with permission from J.H. Prentice, Cheese Rheology, in *Cheese: Chemistry, Physics and Microbiology, Volume 1, General Aspects*, P.F. Fox, ed., © 1987.

Table 13–2. *Source:* Data from van Vliet, 1991 and van Vliet *et al*, 1991.

Figure 13–16. *Source:* Reprinted with permission from F. Shama and P. Sherman, Evaluation of Some Textural Properties of Foods with the Instron Universal Testing Machine, *Journal of Texture Studies*, Vol. 4, pp. 344–353, © 1973, Food and Nutrition Press, Inc.

Figure 13–17. *Source:* Reprinted with permission from J.H. Prentice, Cheese Rheology, in *Cheese: Chemistry, Physics and Microbiology, Volume 1, General Aspects*, P.F. Fox, ed., © 1993.

Figure 13–18. *Source:* Reprinted with permission from J. Culioli and P. Sherman, Evaluation of Gouda Cheese Firmness by Compression Tests, *Journal of Texture Studies*, Vol. 7, pp. 353–372, © 1976, Food and Nutrition Press.

Figure 13–19. *Source:* Reprinted with permission from P. Sherman, Rheological Evaluation of the Textural Properties of Foods, *Progress and*

Trends in Rheology II, pp. 44–53, © 1988, Springer-Verlag New York, Inc.

Figure 13–20. *Source:* Data from Culioli and Sherman, 1976, Vernon Carter and Sherman, 1978 and Prentice *et al*, 1993.

Figure 13–21. *Source:* Reprinted with permission from J. Culioli and P. Sherman, Evaluation of Gouda Cheese Firmness by Compression Tests, *Journal of Texture Studies*, Vol. 7, pp. 353–372, © 1976, Food and Nutrition Press, Inc.

Figure 13–22. *Source:* Reprinted with permission from Chen *et al*, Texture Analysis of Cheese, *Journal of Dairy Science*, Vol. 62, pp. 901–907, © 1979, American Dairy Science Association.

Figure 13–23. *Source:* Reprinted with permission from M.A. Fenelon and T.P. Guinee, Improving the Quality of Low-fat Cheddar Cheese, *Project Report, DPRC No. 4*, © 1999, Dairy Products Research Center, Ireland.

Figure 13–24. *Source:* Reprinted with permission from J. Visser, Factors Affecting the Rheological and Fracture Properties of Hard and Semi-hard Cheese, *Rheological and Fracture Properties of Cheese-Bulletin No. 268*, pp. 49–61, © 1991, International Dairy Federation.

Figure 13–25. *Source:* Reprinted with permission from J. Visser, Factors Affecting the Rheological and Fracture Properties of Hard and Semi-hard Cheese, *Rheological and Fracture Properties of Cheese-Bulletin No. 268*, pp. 49–61, © 1991, International Dairy Federation.

Figure 13–26. *Source:* Reprinted with permission from J. Visser, Factors Affecting the Rheological and Fracture Properties of Hard and Semi-hard Cheese, *Rheological and Fracture Properties of Cheese-Bulletin No. 268*, pp. 49–61, © 1991, International Dairy Federation.

Figure 13–27. *Source:* Reprinted with permission from J. Visser, Factors Affecting the Rheological and Fracture Properties of Hard and Semi-hard Cheese, *Rheological and Fracture Properties of Cheese-Bulletin No. 268*, pp. 49–61, © 1991, International Dairy Federation.

Figure 13–28. *Source:* Guinee *et al*, Milk Protein Standardization by Ultrafiltration for Cheddar Cheese Manufacture, *Journal of Dairy Research*, Vol. 63, pp. 281–293, © 1996. Reprinted with the permission of Cambridge University Press.

Exhibit 13–1. *Source:* Data from Szczesniak, 1963 and Brennan, 1988.

Figure 13–29. *Source:* Reprinted with permission from J.G. Brennan, Texture Perception and Measurement, *Sensory Analysis of Foods*, 2nd Ed., J.R. Piggott, ed., pp. 69–101, © 1988.

Figure 13–30. *Source:* Reprinted with permission from J.G. Brennan, Texture Perception and Measurement, *Sensory Analysis of Foods*, 2nd Ed., J.R. Piggott, ed., pp. 69–101, © 1988.

Exhibit 13–2. *Source:* Reprinted with permission from J.G. Brennan, Texture Perception and Measurement, *Sensory Analysis of Foods*, 2nd Ed., J.R. Piggott, ed., pp. 69–101, © 1988.

Table 13–3. *Source:* Reprinted with permission from Lee *et al*, Evaluation of Cheese Texture, *Journal of Food Science*, Vol. 43, pp. 1600–1605, © 1978, Institute of Food Technologists.

CHAPTER 14

Figure 14–2. *Source:* Data from Gilles and Lawrence, 1973, Fox, 1975, and Pearce and Gilles, 1979.

CHAPTER 16

Figure 16–1. *Source:* Reprinted with permission from T.P. Guinee, P.D. Pudja and N.Y. Farkye, Fresh Acid-curd Cheese Varieties, in *Cheese: Chemistry, Physics and Microbiology, Vol. 2 Major Cheese Groups, 2nd Ed.*, P.F. Fox, ed., p. 364, © 1993, Aspen Publishers, Inc.

Table 16–1. *Source:* Reprinted with permission from T.P. Guinee, P.D. Pudja and N.Y. Farkye, Fresh Acid-curd Cheese Varieties, in *Cheese: Chemistry, Physics and Microbiology, Vol. 2 Major Cheese Groups, 2nd Ed.*, P.F. Fox, ed., p. 367, © 1993, Aspen Publishers, Inc.

Figure 16–2. *Source:* Reprinted with permission from T.P. Guinee, P.D. Pudja and N.Y. Farkye, Fresh Acid-curd Cheese Varieties, in *Cheese: Chemistry, Physics and Microbiology, Vol. 2 Major Cheese Groups, 2nd Ed.*, P.F. Fox, ed., p. 365, © 1993, Aspen Publishers, Inc.

Figure 16–3. *Source:* Reprinted from *International Dairy Journal*, Vol. 40, vanHooydonk *et al*, pH-induced Physico-chemical Changes in Casein Micelles in Milk and Their Effect on Renneting.1.Effect of Acidification on Physico-chemical Properties, pp. 281–296, © 1986, with permission from Elsevier Science.

Figure 16–4. *Source:* D.G. Dalgleish and A.J.R. Law, pH Induced Dissociation of Bovine Casein Micelles.I. Analysis of Liberated Caseins, *Journal of Dairy Research*, Vol. 55, pp. 529–538, © 1988. Reprinted with the permission of Cambridge University Press.

Figure 16–5. *Source:* Reprinted with permission from L.K. Creamer, Water Absorption by Renneted Casein Micelles, *Milchwissenschaft*, Vol. 40, pp. 589–59, © 1985.

Figure 16–6. *Source:* Data from P. McSweeney, T.P. Guinee and M.G. Wilkinson (unpublished results).

Figure 16–10. *Source:* Data from Auty, McSweeney, Guinee and Wilkinson (unpublished results).

Figure 16–11. *Source:* Data from P. McSweeney, T.P. Guinee and M.G. Wilkinson (unpublished results).

CHAPTER 17

Exhibit 17–1 *Source:* Reprinted with permission from P.F. Fox, Cheese: An Overview, *Cheese, Chemistry, Physics and Microbiology, 2nd Ed*, Vol.1, P.F. Fox, ed., pp. 1–36, © 1993, Aspen Publishers, Inc.

Table 17–1 *Source:* Reprinted with permission from *International Dairy Journal*, Vol. 3, L. Berozzi and G. Panari, Cheeses with Appellation d'Origine Controlee (AOC): Factors That Affect Quality, pp. 297–312, © 1993, with permission from Elsevier Science.

Figure 17–6 Courtesy of APV Nordic Cheese, Denmark.

Figure 17–22 *Source:* Reprinted with permission from F.V. Kosikowski and V.V. Mistry, *Cheese and Fermented Milk Foods, 3rd Ed.*, Vols. 1 and 2, © 1997, F.V. Kosikowski, LLC.

Appendix 17-A *Source:* Reprinted with permission from F.V. Kosikowski and V.V. Mistry, *Cheese and Fermented Milk Foods, 3rd Ed.*, Vols. 1 and 2, © 1997, F.V. Kosikowski, LLC.

CHAPTER 18

Table 18–1 *Source:* Reprinted with permission from Fox *et al*, Cheese, Physical, Biochemical and Nutritional Aspects, *Advances in Food Science Nutrition Research*, Vol. 39, pp. 163–329, © 1996, Academic Press, Inc.

Exhibit 18–1 *Source:* Reprinted with permission from Fox *et al*, Cheese, Physical, Biochemical and Nutritional Aspects, *Advances in Food Science Nutrition Research*, Vol. 39, pp. 163–329, © 1996, Academic Press, Inc.

Table 18–2 *Source:* Reprinted with permission from Fox *et al*, Cheese, Physical, Biochemical and Nutritional Aspects, *Advances in Food Science Nutrition Research*, Vol. 39, pp. 163–329, © 1996, Academic Press, Inc.

Table 18–3 *Source:* Reprinted with permission from Fox *et al*, Cheese, Physical, Biochemical and Nutritional Aspects, *Advances in Food Science Nutrition Research*, Vol. 39, pp. 163–329, © 1996, Academic Press, Inc.

Table 18–4 *Source:* Reprinted with permission from Fox *et al*, Cheese, Physical, Biochemical and Nutritional Aspects, *Advances in Food Science Nutrition Research*, Vol. 39, pp. 163–329, © 1996, Academic Press, Inc.

Table 18–5 *Source:* Reprinted with permission from Guinee *et al,* Characteristics of Different Cheeses Used in Pizza Pie, *Australian Journal of Dairy Technology*, Vol. 53, p. 109, © 1998.

Table 18–6 *Source:* Reprinted with permission from Guinee *et al,* Characteristics of Different Cheeses Used in Pizza Pie, *Australian Journal of Dairy Technology*, Vol. 53, p. 109, © 1998.

Figure 18–6 *Source:* Data from Auty and Guinee (unpublished results).

CHAPTER 19

Figure 19–2 *Source:* Data from Guinee *et al* (unpublished results).

Figure 19–3 *Source:* Data from Guinee *et al* (unpublished results).

Figure 19–4 *Source:* Data from Guinee *et al* (unpublished results).

Figure 19–5 *Source:* Data from Guinee *et al* (unpublished results).

Figure 19–6 *Source:* Data from Guinee *et al* (unpublished results).

Figure 19–7 *Source:* Data from Auty and Guinee (unpublished results).

Figure 19–8 *Source:* Reprinted with permission from T.P. Guinee and M.O. Corcoran, Expanded Use of Cheese in Processed Meat Products, *Farm and Food Research*, Vol. 4, No. 1, pp. 25–28, © 1994.

Figure 19–9 *Source:* Reprinted with permission from Guinee *et al*, Functionality of Low Moisture Mozzarella Cheese During Ripening, *Proceedings of the 5th Cheese Symposium*, Cork, T.M. Cogan, ed., © 1997, Teagasc, Ireland.

Figure 19–10 *Source:* Reprinted with permission from Guinee *et al*, Functionality of Low Moisture Mozzarella Cheese During Ripening, *Proceedings of the 5th Cheese Symposium*, Cork, T.M. Cogan, ed., © 1997, Teagasc, Ireland.

Figure 19–11 *Source:* Reprinted with permission from Guinee *et al*, Functionality of Low Moisture Mozzarella Cheese During Ripening, *Proceedings of the 5th Cheese Symposium*, Cork, T.M. Cogan, ed., © 1997, Teagasc, Ireland.

Figure 19–12 *Source:* Reprinted with permission from Guinee *et al*, Functionality of Low Moisture Mozzarella Cheese During Ripening, *Proceedings of the 5th Cheese Symposium*, Cork, T.M. Cogan, ed., © 1997, Teagasc, Ireland.

Figure 19–13 *Source:* Reprinted with permission from Guinee *et al*, Functionality of Low Moisture Mozzarella Cheese During Ripening, *Proceedings of the 5th Cheese Symposium*, Cork, T.M. Cogan, ed., © 1997, Teagasc, Ireland.

Figure 19–14 *Source:* Reprinted with permission from Guinee *et al*, Functionality of Low Moisture Mozzarella Cheese During Ripening, *Proceedings of the 5th Cheese Symposium*, Cork, T.M. Cogan, ed., © 1997, Teagasc, Ireland.

Figure 19–15 *Source:* Reprinted with permission from Guinee *et al*, Functionality of Low Moisture Mozzarella Cheese During Ripening, *Proceedings of the 5th Cheese Symposium*, Cork, T.M. Cogan, ed., © 1997, Teagasc, Ireland.

Table 19–7 *Source:* Reprinted with permission from Guinee *et al*, The Viscosity of Cheese Sauces with Different Starch Systems and Cheese Powders, *International Journal of Dairy Technology*, Vol. 47, pp. 132–138, © 1994, Society of Dairy Technology.

Table 19–8 *Source:* Reprinted with permission from Guinee *et al*, The Viscosity of Cheese Sauces with Different Starch Systems and Cheese Powders, *International Journal of Dairy Technology*, Vol. 47, pp. 132–138, © 1994, Society of Dairy Technology.

CHAPTER 20

Table 20–2 *Source:* Reprinted with permission from Desmasures *et al*, Microbiological Composition of Raw Milk from Selected Farmers in the Camembert Region of Normandy, *Journal of Applied Microbiology*, Vol. 83, pp. 53–58, © 1997, Blackwell Science, Ltd.

Table 20–3 *Source:* Reprinted from EU Guidelines.

Table 20–4 *Source:* Reprinted with permission from Nichols *et al*, The Microbiological Quality of Soft Cheese, *PHLS Microbiological Digest*, Vol. 13, pp. 68–75, © 1996, Public Health Laboratory Services.

Figure 20–1 *Source:* Reprinted with permission from Rea *et al*, Incidence of Pathogenic Bacteria in Raw Milk in Ireland, *Journal of Applied Bacteriology*, Vol. 73, pp. 331–336, © 1992, Blackwell Science, Ltd.

Figure 20–2 *Source:* Reprinted with permission from the Journal of Food Protection. Copyright held by the International Association of Milk, Food and Environmental Sanitarians, Inc., Des Moines, Iowa, USA. C.J. Reitsma and D.R. Henning, Survival of Enterohemmorragic *Escherichia coli* O157:H7 During the Manufacture and Curing of Cheddar Cheese, Vol. 59, pp. 460–464, © 1996; E.T. Ryser and E.H. Marth, Behavior of *Listeria monocytogenes* during the Manufacture and Ripening of Cheddar Cheese, Vol. 50, pp. 7–13, © 1987; and Tuckey *et al*, Relation of Cheesemaking Operations to Survival and Growth of *Staphloccus aureus* in Different Varieties of Cheese, *Journal of Dairy Science*, Vol. 47, pp. 604–611, © 1964.

Figure 20–3 *Source:* Reprinted with permission from H.P. Bachmann and U. Spahr, The Fate of Potentially Pathogenic Bacteria in Swiss Hard and Semi-hard Cheeses Made from Raw Milk, *Journal of Dairy Science*, Vol. 78, pp. 476–483, © 1995, American Dairy Science Association.

Figure 20–4 *Source:* L. Bautista and R.G. Kroll, Survival of Some Non-starter Bacteria in Naturally Ripened and Enzyme-accelerated Cheddar Cheese, *Journal of Dairy Research*, Vol. 55, pp. 597–602, © 1988. Reprinted with the permission of Cambridge University press.

Figure 20–5 *Source:* Reprinted with permission from *the Journal of Food Protection*. Copyright held by the International Association of Milk, Food and Environmental Sanitarians, Inc., Des Moines, Iowa, USA, E.T. Ryser and E.H. Marth, Behavior of *Listeria monocytogenes* during the Manufacture and Ripening of Cheddar Cheese, Vol. 50, pp. 7–13, © 1987.

Figure 20–6 *Source:* Reprinted with permission from C.J. Reitsma and D.R. Henning, Survival of

Enterohemorragic *Escherichia coli* O157:H7 During the Manufacture and Curing of Cheddar Cheese, *Journal of Food Protection*, Vol. 59, pp. 460–464, © 1996, International Association of Milk, Food and Environmental Sanitarians.

Figure 20–7 *Source:* Reprinted with permission from Hargrove *et al*, Factors Affecting Survival of Salmonella in Cheddar and Colby Cheese, *Journal of Food Technology*, Vol. 32, pp. 580–584, © 1969, International Association of Milk, Food and Environmental Sanitarians.

Figure 20–8 *Source:* Reprinted with permission from *the Journal of Food Protection*. Copyright held by the International Association of Milk, Food and Environmental Sanitarians, Inc., Des Moines, Iowa, USA, Frank *et al*, Survival of Enteropathogenic and Non-pathogenic *Escherichia coli* During the Manufacture of Camembert Cheese, Vol. 40, pp. 835–842, © 1977; Rutzinski *et al*, Behaviour of *Enterobacter aerogenes* and *Hafnia* species during the Manufacture and Ripening of Camembert Cheese, Vol. 42, pp. 790–793, © 1979; and E.T. Ryser and E.H. Marth, Behavior of *Listeria monocytogenes* During the Manufacture and Ripening of Cheddar Cheese, Vol. 50, pp. 7–13, © 1987.

Figure 20–9 *Source:* Reprinted with permission from Poullet *et al*, Microbial Study of Casar de Caceres Cheese Throughout Ripening, *Journal of Dairy Research*, Vol. 58, pp. 231–238, © 1991; delPozo *et al*, Changes in the Microflora of LaSerena Ewe's Milk Cheese During Ripening, *Journal of Dairy Research*, Vol. 55, pp. 449–455, © 1985; Cuesta *et al*, Evaluation of the Microbiological and Biochemical Characteristics of Afuga'l Pitu Cheese During Ripening, *Journal of Dairy Science*, Vol. 79, pp. 1693–1698, © 1996; and Gobbetti *et al*, Microbiology and Biochemistry of Pecorino Umbro Cheese During Ripening, *Italian Journal of Food Science*, Vol. 9, pp. 111–126, © 1997.

CHAPTER 21

Exhibit 21–1 *Source:* Reprinted from the U.S. Department of Agriculture and the Department of Health and Human Services, 1995.

Table 21–1 *Source:* Reprinted with permission from Holland *et al*, Milk Products and Eggs, *The Fourth Supplement to McCance and Widdowson's The Composition of Foods, 4th ed.*, © 1989, Royal Society of Chemistry/Ministry of Agriculture, Fisheries and Food, Cambridge, UK. Crown copyright material is adapted/reproduced with the permission of the Controller of Her Majesty's Stationery Office.

Table 21–2 *Source:* Reprinted with permission from Holland *et al*, Milk Products and Eggs, *The Fourth Supplement to McCance and Widdowson's The Composition of Foods, 4th ed.*, © 1989, Royal Society of Chemistry/Ministry of Agriculture, Fisheries and Food, Cambridge, UK. Crown copyright material is adapted/reproduced with the permission of the Controller of Her Majesty's Stationery Office.

Table 21–3 *Source:* Reprinted with permission from Holland *et al*, Milk Products and Eggs, *The Fourth Supplement to McCance and Widdowson's The Composition of Foods, 4th ed.*, © 1989, Royal Society of Chemistry/Ministry of Agriculture, Fisheries and Food, Cambridge, UK. Crown copyright material is adapted/reproduced with the permission of the Controller of Her Majesty's Stationery Office.

Table 21–4 *Source:* Adapted with permission from E. Renner, Nutritional Aspects of Cheese, *Cheese: Chemistry, Physics and Microbiology*, Vol.1, 2nd Edition, P.F. Fox, ed., pp. 345–363, © 1987.

CHAPTER 22

Table 22–1 *Source:* Reprinted with permission from D.M. Mulvihill, Production, Functional Properties and Utilization of Milk Protein Products, *Advanced Dairy Chemistry, Volume 1, Proteins*, P.F. Fox, ed., pp. 369–404, © 1992.

Figure 22–1 *Source:* Reprinted with permission from P.F. Fox and P.H.L. McSweeney, *Dairy Chemistry and Biochemistry*, p. 46, © 1998, Aspen Publishers, Inc.

Figure 22–1 Courtesy of the International Dairy Federation, 1993, Brussels, Belgium.

Figure 22–3 *Source:* Reprinted with permission from P.F. Fox and P.H.L. McSweeney, *Dairy Chemistry and Biochemistry*, p. 51, © 1998, Aspen Publishers, Inc.

Figure 22–4 *Source:* Reprinted with permission from P.F. Fox and P.H.L. McSweeney, *Dairy Chemistry and Biochemistry*, p. 52, © 1998, Aspen Publishers, Inc.

Figure 22–5 Courtesy of the International Dairy Federation, © 1993, Brussels, Belgium.

CHAPTER 23

Figure 23–1 Courtesy of the International Dairy Federation, *Standard 50B, Milk Products, Methods of Sampling*, © 1985, International Dairy Federation, Brussels.

Figure 23–2 Courtesy of the International Dairy Federation, *Standard 50B, Milk Products, Methods of Sampling*, © 1985, International Dairy Federation, Brussels.

Figure 23–3 *Source:* Reprinted with permission from P.H.L. McSweeney and P.F. Fox, Chemical Methods for the Characterization of Proteolysis in Cheese During Ripening, *Le Lait*, Vol. 77, pp. 41–76, © 1997, Editions Scientific Elsevier.

Figure 23–4 *Source:* Reprinted with permission from P.H.L. McSweeney and P.F. Fox, Chemical Methods for the Characterization of Proteolysis in Cheese During Ripening, *Le Lait*, Vol. 77, pp. 41–76, © 1997, Editions Scientific Elsevier.

Figure 23–5 *Source:* Reprinted from *International Dairy Journal*, Vol. 6, O'Shea *et al*, Objective Assessment of Cheddar Cheese Quality, pp. 1135–1147, © 1996, with permission from Elsevier Science.

Figure 23–7 *Source:* Reprinted with permission from P.H.L. McSweeney and P.F. Fox, Chemical Methods for the Characterization of Proteolysis in Cheese During Ripening, *Le Lait*, Vol. 77, pp. 41–76, © 1997, Editions Scientific Elsevier.

Figure 23–8 *Source:* Reprinted with permission from P.H.L. McSweeney and P.F. Fox, Chemical Methods for the Characterization of Proteolysis in Cheese During Ripening, *Le Lait*, Vol. 77, pp. 41–76, © 1997, Editions Scientific Elsevier.

Figure 23–9 *Source:* Reprinted with permission from P.H.L. McSweeney and P.F. Fox, Chemical Methods for the Characterization of Proteolysis in Cheese During Ripening, *Le Lait*, Vol. 77, pp. 41–76, © 1997, Editions Scientific Elsevier.

Exhibit 23–1 Courtesy of C.M. Delahunty and J.M. Murray, *Organoleptic Evaluation of Cheese*, Proceedings of the 5th Cheese Symposium, pp. 90–97, T.M. Cogan, ed., © 1997, Teagasc, Ireland.

Figure 23–10 Courtesy of C.M. Delahunty and J.M. Murray, *Organoleptic Evaluation of Cheese*, Proceedings of the 5th Cheese Symposium, pp. 90–97, T.M. Cogan, ed., © 1997, Teagasc, Ireland.

Index

C

D

E

G

H

M

N

O

P

S

U

V

W

X

Y

Z